tandard Normal Distibrution

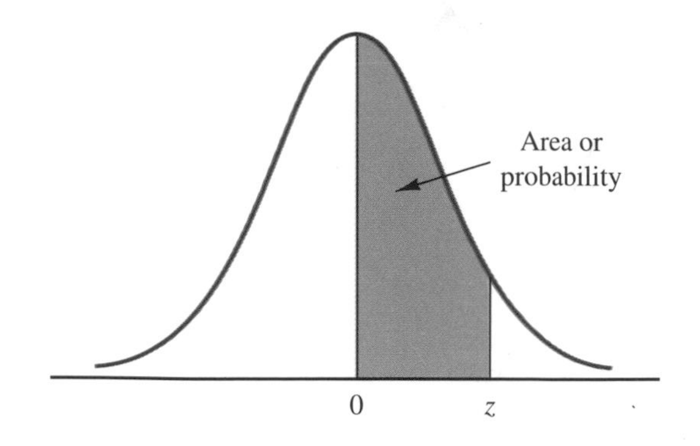

Entries in the table give the area under the curve between the mean and z standard deviations above the mean. For example, for $z = 1.25$ the area under the curve between the mean and z is .3944.

z	.00	.01	.02	.03	.04	.05	.06	.07	.08	.09
.0	.0000	.0040	.0080	.0120	.0160	.0199	.0239	.0279	.0319	.0359
.1	.0398	.0438	.0478	.0517	.0557	.0596	.0636	.0675	.0714	.0753
.2	.0793	.0832	.0871	.0910	.0948	.0987	.1026	.1064	.1103	.1141
.3	.1179	.1217	.1255	.1293	.1331	.1368	.1406	.1443	.1480	.1517
.4	.1554	.1591	.1628	.1664	.1700	.1736	.1772	.1808	.1844	.1879
.5	.1915	.1950	.1985	.2019	.2054	.2088	.2123	.2157	.2190	.2224
.6	.2257	.2291	.2324	.2357	.2389	.2422	.2454	.2486	.2518	.2549
.7	.2580	.2612	.2642	.2673	.2704	.2734	.2764	.2794	.2823	.2852
.8	.2881	.2910	.2939	.2967	.2995	.3023	.3051	.3078	.3106	.3133
.9	.3159	.3186	.3212	.3238	.3264	.3289	.3315	.3340	.3365	.3389
1.0	.3413	.3438	.3461	.3485	.3508	.3531	.3554	.3577	.3599	.3621
1.1	.3643	.3665	.3686	.3708	.3729	.3749	.3770	.3790	.3810	.3830
1.2	.3849	.3869	.3888	.3907	.3925	.3944	.3962	.3980	.3997	.4015
1.3	.4032	.4049	.4066	.4082	.4099	.4115	.4131	.4147	.4162	.4177
1.4	.4192	.4207	.4222	.4236	.4251	.4265	.4279	.4292	.4306	.4319
1.5	.4332	.4345	.4357	.4370	.4382	.4394	.4406	.4418	.4429	.4441
1.6	.4452	.4463	.4474	.4484	.4495	.4505	.4515	.4525	.4535	.4545
1.7	.4554	.4564	.4573	.4582	.4591	.4599	.4608	.4616	.4625	.4633
1.8	.4641	.4649	.4656	.4664	.4671	.4678	.4686	.4693	.4699	.4706
1.9	.4713	.4719	.4726	.4732	.4738	.4744	.4750	.4756	.4761	.4767
2.0	.4772	.4778	.4783	.4788	.4793	.4798	.4803	.4808	.4812	.4817
2.1	.4821	.4826	.4830	.4834	.4838	.4842	.4846	.4850	.4854	.4857
2.2	.4861	.4864	.4868	.4871	.4875	.4878	.4881	.4884	.4887	.4890
2.3	.4893	.4896	.4898	.4901	.4904	.4906	.4909	.4911	.4913	.4916
2.4	.4918	.4920	.4922	.4925	.4927	.4929	.4931	.4932	.4934	.4936
2.5	.4938	.4940	.4941	.4943	.4945	.4946	.4948	.4949	.4951	.4952
2.6	.4953	.4955	.4956	.4957	.4959	.4960	.4961	.4962	.4963	.4964
2.7	.4965	.4966	.4967	.4968	.4969	.4970	.4971	.4972	.4973	.4974
2.8	.4974	.4975	.4976	.4977	.4977	.4978	.4979	.4979	.4980	.4981
2.9	.4981	.4982	.4982	.4983	.4984	.4984	.4985	.4985	.4986	.4986
3.0	.4986	.4987	.4987	.4988	.4988	.4989	.4989	.4989	.4990	.4990

ESSENTIALS OF

Statistics for Business and Economics

ESSENTIALS OF Statistics for Business and Economics

David R. Anderson
University of Cincinnati

Dennis J. Sweeney
University of Cincinnati

Thomas A. Williams
Rochester Institute of Technology

West Publishing Company

Minneapolis/St. Paul
New York
Los Angeles
San Francisco

To Marcia, Cherri, and Robbie

Production Services
Copyediting:
Cheryl Wilms

Text Design:
David J. Farr/ImageSmyth

Composition:
Carlisle Communications

Artwork:
Miyake Illustration

Indexing:
Northwind Editorial Services

Photo Credits:
22/Courtesy of Colgate Palmolive; 61/Andy Sacks/Tony Stone Worldwide; 159/Superstock; 296/David R. Frazier/ Photo Researchers, Inc.; 337/ Courtesy of Fisons Corporation; 388/ Courtesy of United Way of America; 489/ Courtesy of Dow Chemical Company.

West's Commitment to the Environment
In 1906, West Publishing Company began recycling materials left over from the production of books. This began a tradition of efficient and responsible use of resources. Today, 100% of our legal bound volumes are printed on acid-free, recycled paper consisting of 50% new fibers. West recycles nearly 27,700,000 pounds of scrap paper annually—the equivalent of 229,300 trees. Since the 1960s, West has devised ways to capture and recycle waste inks, solvents, oils, and vapors created in the printing process. We also recycle plastics of all kinds, wood, glass, corrugated cardboard, and batteries, and have eliminated the use of polystyrene book packaging. We at West are proud of the longevity and the scope of our commitment to the environment.

West pocket parts and advance sheets are printed on recyclable paper and can be collected and recycled with newspapers. Staples do not have to be removed. Bound volumes can be recycled after removing the cover.

Production, Prepress, Printing, and Binding by West Publishing Company

British Library Cataloguing-in-Publication Data. A catalogue record for this book is available from the British Library

610 Opperman Drive
P.O. Box 64526
St. Paul, MN 55164–0526
1–800–328–9352

Printed in the United States of America

04 03 02 01 00 99 98 97 8 7 6 5 4 3

Library of Congress Cataloging-in-Publication Data

Anderson, David Ray, 1941-
Essentials of statistics for business and economics / David R. Anderson, Dennis J. Sweeney, Thomas A. Williams.
p. cm.
Includes index.
ISBN 0-314-09249-8 (alk. paper)
1. Social sciences—Statistical methods. 2. Commercial statistics. 3. Economics—Statistical methods. 4. Statistics—Problems, exercises, etc. I. Sweeney, Dennis J. II. Williams, Thomas Arthur, 1944-. III. Title.
HA29.A587 1997 IN PROCESS
519.5—dc20 96-33702
CIP

Contents

CHAPTER 13 Statistical Methods for Quality Control 488

The purpose of this book is to give students in business and economics a conceptual introduction to statistics and its many applications. The text is applications oriented and written with the needs of the nonmathematician in mind; the mathematical prerequisite is algebra. The discussion and development of each technique is presented in an application setting, with the statistical results providing insights into decisions and solutions to problems.

We have taken care to use notation that is generally accepted for the topic being covered. Hence, students will find that this text provides good preparation for the study of more advanced statistical material. A bibliography to guide further study is included as an appendix.

Features

Statistics in Practice

To emphasize how companies use statistics, each chapter opens with an application supplied by practitioners from business and economics. Each Statistics in Practice briefly describes an organization and a problem in which the statistical methodology introduced in the chapter was applied; we believe this helps motivate students by placing the chapter material in a real-world context. Procter & Gamble, Polaroid, Monsanto, Xerox, Mead, Dow Chemical, and Colgate-Palmolive are just a few of the companies that have contributed Statistics in Practice features. The table at the end of this preface provides a list of the organizations and applications described.

Examples and Exercises Based on Real Data

To help students understand the wide range of statistical applications, many of the examples and exercises are based on real data and actual statistical studies. Publications such as *The Wall Street Journal, Business Week, USA Today, Fortune, Forbes, Financial World,* and *Barron's* are used to provide referenced applications and exercises that demonstrate uses of statistics in business and economics. The use of real data means that students not only learn about the statistical methodology but also learn about the context of applications encountered in practice.

Methods and Applications Exercises

The end-of-section exercises have been split into two parts: Methods and Applications. The Methods exercises are designed to ensure that students can correctly make the computations required by the methods presented in the section. In the Applications exercises, students are given problems that require applying the material in real-world situations.

Self-Test Exercises

Completely worked-out solutions for exercises identified as self-test exercises are provided in an appendix at the end of the text. Students can work the self-test exercises and immediately check the solution to evaluate their understanding of the concepts presented in the chapter.

Notes & Comments

Notes & Comments at the end of many sections provide additional insights about the statistical methodology and its application. The Notes & Comments include warnings about limitations of the methodology, recommendations for application, brief descriptions of additional technical considerations, and so on.

Computer Software

The text contains numerous examples and discussions of the important role statistical software packages play. In the text we show output obtained using the Minitab statistical software package; the emphasis here is on interpreting the output. Details showing the specific steps required to obtain the Minitab output are provided in the chapter appendix. In this way instructors who use a different statistical software package need not require students to read details that are only appropriate for Minitab. For those instructors using the statistical capabilities available with modern spreadsheet packages, we have included several appendixes describing the use of Excel.

Computer Cases

Many chapters close with computer cases that contain problem scenarios accompanied by modest-sized data sets. Computer solution by Minitab, Excel, The Data Analyst, or another computer package with statistical capabilities is required. Each case outlines a managerial report that the student prepares to summarize statistical results as well as present interpretations and recommendations.

Data Disk

Data sets for text examples, exercises, and computer cases are available on a special Data Disk that can be ordered with the text. The Data Disk is available in a format acceptable to Minitab, a format acceptable to Excel, and in a format acceptable to The Data Analyst.

Illustration of Quality Applications Throughout the Text

The application of statistics for the improvement of quality is demonstrated throughout the text. In most chapters we have included examples and exercises that focus on a wide range of applications that demonstrate ways in which statistics contributes to improving the quality of products and services.

Acknowledgments

We owe a debt to many of our colleagues and friends for their helpful comments and suggestions in the development of this text. The advice that we received from Michael Cicero, Virginia Fisher, Frank Kelly and R. B. Trombley helped us a great deal in making decisions regarding the content of the text and the level of detail to be included. We would also like to recognize the following individuals who have helped us in the past and continue to influence our writing.

Mohammad Ahmadi
Cheryl Asher
Robert Balough
Abdul Basti
Harry Benham
Michael Bernkopf
John Bryant
Peter Bryant
Terri L. Byczkowski
Robert Cochran
Robert Collins
John Cooke
David W. Cravens
George Dery
Gopal Dorai
Edward Fagerlund
Nicholas Farnum
Sharon Fitzgibbons
Jerome Geaun
Paul Guy
Jamshid C. Hasseini
C. Thomas Innis
Ben Isselhardt
Jeffrey Jarrett
Robert Bruce Jones
David Krueger
Martin S. Levy
Thomas McCullough
Bette Midgarden
Glenn Milligan
David Muse
Roger Myerson
Richard O'Connell
Al Palachek
Diane Petersen-Salameh
Tom Pray
Ruby Ramirez
Tom Ryan
James R. Schwenke
Bill Seaver
Alan Smith
Stephen Smith
Suzanne Smith
William E. Stein
Willban Terpening
Ted Tsukahara
Hiroki Tsurumi
Victor Ukpolo
Ebenge Usip
Cindy van Es
Andrew Welki
Ari Wijetunga
Donald Williams
Roy Williams
J.E. Willis
Mustafa Yilmaz
Gary Yoshimoto
Charles Zimmerman
Greg Zimmerman

A special thanks is owed to our associates from business and industry who supplied the Statistics in Practice features. We recognize them individually by a credit line on each Statistics in Practice article. In addition, we acknowledge the cooperation of Minitab. Inc., in permitting the use of Minitab statistical software in this text. Finally, we are also indebted to our editor, Mary C. Schiller, our production editor Amy C. Hanson, and others at West Publishing Company for their editorial counsel and support during the preparation of this text.

David R. Anderson
Dennis J. Sweeney
Thomas A. Williams
August 1996

An Overview of Statistics in Practice Features

Chapter Number	Chapter Title	Organization Featured	Application Topic
1	Data and Statistics	Business Week	Use of Statistics in Business Week
2	Descriptive Statistics I: Tabular and Graphical Methods	Colgate-Palmolive Company	Quality Assurance for Heavy Duty Detergents
3	Descriptive Statistics II: Numerical Methods	Barnes Hospital	Time Spent in Hospice Program
4	Introduction to Probability	Morton International	Evaluation of Customer Service Testing Program
5	Discrete Probability Distributions	Xerox Corporation	Performance Test of an On-line Computerized Publication System
6	Continuous Probability Distributions	Procter & Gamble	Manufacturing Strategy
7	Sampling and Sampling Distributions	Mead Corporation	Estimating the Value of Mead Forest Ownership
8	Interval Estimation	Dollar General Corporation	Sampling and Estimation of LIFO Costs
9	Hypothesis Testing	Harris Corporation	Testing for Defective Plating
10	Comparisons Involving Means	Fisons Corporation	Evaluation of New Drugs
11	Comparisons Involving Proportions	United Way	Determining Community Perceptions of Charities
12	Regression Analysis	Polaroid Corporation	Study of Film Aging
13	Statistical Methods for Quality Control	Dow Chemical U.S.A.	Statistical Process Control

CHAPTER 1

Data and Statistics

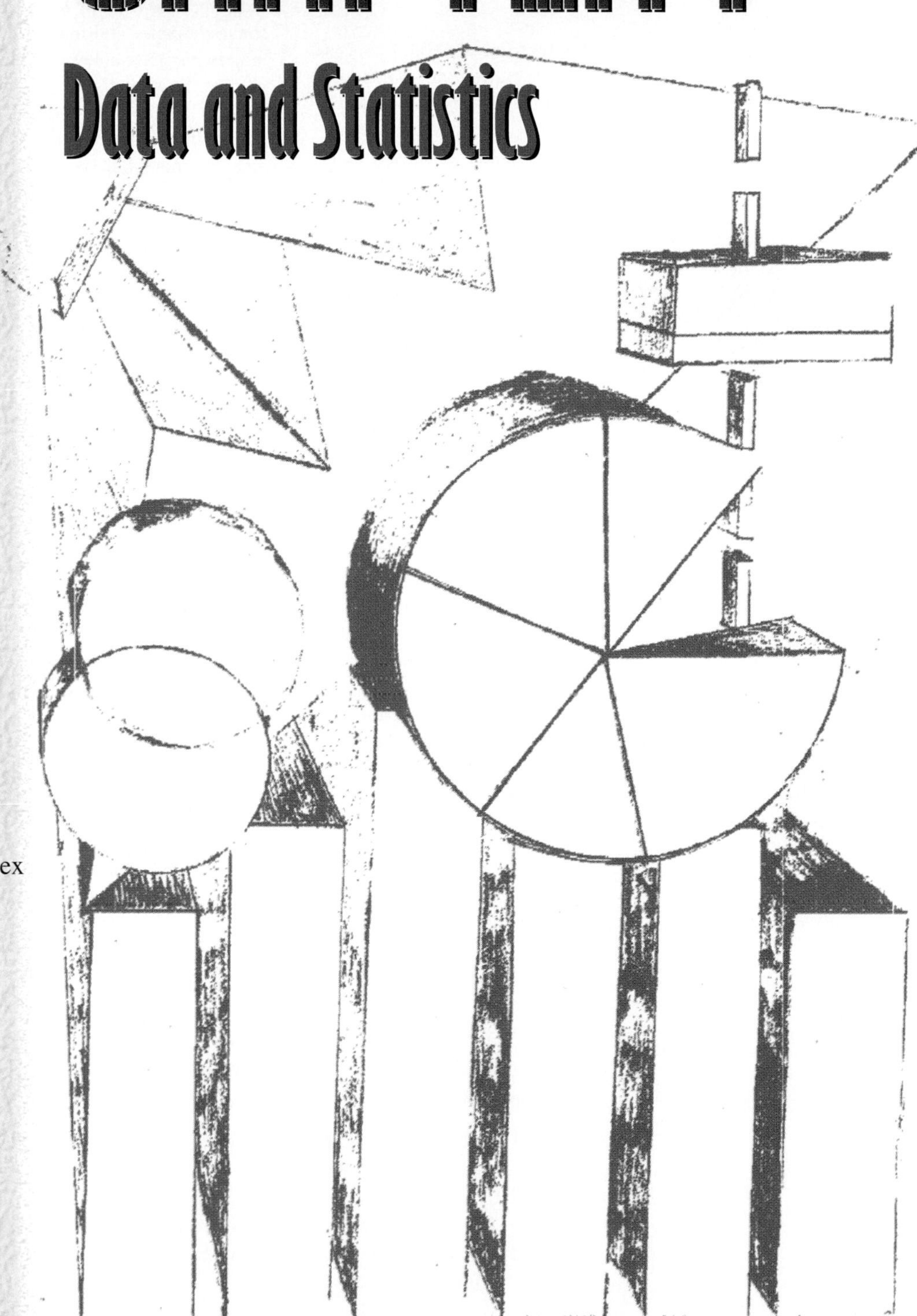

Business Week*

New York, New York

Business Week, published weekly by McGraw-Hill, Inc., is a well-known magazine that offers a variety of articles of interest to the business and economics community. In addition to feature articles on current topics, the magazine includes regular sections on International Business, Economic Analysis, Information Processing, and Science & Technology. The feature articles and regular sections help readers keep abreast of current developments and assess the impact of the developments on the business and economic future.

Most issues of *Business Week* provide an in-depth report on a topic of current interest. For instance, the September 5, 1994 issue contained a special report on database marketing, the October 24, 1994 issue contained a study on the best college and university business schools, and the March 27, 1995 issue contained data on America's most valuable companies and how they rank in terms of market value. Other features that attract reader interest are the annual executive compensation survey and the weekly *Business Week* Index that contains statistics on the state of the economy such as production indexes, stock prices, real estate loan values, and interest rates.

In addition to using statistics and statistical information in its magazine articles, *Business Week* collects and uses statistics to help manage its own business. For example, *Business Week* conducts a survey of its subscribers to learn about their personal profiles, reading habits, shopping practices, lifestyles, and so on. Managers use statistical summaries generated from the survey to provide better service to subscribers and advertisers. For instance, the 1993 U.S. Subscriber Study indicated that 87% of *Business Week* subscribers own a personal computer and that 44% of the subscribers were planning to purchase a new personal computer within the next 12 months. Such statistics alert managers to the fact that subscribers should be interested in articles about personal computers. In addition, the results of the survey are made available to potential advertisers. The high percentage of subscribers indicating an intention to purchase a new personal computer within the next 12 months would be an incentive for personal computer manufacturers to consider advertising in *Business Week.*

In this chapter, we discuss the types of data available for statistical analysis and how the data are obtained. We then introduce descriptive statistics and statistical inference as two ways of converting data into meaningful and easily interpreted statistical information.

Charlene Trentham, Research Manager at Business Week, provided this Statistics in Practice.

Business Week uses business and economic statistics in many of its articles.

Frequently, we see the following kinds of statements in newspaper and magazine articles:

- Sales of new homes are occurring at a rate of 703,000 homes per year (*Business Week,* October 1994).
- Crude oil is averaging $17.37 per barrel (*British Petroleum Annual Report,* 1993).
- The unemployment rate has dropped to 5.9% (*Barron's,* October 10, 1994).
- The Dow Jones Industrial Average closed at 4510.79 (*The Wall Street Journal,* June 19, 1995).

- Stock funds account for 29% of investors' portfolios (*The American Association of Individual Investors Journal,* July 1993).
- Airlines have raised the round-trip discount fare to an average of $270 per trip (*USA Today,* March 30, 1995).

The numerical facts in the preceding statements (703,000, $17.37, 5.9%, 4510.79, 29%, and $270) are called statistics. Thus, in everyday usage, the term "statistics" refers to numerical facts. However, the field or subject of statistics involves much more than numerical facts. In a broad sense, *statistics* is the art and science of collecting, analyzing, presenting, and interpreting data. Particularly in business and economics, a major reason for collecting, analyzing, presenting, and interpreting data is to give managers and decision makers a better understanding of the business and economic environment and thus enable them to make more informed and better decisions. In this text, we emphasize the use of statistics for business and economic decision making.

Chapter 1 begins with some illustrations of the application of statistics in business and economics. A discussion of how data, the raw material of statistics, are acquired and used follows. The use of data in developing descriptive statistics and in making statistical inferences is described in Sections 1.4 and 1.5.

1.1 Applications in Business and Economics

In today's global business and economic environment, vast amounts of statistical information are available. The most successful managers and decision makers are the ones who can understand the information and use it effectively. In this section, we provide examples that illustrate some of the uses of statistics for business and economics.

Accounting

Public accounting firms use statistical sampling procedures when conducting audits for their clients. For instance, suppose an accounting firm wants to determine whether the amount of accounts receivable shown on a client's balance sheet fairly represents the actual amount of accounts receivable. Usually the number of individual accounts receivable is so large that reviewing and validating every account would be too time-consuming and expensive. In such situations, it is common practice for the audit staff to select a subset of the accounts called a sample. After reviewing the accuracy of the sampled accounts, the auditors draw a conclusion as to whether the accounts receivable amount shown on the client's balance sheet is acceptable.

Finance

Financial advisors use a variety of statistical information to guide their investment recommendations. In the case of stocks, the advisors review a variety of financial data including price–earnings ratios and dividend yields. By comparing the information for an individual stock with information about the stock market averages, a financial advisor can begin to draw a conclusion as to whether an individual stock is over- or undervalued. For example, *Barron's* (October 10, 1994) reported that the average price–earnings ratio for the 30 stocks in the Dow Jones Industrial Average was 20.1. On the same day, Philip Morris had a price–earnings ratio of 14. In this case, the statistical information on price–earnings ratios showed that Philip Morris had a lower price in comparison to its earnings than the average for the Dow Jones stocks. A financial advisor therefore might have concluded that Philip Morris was currently underpriced. This and other information about Philip Morris would help the advisor make buy, sell, or hold recommendations for the stock.

Marketing

Electronic scanners at retail checkout counters are being used to collect data for a variety of marketing research applications. For example, data suppliers such as A. C. Nielsen and Information Resources, Inc., purchase point-of-sale scanner data from grocery stores, process the data, and then sell statistical summaries of the data to manufacturers. In 1992, manufacturers spent an average of $310,000 per product category to obtain this type of scanner data (Scanner Data User Survey, Mercer Management Consulting, Inc., August 1993). Manufacturers also purchase data and statistical summaries on promotional activities such as special pricing and the use of in-store displays. Product brand managers can review the scanner statistics and the promotional activity statistics to gain a better understanding of the relationship between promotional activities and sales. Such analyses are helpful in establishing future marketing strategies for the various products.

Production

With today's emphasis on quality, quality control is an important application of statistics in production. A variety of statistical quality control charts are used to monitor the output of a production process. In particular, an x-bar chart is used to monitor the average output. Suppose, for example, that a machine is being used to fill containers with 12 ounces of a well-known soft drink. Periodically, a sample of containers is selected and the average contents of the sample containers determined. This average, or x-bar value, is plotted on an x-bar chart. A plotted value above the chart's upper control limit indicates overfilling and a plotted value below the chart's lower control limit indicates underfilling. Thus, the x-bar chart shows when adjustments are necessary to correct the production process. The process is termed "in control" and allowed to continue as long as the plotted x-bar values are between the chart's upper and lower control limits.

Economics

Economists are frequently asked to provide forecasts about the future of the economy or some aspect of it. They use a variety of statistical information in making such forecasts. For instance, in forecasting inflation rates, economists use statistical information on such indicators as the Producer Price Index, the unemployment rate, and the manufacturing capacity utilization. Often these statistical indicators are entered into computerized forecasting models that predict inflation rates.

Applications of statistics such as those described in this section are an integral part of this text. Such examples provide an overview of the breadth of statistical applications. To supplement these examples, we have asked practitioners in the fields of business and economics to provide chapter-opening Statistics in Practice articles that serve to introduce the material covered in each chapter. The Statistics in Practice applications show the importance of statistics in a wide variety of decision-making situations.

1.2 Data

Data are the facts and figures that are collected, analyzed, and summarized for presentation and interpretation. Together, the data collected in a particular study are referred to as the *data set* for the study. Table 1.1 is a data set for 25 shadow stocks. The term "shadow" is used to indicate that the stocks are for small to medium-size firms that are not followed closely by the major brokerage houses. The data set in Table 1.1 was provided by the American Association of Individual Investors (*AAII Journal,* April 1994).

Elements, Variables, and Observations

The *elements* are the entities on which data are collected. For the data set in Table 1.1, each individual stock is an element. With 25 stocks, there are 25 elements in the data set.

A *variable* is a characteristic of interest for the elements. The data set in Table 1.1 has the following five variables:

TABLE 1.1 A Data Set for 25 Shadow Stocks

Stock	Exchange	Ticker Symbol	Annual Sales ($ Million)	Earnings Per Share ($)	Price-Earnings Ratio
Alcide Corp.	OTC	ALCD	7.4	.52	22.1
ARX Inc.	NYSE	ARX	54.7	.32	14.1
Bowmar	AMEX	BOM	20.7	.10	32.5
Cache Inc.	OTC	CACH	86.6	.25	28.5
CCA Industries	OTC	CCAM	44.3	.32	25.4
Concord Fabrics	AMEX	CIS	197.3	1.54	6.2
DBA Systems	OTC	DBAS	30.6	.38	16.1
Diodes Inc.	AMEX	DIO	26.4	.34	27.6
Gen Magnaplate	OTC	GMCC	10.4	.48	11.7
Harlyn Products	AMEX	HRN	32.0	.40	11.3
Kimmins Environmental	NYSE	KVN	81.9	.21	13.1
Koss Corp.	OTC	KOSS	36.1	.89	14.6
MagneTech Corp.	OTC	MTCC	22.7	.21	28.0
Media Logic Inc.	AMEX	TST	21.2	.79	5.1
Max & Erma's	OTC	MAXE	43.5	.38	23.7
MHI Group Inc.	NYSE	MH	16.7	.95	8.4
Par Technology	NYSE	PTC	81.2	.32	25.0
Penril DataComm	OTC	PNRL	69.6	.26	25.0
Reflectone Inc.	OTC	RFTN	63.1	.86	12.9
Scientific Tech.	OTC	STIZ	17.3	.46	28.8
Stage II Apparel	AMEX	SA	64.1	.23	21.2
Trans Leasing	OTC	TLII	25.5	.13	27.9
Uni-Marts Inc.	AMEX	UNI	336.9	.48	13.0
Western Beef	OTC	BEEF	273.7	.78	12.2
Zygo Corp.	OTC	ZIGO	23.5	.27	28.7

Source: *American Association of Individual Investors Journal,* April 1994.

- *Exchange:* Where the stock is traded—NYSE (New York Stock Exchange), AMEX (American Stock Exchange), and OTC (over the counter).
- *Ticker Symbol:* The abbreviation used to identify the stock on the exchange listing.
- *Annual Sales:* Total sales for the most recent 12 months in millions of dollars.
- *Earnings Per Share:* Net income for the most recent 12 months divided by the number of common shares outstanding.
- *Price-Earnings Ratio:* Market price per share divided by the most recent 12 months' earnings per share.

Data are obtained by collecting measurements on each variable for every element in the study. The set of measurements collected for a particular element is called an *observation.* Referring to Table 1.1, we see that the first element (Alcide Corp.) provides the observation OTC, ALCD, 7.4, .52, and 22.1. With 25 elements, there are 25 observations; with five variables for each observation, there are 125 data values in the data set.

Qualitative and Quantitative Data

The statistical analysis that is appropriate for a particular variable depends on whether the data for the variable are qualitative or quantitative. *Qualitative data* are labels or names used to identify an attribute of each element. For example, referring to the shadow stocks in Table 1.1, we see that the data values for the exchange variable (NYSE, AMEX, and OTC) are labels used to identify where the stocks are traded. Hence, the data are qualitative and exchange is referred to as a qualitative variable. Ticker symbol is also a qualitative variable, with the data values ALCD, ARX, BOM, and so on being the labels used to identify the corresponding stock.

Quantitative data indicate either how much or how many. For example, the data values for annual sales in Table 1.1 are quantitative, with the values 7.4, 54.7, 20.7, and so on indicating how many millions of dollars of sales the company had during the most recent 12-month period. Since the data are quantitative, annual sales is referred to as a quantitative variable. Earnings per share and price–earnings ratio are also quantitative variables.

Quantitative data are always numeric, but qualitative data may be *either numeric or nonnumeric.* For example, to facilitate data collection and prepare the data for easy entry into a computer database, we might decide to use numeric codes for exchange, letting 1 denote a stock traded on the New York Stock exchange, 2 denote a stock traded on the American Stock exchange, and 3 denote a stock traded over the counter. In this case, the numeric values are the labels or codes used to identify where the stock is traded. The data are qualitative and exchange is a qualitative variable even though the data are shown as numeric values. Social security numbers such as 310-22-7924 consist of numeric values. However, social security numbers are qualitative because the data are actually labels that identify particular individuals. Automobile license plate numbers such as CWX802 and product part codes such as A132 are qualitative data, with the combined numeric and nonnumeric entries being the labels used to identify the particular automobile or particular part.

For purposes of statistical analysis, the important distinguishing difference between qualitative and quantitative data is that *ordinary arithmetic operations are meaningful only with quantitative data.* For example, with quantitative data, the data values can be added together and then divided by the number of data values to compute the average value for the data. This average is meaningful and usually easily interpreted. However, when qualitative data are recorded as numeric values, such arithmetic operations provide meaningless results.

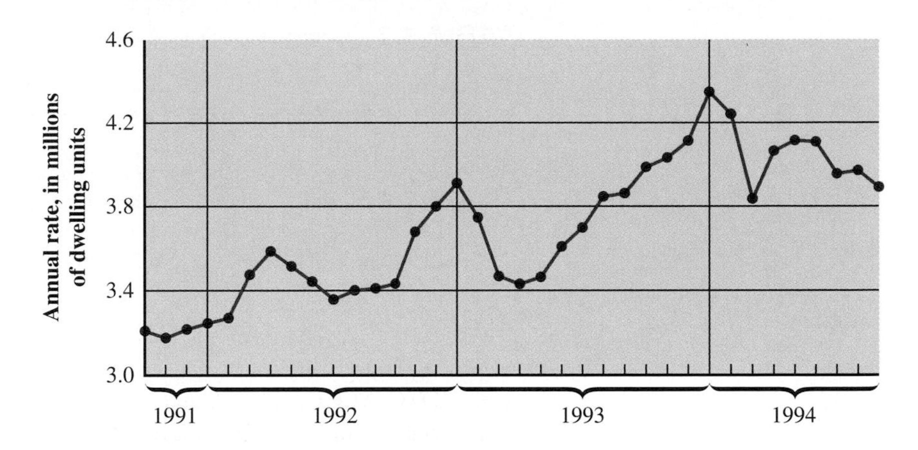

FIGURE 1.1
Time Series Data on Existing Single-Family Home Sales, September 1991 through August 1994 Source: *The Wall Street Journal,* September 2, 1994.

Cross-Sectional and Time Series Data

Cross-sectional data are data collected at the same or approximately the same point in time. The data in Table 1.1 are cross-sectional because they describe the five variables for the 25 shadow stocks at the same point in time, April 1994. *Time series data* are data collected over several time periods. For example, Figure 1.1 is a graph of data for existing single-family home sales over the three-year period from September 1991 to August 1994. There are 36 monthly data values for this time series.

notes and comments

An observation is the set of measurements obtained for each element in a data set. Hence, the number of observations is always the same as the number of elements. The number of measurements obtained on each element is equal to the number of variables. Hence, the total number of data values in a data set is the number of elements multiplied by the number of variables.

1.3 Data Sources

Data can be collected from existing sources or obtained through surveys and experimental studies designed to obtain new data.

Existing Sources

In some cases, data needed for a particular application may already exist within a firm or organization. All companies maintain a variety of databases about their employees, customers, and business operations. Data on employee salaries, ages, and years of service can usually be obtained from internal personnel records. Data on sales, advertising expenditures, distribution costs, inventory levels, and production quantities are generally available from other internal record-keeping systems. Many companies

TABLE 1.2 Examples of Data Available from Internal Company Records

Source	Some of the Data Typically Available
Employee records	Name, address, social security number, salary, number of vacation days, number of sick days, and bonus
Production records	Part or product number, quantity produced, direct labor cost, and materials cost
Inventory records	Part or product number, number of units on hand, reorder level, economic order quantity, and discount schedule
Sales records	Product number, sales volume, sales volume by region, and sales volume by customer type
Credit records	Customer name, address, phone number, credit limit, and accounts receivable balance
Customer profile	Age, gender, income level, household size, address, and preferences

TABLE 1.3 Examples of Data Available from Selected Government Agencies

Government Agency	Some of the Data Available
Bureau of the Census	Population data and its distribution, data on number of households and their distribution, data on household income and its distribution
Federal Reserve Board	Data on the money supply, installment credit, exchange rates, and discount rates
Office of Management and Budget	Data on revenue, expenditures, and debt of the federal government
Department of Commerce	Data on business activity—value of shipments by industry, level of profits by industry, and data on growing and declining industries

also maintain detailed data about their customers. Table 1.2 shows some of the data commonly available from the internal information sources of most companies.

Substantial amounts of business and economic data are available from organizations that specialize in collecting and maintaining data. Companies obtain access to these external data sources through leasing arrangements or by purchase. Dun & Bradstreet and Dow Jones & Company are two firms that provide extensive business database services to clients. A. C. Nielsen and Information Resources, Inc. have built $100-million businesses collecting and processing retail checkout-counter scanner data that they sell to product manufacturers.

Government agencies are another important source of existing data. For instance, the U.S. Department of Labor maintains considerable data on employment rates, wage rates, size of the labor force, and union membership. Table 1.3 lists other selected governmental agencies and some of the data they provide. Data are also available from a variety of industry associations and special-interest organizations. The Travel Industry Association of America maintains travel-related information such as the number of tourists and travel expenditures by state. Such data would be of interest to firms and individuals in the travel industry. The Graduate Management Admission Council maintains data on student characteristics and graduate management education programs. Most of the data from sources such as these are available to qualified users at a modest cost.

Statistical Studies

Sometimes data are not readily available from existing sources. If the data are considered necessary, a statistical study can be conducted to obtain them. Such statistical studies can be classified as either *experimental* or *observational.*

In an experimental study, the variables of interest are first identified. Then one or more factors in the study are controlled so that data can be obtained about how the factors influence the variables. For example, a pharmaceutical firm might be interested in conducting an experiment to learn about how a new drug affects blood pressure. Blood pressure is the variable of interest in the study. The new drug is the factor that influences the blood pressure. To obtain data about the effect of the new drug, a sample of individuals will be selected. The dosage level of the new drug will be controlled, with different groups of individuals being given different dosage levels. Data on blood pressure will be collected for each group. Statistical analysis of the experimental data will help determine how the new drug affects blood pressure.

In nonexperimental, or observational, statistical studies, no attempt is made to control or influence the variables of interest. A survey is perhaps the most common type of observational study. For instance, in a personal interview survey, research questions are first identified. Then a questionnaire is designed and administered to a sample of individuals. Data are obtained about the research variables but no attempt is made to control the factors that influence the variables. Some restaurants use observational studies to obtain data about their customers' opinions of the quality of food, service, atmosphere, and so on. A questionnaire used by the Lobster Pot Restaurant in Redington Shores, Florida, is shown in Figure 1.2. Note that the customers completing the questionnaire are asked to provide ratings for six variables: food, drinks, service, waiter,

FIGURE 1.2
Customer Opinion Questionnaire Used by the Lobster Pot Restaurant, Redington Shores, Florida

The LOBSTER Pot
RESTAURANT

	Excellent	Good	Poor	Comments
Food				
Drinks				
Service				
Waiter				
Captain				
Hostess				

Waiter's Name ______________________

Captain's Name ______________________

Other Comments ______________________

captain, and hostess. The response categories of excellent, good, and poor provide data that enable Lobster Pot's managers to assess the quality of the restaurant's operation.

Managers wanting to use data and statistical analyses as aids to decision making must be aware of the time and cost required to obtain the data. The use of existing data sources is desirable when data must be obtained in a relatively short period of time. If the data are not readily available from an existing source, the additional time and cost involved in obtaining the data must be taken into account. In all cases, the decision maker should consider the contribution of the statistical analysis to the decision-making process. The cost of data acquisition and the subsequent statistical analysis should not exceed the savings generated by using the information to make a better decision.

Data-Acquisition Errors

Managers should always be aware of the possibility of data errors in statistical studies. Using erroneous data could be worse than not using the data and statistical information at all. An error in data acquisition occurs whenever the data value obtained is not equal to the true or actual value that would have been obtained with a correct procedure. Such errors can occur in a number of ways. For example, an interviewer might make a recording error, such as writing the age of a 24-year-old person as 42, or the person answering an interview question might misinterpret the question and make an incorrect response.

Experienced data analysts take great care in collecting and recording data to ensure that errors are not made. Special procedures can be used to check for internal consistency of the data. For instance, such procedures would indicate that the analyst should review data for a respondent who is shown to be 22 years of age but who reports 20 years of work experience. Data analysts also review data for unusually large and small values, called outliers, which are candidates for possible data errors. In Chapter 3, we present some of the methods statisticians use to identify outliers.

The point of this discussion is to alert users of data to the fact that errors can occur during data acquisition. Blindly using any data that happen to be available or using data that were acquired with little care can lead to poor and misleading information. However, taking steps to acquire accurate data can help ensure reliable and valuable decision-making information.

1.4 Descriptive Statistics

Most of the statistical information in newspapers, magazines, reports, and other publications consists of data that are summarized and presented in a form that is easy for the reader to understand. Such summaries of data, which may be tabular, graphical, or numerical, are referred to as *descriptive statistics.*

Tabular and Graphical Approaches

Refer again to the data set in Table 1.1 where 25 shadow stocks are listed. Methods of descriptive statistics can be used to provide summaries of the information in this data set. For example, a tabular summary of the data for the exchange variable is shown in Table 1.4. A graphical summary of the same data is provided by the bar graph in Figure 1.3. The purpose of tabular and graphical summaries such as these is to make the data easier to interpret. Referring to Table 1.4 and Figure 1.3, we can see easily that the majority of the stocks in the data set are traded over the counter. On a percentage basis,

TABLE 1.4 Frequencies and Percentages for the Exchange of 25 Shadow Stocks

Exchange	Frequency	Percent
New York Stock Exchange	4	16
American Stock Exchange	7	28
Over the counter	14	56
Total	25	100

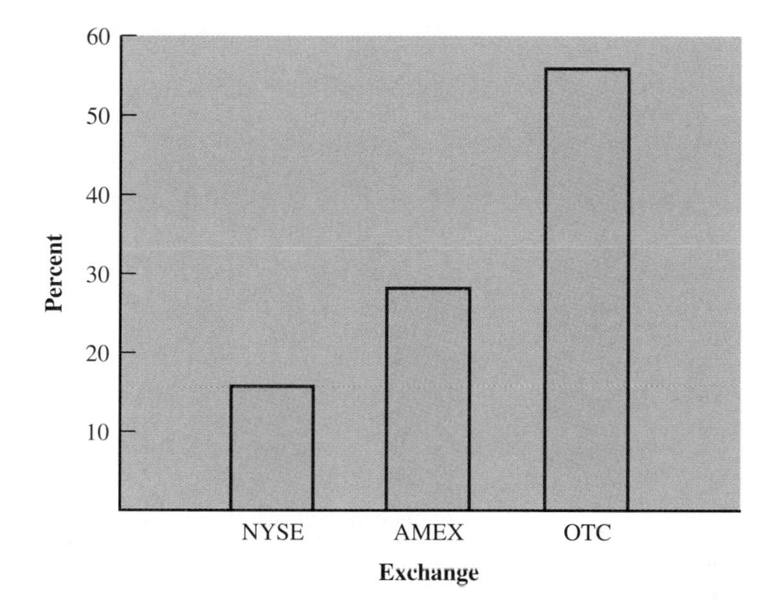

FIGURE 1.3
Bar Graph of Exchange for Shadow Stocks

56% of the stocks are traded over the counter, 28% are traded on the American Stock Exchange, and only 16% are traded on the New York Stock Exchange.

A graphical summary of the data on price–earnings ratio for the stocks in Table 1.1 is provided by the histogram in Figure 1.4. From the histogram, it is easy to see that the price–earnings ratios range from 5 to 35, with the highest concentrations being between 10 and 15 and between 25 and 30.

Numerical Measures and Index Numbers

In addition to tabular and graphical displays, numerical descriptive statistics are used to summarize data. The most common numerical descriptive statistic is the *average* or *mean.* Using the data on annual sales in Table 1.1, we can compute the average annual sales by adding the annual sales for all 25 stocks and dividing the sum by 25. Doing so tells us that the average annual sales for the stocks is $67.5 million. This average is taken as a measure of the central value, or central location, of the data.

Index numbers are numerical descriptive statistics that are widely used in business. For example, each month the U.S. government publishes a variety of indexes designed to help individuals better understand current business and economic conditions. Perhaps the most widely known and cited of these indexes is the Consumer Price Index (CPI). This index, published monthly by the U.S. Bureau of Labor Statistics, is the primary

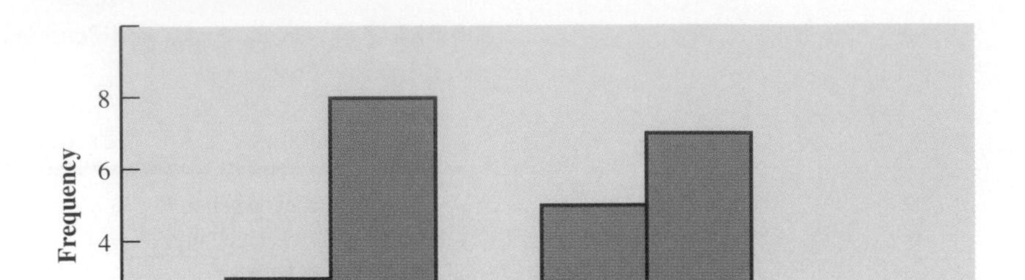

FIGURE 1.4
Price–Earnings Ratio for 25 Shadow Stocks in Table 1.1

measure of the cost of living in the United States. Specifically, the CPI measures changes in price over a period of time. With a given starting point or base period and its associated index of 100, the CPI can be used to compare current period consumer prices with those in the base period. For example, a CPI of 125 reflects the condition that consumer prices as a whole are running approximately 25% above the base period prices.

The Dow Jones averages are indexes that are designed to show price trends and movements on the New York Stock Exchange. The best known of the Dow Jones indexes, referred to as the Dow Jones Industrial Average (DJIA), is a weighted average of the common stock prices of 30 industrial stocks. Other Dow Jones averages are computed for 20 transportation stocks and 15 utility stocks; in addition, a composite average for the 65 industrial, transportation, and utility stocks that make up the three preceding indexes is also provided. The Dow Jones averages are computed and published daily in *The Wall Street Journal* and other financial publications.

Table 1.5 lists the 30 stocks used to compute the DJIA along with their closing prices on November 13, 1995. Although closing prices are listed in *The Wall Street Journal* and other financial publications in fractional form, we have reported the prices in Table 1.5 in their equivalent decimal form; for example, the closing price for American Express, listed as 43⅛ in *The Wall Street Journal,* is shown as 43.125. The DJIA is the sum of the closing prices divided by a number called the divisor. In the Money & Investing section of the November 13, 1995 issue of *The Wall Street Journal,* the divisor was reported to be 0.34599543. The sum of the closing prices is 1684.750. Thus, the DJIA is 1684.750 ÷ 0.34599543 = 4869.28. Whenever the list of 30 companies that make up the list of 30 industrial stocks changes, or whenever a stock split occurs for one of the companies, the divisor is revised to account for these changes. In this way, the DJIA is able to remain an effective measure of overall stock market performance.

notes and comments

Using the closing prices in Table 1.5 we obtained a value for the DJIA of 4869.28. However, the DJIA reported in the November 13, 1995 issue of *The Wall Street Journal* is 4870.37. The slight difference between these two values is due to the fact that the closing prices reported in Table 1.5 are based upon quotations as of 5 P.M. Eastern Time, whereas the value of 4870.37 reported in *The Wall Street Journal* is based upon the prices at the close of trading on the New York Stock Exchange (4 P.M.) In most cases, the effect of after-hour trading on the DJIA is minimal.

TABLE 1.5 Closing Prices on November 13, 1995 for the 30 Stocks Used to Compute the Dow Jones Industrial Average

Company	Closing Price
AT&T	63.000
Allied Signal	44.625
Alcoa	54.125
American Express	43.125
Bethlehem Steel	12.750
Boeing	71.250
Caterpillar	56.625
Chevron	47.625
Coca-Cola	72.750
Disney	59.375
DuPont	61.875
Eastman Kodak	66.375
Exxon	75.500
General Electric	65.625
General Motors	45.625
Goodyear	41.375
IBM	97.375
International Paper	35.250
McDonald's	42.375
Merck	58.250
Minnesota Mining and Manufacturing	59.750
J.P. Morgan	77.125
Philip Morris	87.000
Procter & Gamble	83.000
Sears	38.625
Texaco	67.750
Union Carbide	37.125
United Technologies	90.250
Westinghouse	14.875
Woolworth	14.375

1.5 Statistical Inference

In many situations, statistics are sought for a large group of elements (individuals, stocks, voters, households, products, customers, and so on). Because of time, cost, and other considerations, data are collected from only a small portion of the group. The larger group of elements in a particular study is called the *population* and the smaller group is called the *sample*. Formally, we use the following definitions.

Population

A population is the set of all elements of interest in a particular study.

Sample

A sample is a subset of the population.

TABLE 1.6 Hours Until Burnout for a Sample of 200 Lightbulbs for Norris Electronics

107	73	68	97	76	79	94	59	98	57
54	65	71	70	84	88	62	61	79	98
66	62	79	86	68	74	61	82	65	98
62	116	65	88	64	79	78	79	77	86
74	85	73	80	68	78	89	72	58	69
92	78	88	77	103	88	63	68	88	81
75	90	62	89	71	71	74	70	74	70
65	81	75	62	94	71	85	84	83	63
81	62	79	83	93	61	65	62	92	65
83	70	70	81	77	72	84	67	59	58
78	66	66	94	77	63	66	75	68	76
90	78	71	101	78	43	59	67	61	71
96	75	64	76	72	77	74	65	82	86
66	86	96	89	81	71	85	99	59	92
68	72	77	60	87	84	75	77	51	45
85	67	87	80	84	93	69	76	89	75
83	68	72	67	92	89	82	96	77	102
74	91	76	83	66	68	61	73	72	76
73	77	79	94	63	59	62	71	81	65
73	63	63	89	82	64	85	92	64	73

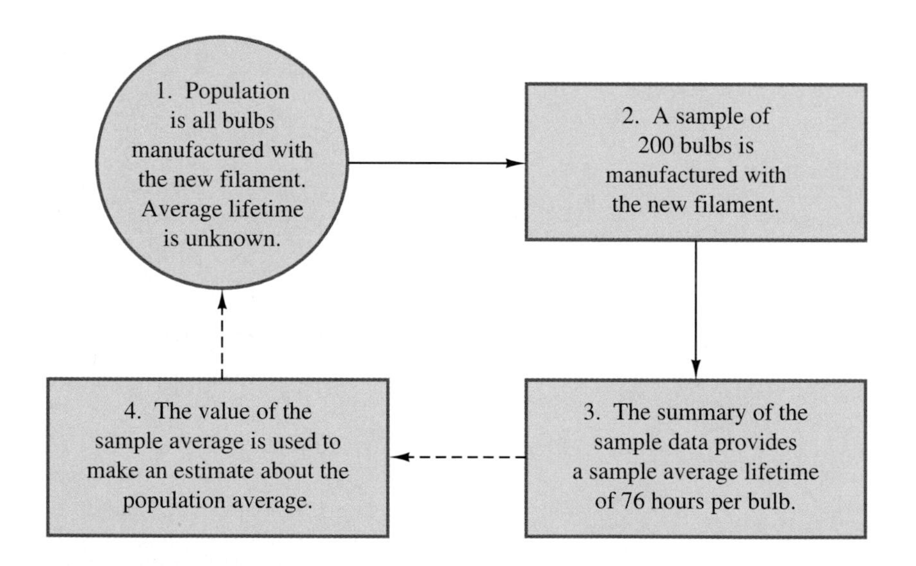

FIGURE 1.5
The Process of Statistical Inference for the Norris Electronics Example

A major contribution of statistics is that data from a sample can be used to make estimates and test hypotheses about the characteristics of a population. This process is referred to as *statistical inference.*

As an example of statistical inference, let us consider the study conducted by Norris Electronics. Norris manufactures a high-intensity lightbulb that is used in a variety of electrical products. In an attempt to increase the useful life of the lightbulb, the product design group has developed a new lightbulb filament. In this case, the population is defined as all lightbulbs that could be produced with the new filament. To evaluate the advantages of the new filament, a sample of 200 new-filament bulbs was manufactured and tested. Data were collected on the number of hours each lightbulb operated before filament burnout. The data from the sample are reported in Table 1.6.

Suppose Norris is interested in using the sample data to make an inference about the average hours of useful life for the population of all lightbulbs that could be produced with the new filament. Adding the 200 values in Table 1.6 and dividing the total by 200 provides the sample average lifetime for the lightbulbs: 76 hours. We can use this sample result to estimate that the average lifetime for the lightbulbs in the population is 76 hours. Figure 1.5 is a graphical summary of the statistical inference process for Norris Electronics.

Whenever statisticians use a sample to make an inference about a characteristic of a population, they provide a statement of the quality, or precision, associated with the inference. For the Norris example, the statistician might state that the estimate of the average lifetime for the population of new lightbulbs is 76 hours with a precision of ±4 hours. Thus, 72 hours to 80 hours is an interval estimate of the average lifetime. The statistician can also state how confident he or she is that the interval from 72 hours to 80 hours contains the population average.

summary

In everyday usage, the term "statistics" refers to numerical facts. However, the field of statistics requires a broader definition of statistics as the art and science of collecting, analyzing, presenting, and interpreting data. Nearly every college student majoring in business or economics is required to take a course in statistics. We began the chapter by describing typical statistical applications for business and economics.

Data are the facts and figures that are collected, analyzed, presented, and interpreted. For purposes of statistical analysis, data are classified as either qualitative or quantitative. Qualitative data consist of labels or names that are used to identify an attribute of an element. Qualitative data may be numeric or nonnumeric. Quantitative data are always numeric and indicate how much or how many for the variable of interest. Ordinary arithmetic operations are meaningful only if the data are *quantitative.* Therefore, statistical computations that are used for quantitative data are not always appropriate for qualitative data.

In Sections 1.4 and 1.5 we introduced the topics of descriptive statistics and statistical inference. Descriptive statistics are the tabular, graphical, and numerical methods used to summarize data. Statistical inference is the process of using data obtained from a sample to make estimates or test claims about the characteristics of a population.

glossary

Data Facts and figures that are collected, analyzed, and interpreted.

Data set All the data collected in a particular study.

Elements The entities on which data are collected.

Variable A characteristic of interest for the elements.

Observation The set of measurements obtained for a single element.

Qualitative data Data that provide labels or names for a characteristic of an element. Qualitative data may be nonnumeric or numeric.

Qualitative variable A variable with qualitative data.

Quantitative data Data that indicate how much or how many of something. Quantitative data are always numeric.

Quantitative variable A variable with quantitative data.

Cross-sectional data Data collected at the same or approximately the same point in time.

Time series data Data collected at several successive points of time.

Descriptive statistics Tabular, graphical, and numerical methods used to summarize data.

Index numbers Numerical descriptive statistics such as the Consumer Price Index (CPI) and the Dow Jones Industrial Average (DJIA).

Population The set of all elements of interest in a particular study.

Sample A subset of the population.

Statistical inference The process of using data obtained from a sample to make estimates or test hypotheses about the characteristics of a population.

exercises

1. Discuss the differences between statistics as numerical facts and statistics as a discipline or field of study.

Self Test

2. The following table shows the chief executive officer (CEO) compensation, industry classification, annual sales, and the CEO compensation versus shareholder return rating data for 10 companies (*Business Week,* April 24, 1994). A CEO compensation versus shareholder return rating of 1 indicates that the company is among the group of companies that have the best ratio of CEO compensation to shareholder return. A rating of 2 indicates that the company is similar to companies that have a very good, but not the best, ratio of CEO compensation to shareholder return. Companies with the worst ratio of CEO compensation to shareholder return have a rating of 5.

Company	CEO Compensation ($1000s)	Industry	Sales ($ millions)	CEO Compensation vs. Shareholder Return Rating
Bankers Trust	8866	Banking	7800	5
Coca-Cola	3654	Beverages	13957	5
General Motors	1375	Automotive	138219	1
Intel	2184	Electronics	8782	3
Motorola	1736	Electronics	16963	4
Readers Digest	1708	Publishing	2821	4
Sears	3095	Retailing	50838	3
Sprint	1692	Telecomm.	11368	3
Walgreen	1145	Retailing	8498	2
Wells Fargo	2125	Banking	4854	2

a. How many elements are in this data set?
b. How many variables are in this data set?
c. Which variables are qualitative and which variables are quantitative?

Self Test

3. Refer to the data in Exercise 2 (*Business Week,* April 24, 1994).

a. Compute the average compensation for the chief executive officers (CEOs).

b. What percentage of the firms are in the banking industry?

c. What percentage of the firms have a value of 3 on the CEO compensation versus shareholder return rating?

Self Test

4. *Fortune* magazine provides data on how the 500 largest U.S. industrial corporations rank in terms of sales and profits. Data for a sample of *Fortune* 500 companies are given in the following table (*Fortune,* April 18, 1994).

Company	Sales ($ millions)	Profit ($ millions)	Industry Code
Coastal	10136	115.1	18
CPC Intl.	6738	454.5	8
Del Monte	1555	(188.0)	8
J. M. Huber	1238	29.7	7
Ivax	645	84.7	19
Northrup	5063	96.0	1
Sealy	683	25.7	10
Unisys	7743	565.4	6
Westvaco	2330	104.3	9
Wrigley	1429	174.9	8

a. How many elements are in this data set?

b. What is the population?

c. Compute the average sales for the sample.

d. Using the results in part (c), what is the estimate of the average sales for the population?

5. Consider the data set in Exercise 4 (*Fortune,* April 18, 1994).

a. How many variables are in the data set?

b. Which of the variables are qualitative and which are quantitative?

c. Compute the average profit for the companies.

d. What percentage of the companies earned a profit over $100 million?

e. What percentage of the companies have an industry code of 8?

6. Columbia House provides CDs, tapes, and records to its mail-order club members. A Columbia House Music Survey conducted in 1992 asked new club members to complete an 11-question survey. Some of the questions asked were:

a. How many albums (CDs, tapes, or records) have you bought in the last 12 months?

b. Are you currently a member of a national mail-order book club? (Yes or No)

c. What is your age?

d. Including yourself, how many people (adults and children) are in your household?

e. What kind of music are you interested in buying? (15 categories were listed, including hard rock, soft rock, adult contemporary, heavy metal, rap, and country.)

Comment on whether each question provides qualitative or quantitative data.

7. The Commerce Department reported that of 1994 applications for the Malcolm Baldrige National Quality Award, 23 were from large manufacturing firms, 18 were from large service firms, and 30 were from small businesses.

a. Is type of business a qualitative or quantitative variable?

b. What percentage of the applications came from small businesses?

8. The Hawaii Visitors Bureau collects data on visitors to Hawaii. The following questions were among 16 asked in a questionnaire handed out to passengers during incoming airline flights in August 1994.

- This trip to Hawaii is my: 1st, 2nd, 3rd, 4th, etc.
- The primary reason for this trip is: (10 categories including vacation, convention, honeymoon)

FIGURE 1.6
Initial Public Offerings Volume of New Equity Issues (in billions of dollars.) Source: *The Wall Street Journal,* October 19, 1994.

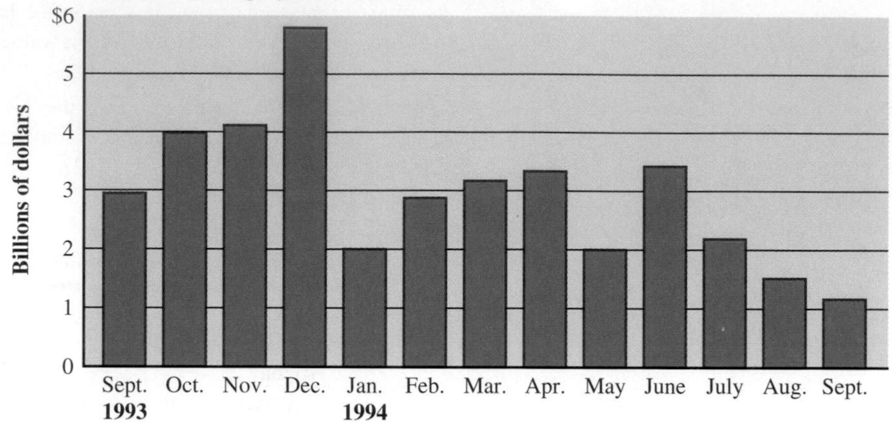

- Where I plan to stay: (11 categories including hotel, apartment, relatives, camping)
- Total days in Hawaii

a. What is the population being studied?
b. Is the use of a questionnaire a good way to reach the population of passengers on incoming airline flights?
c. Comment on each of the preceding four questions in terms of whether it will provide qualitative or quantitative data.

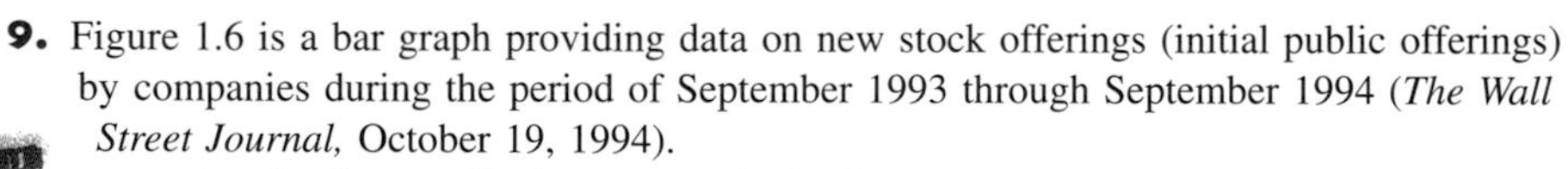

9. Figure 1.6 is a bar graph providing data on new stock offerings (initial public offerings) by companies during the period of September 1993 through September 1994 (*The Wall Street Journal,* October 19, 1994).
a. Are the data qualitative or quantitative?
b. Are they time series or cross-sectional data?
c. What is the variable of interest?
d. Comment on the trend in initial public offerings over time. Would you expect to see an increase or a decrease in October 1994?

10. The following data set provides a snapshot of the recent financial performance of Whirlpool Corporation (*Barrons,* September 26, 1994).

	1990	1991	1992	1993
Earnings Per Share	$1.04	$2.45	$2.90	$3.19
Revenues (billions)	$6.623	$6.770	$7.301	$7.533
Net Income (millions)	$72	$170	$205	$231
Book Value Per Share	$14.20	$8.71	$11.50	$12.31

a. How many variables are there?
b. Are the data qualitative or quantitative?
c. Are they cross-sectional or time series data? Why?

11. The marketing group at your company has come up with a new diet soft drink that it claims will capture a large share of the young adult market.

a. What data would you want to see before deciding to invest substantial funds in introducing the new product into the marketplace?

b. How would you expect the data mentioned in (a) to be obtained?

12. The following table shows the closing prices on May 16, 1995, for the 30 industrial stocks used to compute the Dow Jones industrial Average (DJIA). The divisor for the 30 industrial stocks was 0.36143882 (*The Wall Street Journal,* May 16, 1995).

a. Compute the Dow Jones Industrial Average.

b. Compare your answer to the value for the DJIA obtained using the data in Table 1.5.

Company	Closing Price
AT&T	50.875
Allied Signal	41.625
Alcoa	47.500
American Express	35.375
Bethlehem Steel	14.750
Boeing	54.875
Caterpillar	62.250
Chevron	48.625
Coca-Cola	57.875
Disney	55.125
DuPont	69.000
Eastman Kodak	60.375
Exxon	70.500
General Electric	57.625
General Motors	48.000
Goodyear	41.750
IBM	95.000
International Paper	80.875
McDonald's	37.000
Merck	42.125
Minnesota Mining and Manufacturing	61.375
J.P. Morgan	68.500
Philip Morris	70.875
Procter & Gamble	69.375
Sears	56.625
Texaco	68.500
Union Carbide	31.875
United Technologies	75.000
Westinghouse	14.750
Woolworth	15.625

13. The *Business Week* 1993 U.S. Subscriber Study collected data from a sample of 1597 subscribers. Sixty-six percent of the respondents indicated that their annual income was $50,000 or more and 51% reported having an American Express credit card.

a. What is the population of interest in this study?

b. Is annual income a qualitative or quantitative variable?

c. Is ownership of an American Express card a qualitative or quantitative variable?

d. Does this study involve cross-sectional or time series data?

e. Describe any statistical inferences *Business Week* might make on the basis of the survey.

14. The 1992 Scanner Data User Survey of 52 companies provided the following findings (Mercer Management Consulting, Inc., August 5, 1993):

- 56% percent of users are food manufacturers.
- 12% percent of users are manufacturers of health and beauty aids.
- On a scale of 1 (very dissatisfied) to 5 (very satisfied), the average level of overall satisfaction with scanner data was 3.6.

a. Cite two descriptive statistics.
b. Make an inference of the overall satisfaction in the population of all users of scanner data.
c. Make an inference about the percentage of the population of all scanner data users that are manufacturers of health and beauty aids.

15. A firm is interested in testing the advertising effectiveness of a new television commercial. As part of the test, the commercial is shown on a 6:30 P.M. local news program in Denver, Colorado. Two days later a market research firm conducts a telephone survey to obtain information on recall rates (percentage of viewers who recall seeing the commercial) and impressions of the commercial.
a. What is the population for this study?
b. What is the sample for this study?
c. Why would a sample be used in this situation? Explain.

16. The Nielsen organization conducts weekly surveys of television viewing throughout the United States. The Nielsen statistical ratings indicate the size of the viewing audience for each major network television program. Rankings of the television programs and of the viewing-audience market shares for each network are published each week.
a. What is the Nielsen organization attempting to measure?
b. What is the population?
c. Why would a sample be used for this situation?
d. What kinds of decisions or actions are based on the Nielsen studies?

17. A sample of midterm grades for five students showed the following results: 72, 65, 82, 90, 76. Which of the following statements are correct, and which should be challenged as being too generalized?
a. The average midterm grade for the sample of five students is 77.
b. The average midterm grade for all students who took the exam is 77.
c. An estimate of the average midterm grade for all students who took the exam is 77.
d. More than half of the students who take this exam will score between 70 and 85.
e. If five other students are included in the sample, their grades will be between 65 and 90.

CHAPTER 2

Descriptive Statistics I: Tabular and Graphical Methods

Colgate-Palmolive Company*

New York, New York

The Colgate-Palmolive Company started as a small soap and candle shop in New York City in 1806. Today, Colgate-Palmolive products can be found around the world. International operations exist in 70 countries, and annual sales are in excess of $7 billion. While best known for its traditional line of soaps, detergents, and toothpastes, subsidiary operations include Mennen, Softsoap, Hills Pet Foods, and others.

The Colgate-Palmolive Company uses statistics in its quality assurance program for home laundry detergent products. One concern is customer satisfaction with the quantity of detergent in a carton. Every carton in each size category is filled with the same amount of detergent by weight, but the volume of detergent is affected by the density of the detergent powder. For instance, if the powder density is on the heavy side, a smaller volume of detergent is needed to reach the carton's specified weight. As a result, the carton may appear to be underfilled when opened by the consumer.

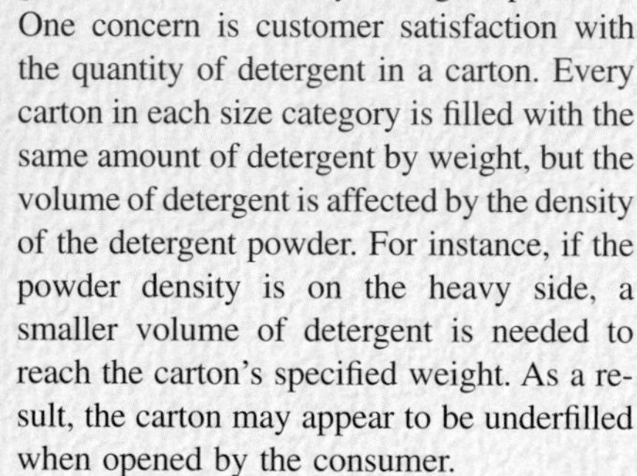

To control the problem of heavy detergent powder, limits are placed on the acceptable range of powder density. Statistical samples are taken periodically and the density of each powder sample is measured. Data summaries are then provided for operating personnel so that corrective action can be taken if necessary to keep the density within the desired quality specifications.

A frequency distribution for the densities of 150 samples taken over a one-week period and a histogram are shown in the accompanying table and figure. Density levels above .40 are unacceptably high. The frequency distribution and histogram show that the operation is meeting its quality guidelines with all of the densities less than or equal to .40. Managers viewing these statistical summaries would be pleased with the quality of the detergent production process.

In this chapter, you will learn about tabular and graphical methods of descriptive statistics such as frequency distributions, bar graphs, histograms, stem-and-leaf displays, dot plots, crosstabulations, and others. The goal of these methods is to summarize data so that they can be easily understood and interpreted.

Frequency Distribution of Density Data

Density	Frequency
.29–.30	30
.31–.32	75
.33–.34	32
.35–.36	9
.37–.38	3
.39–.40	1
Total Samples	150

Histogram of Density Data

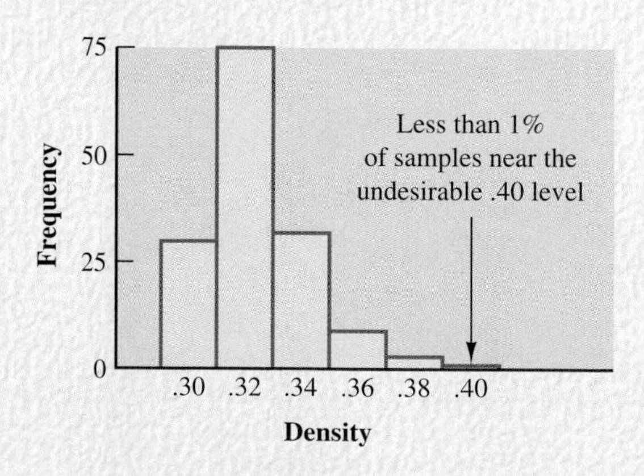

Managers use statistical summaries to help maintain the quality of Colgate's leading products.

William R. Fowle, Manager of Quality Assurance, Colgate-Palmolive Company, provided this Statistics in Practice.

As indicated in Chapter 1, data can be classified as either *qualitative* or *quantitative.* Qualitative data provide labels or names for categories of like items. Quantitative data indicate how much or how many.

The purpose of this chapter is to introduce several tabular and graphical procedures commonly used to summarize both qualitative and quantitative data. Tabular and graphical summaries of data can be found in annual reports, newspaper articles, and research studies. Everyone is exposed to these types of presentations. Hence, it is important to understand how they are prepared and to know how they should be interpreted. We begin with tabular and graphical methods for summarizing qualitative data. Methods for summarizing quantitative data are presented in Section 2.2. Methods for summarizing data involving the relationship between two variables are presented in Section 2.5.

2.1 Summarizing Qualitative Data

Frequency Distribution

We begin the discussion of how tabular and graphical methods can be used to summarize qualitative data with the definition of a *frequency distribution.*

Frequency Distribution

A frequency distribution is a tabular summary of data showing the number of data values in each of several nonoverlapping classes.

The objective in developing a frequency distribution is to provide insights about the data that cannot be quickly obtained by looking only at the original data. To see how frequency distributions can be used with qualitative data, consider the data set in Table 2.1.

TABLE 2.1 Data from a Sample of 50 Computer Purchases

IBM	IBM	Packard Bell
Compaq	IBM	Packard Bell
Apple	Apple	Gateway 2000
Packard Bell	Compaq	Compaq
Gateway 2000	Packard Bell	Compaq
IBM	IBM	Apple
Apple	Compaq	IBM
Packard Bell	Apple	Packard Bell
Packard Bell	Apple	Compaq
Gateway 2000	Compaq	Packard Bell
Compaq	Gateway 2000	Compaq
Compaq	Packard Bell	Packard Bell
Apple	IBM	Compaq
Compaq	IBM	Apple
Apple	Gateway 2000	Packard Bell
Apple	Apple	Apple
Apple	IBM	

What company sells the most personal computers? Dataquest, Inc. reported that, on the basis of third-quarter 1994 sales, Apple, Compaq, Gateway 2000, IBM, and Packard Bell were the top five personal computer vendors (*The Wall Street Journal,* November 15, 1994). Assume that the data in Table 2.1 are from a sample of 50 purchases of personal computers from these five companies. The data shown are qualitative in that each entry is the name of the company supplying the personal computer.

To develop a frequency distribution for these data, we count the number of times each of the five companies appears in the data set. Apple appears 13 times, Compaq appears 12 times, Gateway 2000 appears five times, IBM appears nine times, and Packard Bell appears 11 times. These counts are summarized in the frequency distribution in Table 2.2.

TABLE 2.2 Frequency Distribution of Computer Purchases

Company	Frequency
Apple	13
Compaq	12
Gateway 2000	5
IBM	9
Packard Bell	11
Total	50

The advantage of the frequency distribution is that it provides a data summary that is easier to understand than the original data as shown in Table 2.1. Using the frequency distribution, we see that Apple, Compaq, and Packard Bell are the very close first, second, and third choices for personal computers. IBM and Gateway 2000 lag behind in fourth and fifth places. The frequency distribution provides a data summary showing how the sample of 50 personal computer purchases are distributed across the five vendors.

Relative Frequency and Percent Frequency Distributions

A frequency distribution shows the number of data values in each of several nonoverlapping classes. However, we are often interested in knowing the proportion, or percentage, of the data values in each class. The *relative frequency* of a class is the proportion of the total number of data values belonging to the class. For a data set with n observations, the relative frequency of each class is given by the following formula.

Relative Frequency

$$\text{Relative Frequency of a Class} = \frac{\text{Frequency of the Class}}{n} \tag{2.1}$$

The *percent frequency* of a class is the relative frequency multiplied by 100.

A *relative frequency distribution* is a tabular summary of data showing the relative frequency for each class. A *percent frequency distribution* is a tabular summary of data showing the percent frequency for each class. Using (2.1), we can develop a relative frequency distribution for the personal computer data. In Table 2.3 we see that the relative frequency for Apple is 13/50 = .26, the relative frequency for Compaq is 12/50 = .24, and so on. Computing the relative frequencies for the other computer manufacturers provides the relative frequency distribution in Table 2.3. Multiplying each of the relative frequencies by 100 provides the percent frequency distribution in Table 2.3. From these distributions, we see that on the basis of the sample data, 26% of the purchases were Apple, 24% of the purchases were Compaq, 22% of the purchases were Packard Bell, and so on. We can also note that 26% + 24% + 22% = 72% of the purchases were from the top three manufacturers.

TABLE 2.3 Relative and Percent Frequency Distributions of Computer Purchases

Company	Relative Frequency	Percent Frequency
Apple	.26	26
Compaq	.24	24
Gateway 2000	.10	10
IBM	.18	18
Packard Bell	.22	22
Total	1.00	100

FIGURE 2.1
Bar Graph of Computer Purchases

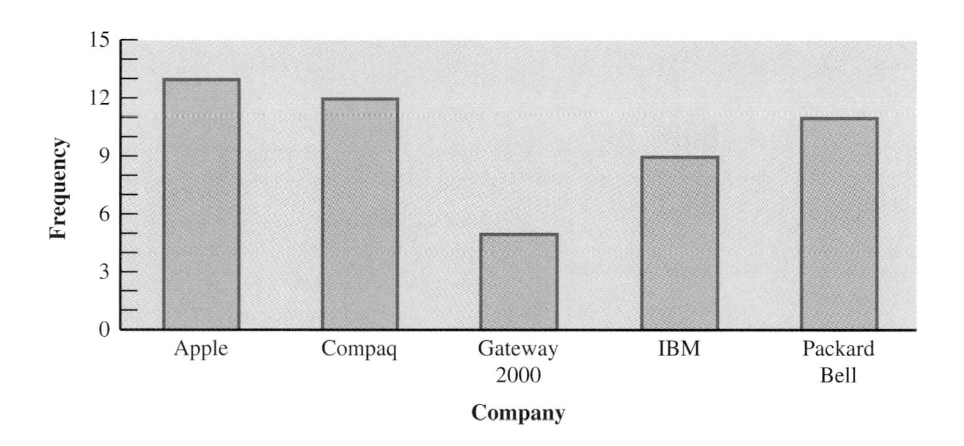

Bar Graphs and Pie Charts

A *bar graph* is a graphical device for depicting qualitative data that have been summarized in a frequency, relative frequency, or percent frequency distribution. On the horizontal axis of the graph, we specify the labels that are used for each of the classes. A frequency, relative frequency, or percent frequency scale can be used for the vertical axis of the graph. Then, using a bar of fixed width drawn above each class label, we extend the height of the bar until we reach the frequency, relative frequency, or percent frequency of the class as indicated by the vertical axis. The bars are separated to emphasize the fact that each class is a separate category. Figure 2.1 is a bar graph of the frequency distribution for the 50 computer purchases. Note how the graphical presentation shows Apple, Compaq, and Packard Bell to be the most preferred brands.

The pie chart is a commonly used graphical device for presenting qualitative data. To draw a pie chart, we first draw a circle; then use the relative frequencies to subdivide the circle into sectors, or parts, that correspond to the relative frequency for each class. For example, since there are 360 degrees in a circle and since Apple has a relative frequency of .26, the sector of the pie chart labeled Apple should consist of $.26 \times 360 = 93.6$ degrees. Similar calculations for the other classes yield the pie chart in Figure 2.2. The numerical values shown for each sector can be frequencies, relative frequencies, or percentages.

FIGURE 2.2
Pie Chart of Computer Purchases

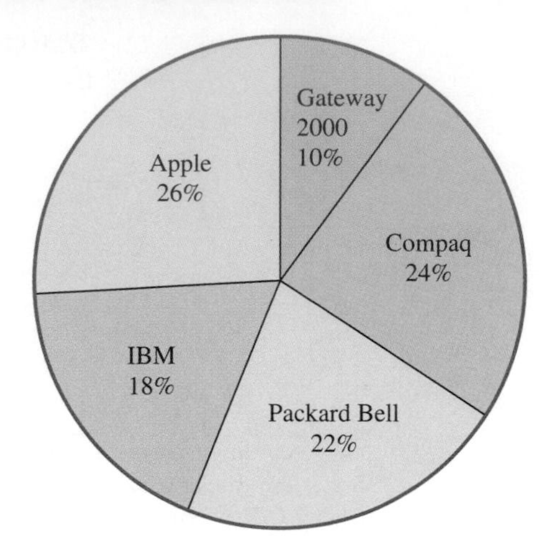

notes and comments

1. Often the number of classes in a frequency distribution is the same as the number of categories found in the data, as is the case for the computer purchase data in this section. The data involve only five computer companies, and a separate frequency distribution class was defined for each one. If the data included all types of personal computers, there would be several models with a small number of purchases. In such cases the categories with smaller frequencies can be grouped into an aggregate class called "other." Most statisticians recommend that from five to 20 classes be used in a frequency distribution; classes with smaller frequencies should normally be grouped.

2. The sum of the frequencies in any frequency distribution always equals the total number of elements in the data set. The sum of the relative frequencies in any relative frequency distribution always equals 1.00 and the sum of the percentages in a percent frequency distribution always equals 100.

exercises

Methods

1. The response to a question has three alternatives: A, B, and C. A sample of 120 responses provides 60 A, 24 B, and 36 C. Show the frequency and relative frequency distributions.

2. The following is a partial relative frequency distribution.

Class	Relative Frequency
A	.22
B	.18
C	.40
D	

a. What is the relative frequency of class D?
b. The total sample size is 200. What is the frequency of class D?
c. Show the frequency distribution.
d. Show the percent frequency distribution.

Self Test

3. A questionnaire provides 58 yes, 42 no, and 20 no-opinion answers.
a. In the construction of a pie chart, how many degrees would be in the section of the pie showing the yes answers?
b. How many degrees would be in the section of the pie showing the no answers?
c. Construct a pie chart.
d. Construct a bar graph.

Applications

4. According to Ward's Automotive Reports, five of the 10 top-selling vehicles in the United States are trucks (*Business Week,* December 5, 1994). The five best-selling trucks in 1994 were Chevy C/K pickup, Dodge Caravan, Ford Explorer, Ford F-Series pickup, and Ford Ranger. Data for a sample of 50 recent purchases follow.

C/K pickup	F-Series pickup	C/K pickup	F-Series pickup	Ford Explorer
Dodge Caravan	C/K pickup	C/K pickup	C/K pickup	Dodge Caravan
Ford Ranger	Dodge Caravan	Ford Ranger	Dodge Caravan	F-Series pickup
C/K pickup	C/K pickup	C/K pickup	F-Series pickup	F-Series pickup
Ford Explorer	F-Series pickup	Ford Ranger	F-Series pickup	Ford Ranger
C/K pickup	F-Series pickup	Dodge Caravan	F-Series pickup	C/K pickup
Ford Ranger	Ford Explorer	F-Series pickup	Ford Explorer	F-Series pickup
F-Series pickup	C/K pickup	Dodge Caravan	C/K pickup	F-Series pickup
Ford Ranger	Ford Explorer	Ford Explorer	Ford Explorer	C/K pickup
F-Series pickup	Ford Ranger	Dodge Caravan	Ford Ranger	Ford Ranger

a. Are these qualitative or quantitative data?
b. Provide frequency and percent frequency distributions for the data.
c. Construct a bar graph and a pie chart for the data.
d. On the basis of the sample, what is America's favorite truck? Is it a close call?

5. Freshmen entering the College of Business at Eastern University were asked to indicate their preferred major. If 55 preferred management, 51 preferred accounting, 28 preferred finance and 80 preferred marketing, summarize the data by constructing:
a. Relative and percent frequency distributions. **b.** A bar graph. **c.** A pie chart.

6. The six best-selling fiction books in November 1994 were *Celestine Prophecy* (C), *Debt of Honor* (D), *Insomnia* (I), *The Lottery Winner* (L), *Politically Correct* (P), and *Wings* (W). (*The Wall Street Journal,* November 23, 1994). Suppose a sample of book purchases in the Houston, Texas, area provided the following data for these six books.

W	D	W	C	I	P	C	W
P	I	W	W	P	C	P	C
L	C	P	C	W	W	P	W
W	C	I	P	L	D	D	
I	P	D	W	C	I	L	
C	D	L	D	I	L	I	

a. Construct frequency and percent frequency distributions for the data.
b. Rank the top six best-selling books.
c. What percentages of the sales are reflected by *Debt of Honor* and *The Lottery Winner?*

7. Leverock's Waterfront Steakhouse in Maderia Beach, Florida, uses a questionnaire to ask customers how they rate the server, food quality, cocktails, prices, and atmosphere at the restaurant. Each characteristic is rated on a scale of outstanding (O), very good (V), good (G), average (A), and poor (P). Use descriptive statistics to summarize the following data collected on food quality. What is your feeling about the food quality ratings at the restaurant?

G	O	V	G	A	O	V	O	V	G	O	V	A
V	O	P	V	O	G	A	O	O	O	G	O	V
V	A	G	O	V	P	V	O	O	G	O	O	V
O	G	A	O	V	O	O	G	V	A	G		

8. Position-by-position data for a sample of 55 members of the Baseball Hall of Fame in Cooperstown, New York, are given below (*Sports Illustrated,* April 6, 1992). Each data item indicates the primary position played by the Hall of Famers: pitcher (P), catcher (H), 1st base (1), 2nd base (2), 3rd base (3), shortstop (S), left field (L), center field (C), and right field (R).

L	P	C	H	2	P	R	1	S	S	1	L	P	R	P
P	P	P	R	C	S	L	R	P	C	C	P	P	R	P
2	3	P	H	L	P	1	C	P	P	P	S	1	L	R
R	1	2	H	S	3	H	2	L	P					

a. Use frequency and relative frequency distributions to summarize the data for the nine positions.
b. What position provides the most Hall of Famers?
c. What position provides the fewest Hall of Famers?
d. What outfield position (L, C, or R) provides the most Hall of Famers?
e. Compare infielders (1, 2, 3, and S) to outfielders (L, C, and R).

9. Employees at Electronics Associates are on a flextime system; they can begin their working day at 7:00, 7:30, 8:00, 8:30, or 9:00 A.M. The following data represent a sample of the starting times selected by the employees.

7:00	8:30	9:00	8:00	7:30	7:30	8:30	8:30	7:30	7:00
8:30	8:30	8:00	8:00	7:30	8:30	7:00	9:00	8:30	8:00

Summarize the data by constructing:

a. A frequency distribution.
b. A percent frequency distribution.
c. A bar graph.
d. A pie chart.
e. What do the summaries tell you about employee preferences in the flextime system?

10. Students in the College of Business Administration at the University of Cincinnati are asked to fill out a course-evaluation questionnaire upon completion of their courses. It consists of a variety of questions that have a five-category response scale. One of the questions follows.

Compared to other courses that you have taken, what is the overall quality of the course you are now completing?

____ Poor ____ Fair ____ Good ____ Very Good ____ Excellent

A sample of 60 students completing a course in business statistics during the spring quarter of 1996 provided the following responses. To aid in computer processing of the questionnaire results, a numeric scale was used with 1 = poor, 2 = fair, 3 = good, 4 = very good, and 5 = excellent.

3	4	4	5	1	5	3	4	5	2	4	5	3	4	4
4	5	5	4	1	4	5	4	2	5	4	2	4	4	4
5	5	3	4	5	5	2	4	3	4	5	4	3	5	4
4	3	5	4	5	4	3	5	3	4	4	3	5	3	3

a. Comment on why these are qualitative data.
b. Provide a frequency distribution and a relative frequency distribution summary of the data.
c. Provide a bar graph and a pie chart summary of the data.
d. On the basis of your summaries, comment on the students' overall evaluation of the course.

2.2 Summarizing Quantitative Data

Frequency Distribution

As defined in Section 2.1, a frequency distribution is a tabular summary of data showing the number of data values in each of several nonoverlapping classes. This definition holds for quantitative as well as qualitative data. However, with quantitative data we have to be more careful in defining the nonoverlapping classes to be used in the frequency distribution.

For example, consider the quantitative data in Table 2.4. These data provide the time in days required to complete year-end audits for a sample of 20 clients of Sanderson and Clifford, a small public accounting firm. The three steps necessary to define the classes for a frequency distribution with quantitative data are:

TABLE 2.4 Year-End Audit Times (in days)

12	14	19	18
15	15	18	17
20	27	22	23
22	21	33	28
14	18	16	13

1. Determine the number of nonoverlapping classes.
2. Determine the width of each class.
3. Determine the class limits.

Let us demonstrate these steps by developing a frequency distribution for the audit-time data in Table 2.4.

NUMBER OF CLASSES Classes are formed by specifying ranges of data values that will be used to group the elements in the data set. As a general guideline, we recommend using between five and 20 classes. Data sets with a larger number of elements usually require a larger number of classes. Data sets with a smaller number of elements can often be summarized easily with as few as five or six classes. The goal is to use enough classes to show the variation in the data, but not so many classes that they contain only a few elements. Since the data set in Table 2.4 is relatively small ($n = 20$), we chose to develop a frequency distribution with five classes.

WIDTH OF THE CLASSES The second step in constructing a frequency distribution for quantitative data is to choose a width for the classes. As a general guideline, we recommend that the width be the same for each class. Thus the choices of the number of classes and the width of the classes are not independent decisions. A larger number of classes means a smaller class width and vice versa. To determine an approximate class width, we begin by identifying the largest and smallest data values in the data set. Then, once the desired number of classes has been specified, we can use the following expression to determine the approximate class width.

$$\text{Approximate Class Width} = \frac{\text{Largest Data Value} - \text{Smallest Data Value}}{\text{Number of Classes}} \tag{2.2}$$

The class width given by (2.2) can be adjusted to a convenient width based on the preference of the person developing the frequency distribution. For example, a computed class width of 9.28 might be adjusted to a class width of 10 simply because 10 is a more convenient class width to use in constructing a frequency distribution.

For the data set involving the year-end audit times, the largest value is 33 and the smallest value is 12. Since we have decided to summarize the data set with five classes, using (2.2) provides an approximate class width of $(33 - 12)/5 = 4.2$. We therefore decided to use a class width of five in the frequency distribution.

In practice, the number of classes and the appropriate class width are determined by trial and error. Once a possible number of classes is chosen, (2.2) is used to find the approximate class width. The process can be repeated for a different number of classes. Ultimately, the analyst uses judgment to determine the combination of number of classes and class width that provides the best means for summarizing the data.

For the audit-time data in Table 2.4, we have decided to use five classes, each with a width of five days, to summarize the data. The next task is to specify the class limits for each of the classes.

CLASS LIMITS The *lower class limit* identifies the smallest possible data value assigned to the class. The *upper class limit* identifies the largest possible data value assigned to the class. Again, the analyst uses judgment, and a variety of acceptable class limits are possible.

For the data in Table 2.4, we defined the class limits as 10–14, 15–19, 20–24, 25–29, and 30–34. The smallest data value, 12, is included in the 10–14 class. The largest data value, 33, is included in the 30–34 class. For the 10–14 class, 10 is the lower class limit and 14 is the upper class limit. The difference between the lower class limits of adjacent classes provides the class width. Using the first two lower class limits of 10 and 15, we see that the class width is $15 - 10 = 5$.

The form of each lower class limit and each upper class limit depends on the number of places to the right of the decimal point for the data. Since the audit times in Table 2.4 are integer, integer class limits of 10–14, 15–19, and so on are acceptable. If the audit times were recorded in tenths of days, such as 12.3, 14.4, 19.3, and so on, the class limits would also be stated in tenths. In that case, class limits of 10.0–14.9, 15.0–19.9, 20.0–24.9, and so on would be appropriate. If the data were in hundredths, which is often the case with dollar-and-cents data, the class limits 10.00–14.99, 15.00–19.99, 20.00–24.99, and so on would be appropriate. Regardless of how the class limits are chosen, they should be defined in such a way that *each data value belongs to one and only one class.* For instance, class limits of 10–15, 15–20, and 20–25 are unacceptable for the audit-time data because the data values of 15 and 20 would be included in two different classes.

TABLE 2.5 Frequency Distribution for the Audit-Time Data

Audit Time (days)	Frequency
10–14	4
15–19	8
20–24	5
25–29	2
30–34	1
Total	20

Once the number of classes, class width, and class limits have been determined, a frequency distribution can be obtained by *counting* the number of data values belonging to each class. For example, the data in Table 2.4 show that four values—12, 14, 14, and 13—belong to the 10–14 class. Thus, the frequency for the 10–14 class is 4. Continuing this counting process for the 15–19, 20–24, 25–29, and 30–34 classes provides the frequency distribution in Table 2.5. Using this frequency distribution, we can observe that:

1. The most frequently occurring audit times are in the class of 15–19 days. Eight of the 20 audit times belong to this class.
2. Only one audit required 30 or more days.

Other relevant observations are possible, depending on the interests of the person viewing the frequency distribution. The value of a frequency distribution is that it provides insights about the data that are not easily obtained by viewing the data in their original unorganized form.

Relative Frequency and Percent Frequency Distributions

We define the relative frequency and percent frequency distributions for quantitative data in the same manner as for qualitative data. First, recall that the relative frequency is simply the proportion of the total number of data values belonging to a class. For a data set having n data values,

TABLE 2.6 Relative and Percent Frequency Distributions for the Audit-Time Data

Audit-Time (days)	Relative Frequency	Percent Frequency
10–14	.20	20
15–19	.40	40
20–24	.25	25
25–29	.10	10
30–34	.05	5
Total	1.00	100

FIGURE 2.3
Dot Plot for the Audit-Time Data

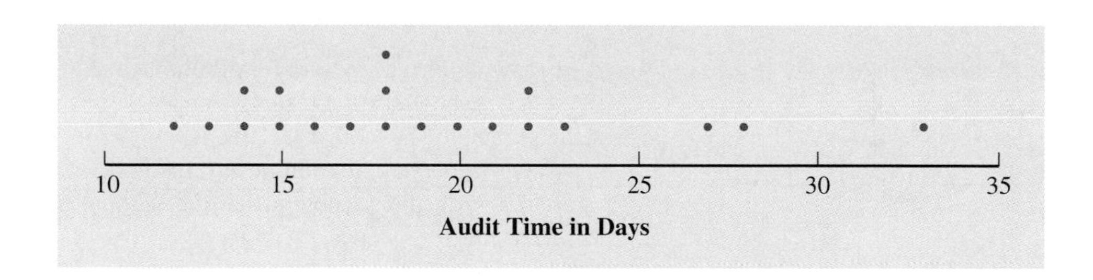

$$\text{Relative Frequency of Class} = \frac{\text{Frequency of the Class}}{n}$$

The percent frequency of a class is the relative frequency multiplied by 100.

Based on the class frequencies in Table 2.5 and with $n = 20$, Table 2.6 shows the relative frequency distribution and percent frequency distribution for the audit-time data. Note that .40 of the audits, or 40%, required from 15 to 19 days. Only .05 of the audits, or 5%, required 30 or more days. Again, additional interpretations and insights can be obtained by using Table 2.6.

Dot Plot

One of the simplest graphical summaries of data is a *dot plot.* A horizontal axis shows the range of values for the data. Then each data value is represented by a dot placed above the axis. Figure 2.3 is the dot plot for the audit-time data in Table 2.4. The three dots located at the value of 18 indicate that 18 occurs three times in the data set. Dot plots show the details of the data and are useful for comparing two or more sets of data.

Histogram

Another common graphical presentation of quantitative data is the *histogram.* This graphical summary can be prepared for data that have been previously summarized in either a frequency, relative frequency, or percent frequency distribution. A histogram is constructed by placing the variable of interest on the horizontal axis and the frequency, relative frequency, or percent frequency on the vertical axis. The frequency, relative frequency, or percent frequency of each class is shown by drawing a rectangle whose base is the class interval on the horizontal axis and whose height is the corresponding frequency, relative frequency, or percent frequency.

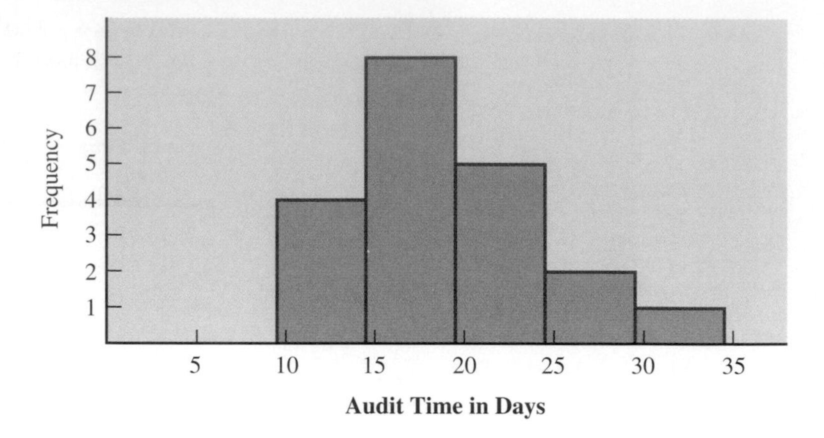

FIGURE 2.4
Histogram for the Audit-Time Data

Figure 2.4 is a histogram for the audit-time data. Note that the class with the greatest frequency is shown by the rectangle appearing above the class of 15–19 days. The height of the rectangle shows that the frequency of this class is 8. A histogram for the relative or percent frequency distribution of this data would look the same as the histogram in Figure 2.4 with the exception that the vertical axis would be labeled with relative or percent frequency values.

As Figure 2.4 shows, the adjacent rectangles of a histogram touch one another. Unlike the bar graph for qualitative data, a histogram has no natural separation between the rectangles of adjacent classes. This is the usual convention for histograms. Since the class limits for the audit-time data are stated as 10–14, 15–19, 20–24, 25–29, and 30–34, there appear to be one-unit intervals of 14 to 15, 19 to 20, 24 to 25, and 29 to 30 between the classes. These spaces are eliminated by drawing the vertical lines of the histogram halfway between the class limits. For example, the vertical lines for the 15–19 class were drawn upward from the values 14.5 and 19.5. Using this procedure for all classes, we drew the vertical lines for the histogram in Figure 2.4 upward from the values 9.5, 14.5, 19.5, 24.5, 29.5, and 34.5. This minor adjustment to eliminate the spaces in a histogram helps show that, even though the data are rounded, all values between the lower limit of the first class and the upper limit of the last class are possible.

Cumulative Distributions

A variation of the frequency distribution that provides another tabular summary of quantitative data is the *cumulative frequency distribution.* The cumulative frequency distribution uses the number of classes, class widths, and class limits that were developed for the frequency distribution. However, rather than showing the frequency of each class, the cumulative frequency distribution shows the number of data values *less than or equal to the upper class limit* of each class. The first two columns of Table 2.7 provide the cumulative frequency distribution for the audit-time data.

To understand how the cumulative frequencies are determined, consider the class with the description "less than or equal to 24." The cumulative frequency for this class is simply the sum of the frequencies for all classes with data values less than or equal to 24. For the frequency distribution in Table 2.5, the sum of the frequencies for classes 10–14, 15–19, and 20–24 indicates that there are $4 + 8 + 5 = 17$ data values less than or equal to 24. Hence, the cumulative frequency for this class is 17. Other observations

TABLE 2.7 Cumulative Frequency, Cumulative Relative Frequency, and Cumulative Percent Frequency Distributions for the Audit-Time Data

Audit Time (days)	Cumulative Frequency	Cumulative Relative Frequency	Cumulative Percent Frequency
Less than or equal to 14	4	.20	20
Less than or equal to 19	12	.60	60
Less than or equal to 24	17	.85	85
Less than or equal to 29	19	.95	95
Less than or equal to 34	20	1.00	100

based on the cumulative frequency distribution in Table 2.7 show that 4 audits were completed in 14 days or less and 19 audits were completed in 29 days or less.

As a final point, we note that a *cumulative relative frequency distribution* shows the proportion and a cumulative percent frequency distribution shows the percentage of data values less than or equal to the upper limit of each class. The cumulative relative frequency distribution can be computed either by summing the relative frequencies in the relative frequency distribution or by dividing the cumulative frequencies by the total number of data values. Using the latter approach, we found the cumulative relative frequencies in column 3 of Table 2.7 by dividing the cumulative frequencies in column 2 by the total number of items ($n = 20$). The cumulative percent frequencies were computed by multiplying the cumulative relative frequencies by 100. The cumulative relative and percent frequency distributions show that .85 of the audits, or 85%, were completed in 24 days or less, .95 of the audits, or 95%, were completed in 29 days or less, and so on.

notes and comments

1. In some applications, we may want to know the *midpoints* of the classes in a frequency distribution for quantitative data. Each class midpoint is simply halfway between the lower and upper class limits. For example, with the class limits of 10–14, 15–19, 20–24, 25–29, and 30–34 in the audit-time example, the five class midpoints would be 12, 17, 22, 27, and 32, respectively.
2. An *open-end* class is one that has only a lower class limit or an upper class limit. For example, in the audit-time data of Table 2.4, suppose two of the audits had taken 58 and 65 days. Rather than continue with the classes of width 5 with class intervals 35–39, 40–44, 45–49, and so on, we could simplify the frequency distribution by showing an open-end class of "35 or more." This class would have a frequency of 2. Most often the open-end class appears at the upper end of the distribution. Sometimes an open-end class appears at the lower end of the distribution, and occasionally such classes appear at both ends.
3. The last entry in a cumulative frequency distribution is always the total number of elements in the data set. The last entry in a cumulative relative frequency distribution is always 1.00 and the last entry in a cumulative percent frequency distribution is always 100.

exercises

Methods

11. Consider the following data.

14	21	23	21	16
19	22	25	16	16
24	24	25	19	16
19	18	19	21	12
16	17	18	23	25
20	23	16	20	19
24	26	15	22	24
20	22	24	22	20

a. Develop a frequency distribution using class limits of 12–14, 15–17, 18–20, 21–23, and 24–26.

b. Develop a relative frequency distribution and a percent frequency distribution using the class limits in part (a).

12. Consider the following frequency distribution.

Class	Frequency
10–19	10
20–29	14
30–39	17
40–49	7
50–59	2

Construct a cumulative frequency distribution and a cumulative relative frequency distribution.

13. Construct a histogram for the data in Exercise 12.

14. Consider the following data.

8.9	10.2	11.5	7.8	10.0	12.2	13.5	14.1	10.0	12.2
6.8	9.5	11.5	11.2	14.9	7.5	10.0	6.0	15.8	11.5

a. Construct a dot plot.
b. Construct a frequency distribution.
c. Construct a percent frequency distribution.

Applications

15. A doctor's office staff has studied the waiting times for patients who arrive at the office with a request for emergency service. The following data were collected over a one-month period (the waiting times are in minutes).

2 5 10 12 4 4 5 17 11 8 9 8 12 21 6 8 7 13 18 3

Use classes of 0–4, 5–9, and so on.

a. Show the frequency distribution.
b. Show the relative frequency distribution.
c. Show the cumulative frequency distribution.
d. Show the cumulative relative frequency distribution.
e. What proportion of patients needing emergency service have a waiting time of nine minutes or less?

16. Data on the 600 largest publicly held U.S. corporations were provided by *Financial World* (December 6, 1994). Return on equity for the most recent 12 months and the five-year average return on equity for 28 insurance companies follow.

RETURN

	Return on Equity (%)			Return on Equity (%)	
Company	12 Month	5-Year Avg.	Company	12 Month	5-Year Avg.
Allstate	6	5	Loews	6	10
Unitron	5	6	General Re	14	16
Aetna	3	5	Geico	15	18
Chubb	13	16	Cigna	10	7
Safeco	12	14	ITT	11	7
Cinti. Fincl.	11	11	Lincoln Natl.	8	10
Aflac	19	16	MGIC Invst.	20	16
Travelers	17	13	UNUM	15	14
Am. Intl. Grp.	14	13	St. Paul	16	9
Torchmark	21	22	Marsh & McL.	28	27
Transamerica	12	9	MBIA	16	14
Jefferson-Pilot	14	12	Aon	15	13
Progressive	24	21	Providian	14	13
American General	13	11	Berkshire	8	6

Source: *Financial World,* December 6, 1994.

a. Develop tabular summaries and a histogram for the 12-month return-on-equity data. Comment on typical returns and their distribution.
b. Develop tabular summaries and a histogram for the data on the five-year average return on equity.
c. Compare returns over the past 12 months with average returns over the last five years.

17. National Airlines accepts flight reservations by telephone. The following data are the call durations (in minutes) for a sample of 20 telephone reservations. Construct the frequency and relative frequency distributions for the data. Also provide a histogram.

2.1	4.8	5.5	10.4	3.3	3.5	4.8	5.8	5.3	5.5
2.8	3.6	5.9	6.6	7.8	10.5	7.5	6.0	4.5	4.8

18. The Roth Young Personnel Service reported that annual salaries for department store assistant managers range from \$28,000 to \$57,000 (*National Business Employment Weekly,* October 16–22, 1994). Assume the following data are a sample of the annual salaries for 40 department store assistant managers (data are in thousands of dollars).

RETAIL

48	35	57	48	52	56	51	44
40	40	50	31	52	37	51	41
47	45	46	42	53	43	44	39
50	50	44	49	45	45	50	42
52	55	46	54	45	41	45	47

a. What are the lowest and highest salaries reported?
b. Use a class width of \$5,000 and prepare tabular summaries of the annual salary data.
c. What proportion of the annual salaries are \$35,000 or less?

d. What percentage of the annual salaries are more than $50,000?
e. Prepare a histogram of the data.

19. The following data are the numbers of units produced by a production employee for the most recent 20 days.

160	170	181	156	176	148	198	179	162	150
162	156	179	178	151	157	154	179	148	156

Summarize the data by constructing:
a. A frequency distribution.
b. A relative frequency distribution.
c. A cumulative frequency distribution.
d. A cumulative relative frequency distribution.

20. *Fortune* magazine conducted a survey to learn about its subscribers in the United States and Canada. One survey question asked the value of subscribers' investment portfolios (stocks, bonds, mutual funds, and certificates of deposit). The following percent frequency distribution was prepared from responses to this question (*Fortune National Subscriber Portrait,* 1994).

Value of Investments	Percent Frequency
Under $25,000	17
$25,000–49,999	9
$50,000–99,999	12
$100,000–249,999	20
$250,000–499,999	13
$500,000–999,999	13
$1,000,000 or over	16
Total	100

a. What percentage of subscribers have investments of less than $100,000?
b. What percentage of subscribers have investments in the $100,000–499,999 range?
c. What percentage of subscribers have investments of $500,000 or more?
d. The percent frequency distribution is based on 816 responses. How many of the respondents reported having investments of $100,000 to $249,999?
e. Estimate the number of respondents reporting investments of less than $100,000.

COMPUTER

21. The personal computer has brought computer convenience and power into the home environment. But just how many hours a week are people actually using their home computers? A study designed to determine the usage of personal computers at home (*U.S. News & World Report,* December 26, 1988) provided the following data in hours per week.

.5	1.2	4.8	10.3	7.0	13.1	16.0	12.7	11.6	5.1
2.2	8.2	.7	9.0	7.8	2.2	1.8	12.8	12.5	14.1
15.5	13.6	12.2	12.5	12.8	13.5	1.3	5.5	5.0	10.8
2.5	3.9	6.5	4.2	8.8	2.8	2.5	14.4	16.0	12.4
2.8	9.5	1.5	10.5	2.2	7.5	10.5	14.1	14.9	.3

Summarize the data by constructing:
a. A frequency distribution (use a class width of three hours).
b. A relative frequency distribution.
c. A histogram.
d. Comment on what the data indicate about personal computer usage at home.

2.3 The Role of the Computer

Computers play an important role in statistical analysis. Several large-scale statistical software packages such as Minitab, SAS, SPSS, and SYSTAT are widely available. They offer extensive data-handling capabilities and numerous statistical analysis routines that can analyze small to very large data sets. Smaller scale statistical software packages, often developed for instructional purposes, are also available. They have limited data-handling capabilities and are intended to be used for smaller data sets. Finally, general-purpose spreadsheet packages such as Microsoft Excel, Lotus 1-2-3, and QuattroPro have statistical analysis capabilities. In all, a variety of options are available for individuals who want to use computers to assist with the analysis and presentation of data.

In this section, we show how computer software packages can assist in the summarization of data. We use Minitab to illustrate the types of output information that can be provided by statistical packages. In the text discussion we focus on the presentation and interpretation of the computer output. In selected chapter appendixes we show the detailed steps necessary to obtain the Minitab output as well as those necessary to obtain similar statistical output with Microsoft Excel spreadsheets. These appendixes are optional and are intended for readers who want to pursue hands-on usage of Minitab and/or Excel.

Let us now examine the Minitab output for the dot plot and histogram of the audit-time data presented in Section 2.2. Minitab uses a worksheet of rows (elements) and columns (variables) to store the data for a particular application. The data can be keyed directly into the worksheet or can be imported from elsewhere, such as from a data disk. We began the session with Minitab by entering the audit-time data from Table 2.4 into a Minitab worksheet. We then instructed Minitab to provide the output shown in Figure 2.5. Panel A is the dot plot of the audit-time data. Note that this output is similar to the dot plot in Figure 2.3. Panel B is the histogram output. Note that the

FIGURE 2.5
Minitab Output for the Audit-Time Data

Panel A: Dot Plot for Audit-Time Data

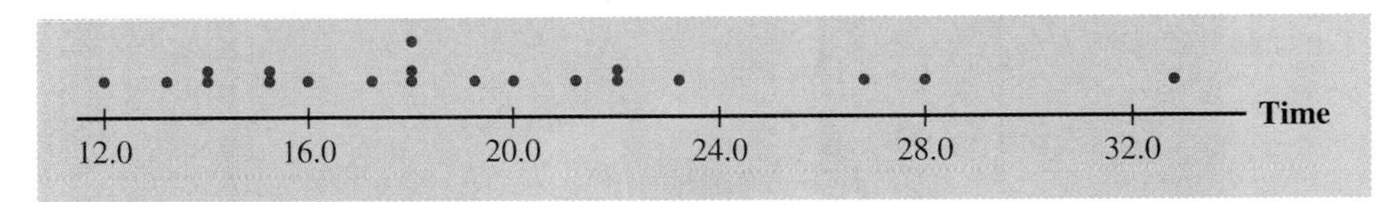

Panel B: Histogram for Audit-Time Data

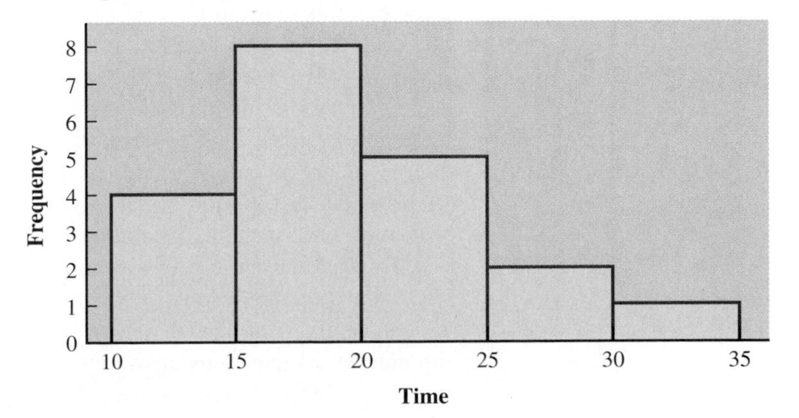

histogram provided by Minitab is essentially identical to the one in Figure 2.4. Appendix 2.1 at the end of the chapter describes the step-by-step procedures used to obtain the Minitab output in Figure 2.5.

notes and comments

This section has described the use of Minitab to summarize the quantitative audit-time data in Table 2.4. Statistical software packages such as Minitab often handle qualitative data by using numeric codes to represent the data. For example, the qualitative data on personal computer purchases in Table 2.1 could have been entered into a Minitab worksheet by using numeric codes such as 1 for Apple, 2 for Compaq, 3 for Gateway 2000, 4 for IBM, and 5 for Packard Bell. With those codes, the Minitab output would appear in the following form.

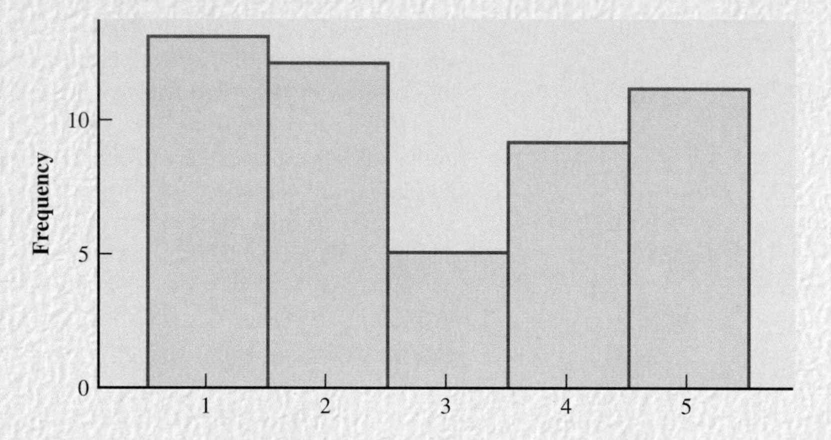

Knowing that the numeric codes are 1 for Apple, 2 for Compaq, and so on, we see that the output provides the bar graph shown previously in Section 2.1.

2.4 Exploratory Data Analysis

The techniques of *exploratory data analysis* consist of simple arithmetic and easy-to-draw pictures that can be used to summarize data quickly. In this section we show how one such technique—referred to as a *stem-and-leaf display*—can be used to rank order data and provide an idea of the shape of the distribution of a set of quantitative data.

One simple method of displaying data is to arrange them in ascending or descending order. This process, referred to as rank ordering of data, provides some degree of organization. However, such an approach provides little insight about the shape of the distribution of data values. A stem-and-leaf display is a device that shows both rank order and shape simultaneously.

To illustrate the use of a stem-and-leaf display, consider the data set in Table 2.8. These data are the results of a 150-question aptitude test given to 50 individuals who were recently interviewed for a position at Haskens Manufacturing. The data indicate the number of questions answered correctly.

TABLE 2.8 Number of Questions Answered Correctly on an Aptitude Test

APTEST

112	72	69	97	107
73	92	76	86	73
126	128	118	127	124
82	104	132	134	83
92	108	96	100	92
115	76	91	102	81
95	141	81	80	106
84	119	113	98	75
68	98	115	106	95
100	85	94	106	119

To develop a stem-and-leaf display for the data in Table 2.8, we first arrange the leading digits of each data item to the left of a vertical line. To the right of the vertical line, we record the last digit for each item as we pass through the scores in the order they were recorded. The last digit for each item is placed on the line corresponding to its first digit.

6	9	8									
7	2	3	6	3	6	5					
8	6	2	3	1	1	0	4	5			
9	7	2	2	6	2	1	5	8	8	5	4
10	7	4	8	0	2	6	6	0	6		
11	2	8	5	9	3	5	9				
12	6	8	7	4							
13	2	4									
14	1										

With this organization of the data, sorting the digits on each line into rank order is simple. Doing so leads to the following stem-and-leaf display of the data.

6	8	9									
7	2	3	3	5	6	6					
8	0	1	1	2	3	4	5	6			
9	1	2	2	2	4	5	5	6	7	8	8
10	0	0	2	4	6	6	6	7	8		
11	2	3	5	5	8	9	9				
12	4	6	7	8							
13	2	4									
14	1										

Each line in this display is referred to as a *stem,* and each digit on the stem is a *leaf.* For example, consider the first line.

6 | 8 9

The meaning attached to this line is that there are two items in the data set whose first digit is six: 68 and 69. Similarly, the second line

7 | 2 3 3 5 6 6

indicates that there are six items whose first digit is seven: 72, 73, 73, 75, 76, and 76. Thus, we see that the data values in this stem-and-leaf display are separated into two parts. The label for each stem is the one or two leading digits of the number (that is, 6, 7, 8, 9, 10, 11, 12, 13, or 14) and the leaf is the single last digit (that is, 0, 1, 2, . . ., 8, 9). The vertical line simply serves to separate the two parts of each number listed.

To focus on the shape indicated by the stem-and-leaf display, let us use a rectangle to depict the "length" of each stem. Doing so, we obtain the following representation.

6	8	9								
7	2	3	3	5	6	6				
8	0	1	1	2	3	4	5	6		
9	1	2	2	2	4	5	5	6	7	8	8
10	0	0	2	4	6	6	6	7	8	
11	2	3	5	5	8	9	9			
12	4	6	7	8						
13	2	4								
14	1									

Rotating this page counterclockwise onto its side provides a picture of the data very similar to that provided by a histogram with classes of 60–69, 70–79, 80–89, and so on.

Although the stem-and-leaf display may appear to offer the same information as a histogram, it has two primary advantages.

1. The stem-and-leaf display is easier to construct.
2. Within a class interval, the stem-and-leaf display provides more information than the histogram because the stem-and-leaf shows the actual data values.

Just as there is no right number of classes in a frequency distribution or histogram, there is no right number of rows or stems in a stem-and-leaf display. If we believe that the original stem-and-leaf display has condensed the data too much, we can easily stretch the display by using two or more stems for each leading digit(s). For example, to use two lines for each leading digit(s), we would place all data values ending in 0, 1, 2, 3, and 4 on one line and all values ending in 5, 6, 7, 8, and 9 on a second line. The following stretched stem-and-leaf display illustrates this approach.

6	
6	8 9
7	2 3 3
7	5 6 6
8	0 1 1 2 3 4
8	5 6
9	1 2 2 2 4
9	5 5 6 7 8 8
10	0 0 2 4
10	6 6 6 7 8
11	2 3
11	5 5 8 9 9
12	4
12	6 7 8
13	2 4
13	
14	1
14	

Note that data 72, 73, and 73 have leaves in the 0–4 range and are shown with the first stem value of 7. The data 75, 76, and 76 have leaves in the 5–9 range and are shown with the second stem value of 7. This stretched stem-and-leaf display is similar to a frequency distribution with intervals of 60–64, 65–69, 70–74, 75–79, and so on.

Figure 2.6 is a portion of the stem-and-leaf output provided by the Minitab computer software package. Note that the output is identical to the stretched stem-and-leaf display we have just discussed.

The preceding example has shown a stem-and-leaf display for data having up to three digits. Stem-and-leaf displays for data with more digits are possible. For example, consider the following data, which show the number of hamburgers sold by a fast-food restaurant for each of 15 weeks.

1852	1644	1766	1888	1912	2044	1812	1790
1679	2008	1565	1852	1967	1954	1733	

A stem-and-leaf display of these data follows.

Leaf Unit = 10

15	6
16	4 7
17	3 6 9
18	1 5 5 8
19	1 5 6
20	0 4

Only the first three digits of each data value are used in the display. The wording "Leaf Unit = 10" indicates that a leaf value of 1 represents 10–19, a leaf value of 2

FIGURE 2.6
Minitab Stem-and-Leaf Display of the Aptitude Test Scores

```
Stem-and-leaf of C1 N = 50
Leaf Unit = 1.0

   6  89
   7  233
   7  566
   8  011234
   8  56
   9  12224
   9  556788
  10  0024
  10  66678
  11  23
  11  55899
  12  4
  12  678
  13  24
  13
  14  1
```

represents 20–29, and so on. For example, the leaf value of 6 on the first line of the display represents a number from 60 to 69, indicating that the data values have four digits and that the data value corresponding to the first line is between 1560 and 1569.

exercises

Methods

22. Construct a stem-and-leaf display for the following data.

70	72	75	64	58	83	80	82
76	75	68	65	57	78	85	72

23. Construct a stem-and-leaf display for the following data.

11.3	9.6	10.4	7.5	8.3	10.5	10.0
9.3	8.1	7.7	7.5	8.4	6.3	8.8

24. Construct a stem-and-leaf display for the following data. Use the first two digits as the stem and the third digit as the leaf.

1161	1206	1478	1300	1604	1725	1361	1422
1221	1378	1623	1426	1557	1730	1706	1689

Applications

25. A psychologist developed a new test of adult intelligence. The test was administered to 20 individuals, and the following data were obtained.

114	99	131	124	117	102	106	127	119	115
98	104	144	151	132	106	125	122	118	118

Construct a stem-and-leaf display for the data.

26. The earnings-per-share data for a sample of 20 companies from the *Fortune* 500 largest U.S. industrial corporations follow (*Fortune,* April 18, 1994):

Company	Earnings per Share ($)	Company	Earnings per Share ($)
Apple Computer	.73	Hewlett-Packard	4.65
Procter & Gamble	−1.11	Sara Lee	1.40
Goodyear	2.64	General Dynamics	14.01
Chiquita Brands	−.99	Compaq Computer	5.45
Hershey Foods	2.15	Sunstrand	3.97
Data General	−1.73	Briggs & Stratton	4.86
Helene Curtis	2.33	Interlake	−1.18
Huffy	−.38	Dell Computer	2.59
Quaker State	.50	Harley-Davidson	−.31
Snap-On Tools	2.02	Zenith Electronics	−3.01

Develop a stem-and-leaf display for the data. Comment on what you learned about the earnings per share for these companies.

JOBSAT

27. In a study of job satisfaction, a series of tests were administered to 50 subjects. The following data were obtained; higher scores represent greater dissatisfaction.

87	76	67	58	92	59	41	50	90	75	80	81	70
73	69	61	88	46	85	97	50	47	81	87	75	60
65	92	77	71	70	74	53	43	61	89	84	83	70
46	84	76	78	64	69	76	78	67	74	64		

Construct a stem-and-leaf display for the data.

PEFORCST

28. Periodically *Barron's* publishes earnings forecasts for the companies listed in the Dow Jones Industrial Average. Shown below are the 1995 forecasts of price–earnings (P/E) ratios for these companies implied by *Barron's* earnings forecasts (*Barron's,* December 12, 1994).

Company	1995 P/E Forecast	Company	1995 P/E Forecast
AT&T	13.6	Alcoa	13.8
Allied Signal	10.4	American Express	9.5
Bethlehem Steel	5.2	Boeing	22.7
Caterpillar	9.4	Chevron	13.5
Coca-Cola	21.6	Disney	16.8
Dupont	11.7	Eastman Kodak	12.8
Exxon	14.7	General Electric	12.0
General Motors	4.7	Goodyear	8.0
IBM	11.3	International Paper	11.5
McDonald's	14.9	Merck	13.8
Minnesota Mining	14.0	J.P. Morgan	8.0
Philip Morris	9.1	Procter & Gamble	15.4
Sears	8.1	Texaco	14.6
Union Carbide	9.8	United Technologies	11.3
Westinghouse	14.8	Woolworth	9.8

a. Develop a stem-and-leaf display for the data.
b. Use the results of the stem-and-leaf display to develop a frequency distribution and percent frequency distribution for the data.

TABLE 2.9 Crosstabulation Format for 300 Los Angeles Restaurants

	Meal Price				
Quality Rating	**$10–19**	**$20–29**	**$30–39**	**$40–49**	**Total**
Good					
Very good					
Excellent					
Total					

2.5 Crosstabulations and Scatter Diagrams

Thus far in this chapter we have focused on tabular and graphical methods that are used to summarize the data for *one variable at a time.* Often a manager or decision maker is interested in tabular and graphical methods that will assist in the understanding of the *relationship between two variables.* Crosstabulation is a tabular method that can be used to summarize the data for two variables simultaneously. A scatter diagram is a graphical method that has a similar purpose.

Let us illustrate the use of a crosstabulation by considering an application. Zagat's Restaurant Review is a service that provides data on restaurants located throughout the world. Data on a variety of variables such as the restaurant's quality rating and typical meal price are reported. Quality rating is a qualitative variable with rating categories of good, very good, and excellent. Meal price is a quantitative variable that generally ranges from $10 to $49. The quality rating and the meal price data were obtained from Compuserve (January 1995) for a sample of 300 restaurants located in the Los Angeles area.

The general format of a crosstabulation for this application is shown in Table 2.9. The left and top margin labels define the classes for the two variables. In the left margin, the row labels (good, very good, and excellent) correspond to the three classes of the quality rating variable. In the top margin, the column labels ($10–19, $20–29, $30–39, and $40–49) correspond to the four classes of the meal price variable. Each restaurant in the sample provides a quality rating and a meal price. Thus, every restaurant in the sample is associated with a cell appearing in one of the rows and one of the columns of the crosstabulation. For example, an Italian restaurant located near Malibu is identified as having a very good quality rating and typical meal price of $32. This restaurant belongs to the cell in row 2 and column 3 of Table 2.9. In constructing a crosstabulation, we simply count the number of restaurants that belong to each of the cells in the crosstabulation table.

Completing the crosstabulation of the data for the sample of 300 restaurants provided the results in Table 2.10. We see that the greatest number of restaurants in the sample (64) have a very good rating and a meal price in the $20–29 range. Only two restaurants have an excellent rating and a meal price in the $10–19 range. Similar interpretations of the other frequencies can be made. In addition, note that the right and bottom margins of the crosstabulation provide the frequency distributions for quality rating and meal price separately. From the frequency distribution in the right margin, we see that data on quality ratings show 84 good restaurants, 150 very good restaurants, and

TABLE 2.10 Crosstabulation of Quality Rating and Meal Price for 300 Los Angeles Restaurants

	Meal Price				
Quality Rating	**$10–19**	**$20–29**	**$30–39**	**$40–49**	**Total**
Good	42	40	2	0	84
Very good	34	64	46	6	150
Excellent	2	14	28	22	66
Total	78	118	76	28	300

TABLE 2.11 Row Percentages for Each Quality Rating Category

	Meal Price				
Quality Rating	**$10–19**	**$20–29**	**$30–39**	**$40–49**	**Total**
Good	50.0	47.6	2.4	0.0	100
Very good	22.7	42.7	30.6	4.0	100
Excellent	3.0	21.2	42.4	33.4	100

66 excellent restaurants. Similarly, the bottom margin shows the frequency distribution for the meal price variable.

The value of a crosstabulation is that it provides insight about the relationship between the variables. From the results in Table 2.10, higher meal prices appear to be associated with the higher quality restaurants, and the lower meal prices appear to be associated with the lower quality restaurants.

Converting the entries in the table into row percentages or column percentages can afford additional insight about the relationship between the variables. For example, the results of dividing each frequency in Table 2.10 by its corresponding row total and expressing the values as percentages are shown in Table 2.11. For the lowest quality category (good), we see that the greatest percentages are for the less expensive restaurants (50.0% have $10–19 meal prices and 47.6% have $20–29 meal prices). For the highest quality category (excellent), we see that the greatest percentages are for the more expensive restaurants (42.4% have $30–39 meal prices and 33.4% have $40–49 meal prices). Thus, we continue to see that the more expensive meals are associated with the higher quality restaurants.

Crosstabulation is widely used for examining the relationship between two variables. In practice, final reports for many statistical surveys include a large number of crosstabulation tables. In the Los Angeles restaurant sample, the crosstabulation is based on one qualitative variable (quality rating) and one quantitative variable (meal price). Crosstabulations can also be developed when both variables are qualitative and when both variables are quantitative.

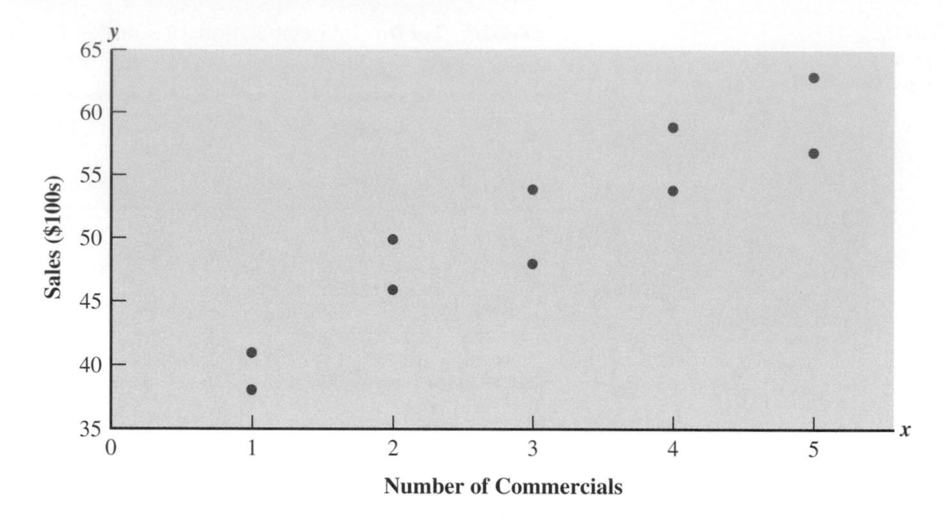

FIGURE 2.7
Scatter Diagram for the Stereo and Sound Equipment Store

TABLE 2.12 Sample Data for the Stereo and Sound Equipment Store

Week	No. of Commercials x	Sales Volume ($100s) y
1	2	50
2	5	57
3	1	41
4	3	54
5	4	54
6	1	38
7	5	63
8	3	48
9	4	59
10	2	46

A scatter diagram is a graphical presentation of the relationship between two quantitative variables. As an illustration of a scatter diagram, consider the situation of a stereo and sound equipment store in San Francisco. On 10 occasions during the past three months, the store has used weekend television commercials to promote sales at its stores. The manager wants to investigate whether there is a relationship between the number of commercials shown and the sales at the store during the following week. Sample data for the 10 weeks with sales in hundreds of dollars are shown in Table 2.12.

Figure 2.7 is the scatter diagram for the data in Table 2.12. The number of commercials (x) is shown on the horizontal axis and the sales (y) is shown on the vertical axis. For week 1, $x = 2$ and $y = 50$. A point with those coordinates is plotted on the scatter diagram. Similar points are plotted for the other nine weeks. The completed scatter diagram in Figure 2.7 indicates a positive relationship between the number of commercials and sales. Higher sales are associated with a higher number of commercials. The relationship is not perfect in that all points are not on a straight line. However, the general pattern of the points suggests that the overall relationship is positive.

Some general scatter diagram patterns and the types of relationships they suggest are shown in Figure 2.8. The top panel depicts a positive relationship similar to the one we saw for the number of commercials and sales example. In the second panel, the scatter diagram shows no apparent relationship between the variables. The third panel depicts a negative relationship where y tends to decrease as x increases.

FIGURE 2.8
Types of Relationships Depicted by Scatter Diagrams

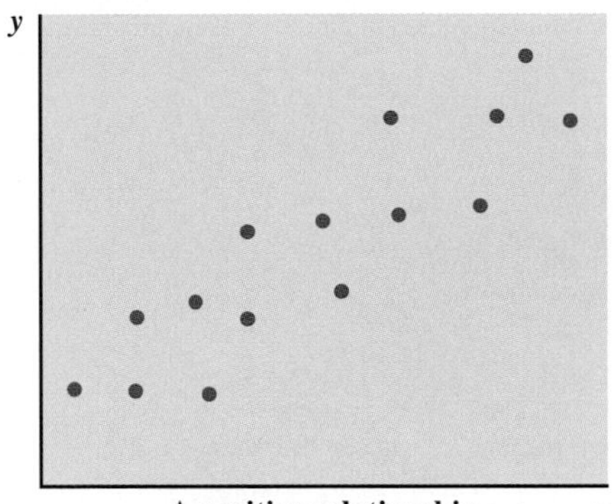

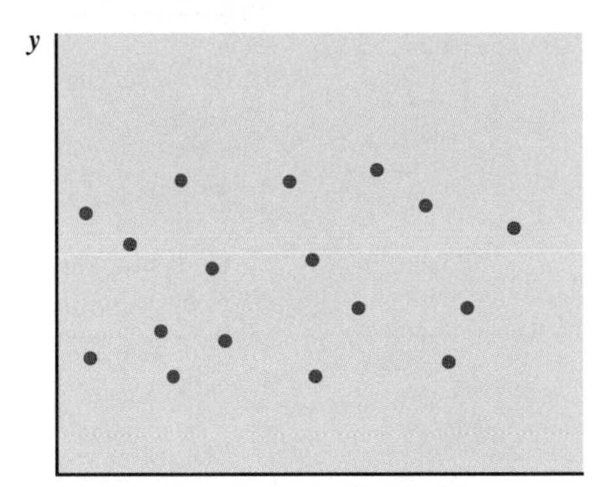

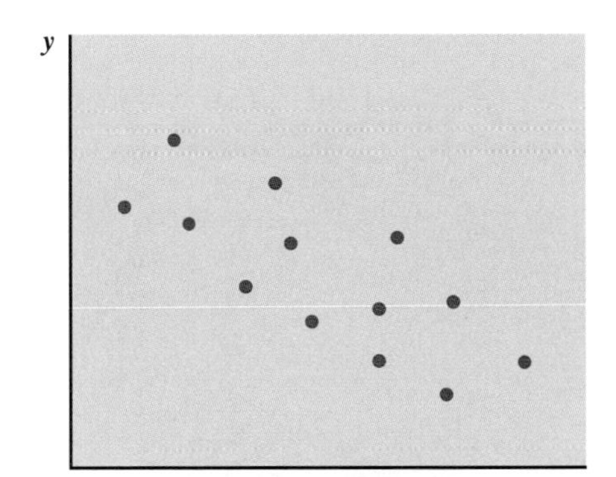

exercises

Methods

29. The following data are for 30 observations on two qualitative variables, x and y. The categories for x are A, B, and C; the categories for y are 1 and 2.

Observation	x	y	Observation	x	y
1	A	1	16	B	2
2	B	1	17	C	1
3	B	1	18	B	1
4	C	2	19	C	1
5	B	1	20	B	1
6	C	2	21	C	2
7	B	1	22	B	1
8	C	2	23	C	2
9	A	1	24	A	1
10	B	1	25	B	1
11	A	1	26	C	2
12	B	1	27	C	2
13	C	2	28	A	1
14	C	2	29	B	1
15	C	2	30	B	2

a. Develop a crosstabulation for the data, using x in the rows and y in the columns.
b. Compute the row percentages.
c. Compute column percentages.
d. What is the relationship, if any, between x and y?

30. The following 20 observations are for two quantitative variables, x and y.

SCATTER

Observation	x	y	Observation	x	y
1	−22	22	11	−37	48
2	−33	49	12	34	−29
3	2	8	13	9	−18
4	29	−16	14	−33	31
5	−13	10	15	20	−16
6	21	−28	16	−3	14
7	−13	27	17	−15	18
8	−23	35	18	12	17
9	14	−5	19	−20	−11
10	3	−3	20	−7	−22

a. Develop a scatter diagram for the relationship between x and y.
b. What is the apparent relationship, if any, between x and y?

Applications

31. Compute column percentages for the restaurant data in Table 2.10. What is the relationship between quality rating and meal price?

32. Shown in Table 2.13 are financial data for a 30-company sample of the largest U.S. companies whose stock is traded over the counter (*Financial World,* September 1, 1994).

a. Prepare a crosstabulation of the data on earnings (rows) and book value (columns). Use classes 0.00–0.99, 1.00–1.99, and 2.00–2.99 for earnings per share and classes 0.00–4.99, 5.00–9.99, 10.00–14.99, and 15.00–15.99 for book value per share.
b. Compute row percentages and comment on any relationship that you see between the variables.

33. Refer to the data in Table 2.13.

a. Prepare a crosstabulation of the data on book value and stock price.
b. Compute column percentages and comment on any relationship that you see between the variables.

TABLE 2.13 Financial Data for a Sample of 30 Companies

FINANCE

Company	Earnings per Share ($)	Book Value per Share ($)	Stock Price ($)
Acme	2.59	15.00	25.50
Aldus	1.41	9.22	29.50
Am. Fed. Bk.	1.26	7.75	11.63
Applebees	.49	3.60	14.75
Banta Corp.	2.16	14.89	32.75
Bob Evans	1.15	7.41	21.50
Cintas Corp.	1.12	5.64	31.00
Comm. Clr. Hse.	.72	5.76	19.00
Cracker Brl.	.89	6.09	23.13
Devon Group	2.01	8.86	19.50
Duracraft	1.69	5.41	40.00
First Alert	.79	3.18	25.75
Food Lion	.26	3.71	5.94
Gentex	.83	3.09	21.00
Gould Pumps	1.12	8.80	20.25
Haggar Corp.	2.80	15.26	29.00
Hubco, Inc.	2.27	11.74	21.13
Info. Res.	.74	8.59	13.75
Irwin Fincl.	2.82	12.08	21.75
Kelly Srvcs.	1.35	11.06	28.75
Lone Star	.62	4.44	20.00
Mark Twain	2.40	13.75	27.50
Micro Sys.	1.03	4.01	28.50
Novell	.98	3.58	16.13
Pacific Phy.	.55	5.00	11.50
Proffitts Inc.	.50	13.39	18.75
Rival	1.44	6.57	20.38
Sybase	1.07	3.81	37.00
Tyson Foods	1.23	17.06	23.88
Zenith Labs	.85	3.63	15.63

Source: *Financial World,* September 1, 1994.

34. Refer to the data in Table 2.13.
   a. Prepare a scatter diagram of the data on earnings per share and book value.
   b. Comment on any relationship that you see between the variables.

35. Refer to the data in Table 2.13.
   a. Prepare a scatter diagram of the data on earnings per share and stock price.
   b. Does there appear to be a relationship between the variables? Comment.

summary

A set of data, even if modest in size, is often difficult to interpret directly in the form in which it is gathered. Tabular and graphical procedures provide means of organizing and summarizing data so that patterns are revealed and the data are more easily interpreted. Frequency distributions, relative frequency distributions, percent frequency distributions,

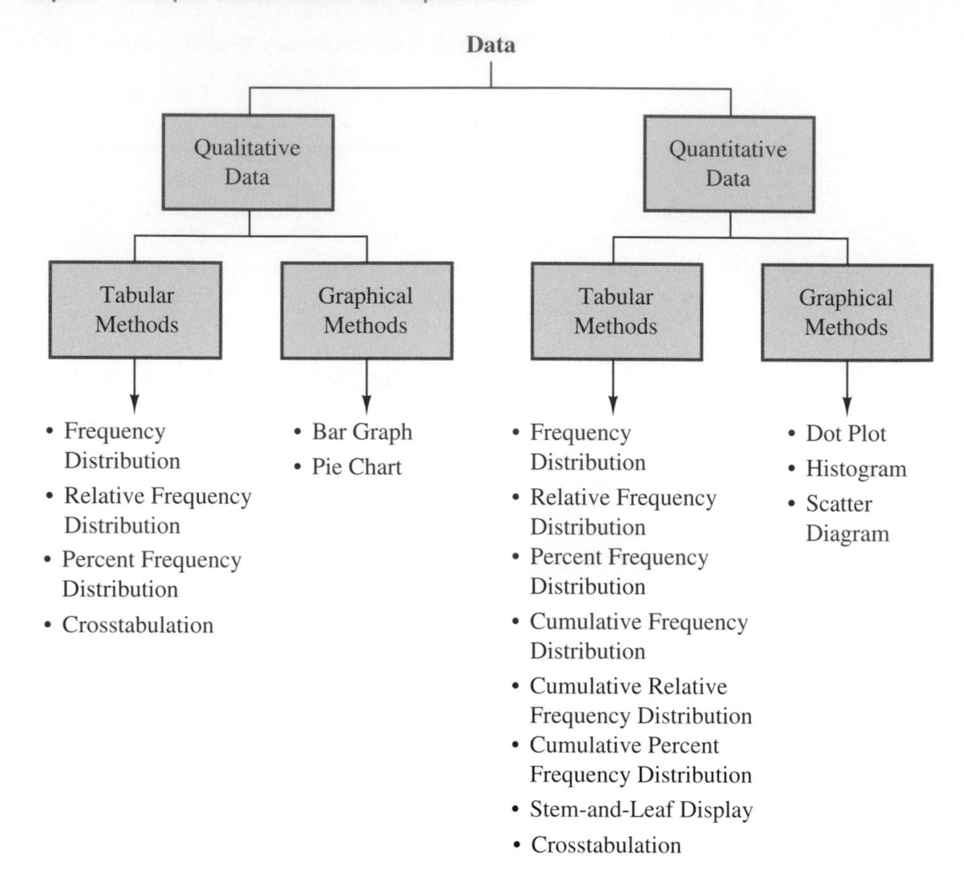

FIGURE 2.9
Tabular and Graphical Procedures for Summarizing Data

bar graphs, and pie charts were presented as tabular and graphical procedures for summarizing qualitative data. Frequency distributions, relative frequency distributions, percent frequency distributions, dot plots, histograms, cumulative frequency distributions, cumulative relative frequency distributions, and cumulative percent frequency distributions were presented as ways of summarizing quantitative data. A stem-and-leaf display was presented as an exploratory data analysis technique that can be used to summarize quantitative data.

Crosstabulation was presented as a tabular method for summarizing data for two variables. The scatter diagram was introduced as a graphical method for showing the relationship between two quantitative variables. Figure 2.9 is a summary of the tabular and graphical methods presented in this chapter.

glossary

Qualitative data Data that provide labels or names for categories of like items.

Quantitative data Data that indicate how much or how many.

Frequency distribution A tabular summary of data showing the number of data values in each of several nonoverlapping classes.

Relative frequency distribution A tabular summary of data showing the relative frequency—that is, the fraction or proportion—of the total number of data values in each of several nonoverlapping classes.

Percent frequency distribution A tabular summary of data showing the percentage of the total number of data values in each of several nonoverlapping classes.

Bar graph A graphical device for depicting the information presented in a frequency distribution, relative frequency distribution, or percent frequency distribution of qualitative data.

Pie chart A graphical device for presenting qualitative data summaries based on subdivision of a circle into sectors that correspond to the relative frequency for each class.

Histogram A graphical presentation of a frequency distribution, relative frequency distribution, or percent frequency distribution of quantitative data constructed by placing the class intervals on the horizontal axis and the frequencies on the vertical axis.

Cumulative frequency distribution A tabular summary of quantitative data showing the number of data values less than or equal to the upper class limit of each class.

Cumulative relative frequency distribution A tabular summary of quantitative data showing the fraction or proportion of data values less than or equal to the upper class limit of each class.

Cumulative percent frequency distribution A tabular summary of quantitative data showing the percentage of data values less than or equal to the upper class limit of each class.

Class midpoint The point in each class that is halfway between the lower and upper class limits.

Stem-and-leaf display An exploratory data analysis technique that simultaneously rank orders quantitative data and provides insight about the shape of the distribution.

Crosstabulation A tabular summary of data for two variables. The classes for one variable are represented by the rows; the classes for the other variable are represented by the columns.

Scatter diagram A graphical means of showing the relationship between two quantitative variables. One variable is shown on the horizontal axis and the other variable is shown on the vertical axis.

key formulas

Relative Frequency

$$\frac{\text{Frequency of the Class}}{n} \tag{2.1}$$

Approximate Class Width

$$\frac{\text{Largest Data Value} - \text{Smallest Data Value}}{\text{Number of Classes}} \tag{2.2}$$

supplementary exercises

36. The Gallup Poll News Service selected a random sample of adults to learn what sports fans select as their favorite sport to watch, in person or on television (*USA Today,* December

12, 1990). The following sample results are consistent with the findings of the Gallup poll. In the data, B is baseball, K is basketball, F is football, I is ice hockey, T is tennis, and O is other sports.

O	F	B	O	B	F	F	K	O	K	F	F	O	F
T	F	F	B	K	F	F	O	F	B	F	O	O	B
I	F	O	B	F	K	B	K	O	O	O	K		

a. Show a frequency distribution.
b. Show a relative frequency distribution. What is the favorite spectator sport?
c. Show a pie chart summary of the data.

37. Each of the *Fortune* 500 companies is classified as belonging to one of several industries (*Fortune,* April 18, 1994). A sample of 20 companies with their corresponding industry classification follows.

Company	Industry Classification	Company	Industry Classification
IBP	Food	Del Monte	Food
Intel	Electronics	McDonnell Douglas	Aerospace
Coca-Cola	Beverage	Morton International	Chemicals
Union Carbide	Chemicals	Quaker Oats	Food
General Electric	Electronics	Pepsico	Beverage
Motorola	Electronics	Maytag	Electronics
Kellogg	Food	Lockheed	Aerospace
Dow Chemical	Chemicals	Pet	Food
Campbell Soup	Food	Westinghouse	Electronics
Ralston Purina	Food	Raychem	Electronics

a. Provide a frequency distribution showing the number of companies in each industry.
b. Provide a percent frequency distribution.
c. Provide a bar graph for the data.

38. An international survey was conducted by the Union Bank of Switzerland to obtain data on the hourly wages of blue- and white-collar workers throughout the world (*Newsweek,* February 17, 1992). Workers in Los Angeles ranked seventh in the world in terms of highest hourly wage. Assume that the following 25 values indicate hourly wages for workers in Los Angeles.

11.50	8.40	11.75	10.05	10.25	8.00	13.65	7.05	9.05
11.90	9.90	6.85	15.35	11.10	14.70	13.15	13.10	6.65
13.10	9.20	9.15	12.05	8.45	5.85	9.80		

a. Construct a frequency distribution using classes of 4.00–5.99, 6.00–7.99, and so on.
b. Construct a relative frequency distribution.
c. Construct cumulative frequency and cumulative relative frequency distributions.
d. Use these distributions to comment on what you have learned about the hourly wages of workers in Los Angeles.

39. The following data represent sales in millions of dollars for 17 companies in the health care services industry (The 1992 *Business Week* 1000).

6099	1709	847	3973	604	2104	166	282	170
233	868	230	225	491	2452	2301	393	

a. Construct a frequency distribution to summarize the data. Use a class width of 1000.
b. Develop a relative frequency distribution for the data.
c. Construct a cumulative frequency distribution for the data.
d. Construct a cumulative relative frequency distribution for the data.
e. Construct a histogram as a graphical representation of the data.

40. The closing prices of 40 common stocks follow (*Investor's Daily,* April 25, 1996).

29⅝	34	43¼	8¾	37⅞	8⅝	7⅝	30⅜	35¼	19⅜
9¼	16½	38	53⅜	16⅝	1¼	48⅜	18	9⅜	9¼
10	37	18	8	28½	24¼	21⅝	18½	33⅝	31⅛
32¼	29⅝	79⅜	11⅜	38⅞	11½	52	14	9	33½

a. Construct frequency and relative frequency distributions for the data.
b. Construct cumulative frequency and cumulative relative frequency distributions for the data.
c. Construct a histogram for the data.
d. Using your summaries, make comments and observations about the price of common stock.

41. Seventy-nine new shadow stocks were reported by the American Association of Individual Investors (*AAII Journal,* April 1992). The term "shadow" indicates stocks for small to medium-size firms not followed closely by the major brokerage houses. Information on where the stock was traded—New York Stock Exchange (NYSE), American Stock Exchange (AMEX), and over the counter (OTC)—the earnings per share, and the price–earnings ratio was provided for the following sample of 15 shadow stocks.

Stock	Exchange	Earnings per Share ($)	Price–Earnings Ratio
Selas Corp of America	AMEX	1.61	7.1
CE Software Holdings	OTC	4.13	7.5
Shult Homes Corp	AMEX	1.05	11.0
Basic American Medical	OTC	1.06	12.0
Titan Corporation	NYSE	.24	18.3
First Team Sports	OTC	.51	18.6
Cooker Restaurant Corp	OTC	.76	43.1
Chempower	OTC	.22	20.5
Benchmark Electronics	AMEX	.62	25.0
Mylex Corp	OTC	.11	37.5
Pharmacy Mgmt Service	OTC	.22	50.0
Arrow Automotive Indus.	AMEX	.23	38.6
U.S. Filter Corp	AMEX	.23	73.4
Sun Sportswear	OTC	.21	28.6
Village Supermarket	OTC	.61	13.5

a. Provide frequency and relative frequency distributions for the exchange data. Where are most shadow stocks listed?
b. Provide frequency and relative frequency distributions for the earnings-per-share and price–earnings ratio data. Use class limits of .00–.49, .50–.99, and so on, for the earnings-per-share data and class limits of 0.0–9.9, 10.0–19.9, and so on for the price–earnings ratio data. What observations and comments can you make about the shadow stocks?

42. The daily high and low temperatures for 24 cities follow (*USA Today,* January 9, 1995).

CITIES

City	High	Low	City	High	Low
Tampa	68	45	Birmingham	58	34
Kansas City	37	28	Minneapolis	17	10
Boise	52	41	Portland	51	48
Los Angeles	66	59	Memphis	54	34
Philadelphia	42	26	Buffalo	26	10
Milwaukee	25	12	Cincinnati	37	24
Chicago	27	16	Charlotte	52	33
Albany	31	9	Boston	38	19
Houston	72	49	Tulsa	54	35
Salt Lake City	51	39	Washington, D.C.	43	28
Miami	75	57	Las Vegas	58	48
Cheyenne	49	35	Detroit	27	13

a. Prepare a stem-and-leaf display for the high temperatures.
b. Prepare a stem-and-leaf display for the low temperatures.
c. Compare the stem-and-leaf displays from (a) and (b) and make some comments about the differences between daily high and low temperatures.
d. Use the stem-and-leaf display from (b) to determine the number of cities having a low temperature of freezing (32°F) or below.
e. Provide frequency distributions for both the high- and low-temperature data.

43. Refer to the data set for high and low temperatures at 24 cities in Exercise 42.
a. Develop a scatter diagram to show the relationship between the two variables, high temperature and low temperature.
b. Comment on the relationship between high and low temperature.

44. Following are financial data for 20 companies in the banking and consumer products industries (*Business Week,* August 15, 1994).

BWDATA

Company	Industry	Price–Earnings Ratio	Profit Margin
Avon Products	Consumer	16	7.2
Bankers Trust	Bank	6	11.1
CoreStates	Bank	15	10.3
Fruit of Loom	Consumer	11	6.1
Mellon Bank	Bank	9	15.4
Liz Claiborne	Consumer	17	3.2
Russell	Consumer	26	5.2
Circuit City	Consumer	16	1.9
State Street	Bank	15	13.5
Banc One	Bank	10	16.7
Maytag	Consumer	19	4.7
First Chicago	Bank	6	15.3
Norwest	Bank	11	13.6
Whirlpool	Consumer	13	4.2
NationsBank	Bank	9	14.6
Wachovia	Bank	11	19.0
Coca-Cola	Consumer	24	17.5
Colgate	Consumer	15	7.5
BankAmerica	Bank	10	14.4
Philip Morris	Consumer	13	9.1

a. Prepare a crosstabulation for the variables industry type and price–earnings ratio. Use industry type as the row labels.
b. Compute the row percentages for your crosstabulation in part (a).
c. What relationship, if any, do you notice between industry type and price–earnings ratio?

45. Refer to the data set in Exercise 44 containing financial data for 20 companies in the banking and consumer products industries.

a. Prepare a scatter diagram to show the relationship between the variables, price–earnings ratio and profit margin.

b. Comment on any relationship that is apparent between the variables.

46. A survey of commercial buildings served by the Cincinnati Gas & Electric Company was concluded in 1992 (CG&E Commercial Building Characteristics Survey, November 25, 1992). One question asked what main heating fuel was used and another asked the year the commercial building was constructed. A partial crosstabulation of the findings follows.

	Fuel Type				
Year Constructed	**Electricity**	**Natural Gas**	**Oil**	**Propane**	**Other**
1973 or before	40	183	12	5	7
1974–1979	24	26	2	2	0
1980–1986	37	38	1	0	6
1987–1991	48	70	2	0	1

a. Complete the crosstabulation by showing the row totals and column totals.

b. Show the frequency distributions for year constructed and for fuel type.

c. Prepare a crosstabulation showing column percentages.

d. Prepare a crosstabulation showing row percentages.

e. Comment on the relationship between year constructed and fuel type.

computer case

Consolidated Foods, Inc.

Consolidated Foods, Inc., operates a chain of supermarkets in New Mexico, Arizona, and California. A recent promotional campaign has advertised the chain's offering of a new credit-card policy whereby Consolidated Foods' customers have the option of paying for their purchases with credit cards such as Visa and MasterCard in addition to the usual options of cash or personal check. The new policy is being implemented on a trial basis with the hope that the credit-card option will encourage customers to make larger purchases.

After the first month of operation, a random sample of 100 customers was selected over a one-week period. Data were collected on the method of payment and how much was spent by each of the 100 customers. The sample data are shown in Table 2.14. Prior to the new credit-card policy, approximately 50% of Consolidated Foods' customers paid in cash and approximately 50% paid by personal check.

Managerial Report

Use the tabular and graphical methods of descriptive statistics to summarize the sample data in Table 2.14. Your report should contain summaries such as the following.

1. A frequency and relative frequency distribution for the method of payment.
2. A bar graph or pie chart for the method of payment.

TABLE 2.14 Purchase Amount and Method of Payment* for a Random Sample of 100 Consolidated Foods Customers

Cash	Personal Check	Credit Card	Cash	Personal Check	Credit Card
$ 7.40	$27.60	$50.30	$ 5.08	$52.87	$69.77
5.15	30.60	33.76	20.48	78.16	48.11
4.75	41.58	25.57	16.28	25.96	
15.10	36.09	46.24	15.57	31.07	
8.81	2.67	46.13	6.93	35.38	
1.85	34.67	14.44	7.17	58.11	
7.41	58.64	43.79	11.54	49.21	
11.77	57.59	19.78	13.09	31.74	
12.07	43.14	52.35	16.69	50.58	
9.00	21.11	52.63	7.02	59.78	
5.98	52.04	57.55	18.09	72.46	
7.88	18.77	27.66	2.44	37.94	
5.91	42.83	44.53	1.09	42.69	
3.65	55.40	26.91	2.96	41.10	
14.28	48.95	55.21	11.17	40.51	
1.27	36.48	54.19	16.38	37.20	
2.87	51.66	22.59	8.85	54.84	
4.34	28.58	53.32	7.22	58.75	
3.31	35.89	26.57		17.87	
15.07	39.55	27.89		69.22	

*The data are based on actual bills and types of payments reported for grocery purchases (*The Wall Street Journal,* April 9, 1992).

CONSOLID

3. Frequency and relative frequency distributions for the amount spent using each method of payment.
4. Histograms and/or stem-and-leaf plots for the amount spent using each method of payment.

What preliminary insights do you have about the amounts spent and method of payment at Consolidated Foods? The data set for this computer case is available in the data file CONSOLID (see Appendix D).

APPENDIX 2.1 Dot Plots and Histograms with Minitab

In this appendix we describe the steps necessary to use Minitab to generate a dot plot and a histogram for the audit-time data in Table 2.4. We first enter the audit-time data into column C1 of a Minitab worksheet. The variable name Time is entered at the top of this column. The following steps generate the computer output shown in Figure 2.5.

STEP 1. Select the **Graph** pull-down menu
STEP 2. Select **Character Graphs**

STEP 3. Select **Dotplot**
STEP 4. When the dialog box appears:
Enter C1 in the **Variables** box
Select **OK**

The dot plot in panel A of Figure 2.5 will then appear.

To obtain the histogram shown in panel B of Figure 2.5, the following steps are necessary.

STEP 1. Select the **Graph** pull-down menu
STEP 2. Select **Histogram**
STEP 3. When the **Histogram** dialog box appears:
Enter C1 in the **Graph variables** box
Select **Bar** under **Display** in the **Data display** box
Select **Graph** under **For each** in the **Data display** box
Select **Options**
STEP 4. When the **Histogram Options** dialog box appears:
Select **Frequency** under **Type of Histogram**
Select **Cutpoint** under **Type of Intervals**
Select **Midpoint/cutpoint positions** under **Definition of Intervals**
Enter **10:35/5** in the **Mid/cutpoint positions** box[1]
Select **OK**
STEP 5. When the Histogram dialog box appears:
Select **OK**

APPENDIX 2.2 Frequency Distributions and Histograms with Spreadsheets

Spreadsheet software packages such as Microsoft Excel, Lotus 1-2-3, and Quattro Pro have the capability of performing many of the statistical methods presented in this text. Generally, the user of the spreadsheet enters data directly into the spreadsheet. Built-in statistical routines, or user-provided formulas, are used to generate the desired statistical information. In this appendix, we show how Excel can be used to construct a frequency distribution and a histogram for the audit-time data in Table 2.4.

Excel provides a worksheet of rows and columns that can be used to enter and store the data. Suppose we enter the 20 observations for the audit-time data set in rows 1 to 20 of column A (see Figure 2.10) and want to develop a frequency distribution and a histogram with five classes: 10–14, 15–19, 20–24, 25–29, and 30–34.

To develop a frequency distribution and histogram, Excel requires the user to identify what are called *bins* for the data. A bin must be identified for each class and its upper limit must be specified. Hence, five bins with upper limits of 14, 19, 24, 29, and 34 are needed for the audit-time data. In constructing the frequency distribution and histogram, Excel provides a count of the number of items with data values *less than or equal to* the upper limit of the first bin (14), a count of the number of items with data values *greater than* the first bin upper limit and *less than or equal to* the second bin

[1]The entry 10:35/5 indicates 10 is the lowest value for the histogram, 35 is the highest value for the histrogram and 5 is the class width.

FIGURE 2.10
Data and Bin Values for the Spreadsheet Summary of the Audit-Time Data

	A	B	C	D
1	12			
2	15		Bin	
3	20		14	
4	22		19	
5	14		24	
6	14		29	
7	15		34	
8	27			
9	21			
10	18			
11	19			
12	18			
13	22			
14	33			
15	16			
16	18			
17	17			
18	23			
19	28			
20	13			

upper limit (19), and so on. With the bin upper limits being equal to the upper class limits, the counts provided by Excel are the class frequencies for the frequency distribution.

Prior to implementing the Excel frequency distribution and/or histogram procedure, we must enter the bin upper limits in *ascending order* in a column or row of the worksheet. Let us select column C as a convenient location and enter the title Bin in row 2. The bin upper limits are then entered in ascending order in rows 3 through 7 of column C. Figure 2.10 is the worksheet with the audit-time data in column A and the bin upper limits in column C.

The following steps describe how to use Excel to produce a frequency distribution and a histogram for the data and worksheet in Figure 2.10.

STEP 1. Select the **Tools** pull-down menu
STEP 2. Choose the **Data Analysis** option
STEP 3. Choose **Histogram** from the list of Analysis Tools
STEP 4. When the dialog box appears:
Enter A1:A20 in the **Input Range** box
Enter C3:C7 in the **Bin Range** box
Select **Output Range**
Enter C10 in the **Output Range** box (this identifies the upper left corner of the section of the worksheet where the frequency distribution will appear)
Select **Chart Output** (this requests a histogram)
Select **OK**

Figure 2.11 is the frequency distribution provided by Excel. The title Bin and the upper limits are shown in cells C10 through C15. The frequencies appear in the corresponding cells of column D. Note that Excel includes a last bin labeled More in

FIGURE 2.11
Frequency Distribution of the Audit-Time Data Constructed by Excel

	A	B	C	D
1	12			
2	15		Bin	
3	20		14	
4	22		19	
5	14		24	
6	14		29	
7	15		34	
8	27			
9	21			
10	18		*Bin*	*Frequency*
11	19		14	4
12	18		19	8
13	22		24	5
14	33		29	2
15	16		34	1
16	18		More	0
17	17			
18	23			
19	28			
20	13			

FIGURE 2.12
A Histogram of the Audit-Time Data Constructed by Excel

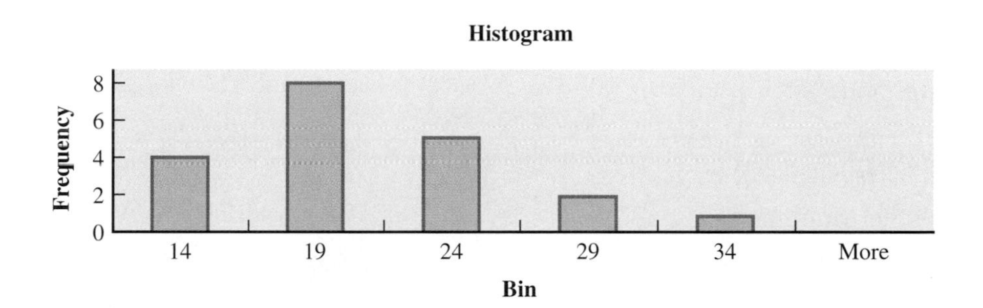

case the data set contains data values greater than the upper limit of the last bin specified by the user. When interpreting the output, the user must remember that the bin values are the upper limits of the corresponding classes. Figure 2.12 is the histogram generated as a result of the preceding steps. We caution that the values shown on the horizontal axis are the upper limits and not class midpoints.

To convert the Excel frequency distribution into an easier-to-read format, the user can replace the title Bin in cell C10 with a descriptive title such as Audit Times. The upper limits in cells C11 through C15 can be replaced by the class limits of 10–14, 15–19, 20–24, 25–29, and 30–34. Finally, the title Total and the sum of the frequencies (50) can be entered in cells C17 and D17 to complete the frequency distribution. Finally, the axis label Bin for the histogram can be replaced by a descriptive title such as Audit Times.

CHAPTER 3

Descriptive Statistics II: Numerical Methods

Barnes Hospital*

St. Louis, Missouri

statistics in practice

Barnes Hospital at the Washington University Medical Center, established in 1914, is the leading provider of health care for the people of St. Louis and neighboring areas. The hospital is nationally recognized as one of the best in the United States. The Hospice Program at Barnes Hospital improves the quality of life for terminally ill patients and their families. The hospice team consists of a medical director, coordinator, RN supervisor, home and inpatient RNs, home health aids, social workers, chaplains, dietitians, trained volunteers, and professionals from other ancillary services as needed. Through the coordinated efforts of the hospice team, patients and families are given the guidance and support necessary to cope with the strains created by serious illness, separation, and death.

In the coordination and administration of the hospice program, monthly reports and quarterly summaries help team members review the ongoing services. Statistical summaries of performance data are used as a basis for planning and implementing policy changes.

For example, data are collected on the length of time patients stay in the hospice program. A sample of 67 patient records showed that the time in the program ranged from one day to 185 days. A frequency distribution was helpful in summarizing and communicating the length-of-stay data. In addition, the following numerical measures of descriptive statistics were used to provide valuable information about the patient time in the program.

Mean:	35.7 days
Median:	17 days
Mode:	1 day

Ms. Paula H. Gianino, Hospice Coordinator at Barnes Hospital, provided this Statistics in Practice.

Interpretation of these statistics shows that the mean, or average, time a patient stays in the program is 35.7 days, or slightly over a month. However, the median shows that half of the patients are in the program 17 days or less and half are in the program 17 days or more. The mode of one day is the most frequent data value and indicates that many patients have a short stay in the program.

Other statistical summaries about the hospice program include the number of admissions, the number of days spent at home versus the number of days in the inpatient unit, the number of discharges from the inpatient unit, and the number of patient deaths at home and in the inpatient unit. These summaries are analyzed according to patient age and Medicare coverage. Overall, descriptive statistics provide valuable information about the hospice services.

In this chapter you will learn how to compute and interpret the statistical measures used by Barnes Hospital. In addition to the mean, median, and mode, you will learn about other descriptive statistics such as range, variance, standard deviation, percentiles, and correlation. These numerical measures will assist in the understanding and interpretation of data.

Barnes Hospital is a leader in health care for all ages.

In Chapter 2 we discussed tabular and graphical methods used to summarize data. These procedures are effective in written reports and as visual aids for presentations to individuals or groups. In this chapter, we present several numerical methods of descriptive statistics that provide additional alternatives for summarizing data.

We start by considering data sets consisting of a single variable. Whenever the data for a single variable, such as age, salary, or the like, have been obtained from a sample of n elements, the data set will contain n items, or data values. The numerical measures of location and dispersion are computed by using the n data values. If there is more than one variable, such numerical measures can be computed separately for each variable. In the two-variable case, we will develop measures of the strength of the relationship between the variables.

Several numerical measures of location, dispersion, and association are introduced. If the measures are computed for data from a sample, they are called *sample statistics.* If the measures are computed for data from a population, they are called *population parameters.*

3.1 Measures of Location

Mean

Perhaps the most important numerical measure of location is the *mean,* or average value, for a variable. The mean provides a measure of central location. It is obtained by adding all the data values and dividing by the number of items. If the data are from a sample, the mean is denoted by $\bar{x}$; if the data are from a population, the mean is denoted by the Greek letter μ.

In specifying statistical formulas, it is customary to denote the value of the first data item by x_1, the value of the second data item by x_2, and so on. In general, the ith data value is denoted by x_i. Stated in this notation, the formula for the sample mean follows.

Sample Mean

$$\bar{x} = \frac{\Sigma x_i}{n} \tag{3.1}$$

The number of items in the sample is denoted by n. In this formula, the numerator is the sum of the n data values. That is,

$$\Sigma x_i = x_1 + x_2 + \cdots + x_n$$

The Greek letter Σ is the summation sign.

To illustrate the computation of a sample mean, let us consider the following class-size data for a sample of five college classes.

46 54 42 46 32

We use the notation x_1, x_2, x_3, x_4, x_5 to represent the number of students in each of the five classes.

$$x_1 = 46 \quad x_2 = 54 \quad x_3 = 42 \quad x_4 = 46 \quad x_5 = 32$$

Hence, to compute the sample mean, we can write

$$\bar{x} = \frac{\Sigma x_i}{n} = \frac{x_1 + x_2 + x_3 + x_4 + x_5}{5} = \frac{46 + 54 + 42 + 46 + 32}{5} = 44$$

For the five classes sampled, the mean class size is 44 students.

Another illustration of the computation of a sample mean is given in the following situation. Suppose that a college placement office sent a questionnaire to a sample of business school graduates requesting information on starting salaries. Table 3.1 shows

TABLE 3.1 Monthly Starting Salaries for a Sample of Business School Graduates

Graduate	Monthly Salary ($)	Graduate	Monthly Salary ($)
1	2350	7	2390
2	2450	8	2630
3	2550	9	2440
4	2380	10	2825
5	2255	11	2420
6	2210	12	2380

the data that have been collected. The mean monthly starting salary for the sample of 12 business college graduates is computed as

$$\bar{x} = \frac{\Sigma x_i}{n} = \frac{x_1 + x_2 + \cdots + x_{12}}{12}$$
$$= \frac{2350 + 2450 + \cdots + 2380}{12}$$
$$= \frac{29{,}280}{12} = 2440$$

Equation (3.1) shows how the mean is computed for a sample with n items. The formula for computing the mean of a population is the same, but we use different notation to indicate that we are working with the entire population. The number of elements in the population is denoted by N and the symbol for the population mean is μ.

Population Mean

$$\mu = \frac{\Sigma x_i}{N} \tag{3.2}$$

Median

The *median* is another measure of central location. The median is the value in the middle when the data are arranged in ascending order (rank ordered from smallest to largest). If there is an odd number of values, the median is the middle value. If there is an even number of values, there is no single middle value. In this case, we follow the convention of defining the median to be the average of the two middle values. For convenience the definition of the median is restated as follows.

Median

If there is an odd number of values, the median is the middle value when the data are arranged in ascending order.

If there is an even number of values, the median is the average of the two middle values when the data are arranged in ascending order.

Let us apply this definition to compute the median class size for the sample of five college classes. Arranging the data in ascending order provides the following rank-ordered list.

32 42 46 46 54

Since $n = 5$ is odd, the median is the middle value in the rank-ordered list. Thus the median class size is 46 students. Even though there are two values of 46, each value is treated as a separate item when we arrange the data in ascending order and determine the median.

Suppose we also compute the median starting salary for the business college graduates. We arrange the data in Table 3.1 in ascending order.

2210 2255 2350 2380 2380 2390 2420 2440 2450 2550 2630 2825

Middle Two Values

Since $n = 12$ is even, we identify the middle two values. The median is the average of these two values.

$$\text{Median} = \frac{2390 + 2420}{2} = 2405$$

Although the mean is the more commonly used measure of central location, there are some situations in which the median is preferred. For instance, suppose that one of the graduates had a starting salary of $10,000 per month (maybe the individual's family owns the company). If we change the highest monthly starting salary in Table 3.1 from $2825 to $10,000 and recompute the mean, the sample mean changes from 2440 to 3038. The median of 2405, however, is unchanged, since 2390 and 2420 are still the middle two values. With the extremely high starting salary included, the median of 2405 provides a better measure of central location than the mean of 3038. We can generalize to say that whenever there are extreme data values, the median is often the preferred measure of central location.

Mode

A third measure of location is the *mode*. The mode is defined as follows.

Mode

The mode is the data value that occurs with greatest frequency.

To illustrate the identification of the mode, consider the sample of five class sizes. The only value that occurs more than once is 46. Since this value, occurring with a frequency of 2, has the greatest frequency, it is the mode. As another illustration, consider the sample of starting salaries for the business school graduates. The only monthly starting salary that occurs more than once is 2380. Since this value has the greatest frequency, it is the mode.

Situations can arise for which the greatest frequency occurs at two or more different values. In these instances more than one mode exists. If the data have exactly two modes, we say that the data are *bimodal.* If data have more than two modes, we say that the data are *multimodal.* In multimodal cases the mode is almost never reported, since listing three or more modes would not be very helpful in describing a location for the data.

The mode is an important measure of location for qualitative data. For example, the qualitative data set in Table 2.1 resulted in the following frequency distribution for personal computer purchases.

Company	**Frequency**
Apple	13
Compaq	12
Gateway 2000	5
IBM	9
Packard Bell	11
Total	50

The mode, or most frequently purchased personal computer, is Apple. For this type of data it obviously makes no sense to speak of the mean or median. The mode provides the information of interest, the most frequently purchased brand of personal computer.

Percentiles

A *percentile* is a measure that locates values in the data that are not necessarily central locations. A percentile provides information about how the data are spread over the interval from the smallest value to the largest value. For data that do not have numerous repeated values, the pth percentile divides the data into two parts. Approximately p percent of the values are less than the pth percentile; approximately $(100 - p)$ percent of the values are greater than the pth percentile. The pth percentile is formally defined as follows.

Percentile

The pth percentile is a value such that *at least* p percent of the data have this value or less and *at least* $(100 - p)$ percent of the data have this value or more.

Admission test scores for colleges and universities are frequently reported in terms of percentiles. For instance, suppose an applicant obtains a raw score of 54 on the verbal portion of an admission test. How this student performed in relation to other students taking the same test may not be readily apparent. However, if the raw score of 54 corresponds to the 70th percentile, we know that approximately 70% of the students had scores lower than this individual's and approximately 30% of the students had scores higher than this individual's.

The following procedure can be used to compute the pth percentile.

Calculating the pth Percentile

STEP 1. Arrange the data in ascending order (rank order from smallest value to largest value).

STEP 2. Compute an index i

$$i = \left(\frac{p}{100}\right)n$$

where p is the percentile of interest and n is the number of items.

STEP 3. (a) If *i is not an integer, round up.* The next integer value *greater* than i denotes the position of the pth percentile.
(b) If *i is an integer,* the pth percentile is the average of the data values in positions i and $i + 1$.

As an illustration of this procedure, let us determine the 85th percentile for the starting salary data in Table 3.1.

STEP 1. Arrange the data in ascending order.

2210 2255 2350 2380 2380 2390 2420 2440 2450 2550 2630 2825

STEP 2.

$$i = \left(\frac{p}{100}\right)n = \left(\frac{85}{100}\right)12 = 10.2$$

STEP 3. Since i is not an integer, *round up.* The position of the 85th percentile is the next integer greater than 10.2, the 11th position.

Returning to the data, we see that the 85th percentile corresponds to the 11th data position, or 2630.

As another illustration of this procedure, let us consider the calculation of the 50th percentile. Applying step 2, we obtain

$$i = \left(\frac{50}{100}\right)12 = 6$$

Since i is an integer, step 3(b) states that the 50th percentile is the average of the sixth and seventh data values; thus the 50th percentile is (2390 + 2420)/2 = 2405. Note that the *50th percentile is also the median.*

Quartiles

It is often desirable to divide data into four parts, with each part containing approximately one-fourth, or 25%, of the data. Figure 3.1 shows data divided into four parts. The division points are referred to as the *quartiles* and are defined as

Q_1 = first quartile, or 25th percentile,

Q_2 = second quartile, or 50th percentile (also the median), and

Q_3 = third quartile, or 75th percentile.

FIGURE 3.1
Location of the Quartiles

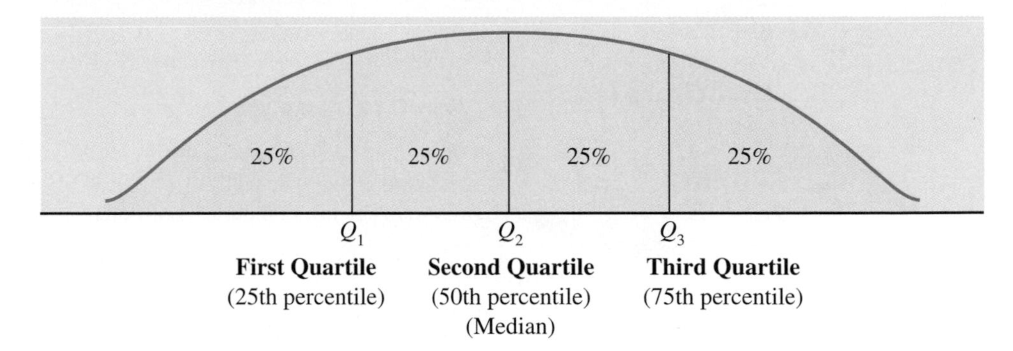

The monthly starting salary data are again arranged in ascending order, Q_2, the second quartile (median), has already been identified as 2405.

2210 2255 2350 2380 2380 2390 2420 2440 2450 2550 2630 2825

The computations of Q_1 and Q_3 require the use of the rule for finding the 25th and 75th percentiles. Those calculations follow.

For Q_1,

$$i = \left(\frac{p}{100}\right)n = \left(\frac{25}{100}\right)12 = 3$$

Since i is an integer, step 3(b) indicates that the first quartile, or 25th percentile, is the average of the third and fourth data values; thus, $Q_1 = (2350 + 2380)/2 = 2365$.

For Q_3,

$$i = \left(\frac{p}{100}\right)n = \left(\frac{75}{100}\right)12 = 9$$

Again, since i is an integer, step 3(b) indicates that the third quartile, or 75th percentile, is the average of the ninth and tenth data values; thus, $Q_3 = (2450 + 2550)/2 = 2500$.

As shown below, the quartiles have divided the data into four parts, with each part consisting of 25% of the data.

2210 2255 2350 | 2380 2380 2390 | 2420 2440 2450 | 2550 2630 2825

$Q_1 = 2365$ $Q_2 = 2405$ (Median) $Q_3 = 2500$

We have defined the quartiles as the 25th, 50th, and 75th percentiles. Thus, we have computed the quartiles in the same way as the other percentiles. However, there are some variations in the conventions used to compute quartiles; the quartiles may vary slightly depending on the convention used. Nevertheless, the objective of all procedures for computing quartiles is to divide data into four equal parts.

exercises

Methods

1. Consider the sample of size 5 with data values of 10, 20, 12, 17, and 16. Compute the mean and median.

2. Consider the sample of size 8 with data values of 27, 25, 20, 15, 30, 34, 28, and 25. Compute the 20th, 25th, 65th, and 75th percentiles.

Applications

3. According to *U.S. News & World Report,* entry-level jobs in accounting paid from $27,000 to $31,000 annually (October 31, 1994). A sample of entry-level salaries follows. Data are in thousands of dollars.

29.6	28.5	28.6	29.4	28.6
31.0	29.2	30.8	28.4	27.5
30.4	29.7	27.2	30.3	29.6
28.0	26.7	28.7	28.6	29.1
28.8	28.5	29.0	27.4	29.9

a. What is the mean entry-level salary?
b. What is the median entry-level salary?
c. What is the mode?
d. What is the first quartile?
e. What is the third quartile?

4. The American Association of Individual Investors conducts an annual survey of discount brokers. Shown below are the commissions charged for a trade of 500 shares at $50 per share for a sample of 20 discount brokers. (*AAII Journal,* January, 1994).

Broker	Commission ($)	Broker	Commission ($)
Peck	100	Pace	75
Aufhauser	32	People's	131
Broker's Ex	160	Pro Value	173
Burke	120	Royal Grimm	75
Schwab	155	Seaport	50
Downstate	90	St. Louis	64
Freeman	145	T. Rowe	134
Kennedy	33	Unified	154
Max Ule	195	White	42
Mongerson	95	Your	55

a. Compute the mean, median, and mode for the commission charged.
b. Compute and interpret the first and third quartiles.

5. *American Demographics* (December 1988) reported that 25 million Americans get up each morning and go to work in their offices at home. The growing use of personal computers is suggested to be one of the reasons more people can operate at-home businesses. The article presented data on the ages of individuals who work at home. Following is a sample of age data for these individuals.

22	58	24	50	29	52	57	31	30	41
44	40	46	29	31	37	32	44	49	29

a. Compute the mean and mode.
b. The median age of the population of all adults is 40.5 years. Use the median age of the preceding data to comment on whether the at-home workers tend to be younger or older than the population of all adults.
c. Compute the first and third quartiles.
d. Compute and interpret the 32nd percentile.

6. A bowler's scores for six games are 182, 168, 184, 190, 170, and 174. Using these data as a sample, compute the following descriptive statistics.

a. Mean **b.** Median **c.** Mode **d.** 75th percentile

7. The American Association of Advertising Agencies records data on nonprogramming minutes per half-hour of prime-time television programming (*U.S. News & World Report,* April 13, 1992). Representative data follow for a sample of prime-time programs on major networks at 8:30 P.M.

6.0	6.6	5.8	7.0	6.3	6.2	7.2	5.7	6.4	7.0
6.5	6.2	6.0	6.5	7.2	7.3	7.6	6.8	6.0	6.2

a. Compute the mean and median.
b. Compute the first and third quartiles.
c. Using the sample mean, find the percentage of viewing time spent on prime-time advertisements, promotions, and credits. What percentage of viewing time is spent on the programs themselves?

8. The *Los Angeles Times* regularly reports the air quality index for various areas of Southern California. Index ratings of 0–50 are considered good, 51–100 moderate, 101–200 unhealthy, 201–275 very unhealthy, and over 275 hazardous. Recent air quality indexes for Pomona were 28, 42, 58, 48, 45, 55, 60, 49, and 50.
a. Compute the mean, median, and mode for the data. Should the Pomona air quality index be considered good?
b. Compute the 25th percentile and 75th percentile for the Pomona air quality data.

9. In automobile mileage and gasoline-consumption testing, 13 automobiles were road tested for 300 miles in both city and country driving conditions. The following data were recorded for miles-per-gallon performance.

City:	16.2	16.7	15.9	14.4	13.2	15.3	16.8	16.0	16.1	15.3	15.2	15.3	16.2
Country:	19.4	20.6	18.3	18.6	19.2	17.4	17.2	18.6	19.0	21.1	19.4	18.5	18.7

Use the mean, median, and mode to make a statement about the difference in performance for city and country driving.

10. Pennsylvania is the nation's fifth largest producer of Christmas trees, with a crop of 1.5 million trees in 1994. Tree prices ranged from $3.50 to $5.50 per foot (*The Philadelphia Inquirer,* December 3, 1994). Suppose the following data represent the price per foot for 16 trees sold in the Philadelphia area.

3.90	4.20	3.90	5.10
4.20	4.50	4.10	5.10
4.30	4.20	4.00	5.20
4.50	4.20	4.50	5.10

a. Compute the mean, median, and mode.
b. Compute the 20th and 90th percentiles.
c. Compute the quartiles.

3.2 Measures of Dispersion

In addition to measures of location, it is often desirable to consider measures of dispersion, or variability, in data. For example, assume that you are a purchasing agent for a large manufacturing firm and that you regularly place orders with two different suppliers. Both suppliers indicate that approximately 10 working days are required to fill your orders. After several months of operation you find that the mean number of days required to fill orders is indeed 10 days for both of the suppliers. The histograms summarizing the number of working days required to fill orders from the suppliers are shown in Figure 3.2. Although the mean number of days is 10 for both suppliers, do the two suppliers have the same degree of reliability in terms of making deliveries on

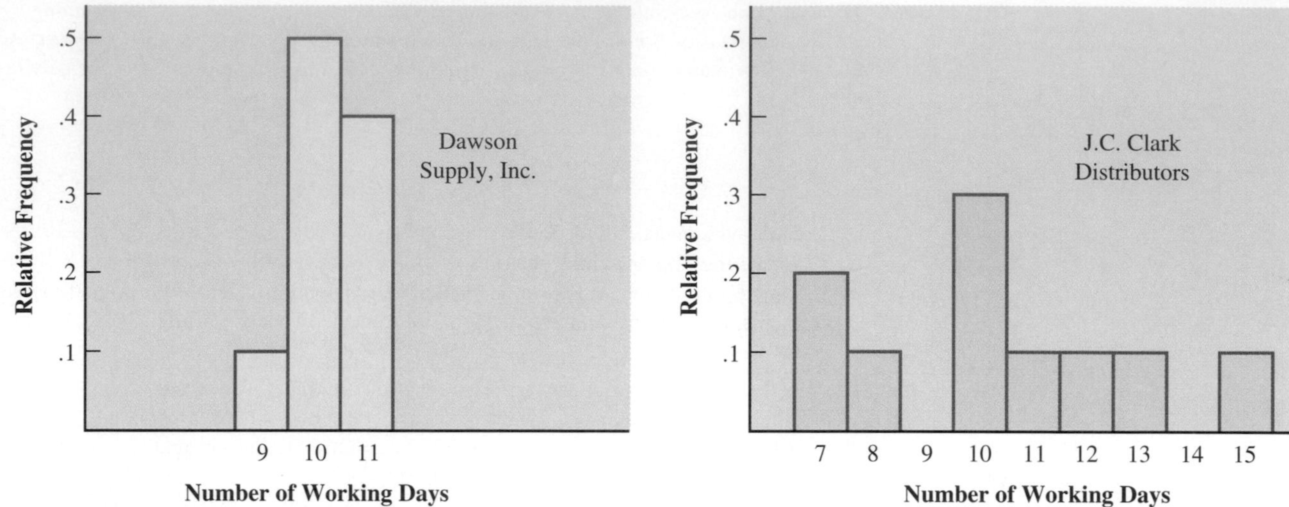

FIGURE 3.2
Historical Data Showing the Number of Days Required to Fill Orders

schedule? Note the dispersion, or variability, in the histograms. Which supplier would you prefer?

For most firms, receiving materials and supplies on schedule is important. The seven- or eight-day deliveries shown for J. C. Clark Distributors might be viewed favorably; however, a few of the slow 13- to 15-day deliveries could be disastrous in terms of keeping a workforce busy and production on schedule. This example illustrates a situation in which the dispersion, or variability, in the delivery times may be an overriding consideration in selecting a supplier. For most purchasing agents, the lower dispersion shown for Dawson Supply, Inc. would make Dawson the more consistent and preferred supplier.

We turn now to a discussion of some commonly used numerical measures of the dispersion, or variability, in data.

Range

Perhaps the simplest measure of dispersion for a data set is the *range*.

Range

$$\text{Range} = \text{Largest Value} - \text{Smallest Value}$$

Let us refer to the data on monthly starting salaries for business school graduates in Table 3.1. The largest starting salary is 2825 and the smallest is 2210. The range is 2825 − 2210 = 615.

Although the range is the easiest of the measures of dispersion to compute, it is seldom used as the only measure of dispersion. The reason is that the range is based on only two data values and is highly influenced by an extremely larger and/or an extremely small data value. Suppose one of the graduates had a starting salary of $10,000.

In this case the range would be 10,000 – 2210 = 7790 rather than 615. This large value for the range would not be very descriptive of the variability in the data, since 11 of the 12 starting salaries are closely grouped between 2210 and 2630.

Interquartile Range

A measure of dispersion that overcomes the dependency on extreme data values is the *interquartile range* (IQR). This measure of dispersion is simply the difference between the third quartile, Q_3, and the first quartile, Q_1. In other words, the interquartile range is the range for the middle 50% of the data.

Interquartile Range

$$IQR = Q_3 - Q_1 \tag{3.3}$$

For the data on monthly starting salaries, the quartiles are $Q_3 = 2500$ and $Q_1 = 2365$. Thus the interquartile range is 2500 – 2365 = 135.

Variance

The *variance* is a measure of dispersion that utilizes all of the data. The variance is based on the difference between each data value and the mean. The difference between each data value x_i and the mean ($\bar{x}$ for a sample, μ for a population) is called a *deviation about the mean.* For a sample, a deviation is written $(x_i - \bar{x})$; for a population, it is written $(x_i - \mu)$. In the computation of the variance, each deviation about the mean is *squared.*

If the data set is a population, the average of the squared deviations is called the *population variance.* The population variance is denoted by the Greek symbol σ^2. For population of size N and with μ denoting the population mean, the definition of the population variance is as follows.

Population Variance

$$\sigma^2 = \frac{\Sigma (x_i - \mu)^2}{N} \tag{3.4}$$

In most statistical applications, the data being analyzed is from a sample of size n. When we compute a sample variance, we are often interested in using it to estimate the population variance σ^2. Although a detailed explanation is beyond the scope of this text, it can be shown that if the sum of the squared deviations about the sample mean is divided by $n - 1$, and not n, the resulting sample variance provides an unbiased estimate of the population variance. For this reason, the *sample variance,* denoted by s^2, is defined as follows.

Sample Variance

$$s^2 = \frac{\Sigma(x_i - \bar{x})^2}{n - 1} \tag{3.5}$$

To illustrate the computation of the sample variance, we use the data on class size for the sample of five college classes. A summary of the data, including the computation of the deviations about the mean and the squared deviations about the mean, is given in Table 3.2. The sum of squared deviations about the mean is $\Sigma(x_i - \bar{x})^2 = 256$. Hence, with $n - 1 = 4$, the sample variance is

$$s^2 = \frac{\Sigma(x_i - \bar{x})^2}{n - 1} = \frac{256}{4} = 64$$

Before moving on, let us note that the units associated with the sample variance often cause confusion. Since the values being summed in the variance calculation, $(x_i - \bar{x})^2$, are squared, the units associated with the sample variance are also *squared.* For instance, the sample variance for the class-size data is $s^2 = 64$ (student)^2. The squared units associated with variance make it difficult to obtain an intuitive understanding and interpretation of the numerical value of the variance. We recommend that you think of the variance as a measure useful in comparing the dispersion in two or more data sets. In a comparison of data sets, the one with the larger variance has the most dispersion. Further interpretation of the value of the variance may not be necessary.

As another illustration of computing a sample variance, consider the starting salaries listed in Table 3.1 for the 12 business school graduates. In Section 3.1, we showed that the sample mean starting salary was 2440. The computation of the sample variance ($s^2 = 27{,}440.91$) is shown in Table 3.3.

Note that in Tables 3.2 and 3.3 we show both the sum of the deviations about the mean and the sum of the squared deviations about the mean. For any data set, the sum of the deviations about the mean will *always equal zero.* Hence, as shown in Tables 3.2

TABLE 3.2 Computation of Deviations and Squared Deviations About the Mean for the Class-Size Data

Number of Students x_i	Sample Mean $\bar{x}$	Deviation About the Mean $(x_i - \bar{x})$	Squared Deviation About the Mean $(x_i - \bar{x})^2$
46	44	2	4
54	44	10	100
42	44	−2	4
46	44	2	4
32	44	−12	144
		0	256
		$\Sigma(x_i - \bar{x})$	$\Sigma(x_i - \bar{x})^2$

TABLE 3.3 Computation of the Sample Variance for the Starting Salary Data

Monthly Salary x_i	Sample Mean $\bar{x}$	Deviation About the Mean $(x_i - \bar{x})$	Squared Deviation About the Mean $(x_i - \bar{x})^2$
2350	2440	−90	8,100
2450	2440	10	100
2550	2440	110	12,100
2380	2440	−60	3,600
2255	2440	−185	34,225
2210	2440	−230	52,900
2390	2440	−50	2,500
2630	2440	190	36,100
2440	2440	0	0
2825	2440	385	148,225
2420	2440	−20	400
2380	2440	−60	3,600
		0	301,850
		$\Sigma(x_i - \bar{x})$	$\Sigma(x_i - \bar{x})^2$

Using (3.5),

$$s^2 = \frac{\Sigma(x_i - \bar{x})^2}{n-1} = \frac{310{,}850}{11} = 27{,}440.91$$

and 3.3, $\Sigma(x_i - \bar{x}) = 0$. This is true because the positive deviations and negative deviations always cancel each other, causing the sum of the deviations about the mean to equal zero.

Standard Deviation

The *standard deviation* is defined to be the positive square root of the variance. Following the notation we adopted for a sample variance and a population variance, we use s to denote the sample standard deviation and σ to denote the population standard deviation. The standard deviation is derived from the variance in the following way.

Standard Deviation

$$\text{Sample Standard Deviation} = s = \sqrt{s^2} \quad (3.6)$$

$$\text{Population Standard Deviation} = \sigma = \sqrt{\sigma^2} \quad (3.7)$$

Recall that the sample variance for the sample of class sizes in five college classes is $s^2 = 64$. Thus the sample standard deviation is $s = \sqrt{64} = 8$. For the data of starting salaries, the sample standard deviation is $s = \sqrt{27{,}440.91} = 165.65$.

What is gained by converting the variance to its corresponding standard deviation? Recall that the units associated with the variance are squared. For example, the sample variance for the starting salary data of business school graduates is $s^2 = 27{,}440.91$ $(\text{dollars})^2$. Since the standard deviation is simply the square root of the variance, the units of the variance, dollars squared, are converted to dollars in the standard deviation. Thus, the standard deviation of the starting salary data is \$165.65. In other words, the standard deviation is measured in the same units as the original data. For this reason the standard deviation is easily compared to the mean and other statistics that are measured in the same units as the original data.

Coefficient of Variation

In some situations we may be interested in a descriptive statistic that indicates how large the standard deviation is in relation to the mean. This measure is called the *coefficient of variation* and is computed as follows.

Coefficient of Variation

$$\frac{\text{Standard Deviation}}{\text{Mean}} \times 100 \tag{3.8}$$

For the class-size data, we found a sample mean of 44 and a sample standard deviation of 8. The coefficient of variation is $(8/44) \times 100 = 18.2$. In words, the coefficient of variation tells us that the standard deviation of the sample is 18.2% of the value of the sample mean. For the starting-salary data with a sample mean of 2440 and a sample standard deviation of 165.65, the coefficient of variation, $(165.65/2440) \times 100 = 6.8$, tells us the standard deviation for this sample is only 6.8% of the value of the sample mean. In general, the coefficient of variation is a useful statistic for comparing the dispersion in data sets having different standard deviations and different means.

notes and comments

1. Rounding the value of the sample mean $\bar{x}$ and the values of the squared deviations $(x_i - \bar{x})^2$ may introduce rounding errors in the computation of the variance and standard deviation. To reduce rounding errors, we recommend carrying at least six significant digits during intermediate calculations. The resulting variance or standard deviation can then be rounded to fewer digits if desired.

2. An alternative formula for the computation of the sample variance is

$$s^2 = \frac{\Sigma x_i^2 - n\bar{x}^2}{n - 1}$$

where $\Sigma x_i^2 = x_1^2 + x_2^2 + \cdots + x_n^2$. Using this formula eases the computational burden slightly and helps reduce rounding errors. Exercise 17 requires use of this alternative formula to compute the sample variance.

exercises

Methods

11. Consider the sample of size 5 with data values of 10, 20, 12, 17, and 16. Compute the range and interquartile range.

12. Consider the sample of size 5 with data values of 10, 20, 12, 17, and 16. Compute the variance and standard deviation.

13. Consider the sample of size 8 with data values of 27, 25, 20, 15, 30, 34, 28, and 25. Compute the range, interquartile range, variance, and standard deviation.

Applications

14. The Hawaii Visitors Bureau collects data on the number of visitors to the islands. The following data are a representative sample of visitors (in thousands) for several days in November 1994 (*The Honolulu Advertiser*, December 28, 1994).

From the mainland, Canada, and Europe:

108.70	112.25	94.01	144.03	162.44	161.61	76.20
102.11	110.87	79.36	129.04	95.16	114.16	121.88

From Asia and the Pacific:

29.89	41.13	40.67	40.41	43.07	24.86
31.61	21.60	27.34	64.57	32.98	41.31

a. Compute the mean and median number of daily visitors from the two sources.
b. Compute the range, the standard deviation, and the coefficient of variation for the two sources of visitors.
c. What comparisons can you make between the numbers of visitors from the two sources?

15. The prices for the population of the 15 basic models of drip coffeemakers follow (*Consumer Reports 1995 Buying Guide*).

Model	Price ($)	Model	Price ($)	Model	Price ($)
Mr. Coffee PR12A	27	Mr. Coffee PR16	25	Braun	60
Krups	50	Mr. Coffee BL110	22	Proctor 42401	35
Proctor 42301	20	Braun	35	Krups	40
Black & Decker 901	22	Bunn	40	Melitta	30
Black & Decker 900	20	West Bend	35	Betty Crocker	19

Compute the range, variance, and standard deviation for this population.

16. The *Los Angeles Times* regularly reports the air quality index for various areas of Southern California. A sample of air quality index values for Pomona provided the following data: 28, 42, 58, 48, 45, 55, 60, 49, and 50.

a. Compute the range and interquartile range.
b. Compute the sample variance and sample standard deviation.
c. A sample of air quality index readings for Anaheim provided a sample mean of 48.5, a sample variance of 136, and a sample standard deviation of 11.66. What comparisons can you make between the air quality in Pomona and that in Anaheim on the basis of these descriptive statistics?

17. The Davis Manufacturing Company has just completed five weeks of operation using a new process that is supposed to increase productivity. The numbers of parts produced each week are 410, 420, 390, 400, and 380. Compute the sample variance and sample standard deviation by using the definition of sample variance (3.5) as well as the alternative formula provided in the Notes and Comments.

18. Assume that the following data are used to construct the histograms of the number of days required to fill orders for Dawson Supply, Inc. and J. C. Clark Distributors (see Figure 3.2).

Dawson Supply Days for Delivery:	11	10	9	10	11	11	10	11	10	10
Clark Distributors Days for Delivery:	8	10	13	7	10	11	10	7	15	12

Use the range and standard deviation to support the previous observation that Dawson Supply provides the more consistent and reliable delivery times.

Self Test

19. A bowler's scores for six games were 182, 168, 184, 190, 170, and 174. Using these data as a sample, compute the following descriptive statistics.

a. Range **b.** Variance
c. Standard deviation **d.** Coefficient of variation

LAWAGES

20. The Union Bank of Switzerland conducted a survey to obtain data on the hourly wages of blue- and white-collar workers throughout the world (*Newsweek,* February 17, 1992). Assume that a sample of 25 workers in the Los Angeles area provided the following data.

11.50	8.40	11.75	10.05	10.25	8.00	13.65	7.05	9.05
11.90	9.90	6.85	15.35	11.10	14.70	13.15	13.10	6.65
13.10	9.20	9.15	12.05	8.45	5.85	9.80		

Provide the following descriptive statistics.

a. Mean **b.** Median
c. Range **d.** Interquartile range
e. Variance **f.** Standard deviation

3.3 Some Uses of the Mean and the Standard Deviation

We have described several measures of location and dispersion for data. The mean is the most widely used measure of location, whereas the standard deviation and variance are the most widely used measures of dispersion. Using only the mean and the standard deviation, we can learn much about a data set.

z-Scores

By using the mean and standard deviation, we can determine the relative location of any data value. Suppose we have a sample of n items, with the values denoted by $x_1, x_2, \ldots, x_n$. In addition, assume that the sample mean, $\bar{x}$, and the sample standard deviation, s, have been computed. Associated with each data value, x_i, is another value called its *z-score.* Equation (3.9) shows how the z-score is computed for data value x_i.

z-Score

$$z_i = \frac{x_i - \bar{x}}{s} \tag{3.9}$$

where

z_i = the z-score for item i

$\bar{x}$ = the sample mean

s = the sample standard deviation

TABLE 3.4 z-Scores for the Class-Size Data

Number of Students (x_i)	Deviation About the Mean ($x_i - \bar{x}$)	z-Score $\left(\frac{x_i - \bar{x}}{s}\right)$
46	2	2/8 = .25
54	10	10/8 = 1.25
42	−2	−2/8 = −.25
46	2	2/8 = .25
32	−12	−12/8 = −1.50

The z-score is often called the *standardized value*. The standardized value or z-score, z_i, can be interpreted as the *number of standard deviations x_i is from the mean $\bar{x}$*. For example, $z_1 = 1.2$ would indicate x_1 is 1.2 standard deviations greater than the sample mean. Similarly, $z_2 = -.5$ would indicate x_2 is .5, or 1/2, standard deviation less than the sample mean. As can be seen from (3.9), z-scores greater than zero occur for data with values greater than the mean, and z-scores less than zero occur for data with values less than the mean. A z-score of zero indicates that the value of the item is equal to the mean.

The z-score for any item can be interpreted as a measure of the relative location of the item in a data set. Items in two different data sets with the same z-score can be said to have the same relative location in terms of being the same number of standard deviations from the mean.

The z-scores for the class-size data are listed in Table 3.4. Recall that the sample mean $\bar{x} = 44$ and sample standard deviation $s = 8$ have been computed previously. The z-score of −1.50 for the fifth item shows it is farthest from the mean; it is 1.50 standard deviations below the mean.

Chebyshev's Theorem

Chebyshev's theorem enables us to make statements about the percentage of the data that must be within a specified number of standard deviations from the mean.

Chebyshev's Theorem

At least $(1 - 1/k^2)$ of the data must be within k standard deviations of the mean, where k is any value greater than 1.

Some of the implications of this theorem, with $k = 2$, 3, and 4 standard deviations, follow.

- At least .75, or 75%, of the data must be within $k = 2$ standard deviations of the mean.
- At least .89, or 89%, of the data must be within $k = 3$ standard deviations of the mean.
- At least .94, or 94%, of the data must be within $k = 4$ standard deviations of the mean.

For an example using Chebyshev's theorem, assume that the midterm test scores for 100 students in a college business statistics course had a mean of 70 and a standard deviation of 5. How many students had test scores between 60 and 80? How many students had test scores between 58 and 82?

For the test scores between 60 and 80, we note that the value 60 is two standard deviations below the mean and the value 80 is two standard deviations above the mean. Using Chebyshev's theorem with $k = 2$, we see that at least .75 or at least 75% of the data must be within two standard deviations of the mean. Thus, at least 75 of the 100 students must have test scores between 60 and 80.

For the test scores between 58 and 82, we see that $(58 - 70)/5 = -2.4$ indicates 58 is 2.4 standard deviations below the mean and that $(82 - 70)/5 = +2.4$ indicates 82 is 2.4 standard deviations above the mean. Applying Chebyshev's theorem with $k = 2.4$, we have

$$\left(1 - \frac{1}{k^2}\right) = \left[1 - \frac{1}{(2.4)^2}\right] = .826$$

Thus, at least 82.6% of the students must have test scores between 58 and 82.

The Empirical Rule

One of the advantages of Chebyshev's theorem is that it applies to any data set regardless of the shape of the distribution of the data. In practical applications, however, it has been found that many data sets have a mound-shaped or bell-shaped distribution like the one shown in Figure 3.3. When the data are believed to approximate this distribution, the *empirical rule* can be used to determine the percentage of data that must be within a specified number of standard deviations of the mean.*

Empirical Rule

For data having a bell-shaped distribution:

- Approximately 68% of the data will be within one standard deviation of the mean.
- Approximately 95% of the data will be within two standard deviations of the mean.
- Almost all of the data will be within three standard deviations of the mean.

For example, liquid-detergent cartons are filled automatically on a production line. The filling weights have a bell-shaped distribution. If the mean filling weight is 16 ounces and the standard deviation is .25 ounces, we can use the empirical rule to draw the following conclusions.

- Approximately 68% of the filled cartons have weights between 15.75 and 16.25 ounces (that is, within one standard deviation of the mean).
- Approximately 95% of the filled cartons have weights between 15.50 and 16.50 ounces (that is, within two standard deviations of the mean).

*The empirical rule is based on the normal probability distribution, which is covered in Chapter 6.

FIGURE 3.3
A Mound-Shaped or Bell-Shaped Distribution

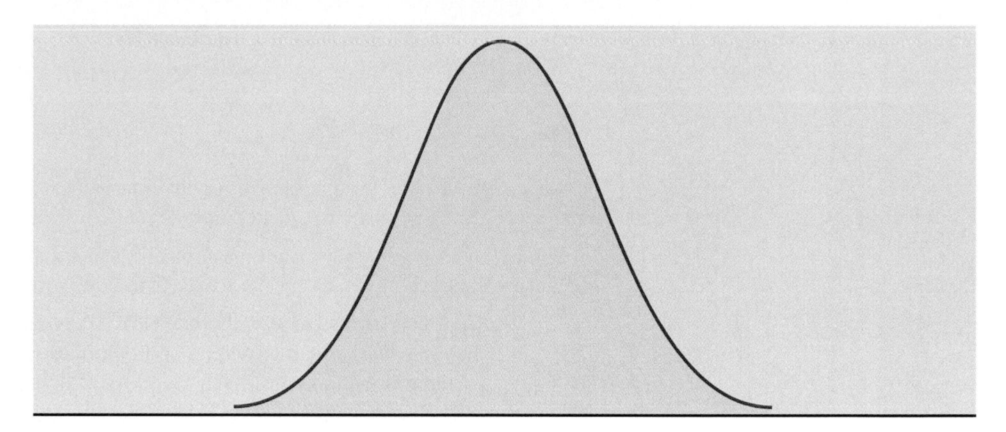

III Almost all filled cartons have weights between 15.25 and 16.75 ounces (that is, within three standard deviations of the mean).

Detecting Outliers

Sometimes a data set will have one or more unusually large or unusually small values. Values such as these are called *outliers.* Experienced statisticians take steps to identify outliers and then review each outlier carefully. An outlier may be data that has been incorrectly recorded. If so, it can be corrected before further analysis. An outlier may also be due to data that was incorrectly included in the data set; if so, it can be removed. Finally, an outlier may just be an unusual data value that has been recorded correctly and does belong in the data set. In such cases it should remain.

Standardized values (z-scores) can be used to help identify outliers. Recall that the empirical rule allows us to conclude that for a bell-shaped distribution, almost all data will be within three standard deviations of the mean. Hence, in using z-scores to identify outliers, we recommend treating any value with a z-score less than -3 or greater than $+3$ as an outlier. Such values can then be reviewed for accuracy and to determine whether or not they belong in the data set.

Refer to the z-scores for the class-size data in Table 3.4. The z-score of -1.50 shows the fifth item is farthest from the mean. However, this standardized value is well within the -3 to $+3$ guideline for outliers. Thus, the z-scores show that outliers are not present in the class-size data.

notes and comments

1. Before analyzing a data set, statisticians usually make a variety of checks to ensure the validity of data. In a large study it is not uncommon for errors to be made in recording data values or in entering the values at a computer. Identifying outliers is one tool used to check the validity of data.
2. Chebyshev's theorem is applicable for any data set and can be used to state the minimum number of items that will be within a certain number of standard deviations of the mean. If the data set is known to be approximately bell-shaped, more can be said. For instance, the empirical rule allows us to say that *approximately* 95% of the items will be within two standard deviations of the mean; Chebyshev's theorem allows us to conclude only that at least 75% of the items will be in that interval.

exercises

Methods

21. Consider the sample of size 5 with data values of 10, 20, 12, 17, and 16. Compute the z-score for each of the five data values.

22. Consider a sample with a mean of 500 and a standard deviation of 100. What is the z-score for each of the data values 520, 650, 500, 450, and 280?

23. Consider a sample with a mean of 30 and a standard deviation of 5. Use Chebyshev's theorem to determine the proportion, or percentage, of the data within each of the following ranges.

a. 20 to 40 **b.** 15 to 45 **c.** 22 to 38 **d.** 18 to 42 **e.** 12 to 48

24. Data that have a bell-shaped distribution have a mean of 30 and a standard deviation of 5. Use the empirical rule to determine the proportion, or percentage, of data within each of the following ranges.

a. 20 to 40 **b.** 15 to 45 **c.** 25 to 35

Applications

25. The Bureau of Economic Analysis at the Department of Commerce reported that 1993 per capita income in California was $21,884 (*USA Today,* August 24, 1994). If the standard deviation is $6,000, what is the z-score for an individual with an annual income of $12,500? What is the z-score of an individual with an annual income of $50,000? Interpret these scores and comment on whether or not either of these annual incomes should be considered an outlier.

26. A sample of 10 men's college basketball scores provided the following winning teams and the numbers of points scored (*USA Today,* January 11, 1995).

Winner	Score	Winner	Score
Boston U.	55	Emory	56
Northeastern	87	Queens College	77
Flagler	89	Millsaps	89
Marquette	70	Wartburg	64
Pepperdine	61	San Francisco	84

a. Compute the mean and standard deviation for the data.
b. In another game, York beat CCNY 108 to 75. Use the z-scores to determine whether the York score should be considered an outlier. Explain.
c. Assume that the distribution of the points scored by winning teams is mound-shaped. Estimate the percentage of all men's college basketball games in which the winning team will score 87 or more points. Estimate the percentage of games in which the winning team will score 46 or less points.

27. According to the Roth Young Personnel Service, salaries for chain store managers range from $30,000 to $62,000 (*National Business Employment Weekly,* October 16–22, 1994). Assume that the following data are the annual salaries for a sample of chain store managers. Data are in $1000s.

33.7	45.4	44.0	47.5	59.6
45.1	37.7	43.9	48.3	53.0
39.5	42.9	51.0	35.6	41.5
49.5	45.4	58.2	55.4	62.3
32.2	45.9	47.6	56.2	56.8
48.8	31.3	51.2	43.2	54.4

a. Compute the mean and standard deviation.
b. A chain store manager in Memphis, Tennessee, earns $28,500 a year. Compute the z-score for this manager and state whether you believe the manager's salary should be considered an outlier.
c. Compute the z-scores for salaries of $30,000, $45,000, $60,000, and $75,000. Should any of them be considered an outlier?

28. The average fuel economy of new cars sold in the United States is 27.5 miles per gallon (*The Wall Street Journal,* April 8, 1992). Assume that the standard deviation is 3.5 miles per gallon.
a. Use Chebyshev's theorem to calculate the percentage of new cars sold with miles-per-gallon ratings between 20.5 and 34.5, between 18.75 and 36.25, and between 17 and 38.
b. If it were reasonable to assume that miles-per-gallon ratings for new cars followed a bell-shaped distribution, what can be said about the percentage of new cars sold with miles-per-gallon ratings between 20.5 and 34.5? Between 17 and 38?

29. IQ scores and birth rates were discussed in an article in the *Atlantic Monthly* (May 1989). IQ scores have a bell-shaped distribution with a mean of 100 and a standard deviation of 15.
a. What percentage of the population should have an IQ score between 85 and 115?
b. What percentage of the population should have an IQ score between 70 and 130?
c. What percentage of the population should have an IQ score of more than 130?
d. A person with an IQ score of more than 145 is considered a genius. Does the empirical rule support this statement? Explain.

3.4 Exploratory Data Analysis

In Chapter 2 we introduced exploratory data analysis. Recall that exploratory data analysis enables us to use simple arithmetic and easy-to-draw pictures to summarize data. In this section we continue exploratory data analysis by considering five-number summaries and box plots.

Five-Number Summary

In a five-number summary, the following five numbers are used to summarize the data.

1. Smallest value
2. First quartile (Q_1)
3. Median
4. Third quartile (Q_3)
5. Largest value

The monthly starting salaries shown in Table 3.1 for a sample of 12 business school graduates follow.

2350	2450	2550	2380	2255	2210
2390	2630	2440	2825	2420	2380

The median of 2405 and the quartiles $Q_1 = 2365$ and $Q_3 = 2500$ were computed in Section 3.1. Reviewing the preceding data shows a smallest value of 2210 and a largest value of 2825. Thus the five-number summary for the salary data is 2210, 2365, 2405, 2500, 2825. Approximately one-fourth, or 25%, of the data values are between adjacent numbers in a five-number summary.

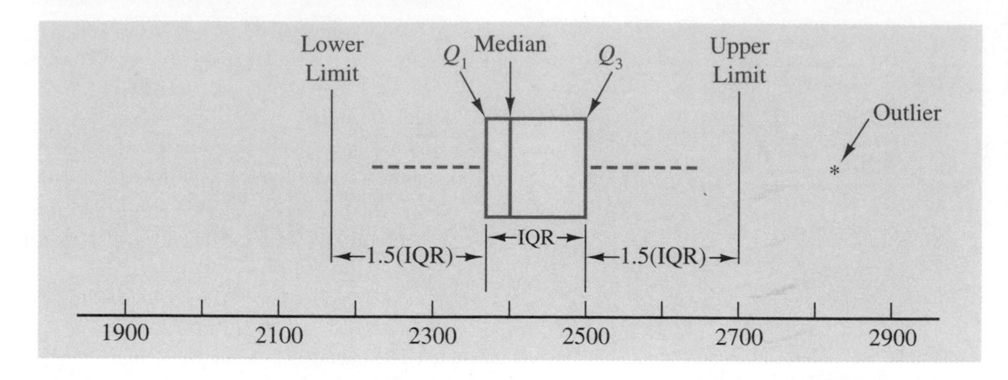

FIGURE 3.4
Box Plot of the Monthly Starting Salaries of Business School Graduates, with Lines Showing the Inner and Outer Fences

Box Plot

Key to the development of a *box plot* is the computation of the median and the quartiles, Q_1 and Q_3. The interquartile range, IQR = $Q_3 - Q_1$, is also used. Figure 3.4 is the box plot for the monthly starting salary data. The steps used to construct the box plot follow.

1. A box is drawn with the ends of the box located at the first and third quartiles. For the salary data, $Q_1 = 2365$ and $Q_3 = 2500$. This box contains the middle 50% of the data.
2. A vertical line is drawn in the box at the location of the median (2405 for the salary data). Thus the median line divides the data into two equal parts.
3. By using the interquartile range, IQR = $Q_3 - Q_1$, limits are located. The limits for the box plot are 1.5(IQR) below Q_1 and 1.5(IQR) above Q_3. For the salary data, IQR = $Q_3 - Q_1 = 2500 - 2365 = 135$. Thus, the limits are $2365 - 1.5(135) = 2162.5$ and $2500 + 1.5(135) = 2702.5$. Data outside these limits are considered *outliers.*
4. The dashed lines in Figure 3.4 are called *whiskers*. The whiskers are drawn from the ends of the box to the smallest and largest data values *inside the limits* computed in step 3. Thus the whiskers end at salary data values of 2210 and 2630.
5. Finally, the location of each outlier is shown with the symbol *. In Figure 3.4 we see one outlier—the data value 2825.

In Figure 3.4 we have included lines showing the location of the limits. These lines were drawn to show how the limits are computed and where they are located for the salary data. Although the limits are always computed, generally they are not drawn on the box plots. Figure 3.5 shows the usual appearance of a box plot for the salary data.

notes and comments

1. When using a box plot, we may or may not identify the same outliers as the ones we select when using z-scores less than -3 and greater than $+3$. However, the objective of both approaches is simply to identify values that should be reviewed to ensure the validity of the data. Outliers identified by either procedure should be reviewed.
2. An advantage of the exploratory data analysis procedures is that they are easy to use; few numerical calculations are necessary. We simply sort the items into ascending order and identify the median and quartiles Q_1 and Q_3 to obtain the five-number summary. The box plot can then easily be determined. It is not necessary to compute the mean and the standard deviation for the data.

FIGURE 3.5
Box Plot of the Monthly Starting Salaries of Business School Graduates

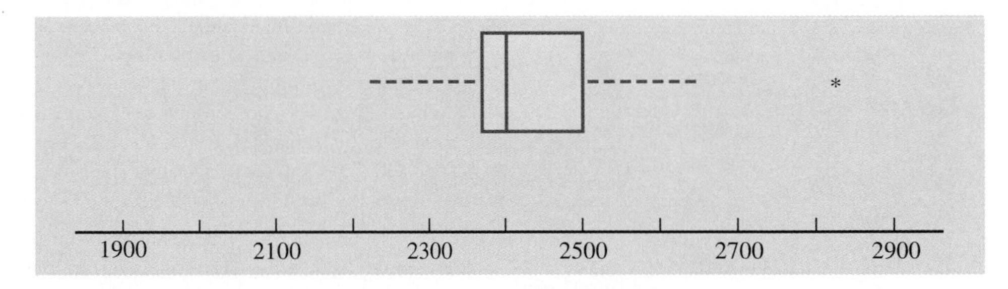

exercises

Methods

30. Consider the sample of size 8 with data values of 27, 25, 20, 15, 30, 34, 28, and 25. Provide the five-number summary for the data and a box plot for the data.

31. Show the five-number summary and the box plot for the following data: 5, 15, 18, 10, 8, 12, 16, 10, 6.

Applications

32. *Fortune* (April 20, 1992) published its annual report on the 500 largest U.S. industrial corporations. Data on the percentage growth in sales for the past 12 months for a sample of 26 companies follow.

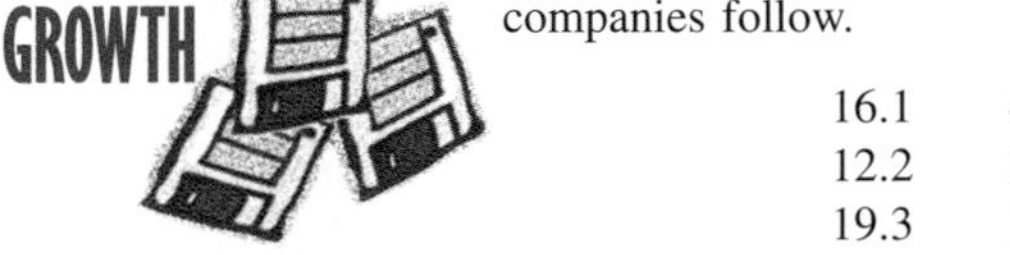

16.1	49.9	23.7	15.6	1.9	10.8	20.4
12.2	22.4	4.9	13.4	6.8	15.8	12.1
19.3	10.0	46.1	27.0	7.0	12.5	6.1
15.6	6.3	16.7	10.1	55.9		

a. Provide a five-number summary.
b. Compute the limits used to identify outliers.
c. Do there appear to be outliers? How would this information be helpful to a financial analyst?
d. Show a box plot.

33. Annual sales, in millions of dollars, for 21 pharmaceutical companies follow (*Business Week*, April 25, 1994).

8408	1374	1872	8879	2459	11413
608	14138	6452	1850	2818	1356
10498	7478	4019	4341	739	2127
3653	5794	8305			

a. Provide a five-number summary.
b. Compute the limits used to identify outliers.
c. Do there appear to be outliers?
d. Johnson & Johnson's sales are the largest in the list at $14,138 million. Suppose a data entry error had been made and the sales had been entered as $41,138 million. Would the method of detecting outliers in part (c) have identified the problem and allowed correction of the data entry error?
e. Show a box plot.

34. *Consumer Reports* provides performance and quality ratings for numerous consumer products. The overall ratings were provided for a sample of 16 midpriced VCRs in the *Consumer Reports 1992 Buying Guide.* The manufacturer brands and overall scores are listed in Table 3.5.

a. Provide the mean and median overall rating.
b. Compute the first and third quartiles.
c. Provide the five-number summary.
d. Similar ratings for camcorders showed a mean of 82.56, a standard deviation of 6.39, and a five-number summary of 75, 77, 82, 86, 93. Compare the *Consumer Reports* ratings data for VCRs and camcorders. Show the box plots for both.
e. Are there any outliers in the VCR data? Explain.

INJURY

35. The Highway Loss Data Institute's Injury and Collision Loss Experience report (September 1988) rates car models on the basis of the number of insurance claims filed after accidents. Index ratings near 100 are considered average. Lower ratings are better, indicating a safer car model. Shown are ratings for 20 midsize cars and 20 small cars

Midsize cars:	81	91	93	127	68	81	60	51	58	75
	100	103	119	82	128	76	68	81	91	82
Small cars:	73	100	127	100	124	103	119	108	109	113
	108	118	103	120	102	122	96	133	80	140

Summarize the data for the midsize and small cars separately.

a. Provide a five-number summary for midsize cars and for small cars.
b. Show the box plots.
c. Make a statement about what your summaries indicate about the safety of midsize cars in comparison to small cars.

TABLE 3.5 Consumer Report Ratings of VCR

Manufacturer	Score
Fisher	77
General Electric	81
Hitachi	89
J. C. Penney	78
JVC	79
Magnavox	80
Montgomery Ward	78
Mitsubishi	90
Panasonic	77
Phillips	73
Quasar	72
Radio Shack	76
RCA	79
Sanyo	75
Sony	86
Toshiba	79

36. Morgan Stanley Capital International in Geneva, Switzerland, provided percent changes in stock markets around the world (*Barrons,* March 30, 1992). Data shown are percent changes over the preceding one-year period.

Country	Percent Change	Country	Percent Change
Australia	29.1	Japan	8.3
Austria	−13.4	Luxembourg	6.7
Belgium	9.3	Netherlands	13.9
Canada	8.3	New Zealand	−12.6
Denmark	15.3	Norway	−15.4
Europe	10.0	Pacific	10.3
Finland	−16.7	Portugal	−7.2
France	15.8	Singapore	22.7
Germany	6.3	Spain	11.6
Hong Kong	42.8	Sweden	9.1
Italy	−4.1	United Kingdom	11.6
Ireland	9.6	United States	27.2

a. What are the mean and median percent changes among these world markets?
b. What are the first and third quartiles?
c. Are there any outliers? Show a box plot.
d. What percentile would you report for the United States?

3.5 Measures of Association Between Two Variables

Thus far we have examined numerical methods that are used to summarize the data for *one variable at a time.* Often a manager or decision maker is interested in the relationship between two variables. In this section we present covariance and correlation as descriptive measures of the relationship between two variables.

We begin by reconsidering the application involving a stereo and sound equipment store in San Francisco as presented in Section 2.5. The store's manager is interested in investigating the relationship between the number of weekend television commercials shown and the sales at the store during the following week. Sample data with sales expressed in hundreds of dollars are provided in Table 3.6. The scatter diagram in Figure 3.6 shows a positive relationship, with higher sales (y) associated with a greater number of commercials (x). In fact, the scatter diagram suggests that a straight line could be used as a linear approximation of the relationship. In the following discussion, we introduce covariance as a descriptive measure of the linear association between two variables.

TABLE 3.6 Sample Data for the Stereo and Sound Equipment Store

Week	Number of Commercials x	Sales Volume ($100s) y
1	2	50
2	5	57
3	1	41
4	3	54
5	4	54
6	1	38
7	5	63
8	3	48
9	4	59
10	2	46

Covariance

For a sample of n elements with the corresponding pairs of data values $x_1\,y_1$, $x_2\,y_2$, and so on, the *sample covariance* is defined by the following equation.

Sample Covariance

$$s_{xy} = \frac{\Sigma(x_i - \bar{x})(y_i - \bar{y})}{n - 1} \quad (3.10)$$

In this formula each x_i value is paired with a y_i value. We then sum the products obtained by multiplying the deviation of each x_i from its sample mean $\bar{x}$ by the deviation of the corresponding y_i from its sample mean $\bar{y}$; this sum is then divided by $n - 1$.

To measure the strength of the linear relationship between the number of commercials x and the sales volume y in the stereo and sound equipment store problem, we use (3.10) to compute the sample covariance. The calculations in Table 3.7 show the computation of $\Sigma(x_i - \bar{x})(y_i - \bar{y})$. Note that $\bar{x} = 30/10 = 3$ and $\bar{y} = 510/10 = 51$. Using (3.10), we obtain a sample covariance of

$$s_{xy} = \frac{\Sigma(x_i - \bar{x})(y_i - \bar{y})}{n - 1} = \frac{99}{9} = 11$$

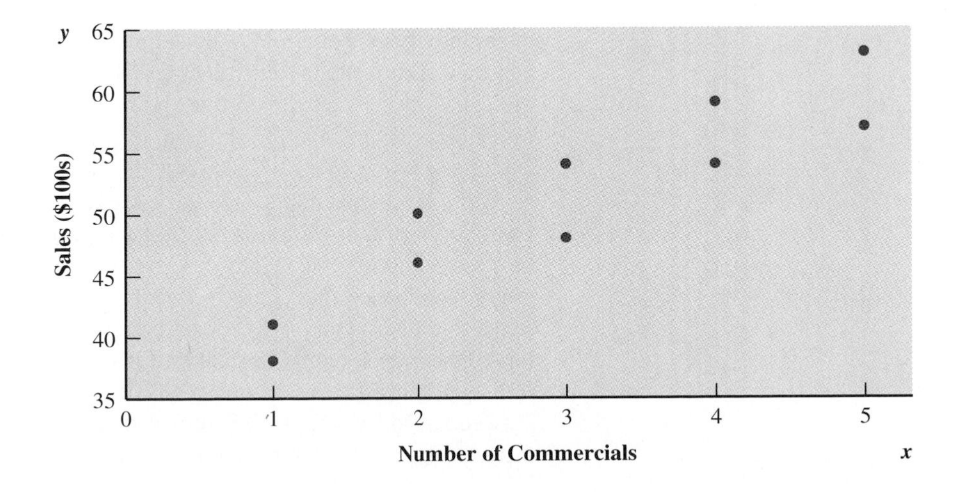

FIGURE 3.6
Scatter Diagram for the Stereo and Sound Equipment Store

TABLE 3.7 Calculations for the Sample Covariance

	x_i	y_i	$x_i - \bar{x}$	$y_i - \bar{y}$	$(x_i - \bar{x})(y_i - \bar{y})$
	2	50	−1	−1	1
	5	57	2	6	12
	1	41	−2	−10	20
	3	54	0	3	0
	4	54	1	3	3
	1	38	−2	−13	26
	5	63	2	12	24
	3	48	0	−3	0
	4	59	1	8	8
	2	46	−1	−5	5
Totals	30	510	0	0	99

The formula for computing the covariance of a population of size N is similar to (3.10), but we use different notation to indicate that we are working with the entire population.

Population Covariance

$$\sigma_{xy} = \frac{\Sigma(x_i - \mu_x)(y_i - \mu_y)}{N} \tag{3.11}$$

In (3.11) we use the notation μ_x for the population mean of the variable x and μ_y for the population mean of the variable y. The population covariance σ_{xy} is defined for a population of size N.

Interpretation of the Covariance

To aid in the interpretation of the *covariance,* consider Figure 3.7. It is the same as the scatter diagram of Figure 3.6 with a vertical line at $x = 3$ (the value of $\bar{x}$) and a horizontal line at $y = 51$ (the value of $\bar{y}$). Four quadrants have been identified on the graph. Points in quadrant I correspond to x_i values greater than $\bar{x}$ and y_i values greater than $\bar{y}$, points in quadrant II correspond to x_i values less than $\bar{x}$ and y_i values greater than $\bar{y}$, and so on. Thus, the value of $(x_i - \bar{x})(y_i - \bar{y})$ must be positive for points in quadrant I, negative for points in quadrant II, positive for points in quadrant III, and negative for points in quadrant IV.

If the value of s_{xy} is positive, the points that have had the greatest influence on s_{xy} must be in quadrants I and III. Hence, a positive value for s_{xy} is indicative of a positive linear association between x and y; that is, as the value of x increases, the value of y increases. If the value of s_{xy} is negative, however, the points that have had the greatest influence on s_{xy} are in quadrants II and IV. Hence, a negative value for s_{xy} is indicative of a negative linear association between x and y; that is, as the value of x increases, the value of y decreases. Finally, if the points are evenly distributed across all four quadrants, the value of s_{xy} will be close to zero, indicating no linear association between x and y. Figure 3.8 shows the values of s_{xy} that can be expected with three different types of scatter diagrams.

FIGURE 3.7
Partitioned Scatter Diagram for the Stereo and Sound Equipment Store

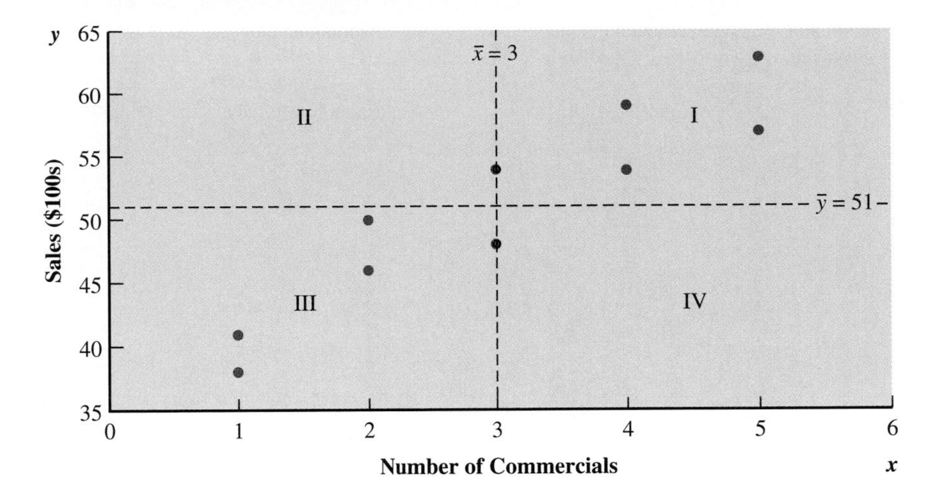

Referring again to Figure 3.7, we see that the scatter diagram for the stereo and sound equipment store follows the pattern in the top panel of Figure 3.8. As we should expect, the value of the sample covariance is positive with $s_{xy} = 11$.

From the preceding discussion, it might appear that a large positive value for the covariance is indicative of a strong positive linear relationship and that a large negative value is indicative of a strong negative linear relationship. However, one problem with using covariance as a measure of the strength of the linear relationship is that the value we obtain for the covariance depends on the units of measurement for x and y. For example, suppose we are interested in the relationship between height x and weight y for individuals. Clearly the strength of the relationship should be the same whether we measure height in feet or inches. When height is measured in inches, however, we get much larger numerical values for $(x_i - \bar{x})$ than we get when it is measured in feet. Thus, with height measured in inches, we would obtain a larger value for the numerator $\Sigma(x_i - \bar{x})(y_i - \bar{y})$—and hence a larger covariance—when in fact there is no difference in the relationship. A measure of the relationship between two variables that avoids this difficulty is the *correlation coefficient.*

Correlation Coefficient

Equation (3.12) shows that the Pearson product moment correlation coefficient for sample data (commonly referred to more simply as the *sample correlation coefficient*) is computed by dividing the sample covariance by the product of the standard deviation of x and the standard deviation of y.

Pearson Product Moment Correlation Coefficient: Sample Data

$$r_{xy} = \frac{s_{xy}}{s_x s_y} \tag{3.12}$$

where

r_{xy} = sample correlation coefficient
s_{xy} = sample covariance
s_x = sample standard deviation of x
s_y = sample standard deviation of y

FIGURE 3.8
Interpretation of Sample Covariance

s_{xy} **Positive:**
(x and y are positively linearly related)

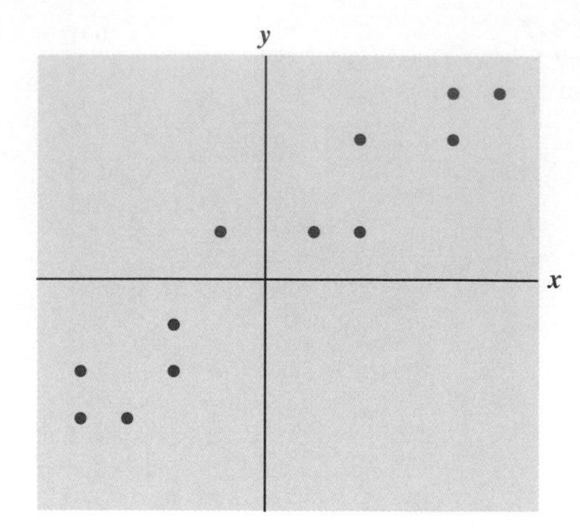

s_{xy} **Approximately 0:**
(x and y are not linearly related)

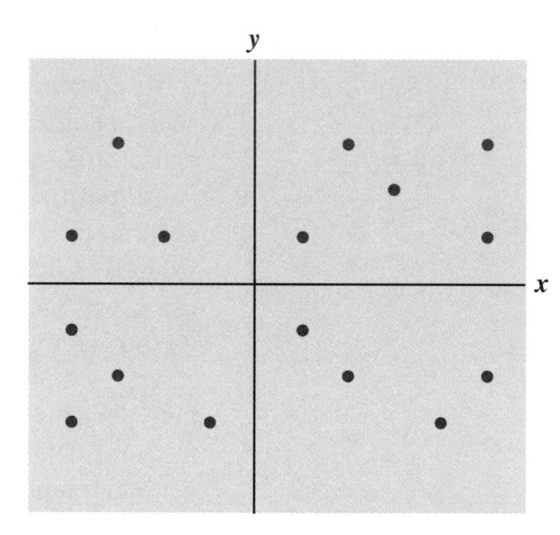

s_{xy} **Negative:**
(x and y are negatively linearly related)

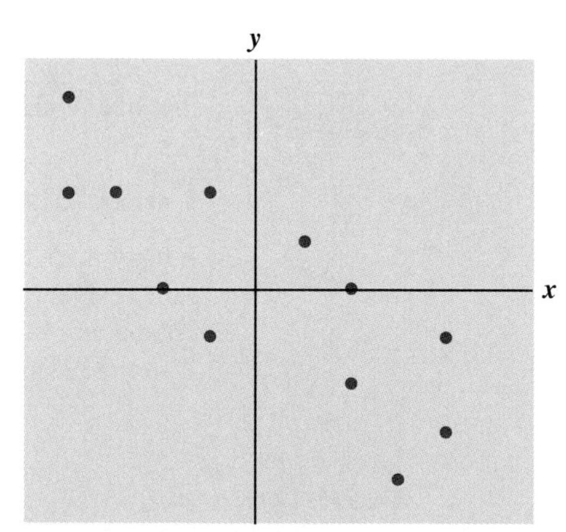

TABLE 3.8 Computations for Using the Alternate Formula to Compute r_{xy}

	x_i	y_i	$x_i y_i$	x_i^2	y_i^2
	2	50	100	4	2500
	5	57	285	25	3249
	1	41	41	1	1681
	3	54	162	9	2916
	4	54	216	16	2916
	1	38	38	1	1444
	5	63	315	25	3969
	3	48	144	9	2304
	4	59	236	16	3481
	2	46	92	4	2116
Totals	30	510	1629	110	26576

Let us now compute the sample correlation coefficient for the stereo and sound equipment store. Using the data in Table 3.6, we can compute the sample standard deviations for the two variables.

$$s_x = \sqrt{\frac{\Sigma(x_i - \bar{x})^2}{n - 1}} = \sqrt{\frac{20}{9}} = 1.4907$$

$$s_y = \sqrt{\frac{\Sigma(y_i - \bar{y})^2}{n - 1}} = \sqrt{\frac{566}{9}} = 7.9303$$

Now, since $s_{xy} = 11$, we have a sample correlation coefficient of

$$r_{xy} = \frac{s_{xy}}{s_x s_y} = \frac{11}{(1.4907)(7.9303)} = +.93$$

When a calculator is used to compute the sample correlation coefficient, the formula (3.13) may be preferred because it does not require the computation of each $(x_i - \bar{x})$ and $(y_i - \bar{y})$ deviation.

Pearson Product Moment Correlation Coefficient: Sample Data, Alternate Formula

$$r_{xy} = \frac{\Sigma x_i y_i - (\Sigma x_i \Sigma y_i)/n}{\sqrt{\Sigma x_i^2 - (\Sigma x_i)^2/n}\sqrt{\Sigma y_i^2 - (\Sigma y_i)^2/n}} \tag{3.13}$$

Algebraically, (3.12) and (3.13) are equivalent. In Table 3.8 we provide the calculations needed to use (3.13) to compute the correlation coefficient for the stereo and sound equipment store data. Based on these results, we obtain

$$r_{xy} = \frac{1629 - (30)(510)/10}{\sqrt{110 - (30)^2/10}\sqrt{26576 - (510)^2/10}} = \frac{99}{\sqrt{20}\sqrt{566}} = +.93$$

FIGURE 3.9
Scatter Diagram Depicting a Perfect Positive Linear Relationship

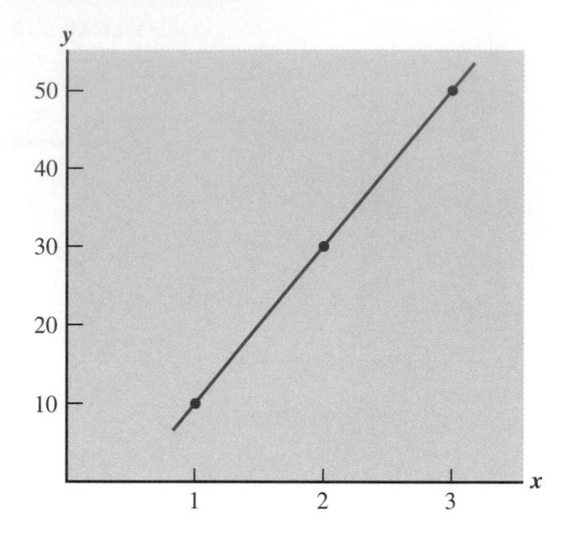

Thus we see that the value for r_{xy} is the same whether we use (3.12) or (3.13).

The formula for computing the correlation coefficient for a population, denoted by the Greek letter ρ_{xy} (rho, pronounced "row"), follows.

Pearson Product Moment Correlation Coefficient: Population Data

$$\rho_{xy} = \frac{\sigma_{xy}}{\sigma_x \sigma_y} \tag{3.14}$$

where

ρ_{xy} = population correlation coefficient

σ_{xy} = population covariance

σ_x = population standard deviation for x

σ_y = population standard deviation for y

The sample correlation coefficient r_{xy} is an estimate of the population correlation coefficient ρ_{xy}.

Interpretation of the Correlation Coefficient

First let us consider a simple example that illustrates the concept of a perfect positive linear relationship. The scatter diagram in Figure 3.9 depicts the relationship between the following $n = 3$ pairs of points.

x_i	1	2	3
y_i	10	30	50

The straight line drawn through each of the three points shows that there is a perfect linear relationship between the two variables x and y. The calculations needed to compute r_{xy} are shown in Table 3.9. Using (3.13), we obtain

TABLE 3.9 Calculations for Computing r_{xy} for the Relationship in Figure 3.9

	x_i	y_i	$x_i y_i$	x_i^2	y_i^2
	1	10	10	1	100
	2	30	60	4	900
	3	50	150	9	2500
Totals	6	90	220	14	3500

$$r_{xy} = \frac{\Sigma x_i y_i - (\Sigma x_i \Sigma y_i)/n}{\sqrt{\Sigma x_i^2 - (\Sigma x_i)^2/n}\sqrt{\Sigma y_i^2 - (\Sigma y_i)^2/n}} = \frac{220 - 6(90)/3}{\sqrt{14 - (6)^2/3}\sqrt{3500 - (90)^2/3}}$$

$$= \frac{40}{\sqrt{2}\sqrt{800}} = 1$$

Thus, we see that the value of the sample correlation coefficient for this data set is 1.

In general, it can be shown that if all the points in a data set are on a straight line having positive slope, the value of the sample correlation coefficients is +1; that is, a sample correlation coefficient of +1 corresponds to a perfect positive linear relationship between x and y. Moreover, if the points in the data set are on a straight line having negative slope, the value of the sample correlation coefficient is −1; that is, a sample correlation coefficient of −1 corresponds to a perfect negative linear relationship between x and y.

Let us now suppose that for a certain data set there is a positive linear relationship between x and y but that the relationship is not perfect. The value of r_{xy} will be less than 1, indicating that the points in the scatter diagram are not all on a straight line. As the points in a data set deviate more and more from a perfect positive linear relationship, the value of r_{xy} becomes smaller and smaller. A value of r_{xy} equal to zero indicates no linear relationship between x and y and values of r_{xy} near zero indicate a weak linear relationship.

For the data set involving the stereo and sound equipment store, recall that r_{xy} = +.93. Therefore, we conclude that there is a positive linear relationship between the number of commercials and sales. More specifically, an increase in the number of commercials is associated with an increase in sales.

exercises

Self Test

Methods

37. Five observations taken for two variables follow.

x_i	4	6	11	3	16
y_i	50	50	40	60	30

a. Develop a scatter diagram with x on the horizontal axis.
b. What does the scatter diagram indicate about the relationship between the two variables?
c. Compute and interpret the sample covariance for the data.
d. Compute and interpret the sample correlation coefficient for the data.

TABLE 3.10 Book Value and Dividends per Share for 15 Utility Stocks

UTILITY

Company	Book Value ($)	Annual Dividend ($)	Company	Book Value ($)	Annual Dividend ($)
Am Elec	22.44	2.40	Centerior	12.14	.80
Con Ed	20.89	2.98	Cons N Gas	23.31	1.94
Detroit Ed	22.09	2.06	Houston Ind	16.23	3.00
Niag Moh	14.48	1.09	NorAm Enrgy	.56	.28
Pac G&E	20.73	1.96	Panh East	.84	.84
Peco	19.25	1.55	Peoples En	18.05	1.80
Pub Sv Ent	20.37	2.16	SCEcorp	12.45	1.21
UnicomCp	26.43	1.60			

Source: *Barron's,* January 2, 1995.

38. Five observations taken for two variables follow.

x_i	6	11	15	21	27
y_i	6	9	6	17	12

a. Develop a scatter diagram for these data.
b. What does the scatter diagram indicate about the relationship between x and y?
c. Compute and interpret the sample covariance for the data.
d. Compute and interpret the sample correlation coefficient for the data.

Applications

39. A high school guidance counselor collected the following data about the grade point averages (GPA) and the SAT mathematics test scores for six seniors.

GPA	2.7	3.5	3.7	3.3	3.6	3.0
SAT	450	560	700	620	640	570

a. Develop a scatter diagram for the data with GPA on the horizontal axis.
b. Does there appear to be any relationship between the GPA and the SAT mathematics test score? Explain.
c. Compute and interpret the sample covariance for the data.
d. Compute the sample correlation coefficient for the data. What does this value tell us about the relationship between the two variables?

40. A department of transportation's study on driving speed and miles per gallon for midsize automobiles resulted in the following data.

Driving Speed	30	50	40	55	30	25	60	25	50	55
Miles per gallon	28	25	25	23	30	32	21	35	26	25

Compute and interpret the sample correlation coefficient for these data.

41. A sociologist collected data on the ages of wives and husbands when they married.

Wife's Age	22	27	25	32	34	25
Husband's Age	24	33	28	30	40	25

a. Develop a scatter diagram for these data with the wife's age on the horizontal axis.
b. Does there appear to be a linear association? Explain.
c. Compute and interpret the sample correlation coefficient for these data.

42. Table 3.10 gives the book value per share and the annual dividend for 15 utility stocks (*Barrons,* January 2, 1995).

a. Develop a scatter diagram with book value on the horizontal axis.
b. Compute and interpret the sample correlation coefficient.

43. The daily high and low temperatures for 24 cities follow (*USA Today,* January 9, 1995).

City	High	Low	City	High	Low
Tampa	68	45	Birmingham	58	34
Kansas City	37	28	Minneapolis	17	10
Boise	52	41	Portland	51	48
Los Angeles	66	59	Memphis	54	34
Philadelphia	42	26	Buffalo	26	10
Milwaukee	25	12	Cincinnati	37	24
Chicago	27	16	Charlotte	52	33
Albany	31	9	Boston	38	19
Houston	72	49	Tulsa	54	35
Salt Lake City	51	39	Washington, D.C.	43	28
Miami	75	57	Las Vegas	58	48
Cheyenne	49	35	Detroit	27	13

What is the correlation between the high and low temperatures?

3.6 The Role of the Computer

Computer software packages can provide the descriptive statistics presented in this chapter. After the data have been entered into the computer, a few simple commands can be used to generate the desired output. In this section, we examine the descriptive statistics provided by the Minitab statistical software package; the instructions necessary

FIGURE 3.10
Descriptive Statistics and Box Plot Provided by Minitab

Panel A: Descriptive Statistics

N	Mean	Median	TrMean	StDev	SEMean
12	2440.0	2405.0	2424.5	165.7	47.8

Min	Max	Q1	Q3
2210.0	2825.0	2357.5	2525.0

Panel B: Box Plot

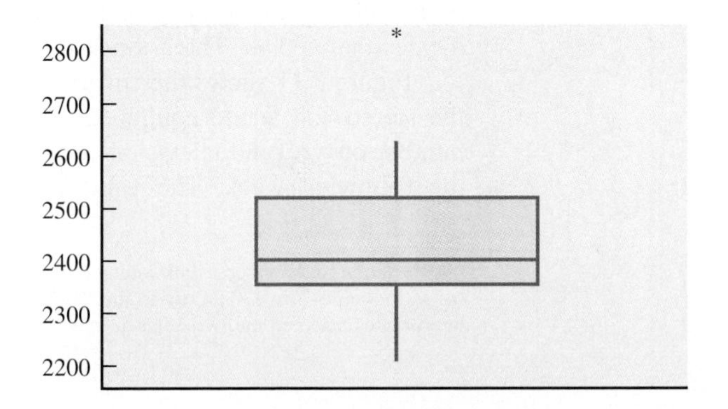

to generate the output are given in Appendix 3.1. Appendix 3.2 shows how an Excel spreadsheet package can be used to provide similar information.

Table 3.1 lists the starting salaries for 12 business school graduates. Panel A of Figure 3.10 shows the descriptive statistics obtained by using Minitab to summarize these data. Definitions of the headings in Panel A follow.

N	number of data values
Mean	mean
Median	median
StDev	standard deviation
Min	minimum data value
Max	maximum data value
Q1	first quartile
Q3	third quartile

We have not discussed the numerical measures labelled TrMean and SEMean. TrMean refers to the *trimmed mean.* The trimmed mean indicates the central location of the data after removing the effect of the smallest and largest data values in the data set. Minitab provides the 5% trimmed mean, the smallest 5% of the data values and the largest 5% of the data values are removed. The 5% trimmed mean is found by computing the mean for the middle 90% of the data. SEMean, which is the *standard error of the mean,* is computed by dividing the standard deviation by the square root of N. The interpretation and use of this measure are discussed in Chapter 7 when we introduce the topics of sampling and sampling distributions.

Although the numerical measures of range, interquartile range, variance, and coefficient of variation do not appear on the Minitab output, these values can be easily computed if desired from the results in Figure 3.10 by the following formulas.

$$\text{Range} = \text{Max} - \text{Min}$$
$$\text{IQR} = \text{Q3} - \text{Q1}$$
$$\text{Variance} = (\text{StDev})^2$$
$$\text{Coefficient of Variation} = (\text{StDev}/\text{Mean}) \times 100$$

Finally, note that Minitab's quartiles $Q_1 = 2357.5$ and $Q_3 = 2525$ provide slightly different values than the quartiles $Q_1 = 2365$ and $Q_3 = 2500$ that we computed in Section 3.1. The reason is that different conventions* can be used to identify the quartiles. Hence, the values of Q_1 and Q_3 provided by one convention may not be identical to the values of Q_1 and Q_3 provided by another convention. Any differences tend to be negligible, however, and the results provided should not mislead the user in making the usual interpretations associated with quartiles.

Panel B of Figure 3.10 is a box plot provided by Minitab. The box drawn from the first to the third quartiles contains the middle 50% of the data. The line within the box locates the median. The asterisk indicates an outlier at 2825.

Figure 3.11 shows the covariance and correlation output that Minitab provided for the stereo and sound equipment store data in Table 3.6. The variable x denotes the number of weekend television commercials and the variable y denotes the sales during the following week. The value in column x and row y is the sample covariance

---

*With the n data values rank ordered from smallest to largest, Minitab uses the positions given by $(n + 1)/4$ and $3(n + 1)/4$ to locate Q_1 and Q_3, respectively. When a position is fractional, Minitab interpolates between the two adjacent rank-ordered data values to determine the corresponding quartile.

FIGURE 3.11
Covariance and Correlation Provided by Minitab for the Number of Commercials (x) and Sales (y) Data

```
                x           y
x         2.22222
y        11.00000    62.88889

Correlation of x and y = 0.930
```

$s_{xy} = 11$ as computed in Section 3.5. The value in column x and row x is the sample variance for the number of commercials and the value in column y and row y is the sample variance for sales. The sample correlation coefficient $r_{xy} = +.93$ is shown at the bottom of the output.

3.7 The Weighted Mean and Working with Grouped Data

In Section 3.1, we presented the mean as one of the most important measures of descriptive statistics. The formula for the mean of a sample with n data values $x_1, x_2, x_3, \ldots, x_n$ is restated as follows.

$$\bar{x} = \frac{\Sigma x_i}{n} = \frac{x_1 + x_2 + x_3 + \ldots + x_n}{n} \tag{3.15}$$

In this formula, each data value x_i is given an equal importance or weight. Although this practice is most common, in some instances, the mean is computed by giving each data value a weight that reflects its importance. A mean computed in this manner is referred to as a *weighted mean.*

The Weighted Mean

The weighted mean is computed as follows:

The Weighted Mean

$$\bar{x}_{wt} = \frac{\Sigma w_i x_i}{\Sigma w_i} \tag{3.16}$$

where

$\bar{x}_{wt}$ = the weighted mean
x_i = data value i
w_i = weight for data value i

When the data are from a sample, (3.16) provides the weighted sample mean. When the data are from a population, μ_{wt} replaces $\bar{x}_{wt}$ and (3.16) provides the weighted population mean.

As an example of the need for a weighted mean, consider the following sample of five purchases of a raw material over the past three months.

Purchase	Cost per Pound	Number of Pounds
1	\$3.00	1200
2	\$3.40	500
3	\$2.80	2500
4	\$2.90	1000
5	\$3.25	800

Note that the cost per pound has varied from \$2.80 to \$3.40 and the quantity purchased has varied from 500 to 2500 pounds. Suppose that a manager has asked for information about the mean cost per pound of the raw material. Since the quantities ordered vary, we must use the formula for a weighted mean. The five cost-per-pound data values are $x_1 = 3.00$, $x_2 = 3.40$, $x_3 = 2.80$, $x_4 = 2.90$, and $x_5 = 3.25$. The mean cost per pound is found by weighting each cost by its corresponding quantity. For this example, the weights are $w_1 = 1200$, $w_2 = 500$, $w_3 = 2500$, $w_4 = 1000$, and $w_5 = 800$. Using (3.16), the weighted mean is calculated as follows:

$$\bar{x}_{wt} = \frac{1200\,(3.00) + 500(\,3.40) + 2500(2.80) + 1000(\,2.90) + 800(3.25)}{1200 + 500 + 2500 + 1000 + 800}$$

$$= \frac{17{,}800}{6{,}000} = 2.967$$

Thus, the weighted mean computation shows that the mean cost per pound for the raw material is \$2.967. Note that using (3.15) rather than the weighted mean would have provided misleading results. In this case, the mean of the five cost-per-pound values is $(3.00 + 3.40 + 2.80 + 2.90 + 3.25)/5 = 15.35/5 = \$3.07$, which overstates the actual mean cost per pound.

The choice of weights for a particular weighted mean computation depends upon the application. An example that is well known to college students is the computation of a grade point average (GPA). In this computation, the data values generally used are 4 for an A grade, 3 for a B grade, 2 for a C grade, 1 for a D grade, and 0 for an F grade. The weights are the number of credits hours earned for each grade. Exercise 46 provides an example of this weighted mean computation. In other weighted mean computations, quantities such as pounds, dollars, and/or volume are frequently used as weights. In any case, when data values vary in importance, the analyst must choose the weight that best reflects the importance of each data value in the determination of the mean.

Grouped Data

In most cases, measures of location and dispersion are computed by using the individual data values. Sometimes, however, we have data in only a grouped or frequency distribution form. In the following discussion, we show how the weighted mean computation can be used to obtain approximations of the mean, variance, and standard deviation for the grouped data.

In Section 2.2 we provided a frequency distribution of the time in days required to complete year-end audits for the public accounting firms of Sanderson and Clifford. The frequency distribution of audit times based on a sample of 20 clients is shown again in Table 3.11. Based on this frequency distribution, what is the sample mean audit time?

TABLE 3.11 Frequency Distribution of Audit Times

Audit Time (Days)	Frequency
10–14	4
15–19	8
20–24	5
25–29	2
30–34	1
Total	20

TABLE 3.12 Computation of the Sample Mean Audit Time for Grouped Data

Audit Time (Days)	Class Midpoint M_i	Frequency f_i	$f_i M_i$
10–14	12	4	48
15–19	17	8	136
20–24	22	5	110
25–29	27	2	54
30–34	32	1	32
		20	380
			$\Sigma f_i M_i$

$$\text{Sample mean } \bar{x} = \frac{\Sigma f_i M_i}{n} = \frac{380}{20} = 19 \text{ days}$$

To compute the mean, we treat the midpoint of each class as though it were the mean of all items in the class. Let M_i denote the midpoint for class i and let f_i denote the frequency of class i. The weighted mean formula (3.16) is then used with the data values noted as M_i and the weights given by the frequencies f_i. In this case, the denominator of (3.16) is the sum of the frequencies which is the sample size n. That is, $\Sigma f_i = n$. Thus, the equation for the sample mean for grouped data is as follows.

Sample Mean for Grouped Data

$$\bar{x} = \frac{\Sigma f_i M_i}{n} \tag{3.17}$$

where

M_i = the midpoint for class i
f_i = the frequency for class i
n = the sample size

With the class midpoints, M_i, halfway between the class limits, the first class of 10–14 in Table 3.11 has a midpoint at $(10 + 14)/2 = 12$. The five class midpoints and the weighted mean computation for the audit-time data are summarized in Table 3.12. As can be seen, the sample mean audit time is 19 days.

To compute the variance for grouped data, we use a slightly altered version of the formula for the variance provided in (3.5). In (3.5), the squared deviations of the data values about the sample mean $\bar{x}$ were written $(x_i - \bar{x})^2$. However, with grouped data, the individual data values, x_i, are not known. In this case, we treat the class midpoint, M_i, as being a representative value for the x_i values in the corresponding class. Thus the squared deviations about the sample mean, $(x_i - \bar{x})^2$, are replaced by $(M_i - \bar{x})^2$. Then, just as we did with the sample mean calculations for grouped data, we weight each value by the frequency of the class, f_i. The sum of the squared deviations about the mean for all the data is approximated by $\Sigma f_i (M_i - \bar{x})^2$. The term $n-1$ rather than n appears in the denominator in order to make the sample variance the estimate of the population variance σ^2. Thus, the following formula is used to obtain the sample variance for grouped data.

Sample Variance for Grouped Data

$$s^2 = \frac{\Sigma f_i(M_i - \bar{x})^2}{n - 1} \tag{3.18}$$

The calculation of the sample variance for audit times based on the grouped data from Table 3.11 is shown in Table 3.13.

The standard deviation for grouped data is simply the square root of the variance for grouped data. For the audit-time data, the sample standard deviation is $s = \sqrt{30} = 5.48$.

Before closing this section on computing measures of location and dispersion for grouped data, we note that the formulas (3.17) and (3.18) are for a sample. Population summary measures are computed similarly. The grouped data formulas for a population mean and variance follow.

Population Mean for Grouped Data

$$\mu = \frac{\Sigma f_i M_i}{N} \tag{3.19}$$

Population Variance for Grouped Data

$$\sigma^2 = \frac{\Sigma f_i(M_i - \mu)^2}{N} \tag{3.20}$$

notes and comments

1. An alternative formula for the computation of the sample variance for grouped data is

$$s^2 = \frac{\Sigma f_i M_i^2 - n\bar{x}^2}{n - 1}$$

where $\Sigma f_i M_i^2 = f_1 M_1^2 + f_2 M_2^2 + \cdots + f_k M_k^2$ and k is the number of classes used to group the data. Using this formula may ease the computations slightly.

2. In computing descriptive statistics for grouped data, the class midpoints are used to approximate the data values in each class. As a result, the descriptive statistics for grouped data are approximations of the descriptive statistics that would result from using the original data directly. We therefore recommend computing descriptive statistics from the original data rather than from grouped data whenever possible.

TABLE 3.13 Computation of the Sample Variance of Audit Times for Grouped Data (Sample Mean $\bar{x}$ = 19)

Audit Time (Days)	Class Midpoint M_i	Frequency f_i	Deviation $(M_i - \bar{x})$	Squared Deviation $(M_i - \bar{x})^2$	$f_i(M_i - \bar{x})^2$
10–14	12	4	−7	49	196
15–19	17	8	−2	4	32
20–24	22	5	3	9	45
25–29	27	2	8	64	128
30–34	32	1	13	169	169
		20			570
					$\Sigma f_i (M_i - \bar{x})^2$

$$\text{Sample variance } s^2 = \frac{\Sigma f_i (M_i - \bar{x})^2}{n - 1} = \frac{570}{19} = 30$$

exercises

Methods

44. Consider the following data and corresponding weights.

x_i	Weight (w_i)
3.2	6
2.0	3
2.5	2
5.0	8

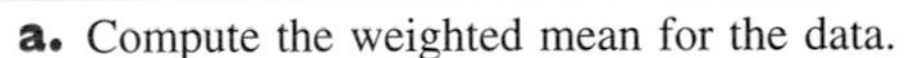

a. Compute the weighted mean for the data.

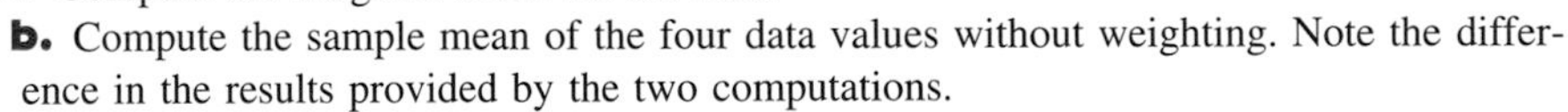

b. Compute the sample mean of the four data values without weighting. Note the difference in the results provided by the two computations.

Self Test

45. Consider the sample data in the following frequency distribution.

Class	Midpoint	Frequency
3–7	5	4
8–12	10	7
13–17	15	9
18–22	20	5

a. Compute the sample mean.

b. Compute the sample variance and sample standard deviation.

Applications

46. The grade point average for college students is based on a weighted mean computation. For most colleges, the grades are given the following data values: A (4), B (3), C (2), D (1), and F (0). After 60 credit hours of course work, a student at State University has earned 9 credit hours of A, 15 credit hours of B, 33 credit hours of C, and 3 credit hours of D.

Self Test

a. Compute the student's grade point average.

b. Students at State University must have a 2.5 grade point average for their first 60 credit hours of course work in order to be admitted to the business college. Will this student be admitted?

47. The dividend yield for a stock is the percentage of the amount invested in a stock that will be paid as an annual dividend to the stockholder. A sample of eight stocks held by Innis Investments had the dividend yields shown below (*The Wall Street Journal*, December 31, 1994). The amount Innis has invested in each stock is also shown. What is the mean dividend yield for the portfolio?

Company	Dividend Yield	Amount Invested ($)
Apple Computer	1.23	37,830
Chevron Corp.	4.15	27,667
Eastman Kodak	3.35	31,037
Exxon Corp.	4.94	27,336
Merck & Co.	3.15	37,553
Franklin Resources	1.12	17,812
Sears	3.48	32,660
Woolworth	4.00	17,775

48. A service station has recorded the following frequency distribution for the number of gallons of gasoline sold per car in a sample of 680 cars.

Gasoline (gallons)	Frequency
0–4	74
5–9	192
10–14	280
15–19	105
20–24	23
25–29	6
Total	680

Compute the mean, variance, and standard deviation for these grouped data. If the service station expects to service about 120 cars on a given day, what is an estimate of the total number of gallons of gasoline that will be sold?

49. In a survey of subscribers to *Fortune* magazine, the following question was asked: "How many of the last four issues have you read or looked through?" The following frequency distribution summarizes 500 responses (*Fortune* National Subscriber Portrait, 1994).

Number Read	Frequency
0	15
1	10
2	40
3	85
4	350
	Total 500

a. What is the mean number of issues read by a *Fortune* subscriber?
b. What is the standard deviation of the number of issues read?

summary

In this chapter we introduced several descriptive statistics that can be used to summarize the location and dispersion of data. Unlike the tabular and graphical procedures, the measures introduced in this chapter summarize the data in terms of numerical values. When the numerical values obtained are for a sample, they are called sample statistics. When the numerical values obtained are for a population, they are called population parameters. Some of the notation used for sample statistics and population parameters follow.

	Sample Statistic	Population Parameter
Mean	$\bar{x}$	μ
Variance	s^2	σ^2
Standard deviation	s	σ

As measures of location, we defined the mean, median, and mode. Then the concept of percentiles was used to describe the location of other values in the data set. Next, we presented the range, interquartile range, variance, standard deviation, and coefficient of variation as measures of variability or dispersion. We then described how the mean and standard deviation could be used together, applying the empirical rule and Chebyshev's theorem, to provide more information about the distribution of data.

A discussion of two exploratory data analysis techniques was included in Section 3.4. Specifically, we showed how to develop a five-number summary and a box plot to provide simultaneous information about the location, dispersion, and shape of the distribution. In Section 3.5 we introduced covariance and the correlation coefficient as measures of association between two variables.

Output from the software package Minitab was used to illustrate how statistical computing systems can support the analysis and summarization of data. Finally, we described how the mean, variance, and standard deviation could be computed for grouped data. However, we recommend using the measures based on the individual data values unless the grouped format is the only one in which the data are available.

glossary

Population parameter A numerical value used as a summary measure for a population of data (e.g., the population mean, μ, the population variance, σ^2, and the population standard deviation, σ).

Sample statistic A numerical value used as a summary measure for a sample (e.g., the sample mean, $\bar{x}$, the sample variance, s^2, and the sample standard deviation, s).

Mean A measure of central location for a data set. It is computed by summing all the data values and dividing by the number of items.

Trimmed mean The mean of the data remaining after α percent of the smallest and α percent of the largest items have been removed. The purpose of a trimmed mean is to provide a measure of central location after elimination of the effect of extremely large and extremely small data values.

Median A measure of central location. It is the value that splits the data into two equal groups, one with values greater than or equal to the median and one with values less than or equal to the median.

Mode A measure of location, defined as the most frequently occurring data value.

Percentile A value such that at least p percent of the data are less than or equal to this value and at least $(100 - p)$ percent of the data are greater than or equal to this value. The 50th percentile is the median.

Quartiles The 25th, 50th, and 75th percentiles referred to as the first quartile, the second quartile (median), and third quartile, respectively. The quartiles can be used to divide the data set into four parts, with each part containing approximately 25% of the data.

Range A measure of dispersion, defined to be the largest value minus the smallest value.

Interquartile range (IQR) A measure of dispersion, defined to be the difference between the third and first quartiles.

Variance A measure of dispersion, based on the squared deviations of the data values about the mean.

Standard deviation A measure of dispersion, found by taking the positive square root of the variance.

Coefficient of variation A measure of relative dispersion, found by dividing the standard deviation by the mean and multiplying by 100.

z-Score A value found by dividing the deviation about the mean $(x_i - \bar{x})$ by the standard deviation s. A z-score is referred to as a standardized value and denotes the number of standard deviations a data value x_i is from the mean.

Chebyshev's theorem A theorem that can be used to make statements about the percentage of data that must be within a specified number of standard deviations from the mean.

Empirical rule A rule that states the percentages of data that are within one, two, and three standard deviations from the mean for mound-shaped, or bell-shaped, distributions.

Outlier An unusually small or unusually large data value.

Five-number summary An exploratory data analysis technique that uses the following five numbers to summarize the data: smallest value, first quartile, median, third quartile, and largest value.

Box plot A graphical summary of data based on the five-number summary. A box, drawn from the first to the third quartiles, shows the location of the middle 50% of the data. Dashed lines, called whiskers, extending from the ends of the box, show the location of data greater than the third quartile and data less than the first quartile. The locations of any outliers are also noted.

Covariance A numerical measure of linear association between two variables. Positive values indicate a positive relationship; negative values indicate a negative relationship.

Correlation coefficient A numerical measure of linear association between two variables that has values between −1 and +1. Values near +1 indicate a strong positive linear relationship, values near −1 indicate a strong negative linear relationship, and values near zero indicate lack of a linear relationship.

Weighted mean A mean computed by weighting each data value by its relative importance.

Grouped data Data available in class intervals as summarized by a frequency distribution. Individual values of the original data are not recorded.

key formulas

Sample Mean

$$\bar{x} = \frac{\Sigma x_i}{n} \quad (3.1)$$

Population Mean

$$\mu = \frac{\Sigma x_i}{N} \quad (3.2)$$

Interquartile Range

$$\text{IQR} = Q_3 - Q_1 \quad (3.3)$$

Population Variance

$$\sigma^2 = \frac{\Sigma(x_i - \mu)^2}{N} \quad (3.4)$$

Sample Variance

$$s^2 = \frac{\Sigma(x_i - \bar{x})^2}{n - 1} \quad (3.5)$$

Standard Deviation

$$\text{Sample Standard Deviation} = s = \sqrt{s^2} \tag{3.6}$$

$$\text{Population Standard Deviation} = \sigma = \sqrt{\sigma^2} \tag{3.7}$$

Coefficient of Variation

$$\left(\frac{\text{Standard Deviation}}{\text{Mean}}\right) \times 100 \tag{3.8}$$

***z*-Score**

$$z_i = \frac{x_i - \bar{x}}{s} \tag{3.9}$$

Sample Covariance

$$s_{xy} = \frac{\Sigma(x_i - \bar{x})(y_i - \bar{y})}{n - 1} \tag{3.10}$$

Pearson Product Moment Correlation Coefficient: Sample Data

$$r_{xy} = \frac{s_{xy}}{s_x s_y} \tag{3.12}$$

Pearson Product Moment Correlation Coefficient: Sample Data, Alternate Formula

$$r_{xy} = \frac{\Sigma x_i y_i - (\Sigma x_i \Sigma y_i)/n}{\sqrt{\Sigma x_i^2 - (\Sigma x_i)^2/n}\,\sqrt{\Sigma y_i^2 - (\Sigma y_i)^2/n}} \tag{3.13}$$

Weighted Mean

$$\bar{x}_{wt} = \frac{\Sigma w_i x_i}{\Sigma w_i} \tag{3.16}$$

Sample Mean for Grouped Data

$$\bar{x} = \frac{\Sigma f_i M_i}{n} \tag{3.17}$$

Sample Variance for Grouped Data

$$s^2 = \frac{\Sigma f_i (M_i - \bar{x})^2}{n - 1} \tag{3.18}$$

Population Mean for Grouped Data

$$\mu = \frac{\Sigma f_i M_i}{N} \tag{3.19}$$

Population Variance for Grouped Data

$$\sigma^2 = \frac{\Sigma f_i (M_i - \mu)^2}{N} \tag{3.20}$$

supplementary exercises

PRICES

50. A sample of economists predicted what would happen to consumer prices during 1992 (*Business Week,* December 30, 1991). Their predictions about the percentage increase in consumer prices follow.

4.6	3.4	2.3	2.9	2.7	3.2	2.9
3.7	3.7	3.3	3.7	2.9	3.3	2.9
3.6	3.7	3.3	3.5	3.4	2.7	3.3
2.9	3.0	3.0	1.8			

a. Compute the mean, median, and mode.
b. Compute the first and third quartiles.
c. Compute the range and interquartile range.
d. Compute the variance and standard deviation.
e. Are there any outliers?

51. Following is a sample of yields for 10 bonds provided by Lehman Brothers (*Barron's,* January 23, 1995).

Issuer	Yield (%)	Issuer	Yield (%)
American Standard	10.59	Comcast	10.78
Inland Steel	9.92	Kroger A	9.54
Owens Illinois	10.64	Northwest Steel & Wire	11.29
Rogers Cantel	10.74	Stone Container	11.37
Westpoint Stevens	10.24	Unisys	10.58

Compute the following descriptive statistics for the data.

a. Mean **b.** Median **c.** Mode
d. 25th percentile **e.** Range **f.** Interquartile range
g. Variance **h.** Standard deviation **i.** Coefficient of variation

52. *Time* (January 9, 1989) published an article on the academic ability of college athletes. The article noted that some of the most successful athletic programs (citing the University of Notre Dame and Duke University) have athletes with very good college board scores. Assume that the following sample data are typical of college board scores for Notre Dame football players.

1100 970 1000 1250 880 790 1300 1050 900 950 1120

a. Compute the mean, median, and mode.
b. Compute the range and interquartile range.
c. Compute the variance and standard deviation.
d. Using z-scores, state whether or not there are any outliers.

53. A sample of 10 stocks on the New York Stock Exchange (*The Wall Street Journal,* April 28, 1995) has the following price-earnings ratios: 9, 4, 6, 7, 3, 11, 4, 6, 4, 7. Using these data, compute the mean, median, mode, range, variance, and standard deviation.

54. Public transportation and the automobile are two methods an employee can use to travel to work each day. Samples of times recorded for each method are shown. Times are in minutes.

Public Transportation:	28	29	32	37	33	25	29	32	41	34
Automobile:	29	31	33	32	34	30	31	32	35	33

a. Compute the sample mean time to travel to work for each method.
b. Compute the sample standard deviation for each method.
c. On the basis of your results from (a) and (b), which method of transportation should be preferred? Explain.
d. Develop a box plot for each method. Does a comparison of the box plots support your conclusion in (c)?

55. The median income and the median home price for a sample of six cities follow (*Who's Buying Homes in America,* Chicago Title and Trust Company, 1994). Data are in thousands of dollars.

City	Median Income	Median Home Price
Atlanta, Georgia	$65.2	$120.2
Cleveland, Ohio	49.8	92.7
Denver, Colorado	53.8	111.7
Dallas, Texas	62.7	104.7
Orlando, Florida	50.9	98.5
Minneapolis, Minnesota	53.1	105.8

a. What is the value of the sample covariance? Does it indicate a positive or negative relationship?
b. What is the sample correlation coefficient?

56. *Road & Track,* October 1994 provided the following sample of the tire ratings and load-carrying capacity of automobiles tires.
a. Develop a scatter diagram for the data with tire rating on the x axis.
b. What is the sample correlation coefficient and what does it tell you about the relationship between tire rating and load-carrying capacity?

Tire Rating	Load-carrying Capacity
75	853
82	1047
85	1135
87	1201
88	1235
91	1356
92	1389
93	1433
105	2039

57. The days to maturity for a sample of five money-market funds are shown below. The dollar amounts invested in the funds are provided. Use the weighted mean to determine the mean number of days to maturity for dollars invested in these five money-market funds.

Days to Maturity	Dollar Value ($million)
20	20
12	30
7	10
5	15
6	10

58. A forecasting technique referred to as moving averages uses the average or mean of the most recent n periods to forecast the next value for time series data. With a three-period moving average, the most recent three periods of data are used in the forecast computation. Consider a product with the following demand for the first three months of the current year: January (800 units), February (750 units), and March (900 units).

a. What is the three-month moving average forecast for April?

b. A variation of this forecasting technique is called weighted moving averages. The weighting allows the more recent time series data to receive more weight or more importance in the computation of the forecast. For example, a weighted three-month moving average might give a weight of 3 to data one month old, a weight of 2 to data two months old, and a weight of 1 to data three months old. Use the data above to provide a three-month weighted moving average forecast for April.

59. Automobiles traveling on the New York State Thruway are checked for speed by a state police radar system. A frequency distribution of speeds is shown in the following table.

Speed (Miles per Hour)	Frequency
45–49	10
50–54	40
55–59	150
60–64	175
65–69	75
70–74	15
75–79	10
Total	475

a. What is the mean speed of the automobiles traveling on the New York State Thruway?

b. Compute the variance and the standard deviation.

DUKE

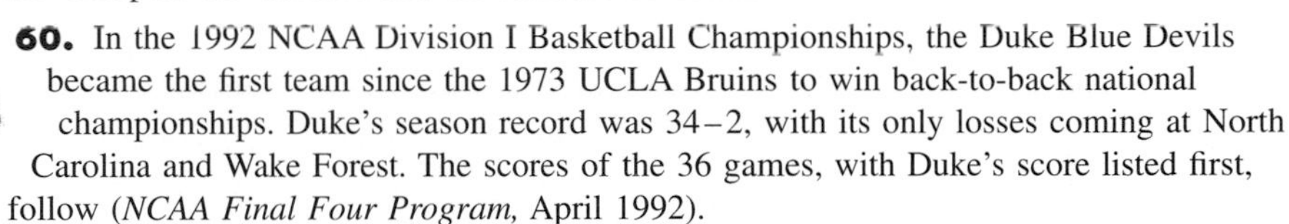

60. In the 1992 NCAA Division I Basketball Championships, the Duke Blue Devils became the first team since the 1973 UCLA Bruins to win back-to-back national championships. Duke's season record was 34–2, with its only losses coming at North Carolina and Wake Forest. The scores of the 36 games, with Duke's score listed first, follow (*NCAA Final Four Program,* April 1992).

Opponent	Score	Opponent	Score
East Carolina	103–75	Louisiana State	77–67
Harvard	118–65	Georgia Tech	71–62
St. John's	91–81	North Carolina State	71–63
Canisius	96–60	Maryland	91–89
Michigan	88–85	Wake Forest	68–72
William & Mary	97–61	Virginia	76–67
Virginia	68–62	UCLA	75–65
Florida State	86–70	Clemson	98–97
Maryland	83–66	North Carolina	89–77
Georgia Tech	97–84	Maryland	94–87
North Carolina State	110–75	Georgia Tech	89–76
N.C.—Charlotte	104–82	North Carolina	94–74
Boston University	95–85	Campbell	82–56
Wake Forest	84–68	Iowa	75–62
Clemson	112–73	Seton Hall	81–69
Florida State	75–62	Kentucky	104–103
Notre Dame	100–71	Indiana	81–78
North Carolina	73–75	Michigan	71–51

a. Compute the mean and median scores for Duke and its opponents.
b. Compute the range and interquartile range for Duke and its opponents.
c. Provide box plots for Duke and its opponents.
d. Comment on what you learned.

computer case I

Consolidated Foods, Inc.

Consolidated Foods, Inc., operates a chain of supermarkets in New Mexico, Arizona, and California. (See Computer Case, Chapter 2). Data in Table 3.14 show the dollar amounts and method of payment for a sample of 100 customers. Consolidated's managers requested the sample be taken to learn about payment practices of the store's customers. In particular, managers were interested in learning about how a new credit-card payment option was related to the customers' purchase amounts.

Managerial Report

Use the methods of descriptive statistics presented in Chapter 3 to summarize the sample data. Provide summaries of the dollar purchase amounts for cash customers,

TABLE 3.14 Purchase Amount and Method of Payment for a Random Sample of 100 Consolidated Foods Customers

CONSOLID

Cash	Personal Check	Credit Card	Cash	Personal Check	Credit Card
$ 7.40	$27.60	$50.30	$ 5.08	$52.87	$69.77
5.15	30.60	33.76	20.48	78.16	48.11
4.75	41.58	25.57	16.28	25.96	
15.10	36.09	46.24	15.57	31.07	
8.81	2.67	46.13	6.93	35.38	
1.85	34.67	14.44	7.17	58.11	
7.41	58.64	43.79	11.54	49.21	
11.77	57.59	19.78	13.09	31.74	
12.07	43.14	52.35	16.69	50.58	
9.00	21.11	52.63	7.02	59.78	
5.98	52.04	57.55	18.09	72.46	
7.88	18.77	27.66	2.44	37.94	
5.91	42.83	44.53	1.09	42.69	
3.65	55.40	26.91	2.96	41.10	
14.28	48.95	55.21	11.17	40.51	
1.27	36.48	54.19	16.38	37.20	
2.87	51.66	22.59	8.85	54.84	
4.34	28.58	53.32	7.22	58.75	
3.31	35.89	26.57		17.87	
15.07	39.55	27.89		69.22	

personal-check customers, and credit-card customers separately. Your report should contain the following summaries and discussions.

1. A comparison and interpretation of means and medians.
2. A comparison and interpretation of measures of dispersion such as the range and standard deviation.
3. The identification and interpretation of the five-number summaries for each method of payment.
4. Box plots for each method of payment.

Use the summary section of your report to provide a discussion of what you have learned about the method of payment and the amounts of payments for Consolidated Foods' customers. The data set for this computer case are in the data file CONSOLID.

computer case 2

National Health Care Association

The National Health Care Association is concerned about the shortage of nurses the health care profession is projecting for the future. To learn the current degree of job satisfaction among nurses, the association has sponsored a study of hospital nurses throughout the country. As part of this study, a sample of 50 nurses were asked to indicate their degree of satisfaction in their work, their pay, and their opportunities for promotion. Each of the three aspects of satisfaction was measured on a scale from 0 to 100, with larger values indicating higher degrees of satisfaction. The data in Table 3.15 were collected and are available in the data set HEALTH1.

In addition, the sample data in Table 3.15 were broken down by the types of hospitals employing the nurses. The types of hospitals considered were private (P), Veterans Administration (VA), and university (U). The data in Table 3.16 are available in the data set HEALTH2.

Managerial Report

Use methods of descriptive statistics to summarize the data. Present the summaries that will be beneficial in communicating the results to others. Discuss your findings. Specifically, comment on the following questions.

1. On the basis of the entire data set and the three job-satisfaction variables, what aspect of the job is most satisfying for the nurses? What appears to be the least satisfying? In what area(s), if any, do you feel improvements should be made? Discuss.
2. On the basis of descriptive measures of dispersion, what measure of job satisfaction appears to generate the greatest difference of opinion among the nurses? Explain.
3. What can be learned about the types of hospitals? Does any particular type of hospital seem to have better levels of job satisfaction than the other types? Do your results suggest any recommendations for learning about and/or improving job satisfaction? Discuss.
4. What additional descriptive statistics and insights can you use to learn about and possibly improve job satisfaction?

TABLE 3.15 Work, Pay, and Promotion Job-Satisfaction Scores for a Sample of 50 Nurses

HEALTH1

Work	Pay	Promotion	Work	Pay	Promotion
71	49	58	72	76	37
84	53	63	71	25	74
84	74	37	69	47	16
87	66	49	90	56	23
72	59	79	84	28	62
72	37	86	86	37	59
72	57	40	70	38	54
63	48	78	86	72	72
84	60	29	87	51	57
90	62	66	77	90	51
73	56	55	71	36	55
94	60	52	75	53	92
84	42	66	74	59	82
85	56	64	76	51	54
88	55	52	95	66	52
74	70	51	89	66	62
71	45	68	85	57	67
88	49	42	65	42	68
90	27	67	82	37	54
85	89	46	82	60	56
79	59	41	89	80	64
72	60	45	74	47	63
88	36	47	82	49	91
77	60	75	90	76	70
64	43	61	78	52	72

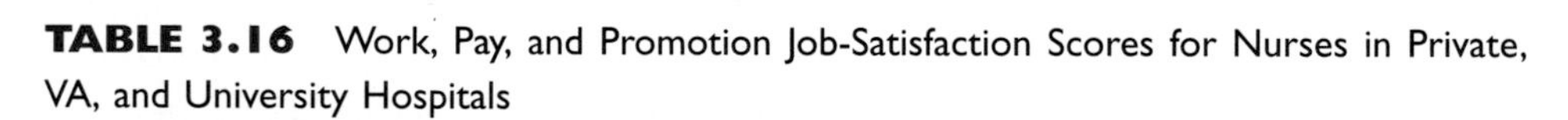

TABLE 3.16 Work, Pay, and Promotion Job-Satisfaction Scores for Nurses in Private, VA, and University Hospitals

HEALTH2

Private Hospitals			VA Hospitals			University Hospitals		
Work	Pay	Promotion	Work	Pay	Promotion	Work	Pay	Promotion
72	57	40	71	49	58	84	53	63
90	62	66	84	74	37	87	66	49
84	42	66	72	37	86	72	59	79
85	56	64	63	48	78	88	55	52
71	45	68	84	60	29	74	70	51
88	49	42	73	56	55	85	89	46
72	60	45	94	60	52	79	59	41
88	36	47	90	27	67	69	47	16
77	60	75	72	76	37	90	56	23
64	43	61	86	37	59	77	90	51
71	25	74	86	72	72	71	36	55
84	28	62	95	66	52	75	53	92
70	38	54	65	42	68	76	51	54
87	51	57	82	37	54	89	80	64
74	59	82	82	60	56			
89	66	62	90	76	70			
85	57	67	78	52	72			
74	47	63						
82	49	91						

APPENDIX 3.1 Descriptive Statistics with Minitab

In this appendix, we describe the steps necessary to use Minitab to generate the descriptive statistics in Figures 3.10 and 3.11 of Section 3.6. The descriptive statistics in Figure 3.10 are for the data on starting salaries of 12 business school graduates (see Table 3.1). After the data are entered into column C1 of a Minitab worksheet, the following steps generate the computer output shown in panel A of Figure 3.10.

STEP 1: Select the **Stat** pull-down menu
STEP 2: Select the **Basic Statistics** pull-down menu
STEP 3: Select the **Descriptive Statistics** option
STEP 4: When the dialog box appears:
Enter C1 in the **Variables** box
Select **OK**

The following steps generate the box plot shown in panel B of Figure 3.10.

STEP 1: Select the **Graph** pull-down menu
STEP 2: Select **Boxplot**
STEP 3: When the dialog box appears:
Enter C1 under **Y** in the **Graph variables** box
Select **OK**

The covariance and correlation output in Figure 3.11 are for the advertising and sales data at the stereo and sound equipment store (see Table 3.6). After entering the data for the number of commercials into column C1 and the data for sales into column C2 of the Minitab worksheet, we named the columns x and y, respectively. The steps necessary to generate the covariance output in the first three rows of Figure 3.11 follow.

STEP 1: Select the **Stat** pull-down menu
STEP 2: Select the **Basic Statistics** pull-down menu
STEP 3: Select the **Covariance** option
STEP 4: When the dialog box appears:
Enter C1 C2 in the **Variables** box
Select **OK**

After selection of **OK,** the variances and covariance appear as shown in the figure.

To obtain the correlation coefficient in the last line of Figure 3.11, only one change is necessary in the steps for obtaining the covariance. In step 3, the **Correlation** option is selected.

APPENDIX 3.2 Descriptive Statistics with Spreadsheets

Spreadsheet software packages can be used to generate the descriptive statistics discussed in this chapter. In this appendix, we show how Excel can be used to generate several of the measures of location and dispersion for a single variable and to generate the covariance and correlation coefficient as measures of association between two variables.

FIGURE 3.12
Monthly Starting Salary Data in Column A of Spreadsheet

	A	B	C	D	E
1	2350				
2	2450				
3	2550				
4	2380				
5	2255				
6	2210				
7	2390				
8	2630				
9	2440				
10	2825				
11	2420				
12	2380				

FIGURE 3.13
Descriptive Statistics for Monthly Starting Salaries Provided by Excel

	A	B	C	D
1	2350		*Column 1*	
2	2450			
3	2550		Mean	2440
4	2380		Standard Error	47.8198957
5	2255		Median	2405
6	2210		Mode	2380
7	2390		Standard Deviation	165.652978
8	2630		Sample Variance	27440.9091
9	2440		Kurtosis	1.71888364
10	2825		Skewness	1.09110869
11	2420		Range	615
12	2380		Minimum	2210
13			Maximum	2825
14			Sum	29280
15			Count	12
16			Confidence Level (95.0%)	105.250934

The starting-salary data from Table 3.1 have been entered into cells A1 through A12 of the spreadsheet in Figure 3.12. The following steps describe how to use Excel to generate descriptive statistics for these data.

STEP 1: Select the **Tools** pull-down menu
STEP 2: Choose the **Data Analysis** option
STEP 3: Choose **Descriptive Statistics** from the list of Analysis Tools
STEP 4: When the dialog box appears:
Enter A1:A12 in the **Input Range** box
Select **Output Range**
Enter C1 in the **Output Range** box (This identifies the upper left corner of the section of the worksheet where the descriptive statistics will appear.)
Select **Summary Statistics**
Select **OK**

Figure 3.13 shows the descriptive statistics provided by Excel. The highlighted rows contain the descriptive statistics covered in this chapter. The other items are either covered subsequently in the text or covered in more advanced texts.

FIGURE 3.14
Advertising and Sales Data for Stereo and Sound Equipment Store

	A	B	C	D	E
1	2	50			
2	5	57			
3	1	41			
4	3	54			
5	4	54			
6	1	38			
7	5	63			
8	3	48			
9	4	59			
10	2	46			

	A	B	C	D	E	F
1	2	50			*Column 1*	*Column 2*
2	5	57		Column 1	2.22222222	
3	1	41		Column 2	11	62.8888889
4	3	54				
5	4	54				
6	1	38			*Column 1*	*Column 2*
7	5	63		Column 1	1	
8	3	48		Column 2	0.93049058	1
9	4	59				
10	2	46				

FIGURE 3.15
Covariance and Correlation Coefficient for Advertising and Sales Data as Computed by Excel

We next illustrate the computation of the covariance and correlation coefficient, using the advertising and sales data for the stereo and sound equipment store (see Table 3.6). Figure 3.14 is an Excel spreadsheet with the data on the number of commercials in column A and the data on sales in column B. The steps necessary to generate the covariance for these data follow.

STEP 1: Select the **Tools** pull-down menu
STEP 2: Choose the **Data Analysis** option
STEP 3: Choose **Covariance** from the list of Analysis Tools
STEP 4: When the dialog box appears:
Enter A1:B10 in the **Input Range** box
Select **Output Range**
Enter D1 in the **Output Range** box
Select **OK**

Figure 3.15 shows the output of the covariance procedure in rows 1 to 3 of columns D, E, and F. The entry in cell E3 is the covariance between the two variables, $s_{xy} = 11$. The entry in cell E2 is the variance for the number of commercials and the entry in cell F3 is the variance for sales.

The steps necessary to compute the correlation coefficient from the spreadsheet in Figure 3.14 follow.

STEP 1: Select the **Tools** pull-down menu
STEP 2: Choose the **Data Analysis** option
STEP 3: Choose **Correlation** from the list of Analysis Tools
STEP 4: When the dialog box appears:
Enter A1:B10 in the **Input Range** box
Select **Output Range**
Enter D6 in the **Output Range** box
Select **OK**

The output generated by Excel begins in cell D6 of Figure 3.15. The entry in cell E8 is the correlation coefficient, $r_{xy} = +.93$; note that it is the same value we computed in Section 3.5. The entry in cell E7 is the correlation coefficient of the first variable (number of commercials) with itself and the entry in cell F8 is the correlation coefficient of the second variable (sales) with itself. Since a variable is always perfectly correlated with itself, these values are always equal to 1.

CHAPTER 4

Introduction to Probability

Morton International*

Chicago, Illinois

Morton International is a company with businesses in salt, household products, rocket motors, and specialty chemicals. Carstab Corporation, a subsidiary of Morton International, produces specialty chemicals and offers a variety of chemicals designed to meet the unique specifications of its customers. For one particular customer, Carstab produced an expensive catalyst used in chemical processing. Some, but not all, of the lots produced by Carstab met the customer's specifications for the product.

Carstab's customer agreed to test each lot after receiving it and determine whether the catalyst would perform the desired function. Lots that did not pass the customer's test would be returned to Carstab. Over time, Carstab found that the customer was accepting 60% of the lots and returning 40%. In probability terms, this meant that each Carstab shipment to the customer had a .60 probability of being accepted and a .40 probability of being returned.

Neither Carstab nor its customer was pleased with these results. In an effort to improve service, Carstab explored the possibility of duplicating the customer's test prior to shipment. However, the high cost of the special testing equipment made that alternative infeasible. Carstab's chemists then proposed a new, relatively low-cost test designed to indicate whether a lot would pass the customer's test. The probability question of interest was: What is the probability that a lot will pass the customer's test if it has passed the new Carstab test?

A sample of lots was produced and subjected to the new Carstab test. Only lots that passed the new test were sent to the customer. Probability analysis of the data indicated that if a lot passed the Carstab test, it had a .909 probability of passing the customer's test and being accepted. Alternatively, if a lot passed the Carstab test, it had only a .091 probability of being returned. The probability analysis provided key supporting evidence for the adoption and implementation of the new testing procedure at Carstab. The new test resulted in an immediate improvement in customer service and a substantial reduction in shipping and handling costs for returned lots.

The probability of a lot being accepted by the customer after passing the new Carstab test is called a conditional probability. In this chapter, you will learn how to compute this and other probabilities that are helpful in decision making.

Morton Salt: "When It Rains It Pours".

**The authors are indebted to Michael Haskell of Morton International for providing this Statistics in Practice.*

Business decisions are often based on an analysis of uncertainties such as the following:

1. What are the chances that sales will decrease if we increase prices?
2. What is the likelihood a new assembly method will increase productivity?
3. How likely is it that the project will be finished on time?
4. What are the odds in favor of a new investment being profitable?

Probability is a numerical measure of the likelihood that an event will occur. Thus, probabilities could be used as measures of the degree of uncertainty associated with the four events previously listed. If probabilities were available, we could determine the likelihood of each event occurring.

FIGURE 4.1
Probability as a Numerical Measure of the Likelihood of Occurrence

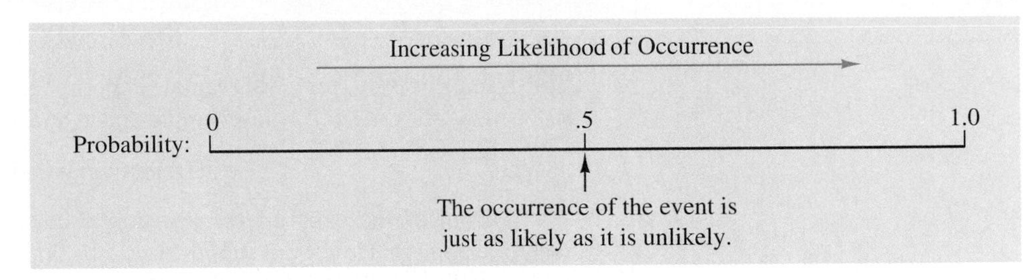

Probability values are always assigned on a scale from 0 to 1. A probability near zero indicates an event is very unlikely to occur; a probability near 1 indicates an event is almost certain to occur. Other probabilities between 0 and 1 represent degrees of likelihood that an event will occur. For example, if we consider the event "rain tomorrow," we understand that when the weather report indicates " a near-zero probability of rain," there is almost no chance of rain. However, if a .90 probability of rain is reported, we know that rain is likely to occur. A .50 probability indicates that rain is just as likely to occur as not. Figure 4.1 depicts the view of probability as a numerical measure of the likelihood of an event occurring.

4.1 Experiments, Counting Rules, and Assigning Probabilities

In discussing probability, we define an *experiment* to be any process that generates well-defined outcomes. On any single repetition of an experiment, one and only one of the possible experimental outcomes will occur. Several examples of experiments and their associated outcomes follow.

Experiment	Experimental Outcomes
Toss a coin	Head, Tail
Select a part for inspection	Defective, nondefective
Conduct a sales call	Purchase, no purchase
Roll a die	1, 2, 3, 4, 5, 6
Play a football game	Win, lose, tie

When we have specified all possible experimental outcomes, we have identified the *sample space* for an experiment.

Sample Space

The sample space for an experiment is the set of all experimental outcomes.

An experimental outcome is also called a *sample point* to identiify it as an element of the sample space.

Consider the first experiment in the table above—tossing a coin. The experimental outcomes (sample points) are determined by the upward face of the coin—a head or a tail. If we let S denote the sample space, we can use the following notation to describe the sample space.

$$S = \{\text{Head, Tail}\}$$

The sample space for the second experiment in the table—selecting a part for inspection—has the following sample space and sample points.

$$S = \{\text{Defective, Nondefective}\}$$

In both of the experiments just described, there are two experimental outcomes (sample points). However, suppose we consider the fourth experiment listed in the table—rolling a die. With the experimental outcomes defined as the number of dots appearing on the upward face of the die, there are six points in the sample space for this experiment.

$$S = \{1, 2, 3, 4, 5, 6\}$$

Counting Rules, Combinations, and Permutations

Being able to identify and count the experimental outcomes is a necessary step in assigning probabilities. We now discuss three counting rules that are useful.

MULTIPLE STEP EXPERIMENTS The first counting rule is for multiple-step experiments. Consider the experiment of tossing two coins. Let the experimental outcomes be defined in terms of the pattern of heads and tails appearing on the upward faces of the two coins. How many experimental outcomes are possible for this experiment? This can be thought of as a two-step experiment in which step 1 is the tossing of the first coin and step 2 is the tossing of the second coin. If we use H to denote a head and T to denote a tail, (H, H) indicates the experimental outcome with a head on the first coin and a head on the second coin. Continuing this notation, we can describe the sample space (S) for this coin-tossing experiment as follows:

$$S = \{(H, H), (H, T), (T, H), (T, T)\}$$

Thus, we see that there are four experimental outcomes. In this case, it is not difficult to list all of the experimental outcomes.

The counting rule for multiple-step experiments makes it possible to determine the number of experimental outcomes without listing them.

A Counting Rule for Multiple-Step Experiments

If an experiment can be described as a sequence of k steps in which there are n_1 possible outcomes on the first step, n_2 possible outcomes on the second step, and so on, then the total number of experimental outcomes is given by $(n_1)(n_2) \ldots (n_k)$.

Viewing the experiment of tossing two coins as a sequence of first tossing one coin ($n_1 = 2$) and then tossing the other coin ($n_2 = 2$), we can see from the counting rule that there must be $(2)(2) = 4$ distinct experimental outcomes. As shown above, they are $S = \{(H, H), (H, T), (T, H), (T, T)\}$. The number of experimental outcomes in an experiment involving tossing six coins is $(2)(2)(2)(2)(2)(2) = 64$.

A graphical device that is helpful in visualizing an experiment and enumerating outcomes in a multiple-step experiment is a *tree diagram.* Figure 4.2 shows a tree

FIGURE 4.2
Tree Diagram for the Experiment of Tossing Two Coins

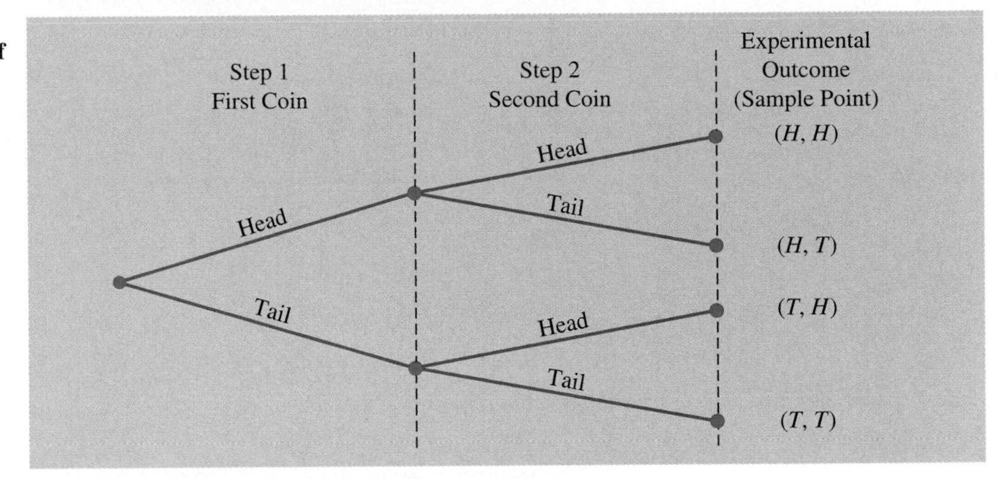

diagram for the experiment of tossing two coins. The sequence of steps is depicted by moving from left to right through the tree. Step 1 corresponds to tossing the first coin, and there are two branches corresponding to the two possible outcomes. Step 2 corresponds to tossing the second coin, and for each possible outcome at step 1, there are two branches corresponding to the two possible outcomes at step 2. Finally, each of the points on the right end of the tree corresponds to an experimental outcome. Each path through the tree from the leftmost node to one of the nodes at the right side of the tree corresponds to a unique sequence of outcomes.

Let us now see how the counting rule for multiple-step experiments can be used in the analysis of a capacity expansion project faced by the Kentucky Power & Light Company (KP&L). The project KP&L is starting to work on is designed to increase the generating capacity of one of its plants in northern Kentucky. The project is divided into two sequential stages or steps: stage 1 (design) and stage 2 (construction). While each stage will be scheduled and controlled as closely as possible, management cannot predict beforehand the exact time required to complete each stage of the project. An analysis of similar construction projects has shown completion times for the design stage of 2, 3, or 4 months and completion times for the construction stage of 6, 7, or 8 months. In addition, because of the critical need for additional electrical power, management has set a goal of 10 months for the completion of the entire project.

Since there are three possible completion times for the design stage (step 1) and three possible completion times for the construction stage (step 2), the counting rule for multiple-step experiments can be applied here to determine that there is a total of $(3)(3) = 9$ experimental outcomes. To describe the experimental outcomes, we will use a two-number notation; for instance, (2, 6) will indicate that the design stage is completed in 2 months and the construction stage is completed in 6 months. This experimental outcome results in a total of $2 + 6 = 8$ months to complete the entire project. Table 4.1 summarizes the nine experimental outcomes for the KP&L problem. The tree diagram in Figure 4.3 shows how the nine outcomes (sample points) occur.

The counting rule and tree diagram have been used to help the project manager identify the experimental outcomes and determine the possible project completion times. From the information in Figure 4.3, we see that the project will be completed in from 8 to 12 months, with six of the nine experimental outcomes providing the desired completion time of 10 months or less. While it has been helpful to identify the experimental outcomes, we will need to consider how probability values can be

TABLE 4.1 Listing of Experimental Outcomes (Sample Points) for the KP&L Problem

Completion Time (months)			
Stage 1 (Design)	Stage 2 (Construction)	Notation for Experimental Outcome	Total Project Completion Time (months)
2	6	(2, 6)	8
2	7	(2, 7)	9
2	8	(2, 8)	10
3	6	(3, 6)	9
3	7	(3, 7)	10
3	8	(3, 8)	11
4	6	(4, 6)	10
4	7	(4, 7)	11
4	8	(4, 8)	12

FIGURE 4.3
Tree Diagram for the KP&L Project

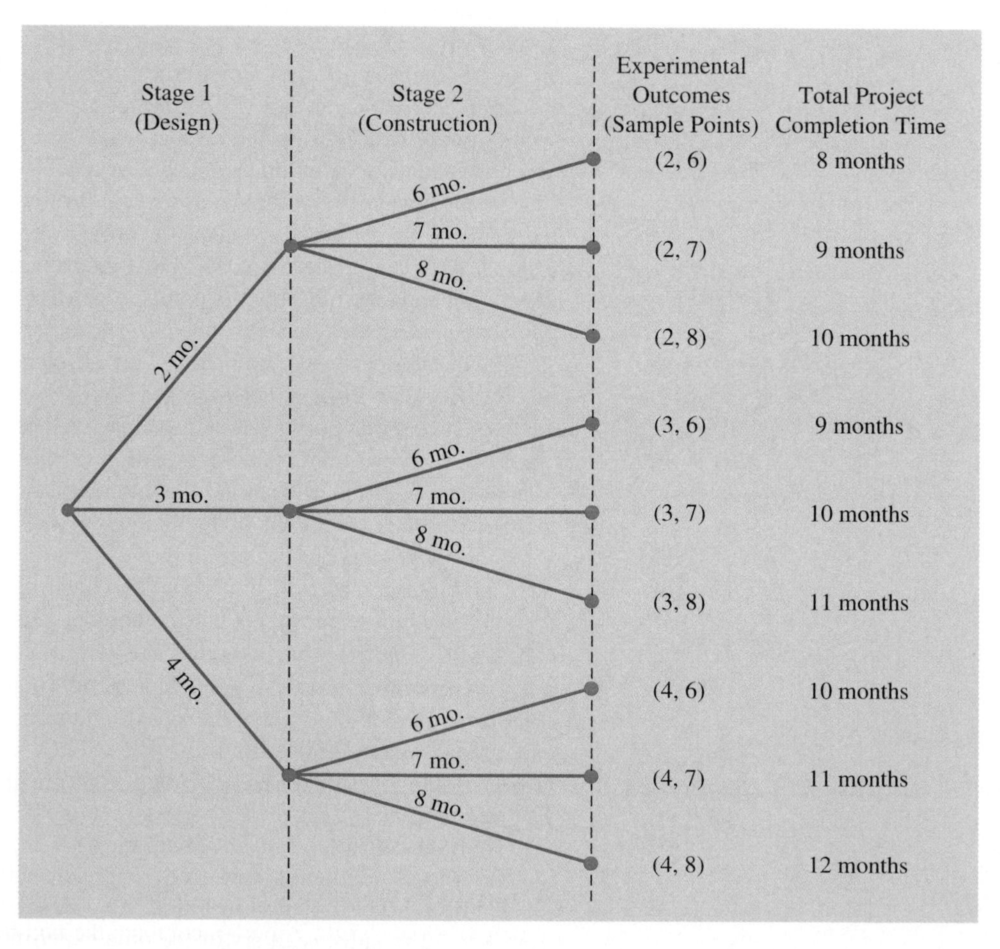

assigned to the experimental outcomes before making an assessment of the probability that the project will be completed within the desired 10 months.

COMBINATIONS A second counting rule that is often useful allows one to count the number of experimental outcomes when n objects are to be selected from a set of N objects. It is called the counting rule for combinations.

Counting Rule for Combinations

The number of combinations of N objects taken n at a time is

$$C_n^N = \binom{N}{n} = \frac{N!}{n!(N-n)!} \tag{4.1}$$

where

$$N! = N(N-1)(N-2)\ldots(2)(1)$$
$$n! = n(n-1)(n-2)\ldots(2)(1)$$

and

$$0! = 1.$$

The notation ! means *factorial;* for example, 5 factorial is $5! = (5)(4)(3)(2)(1) = 120$. By definition, 0! is equal to 1.

As an illustration of the counting rule for combinations, consider a quality control procedure where an inspector randomly selects two of five parts to test for defects. In a group of five parts, how many combinations of two parts can be selected? The counting rule in (4.1) shows that with $N = 5$ and $n = 2$, we have

$$C_2^5 = \binom{5}{2} = \frac{5!}{2!(5-2)!} = \frac{(5)(4)(3)(2)(1)}{(2)(1)(3)(2)(1)} = \frac{120}{12} = 10$$

Thus, there are 10 outcomes for the experiment of randomly selecting two parts from a group of five. If we label the five parts as A, B, C, D, and E, the 10 combinations or experimental outcomes can be identified as AB, AC, AD, AE, BC, BD, BE, CD, CE, and DE.

As another example, consider that the Ohio lottery system uses the random selection of six integers from a group of 47 to determine the weekly lottery winner. The counting rule for combinations, equation (4.1), can be used to determine the number of ways six different integers can be selected from a group of 47.

$$\binom{47}{6} = \frac{47!}{6!(47-6)!} = \frac{47!}{6!41!} = \frac{(47)(46)(45)(44)(43)(42)}{(6)(5)(4)(3)(2)(1)} = 10{,}737{,}573$$

The counting rule for combinations tells us that there are more than 10 million experimental outcomes in the lottery drawing. An individual who buys a lottery ticket has 1 chance in 10,737,573 of winning.

PERMUTATIONS A third counting rule that is sometimes useful is the counting rule for permutations. It allows one to compute the number of experimental outcomes when n objects are to be selected from a set of N objects where the order of selection is important. The same n objects selected in a different order is considered a different experimental outcome.

Counting Rule for Permutations

The number of permutations of N objects taken n at a time is given by

$$P_n^N = n!\binom{N}{n} = \frac{N!}{(N-n)!} \tag{4.2}$$

The counting rule for permutations is closely related to the one for combinations; however, there are more permutations than combinations for the same number of objects. This is because for every selection of n objects there are $n!$ different ways to order them.

As an example, consider again the quality control process in which an inspector selects two of five parts to inspect for defects. How many permutations may be selected? The counting rule in equation (4.2) shows that with $N = 5$ and $n = 2$, we have

$$P_2^5 = \frac{5!}{(5-2)!} = \frac{5!}{3!} = \frac{(5)(4)(3)(2)(1)}{(3)(2)(1)} = (5)(4) = 20$$

Thus, there are 20 outcomes for the experiment of randomly selecting two parts from a group of five when the order of selection must be taken into account. If we label the parts A, B, C, D, and E, the 20 permutations are AB, BA, AC, CA, AD, DA, AE, EA, BC, CB, BD, DB, BE, EB, CD, DC, CE, EC, DE, and ED.

Assigning Probabilities

Now let us see how probabilities can be assigned to experimental outcomes. The three approaches most frequently used are the classical, relative frequency, and subjective methods. Regardless of the method used, the probabilities assigned must satisfy two basic requirements.

Basic Requirements for Assigning Probabilities

1. The probability assigned to each experimental outcome must be between 0 and 1, inclusively. If we let E_i denote the ith experimental outcome and $P(E_i)$ be its probability, then this requirement can be written as

$$0 \leq P(E_i) \leq 1 \text{ for all } i \tag{4.3}$$

2. The sum of the probabilities for all the experimental outcomes must equal 1. If there are n experimental outcomes, this requirement can be written as

$$P(E_1) + P(E_2) + \cdots + P(E_n) = 1 \tag{4.4}$$

The *classical method* of assigning probabilities is appropriate when all the experimental outcomes are equally likely. If there are n experimental outcomes, a probability of $1/n$ is assigned to each experimental outcome. When using this approach, the two basic requirements for assigning probabilities are automatically satisfied.

For an example, consider the experiment of tossing a fair coin, the two experimental outcomes—head and tail—are equally likely. Since one of the two equally likely outcomes is a head, the probability of observing a head is ½, or .50. Similarly, the probability of observing a tail is also ½, or .50.

As an another example, consider the experiment of rolling a die. It would seem reasonable to conclude that the six possible outcomes are equally likely, and hence each outcome is assigned a probability of ⅙. If $P(1)$ denotes the probability that one dot appears on the upward face of the die, then $P(1) = 1/6$. Similarly, $P(2)=1/6$. $P(3)=1/6$, $P(4)=1/6$, $P(5) = 1/6$, and $P(6)=1/6$. Note that requirements (4.3) and (4.4) are both satisfied since each of the probabilities is greater than or equal to zero and they sum to one.

The *relative frequency method* of assigning probabilities is appropriate when data are available to estimate the proportion of the time the experimental outcome will occur when the experiment is repeated a large number of times. As an example consider a study of waiting times in the X-ray department for a local hospital. The number of patients waiting for service at 9:00 A.M. was recorded for 20 successive days. The following results were obtained.

Number Waiting	Number of Days Outcome Occurred
0	2
1	5
2	6
3	4
4	3
Total	20

These data show that on 2 of the 20 days, zero patients were waiting for service; on 5 of the days, one patient was waiting for service; and so on. Using the relative frequency method, we would assign a probability of $2/20 = .10$ to the experimental outcome of zero patients waiting for service, $5/20 = .25$ to the experimental outcome of one patient waiting, $6/20 = .30$ to two patients waiting, $4/20 = .20$ to three patients waiting, and $3/20 = .15$ to four patients waiting.

As with the classical method, the two basic requirements of (4.3) and (4.4) are automatically satisfied when the relative frequency method is used.

The *subjective method* of assigning probabilities is most appropriate when it is unrealistic to assume that the experimental outcomes are equally likely and when little relevant data are available. When the subjective method is used to assign probabilities to the experimental outcomes, we may use any information available, such as our experience or intuition. After considering all available information, a probability value that expresses our *degree of belief* that the experimental outcome will occur is specified. Since subjective probability expresses a person's degree of belief, it is personal. Using the subjective method, different people can be expected to assign different probabilities to the same experimental outcome.

When using the subjective probability assignment method, extra care must be taken to ensure that requirements (4.3) and (4.4) are satisfied. Regardless of a person's degree of belief, the probability value assigned to each experimental outcome must be between 0 and 1, inclusive, and the sum of all the experimental outcome probabilities must equal one.

TABLE 4.2 Completion Results for 40 KP&L Projects

Completion Time (months)			Number of Past Projects Having These Completion Times
Stage 1	Stage 2	Sample Point	
2	6	(2, 6)	6
2	7	(2, 7)	6
2	8	(2, 8)	2
3	6	(3, 6)	4
3	7	(3, 7)	8
3	8	(3, 8)	2
4	6	(4, 6)	2
4	7	(4, 7)	4
4	8	(4, 8)	6
		Total	40

Consider the case in which Tom and Judy Elsbernd have just made an offer to purchase a house. Two outcomes are possible:

$$E_1 = \text{their offer is accepted}$$
$$E_2 = \text{their offer is rejected}$$

Judy believes that the probability their offer will be accepted is .8; thus, Judy would set $P(E_1) = .8$ and $P(E_2) = .2$. Tom, however, believes that the probability that their offer will be accepted is .6; hence, Tom would set $P(E_1) = .6$ and $P(E_2) = .4$. Note that Tom's probability estimate for E_1 reflects the fact that he is a bit more pessimistic than Judy is about their offer being accepted.

Both Judy and Tom have assigned probabilities that satisfy the two basic requirements. The fact that their probability estimates are different emphasizes the personal nature of the subjective method.

Even in business situations where either the classical or the relative frequency approach can be applied, managers may want to provide subjective probability estimates. In such cases, the best probability estimates often are obtained by combining the estimates from the classical or relative frequency approach with subjective probability estimates.

Probabilities for the KP&L Project

To perform further analysis on the KP&L project, we must develop probabilities for each of the nine experimental outcomes listed in Table 4.1. On the basis of experience and judgment, management concluded that the experimental outcomes were not equally likely. Hence, the classical method of assigning probabilities could not be used. Management then decided to conduct a study of the completion times for similar projects undertaken by KP&L over the past three years. The results of a study of 40 similar projects are summarized in Table 4.2.

After reviewing the results of the study, management decided to employ the relative frequency method of assigning probabilities. Management could have provided subjective probability estimates, but felt that the current project was quite similar to the 40 previous projects. Thus, the relative frequency method was judged best.

TABLE 4.3 Probability Assignments for the KP&L Problem Based on the Relative Frequency Method

Sample Point	Project Completion Time	Probability of Sample Point
(2, 6)	8 months	$P(2, 6) = 6/40 = .15$
(2, 7)	9 months	$P(2, 7) = 6/40 = .15$
(2, 8)	10 months	$P(2, 8) = 2/40 = .05$
(3, 6)	9 months	$P(3, 6) = 4/40 = .10$
(3, 7)	10 months	$P(3, 7) = 8/40 = .20$
(3, 8)	11 months	$P(3, 8) = 2/40 = .05$
(4, 6)	10 months	$P(4, 6) = 2/40 = .05$
(4, 7)	11 months	$P(4, 7) = 4/40 = .10$
(4, 8)	12 months	$P(4, 8) = 6/40 = .15$
		Total 1.00

In using the data in Table 4.2 to compute probabilities, we note that outcome (2, 6)—stage 1 completed in 2 months and stage 2 completed in 6 months—occurred six times in the 40 projects. We can use the relative frequency method to assign a probability of 6/40 = .15 to this outcome. Similarly, outcome (2, 7) also occurred in six of the 40 projects, providing a 6/40 = .15 probability. Continuing in this manner, we obtain the probability assignments for the sample points of the KP&L project shown in Table 4.3. Note that $P(2, 6)$ represents the probability of the sample point (2, 6), $P(2, 7)$ represents the probability of the sample point (2, 7), and so on.

notes and comments

1. In statistics, the notion of an experiment is somewhat different from the notion of an experiment in the physical sciences. In the physical sciences, an experiment is usually conducted in a laboratory or a controlled environment in order to learn about a scientific occurrence. In statistical experiments, the outcomes are determined by probability. Even though the experiment is repeated in exactly the same way, an entirely different outcome may occur. Because of this influence of probability on the outcome, the experiments of statistics are sometimes called *random experiments.*

2. When drawing a random sample without replacement from a population of size N, the counting rule for combinations is used to find the number of different samples of size n that can be selected.

exercises

Methods

1. An experiment has three steps with three outcomes possible for the first step, two outcomes possible for the second step, and four outcomes possible for the third step. How many experimental outcomes exist for the entire experiment?

Self Test

2. How many ways can three items be selected from a group of six items? Use the letters A, B, C, D, E, and F to identify the items, and list each of the different combinations of three items.

3. How many permutations of three items can be selected from a group of six? Use the letters A, B, C, D, E, and F to identify the items, and list each of the permutations when the three items (B, D, F) are selected.

4. Consider the experiment of tossing a coin three times.

a. Develop a tree diagram for the experiment.
b. List the experimental outcomes.
c. What is the probability for each outcome?

5. Suppose an experiment has five equally likely outcomes: E_1, E_2, E_3, E_4, E_5. Assign probabilities to each outcome and show that conditions (4.3) and (4.4) are satisfied. What method did you use?

Self Test

6. An experiment with three outcomes has been repeated 50 times and it was learned that E_1 occurred 20 times, E_2 occurred 13 times, and E_3 occurred 17 times. Assign probabilities to the outcomes. What method did you use?

7. A decision maker has subjectively assigned the following probabilities to the four outcomes of an experiment: $P(E_1)=.10$, $P(E_2) = .15$, $P(E_3) = .40$, and $P(E_4) = .20$. Are these valid probability assignments? Check to see if (4.3) and (4.4) are satisfied.

Applications

8. In the city of Milford, applications for zoning changes go through a two-step process: a review by the planning commission and a final decision by the city council. At step 1 the planning commission will review the zoning change request and make a positive or negative recommendation concerning the change. At step 2 the city council will review the planning commission's recommendation and then vote to approve or to disapprove the zoning change. An application for a zoning change has just been submitted by the developer of an apartment complex. Consider the application process as an experiment.

a. How many sample points are there for this experiment? List the sample points.
b. Construct a tree diagram for the experiment.

Self Test

9. Simple random sampling uses a sample of size n from a population of size N to obtain data that can be used to make inferences about the characteristics of a population. Suppose we have a population of 50 bank accounts and want to take a random sample of four accounts in order to learn about the population. How many different random samples of four accounts are possible?

Self Test

10. In a survey of new matriculants to MBA programs ("School Selection by Students," *GMAC Occasional Papers,* Stolzenberg and Giarrusso, March 1988), the following data were obtained on the marital status of the students.

Marital Status	Frequency
Never married	1106
Married	826
Other (separated, widowed, divorced)	106
Total	2038

Consider the experiment of interviewing a new MBA student and recording her or his marital status. Show your probability assignments.

11. Strom Construction has made bids on two contracts. The owner has identified the possible outcomes and subjectively assigned the following probabilities.

Experimental Outcome	Obtain Contract 1	Obtain Contract 2	Probability
1	Yes	Yes	.15
2	Yes	No	.15
3	No	Yes	.30
4	No	No	.25

a. Are these valid probability assignments? Why or why not?
b. What would have to be done to make the probability assignments valid?

12. Faced with the question of determining the probability of obtaining either 0 heads, 1 head, or 2 heads when flipping a coin twice, an individual argued that, since it seems reasonable to treat the outcomes as equally likely, the probability of each event is ⅓. Do you agree? Explain.

13. A company that manufactures toothpaste is studying five different package designs. Assuming that one design is just as likely to be selected by a consumer as any other design, what selection probability would you assign to each of the package designs? In an actual experiment, 100 consumers were asked to pick the design they preferred. The following data were obtained. Do the data appear to confirm the belief that one design is just as likely to be selected as another? Explain.

Design	Number of Times Preferred
1	5
2	15
3	30
4	40
5	10

4.2 Events and Their Probabilities

Until now we have used the term "event" much as it would be used in everyday language. We must now introduce the formal definition of an *event* as it relates to probability.

Event

An *event* is a collection of sample points.

For an example, let us return to the KP&L problem and assume that the project manager is interested in the event that the entire project can be completed in 10 months or less. Referring to Table 4.3, we see that six sample points—(2, 6), (2, 7), (2, 8), (3, 6), (3, 7), and (4, 6)—provide a project completion time of 10 months or less. Let C denote the event that the project is completed in 10 months or less; we write

$$C = \{(2, 6), (2, 7), (2, 8), (3, 6), (3, 7), (4, 6)\}.$$

Event C is said to occur if *any one* of the six sample points shown above appears as the experimental outcome.

Other events that might be of interest to KP&L management include the following.

L = the event that the project is completed in *less* than 10 months

M = the event that the project is completed in *more* than 10 months

Using the information in Table 4.3, we see that these events consist of the following sample points.

$$L = \{(2, 6), (2, 7), (3, 6)\}$$

$$M = \{(3, 8), (4, 7), (4, 8)\}$$

A variety of additional events can be defined for the KP&L problem, but in each case the event must be identified as a collection of sample points for the experiment.

Given the probabilities of the sample points shown in Table 4.3, we can use the following definition to compute the probability of any event that KP&L management might want to consider.

Probability of an Event

The probability of any event is equal to the sum of the probabilities of the sample points in the event.

Using this definition, we calculate the probability of a particular event by adding the probabilities of the sample points (experimental outcomes) that make up the event. We can now compute the probability that the project will take 10 months or less to complete. Since this event is given by $C = \{(2, 6), (2, 7), (2, 8), (3, 6), (3, 7), (4, 6)\}$, the probability ($P$) of event C is shown by

$$P(C) = P(2, 6) + P(2, 7) + P(2, 8) + P(3, 6) + P(3, 7) + P(4, 6)$$

Refer to the sample point probabilities in Table 4.3; we have

$$P(C) = .15 + .15 + .05 + .10 + .20 + .05 = .70$$

Similarly, since the event that the project is completed in less than 10 months is given by $L = \{(2, 6), (2, 7), (3, 6)\}$, the probability of this event is given by

$$\begin{aligned} P(L) &= P(2, 6) + P(2, 7) + P(3, 6) \\ &= .15 + .15 + .10 = .40 \end{aligned}$$

Finally, for the event that the project is completed in more than 10 months, we have $M = \{(3, 8), (4, 7), (4, 8)\}$ and thus

$$\begin{aligned} P(M) &= P(3, 8) + P(4, 7) + P(4, 8) \\ &= .05 + .10 + .15 = .30 \end{aligned}$$

Using these probability results, we can now tell KP&L management that there is a .70 probability that the project will be completed in 10 months or less, a .40 probability that the project will be completed in less than 10 months, and a .30 probability that the project will be completed in more than 10 months. This procedure of computing event probabilities can be repeated for any event of interest to the KP&L management.

Any time that we can identify all the sample points of an experiment and assign the corresponding sample point probabilities, we can use the definition to compute the probability of an event. However, in many experiments the number of sample points is large and the identification of the sample points, as well as the determination of their associated probabilities, is extremely cumbersome, if not impossible. In the remaining sections of this chapter, we present some basic probability relationships that can be used to compute the probability of an event without knowledge of all sample point probabilities.

notes and comments

1. The sample space, S, is an event. Since it contains all the experimental outcomes, it has a probability of 1; that is, $P(S) = 1$.
2. When the classical method is used to assign probabilities, the assumption is that the experimental outcomes are equally likely. In such cases, the probability of an event can be computed by counting the number of experimental outcomes in the event and dividing the result by the total number of experimental outcomes.

exercises

Methods

14. An experiment has four equally likely outcomes.

a. What is the probability that E_2 occurs?
b. What is the probability that any two of the outcomes occur (e.g., E_1 or E_3)?
c. What is the probability that any three of the outcomes occur (e.g., E_1 or E_2 or E_4)?

Self Test

15. Consider the experiment of selecting a card from a deck of 52 cards. Each card corresponds to a sample point with a 1/52 probability.

a. List the sample points in the event an ace is selected.
b. List the sample points in the event a club is selected.
c. List the sample points in the event a face card (jack, queen, or king) is selected.
d. Find the probabilities associated with each of the events in (a), (b), and (c).

16. Consider the experiment of rolling a pair of dice. Suppose that we are interested in the sum of the face values showing on the dice.

a. How many sample points are possible? (Hint: Use the counting rule for multiple-step experiments.)
b. List the sample points.
c. What is the probability of obtaining a value of 7?
d. What is the probability of obtaining a value of 9 or greater?
e. Since there are six possible even values (2, 4, 6, 8, 10, and 12) and only five possible odd values (3, 5, 7, 9, and 11), the dice should show even values more often than odd values. Do you agree with this statement? Explain.
f. What method did you use to assign the probabilities requested above?

Applications

Self Test

17. Refer to the KP&L sample points and sample point probabilities in Table 4.3.

a. The design stage (stage 1) will run over budget if it takes 4 months to complete. List the sample points in the event the design stage is over budget.

b. What is the probability that the design stage is over budget?

c. The construction stage (stage 2) will run over budget if it takes 8 months to complete. List the sample points in the event the construction stage is over budget.

d. What is the probability that the construction stage is over budget?

e. What is the probability that both stages are over budget?

18. Suppose that a manager of a large apartment complex provides the following subjective probability estimates about the number of vacancies that will exist next month.

Vacancies	Probability
0	.05
1	.15
2	.35
3	.25
4	.10
5	.10

List the sample points in each of the following events and provide the probability of the event.

a. No vacancies. **b.** At least four vacancies. **c.** Two or fewer vacancies.

19. The manager of a furniture store sells from 0 to 4 china hutches each week. On the basis of past experience, the following probabilities are assigned to sales of 0, 1, 2, 3, or 4 hutches: $P(0) = .08$; $P(1) = .18$; $P(2) = .32$; $P(3) = .30$; and $P(4) = .12$.

a. Are these valid probability assignments? Why or why not?

b. Let A be the event that 2 or fewer are sold in one week. Find $P(A)$.

c. Let B be the event that 4 or more are sold in one week. Find $P(B)$.

20. The 1992 *Wall Street Journal* Subscriber Study revealed characteristics of the *Journal*'s subscribers, including business responsibilities, investment activities, life-style characteristics, and personal affluence. We show the total value of stock owned by the 1317 respondents.

Amount ($)	Number
Less than $15,000	216
$15,000–49,999	232
$50,000–99,999	200
$100,000–299,999	316
$300,000 or more	353

Suppose a subscriber is selected at random. What are the probabilities of the following events?

a. Let A be the event that the total value of stock owned is at least $50,000 but less than $100,000. Find $P(A)$.

b. Let B be the event that the total value of stock owned is less than $50,000. Find $P(B)$.

c. Let C be the event that the total value of stock owned is $100,000 or more. Find $P(C)$.

21. A survey of 50 students at Tarpon Springs College about the number of extracurricular activities resulted in the data shown.

Number of Activities	Frequency
0	8
1	20
2	12
3	6
4	3
5	1

a. Let A be the event that a student participates in at least 1 activity. Find $P(A)$.
b. Let B be the event that a student participates in 3 or more activities. Find $P(B)$.
c. What is the probability that a student participates in exactly 2 activities?

4.3 Some Basic Relationships of Probability

Complement of an Event

Given an event A, the *complement* of A is defined to be the event consisting of all sample points that are *not* in A. The complement of A is denoted by A^c. Figure 4.4 is a diagram, known as a *Venn diagram,* which illustrates the concept of a complement. The rectangular area represents the sample space for the experiment and as such contains all possible sample points. The circle represents event A and contains only the sample points that belong to A. The blue shaded region of the rectangle contains all sample points not in event A, and is by definition the complement of A.

In any probability application, either event A or its complement A^c must occur. Therefore, we have

$$P(A) + P(A^c) = 1$$

Solving for $P(A)$, we obtain the following result.

Computing Probability Using the Complement

$$P(A) = 1 - P(A^c) \tag{4.5}$$

FIGURE 4.4
Complement of Event A

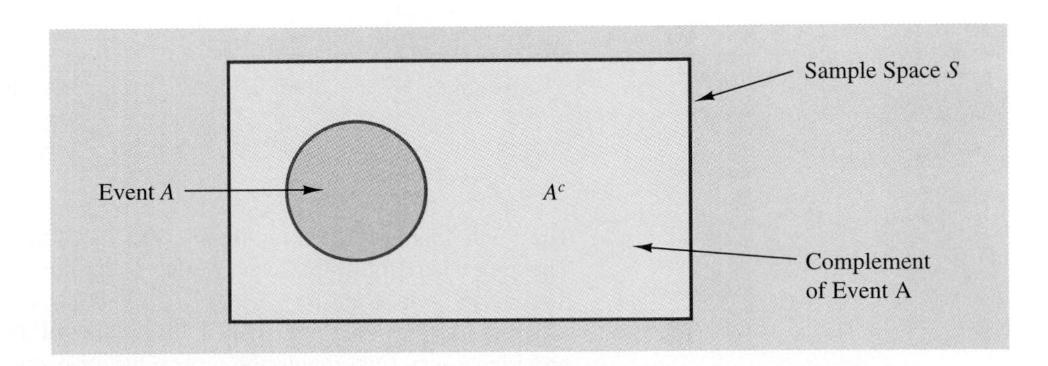

Equation (4.5) shows that the probability of an event A can be computed easily if the probability of its complement, $P(A^c)$, is known.

As an example, consider the case of a sales manager who, after reviewing sales reports, states that 80% of new customer contacts result in no sale. By allowing A to denote the event of a sale and A^c to denote the event of no sale, the manager is stating that $P(A^c) = .80$. Using (4.5), we see that

$$P(A) = 1 - P(A^c) = 1 - .80 = .20$$

We can conclude that there is a .20 probability that a sale will be made on a new customer contact.

In another example, a purchasing agent states that there is a .90 probability that a supplier will send a shipment that is free of defective parts. Using the complement, we can conclude that there is a $1 - .90 = .10$ probability that the shipment will contain defective parts.

Addition Law

The addition law is helpful when we have two events and are interested in knowing the probability that at least one of the events occurs. That is, with events A and B we are interested in knowing the probability that event A or event B or both occur.

Before we present the addition law, we need to discuss two concepts related to the combination of events: the *union* of events and the *intersection* of events. Given two events A and B, the union of A and B is defined as follows.

Union of Two Events

The *union* of A and B is the event containing *all* sample points belonging to A *or* B *or both*. The union is denoted by $A \cup B$.

The Venn diagram in Figure 4.5 depicts the union of events A and B. Note that the two circles contains all the sample points in event A as well as all the sample points in event B. The fact that the circles overlap indicates that some sample points are contained in both A and B.

The definition of the intersection of two events A and B follows.

Intersection of Two Events

Given two events A and B, the *intersection* of A and B is the event containing the sample points belonging to *both* A *and* B. The intersection is denoted by $A \cap B$.

The Venn diagram depicting the intersection of the two events is shown in Figure 4.6. The area where the two circles overlap is the intersection; it contains the sample points that are in both A and B.

Let us now continue with a discussion of the addition law. The addition law provides a way to compute the probability of event A or B or both A and B occurring.

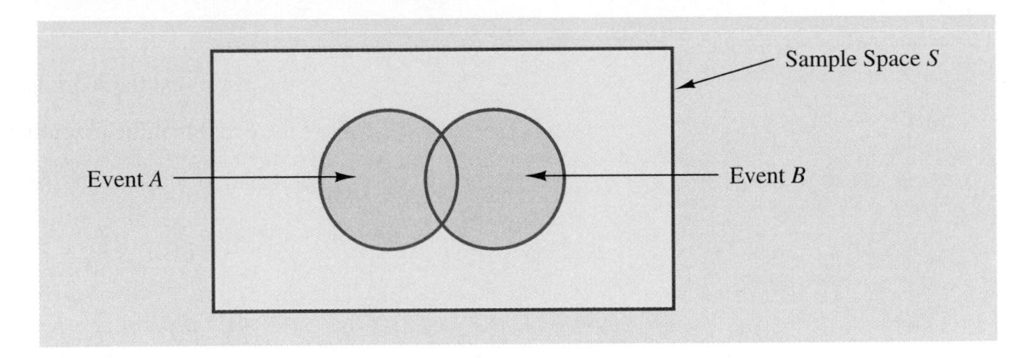

FIGURE 4.5
Union of Events A and B

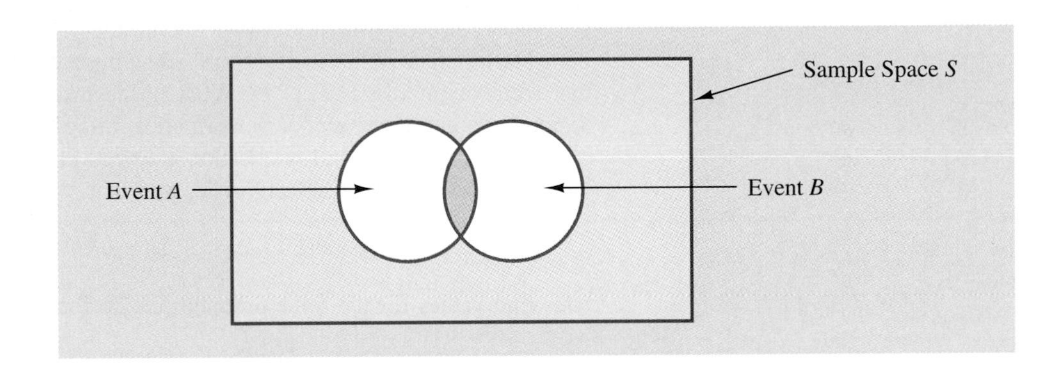

FIGURE 4.6
Intersection of Events A and B

In other words, the addition law is used to compute the probability of the union of two events, $A \cup B$. The addition law is written as follows.

Addition Law

$$P(A \cup B) = P(A) + P(B) - P(A \cap B) \tag{4.6}$$

To grasp the addition law intuitively, note that the first two terms in the addition law, $P(A) + P(B)$, account for all the sample points in $A \cup B$. However, since the sample points in the intersection $A \cap B$ are in both A and B, when we compute $P(A) + P(B)$, we are in effect counting each of the sample points in $A \cap B$ twice. We correct for this by subtracting $P(A \cap B)$.

As an example of an application of the addition law, let us consider the case of a small assembly plant with 50 employees. Each worker is expected to complete work assignments on time and in such a way that the assembled product will pass a final inspection. On occasion, some of the workers fail to meet the performance standards by completing work late and/or assembling defective products. At the end of a performance evaluation period, the production manager found that 5 of the 50 workers had completed work late, 6 of the 50 workers had assembled defective products, and 2 of the 50 workers had both completed work late *and* assembled defective products.

Let

L = the event that the work is completed late

D = the event that the assembled product is defective

The relative frequency information leads to the following probabilities.

$$P(L) = \frac{5}{50} = .10$$

$$P(D) = \frac{6}{50} = .12$$

$$P(L \cap D) = \frac{2}{50} = .04$$

After reviewing the performance data, the production manager decided to assign a poor performance rating to any employee whose work was either late or defective; thus the event of interest is $L \cup D$. What is the probability that the production manager assigned an employee a poor performance rating?

Note that the probability question is about the union of two events. Specifically, we want to know $P(L \cup D)$. Using (4.6), we have

$$P(L \cup D) = P(L) + P(D) - P(L \cap D)$$

Knowing values for the three probabilities on the right side of this expression, we can write

$$P(L \cup D) = .10 + .12 - .04 = .18$$

This tells us that there is a .18 probability that an employee received a poor performance rating.

As another example of the addition law, consider a recent study conducted by the personnel manager of a major computer software company. It was found that 30% of the employees who left the firm within two years did so primarily because they were dissatisfied with their salary, 20% left because they were dissatisfied with their work assignments, and 12% of the former employees indicated dissatisfaction with *both* their salary and their work assignments. What is the probability that an employee who leaves within two years does so because of dissatisfaction with salary, dissatisfaction with the work assignment, or both?

Let

S = the event that the employee leaves because of salary

W = the event that the employee leaves because of work assignment

We have $P(S) = .30$, $P(W) = .20$, and $P(S \cap W) = .12$. Using (4.6), the addition law, we have

$$P(S \cup W) = P(S) + P(W) - P(S \cap W) = .30 + .20 - .12 = .38$$

We find that there is a .38 probability that an employee leaves for salary or work assignment reasons.

Before we conclude our discussion of the addition law, let us consider a special case that arises for *mutually exclusive events.*

FIGURE 4.7
Mutually Exclusive Events

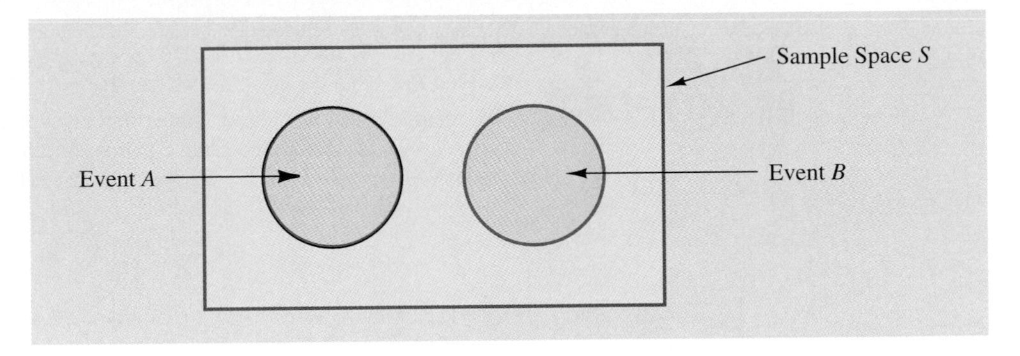

Mutually Exclusive Events

Two events are said to be *mutually exclusive* if the events have no sample points in common.

That is, events A and B are mutually exclusive if, when one event occurs, the other cannot occur. Thus, a requirement for A and B to be mutually exclusive is that their intersection must contain no sample points. The Venn diagram depicting two mutually exclusive events A and B is shown in Figure 4.7. In this case $P(A \cap B) = 0$ and the addition law can be written as follows.

Addition Law for Mutually Exclusive Events

$$P(A \cup B) = P(A) + P(B)$$

exercises

Methods

22. Suppose that we have a sample space with five equally likely experimental outcomes: E_1, E_2, E_3, E_4, E_5.
Let

$$A = \{E_1, E_2\}$$
$$B = \{E_3, E_4\}$$
$$C = \{E_2, E_3, E_5\}$$

a. Find $P(A)$, $P(B)$, and $P(C)$.
b. Find $P(A \cup B)$. Are A and B mutually exclusive?

c. Find A^c, C^c, $P(A^c)$, and $P(C^c)$.
d. Find $A \cup B^c$ and $P(A \cup B^c)$.
e. Find $P(B \cup C)$.

Self Test

23. Suppose that we have a sample space $S = \{E_1, E_2, E_3, E_4, E_5, E_6, E_7\}$, where $E_1, E_2, \ldots, E_7$ denote the sample points. The following probability assignments apply: $P(E_1) = .05$, $P(E_2) = .20$, $P(E_3) = .20$, $P(E_4) = .25$, $P(E_5) = .15$, $P(E_6) = .10$, and $P(E_7) = .05$. Let

$$A = \{E_1, E_4, E_6\}$$

$$B = \{E_2, E_4, E_7\}$$

$$C = \{E_2, E_3, E_5, E_7\}$$

a. Find $P(A)$, $P(B)$, and $P(C)$.
b. Find $A \cup B$ and $P(A \cup B)$.
c. Find $A \cap B$ and $P(A \cap B)$.
d. Are events A and C mutually exclusive?
e. Find B^c and $P(B^c)$.

Applications

24. The public accounting firm Grant Thornton conducted a survey to see how executives felt about the 1992 recession and the potential for recovery (*Journal of Accountancy,* February 1992). The probability that an executive indicated a recession existed was .74. If we were to choose one of the executives, what is the probability that she or he would indicate a recession does not exist?

25. A survey of benefits for 254 corporate executives (*Business Week,* October 24, 1994) showed that 155 executives were provided mobile phones, 152 were provided club memberships, and 110 were provided both mobile phones and club memberships as perks associated with their position.

a. Let M be the event of having a mobile phone and C be the event of having a club membership. Find the following probabilities: $P(M)$, $P(C)$, and $P(M \cap C)$.
b. Use the probabilities in part (a) to compute the probability that a corporate executive has at least one of the two perks.
c. What is the probability that a corporate executive does not have either of these perks?

26. A *U.S. News/UCLA* survey of 867 entertainment leaders in Hollywood studied how the entertainment industry views itself in terms of the amount of violence on television and the general quality of television programming (*U.S. News & World Report,* May 9, 1994). Results showed that 624 leaders felt the amount of violent programming had increased in the last 10 years, 390 felt the quality of programming had decreased over the same 10 years, and 234 leaders responded both that the amount of violent programming had increased and that the quality of programming had decreased.

a. Letting V be the event that the amount of violent programming has increased and Q be the event that the quality of programming has decreased, compute the following probabilities: $P(V)$, $P(Q)$, and $P(V \cap Q)$.
b. Use the probabilities in part (a) to find the probability that a leader made at least one of the following two comments: the amount of violent programming has increased or the quality of programming has decreased.
c. What is the probability that a leader did not agree with either of the two comments?

27. A survey of subscribers to *Forbes* showed that 72% have investments in money market funds and 36.4% have investments in certificates of deposit (CDs) (*Forbes* 1993 Subscriber Study). If 20% have investments in both money market funds and CDs, what is the probability that a subscriber has investments in either money market funds or CDs? What is the probability that a subscriber does not have investments in either money market funds or CDs?

Self Test

28. The survey of subscribers to *Forbes* showed that 45.8% rented a car during the past 12 months for business reasons, 54% rented a car during the past 12 months for personal reasons, and 30% rented a car during the past 12 months for both business and personal reasons (*Forbes* 1993 Subscriber Study).

a. What is the probability that a subscriber rented a car during the past 12 months for business or personal reasons?

b. What is the probability that a subscriber did not rent a car during the past 12 months for either business or personal reasons?

29. Let

A = the event that a person runs five miles or more per week
B = the event that a person dies of heart disease
C = the event that a person dies of cancer

Further, suppose that $P(A) = .01$, $P(B) = .25$, and $P(C) = .20$.

a. Are events A and B mutually exclusive? Can you find $P(A \cap B)$?

b. Are events B and C mutually exclusive? Find the probability that a person dies of heart disease or cancer.

c. Find the probability that a person dies from causes other than cancer.

4.4 Conditional Probability

Often, the probability of an event is influenced by whether a related event has occurred. Suppose we have an event A with probability $P(A)$. If we obtain new information and learn that a related event, denoted by B, has occurred, we will want to take advantage of this information in calculating a new probability for event A. This new probability of event A is written $P(A \mid B)$. The notation $\mid$ is used to denote the fact that we are considering the probability of event A *given* the condition that event B has occurred. Hence, the notation $P(A \mid B)$ is read "the probability of A given B."

As an illustration of the application of *conditional probability,* consider the situation of the promotion status of male and female officers of a major metropolitan police force in the eastern United States. The police force consists of 1200 officers, 960 men and 240 women. Over the past two years, 324 officers on the police force have been awarded promotions. The specific breakdown of promotions for male and female officers is shown in Table 4.4.

After reviewing the promotion record, a committee of female officers raised a discrimination case on the basis that 288 male officers had received promotions but only 36 female officers had received promotions. The police administration argued that the relatively low number of promotions for female officers was due not to discrimination, but to the fact that there are relatively few female officers on the police force. Let us show how conditional probability could be used to analyze the discrimination charge.

Let

M = event an officer is a man

W = event an officer is a woman

A = event an officer is promoted

A^c = event an officer is not promoted

Dividing the data values in Table 4.4 by the total of 1200 officers enables us to summarize the available information with the following probability values.

Table 4.4 Promotion Status of Police Officers over the Past Two Years

	Men	Women	Totals
Promoted	288	36	324
Not Promoted	672	204	876
Totals	960	240	1200

Table 5.5 Joint Probability Table for Promotions

Joint probabilites appear in the body of the table

	Men (M)	Women (W)	Totals
Promoted (A)	.24	.03	.27
Not Promoted (A^c)	.56	.17	.73
Totals	.80	.20	1.00

Marginal probabilities appear in the margins of the table

$P(M \cap A) = 288/1200 = .24 =$ probability that a randomly selected officer is a man *and* is promoted

$P(M \cap A^c) = 672/1200 = .56 =$ probability that a randomly selected officer is a man *and* is not promoted

$P(W \cap A) = 36/1200 = .03 =$ probability that a randomly selected officer is a woman *and* is promoted

$P(W \cap A^c) = 204/1200 = .17 =$ probability that a randomly selected officer is a woman *and* is not promoted

Since each of these values gives the probability of the intersection of two events, the probabilities are called *joint probabilities.* Table 4.5, which provides a summary of the probability information for the police officer promotion situation, is referred to as a *joint probability table.*

The values in the margins of the joint probability table provide the probabilities of each event separately. That is, $P(M) = .80$, $P(W) = .20$, $P(A) = .27$, and $P(A^c) = .73$. These probabilities are referred to as *marginal probabilities* because of their location in the margins of the joint probability table. We note that the marginal probabilities are found by summing the joint probabilities in the corresponding row or column of the joint probability table. For instance, the marginal probability of being promoted is $P(A) = P(M \cap A) + P(W \cap A) = .24 + .03 = .27$. From the marginal probabilities, we see that 80% of the force is male, 20% of the force is female, 27% of all officers received promotions, and 73% were not promoted.

Let us begin the conditional probability analysis by computing the probability that an officer is promoted given that the officer is a man. In conditional probability notation, we are attempting to determine $P(A \mid M)$. To calculate $P(A \mid M)$, we first realize that this notation simply means that we are considering the probability of the event A (promotion) given that the condition designated as event M (the officer is a man) is known to exist. Thus $P(A \mid M)$ tells us that we are now concerned only with the promotion status of the 960 male officers. Since 288 of the 960 male officers received promotions, the probability of being promoted given that the officer is a man is 288/960 = .30. In other words, given that an officer is a man, there has been a 30% chance of receiving a promotion over the past two years.

The above procedure was easy to apply because the values in Table 4.4 show the number of officers in each category. We now want to demonstrate how conditional probabilities such as $P(A \mid M)$ can be computed directly from event probabilities rather than the frequency data of Table 4.4.

We have shown that $P(A \mid M) = 288/960 = .30$. Let us now divide both the numerator and denominator of this fraction by 1200, the total number of officers in the study.

$$P(A \mid M) = \frac{288}{960} = \frac{288/1200}{960/1200} = \frac{.24}{.80} = .30$$

We now see that the conditional probability $P(A \mid M)$ can be computed as .24/.80. Refer to the joint probability table (Table 4.5). Note in particular that .24 is the joint probability of A and M; that is, $P(A \cap M) = .24$. Also note that .80 is the marginal probability that a randomly selected officer is a man; that is, $P(M) = .80$. Thus, the conditional probability $P(A \mid M)$ can be computed as the ratio of the joint probability $P(A \cap M)$ to the marginal probability $P(M)$.

$$P(A \mid M) = \frac{P(A \cap M)}{P(M)} = \frac{.24}{.80} = .30$$

The fact that conditional probabilities can be computed as the ratio of a joint probability to a marginal probability provides the following general formula for conditional probability calculations for two events A and B.

Conditional Probability

$$P(A \mid B) = \frac{P(A \cap B)}{P(B)} \tag{4.7}$$

or

$$P(B \mid A) = \frac{P(A \cap B)}{P(A)} \tag{4.8}$$

The Venn diagram in Figure 4.8 is helpful in obtaining an intuitive understanding of conditional probability. The circle on the right shows that event B has occurred; the portion of the circle that overlaps with event A denotes the event $(A \cap B)$. We know that once event B has occurred, the only way that we can also observe event A is for the event $(A \cap B)$ to occur. Thus, the ratio $P(A \cap B)/P(B)$ provides the conditional probability that we will observe event A given event B has already occurred.

FIGURE 4.8
Conditional Probability

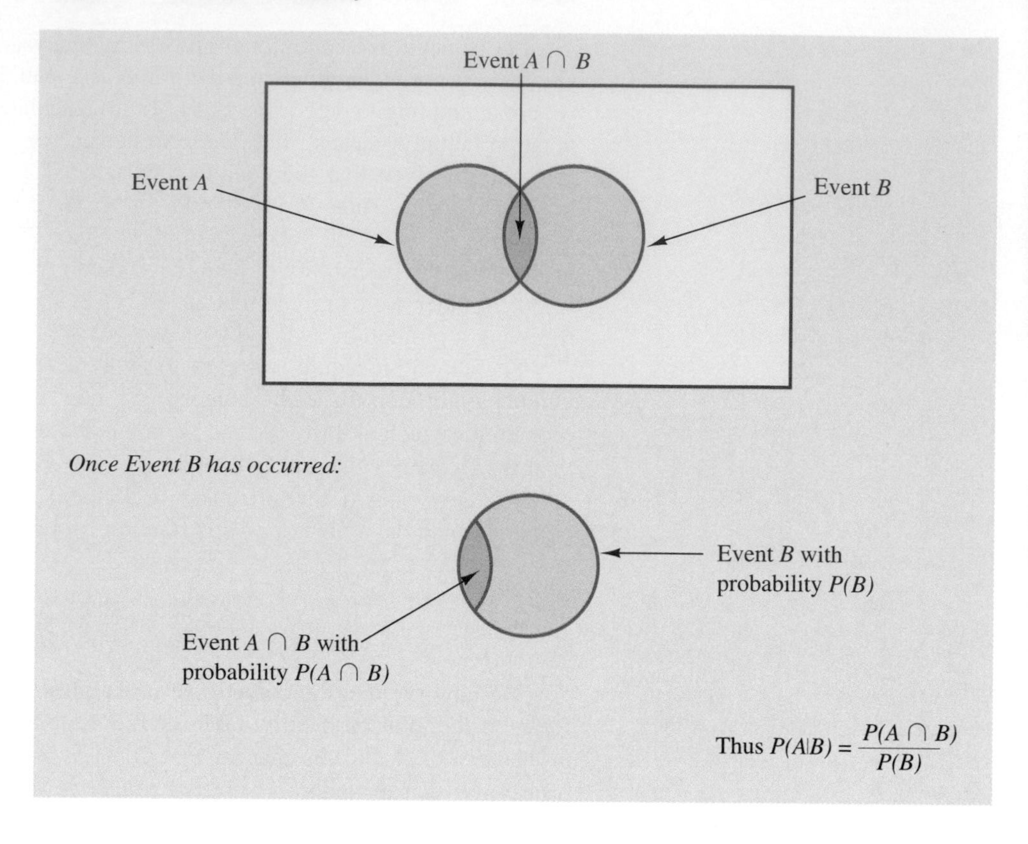

Let us return to the issue of discrimination against the female officers. The marginal probability in row 1 of Table 4.5 shows that the probability of promotion of an officer is $P(A) = .27$ (regardless of whether that officer is male or female). However, the critical issue in the discrimination case involves the two conditional probabilities $P(A \mid M)$ and $P(A \mid W)$. That is, what is the probability of a promotion *given* that the officer is a man, and what is the probability of a promotion *given* that the officer is a woman? If these two probabilities are equal, there is no basis for a discrimination argument because the chances of a promotion are the same for male and female officers. However, if the two conditional probabilities differ, there will be support for the position that male and female officers are treated differently in promotion decisions.

We have already determined that $P(A \mid M) = .30$. Let us now use the probability values in Table 4.5 and the basic relationship of conditional probability (4.7) to compute the probability that an officer is promoted given that the officer is a woman; that is, $P(A \mid W)$. Using (4.7), we obtain

$$P(A \mid W) = \frac{P(A \cap W)}{P(W)} = \frac{.03}{.20} = .15$$

What conclusions do you draw? The probability of a promotion given that the officer is a man is .30, twice the .15 probability of a promotion given that the officer is a woman. While the use of conditional probability does not in itself prove that discrimination exists in this case, the conditional probability values support the argument presented by the female officers.

Independent Events

In the preceding illustration, $P(A) = .27$, $P(A \mid M) = .30$, and $P(A \mid W) = .15$. We see that the probability of a promotion (event A) is affected or influenced by whether the officer is a man or a woman. Particularly, since $P(A \mid M) \neq P(A)$, we would say that events A and M are *dependent* events. That is, the probability of event A (promotion) is altered or affected by knowing event M (the officer is a man) exists. Similarly, with $P(A \mid W) \neq P(A)$, we would say that events A and W are *dependent* events. However, if the probability of event A is not changed by the existence of event M—that is, $P(A \mid M) = P(A)$—we would say that events A and M are *independent* events. This leads to the following definition of the independence of two events.

Independent Events

Two events A and B are independent if

$$P(A \mid B) = P(A) \tag{4.9}$$

or

$$P(B \mid A) = P(B). \tag{4.10}$$

Otherwise, the events are dependent.

Multiplication Law

Whereas the addition law of probability is used to compute the probability of a union of two events, the multiplication law is used to compute the probability of an intersection of two events. The multiplication law is based on the definition of conditional probability. Using (4.7) and (4.8) and solving for $P(A \cap B)$, we obtain the *multiplication law.*

Multiplication Law

$$P(A \cap B) = P(B)P(A \mid B) \tag{4.11}$$

or

$$P(A \cap B) = P(A)P(B \mid A) \tag{4.12}$$

To illustrate the use of the multiplication law, consider a newspaper circulation department where it is known that 84% of the households in a particular neighborhood subscribe to the daily edition of the paper. If we let D denote the event that a household subscribes to the daily edition, $P(D) = .84$. In addition, it is known that the probability that a household who already holds a daily subscription also subscribes to the Sunday edition (event S) is .75; that is, $P(S \mid D) = .75$. What is the probability that a household subscribes to both the Sunday and daily editions of the newspaper? Using the multiplication law, we compute the desired $P(S \cap D)$ as

$$P(S \cap D) = P(D)P(S \mid D) = .84(.75) = .63.$$

We now know that 63% of the households subscribe to both the Sunday and daily editions.

Before concluding this section, let us consider the special case of the multiplication law when the events involved are independent. Recall that we defined independent events to exist whenever $P(A \mid B) = P(A)$ or $P(B \mid A) = P(B)$. Hence, using (4.11) and (4.12) for the special case of independent events, we obtain the following multiplication law.

Multiplication Law for Independent Events

$$P(A \cap B) = P(A)P(B) \tag{4.13}$$

To compute the probability of the intersection of two independent events, we simply multiply the corresponding probabilities. Note that the multiplication law for independent events provides another way to determine whether A and B are independent. That is, if $P(A \cap B) = P(A)P(B)$, then A and B are independent; if $P(A \cap B) \neq P(A)P(B)$, then A and B are dependent.

As an application of the multiplication law for independent events, consider the situation of a service station manager who knows from past experience that 80% of the customers use a credit card when they purchase gasoline. What is the probability that the next two customers purchasing gasoline will each use a credit card? If we let

A = the event that the first customer uses a credit card
B = the event that the second customer uses a credit card

then the event of interest is $A \cap B$. Given no other information, we can reasonably assume that A and B are independent events. Thus,

$$P(A \cap B) = P(A)P(B) = (.80)(.80) = .64$$

notes and comments

Do not confuse the notion of mutually exclusive events with that of independent events. Two events with nonzero probabilities cannot be both mutually exclusive and independent. If one mutually exclusive event is known to occur, the probability of the other occurring is reduced to zero. They are therefore dependent.

exercises

Methods

30. Suppose that we have two events, A and B, with $P(A) = .50$, $P(B) = .60$, and $P(A \cap B) = .40$.

a. Find $P(A \mid B)$. **b.** Find $P(B \mid A)$.
c. Are A and B independent? Why or why not?

31. Assume that we have two events, A and B, that are mutually exclusive. Assume further that we know $P(A) = .30$ and $P(B) = .40$.

a. What is $P(A \cap B)$?
b. What is $P(A \mid B)$?
c. A student in statistics argues that the concepts of mutually exclusive events and independent events are really the same, and that if events are mutually exclusive they must be independent. Do you agree with this statement? Use the probability information in this problem to justify your answer.
d. What general conclusion would you make about mutually exclusive and independent events given the results of this problem?

Applications

32. A Daytona Beach nightclub has the following data on the age and marital status of 140 customers.

		Marital Status	
		Single	*Married*
Age	*Under 30*	77	14
	30 or Over	28	21

a. Develop a joint probability table for these data.
b. Use the marginal probabilities to comment on the age of customers attending the club.
c. Use the marginal probabilities to comment on the marital status of customers attending the club.
d. What is the probability of finding a customer who is single and under the age of 30?
e. If a customer is under 30, what is the probability that he or she is single?
f. Is marital status independent of age? Explain, using probabilities.

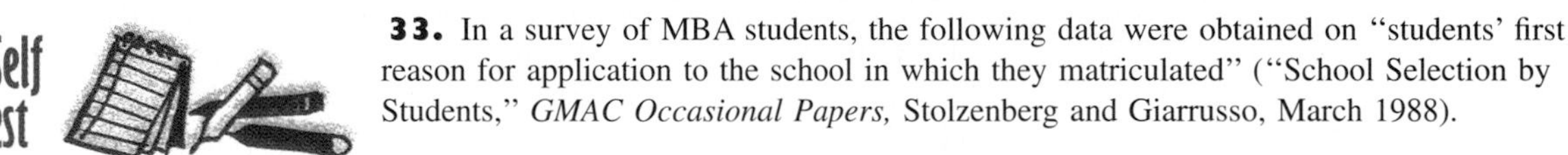

33. In a survey of MBA students, the following data were obtained on "students' first reason for application to the school in which they matriculated" ("School Selection by Students," *GMAC Occasional Papers,* Stolzenberg and Giarrusso, March 1988).

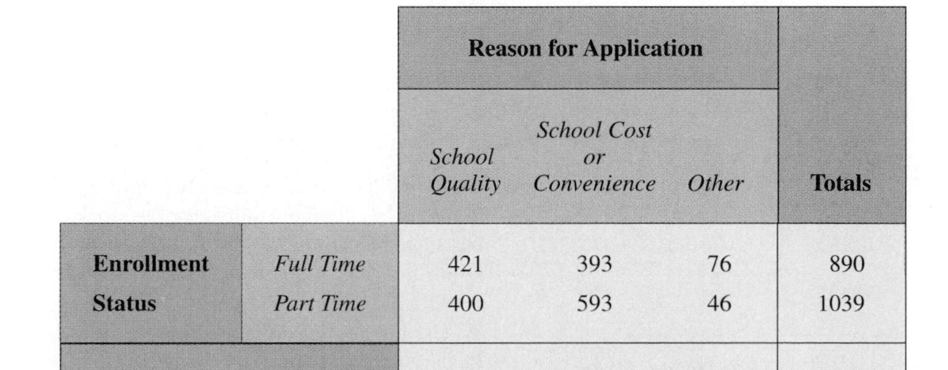

		Reason for Application			
		School Quality	*School Cost or Convenience*	*Other*	**Totals**
Enrollment Status	*Full Time*	421	393	76	890
	Part Time	400	593	46	1039
	Totals	821	986	122	1929

a. Develop a joint probability table for these data.
b. Use the marginal probabilities of school quality, cost/convenience, and other to comment on the most important reason for choosing a school.
c. If a student goes full time, what is the probability that school quality is the first reason for choosing a school?
d. If a student goes part time, what is the probability that school quality is the first reason for choosing a school?
e. Let A be the event that a student is full time and let B be the event that the student lists school quality as the first reason for applying. Are events A and B independent? Justify your answer.

34. *The* 1992 *Wall Street Journal* Subscriber Study provided data on automobiles in each subscriber's household. Data for 1900 respondents follow.

		Do You Have a U.S. Car?		
		Yes	*No*	**Totals**
Do You Have a Foreign Car?	*Yes*	734	430	1164
	No	701	35	736
	Totals	1435	465	1900

a. Show the joint probability table for these data.
b. Use the marginal probabilities to compare U.S. and foreign car preferences among subscribers.
c. What is the probability that a household has both a U.S. and a foreign car?
d. What is the probability that a household has a car, U.S. or foreign?
e. If a household has a U.S. car, what is the probability that it also has a foreign car?
f. If a household has a foreign car, what is the probability that it also has a U.S. car?
g. Are having a U.S. car and having a foreign car independent events? Explain.

35. Some investment analysts believe the January performance of the stock market is an indicator of how the market will perform during the coming year (*USA Today,* January 10, 1995). Historical data suggest that if the stock market rises in January, the outlook is good for stocks during the coming year. Suppose an investment analyst provides the following probability estimates.

- The probability that the stock market will be up for January is .70.
- The probability that the stock market will be up for the year is .80.
- The probability that the stock market will be up for January and up for the year is .63.

a. Given that the stock market is up for January, use the investment analyst's estimates to determine the probability that the stock market will be up for the year.
b. Suppose the probability that the stock market will both not be up for January and be up for the year is .17. If the stock market is not up for January, what is the probability that it will be up for the year?
c. Use the above conditional probabilities to comment on the use of the stock market's January performance as an indicator of its performance for the year.
d. Do the probabilities suggest that the stock market's January performance and its annual performance are independent or dependent events? Explain.

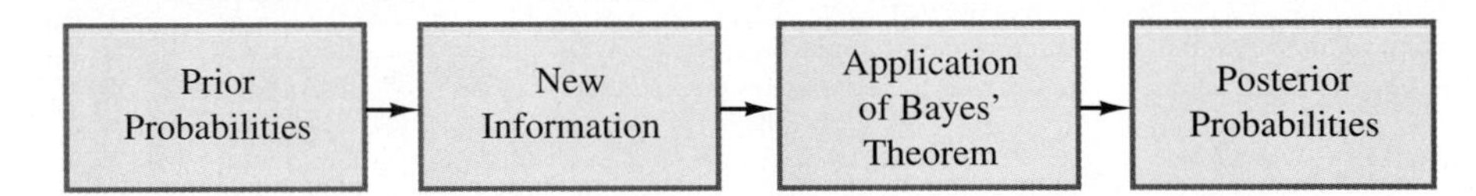

FIGURE 4.9
Probability Revision Using Bayes' Theorem

36. The Grant Thornton public accounting firm conducted a survey to see how executives felt about the 1992 recession and recovery potential (*Journal of Accountancy,* February 1992). Results showed that the probability of an executive indicating the existence of a recession was .74. The probability of an executive stating both that a recession existed and that a recovery would occur within six months was .41.

a. Given that an executive indicated the existence of a recession, what is the probability that the executive felt that a recovery would occur within six months?

b. Assume that if an executive denied the existence of a recession, he or she also believed that a recovery had already begun. Construct a joint probability table for recession—no recession, and recovery—no recovery. Executives who feel a recovery is already underway and those who feel one will occur in six months should be put in the same category for this table.

c. Use the joint probability table in part (b) to find the marginal probability of a recovery.

37. A purchasing agent has placed rush orders for a particular raw material with two different suppliers, A and B. If neither order arrives in four days, the production process must be shut down until at least one of the orders arrives. The probability that supplier A can deliver the material in four days is .55. The probability that supplier B can deliver the material in four days is .35.

a. What is the probability that both suppliers will deliver the material in four days? Since two separate suppliers are involved, we are willing to assume independence.

b. What is the probability that at least one supplier will deliver the material in four days?

c. What is the probability that the production process will be shut down in four days because of a shortage of raw material (that is, both orders are late)?

38. In a 1992 study of the consumer's view of the economy, the probability that a consumer would buy a house during the year was .033 and the probability that a consumer would buy a car during the year was .168 (*U.S. News & World Report,* April 13, 1992). Assume that there was only a .004 probability that a consumer would buy a house and a car during the year.

a. What is the probability that a consumer would buy either a car or a house during the year?

b. What is the probability that a consumer would buy a car during the year given that the consumer purchased a house during the year?

c. Are buying a car and buying a house independent events? Explain.

4.5 Bayes' Theorem

In the discussion of conditional probability, we indicated that revising probabilities when new information is obtained is an important phase of probability analysis. Often, we begin the analysis with initial or *prior* probability estimates for specific events of interest. Then, from sources such as a sample, a special report, or a product test, we obtain some additional information about the events. Given this new information, we update the prior probability values by calculating revised probabilities, referred to as *posterior probabilities. Bayes' theorem* provides a means for making these probability calculations. The steps in this probability revision process are shown in Figure 4.9.

As an application of Bayes' theorem, consider a manufacturing firm that receives shipments of parts from two different suppliers. Let A_1 denote the event that a part is from supplier 1 and A_2 denote the event that a part is from supplier 2. Currently, 65%

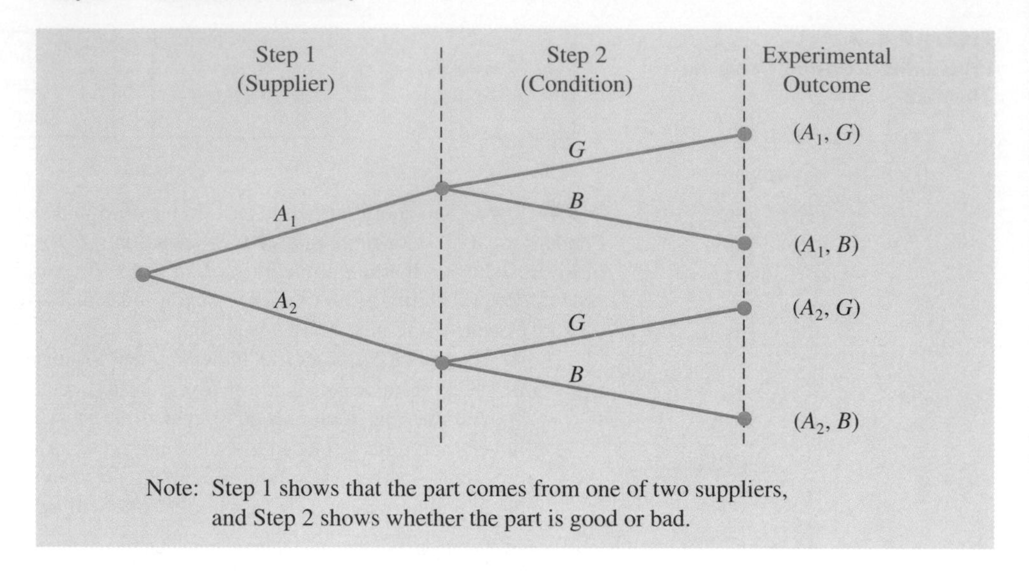

FIGURE 4.10
Two-Step Tree Diagram

Table 4.6 Historical Quality Levels of Two Suppliers

	Percentage Good Parts	Percentage Bad Parts
Supplier 1	98	2
Supplier 2	95	5

of the parts purchased by the company are from supplier 1 and the remaining 35% are from supplier 2. Hence, if a part is selected at random, we would assign the prior probabilities $P(A_1) = .65$ and $P(A_2) = .35$.

The quality of the purchased parts varies with the source of supply. Historical data suggest that the quality ratings of the two suppliers are as shown in Table 4.6. If we let G denote the event that a part is good and B denote the event that a part is bad, the information in Table 4.6 provides the following conditional probability values.

$$P(G \mid A_1) = .98 \quad P(B \mid A_1) = .02$$

$$P(G \mid A_2) = .95 \quad P(B \mid A_2) = .05$$

The tree diagram in Figure 4.10 depicts the process of the firm receiving a part from one of the two suppliers and then discovering that the part is good or bad as a two-step experiment. We see that there are four experimental outcomes; two correspond to the part being good and two correspond to the part being bad.

Each of the experimental outcomes is the intersection of two events, so we can use the multiplication rule to compute the probabilities. For instance,

$$P(A_1, G) = P(A_1 \cap G) = P(A_1)P(G \mid A_1)$$

FIGURE 4.11
Probability Tree for Two-Supplier Example

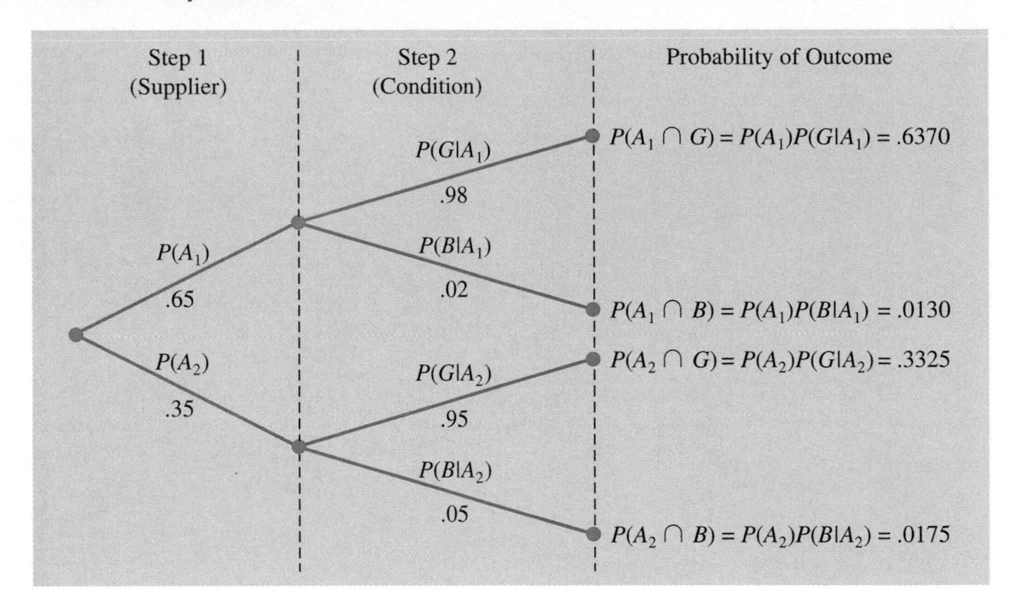

The process of computing these joint probabilities can be depicted in what is called a probability tree (see Figure 4.11). From left to right through the tree, the probabilities for each branch at step 1 are prior probabilities and the probabilities for each branch at step 2 are conditional probabilities. To find the probabilities of each experimental outcome, we simply multiply the probabilities on the branches leading to the outcome. Each of these joint probabilities is shown in Figure 4.11 along with the known probabilities for each branch.

Suppose now that the parts from the two suppliers are used in the firm's manufacturing process and that a machine breaks down because it attempts to process a bad part. Given the information that the part is bad, what is the probability that it came from supplier 1 and what is the probability that it came from supplier 2? With the information in the probability tree (Figure 4.11), Bayes' theorem can be used to answer these questions.

Letting B denote the event that the part is bad, we are looking for the posterior probabilities $P(A_1 \mid B)$ and $P(A_2 \mid B)$. From the law of conditional probability, we know that

$$P(A_1 \mid B) = \frac{P(A_1 \cap B)}{P(B)} \tag{4.14}$$

Referring to the probability tree, we see that

$$P(A_1 \cap B) = P(A_1)P(B \mid A_1) \tag{4.15}$$

To find $P(B)$, we note that there are only two ways event B can occur: $(A_1 \cap B)$ and $(A_2 \cap B)$. Therefore, we have

$$\begin{aligned} P(B) &= P(A_1 \cap B) + P(A_2 \cap B) \\ &= P(A_1)P(B \mid A_1) + P(A_2)P(B \mid A_2) \end{aligned} \tag{4.16}$$

Substituting from (4.15) and (4.16) into (4.14) and writing a similar result for $P(A_2 \mid B)$, we obtain Bayes' theorem for the case of two events.

Bayes' Theorem (Two-Event Case)

$$P(A_1 \mid B) = \frac{P(A_1)P(B \mid A_1)}{P(A_1)P(B \mid A_1) + P(A_2)P(B \mid A_2)} \tag{4.17}$$

$$P(A_2 \mid B) = \frac{P(A_2)P(B \mid A_2)}{P(A_1)P(B \mid A_1) + P(A_2)P(B \mid A_2)} \tag{4.18}$$

Using (4.17) and the probability values provided in the example, we have

$$\begin{aligned} P(A_1 \mid B) &= \frac{P(A_1)P(B \mid A_1)}{P(A_1)P(B \mid A_1) + P(A_2)P(B \mid A_2)} \\ &= \frac{(.65)(.02)}{(.65)(.02) + (.35)(.05)} = \frac{.0130}{.0130 + .0175} \\ &= \frac{.0130}{.0305} = .4262 \end{aligned}$$

In addition, using (4.18), we find $P(A_2 \mid B)$.

$$\begin{aligned} P(A_2 \mid B) &= \frac{(.35)(.05)}{(.65)(.02) + (.35)(.05)} \\ &= \frac{.0175}{.0130 + .0175} = \frac{.0175}{.0305} = .5738 \end{aligned}$$

Note that in this application we started with a probability of .65 that a part selected at random was from supplier 1. However, given information that the part is bad, the probability that the part is from supplier 1 drops to .4262. In fact, if the part is bad, there is a better than 50–50 chance that the part came from supplier 2; that is, $P(A_2 \mid B) = .5738$.

Bayes' theorem is applicable when the events for which we want to compute posterior probabilities are mutually exclusive and their union is the entire sample space.* Bayes' theorem can be extended to the case where there are n mutually exclusive events $A_1, A_2, \ldots, A_n$ whose union is the entire sample space. In such a case, Bayes' theorem for the computation of any posterior probability $P(A_i \mid B)$ has the following form.

Bayes' Theorem

$$P(A_i \mid B) = \frac{P(A_i)P(B \mid A_i)}{P(A_1)P(B \mid A_1) + P(A_2)P(B \mid A_2) + \cdots + P(A_n)P(B \mid A_n)} \tag{4.19}$$

With prior probabilities $P(A_1), P(A_2), \ldots, P(A_n)$ and the appropriate conditional probabilities $P(B \mid A_1), P(B \mid A_2), \ldots, P(B \mid A_n)$, Equation (4.19) can be used to compute the posterior probability of the events $A_1, A_2, \ldots, A_n$.

*If the union of events is the entire sample space, the events are said to be *collectively exhaustive.*

TABLE 4.7 Summary of Bayes' Theorem Calculations for the Two-Supplier Problem

(1) Events A_i	(2) Prior Probabilities $P(A_i)$	(3) Conditional Probabilities $P(B \mid A_i)$	(4) Joint Probabilities $P(A_i \cap B)$	(5) Posterior Probabilities $P(A_i \mid B)$
A_1	.65	.02	.0130	.0130/.0305 = .4262
A_2	.35	.05	.0175	.0175/.0305 = .5738
	1.00		$P(B)$ = .0305	1.0000

The Tabular Approach

A tabular approach is helpful in conducting the Bayes' theorem calculations. Such an approach is shown in Table 4.7 for the parts supplier problem. The computations shown there are done in the following steps.

STEP 1: Prepare the following three columns:

Column 1—The mutually exclusive events for which posterior probabilities are desired.

Column 2—The prior probabilities for the events.

Column 3—The conditional probabilities of the new information *given* each event.

STEP 2: In column 4, compute the joint probabilities for each event and the new information B by using the multiplication law. These joint probabilities are found by multiplying the prior probabilities in column 2 by the corresponding conditional probabilities in column 3; that is, $P(A_i \cap B) = P(A_i)P(B \mid A_i)$.

STEP 3: Sum the joint probabilities in column 4. The sum is the probability of the new information, $P(B)$. Thus we see that in the above example there is a .0130 probability of a bad part from supplier 1 and there is a .0175 probability of a bad part from supplier 2. Since these are the only two ways in which a bad part can be obtained, the sum .0130 + .0175 shows that there is an overall probability of .0305 of finding a bad part from the combined shipments of the two suppliers.

STEP 4: In column 5, compute the posterior probabilities using the basic relationship of conditional probability,

$$P(A_i \mid B) = \frac{P(A_i \cap B)}{P(B)}$$

Note that the joint probabilities $P(A_i \cap B)$ are in column 4 and the probability $P(B)$ is the sum of column 4.

notes and comments

1. Bayes' theorem is used extensively in decision analysis. The prior probabilities are often subjective estimates provided by a decision maker. Sample information is obtained and posterior probabilities are computed for use in developing a decision strategy.

2. An event and its complement are mutually exclusive, and their union is the entire sample space. Thus, Bayes' theorem is always applicable for computing posterior probabilities of an event and its complement.

exercises

Methods

39. The prior probabilities for events A_1 and A_2 are $P(A_1) = .40$ and $P(A_2) = .60$. It is also known that $P(A_1 \cap A_2) = 0$. Suppose $P(B \mid A_1) = .20$ and $P(B \mid A_2) = .05$.

a. Are A_1 and A_2 mutually exclusive? Why or why not?

b. Compute $P(A_1 \cap B)$ and $P(A_2 \cap B)$.

c. Compute $P(B)$.

d. Apply Bayes' theorem to compute $P(A_1 \mid B)$ and $P(A_2 \mid B)$.

40. The prior probabilities for events A_1, A_2, and A_3 are $P(A_1) = .20$, $P(A_2) = .50$, and $P(A_3) = .30$. The conditional probabilities of event B given A_1, A_2, and A_3 are $P(B \mid A_1) = .50$, $P(B \mid A_2) = .40$, and $P(B \mid A_3) = .30$.

a. Compute $P(B \cap A_1)$, $P(B \cap A_2)$, and $P(B \cap A_3)$.

b. Apply Bayes' theorem, Equation (4.19), to compute the posterior probability $P(A_2 \mid B)$.

c. Use the tabular approach to applying Bayes' theorem to compute $P(A_1 \mid B)$, $P(A_2 \mid B)$, and $P(A_3 \mid B)$.

Applications

41. A consulting firm has submitted a bid for a large research project. The firm's management initially felt there was a 50–50 chance of getting the project. However, the agency to which the bid was submitted has subsequently requested additional information on the bid. Past experience indicates that on 75% of the successful bids and 40% of the unsuccessful bids the agency requested additional information.

a. What is the prior probability of the bid being successful (that is, prior to the request for additional information)?

b. What is the conditional probability of a request for additional information given that the bid will ultimately be successful?

c. Compute a posterior probability that the bid will be successful given that a request for additional information has been received.

42. A local bank is reviewing its credit-card policy with a view toward recalling some of its credit cards. In the past approximately 5% of cardholders have defaulted and the bank has been unable to collect the outstanding balance. Hence, management has established a prior probability of .05 that any particular cardholder will default. The bank has further found that the probability of missing one or more monthly payments is .20 for customers who do not default. Of course, the probability of missing one or more payments for those who default is 1.

a. Given that a customer has missed a monthly payment, compute the posterior probability that the customer will default.

b. The bank would like to recall its card if the probability that a customer will default is greater than .20. Should the bank recall its card if the customer misses a monthly payment? Why or why not?

43. *The Book of Risks* (1994) contains probability information about the chances people take in everyday activities. For example, the probability of a man having a motor vehicle accident during a one-year period is reported to be twice as great as the probability of a woman having a motor vehicle accident during a one-year period. Indicated probabilities are .113 for men and .057 for women. Suppose that 55% of the drivers in Lucas County are men. In filling out a driving history questionnaire, a person from Lucas County indicates involvement in a motor vehicle accident during the past year. What is the probability that the person is a woman?

44. A city has a professional basketball team playing at home and a professional hockey team playing away on the same night. According to probabilities for professional sports published in

Chance (Fall 1992), a professional basketball team has a .641 probability of winning a home game and a professional hockey team has a .462 probability of winning an away game. Historically, when both teams play on the same night, the chance that the next morning's leading sports story will be about the basketball game is 60% and the chance that it will be about the hockey game is 40%. Suppose that on the morning after these games the newspaper's leading sports story begins with the headline "We Win!!" What is the probability that the story is about the basketball team?

45. *M.D. Computing* (May, 1991) describes the use of Bayes' theorem and the use of conditional probability in medical diagnosis. Prior probabilities of diseases are based on the physician's assessment of such things as geographical location, seasonal influence, occurrence of epidemics, and so forth. Assume that a patient is believed to have one of two diseases, denoted D_1 and D_2, with $P(D_1) = .60$ and $P(D_2) = .40$ and that medical research has shown there is a probability associated with each symptom that may accompany the diseases. Suppose that, given diseases D_1 and D_2, the probabilities that the patient will have symptoms S_1, S_2, or S_3 are as follows.

		Symptoms		
		S_1	S_2	S_3
Disease	D_1	.15	.10	.15 ← $P(S_3 \mid D_1)$
	D_2	.80	.15	.03

After a certain symptom is found to be present, the medical diagnosis may be aided by finding the revised probabilities of each particular disease. Compute the posterior probabilities of each disease given the following medical findings.

a. The patient has symptom S_1.
b. The patient has symptom S_2.
c. The patient has symptom S_3.
d. For the patient with symptom S_1 in (a), suppose we also find symptom S_2. What are the revised probabilities of D_1 and D_2?

summary

In this chapter we introduced basic probability concepts and illustrated how probability analysis can be used to provide helpful information for decision making. We described how probability can be interpreted as a numerical measure of the likelihood that an event will occur. In addition, we saw that the probability of an event can be computed either by summing the probabilities of the experimental outcomes (sample points) comprising the event or by using the relationships established by the addition, conditional probability, and multiplication laws of probability. For cases in which additional information is available, we showed how Bayes' theorem can be used to obtain revised or posterior probabilities.

glossary

Probability A numerical measure of the likelihood that an event will occur.

Experiment Any process that generates well-defined outcomes.

Sample space The set of all possible sample points (experimental outcomes).

Sample points The individual outcomes of an experiment.

Tree diagram A graphical device helpful in defining sample points of an experiment involving multiple steps.

Basic requirements of probability Two requirements that restrict the manner in which probability assignments can be made:

a. For each experimental outcome E_i we must have $0 \leq P(E_i) \leq 1$.
b. Considering all experimental outcomes, we must have $\Sigma P(E_i) = 1$.

Classical method A method of assigning probabilities which assumes that the experimental outcomes are equally likely.

Relative frequency method A method of assigning probabilities on the basis of experimentation or historical data.

Subjective method A method of assigning probabilities on the basis of judgment.

Event A collection of sample points.

Complement of event *A* The event containing all sample points that are not in A.

Venn diagram A graphical device for representing symbolically the sample space and operations involving events.

Union of events *A* and *B* The event containing all sample points that are in A, in B, or in both. The union is denoted $A \cup B$.

Intersection of *A* and *B* The event containing all sample points that are in both A and B. The intersection is denoted $A \cap B$.

Addition law A probability law used to compute the probability of a union, $P(A \cup B)$. It is $P(A \cup B) = P(A) + P(B) - P(A \cap B)$. For mutually exclusive events, since $P(A \cap B) = 0$, it reduces to $P(A \cup B) = P(A) + P(B)$.

Mutually exclusive events Events that have no sample points in common; that is, $A \cap B$ is empty and $P(A \cap B) = 0$.

Conditional probability The probability of an event given that another event has occurred. The conditional probability of A given B is $P(A \mid B) = P(A \cap B)/P(B)$.

Independent events Two events A and B where $P(A \mid B) = P(A)$ or $P(B \mid A) = P(B)$; that is, the events have no influence on each other.

Multiplication law A probability law used to compute the probability of an intersection, $P(A \cap B)$. It is $P(A \cap B) = P(A)P(B \mid A)$ or $P(A \cap B) = P(B)P(A \mid B)$. For independent events it reduces to $P(A \cap B) = P(A)P(B)$.

Prior probabilities Initial estimates of the probabilities of events.

Posterior probabilities Revised probabilities of events based on additional information.

Bayes' theorem A method used to compute posterior probabilities.

key formulas

Counting Rule for Combinations

$$C_n^N = \binom{N}{n} = \frac{N!}{n!(N-n)!} \tag{4.1}$$

Counting Rule for Permutations

$$P_n^N = n!\binom{N}{n} = \frac{N!}{(N-n)!} \tag{4.2}$$

Computing Probability Using the Complement

$$P(A) = 1 - P(A^c) \tag{4.5}$$

Addition Law

$$P(A \cup B) = P(A) + P(B) - P(A \cap B) \tag{4.6}$$

Conditional Probability

$$P(A \mid B) = \frac{P(A \cap B)}{P(B)} \tag{4.7}$$

$$P(B \mid A) = \frac{P(A \cap B)}{P(A)} \tag{4.8}$$

Multiplication Law

$$P(A \cap B) = P(B)P(A \mid B) \tag{4.11}$$

$$P(A \cap B) = P(A)P(B \mid A) \tag{4.12}$$

Multiplication Law for Independent Events

$$P(A \cap B) = P(A)P(B) \tag{4.13}$$

Bayes' Theorem

$$P(A_i \mid B) = \frac{P(A_i)P(B \mid A_i)}{P(A_1)P(B \mid A_1) + P(A_2)P(B \mid A_2) + \cdots + P(A_n)P(B \mid A_n)} \tag{4.19}$$

supplementary exercises

46. The long-distance calling market is shared by AT&T, MCI, and Sprint (*Business Week,* February 20, 1995). Suppose a survey of 200 small businesses finds 122 AT&T users, 38 MCI users, 20 Sprint users, and 20 users of other long-distance systems.

Table 4.8 Number of Schools Applied to by MBA Students

Number of Schools	Number of Students
1	1,230
2	304
3	184
4	118
5	78
6	51
7	25
8	13
9	20
10	8
11	9
12	6
Total	2,046

a. Consider the experiment of observing the supplier of long-distance service for a randomly selected small business. How many experimental outcomes are possible?
b. Use the sample data to assign probabilities to the experimental outcomes.

47. A financial manager has just made two new investments—one in the oil industry and one in municipal bonds. After a one-year period, each of the investments will be classified as either successful or unsuccessful. Consider the making of the two investments as an experiment.
a. How many sample points exist for this experiment?
b. Show a tree diagram and list the sample points.
c. Let O = the event that the oil investment is successful and M = the event that the municipal bond investment is successful. List the sample points in O and in M.
d. List the sample points in the union of the events $(O \cup M)$.
e. List the sample points in the intersection of the events $(O \cap M)$.
f. Are events O and M mutually exclusive? Explain.

48. A survey of 1174 *Forbes* subscribers found the following information about corporate credit cards (*Forbes* 1993 Subscriber Study).

Type of Card	Number of Subscribers
AT&T	195
American Express	406
Diner's Club	30
MasterCard	109
Visa	157
Other	50

a. If 629 subscribers indicated they have at least one corporate credit card, what is the probability that a *Forbes* subscriber does not have a corporate credit card?
b. What is the probability that a subscriber holds a corporate American Express card?
c. Suppose it is known that 90 subscribers hold both MasterCard and Visa corporate credit cards. What is the probability that a subscriber holds one or both?

49. A survey of new matriculants to MBA programs was conducted by the Graduate Management Admissions Council during 1985 ("School Selection by Students," *GMAC Occasional Papers,* Stolzenberg and Giarrusso, March 1988). Table 4.8 shows the number of schools to which students applied.
a. Use these data to assign probabilities to the number of schools to which a randomly selected MBA student applied.
b. What is the probability that the student applied to only one school?
c. What is the probability that the student applied to three or more schools?
d. What is the probability that the student applied to more than six schools?

50. A telephone survey was used to determine viewer response to a new television show. The following data were obtained.

Rating	Frequency
Poor	4
Below average	8
Average	11
Above average	14
Excellent	13

a. What is the probability that a randomly selected viewer rated the new show as average or better?
b. What is the probability that a randomly selected viewer rated the new show below average or worse?

51. A bank has observed that credit-card account balances have been growing over the past year. A sample of 200 customer accounts resulted in the data shown.

Amount owed ($)	Frequency
0–99	62
100–199	46
200–299	24
300–399	30
400–499	26
500 and over	12

a. Let A be the event that a customer's balance is less than $200. Find $P(A)$.
b. Let B be the event that a customer's balance is $300 or more. Find $P(B)$.

52. The GMAC MBA new-matriculants survey provided the following data for 2018 students.

		Applied to More Than One School	
		Yes	*No*
Age Group	*23 and under*	207	201
	24–26	299	379
	27–30	185	268
	31–35	66	193
	36 and over	51	169

a. For a randomly selected MBA student, prepare a joint probability table for the experiment consisting of observing the student's age and the number of schools to which the student applied.
b. What is the probability that an applicant was 23 or under?
c. What is the probability that an applicant was older than 26?
d. What is the probability that an applicant applied to more than one school?

53. Refer again to the data from the GMAC new-matriculants survey in Exercise 52.
a. Given that a person applied to more than one school, what is the probability that the person was 24–26 years old?
b. Given that a person is in the 36-and-over age group, what is the probability that the person applied to more than one school?
c. What is the probability that a person is 24–26 years old *or* applied to more than one school?

d. Suppose a person is known to have applied to only one school. What is the probability that the person was 31 or more years old?
e. Is the number of schools applied to independent of age? Explain.

54. A William M. Mercer Inc. survey provided the following data on salary and bonus compensation for 105 CEOs of technology and financial corporations (*The Wall Street Journal,* April 13, 1994). Total compensation is shown in millions of dollars.

		Total Compensation			
		Under \$1M	*\$1M–\$2M*	*Over \$2M*	**Total**
Corporation	*Technology*	17	21	7	45
	Financial	12	31	17	60
	Total	29	52	24	105

a. Show the joint probability table for the data.
b. Use the marginal probabilities to comment on the most probable of the three compensation ranges.
c. Let T represent technology, F represent financial, and $2M$ represent total compensation over \$2 million. Find $P(2M)$. Then compute the conditional probabilities $P(2M \mid T)$ and $P(2M \mid F)$. What conclusion can you draw about the compensation levels for technology and financial CEOs?
d. Is compensation independent of the type of corporation? Explain.

55. A large consumer goods company has been running a television advertisement for one of its soap products. A survey was conducted. On the basis of this survey, probabilities were assigned to the following events.

B = individual purchased the product

S = individual recalls seeing the advertisement

$B \cap S$ = individual purchased the product and recalls seeing the advertisement

The probabilities assigned were $P(B) = .20$, $P(S) = .40$, and $P(B \cap S) = .12$. The following problems relate to this situation.
a. What is the probability of an individual's purchasing the product given that the individual recalls seeing the advertisement? Does seeing the advertisement increase the probability that the individual will purchase the product? As a decision maker, would you recommend continuing the advertisement (assuming that the cost is reasonable)?
b. Assume that individuals who do not purchase the company's soap product buy from its competitors. What would be your estimate of the company's market share? Would you expect that continuing the advertisement will increase the company's market share? Why or why not?
c. The company has also tested another advertisement and assigned it values of $P(S) = .30$ and $P(B \cap S) = .10$. What is $P(B \mid S)$ for this other advertisement? Which advertisement seems to have had the bigger effect on customer purchases?

56. In the evaluation of a sales training program, a firm found that of 50 salespersons making a bonus last year, 20 had attended a special sales training program. The firm has 200

salespersons. Let B = the event that a salesperson makes a bonus and S = the event that a salesperson attends the sales training program.

a. Find $P(B)$, $P(S \mid B)$, and $P(S \cap B)$.

b. Assume that 40% of the salespersons have attended the training program. What is the probability that a salesperson makes a bonus given that the salesperson attended the sales training program, $P(B \mid S)$?

c. If the firm evaluates the training program in terms of its effect on the probability of a salesperson's making a bonus, what is your evaluation of the training program? Comment on whether B and S are dependent or independent events.

57. A company has studied the number of lost-time accidents occurring at its Brownsville, Texas, plant. Historical records show that 6% of the employees had lost-time accidents last year. Management believes that a special safety program will reduce such accidents to 5% during the current year. In addition, it estimates that 15% of employees who had lost-time accidents last year will have a lost-time accident during the current year.

a. What percentage of the employees will have lost-time accidents in both years?

b. What percentage of the employees will have at least one lost-time accident over the two-year period?

58. The Dallas IRS auditing staff is concerned with identifying potentially fraudulent tax returns. From past experience they believe that the probability of finding a fraudulent return given that the return contains deductions for contributions exceeding the IRS standard is .20. Given that the deductions for contributions do not exceed the IRS standard, the probability of a fraudulent return decreases to .02. If 8% of all returns exceed the IRS standard for deductions due to contributions, what is the best estimate of the percentage of fraudulent returns?

59. An oil company has purchased an option on land in Alaska. Preliminary geologic studies have assigned the following prior probabilities.

$$P(\text{high-quality oil}) = .50$$

$$P(\text{medium-quality oil}) = .20$$

$$P(\text{no oil}) = .30$$

a. What is the probability of finding oil?

b. After 200 feet of drilling on the first well, a soil test is taken. The probabilities of finding the particular type of soil identified by the test follow.

$$P(\text{soil} \mid \text{high-quality oil}) = .20$$

$$P(\text{soil} \mid \text{medium-quality oil}) = .80$$

$$P(\text{soil} \mid \text{no oil}) = .20$$

How should the firm interpret the soil test? What are the revised probabilities, and what is the new probability of finding oil?

60. A Bayesian approach can be used to revise probabilities that a prospect field will produce oil (*Oil & Gas Journal,* January 11, 1988). In one case, geological assessment indicates a 25% chance that the field will produce oil. Further, there is an 80% chance that a particular well will strike oil given that oil is present in the prospect field.

a. Suppose that one well is drilled on the field and it comes up dry. What is the probability that the prospect field will produce oil?

b. If two wells come up dry, what is the probability that the field will produce oil?

c. The oil company would like to keep looking as long as the chances of finding oil are greater than 1%. How many dry wells must be drilled before the field will be abandoned?

CHAPTER 5

Discrete Probability Distributions

Xerox Corporation*

Stamford, Connecticut

Xerox Corporation is a worldwide leader in information products and services. Almost everyone is familiar with Xerox copying machines, but the company is involved in many other businesses also. For example, Xerox's Multinational Documentation and Training Services (MD&TS) group provides cost-effective, high-quality communication services including documentation, training, translation, and publishing in a variety of languages.

Professional writers and translators working for MD&TS use an online computerized publication system. Several individuals can access the system simultaneously. Occasionally, during peak demand periods, one or more individuals may be denied access to the system. In considering design alternatives for the online computerized publication system, management was interested in estimating the probability that a user would be denied access.

statistics in practice

A computer simulation model was developed to evaluate the various designs. A key input to the simulation model was the length of time a user is on the system in each session. From observation of MD&TS users, the time per session was found to range from 10 minutes to 90 minutes. The probability distribution for the time per session, with x denoting the time per session in minutes, is shown. The column labeled $f(x)$ shows probabilities of .05 for a 10-minute session, .06 for a 20-minute session, and so on. The highest probability, .25, is for a session lasting 50 minutes.

In probability terminology, session duration is referred to as a random variable and the table shown is referred to as the probability distribution for the random variable. This probability distribution and the probability distribution for the length of time between sessions were key inputs to the computer simulation model. The simulation results based on these probability distributions helped management select a computerized publication system design that ensured a near-zero probability of a user being denied access to the system.

The authors are indebted to Soterios M. Flouris for providing this Statistics in Practice.

Probability distributions such as the ones used by Xerox are the topic of this chapter. You will also learn about some special probability distributions that are widely available and the situations in which they are applicable.

Probability Distribution for Session Duration

x (minutes)	Probability $f(x)$
10	.05
20	.06
30	.08
40	.20
50	.25
60	.20
70	.08
80	.06
90	.02

The actual probability distribution used in the simulation study was modified to protect proprietary information and to simplify the discussion.

Individual workstations provide access to an online computerized publication system.

In this chapter we continue the study of probability by introducing the concepts of random variables and probability distributions. The focus of this chapter is discrete probability distributions. Two special discrete probability distributions—the binomial and Poisson—are covered.

5.1 Random Variables

In Chapter 4 we defined the concept of an experiment and its associated experimental outcomes. A random variable provides a means for describing experimental outcomes by numerical values. The definition of a random variable follows.

Random Variable

A *random variable* is a numerical description of the outcome of an experiment.

In effect, a random variable associates a numerical value with each possible outcome. The particular numerical value of the random variable depends on the outcome of the experiment. A random variable can be classified as being either *discrete* or *continuous* depending on the numerical values it assumes.

Discrete Random Variables

A random variable that may assume either a finite number of values or an infinite sequence of values such as 0, 1, 2, . . ., is referred to as a *discrete random variable.* For example, consider the experiment of an accountant taking the certified public accountant (CPA) examination. The examination has four parts. We can define the discrete random variable as x = the number of parts of the CPA examination passed. This discrete random variable may assume the finite number of values 0, 1, 2, 3, or 4.

As another example of a discrete random variable, consider the experiment of cars arriving at a tollbooth. The random variable of interest is x = the number of cars arriving during a one-day period. The possible values for x come from the sequence of integers 0, 1, 2, and so on. Hence, x is a discrete random variable assuming one of the values in this infinite sequence.

Although many experiments have outcomes that are naturally described by numerical values, others do not. For example, a survey question might ask an individual to recall the message in a recent television commercial. There would be two experimental outcomes: the individual cannot recall the message and the individual can recall the message. We can still describe these experimental outcomes numerically by defining the discrete random variable x as follows: let $x = 0$ if the individual cannot recall the message and $x = 1$ if the individual can recall the message. The numerical values for this random variable are arbitrary (we could have used 5 and 10), but they are acceptable in terms of the definition of a random variable—namely, x is a random variable because it provides a numerical description of the outcome of the experiment.

Table 5.1 provides some additional examples of discrete random variables. Note that in each example the discrete random variable assumes a finite number of values or an infinite sequence of values such as 0, 1, 2, ... Discrete random variables such as these are discussed in detail in this chapter.

TABLE 5.1 Examples of Discrete Random Variables

Experiment	Random Variable (x)	Possible Values for the Random Variable
Contact five customers	Number of customers who place an order	0, 1, 2, 3, 4, 5
Inspect a shipment of 50 radios	Number of defective radios	0, 1, 2, . . ., 49, 50
Operate a restaurant for one day	Number of customers	0, 1, 2, 3, . . .
Sell an automobile	Gender of the customer	0 if male; 1 if female

TABLE 5.2 Examples of Continuous Random Variables

Experiment	Random Variable (x)	Possible Values for the Random Variable
Operate a bank	Time between customer arrivals in minutes	$x \geq 0$
Fill a soft drink can (max = 12.1 ounces)	Number of ounces	$0 \leq x \leq 12.1$
Work on a project to construct a new library	Percentage of project complete after six months	$0 \leq x \leq 100$
Test a new chemical process	Temperature when the desired reaction takes place (min 150 F; max 212 F)	$150 \leq x \leq 212$

Continuous Random Variables

A random variable that may assume any numerical value in an interval or collection of intervals is called a *continuous random variable.* Experimental outcomes that are based on measurement scales such as time, weight, distance, and temperature can be described by continuous random variables. For example, consider an experiment of monitoring incoming telephone calls to the claims office of a major insurance company. Suppose the random variable of interest is x = the time between consecutive incoming calls in minutes. This random variable may assume any value in the interval $x \geq 0$. Actually, an infinite number of values are possible for x, including values such as 1.26 minutes, 2.751 minutes, 4.3333 minutes, and so on. As another example, consider a 90-mile section of interstate highway I-75 north of Atlanta, Georgia. For an emergency ambulance service located in Atlanta, we might define the random variable as x = the location of the next traffic accident along this section of I-75. In this case, x would be a continuous random variable assuming any value in the interval $0 \leq x \leq 90$. Additional examples of continuous random variables are listed in Table 5.2. Note that each example describes a random variable that may assume any value in an interval of values. Continuous random variables and their probability distributions will be the topic of Chapter 6.

notes and comments

One way to determine whether a random variable is discrete or continuous is to think of the values of the random variable as points on a line segment. Choose two points representing values of the random variable. If the entire line segment between the two points also represents possible values for the random variable, the random variable is continuous.

exercises

Methods

1. Consider the experiment of tossing a coin twice.
a. List the experimental outcomes.
b. Define a random variable that represents the number of heads occurring on the two tosses.
c. Show what value the random variable would assume for each of the experimental outcomes.
d. Is this random variable discrete or continuous?

2. Consider the experiment of a worker assembling a product, and record how long it takes.
a. Define a random variable that represents the time in minutes required to assemble the product.
b. What values may the random variable assume?
c. Is the random variable discrete or continuous?

Applications

3. Three students have interviews scheduled for summer employment at the Brookwood Institute. In each case the result of the interview will be that a position is either offered or not offered. Experimental outcomes are defined in terms of the results of the three interviews.
a. List the experimental outcomes.
b. Define a random variable that represents the number of offers made. Is this a discrete or continuous random variable?
c. Show the value of the random variable for each of the experimental outcomes.

4. Home mortgage rates were listed for 12 Florida lending institutions (*The Tampa Tribune*, February 25, 1995). Assume that the random variable of interest is the number of lending institutions in this group that offer a 30-year fixed rate of 8.5% or less. What values may this random variable assume?

5. To perform a certain type of blood analysis, lab technicians must perform two procedures. The first procedure requires either 1 or 2 separate steps and the second procedure requires either 1, 2, or 3 steps.
a. List the experimental outcomes associated with performing an analysis.
b. If the random variable of interest is the total number of steps required to do the complete analysis, show what value the random variable will assume for each of the experimental outcomes.

6. Listed is a series of experiments and associated random variables. In each case, identify the values that the random variable can assume and state whether the random variable is discrete or continuous.

Experiment	Random Variable (x)
a. Take a 20-question examination	Number of questions answered correctly
b. Observe cars arriving at a tollbooth for one hour	Number of cars arriving at tollbooth
c. Audit 50 tax returns	Number of returns containing errors
d. Observe an employee's work	Number of nonproductive hours in an eight-hour work day
e. Weigh a shipment of goods	Number of pounds

5.2 Discrete Probability Distributions

The *probability distribution* for a random variable describes how probabilities are distributed over the values of the random variable. For a discrete random variable x, the probability distribution is defined by a *probability function,* denoted by $f(x)$. The probability function provides the probability for each value of the random variable.

As an illustration of a discrete random variable and its probability distribution, consider the sales of automobiles at DiCarlo Motors in Saratoga, New York. Over the past 300 days of operation, sales data show 54 days with no automobiles sold, 117 days with 1 automobile sold, 72 days with 2 automobiles sold, 42 days with 3 automobiles sold, 12 days with 4 automobiles sold, and 3 days with 5 automobiles sold. Suppose we consider the experiment of selecting a day of operation at DiCarlo Motors. We define the random variable of interest as x = the number of automobiles sold during a day. From historical data, we know x is a discrete random variable that can assume the values 0, 1, 2, 3, 4, or 5. In probability function notation, $f(0)$ provides the probability of 0 automobiles sold, $f(1)$ provides the probability of 1 automobile sold, and so on. Since historical data show 54 of 300 days with 0 automobiles sold, we assign the value 54/300 = .18 to $f(0)$, indicating that the probability of 0 automobiles being sold during a day is .18. Similarly, since 117 of 300 days had 1 automobile sold, we assign the value 117/300 = .39 to $f(1)$, indicating that the probability of exactly 1 automobile being sold during a day is .39. Continuing in this way for the other values of the random variable, we compute the values for $f(2)$, $f(3)$, $f(4)$, and $f(5)$ as shown in Table 5.3, the probability distribution for the number of automobiles sold during a day at DiCarlo Motors.

TABLE 5.3
Probability Distribution for the Number of Automobiles Sold During a Day at DiCarlo Motors

x	$f(x)$
0	.18
1	.39
2	.24
3	.14
4	.04
5	.01
Total	1.00

A primary advantage of defining a random variable and its probability distribution is that once the probability distribution is known, it is relatively easy to determine the probability of a variety of events that may be of interest to a decision maker. For example, using the probability distribution for DiCarlo Motors as shown in Table 5.3, we see that the most probable number of automobiles sold during a day is 1 with a probability of $f(1) = .39$. In addition, there is an $f(3) + f(4) + f(5) = .14 + .04 + .01 = .19$ probability of selling 3 or more automobiles during a day. These probabilities, plus others the decision maker may ask about, provide information that can help the decision maker understand the process of selling automobiles at DiCarlo Motors.

In the development of a probability function for any discrete random variable, the following two conditions must be satisfied.

Required Conditions for a Discrete Probability Function

$$f(x) \geq 0 \tag{5.1}$$

$$\Sigma f(x) = 1 \tag{5.2}$$

FIGURE 5.1
Graphical Representation of the Probability Distribution for Automobile Sales at DiCarlo Motors

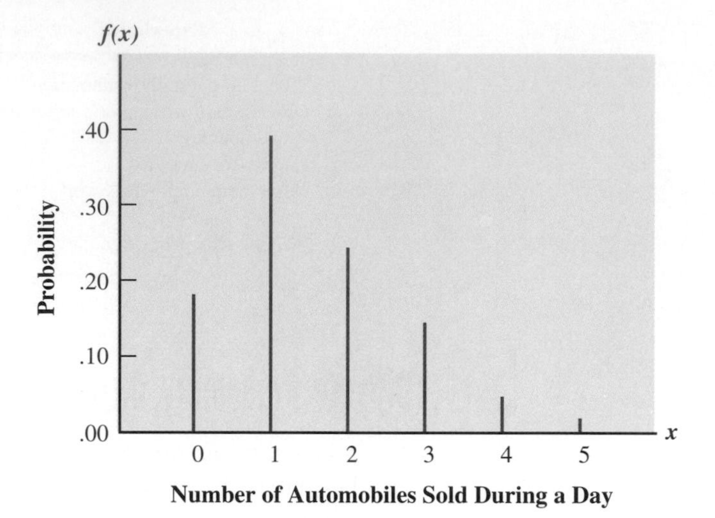

Table 5.3 shows that the probabilities for the random variable x satisfy condition (5.1); $f(x)$ is greater than or equal to 0 for all values of x. In addition, the probabilities sum to 1 so (5.2) is satisfied. Thus, the DiCarlo Motors probability function is a valid discrete probability function.

We can also present probability distributions graphically. In Figure 5.1 the values of the random variable x representing daily sales at DiCarlo Motors are shown on the horizontal axis and the probability associated with these values is shown on the vertical axis.

In addition to tables and graphs, a formula that gives $f(x)$ for every value of x can provide the probability distribution for some discrete random variables. For example, consider the random variable x with the following probability distribution.

x	$f(x)$
1	1/10
2	2/10
3	3/10
4	4/10

This probability distribution can also be defined by the formula

$$f(x) = \frac{x}{10} \qquad \text{for } x = 1, 2, 3, \text{ or } 4$$

Evaluating $f(x)$ for a given value of the random variable will provide the associated probability. For example, using the above probability function, we see that $f(2) = 2/10$ provides the probability that the random variable assumes a value of 2.

The more widely used discrete probability distributions generally are specified by formulas. Two important cases are the binomial and Poisson probability distributions; they are discussed later in the chapter.

exercises

Methods

7. The probability distribution of the random variable x is shown.

x	$f(x)$
20	.20
25	.15
30	.25
35	.40
Total	1.00

a. Is this a proper probability distribution? Check to see that (5.1) and (5.2) are satisfied.
b. What is the probability that $x = 30$?
c. What is the probability that x is less than or equal to 25?
d. What is the probability that x is greater than 30?

Applications

8. The following data were collected by counting the number of operating rooms in use at Tampa General Hospital during a 20-day period: On 3 of the days only 1 operating room was used, on 5 of the days 2 were used, on 8 of the days 3 were used, and on 4 days all 4 of the hospital's operating rooms were used.
a. Use the relative frequency approach to construct a probability distribution for the number of operating rooms in use on any given day.
b. Draw a graph of the probability distribution.
c. Show that your probability distribution satisfies the required conditions for a valid discrete probability distribution.

9. Information on 3731 subscribers to *The Wall Street Journal* includes the following data on household members (*The Wall Street Journal* Subscriber Study, 1992).

Number of Household Members	Frequency
1	474
2	1664
3	627
4	522
5	444

Let x be a random variable indicating the number of household members.
a. Use the data to develop a probability distribution for x. Specify the values for the random variable and the corresponding values for the probability function $f(x)$.
b. Draw a graph of the probability distribution.
c. Show that the probability distribution satisfies (5.1) and (5.2).

10. QA Properties is considering making an offer to purchase an apartment building. Management has subjectively assessed a probability distribution for x, the purchase price, as shown below.

x	$f(x)$
$148,000	.20
$150,000	.40
$152,000	.40

a. Determine whether this is a proper probability distribution by checking (5.1) and (5.2).
b. What is the probability that the apartment house can be purchased for $150,000 or less?

11. The cleaning and changeover operation for a production system requires 1, 2, 3, or 4 hours, depending on the specific product that will begin production. Let x be a random variable indicating the time in hours required to make a changeover. The following probability function can be used to compute the probability associated with any changeover time x.

$$f(x) = \frac{x}{10} \qquad \text{for } x = 1, 2, 3, \text{ or } 4$$

a. Show that the probability function meets the required conditions of (5.1) and (5.2).
b. What is the probability that a changeover will take 2 hours?
c. What is the probability that a changeover will take more than 2 hours?
d. Graph the probability distribution for changeover time.

12. The director of admissions at Lakeville Community College has subjectively assessed a probability distribution for x, the number of entering students.

x	$f(x)$
1000	.15
1100	.20
1200	.30
1300	.25
1400	.10

a. Is this a valid probability distribution?
b. What is the probability that there will be 1200 or fewer entering students?

13. A psychologist has determined that the number of hours required to obtain the trust of a new patient is either 1, 2, or 3. Let x be a random variable indicating the time in hours required to gain the patient's trust. The following probability function has been proposed.

$$f(x) = \frac{x}{6} \qquad \text{for } x = 1, 2, \text{ or } 3$$

a. Is this a valid probability function? Explain.
b. What is the probability that it takes exactly 2 hours to gain the patient's trust?
c. What is the probability that it takes at least 2 hours to gain the patient's trust?

14. The following is a partial probability distribution for the MRA Company's projected profits (x = profit in $1000s) for the first year of operation (the negative value denotes a loss).

x	$f(x)$
−100	.10
0	.20
50	.30
100	.25
150	.10
200	

a. What is the proper value for $f(200)$? What is your interpretation of this value?
b. What is the probability that MRA will be profitable?
c. What is the probability that MRA will make at least $100,000?

5.3 Expected Value and Variance

Expected Value

The *expected value,* or mean, of a random variable is a measure of the central location for the random variable. The mathematical expression for the expected value of a discrete random variable x follows.

Expected Value of a Discrete Random Variable

$$E(x) = \mu = \Sigma x f(x) \tag{5.3}$$

TABLE 5.4 Calculation of the Expected Value for the Number of Automobiles Sold During a Day at DiCarlo Motors

x	$f(x)$	$xf(x)$
0	.18	0(.18) = .00
1	.39	1(.39) = .39
2	.24	2(.24) = .48
3	.14	3(.14) = .42
4	.04	4(.04) = .16
5	.01	5(.01) = .05
		1.50

$E(x) = \mu = \Sigma\, xf(x)$

Both the notations $E(x)$ and μ can be used to denote the expected value of a random variable.

Equation (5.3) shows that to compute the expected value of a discrete random variable, we must multiply each value of the random variable by the corresponding probability $f(x)$ and then add the resulting products. Using the DiCarlo Motors automobile sales example from Section 5.2, we show the calculation of the expected value for the number of automobiles sold during a day in Table 5.4. The sum of the entries in the $xf(x)$ column shows that the expected value is 1.50 automobiles per day. We therefore know that although sales of 0, 1, 2, 3, 4, or 5 automobiles are possible on any one day, over time DiCarlo can anticipate selling an average of 1.50 automobiles per day. Assuming 30 days of operation during a month, we can use the expected value of 1.50 to anticipate average monthly sales of 30(1.50) = 45 automobiles.

Variance

While the expected value provides the mean value for the random variable, we often need a measure of dispersion, or variability. Just as we used the variance in Chapter 3 to summarize the dispersion in a data set, we now use variance to summarize the variability in the values of a random variable. The mathematical expression for the variance of a discrete random variable follows.

TABLE 5.5 Calculation of the Variance for the Number of Automobiles Sold During a Day at DiCarlo Motors

x	$x-\mu$	$(x-\mu)^2$	$f(x)$	$(x-\mu)^2f(x)$
0	0−1.50 = −1.50	2.25	.18	2.25(.18) = .4050
1	1−1.50 = −.50	.25	.39	.25(.39) = .0975
2	2−1.50 = .50	.25	.24	.25(.24) = .0600
3	3−1.50 = 1.50	2.25	.14	2.25(.14) = .3150
4	4−1.50 = 2.50	6.25	.04	6.25(.04) = .2500
5	5−1.50 = 3.50	12.25	.01	12.25(.01) = .1225
				1.2500

$\sigma^2 = \Sigma(x-\mu)^2 f(x)$

Variance of a Discrete Random Variable

$$\text{Var}(x) = \sigma^2 = \Sigma(x - \mu)^2 f(x) \tag{5.4}$$

As (5.4) shows, an essential part of the variance formula is the deviation, $x - \mu$, which measures how far a particular value of the random variable is from the expected value or mean, μ. In computing the variance of a random variable, the deviations are squared and then weighted by the corresponding value of the probability function. The sum of these weighted squared deviations for all values of the random variable is referred to as the *variance*. The notations Var(x) and σ^2 are both used to denote the variance of a random variable.

The calculation of the variance for the probability distribution of the number of automobiles sold during a day at DiCarlo Motors is summarized in Table 5.5. We see that the variance is 1.25. The *standard deviation,* σ, is defined as the positive square root of the variance. Thus, the standard deviation for the number of automobiles sold during a day is

$$\sigma = \sqrt{1.25} = 1.118$$

The standard deviation is measured in the same units as the random variable ($\sigma = 1.118$ automobiles) and therefore is often preferred in describing the variability of a random variable. The variance σ^2 is measured in squared units and is thus more difficult to interpret.

exercises

Methods

15. The following is a probability distribution for the random variable x.

x	$f(x)$
3	.25
6	.50
9	.25
Total	1.00

a. Compute $E(x)$, the expected value of x.
b. Compute σ^2, the variance of x.
c. Compute σ, the standard deviation of x.

Self Test

16. The following is a probability distribution for the random variable y.

y	$f(y)$
2	.20
4	.30
7	.40
8	.10
Total	1.00

a. Compute $E(y)$.
b. Compute Var(y) and σ.

Applications

17. A volunteer ambulance service handles from 0 to 5 service calls on any given day. The probability distribution for the number of service calls is shown.

Number of Service Calls	Probability
0	.10
1	.15
2	.30
3	.20
4	.15
5	.10

a. What is the expected number of service calls?
b. What is the variance in the number of service calls? What is the standard deviation?

18. *The 1994 Statistical Abstract of the United States* shows that the average number of television sets per household is 2.2. Assume that the probability distribution for the number of television sets per household is as shown.

x	$f(x)$
0	.02
1	.24
2	.42
3	.20
4	.08
5	.04

a. Compute the expected value of the number of television sets per household and compare it with the average reported in the *Statistical Abstract.*
b. What are the variance and standard deviation of the number of television sets per household?

19. The actual shooting records of the 1992 NCAA championship final four teams (*NCAA Final Four Program,* April 1992) showed the probability of making a 2-point basket was .50 and the probability of making a 3-point basket was .39.
a. What is the expected value of a 2-point shot for these teams?
b. What is the expected value of a 3-point shot for these teams?
c. Since the probability of making a 2-point basket is greater than the probability of making a 3-point basket, why do coaches allow some players to shoot the 3-point shot if they have the opportunity? Use expected value to explain your answer.

20. The probability distribution for damage claims paid by the Newton Automobile Insurance Company on collision insurance is shown.

Payment ($)	Probability
0	.90
400	.04
1000	.03
2000	.01
4000	.01
6000	.01

a. Use the expected collision payment to determine the collision insurance premium that would enable the company to break even.
b. The insurance company charges an annual rate of $260 for the collision coverage. What is the expected value of the collision policy for a policyholder? (Hint: It is the expected payments from the company minus the cost of coverage.) Why does the policyholder purchase a collision policy with this expected value?

21. The *Forbes* 1993 Subscriber Study and the *Fortune* 1994 National Subscriber Portrait reported the following probability distributions for the number of vehicles per subscriber household

x	*Forbes* $f(x)$	*Fortune* $f(x)$
0	.045	.028
1	.230	.165
2	.449	.489
3	.169	.185
4	.107	.133

a. What is the expected value of the number of vehicles per household for each subscriber group?
b. What is the variance of the number of vehicles per household for each subscriber group?
c. Using your answers to parts (a) and (b), what comparisons can you make about the number of vehicles per household for *Forbes* and *Fortune* subscribers?

22. The J. R. Ryland Computer Company is considering a plant expansion that will enable the company to begin production of a new computer product. The company's president must determine whether to make the expansion a medium- or large-scale project. An uncertainty is the demand for the new product, which for planning purposes may be low demand, medium demand, or high demand. The probability estimates for demand are .20, .50, and .30, respectively. Letting x indicate the annual profit in $1000s, the firm's planners have developed the following profit forecasts for the medium- and large-scale expansion projects.

		Medium-Scale Expansion Profits		Large-Scale Expansion Profits	
		x	$f(x)$	y	$f(y)$
Demand	*Low*	50	.20	0	.20
	Medium	150	.50	100	.50
	High	200	.30	300	.30

a. Compute the expected value for the profit associated with the two expansion alternatives. Which decision is preferred for the objective of maximizing the expected profit?
b. Compute the variance for the profit associated with the two expansion alternatives. Which decision is preferred for the objective of minimizing the risk or uncertainty?

5.4 The Binomial Probability Distribution

The *binomial probability distribution* is a discrete probability distribution that has many applications. It is associated with a multiple-step experiment that we call the binomial experiment.

A Binomial Experiment

A *binomial experiment* has the following four properties.

Properties of a Binomial Experiment

1. The experiment consists of a sequence of n identical trials.
2. Two outcomes are possible on each trial. We refer to one as a *success* and the other as a *failure.*
3. The probability of a success, denoted by p, does not change from trial to trial. Consequently, the probability of a failure, denoted by $1 - p$, does not change from trial to trial.
4. The trials are independent.

If properties 2, 3, and 4 are present, we say the trials are generated by a Bernoulli process. If, in addition, property 1 is present, we say we have a *binomial experiment.* Figure 5.2 depicts one possible sequence of outcomes of a binomial experiment involving eight trials.

In a binomial experiment, our interest is in the *number of successes occurring in the n trials.* If we let x denote the number of successes occurring in the n trials, we see that x can assume the values of 0, 1, 2, 3, . . . , n. Since the number of values is finite, x is a *discrete* random variable. The probability distribution associated with this random variable is called the *binomial probability distribution.*

For example, consider the experiment of tossing a coin five times and on each toss observing whether the coin lands with a head or a tail on its upward face. Suppose we

FIGURE 5.2
Diagram of an Eight-Trial Binomial Experiment

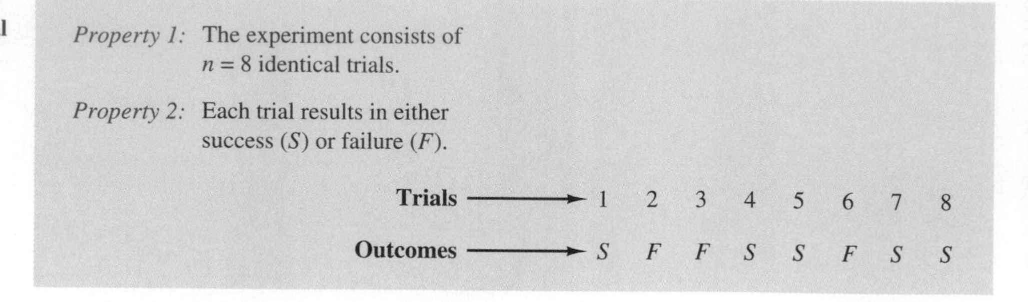

are interested in counting the number of heads appearing over the five tosses. Does this experiment have the properties of a binomial experiment? What is the random variable of interest? Note that:

1. The experiment consists of five identical trials, where each trial involves the tossing of one coin.
2. Two outcomes are possible for each trial: a head and a tail. We can designate head a success and tail a failure.
3. The probability of a head and the probability of a tail are the same for each trial, with $p = .5$ and $1 - p = .5$.
4. The trials or tosses are independent, since the outcome on any one trial is not affected by what happens on other trials or tosses.

Thus, the properties of a binomial experiment are satisfied. The random variable of interest is x = the number of heads appearing in the five trials. In this case, x can assume the values of 0, 1, 2, 3, 4, or 5.

As another example, consider an insurance salesperson who visits 10 randomly selected families. The outcome associated with each visit is classified as a success if the family purchases an insurance policy and a failure if the family does not. From past experience, the salesperson knows the probability that a randomly selected family will purchase an insurance policy is .10. Checking the properties of a binomial experiment, we observe that:

1. The experiment consists of 10 identical trials, where each trial involves contacting one family.
2. Two outcomes are possible on each trial: the family purchases a policy (success) or the family does not purchase a policy (failure).
3. The probabilities of a purchase and a nonpurchase are assumed to be the same for each sales call, with $p = .10$ and $1 - p = .90$.
4. The trials are independent since the decision of any one family does not influence the decision of any other family.

Since the four assumptions are satisfied, this is a binomial experiment. The random variable of interest is the number of sales obtained in contacting the 10 families. In this case, x can assume the values of 0, 1, 2, 3, 4, 5, 6, 7, 8, 9, or 10.

Property 3 of the binomial experiment is called the *stationarity assumption* and is sometimes confused with property 4, independence of trials. To see how they differ, consider again the case of the salesperson calling on families to sell insurance policies. If, as the day wore on, the salesperson got tired and lost enthusiasm, the probability of success (selling a policy) might drop to .05, for example, by the tenth call. In such a case, property 3 (stationarity) would not be satisfied, and we would not have a binomial experiment. This would be true even if property 4 held—that is, the purchase decisions of each family were made independently.

FIGURE 5.3
Tree Diagram for the Martin Clothing Store Problem

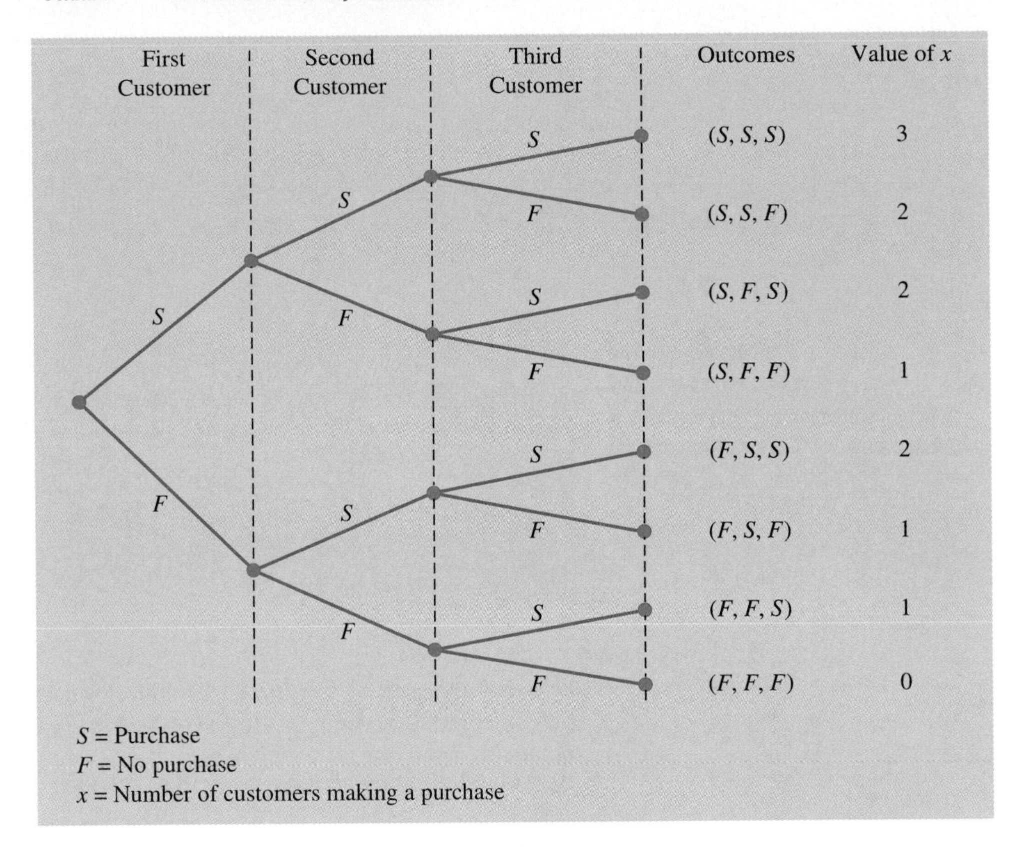

In applications involving binomial experiments, a special mathematical formula, called the *binomial probability function,* can be used to compute the probability of x successes in the n trials. Using probability concepts introduced in Chapter 4, we will show in the context of an illustrative problem how the formula can be developed.

Martin Clothing Store Problem

Let us consider the purchase decisions of the next three customers who enter the Martin Clothing Store. On the basis of past experience, the store manager estimates the probability that any one customer will make a purchase is .30. What is the probability that two of the next three customers will make a purchase?

Using a tree diagram (Figure 5.3), we can see that the experiment of observing the three customers each making a purchase decision has eight possible outcomes. Using S to denote success (a purchase) and F to denote failure (no purchase), we are interested in experimental outcomes involving two successes in the three trials (two purchases). Next, let us verify that the experiment involving the sequence of three purchase decisions can be viewed as a binomial experiment. Checking the four requirements for a binomial experiment, we note that:

1. The experiment can be described as a sequence of three identical trials, one trial for each of the three customers who will enter the store.
2. Two outcomes—the customer makes a purchase (success) or the customer does not make a purchase (failure)—are possible for each trial.
3. The probability that the customer will make a purchase (.30) or will not make a purchase (.70) is assumed to be the same for all customers.

4. The purchase decision of each customer is independent of the decisions of the other customers.

Hence, the properties of a binomial experiment are present.

The number of experimental outcomes resulting in exactly x successes in n trials can be computed from the following formula.*

Number of Experimental Outcomes Providing Exactly x Successes in n Trials

$$\binom{n}{x} = \frac{n!}{x!(n-x)!} \tag{5.5}$$

where

$$n! = n(n-1)(n-2)\ldots(2)(1) \tag{5.6}$$

and

$$0! = 1$$

Now let us return to the Martin Clothing Store experiment involving three customer purchase decisions. Equation (5.5) can be used to determine the number of experimental outcomes involving two purchases; that is, the number of ways of obtaining $x = 2$ successes in the $n = 3$ trials. From (5.5) we have

$$\binom{n}{x} = \binom{3}{2} = \frac{3!}{2!(3-2)!} = \frac{(3)(2)(1)}{(2)(1)(1)} = \frac{6}{2} = 3$$

Formula (5.5) shows that three of the outcomes yield two successes. From Figure 5.3 we see these three outcomes are denoted by *SSF, SFS,* and *FSS.*

Using (5.5) to determine how many experimental outcomes have three successes (purchases) in the three trials, we obtain

$$\binom{n}{x} = \binom{3}{3} = \frac{3!}{3!(3-3)!} = \frac{3!}{3!0!} = \frac{(3)(2)(1)}{(3)(2)(1)(1)} = \frac{6}{6} = 1$$

From Figure 5.3 we see that the one experimental outcome with three successes is identified by *SSS.*

We know that (5.5) can be used to determine the number of experimental outcomes that result in x successes. But, if we are to determine the probability of x successes in n trials, we must also know the probability associated with each of these experimental outcomes. Since the trials of a binomial experiment are independent, we can simply multiply the probabilities associated with each trial outcome to find the probability of a particular sequence of outcomes.

The probability of purchases by the first two customers and no purchase by the third customer is given by

$$pp(1-p)$$

*This is the formula introduced in Chapter 4 to determine the number of combinations of n objects selected x at a time. For the binomial experiment, this combinatorial formula provides the number of experimental outcomes (sequences of n trials) resulting in x successes.

With a .30 probability of a purchase on any one trial, the probability of a purchase on the first two trials and no purchase on the third is given by

$$(.30)(.30)(.70) = (.30)^2(.70) = .063$$

Two other sequences of outcomes result in two successes and one failure. The probabilities for all three sequences involving two successes are shown below.

Trial Outcomes				
1st Customer	**2nd Customer**	**3rd Customer**	**Success-Failure Notation**	**Probability of Experimental Outcome**
Purchase	Purchase	No purchase	*SSF*	$pp(1-p) = p^2(1-p)$ $= (.30)^2(.70) = .063$
Purchase	No purchase	Purchase	*SFS*	$p(1-p)p = p^2(1-p)$ $= (.30)^2(.70) = .063$
No purchase	Purchase	Purchase	*FSS*	$(1-p)pp = p^2(1-p)$ $= (.30)^2(.70) = .063$

Observe that all three outcomes with two successes have exactly the same probability. This observation holds in general. In any binomial experiment, all sequences of trial outcomes yielding x successes in n trials have the *same probability* of occurrence. The probability of each sequence of trials yielding x successes in n trials follows.

$$\begin{array}{l}\text{Probability of a Particular} \\ \text{Sequence of Trial Outcomes} \\ \text{with } x \text{ Successes in } n \text{ Trials}\end{array} = p^x(1-p)^{(n-x)} \qquad (5.7)$$

For the Martin Clothing Store, this formula shows that any outcome with two successes has a probability of $p^2(1-p)^{(3-2)} = p^2(1-p)^1 = (.30)^2(.70)^1 = .063$, as shown.

Since (5.5) shows the number of outcomes in a binomial experiment with x successes and (5.7) gives the probability for each sequence involving x successes, we combine (5.5) and (5.7) to obtain the following *binomial probability function.*

Binomial Probability Function

$$f(x) = \binom{n}{x} p^x(1-p)^{(n-x)} \qquad (5.8)$$

where

$$\begin{aligned} f(x) &= \text{the probability of } x \text{ successes in } n \text{ trials} \\ n &= \text{the number of trials} \\ \binom{n}{x} &= \frac{n!}{x!(n-x)!} \\ p &= \text{the probability of a success on any one trial} \\ (1-p) &= \text{the probability of a failure on any one trial} \end{aligned}$$

FIGURE 5.4
Graphical Representation of the Probability Distribution for the Martin Clothing Store Problem

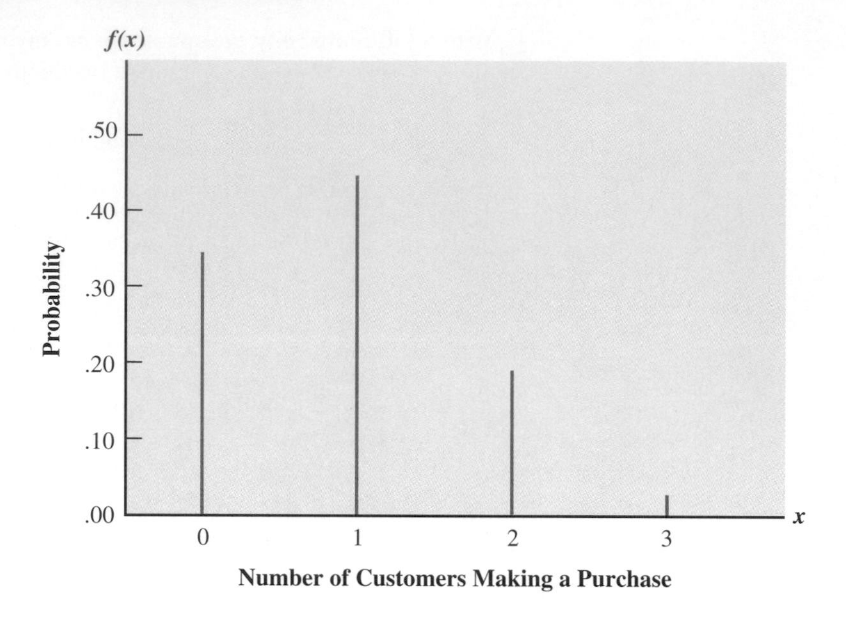

In the Martin Clothing Store example, let us compute the probability that no customer makes a purchase, exactly one customer makes a purchase, exactly two customers make a purchase, and all three customers make a purchase. The calculations are summarized in Table 5.6, which gives the probability distribution of number of customers purchasing. Figure 5.4 is a graph of this probability distribution.

TABLE 5.6 Probability Distribution for the Number of Customers Making a Purchase

x	$f(x)$
0	$\frac{3!}{0!3!}(.30)^0(.70)^3 = .343$
1	$\frac{3!}{1!2!}(.30)^1(.70)^2 = .441$
2	$\frac{3!}{2!1!}(.30)^2(.70)^1 = .189$
3	$\frac{3!}{3!0!}(.30)^3(.70)^0 = .027$
	1.000

The binomial probability function can be applied to *any* binomial experiment. If we are satisfied that a situation has the properties of a binomial experiment and if we know the values of n, p, and $(1 - p)$, we can use (5.8) to compute the probability of x successes in the n trials.

If we consider variations of the Martin experiment, such as 10 customers rather than three entering the store, the binomial probability function given by (5.8) is still applicable. For example, the probability of making exactly four sales to 10 potential customers entering the store is

$$f(4) = \frac{10!}{4!6!}(.30)^4(.70)^6 = .2001$$

This is a binomial experiment with $n = 10$, $x = 4$, and $p = .30$.

Using Tables of Binomial Probabilities

Tables have been developed that give the probability of x successes in n trials for a binomial experiment. The tables are generally easy to use and quicker than (5.8). A table of binomial probabilities is provided as Table 5 of Appendix B. A portion of this table is given in Table 5.7. To use this table, we must specify the values of n, p, and x for the binomial experiment of interest. In the example at the top of Table 5.7, we see that the probability of $x = 3$ successes in a binomial experiment with $n = 10$ and $p = .40$ is .2150. You can use (5.8) to verify that you would obtain the same answer if you used the binomial probability function directly.

TABLE 5.7 Selected Values from the Binomial Probability Table Example: $n = 10$, $x = 3$, $p = .40$; $f(3) = .2150$

						p					
n	x	.05	.10	.15	.20	.25	.30	.35	.40	.45	.50
9	0	.6302	.3874	.2316	.1342	.0751	.0404	.0207	.0101	.0046	.0020
	1	.2985	.3874	.3679	.3020	.2253	.1556	.1004	.0605	.0339	.0176
	2	.0629	.1722	.2597	.3020	.3003	.2668	.2162	.1612	.1110	.0703
	3	.0077	.0446	.1069	.1762	.2336	.2668	.2716	.2508	.2119	.1641
	4	.0006	.0074	.0283	.0661	.1168	.1715	.2194	.2508	.2600	.2461
	5	.0000	.0008	.0050	.0165	.0389	.0735	.1181	.1672	.2128	.2461
	6	.0000	.0001	.0006	.0028	.0087	.0210	.0424	.0743	.1160	.1641
	7	.0000	.0000	.0000	.0003	.0012	.0039	.0098	.0212	.0407	.0703
	8	.0000	.0000	.0000	.0000	.0001	.0004	.0013	.0035	.0083	.0176
	9	.0000	.0000	.0000	.0000	.0000	.0000	.0001	.0003	.0008	.0020
10	0	.5987	.3487	.1969	.1074	.0563	.0282	.0135	.0060	.0025	.0010
	1	.3151	.3874	.3474	.2684	.1877	.1211	.0725	.0403	.0207	.0098
	2	.0746	.1937	.2759	.3020	.2816	.2335	.1757	.1209	.0763	.0439
	3	.0105	.0574	.1298	.2013	.2503	.2668	.2522	**.2150**	.1665	.1172
	4	.0010	.0112	.0401	.0881	.1460	.2001	.2377	.2508	.2384	.2051
	5	.0001	.0015	.0085	.0264	.0584	.1029	.1536	.2007	.2340	.2461
	6	.0000	.0001	.0012	.0055	.0162	.0368	.0689	.1115	.1596	.2051
	7	.0000	.0000	.0001	.0008	.0031	.0090	.0212	.0425	.0746	.1172
	8	.0000	.0000	.0000	.0001	.0004	.0014	.0043	.0106	.0229	.0439
	9	.0000	.0000	.0000	.0000	.0000	.0001	.0005	.0016	.0042	.0098
	10	.0000	.0000	.0000	.0000	.0000	.0000	.0000	.0001	.0003	.0010

Now let us use this table to verify the probability of four successes in 10 trials for the Martin Clothing Store problem. Note that the value of $f(4) = .2001$ can be read directly from the table of binomial probabilities, with $n = 10$, $x = 4$, and $p = .30$.

While the tables of binomial probabilities are relatively easy to use, it is impossible to have tables that show all possible values of n and p that might be encountered in a binomial experiment. However, with today's calculators, using (5.8) to calculate the desired probability is not difficult, especially if the number of trials is not large. In the exercises, you should practice using (5.8) to compute the binomial probabilities unless the problem specifically requests that you use the binomial probability table.

Using the Computer to Obtain Binomial Probabilities

Statistical software packages such as Minitab provide a capability for computing binomial probabilities. Consider the Martin Clothing Store example with $n = 10$ and $p = .30$. To generate probabilities, the user first must enter the values of the random variable x for which probabilities are desired into a column of the worksheet. We entered the values 0, 1, 2, . . . , 10 into column 1 of a Minitab worksheet to generate the entire binomial probability distribution for the Martin Clothing Store example.

Selecting the binomial probability distribution option of Minitab provided the output shown in Figure 5.5. The first column contains the values of the random variable and the second column contains the associated probabilities. Note that the probability of

FIGURE 5.5
Minitab Output Showing Binomial Probabilities for the Martin Clothing Store Problem

x	P(X = x)
0.00	0.0282
1.00	0.1211
2.00	0.2335
3.00	0.2668
4.00	0.2001
5.00	0.1029
6.00	0.0368
7.00	0.0090
8.00	0.0014
9.00	0.0001
10.00	0.0000

four successes is .2001, as we found previously using (5.8) and Table 5.7. Computer packages can easily generate binomial probabilities for problems with any values for n and p. The appendixes at the end of this chapter show how spreadsheet packages can be used to generate probabilities. Appendix 5.1 gives the step-by-step procedure for using Minitab to determine the binomial probabilities for the Martin Clothing Store problem. Appendix 5.2 describes how Excel can be used to compute probabilities for any binomial probability distribution.

The Expected Value and Variance for the Binomial Probability Distribution

In Section 5.3 we provided formulas for computing the expected value and variance of a discrete random variable. In the special case where the random variable has a binomial probability distribution with a known number of trials n and a known probability of success p, the general formulas for the expected value and variance can be simplified. The results follow.

Expected Value and Variance for the Binomial Probability Distribution

$$E(x) = \mu = np \tag{5.9}$$

$$\text{Var}(x) = \sigma^2 = np(1 - p) \tag{5.10}$$

For the Martin Clothing Store problem with three customers, we can use (5.9) to compute the expected number of customers making a purchase.

$$E(x) = np = 3(.30) = .9$$

Suppose that for the next month the Martin Clothing Store forecasts 1000 customers will enter the store. What is the expected number of customers who will make a purchase? The answer is $\mu = np = (1000)(.3) = 300$. Thus, to increase the expected number of sales, Martin's must induce more customers to enter the store and/or somehow increase the probability that any individual customer will make a purchase after entering.

For the Martin Clothing Store problem with three customers, we see that the variance and standard deviation for the number of customers making a purchase are

$$\sigma^2 = np(1 - p) = 3(.3)(.7) = .63$$
$$\sigma = \sqrt{.63} = .79$$

For the next 1000 customers entering the store, the variance and standard deviation for the number of customers making a purchase are

$$\sigma^2 = np(1 - p) = 1000(.3)(.7) = 210$$
$$\sigma = \sqrt{210} = 14.49$$

notes and comments

1. Some binomial tables show values of p only up to and including $p = .50$. It would appear that such tables cannot be used when the probability of success exceeds $p = .50$. However, they can be used by noting that the probability of $n - x$ failures is also the probability of x successes. When the probability of success is greater than $p = .50$, one can compute the probability of $n - x$ failures instead. The probability of failure, $1 - p$, will be less than .50 when $p > .50$.
2. Some sources present binomial tables in a cumulative form. In using such tables, one must subtract to find the probability of x successes in n trials. For example, $f(2) = P(x \leq 2) - P(x \leq 1)$. Our tables provide these probabilities directly. To compute cumulative probabilities using our tables, one simply sums the individual probabilities. For example, to compute $P(x \leq 2)$ using our tables, we sum $f(0) + f(1) + f(2)$.

exercises

Methods

Self Test

23. Consider a binomial experiment with two trials and $p = .4$.
a. Draw a tree diagram showing this as a two-trial experiment (see Figure 5.3).
b. Compute the probability of one success, $f(1)$.
c. Compute $f(0)$.
d. Compute $f(2)$.
e. Find the probability of at least one success.
f. Find the expected value, variance, and standard deviation.

24. Consider a binomial experiment with $n = 10$ and $p = .10$. Use the binomial tables (Table 5 of Appendix B) to answer (a) through (d).
a. Find $f(0)$. **b.** Find $f(2)$. **c.** Find $P(x \leq 2)$.
d. Find $P(x \geq 1)$. **e.** Find $E(x)$. **f.** Find $\text{Var}(x)$ and σ.

25. Consider a binomial experiment with $n = 20$ and $p = .70$. Use the binomial tables (Table 5 of Appendix B) to answer (a) through (d).
a. Find $f(12)$. **b.** Find $f(16)$. **c.** Find $P(x \geq 16)$.
d. Find $P(x \leq 15)$. **e.** Find $E(x)$. **f.** Find $\text{Var}(x)$ and σ.

Applications

26. The greatest number of complaints by owners of two-year-old automobiles pertain to electrical system performance (*Consumer Reports 1995 Buyer's Guide*). Assume that an annual questionnaire sent to owners of over 300 makes and models of automobiles reveals that 10% of the owners of two-year-old automobiles found trouble spots in the electrical system that included the starter, alternator, battery, switch controls, instruments, wiring, lights, and radio.

a. What is the probability that a sample of 12 owners of two-year-old automobiles will find exactly two owners with electrical system problems?

b. What is the probability that a sample of 12 owners of two-year-old automobiles will find at least two owners with electrical system problems?

c. What is the probability that a sample of 20 owners of two-year-old automobiles will find at least one owner with an electrical system problem?

27. *The American Almanac of Jobs and Salaries, 1994–95*, reported that 25% of accountants are employed in public accounting. Assume that this percentage applies to a group of 15 college graduates just entering the accounting profession. What is the probability that at least three graduates will be employed in public accounting?

Self Test

28. When a new machine is functioning properly, only 3% of the items produced are defective. Assume that we will randomly select two parts produced on the machine and that we are interested in the number of defective parts found.

a. Describe the conditions under which this situation would be a binomial experiment.

b. Draw a tree diagram similar to Figure 5.3 showing this as a two-trial experiment.

c. How many experimental outcomes result in exactly one defect being found?

d. Compute the probabilities associated with finding no defects, exactly one defect, and two defects.

29. Five percent of American truck drivers are women (*Statistical Abstract of the United States,* 1994). Suppose 10 truck drivers are selected randomly to be interviewed about quality of work conditions.

a. Is the selection of the 10 drivers a binomial experiment? Explain.

b. What is the probability that two of the drivers will be women?

c. What is the probability that none will be women?

d. What is the probability that at least one will be a woman?

30. An American Association of Individual Investors survey found that 23% of AAII members had purchased shares of stock directly through an initial public offering (IPO) (*AAII Journal,* July 1994). In a sample of 12 AAII members,

a. What is the probability exactly three members have purchased IPOs?

b. What is the probability at least one member has purchased IPOs?

c. What is the probability that two or more members have purchased IPOs?

31. A university has found that 20% of its students withdraw without completing the introductory statistics course. Assume that 20 students have registered for the course this quarter.

a. What is the probability that two or fewer will withdraw?

b. What is the probability that exactly four will withdraw?

c. What is the probability that more than three will withdraw?

d. What is the expected number of withdrawals?

32. For the special case of a binomial random variable, we stated that the variance measure could be computed from the formula $\sigma^2 = np(1 - p)$. For the Martin Clothing Store problem data in Table 5.6, we found $\sigma^2 = np(1 - p) = .63$. Use the general definition of variance for a discrete random variable, Equation (5.4), and the data in Table 5.6 to verify that the variance is in fact .63.

33. Of the next-day express mailings handled by the U.S. Postal Service, 85% are actually received by the addressee one day after the mailing. What is the expected value and variance for the number of one-day deliveries in a group of 250 express mailings?

5.5 The Poisson Probability Distribution

In this section we consider a discrete random variable that is often useful in describing the number of occurrences over a specified interval of time or space. For example, the random variable of interest might be the number of arrivals at a car wash in one hour, the number of repairs needed in 10 miles of highway, or the number of leaks in 100 miles of pipeline. If the following three properties are satisfied, the number of occurrences is a random variable described by the *Poisson probability function.*

Properties of a Poisson Experiment

1. The experiment consists of observing the number of occurrences, x, in a continuous interval.
2. The probability of an occurrence is the same for any two intervals of equal length.
3. The occurrence or nonoccurrence in any interval is independent of the occurrence or nonoccurrence in any other interval.

The Poisson probability function is given by (5.11).

Poisson Probability Function

$$f(x) = \frac{\mu^x e^{-\mu}}{x!} \tag{5.11}$$

where

$f(x)$ = the probability of x occurrences in an interval

μ = expected value or mean number of occurrences in an interval

$e = 2.71828$

Before we consider a specific example to see how the Poisson distribution can be applied, note that there is no upper limit on x, the number of occurrences. It is a discrete random variable that may assume an infinite sequence of values ($x = 0, 1, 2, \ldots$). The Poisson random variable has no upper limit.

An Example Involving Time Intervals

Suppose that we are interested in the number of arrivals at the drive-in teller window of a bank during a 15-minute period on weekday mornings. If we can assume that the probability of a car arriving is the same for any two time periods of equal length and that the arrival or nonarrival of a car in any time period is independent of the arrival or nonarrival in any other time period, the Poisson probability function is applicable. Suppose these assumptions are satisfied and an analysis of historical data shows that the average number of cars arriving in a 15-minute period of time is 10; then the following probability function applies.

TABLE 5.8 Selected Values from the Poisson Probability Tables Example: $\mu = 10$, $x = 5$; $f(5) = .0378$

	μ									
x	9.1	9.2	9.3	9.4	9.5	9.6	9.7	9.8	9.9	10
0	.0001	.0001	.0001	.0001	.0001	.0001	.0001	.0001	.0001	.0000
1	.0010	.0009	.0009	.0008	.0007	.0007	.0006	.0005	.0005	.0005
2	.0046	.0043	.0040	.0037	.0034	.0031	.0029	.0027	.0025	.0023
3	.0140	.0131	.0123	.0115	.0107	.0100	.0093	.0087	.0081	.0076
4	.0319	.0302	.0285	.0269	.0254	.0240	.0226	.0213	.0201	.0189
5	.0581	.0555	.0530	.0506	.0483	.0460	.0439	.0418	.0398	**.0378**
6	.0881	.0851	.0822	.0793	.0764	.0736	.0709	.0682	.0656	.0631
7	.1145	.1118	.1091	.1064	.1037	.1010	.0982	.0955	.0928	.0901
8	.1302	.1286	.1269	.1251	.1232	.1212	.1191	.1170	.1148	.1126
9	.1317	.1315	.1311	.1306	.1300	.1293	.1284	.1274	.1263	.1251
10	.1198	.1210	.1219	.1228	.1235	.1241	.1245	.1249	.1250	.1251
11	.0991	.1012	.1031	.1049	.1067	.1083	.1098	.1112	.1125	.1137
12	.0752	.0776	.0799	.0822	.0844	.0866	.0888	.0908	.0928	.0948
13	.0526	.0549	.0572	.0594	.0617	.0640	.0662	.0685	.0707	.0729
14	.0342	.0361	.0380	.0399	.0419	.0439	.0459	.0479	.0500	.0521
15	.0208	.0221	.0235	.0250	.0265	.0281	.0297	.0313	.0330	.0347
16	.0118	.0127	.0137	.0147	.0157	.0168	.0180	.0192	.0204	.0217
17	.0063	.0069	.0075	.0081	.0088	.0095	.0103	.0111	.0119	.0128
18	.0032	.0035	.0039	.0042	.0046	.0051	.0055	.0060	.0065	.0071
19	.0015	.0017	.0019	.0021	.0023	.0026	.0028	.0031	.0034	.0037
20	.0007	.0008	.0009	.0010	.0011	.0012	.0014	.0015	.0017	.0019
21	.0003	.0003	.0004	.0004	.0005	.0006	.0006	.0007	.0008	.0009
22	.0001	.0001	.0002	.0002	.0002	.0002	.0003	.0003	.0004	.0004
23	.0000	.0001	.0001	.0001	.0001	.0001	.0001	.0001	.0002	.0002
24	.0000	.0000	.0000	.0000	.0000	.0000	.0000	.0001	.0001	.0001

$$f(x) = \frac{10^x e^{-10}}{x!}$$

The random variable here is x = number of cars arriving in any 15-minute period.

If management wanted to know the probability of exactly five arrivals in 15 minutes, we would set $x = 5$ and thus obtain*

$$\text{Probability of Exactly 5 Arrivals in 15 Minutes} = f(5) = \frac{10^5 e^{-10}}{5!} = .0378$$

Although the above probability was determined by evaluating the probability function with $\mu = 10$ and $x = 5$, it is often easier to refer to tables for the Poisson probability distribution. These tables provide probabilities for specific values of x and μ. We have included such a table as Table 7 of Appendix B. For convenience we have reproduced a portion of this table as Table 5.8. Note that to use the table of Poisson probabilities, we need know only the values of x and μ. From Table 5.8 we see that the probability of five

*Values of $e^{-\mu}$ can be found in Table 6 of Appendix B. Most calculators also provide these values.

arrivals in a 15-minute period is found by locating the value in the row of the table corresponding to $x = 5$ and the column of the table corresponding to $\mu = 10$. Hence, we obtain $f(5) = .0378$.

This illustration involves computing the probability of five arrivals in a 15-minute period, but other time periods can be used. Suppose we want to compute the probability of one arrival in a three-minute period. Since 10 is the expected number of arrivals in a 15-minute period, we see that $10/15 = 2/3$ is the expected number of arrivals in a one-minute period and that (2/3) (3 minutes) = 2 is the expected number of arrivals in a three-minute period. Thus, the probability of x arrivals in a three-minute time period with $\mu = 2$ is given by the following Poisson probability function.

$$f(x) = \frac{2^x e^{-2}}{x!}$$

To find the probability of one arrival in a three-minute period, we can either use Table 7 in Appendix B or compute the probability directly.

$$\begin{matrix}\text{Probability of Exactly} \\ \text{1 Arrival in 3 Minutes}\end{matrix} = f(1) = \frac{2^1 e^{-2}}{1!} = .2707$$

An Example Involving Length or Distance Intervals

Let us illustrate another application where the Poisson probability distribution is useful. Suppose we are concerned with the occurrence of major defects in a section of highway one month after resurfacing. We will assume that the probability of a defect in this section of highway is the same for any two intervals of equal length and that the occurrence or nonoccurrence of a defect in any one interval is independent of the occurrence or nonoccurrence of a defect in any other interval. Hence, the Poisson probability distribution can be applied.

Suppose we learn that major defects one month after resurfacing occur at the average rate of two per mile. Let us find the probability that there will be no major defects in a particular three-mile section of the highway. Since we are interested in an interval with a length of three miles, $\mu = (2 \text{ defects/mile})(3 \text{ miles}) = 6$ represents the expected number of major defects over the three-mile section of highway. By using (5.11) or Table 7 in Appendix B, we see that the probability of no major defects is .0025. Thus, it is very unlikely that there will be no major defects in the three-mile section. In fact, there is a $1 - .0025 = .9975$ probability of at least one major defect in the highway section.

Poisson Approximation of the Binomial Probability Distribution

The Poisson probability distribution can be used as an approximation of the binomial probability distribution when p, the probability of success, is small and n, the number of trials, is large. Simply set $\mu = np$ and use the Poisson tables. As a rule of thumb, the approximation will be good whenever $p \leq .05$ and $n \geq 20$.

As an example, suppose we want to compute the binomial probability of $x = 3$ successes in $n = 250$ trials with $p = .01$. To use the Poisson approximation, we set $\mu = np = 250(.01) = 2.5$. Referring to the Poisson probability tables (Table 7 in Appendix B), we find $f(3) = .2138$. So, we would approximate the binomial probability of three successes as $f(3) = .2138$.

exercises

Methods

34. Consider a Poisson probability distribution with $\mu = 3$.

a. Write the appropriate Poisson probability function.

b. Find $f(2)$. **c.** Find $f(1)$. **d.** Find $P(x \geq 2)$.

35. Consider a Poisson probability distribution with an average number of occurrences per time period of two.

a. Write the appropriate Poisson probability function.

b. What is the average number of occurrences in three time periods?

c. Write the appropriate Poisson probability function to determine the probability of x occurrences in three time periods.

d. Find the probability of two occurrences in one time period.

e. Find the probability of six occurrences in three time periods.

f. Find the probability of five occurrences in two time periods.

Applications

36. Phone calls arrive at the rate of 48 per hour at the reservation desk for Regional Airways.

a. Find the probability of receiving three calls in a five-minute interval of time.

b. Find the probability of receiving exactly 10 calls in 15 minutes.

c. Suppose no calls are currently on hold. If the agent takes five minutes to complete the current call, how many callers do you expect to be waiting by that time? What is the probability that none will be waiting?

d. If no calls are currently being processed, what is the probability that the agent can take three minutes for personal time without being interrupted?

37. During the period of time phone-in reservations are being taken at a local university, calls come in at the rate of one every two minutes.

a. What is the expected number of calls in one hour?

b. What is the probability of three calls in five minutes?

c. What is the probability of no calls in a five-minute period?

38. The mean number of times per month that *Forbes* magazine subscribers entertain for business reasons is 3.9 (*Forbes* 1993 Subscriber Study). Assume that the number of times a subscriber entertains for business reasons follows the assumptions of a Poisson experiment and answer the following questions.

a. What is the probability that a subscriber entertains for business reasons exactly three times during a one-month period?

b. What is the mean number of times a subscriber entertains for business reasons over a two-month period?

c. What is the probability that a subscriber entertains for business reasons at least once during a two-month period?

39. Airline passengers arrive randomly and independently at the passenger-screening facility at a major international airport. The mean arrival rate is 10 passengers per minute.

a. What is the probability of no arrivals in a one-minute period?

b. What is the probability that three or fewer passengers arrive in a one-minute period?

c. What is the probability of no arrivals in a 15-second period?

d. What is the probability of at least one arrival in a 15-second period?

40. Investment activities for subscribers to *The Wall Street Journal* show that the average number of stock transactions per year is 12 (*The Wall Street Journal* Subscriber Study, 1992). Assume that a particular investor makes transactions at this rate. Furthermore, assume that the probability of a transaction for this investor is the same for any two months and transactions in one month are independent of transactions in any other month. Answer the following questions.

a. What is the mean number of transactions per month?
b. What is the probability of no stock transactions during a month?
c. What is the probability of exactly one stock transaction during a month?
d. What is the probability of more than one stock transaction during a month?

41. A survey found that only 2% of investors believed money-market funds are not safe investments (*Business Week,* August 15, 1994). In a sample of 100 investors, what is the probability that

a. exactly two investors indicate money-market funds are not safe investments.
b. at least two investors indicate money-market funds are not safe investments.

summary

The concept of a random variable was introduced to provide a numerical description of the outcome of an experiment. We saw that the probability distribution for a random variable describes how the probabilities are distributed over the values the random variable can assume. For any discrete random variable x, the probability distribution is defined by a probability function, denoted by $f(x)$, which provides the probability associated with each value of the random variable. Once the probability function has been defined, we can compute the expected value and the variance for the random variable.

The binomial probability distribution can be used to determine the probability of x successes in n trials whenever the experiment has the following properties:

1. The experiment consists of a sequence of n identical trials.
2. Two outcomes are possible on each trial, one called success and the other failure.
3. The probability of a success p does not change from trial to trial. Consequently, the probability of failure, $1-p$, does not change from trial to trial.
4. The trials are independent.

When the four conditions hold, a binomial probability function, or a table of binomial probabilities, can be used to determine the probability of x successes in n trials. Formulas were also presented for the mean and variance of the binomial probability distribution.

The Poisson probability distribution is used when it is desirable to determine the probability of obtaining x occurrences over an interval of time or space. The following assumptions are necessary for the Poisson distribution to be applicable.

1. The experiment consists of observing the number of occurrences, x, in a continuous interval.
2. The probability of an occurrence of the event is the same for any two intervals of equal length.
3. The occurrence or nonoccurrence of the event in any interval is independent of the occurrence or nonoccurrence of the event in any other interval.

glossary

Random variable A numerical description of the outcome of an experiment.

Discrete random variable A random variable that can assume only a finite or infinite sequence of values.

Continuous random variable A random variable that may assume any value in an interval or collection of intervals.

Probability distribution A description of how the probabilities are distributed over the values the random variable can assume.

Probability function A function, denoted by $f(x)$, that provides the probability that x takes a particular value for a discrete random variable.

Expected value A measure of the mean, or central location, of a random variable.

Variance A measure of the dispersion, or variability, of a random variable.

Standard deviation The positive square root of the variance.

Binomial experiment A probability experiment having the four properties stated in Section 5.4.

Binomial probability distribution A probability distribution showing the probability of x successes in n trials of a binomial experiment.

Binomial probability function The function used to compute probabilities in a binomial experiment.

Poisson probability distribution A probability distribution showing the probability of x occurrences of an event over a specified interval of time or space.

Poisson probability function The function used to compute Poisson probabilities.

key formulas

Expected Value of a Discrete Random Variable

$$E(x) = \mu = \Sigma x f(x) \tag{5.3}$$

Variance of a Discrete Random Variable

$$\text{Var}(x) = \sigma^2 = \Sigma(x - \mu)^2 f(x) \tag{5.4}$$

Number of Experimental Outcomes Providing Exactly x Successes in n Trials

$$\binom{n}{x} = \frac{n!}{x!(n-x)!} \tag{5.5}$$

Binomial Probability Function

$$f(x) = \binom{n}{x} p^x (1 - p)^{(n - x)} \tag{5.8}$$

Expected Value for the Binomial Probability Distribution

$$E(x) = \mu = np \tag{5.9}$$

Variance for the Binomial Probability Distribution

$$\text{Var}(x) = \sigma^2 = np(1 - p) \qquad (5.10)$$

Poisson Probability Function

$$f(x) = \frac{\mu^x e^{-\mu}}{x!} \qquad (5.11)$$

supplementary exercises

42. The number of weekly lost-time injuries at a particular plant (x) has the probability distribution shown.

x	$f(x)$
0	.05
1	.20
2	.40
3	.20
4	.15

a. Compute the expected value.
b. Compute the variance.

43. Assume that the plant in Exercise 42 initiated a safety training program and that the following numbers of lost-time injuries occurred during the 20 weeks after the training program.

Number of Injuries	Number of Weeks
0	2
1	8
2	6
3	3
4	1
Total	20

a. Construct a probability distribution for weekly lost-time injuries based on these data.
b. Compute the expected value and the variance and use both to evaluate the effectiveness of the safety training program.

44. The budgeting process for a midwestern college resulted in expense forecasts for the coming year (in $1,000,000s) of $9, $10, $11, $12, and $13. The actual expenses are unknown, but the following respective probabilities are assigned: .3, .2, .25, .05, and .2.
a. Show the probability distribution for the expense forecast.
b. What is the expected value of the expenses for the coming year?
c. What is the variance in the expenses for the coming year?
d. If income projections for the year are estimated at $12 million, comment on the financial position of the college.

45. A study conducted at the University of Southern California investigated the use of music in television commercial messages (*New York,* March 23, 1992). The study found 42% of commercials include music. Consider a sample of 12 commercials.

a. What is the probability that exactly six of the commercials include music?
b. What is the probability that exactly three of the commercials include music?
c. What is the probability that at least three of the commercials include music?

46. A Louis Harris & Associates survey of senior executives revealed that 89% were optimistic about the outlook for the U.S. economy in 1995 (*Business Week,* January 9, 1995). Answer the following questions for a sample of 15 senior executives.

a. What is the probability that exactly 13 of the senior executives were optimistic about the U.S. economy?
b. What is the probability that at least 13 of the senior executives were optimistic about the U.S. economy?
c. What is the probability that exactly three of the senior executives were *not optimistic* about the U.S. economy?

47. According to the *1994 Statistical Abstract of the United States,* 42% of all households have guns. Assume a random selection of 10 households. What is the probability that:

a. Exactly four households have guns.
b. No households have guns.
c. Eight or more households have guns.

48. Many companies use a quality-control technique called "acceptance sampling" to monitor incoming shipments of parts, raw materials, and so on. In the electronics industry, component parts are commonly shipped from suppliers in large lots. Inspection of a sample of n components can be viewed as the n trials of a binomial experiment. The outcome for each component tested (trial) will be that the component is good or defective. Reynolds Electronics accepts a lot from a particular supplier if the defective components in the lot do not exceed 1%. Suppose a random sample of five items from a recent shipment has been tested.

a. Assume that 1% of the shipment is defective. Compute the probability that no items in the sample are defective.
b. Assume that 1% of the shipment is defective. Compute the probability that exactly one item in the sample is defective.
c. What is the probability of observing one or more defective items in the sample if 1% of the shipment is defective?
d. Would you feel comfortable accepting the shipment if one item were found to be defective? Why or why not?

49. Cars arrive at a car wash randomly and independently; the probability of an arrival is the same for any two time intervals of equal length. The mean arrival rate is 15 cars per hour. What is the probability that 20 or more cars will arrive during any given hour of operation?

50. A new automated production process has had an average of 1.5 breakdowns per day. Because of the cost associated with a breakdown, management is concerned about the possibility of having three or more breakdowns during a day. Assume that breakdowns occur randomly, that the probability of a breakdown is the same for any two time intervals of equal length, and that breakdowns in one period are independent of breakdowns in other periods. What is the probability of having three or more breakdowns during a day?

51. A regional director responsible for business development in the state of Pennsylvania is concerned about the number of small business failures. If the mean number of small business failures per month is 10, what is the probability that exactly four small businesses will fail during a given month? Assume that the probability of a failure is the same for any two months and that the occurrence or nonoccurrence of a failure in any month is independent of failures in any other month.

52. Customer arrivals at a bank are random and independent; the probability of an arrival in any one-minute period is the same as the probability of an arrival in any other one-minute period. Answer the following questions, assuming a mean arrival rate of three customers per minute.

a. What is the probability of exactly three arrivals in a one-minute period?
b. What is the probability of at least three arrivals in a one-minute period?

APPENDIX 5.1 Discrete Probability Distributions with Minitab

Statistical packages such as Minitab offer a relatively easy and efficient procedure for computing binomial probabilities. In this appendix, we show the step-by-step procedure for determining the binomial probabilities for the Martin Clothing Store problem in Section 5.4. Recall that the desired binomial probability is based on $n = 10$, $x = 4$, and $p = .30$. Before beginning the Minitab routine, the user must enter the desired values of the random variable x into a column of the worksheet. We entered the values 0, 1, 2, . . . , 10 in column 1 (see Figure 5.5) to generate the entire binomial probability distribution. The Minitab steps to obtain the desired binomial probabilities follow.

STEP 1. Select the **Calc** pull-down menu.
STEP 2. Select the **Probability Distributions** pull-down menu.
STEP 3. Select the **Binomial** option.
STEP 4. When the dialog box appears:
Select **Probability**
Enter 10 in the **Number of trials** box,
Enter .3 in the **Probability of success** box, and
Enter C1 in the **Input column** box.
Select **OK** to produce the binomial probabilities.

The Minitab output with the binomial probabilities is shown in Figure 5.5.

Minitab provides Poisson probabilities in a similar manner. The only differences are in Step 3, where the **Poisson** option would be selected, and Step 4, where the **Mean** would be entered rather than the number of trials and the probability of success.

APPENDIX 5.2 Discrete Probability Distributions with Spreadsheets

Excel has the capability of computing probabilities for several discrete probability distributions including the binomial and Poisson. In this appendix, we describe how Excel can be used to compute the probabilities for any binomial probability distribution. The procedures for the Poisson probability distribution are similar to the one we describe for the binomial probability distribution.

Let us return to the Martin Clothing Store problem, where the binomial probabilities of interest are based on a binomial experiment with $n = 10$ and $p = .30$. Let us assume that the user is interested in the probability of $x = 4$ successes in the 10 trials. The following steps describe how to use Excel to produce the desired binomial probability.

STEP 1: Select a cell in the worksheet where you want the binomial probability to appear.
STEP 2: Select the **Insert** pull-down menu.
STEP 3: Choose the **Function** option.

STEP 4: When the Function Wizard–Step 1 of 2 dialog box appears:
Choose **Statistical** from the **Function Category** box,
Choose **BINOMDIST** from the **Function Name** box,
Select **Next>**.

STEP 5: When the Function Wizard–Step 2 of 2 dialog box appears:
Enter **4** in the **numbers** box (the value of x),
Enter **10** in the **trials** box (the value of n),
Enter **.30** in the **probability** box (the value of p), and
Enter false in the **cumulative** box.*

Note: At this point the desired binomial probability of .2001 is automatically computed and appears in the **Value** box in the upper right corner of the dialog box.

Select **Finish** and the binomial probability will appear in the worksheet cell requested in Step 1.

If the user wants other binomial probabilities, there are ways of obtaining the information without repeating the steps for each probability desired. Perhaps the easiest alternative is to stay in Step 5. After the four entries have been made and the first probability appears in the value box, simply return to the numbers box and insert a new value of x. The new probability will appear in the value box. Repeated changes can be made in the dialog box, including changes to the trials, probability, and/or cumulative boxes. For each change, the desired probability will appear in the value box. When **Finish** is selected, only the last binomial probability will be placed in the worksheet.

If the user wants to insert multiple binomial probabilities into the worksheet, the desired values of x should be entered into the worksheet first. Then, in Step 5, the user must enter the cell location of one of the values of x in the numbers box. After completing the steps for one binomial probability, individuals experienced with Excel can use Excel's Copy command to copy the binomial function into the cells where the other binomial probabilities are to appear.

The Excel procedure for generating Poisson probabilities is similar to the procedure already described. Step 4 can be used to select the **POISSON** function name. The dialog box in Step 5 will guide the user through the input values required to compute the desired probabilities.

---

*Placing false in the cumulative box provides the probability of exactly four successes. Placing true in this box provides the cumulative probability of four *or fewer* successes.

CHAPTER 6

Continuous Probability Distributions

Procter & Gamble*

Cincinnati, Ohio

Procter & Gamble (P&G) is in the consumer-products business worldwide. P&G produces and markets such products as detergents, disposable diapers, over-the-counter pharmaceuticals, dentifrices, bar soaps, mouthwashes, and paper towels. It has the leading brand in more categories than any other consumer-products company.

As a leader in the application of statistical methods to decision making, P&G employs people with diverse academic backgrounds: engineering, statistics, operations research, and business. The major quantitative technologies for which these people provide support are probabilistic decision and risk analysis, advanced simulation, quality improvement, and quantitative methods (e.g., linear programming, regression analysis, probability analysis).

The Industrial Chemicals Division of P&G is a major supplier of fatty alcohols derived from natural substances such as coconut oil and from petroleum-based derivatives. The division wanted to know the economic risks and opportunities of expanding its fatty-alcohol production facilities, and P&G's experts in probabilistic decision and risk analysis were called in to help. After structuring and modeling the problem, they determined that the key to profitability was the cost difference between the petroleum- and coconut-based raw materials. Future costs were unknown, but the analysts were able to represent them with the following continuous random variables.

x = the coconut oil price per pound of fatty alcohol

and

y = the petroleum raw material price per pound of fatty alcohol

Since the key to profitability was the difference between these two random variables, a third random variable, $d = x - y$, was used in the analysis. Experts were interviewed to determine the probability distribution for x and y. In turn, this information was used to develop a continuous probability distribution for the difference d. The continuous probability distribution showed that there was a .90 probability that the price difference would be \$.0655 or less and that there was a .50 probability that the price difference would be \$.035 or less. In addition, there was only a .10 probability that the price difference would be \$.0045 or less.†

The Industrial Chemicals Division thought that being able to quantify the impact of raw material price differences was key to reaching a consensus. The probabilities obtained were used in a sensitivity analysis of the raw material price difference. The analysis yielded sufficient insight to form the basis for a recommendation to management.

The use of continuous random variables and their probability distributions was helpful to P&G in analyzing the economic risks associated with its fatty-alcohol production. In this chapter, you will gain an understanding of continuous random variables and their probability distributions including one of the most important probability distributions in statistics, the normal distribution.

Two of Procter & Gamble's many well-known products.

The authors are indebted to Mr. Joel Kahn of Procter & Gamble for providing this Statistics in Practice.

† *The price differences stated here have been modified to protect proprietary data.*

In the preceding chapter we discussed discrete random variables and their probability distributions. In this chapter we turn to the study of continuous random variables. Specifically, we discuss three continuous probability distributions: the uniform, the normal, and the exponential.

To understand the difference between discrete and continuous random variables, first recall that for a discrete random variable we can compute the probability of the random variable assuming a particular value. For continuous random variables, the situation is much different. A continuous random variable may assume any value in an interval on the real line or in a collection of intervals. Since any interval contains an infinite number of values, it is not possible to talk about the probability that the random variable will assume a specific value; instead, we must think in terms of the probability that a continuous random variable will assume a value within a given interval.

In the discussion of discrete probability distributions, we introduced the concept of a probability function $f(x)$. In the continuous case, the counterpart of the probability function is the *probability density function,* also denoted by $f(x)$. For a continuous random variable, the probability density function provides the value of the function at any particular value of x; it does not directly provide the probability of the random variable assuming some specific value. However, the area under the graph of $f(x)$ corresponding to a given interval provides the probability that the continuous random variable will assume a value in that interval. In Section 6.1 we demonstrate these concepts for a continuous random variable that has a uniform probability distribution.

Much of the chapter is devoted to describing and showing applications of the normal probability distribution. The normal probability distribution is of major importance; it is used extensively in statistical inference. The normal distribution can also be used as an approximation to the discrete binomial distribution, as we show in Section 6.3. The chapter closes with a discussion of the exponential probability distribution.

6.1 Uniform Probability Distribution

Consider the random variable x that represents the flight time of an airplane traveling from Chicago to New York. Suppose the flight time can be any value in the interval from 120 minutes to 140 minutes. Since the random variable x can assume any value in that interval, x is a continuous rather than a discrete random variable. Let us assume that sufficient actual flight data are available to conclude that the probability of a flight time within any one-minute interval is the same as the probability of a flight time within any other one-minute interval from 120 to 140 minutes. With every one-minute interval being equally likely, the random variable x is said to have a *uniform probability distribution.* The *probability density function,* which defines the uniform probability distribution for the flight time random variable, is

$$f(x) = \begin{cases} 1/20 & \text{for } 120 \le x \le 140 \\ 0 & \text{elsewhere} \end{cases}$$

Figure 6.1 is a graph of this probability density function. In general, the uniform probability density function for a random variable x is found by using the following formula.

Uniform Probability Density Function

$$f(x) = \begin{cases} \dfrac{1}{b-a} & \text{for } a \le x \le b \\ 0 & \text{elsewhere} \end{cases} \qquad (6.1)$$

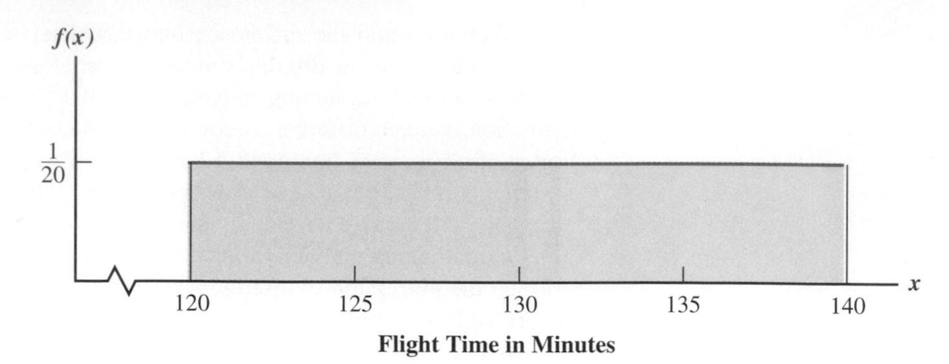

FIGURE 6.1
Uniform Probability Density Function for Flight Time

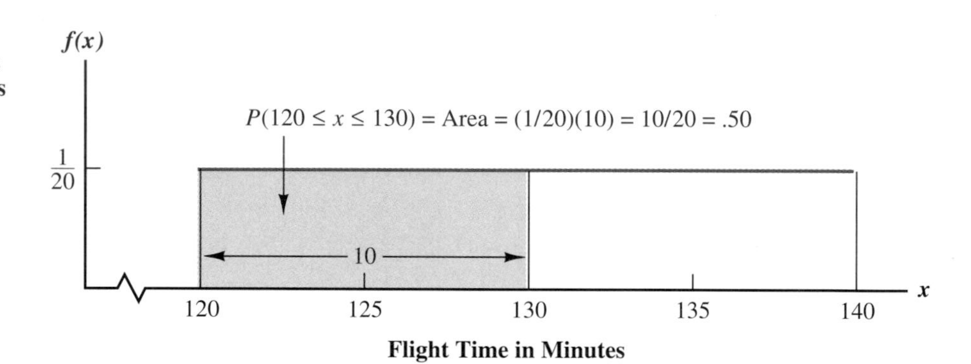

FIGURE 6.2
Area Provides Probability of Flight Time Between 120 and 130 Minutes

In the flight-time example, $a = 120$ and $b = 140$.

The graph of the probability density function $f(x)$ provides the height or value of the function at any particular value of x. Note that for a *uniform* probability density function, the height of the function is the same for each value of x. For example, in the flight-time example, $f(x) = 1/20$ for all values of x between 120 and 140. In general, the probability density function $f(x)$, unlike the probability function for a discrete random variable, *does not represent probability.* Rather, it simply provides *the height of the function at any particular value of x.*

For a continuous random variable, we consider probability only in terms of the likelihood that a random variable has a value within a *specified interval.* In the flight-time example, an acceptable probability question is: What is the probability that the flight time is between 120 and 130 minutes? That is, what is $P(120 \le x \le 130)$? Since the flight time must be between 120 and 140 minutes and since the probability is described as being uniform over this interval, we feel comfortable saying $P(120 \le x \le 130) = .50$. In the following subsection we show that this probability can be computed as the area under the graph of $f(x)$ from 120 to 130.

Area as a Measure of Probability

Let us make an observation about the graph in Figure 6.2. Consider the *area* under the graph of $f(x)$ in the interval from 120 to 130. The region is rectangular, and the area of a rectangle is simply the width multiplied by the height. With the width of the interval equal to $130 - 120 = 10$ and the height equal to the value of the probability density function $f(x) = 1/20$, we have area = width × height = $10(1/20) = 10/20 = .50$.

What observation can you make about the area under the graph of $f(x)$ and probability? They are identical! Indeed, this is true for all continuous random variables. Once a probability density function $f(x)$ has been identified, the probability that x assumes a value between some lower value x_1 and some higher value x_2 can be found by computing the *area* under the graph of $f(x)$ over the interval x_1 to x_2 .

Once we have the appropriate probability distribution and accept the interpretation of area as probability, we can answer any number of probability questions. For example, what is the probability of a flight time between 128 and 136 minutes? The width of the interval is $136 - 128 = 8$. With the uniform height of $1/20$, we see that $P(128 \leq x \leq 136) = 8(1/20) = .40$.

Note that $P(120 \leq x \leq 140) = 20(1/20) = 1$. That is, the total area under the graph of $f(x)$ is equal to 1. This property holds for all continuous probability distributions and is the analog of the condition that the sum of the probabilities must equal 1 for a discrete probability function. For a continuous probability density function, we must also require that $f(x) \geq 0$ for all values of x. This is the analog of the requirement that $f(x) \geq 0$ for discrete probability functions.

Two major differences stand out between the treatment of continuous random variables and the treatment of their discrete counterparts.

1. We no longer talk about the probability of the random variable assuming a particular value. Instead, we talk about the probability of the random variable assuming a value within some given interval.
2. The probability of the random variable assuming a value within some given interval from x_1 to x_2 is defined to be the area under the graph of the probability density function between x_1 and x_2. *This implies that the probability that a continuous random variable assumes any particular value exactly is zero, since the area under the graph of $f(x)$ at a single point is zero.*

The calculation of the expected value and variance for a continuous random variable is analogous to that for a discrete random variable. However, since the computational procedure involves integral calculus, we leave the derivation of the appropriate formulas to more advanced texts.

For the uniform continuous probability distribution introduced in this section, the formulas for the expected value and variance are

$$E(x) = \frac{a + b}{2}$$

$$\text{Var}(x) = \frac{(b - a)^2}{12}$$

In these formulas, a is the smallest value and b is the largest value that the random variable may assume.

Applying these formulas to the uniform probability distribution for flight times from Chicago to New York, we obtain

$$E(x) = \frac{(120 + 140)}{2} = 130$$

$$\text{Var}(x) = \frac{(140 - 120)^2}{12} = 33.33$$

The standard deviation of flight times can be found by taking the square root of the variance. Thus, $\sigma = 5.77$ minutes.

notes and comments

1. Since for a continuous random variable the probability of any particular value is zero, we have $P(a \leq x \leq b) = P(a < x < b)$. This equation shows that the probability of a random variable assuming a value in any interval is the same whether or not the endpoints are included.

2. To see more clearly why the height of a probability density function is not a probability, think about a random variable with the following uniform probability distribution.

$$f(x) = \begin{cases} 2 & \text{for } 0 \leq x \leq .5 \\ 0 & \text{elsewhere} \end{cases}$$

The height of the probability density function is 2 for values of x between 0 and .5. However, we know probabilities can never be greater than 1.

exercises

Methods

1. The random variable x is known to be uniformly distributed between 1.0 and 1.5.

a. Show the graph of the probability density function.
b. Find $P(x = 1.25)$.
c. Find $P(1.0 \leq x \leq 1.25)$.
d. Find $P(1.20 < x < 1.5)$.

2. The random variable x is known to be uniformly distributed between 10 and 20.

a. Show the graph of the probability density function.
b. Find $P(x < 15)$.
c. Find $P(12 \leq x \leq 18)$.
d. Find $E(x)$.
e. Find $\text{Var}(x)$.

Applications

3. Delta Airlines quotes a flight time of 1 hour, 52 minutes for its flights from Cincinnati to Tampa. Suppose we believe that actual flight times are uniformly distributed between the quoted time and 2 hours, 10 minutes.

a. Show the graph of the probability density function for flight times.
b. What is the probability that the flight will be no more than five minutes late?
c. What is the probability that the flight will be more than 10 minutes late?
d. What is the expected flight time?

4. Most computer languages have a function that can be used to generate random numbers. In Microsoft's QuickBASIC, the RND function can be used to generate random numbers between 0 and 1. If we let x denote the random number generated, then x is a continuous random variable with the following probability density function.

$$f(x) = \begin{cases} 1 & \text{for } 0 \leq x \leq 1 \\ 0 & \text{elsewhere} \end{cases}$$

a. Graph the probability density function.
b. What is the probability of generating a random number between .25 and .75?
c. What is the probability of generating a random number with a value less than or equal to .30?
d. What is the probability of generating a random number with a value greater than .60?

5. The driving distance for the top golfers on the PGA tour is between 270 and 280 yards (*Golf World,* March 25, 1994). Assume a particular golfer routinely hits drives in this range and that the distance is uniform over the interval from 270 to 280 yards.
a. Give a mathematical expression for the probability density function.
b. What is the probability that the driving distance will be less than 274 yards?
c. What is the probability that the driving distance will be between 272 and 277 yards?

6. The label on a bottle of liquid detergent shows contents to be 12 ounces per bottle. The production operation fills the bottle uniformly according to the following probability density function.

$$f(x) = \begin{cases} 8 & \text{for } 11.975 \leq x \leq 12.10 \\ 0 & \text{elsewhere} \end{cases}$$

a. What is the probability that a bottle will be filled with between 12 and 12.05 ounces?
b. What is the probability that a bottle will be filled with 12.02 or more ounces?
c. Quality control accepts production that is within .02 ounces of the number of ounces shown on the container label. What is the probability that a bottle of this liquid detergent will fail to meet the quality control standard?

7. Suppose we are interested in bidding on a piece of land and we know there is one other bidder.* The seller has announced that the highest bid in excess of $10,000 will be accepted. Assume that the competitor's bid x is a random variable that is uniformly distributed between $10,000 and $15,000.
a. Suppose you bid $12,000. What is the probability that your bid will be accepted?
b. Suppose you bid $14,000. What is the probability that your bid will be accepted?
c. What amount should you bid to maximize the probability that you get the property?
d. Suppose you know someone who is willing to pay you $16,000 for the property. Would you consider bidding less than the amount in (c)? Why or why not?

6.2 Normal Probability Distribution

Perhaps the most important probability distribution for describing a continuous random variable is the *normal probability distribution.* The normal probability distribution has been used in a wide variety of practical applications in which the random variables are heights and weights of people, IQ scores, scientific measurements, amounts of rainfall, and so on. Use of this probability distribution requires that the random variable be continuous. However, as we shall see in Section 6.3, a continuous normal random variable can be used as an approximation in situations involving discrete random variables.

Normal Curve

The form, or shape, of the normal probability distribution is illustrated by the bell-shaped curve in Figure 6.3. The probability density function that defines the bell-shaped curve of the normal probability distribution follows.

*This exercise is based on a problem suggested to us by Professor Roger Myerson of Northwestern University.

FIGURE 6.3
Bell-Shaped Curve for the Normal Probability Distribution

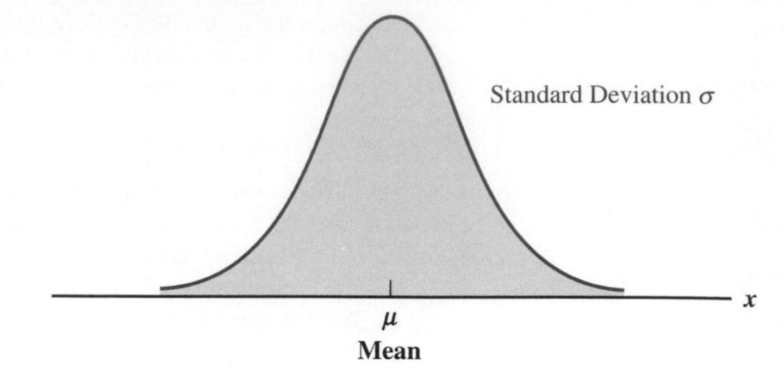

Normal Probability Density Function

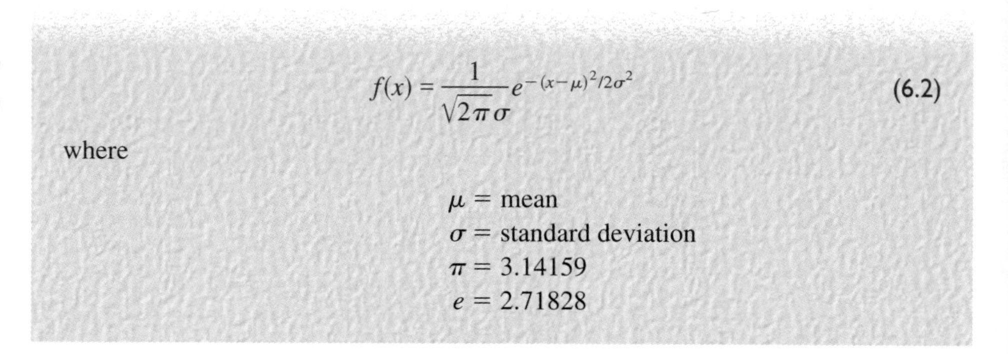

$$f(x) = \frac{1}{\sqrt{2\pi}\sigma} e^{-(x-\mu)^2/2\sigma^2} \tag{6.2}$$

where

μ = mean
σ = standard deviation
π = 3.14159
e = 2.71828

We make some observations about the characteristics of the normal probability distribution.

1. There is an entire family of normal probability distributions, with each specific normal distribution being differentiated by its mean μ and its standard deviation σ.
2. The highest point on the normal curve is at the mean, which is also the median and mode of the distribution.
3. The mean of the distribution can be any numerical value: negative, zero, or positive. Three normal curves with the same standard deviation but three different means (−10, 0, and 20) are shown below.

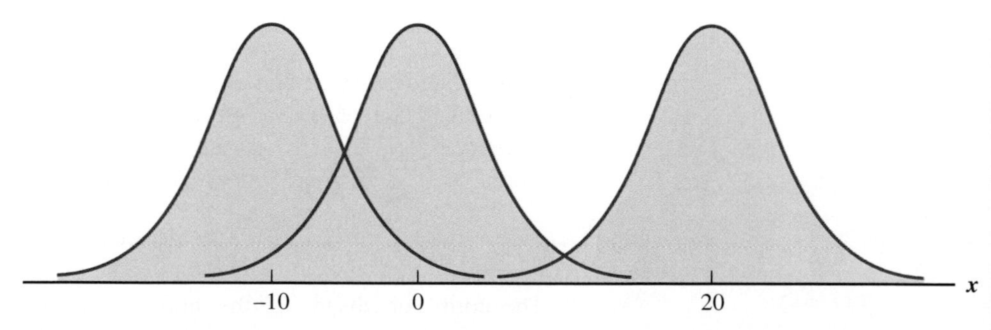

4. The normal probability distribution is symmetric, with the shape of the curve to the left of the mean a mirror image of the shape of the curve to the right of the mean. The tails of the curve extend to infinity in both directions and theoretically never touch the horizontal axis.

FIGURE 6.4
Areas Under the Curve for Any Normal Probability Distribution

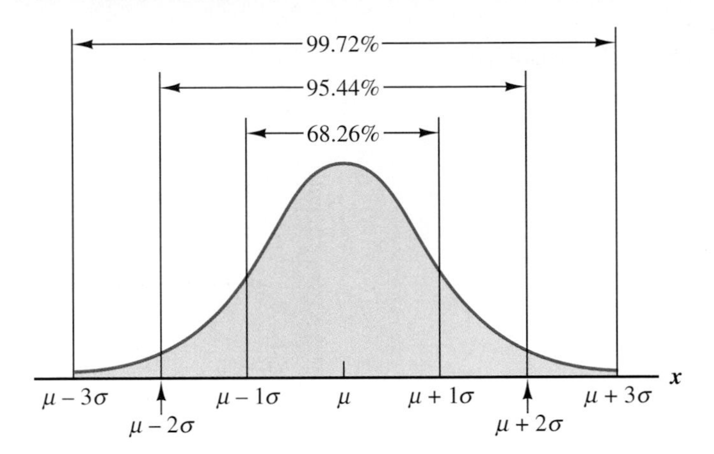

5. The standard deviation determines the width of the curve. Larger values of the standard deviation result in wider, flatter curves, showing more dispersion in the data. Two normal distributions with the same mean but with different standard deviations are shown below.

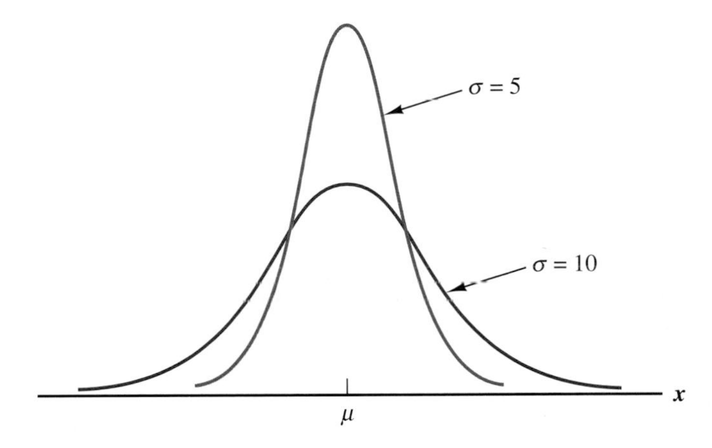

6. The total area under the curve for the normal probability distribution is 1. (This is true for all continuous probability distributions.)

7. Probabilities for the normal random variable are given by areas under the curve. Probabilities for some commonly used intervals are:

a. 68.26% of the time, a normal random variable assumes a value within plus or minus one standard deviation of its mean.

b. 95.44% of the time, a normal random variable assumes a value within plus or minus two standard deviations of its mean.

c. 99.72% of the time, a normal random variable assumes a value within plus or minus three standard deviations of its mean. Figure 6.4 shows properties (a), (b), and (c) graphically.

Standard Normal Probability Distribution

A random variable that has a normal distribution with a mean of zero and a standard deviation of one is said to have a *standard normal probability distribution.* The letter z

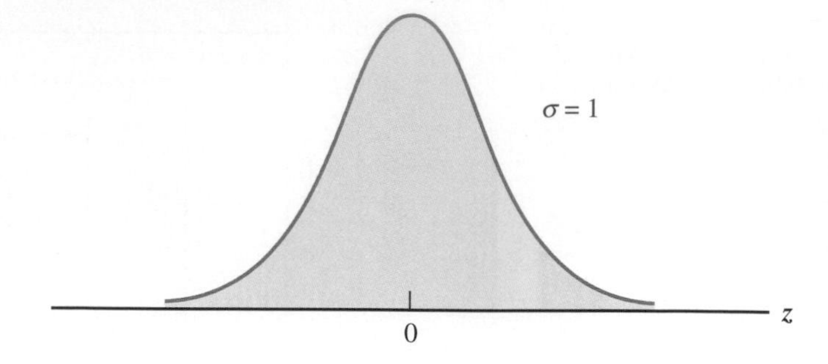

FIGURE 6.5
The Standard Normal Probability Distribution

is commonly used to designate this particular normal random variable. Figure 6.5 is the graph of the standard normal probability distribution. It has the same general appearance as other normal distributions, but with the special properties of $\mu = 0$ and $\sigma = 1$.

As with other continuous random variables, probability calculations with any normal probability distribution are made by computing areas under the graph of the probability density function. Thus, to find the probability that a normal random variable is within any specific interval, we must compute the area under the normal curve over that interval. For the standard normal probability distribution, areas under the normal curve have been computed and are available in tables that can be used in computing probabilities. Table 6.1 is such a table; it is also available as Table 1 of Appendix B and inside the front cover of this text.

To see how the table of areas under the curve for the standard normal probability distribution (Table 6.1) can be used to find probabilities, let us consider some examples. Later, we will see how this same table can be used to compute probabilities for any normal distribution. To begin, let us see how we can compute the probability that the z value for the standard normal random variable will be between .00 and 1.00, that is, $P(.00 \leq z \leq 1.00)$. The shaded region in the following graph shows this area or probability.

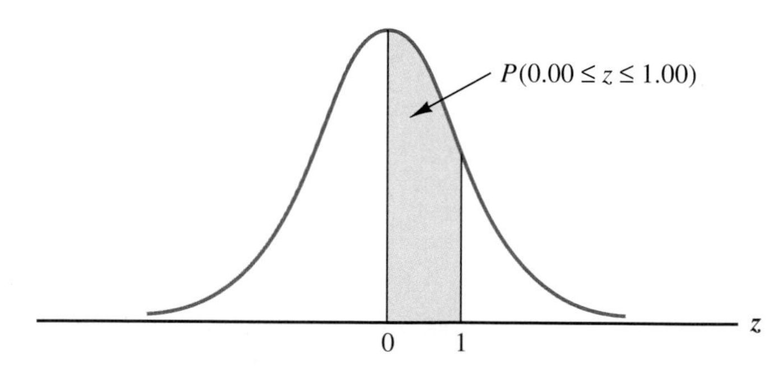

The entries in Table 6.1 give the area under the standard normal curve between the mean, $z = 0$, and a specified positive value of z (see the graph at the top of the table). In this case we are interested in the area between $z = 0$ and $z = 1.00$. Thus, we must find the entry in the table corresponding to $z = 1.00$. To do this, we first find 1.0 in the left column of the table and then find .00 in the top row of the table. By looking in the body of the table, we find that the 1.0 row and the .00 column intersect at the value of .3413. We have found the desired probability: $P(.00 \leq z \leq 1.00) = .3413$. A portion of Table 6.1 showing these steps follows.

TABLE 6.1 Areas, or Probabilities, for the Standard Normal Distribution

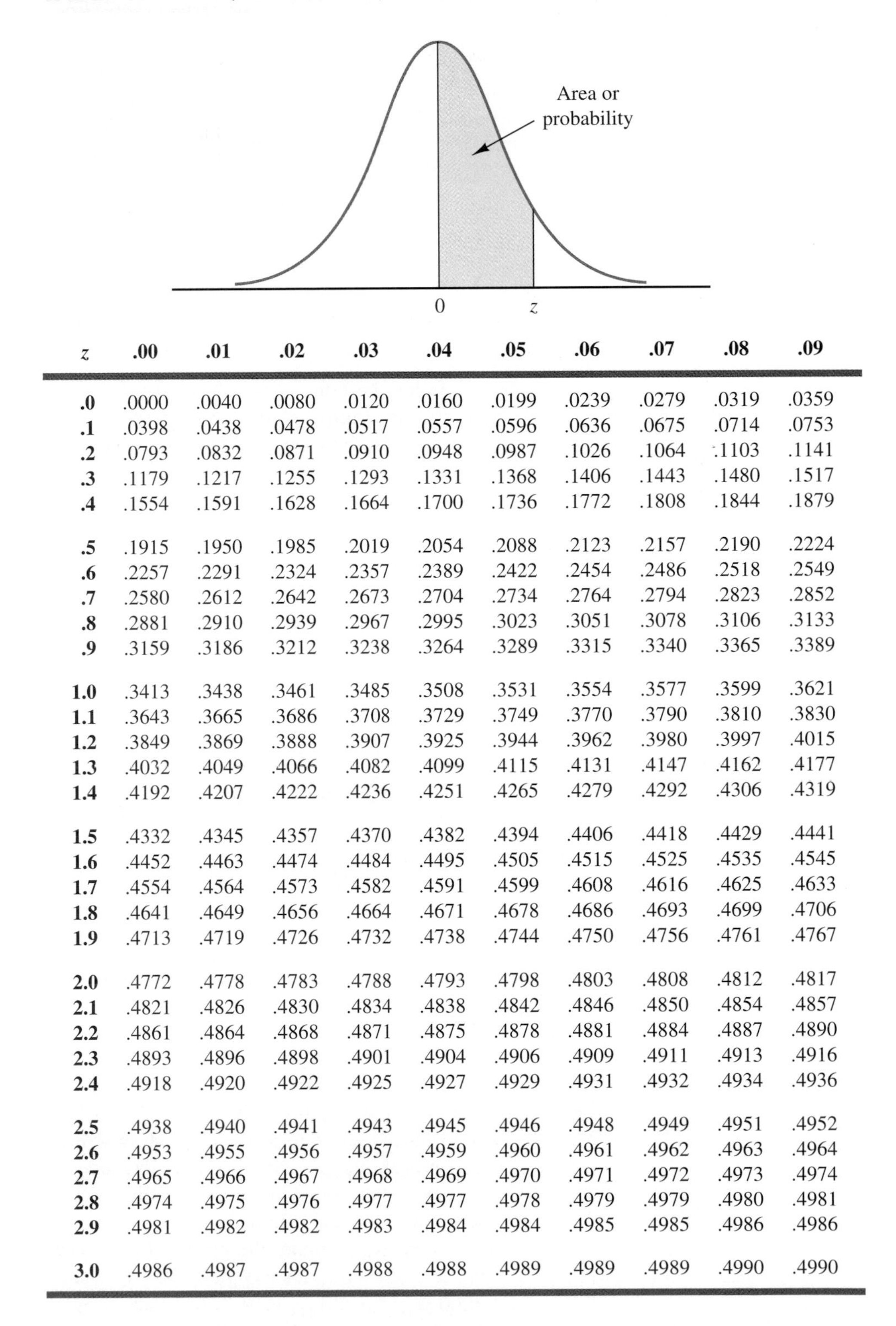

z	.00	.01	.02	.03	.04	.05	.06	.07	.08	.09
.0	.0000	.0040	.0080	.0120	.0160	.0199	.0239	.0279	.0319	.0359
.1	.0398	.0438	.0478	.0517	.0557	.0596	.0636	.0675	.0714	.0753
.2	.0793	.0832	.0871	.0910	.0948	.0987	.1026	.1064	.1103	.1141
.3	.1179	.1217	.1255	.1293	.1331	.1368	.1406	.1443	.1480	.1517
.4	.1554	.1591	.1628	.1664	.1700	.1736	.1772	.1808	.1844	.1879
.5	.1915	.1950	.1985	.2019	.2054	.2088	.2123	.2157	.2190	.2224
.6	.2257	.2291	.2324	.2357	.2389	.2422	.2454	.2486	.2518	.2549
.7	.2580	.2612	.2642	.2673	.2704	.2734	.2764	.2794	.2823	.2852
.8	.2881	.2910	.2939	.2967	.2995	.3023	.3051	.3078	.3106	.3133
.9	.3159	.3186	.3212	.3238	.3264	.3289	.3315	.3340	.3365	.3389
1.0	.3413	.3438	.3461	.3485	.3508	.3531	.3554	.3577	.3599	.3621
1.1	.3643	.3665	.3686	.3708	.3729	.3749	.3770	.3790	.3810	.3830
1.2	.3849	.3869	.3888	.3907	.3925	.3944	.3962	.3980	.3997	.4015
1.3	.4032	.4049	.4066	.4082	.4099	.4115	.4131	.4147	.4162	.4177
1.4	.4192	.4207	.4222	.4236	.4251	.4265	.4279	.4292	.4306	.4319
1.5	.4332	.4345	.4357	.4370	.4382	.4394	.4406	.4418	.4429	.4441
1.6	.4452	.4463	.4474	.4484	.4495	.4505	.4515	.4525	.4535	.4545
1.7	.4554	.4564	.4573	.4582	.4591	.4599	.4608	.4616	.4625	.4633
1.8	.4641	.4649	.4656	.4664	.4671	.4678	.4686	.4693	.4699	.4706
1.9	.4713	.4719	.4726	.4732	.4738	.4744	.4750	.4756	.4761	.4767
2.0	.4772	.4778	.4783	.4788	.4793	.4798	.4803	.4808	.4812	.4817
2.1	.4821	.4826	.4830	.4834	.4838	.4842	.4846	.4850	.4854	.4857
2.2	.4861	.4864	.4868	.4871	.4875	.4878	.4881	.4884	.4887	.4890
2.3	.4893	.4896	.4898	.4901	.4904	.4906	.4909	.4911	.4913	.4916
2.4	.4918	.4920	.4922	.4925	.4927	.4929	.4931	.4932	.4934	.4936
2.5	.4938	.4940	.4941	.4943	.4945	.4946	.4948	.4949	.4951	.4952
2.6	.4953	.4955	.4956	.4957	.4959	.4960	.4961	.4962	.4963	.4964
2.7	.4965	.4966	.4967	.4968	.4969	.4970	.4971	.4972	.4973	.4974
2.8	.4974	.4975	.4976	.4977	.4977	.4978	.4979	.4979	.4980	.4981
2.9	.4981	.4982	.4982	.4983	.4984	.4984	.4985	.4985	.4986	.4986
3.0	.4986	.4987	.4987	.4988	.4988	.4989	.4989	.4989	.4990	.4990

z	.00	.01	.02
.			
.			
.			
.9	.3159	.3186	.3212
1.0	.3413	.3438	.3461
1.1	.3643	.3665	.3686
1.2	.3849	.3869	.3888
.			
.			
.			

$P(.00 \le z \le 1.00)$

Using the same approach, we can find $P(.00 \le z \le 1.25)$. By first locating the 1.2 row and then moving across to the .05 column, we find $P(.00 \le z \le 1.25) = .3944$.

As another example of the use of the table of areas for the standard normal distribution, we compute the probability of obtaining a z value between $z = -1.00$ and $z = 1.00$; that is, $P(-1.00 \le z \le 1.00)$. Note that we have already used Table 6.1 to show that the probability of a z value between $z = .00$ and $z = 1.00$ is .3413, and recall that the normal probability distribution is *symmetric*. Thus, the probability of a z value between $z = .00$ and $z = -1.00$ is the *same* as the probability of a z value between $z = .00$ and $z = +1.00$. Hence, the probability of a z value between $z = -1.00$ and $z = +1.00$ is

$$P(-1.00 \le z \le .00) + P(.00 \le z \le 1.00) = .3413 + .3413 = .6826.$$

This area is shown graphically in the following figure.

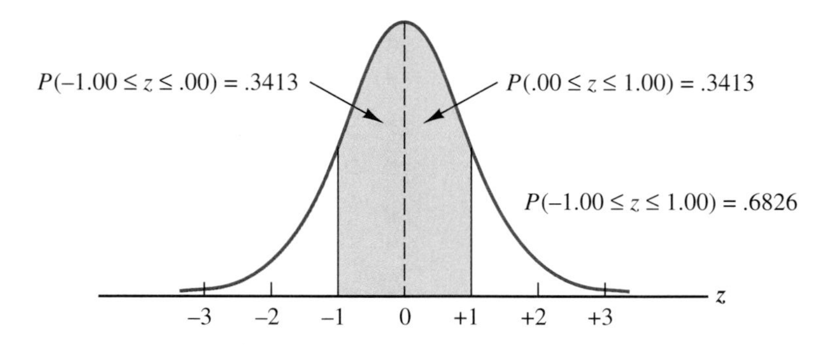

In a similar manner, we can use the values in Table 6.1 to show that the probability of a z value between −2.00 and +2.00 is .4772 + .4772 = .9544 and that the probability of a z value between −3.00 and +3.00 is .4986 + .4986 = .9972. Since we know that the total probability or total area under the curve for any continuous random variable must be 1.0000, the probability .9972 tells us that the value of z will almost always be between −3.00 and +3.00.

Next, we compute the probability of obtaining a z value of at least 1.58; that is, $P(z \ge 1.58)$. First, we use the $z = 1.5$ row and the .08 column of Table 6.1 to find that $P(.00 \le z \le 1.58) = .4429$. Now, since the normal probability distribution is symmetric and the total area under the curve equals 1, we know that 50% of the area must be above the mean (i.e., $z = 0$) and 50% of the area must be below the mean. Since .4429 is the area between the mean and $z = 1.58$, the area or probability corresponding to $z \geq 1.58$ must be $.5000 - .4429 = .0571$. This probability is shown in the following figure.

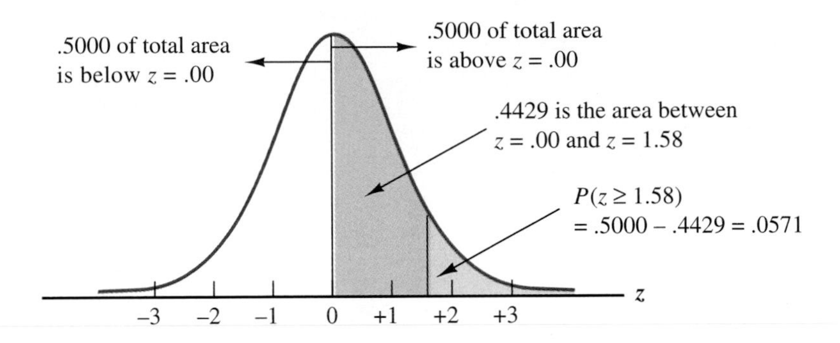

As another illustration, consider the probability that the random variable z assumes a value of $-.50$ or larger; that is, $P(z \geq -.50)$. To make this computation, we note that the probability we are seeking can be written as the sum of two probabilities: $P(z \geq -.50) = P(-.50 \leq z \leq .00) + P(z \geq 0.00)$. We have previously seen that $P(z \geq .00) = .50$. Also, we know that, since the normal distribution is symmetric, $P(-.50 \leq z \leq .00) = P(.00 \leq z \leq .50)$. Referring to Table 6.1, we find that $P(.00 \leq z \leq .50) = .1915$. Therefore $P(z \geq -.50) = .1915 + .5000 = .6915$. The following graph shows this area.

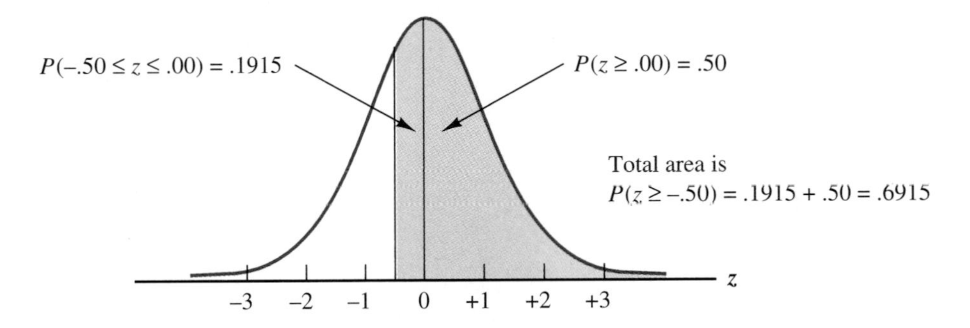

Next, we compute the probability of obtaining a z value between 1.00 and 1.58; that is, $P(1.00 \leq z \leq 1.58)$. From our previous examples, we know that there is a .3413 probability of a z value between $z = 0.00$ and $z = 1.00$ and that there is a .4429 probability of a z value between $z = 0.00$ and $z = 1.58$. Hence, there must be a $.4429 - .3413 = .1016$ probability of a z value between $z = 1.00$ and $z = 1.58$. Thus, $P(1.00 \leq z \leq 1.58) = .1016$. This situation is shown graphically in the following figure.

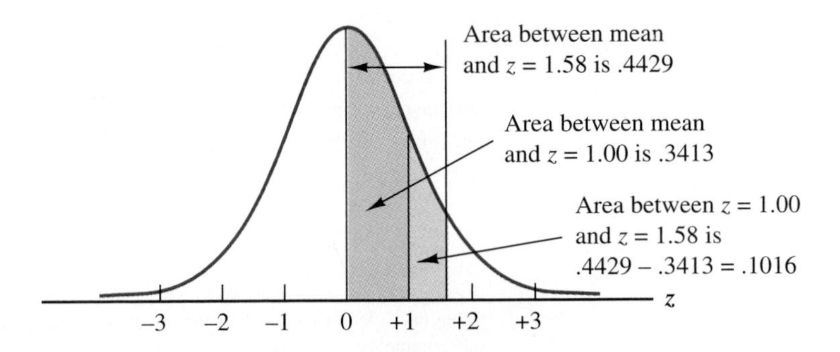

As a final illustration, let us find a z value such that the probability of obtaining a larger z value is only .10. The following figure shows this situation graphically.

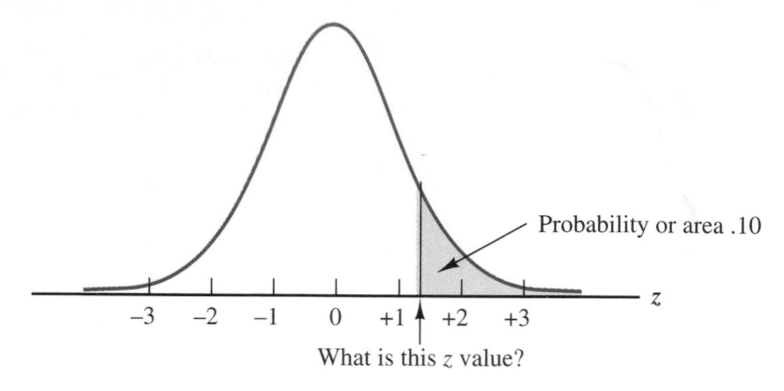

This problem is the reverse of the preceding examples. Previously we specified the z value of interest and then found the corresponding probability, or area. In this example, we are given the probability, or area, and asked to find the corresponding z value. To do so we use the table of areas for the standard normal probability distribution (Table 6.1) somewhat differently.

Recall that the body of Table 6.1 gives the area under the curve between the mean and a particular z value. We have been given the information that the area in the upper tail of the curve is .10. Hence, we must determine how much of the area is between the mean and the z value of interest. Since we know .5000 of the area is above the mean, .5000 − .1000 = .4000 must be the area under the curve *between* the mean and the desired z value. Scanning the body of the table, we find .3997 as the probability value closest to .4000. The section of the table providing this result follows.

z	**.06**	**.07**	**.08**	**.09**
.				
.				
.				
1.0	.3554	.3577	.3599	.3621
1.1	.3770	.3790	.3810	.3830
1.2	.3962	.3980	.3997	.4015
1.3	.4131	.4147	.4162	.4177
1.4	.4279	.4292	.4306	.4319
.				
.	Area value in body of table closest to .4000			
.				

Reading the z value from the left column and the top row of the table, we find that the corresponding z value is 1.28. Thus, there will be an area of approximately .4000 (actually .3997) between the mean and z = 1.28.* In terms of the question originally asked, there is an approximately .10 probability of a z value larger than 1.28.

*We could use interpolation in the body of the table to get a better approximation of the z value that corresponds to an area of .4000. Doing so to provide one more decimal place of accuracy would yield a z value of 1.282. However, in most practical situations, sufficient accuracy is obtained by simply using the table value closest to the desired probability.

The examples illustrate that the table of areas for the standard normal probability distribution can be used to find probabilities associated with values of the standard normal random variable z. Two types of questions can be asked. The first type of question specifies a value, or values, for z and asks us to use the table to determine the corresponding areas, or probabilities. The second type of question provides an area, or probability, and asks us to use the table to determine the corresponding z value. Thus, we need to be flexible in using the standard normal probability table to answer the desired probability question. In most cases, sketching a graph of the standard normal probability distribution and shading the appropriate area or probability helps to visualize the situation and aids in determining the correct answer.

Computing Probabilities for Any Normal Probability Distribution

The reason for discussing the standard normal distribution so extensively is that probabilities for all normal distributions are computed by using the standard normal distribution. That is, when we have a normal distribution with any mean μ and any standard deviation σ, we answer probability questions about the distribution by first converting to the standard normal distribution. Then we can use Table 6.1 and the appropriate z values to find the desired probabilities. The formula used to convert any normal random variable x with mean μ and standard deviation σ to the standard normal distribution follows.

Converting to the Standard Normal Distribution

$$z = \frac{x - \mu}{\sigma} \tag{6.3}$$

A value of x equal to its mean μ results in $z = (\mu - \mu)/\sigma = 0$. Thus, we see that a value of x equal to its mean μ corresponds to a value of z at its mean 0. Now suppose that x is one standard deviation above its mean; that is, $x = \mu + \sigma$. Applying (6.3), we see that the corresponding z value is $z = [(\mu + \sigma) - \mu]/\sigma = \sigma/\sigma = 1$. Thus, a value that is one standard deviation above its mean yields $z = 1$. In other words, we can interpret z as *the number of standard deviations that the normal random variable x is from its mean μ.*

To see how this conversion enables us to compute probabilities for any normal distribution, suppose we have a normal distribution with $\mu = 10$ and $\sigma = 2$. What is the probability that the random variable x is between 10 and 14? Using (6.3) we see that at $x = 10$, $z = (x - \mu)/\sigma = (10 - 10)/2 = 0$ and that at $x = 14$, $z = (14 - 10)/2 = 4/2 = 2$. Thus, the answer to our question about the probability of x being between 10 and 14 is given by the equivalent probability that z is between 0 and 2 for the standard normal distribution. In other words, the probability that we are seeking is the probability that the random variable x is between its mean and two standard deviations above the mean. Using $z = 2.00$ and Table 6.1, we see that the probability is .4772. Hence the probability that x is between 10 and 14 is .4772.

Grear Tire Company Problem

Let us look at an application of the use of the normal probability distribution. Suppose the Grear Tire Company has just developed a new steel-belted radial tire that will be

FIGURE 6.6
Grear Tire Company Mileage Distribution

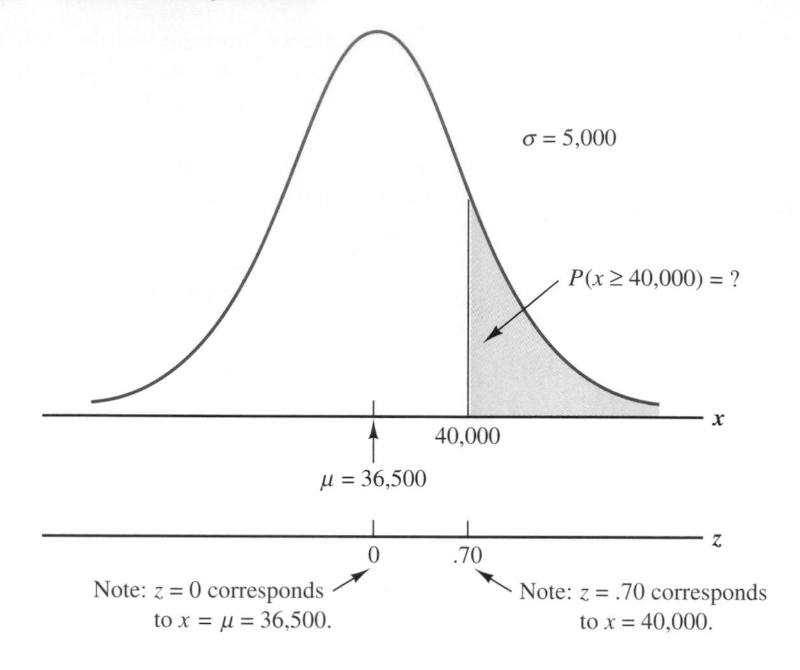

sold through a national chain of discount stores. Since the tire is a new product, Grear's managers believe that the mileage guarantee offered with the tire will be an important factor in the acceptance of the product. Before finalizing the tire mileage guarantee policy, Grear's managers want probability information about the number of miles the tires will last.

From actual road tests with the tires, Grear's engineering group has estimated the mean tire mileage at $\mu = 36{,}500$ miles and the standard deviation at $\sigma = 5000$. In addition, the data collected indicate that a normal distribution is a reasonable assumption. What percentage of the tires can be expected to last more than 40,000 miles? In other words, what is the probability that the tire mileage will exceed 40,000? This question can be answered by finding the area of the shaded region in Figure 6.6.

At $x = 40{,}000$, we have

$$z = \frac{x - \mu}{\sigma} = \frac{40{,}000 - 36{,}500}{5000} = \frac{3500}{5000} = .70$$

Refer now to the bottom of Figure 6.6. We see that a value of $x = 40{,}000$ on the Grear Tire normal distribution corresponds to a value of $z = .70$ on the standard normal distribution. Using Table 6.1, we see that the area between the mean and $z = .70$ is .2580. Referring again to Figure 6.6, we see that the area between $x = 36{,}500$ and $x = 40{,}000$ on the Grear Tire normal distribution is also .2580. Thus, .5000 – .2580 = .2420 is the probability that x will exceed 40,000. We can conclude that about 24.2% of the tires will exceed 40,000 in mileage.

Let us now assume that Grear is considering a guarantee that will provide a discount on replacement tires if the original tires do not exceed the mileage stated in the guarantee. What should the guarantee mileage be if Grear wants no more than 10% of the tires to be eligible for the discount guarantee? This question is interpreted graphically in Figure 6.7.

According to Figure 6.7, 40% of the area must be between the mean and the unknown guarantee mileage. We look up .4000 in the body of Table 6.1 and see that this area is at approximately 1.28 standard deviations *below the mean.* That is, $z = -1.28$ is

FIGURE 6.7
Grear's Discount Guarantee

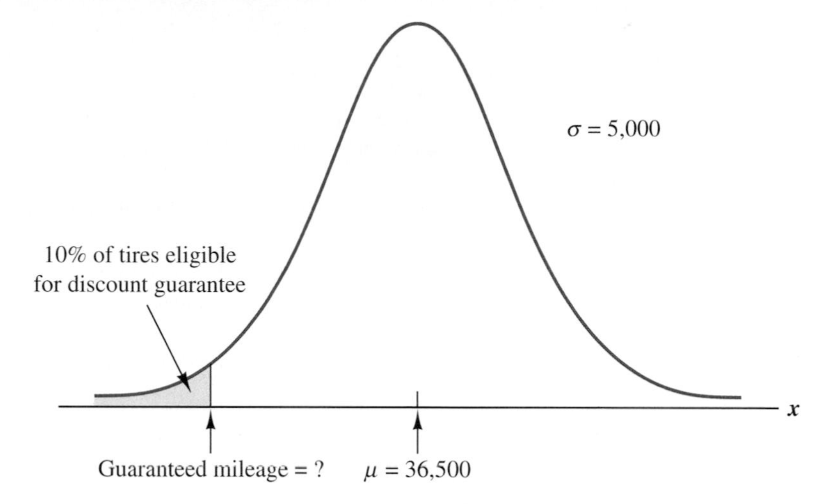

the value of the standard normal random variable corresponding to the desired mileage guarantee on the Grear Tire normal distribution. To find the mileage x corresponding to $z = -1.28$, we have

$$z = \frac{x - \mu}{\sigma} = -1.28$$

$$x - \mu = -1.28\sigma$$

$$x = \mu - 1.28\sigma$$

or, with $\mu = 36{,}500$ and $\sigma = 5000$,

$$x = 36{,}500 - 1.28(5000) = 30{,}100$$

Thus, a guarantee of 30,100 miles will meet the requirement that approximately 10% of the tires will be eligible for the guarantee. Perhaps, with this information, the firm will set its tire mileage guarantee at 30,000 miles.

Again, we see the important role that probability distributions play in providing decision-making information. Namely, once a probability distribution is established for a particular application, it can be used quickly and easily to obtain probability information about the problem. Probability does not make a decision recommendation directly, but it provides information that helps the decision maker better understand the risks and uncertainties associated with the problem. Ultimately, this information may assist the decision maker in reaching a good decision.

exercises

Methods

8. Using Figure 6.4 as a guide, sketch a normal curve for a random variable x that has a mean of $\mu = 100$ and a standard deviation of $\sigma = 10$. Label the horizontal axis with values of 70, 80, 90, 100, 110, 120, and 130.

9. A random variable is normally distributed with a mean of $\mu = 50$ and a standard deviation of $\sigma = 5$.

a. Sketch a normal curve for this random variable. Label the horizontal axis with values of 35, 40, 45, 50, 55, 60, and 65 minutes. Figure 6.4 shows that the normal curve almost touches the horizontal axis at three standard deviations below and at three standard deviations above the mean (in this case at 35 and 65).

b. What is the probability that the random variable will assume a value between 45 and 55?

c. What is the probability that the random variable will assume a value between 40 and 60?

10. Given that z is a standard normal random variable, sketch the standard normal curve. Label the horizontal axis at values of −3, −2, −1, 0, 1, 2, and 3. Then use the table of probabilities for the standard normal distribution to compute the following probabilities.

a. $P(0 \le z \le 1)$ **b.** $P(0 \le z \le 1.5)$ **c.** $P(0 < z < 2)$ **d.** $P(0 < z < 2.5)$

11. Given that z is a standard normal random variable, compute the following probabilities.

a. $P(-1 \le z \le 0)$ **b.** $P(-1.5 \le z \le 0)$ **c.** $P(-2 < z < 0)$
d. $P(-2.5 \le z \le 0)$ **e.** $P(-3 < z \le 0)$

12. Given that z is a standard normal random variable, compute the following probabilities.

a. $P(0 \le z \le .83)$ **b.** $P(-1.57 \le z \le 0)$ **c.** $P(z > .44)$
d. $P(z \ge -.23)$ **e.** $P(z < 1.20)$ **f.** $P(z \le -.71)$

13. Given that z is a standard normal random variable, compute the following probabilities.

a. $P(-1.98 \le z \le .49)$ **b.** $P(.52 \le z \le 1.22)$ **c.** $P(-1.75 \le z \le -1.04)$

14. Given that z is a standard normal random variable, find z for each situation.

a. The area between 0 and z is .4750.
b. The area between 0 and z is .2291.
c. The area to the right of z is .1314.
d. The area to the left of z is .6700.

Self Test

15. Given that z is a standard normal random variable, find z for each situation.

a. The area to the left of z is .2119.
b. The area between $-z$ and z is .9030.
c. The area between $-z$ and z is .2052.
d. The area to the left of z is .9948.
e. The area to the right of z is .6915.

16. Given that z is a standard normal random variable, find z for each situation.

a. The area to the right of z is .01.
b. The area to the right of z is .025.
c. The area to the right of z is .05.
d. The area to the right of z is .10.

Applications

17. The demand for a new product is assumed to be normally distributed with $\mu = 200$ and $\sigma = 40$. Letting x be the number of units demanded, find the following probabilities.

a. $P(180 \le x \le 220)$ **b.** $P(x \ge 250)$ **c.** $P(x \le 100)$ **d.** $P(225 \le x \le 250)$

18. The mean cost for employee alcohol rehabilitation programs involving hospitalization is $10,000 (*USA Today,* September 12, 1991). Assume the rehabilitation program cost has a normal probability distribution with a standard deviation of $2200. Answer the following questions.

a. What is the probability that a rehabilitation program will cost at least $12,000?
b. What is the probability that a rehabilitation program will cost at least $6000?
c. What is the cost range for the most expensive 10% of the rehabilitation programs?

19. College presidents receive a housing provision that averages $26,234 annually (*USA Today,* April 18, 1994). Assume that a normal distribution applies and that the standard deviation is $5000.

a. What percentage of college presidents receive an annual housing provision exceeding \$35,000 per year?

b. What percentage of college presidents receive an annual housing provision less than \$20,000 per year?

c. What is the annual housing provision for the 10% of the college presidents receiving the largest provision?

20. Miami University reported admission statistics for 3339 students who were admitted as freshmen for the fall semester of 1991. Of these students, 1590 had taken the Scholastic Aptitude Test (SAT). Assume the SAT verbal test scores were normally distributed with a mean of 530 and a standard deviation of 70.

a. What percentage of students were admitted with SAT verbal scores between 500 and 600?

b. What percentage of students were admitted with SAT verbal scores of 600 or more?

c. What percentage of students were admitted with SAT verbal scores of 480 or less?

21. Mensa is the international high-IQ society. To be a Mensa member, a person must have an IQ of 132 or higher (*USA Today,* February 13, 1992). If IQ scores are normally distributed with a mean of 100 and a standard deviation of 15, what percentage of the population qualifies for membership in Mensa?

22. Drivers who are members of the Teamsters Union earn an average of \$17.15 per hour (*U.S. News & World Report,* April 11, 1994). Assume that available data indicate wages are normally distributed with a standard deviation of \$2.25.

a. What is the probability that wages are between \$15.00 and \$20.00 per hour?

b. What is the hourly wage of the highest paid 15% of the Teamster drivers?

c. What is the probability that wages are less than \$12.00 per hour?

23. The time needed to complete a final examination in a particular college course is normally distributed with a mean of 80 minutes and a standard deviation of 10 minutes. Answer the following questions.

a. What is the probability of completing the exam in one hour or less?

b. What is the probability that a student will complete the exam in more than 60 minutes but less than 75 minutes?

c. Assume that the class has 60 students and that the examination period is 90 minutes in length. How many students do you expect will be unable to complete the exam in the allotted time?

24. The average age for a person getting married for the first time is 26 years (*U.S. News & World Report,* June 6, 1994). Assume the ages for first marriages have a normal distribution with a standard deviation of four years.

a. What is the probability that a person getting married for the first time is younger than 23 years of age?

b. What is the probability that a person getting married for the first time is in his or her twenties?

c. 90% of people getting married for the first time get married before what age?

25. *Team Marketing Report,* a sports-business newsletter, estimates that the average total cost for a family of four to attend a 1994 major league baseball game was \$95.80 (*The Wall Street Journal,* April 5, 1994). Assume that a normal distribution applies and that the standard deviation is \$10.00.

a. What is the probability that the cost will exceed \$100.00?

b. What is the probability that a family of four will spend \$75.00 or less?

c. What is the probability that the cost will be between \$85.00 and \$100.00?

6.3 Normal Approximation of Binomial Probabilities

In Section 5.4 we presented the binomial probability distribution. Recall that a binomial experiment consists of a sequence of n identical independent trials with each trial having

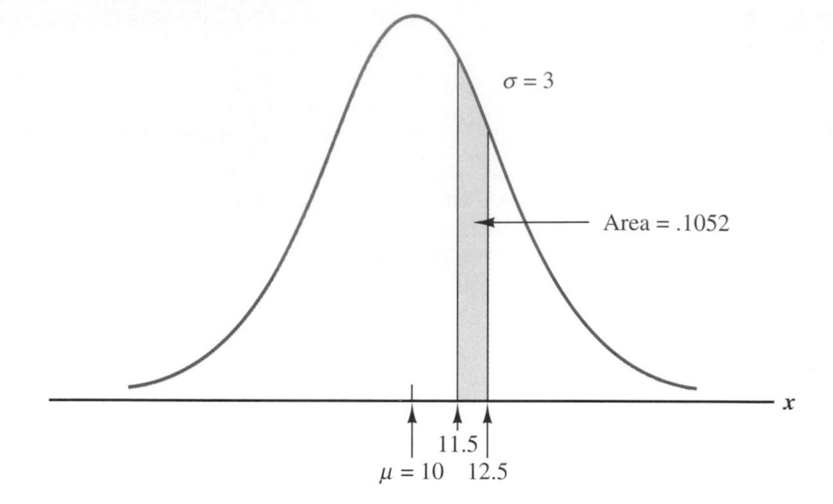

FIGURE 6.8
Normal Approximation to a Binomial Probability Distribution With n = 100 and p = .10 Showing the Probability of 12 Errors

two possible outcomes, a success or a failure. The probability of a success on a trial is the same for all trials and is denoted by p. The binomial random variable is the number of successes in the n trials, and probability questions pertain to the probability of x success in the n trials.

When the number of trials becomes large, evaluating the binomial probability function by hand or with a calculator is difficult. In addition, the binomial tables in Appendix B do not include values of n greater than 20. Hence, when we encounter a binomial probability distribution problem with a large number of trials, we may want to approximate the binomial probability distribution. In cases where the number of trials is greater than 20, $np \geq 5$, and $n(1 - p) \geq 5$, the normal probability distribution provides an easy-to-use approximation of binomial probabilities.

When using the normal approximation to the binomial, we set $\mu = np$ and $\sigma = \sqrt{np(1-p)}$ in the definition of the normal curve. Let us illustrate the normal approximation to the binomial by supposing that a particular company has a history of making errors in 10% of its invoices. A sample of 100 invoices has been taken, and we want to compute the probability that 12 invoices contain errors. That is, we want to find the binomial probability of 12 successes in 100 trials.

In applying the normal approximation to the binomial, we set $\mu = np = (100)(.1) = 10$ and $\sigma = \sqrt{np(1-p)} = \sqrt{(100)(.1)(.9)} = 3$. A normal distribution with $\mu = 10$ and $\sigma = 3$ is shown in Figure 6.8.

Recall that, with a continuous probability distribution, probabilities are computed as areas under the probability density function. As a result, the probability of any single value for the random variable is zero. To approximate the binomial probability of 12 successes, we must compute the area under the corresponding normal curve between 11.5 and 12.5. The .5 that we add and subtract from 12 is called a *continuity correction factor.* It is introduced because a continuous distribution is being used to approximate a discrete distribution. Thus, $P(x = 12)$ for the *discrete* binomial distribution is approximated by $P(11.5 \leq x \leq 12.5)$ for the *continuous* normal distribution.

Converting to the standard normal distribution to compute $P(11.5 \leq x \leq 12.5)$, we have

$$z = \frac{x - \mu}{\sigma} = \frac{12.5 - 10.0}{3} = .83 \qquad \text{at } x = 12.5$$

and

FIGURE 6.9
Normal Approximation to a Binomial Probability Distribution With n = 100 and p = .10 Showing the Probability of 13 or Fewer Errors

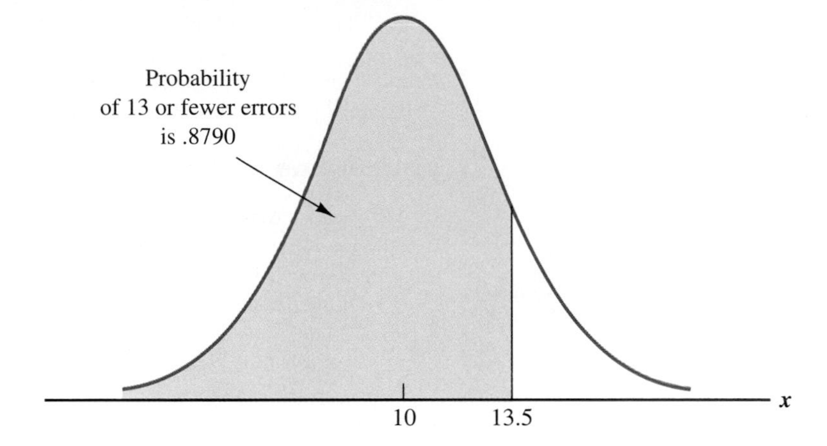

$$z = \frac{x - \mu}{\sigma} = \frac{11.5 - 10.0}{3} = .50 \qquad \text{at } x = 11.5$$

From Table 6.1 we find that the area under the curve between 10 and 12.5 is .2967. Similarly, the area under the curve between 10 and 11.5 is .1915. Therefore, the area between 11.5 and 12.5 is .2967 − .1915 = .1052. The normal approximation to the probability of 12 successes in 100 trials is .1052 (see Figure 6.8).

For another illustration, suppose we want to compute the probability of 13 or fewer errors in the sample of 100 invoices. Figure 6.9 shows the area under the normal curve that approximates this probability. Note that the use of the continuity correction factor results in the value of 13.5 being used to compute the desired probability. The z value corresponding to x = 13.5 is

$$z = \frac{13.5 - 10.0}{3.0} = 1.17$$

Table 6.1 shows that the area under the standard normal curve between 0 and 1.17 is .3790. The area under the normal curve approximating the probability of 13 or fewer errors is given by the shaded portion of the graph in Figure 6.9. The probability is .3790 + .5000 = .8790.

exercises

Methods

26. A binomial probability distribution has p = .20 and n = 100.

a. What is the mean and standard deviation?

b. Is this a situation in which binomial probabilities can be approximated by the normal probability distribution? Explain.

c. What is the probability of exactly 24 successes?

d. What is the probability of 18 to 22 successes?

e. What is the probability of 15 or fewer successes?

27. Assume a binomial probability distribution has p = .60 and n = 200.

a. What is the mean and standard deviation?

b. Is this a situation in which binomial probabilities can be approximated by the normal probability distribution? Explain.

c. What is the probability of between 100 and 110 successes?
d. What is the probability of 130 or more successes?
e. What is the advantage of using the normal probability distribution to approximate the binomial probabilities? Use part (d) to explain the advantage.

Applications

28. A *Consumer Reports* survey listed Saturn, Infiniti, and Lexus automobile dealers as the top three in customer service (*Consumer Reports,* April 1994). Saturn ranked number one, with only 4% of the Saturn customers citing some form of dissatisfaction with the dealer. Answer the following questions about a group of 250 Saturn customers.
a. What is the probability that 12 or fewer customers will have some form of dissatisfaction with the dealer?
b. What is the probability that five or more customers will have some form of dissatisfaction with the dealer?
c. What is the probability that eight customers will have some form of dissatisfaction with the dealer?

29. The true unemployment rate is 7% (*Business Week,* November 7, 1994). Assume that 100 employable people are selected randomly.
a. What is the expected number who are unemployed?
b. What is the variance and standard deviation of the number who are unemployed?
c. What is the probability that exactly nine are unemployed?
d. What is the probability that at least five are unemployed?

30. Homes in Chicago, Illinois, in the price range \$95,000–\$130,000 are on the market an average of 70 days prior to sale (*U.S. News & World Report,* April 6, 1992). Assume the distribution of days on the market is normal with a standard deviation of 25 days.
a. What is the probability that a house will be on the market 100 days or more?
b. What is the probability that a house will sell during the second month it is on the market? That is, $P(31 \leq x \leq 60)$?
c. For how many days are the fastest selling 20% on the market?

31. A Myrtle Beach resort hotel has 120 rooms. In the spring months, hotel room occupancy is approximately 75%. Use the normal approximation to the binomial distribution to answer the following questions.
a. What is the probability that at least half of the rooms are occupied on a given day?
b. What is the probability that 100 or more rooms are occupied on a given day?
c. What is the probability that 80 or fewer rooms are occupied on a given day?

32. It is known that 30% of all customers of a major national charge card pay their bills in full before any interest charges are incurred. Use the normal approximation to the binomial distribution to answer the following questions for a group of 150 credit-card holders.
a. What is the probability that between 40 and 60 customers pay their account balances before any interest charges are incurred? That is, find $P(40 \leq x \leq 60)$.
b. What is the probability that 30 or fewer customers pay their account balances before any interest charges are incurred?

6.4 Exponential Probability Distribution

A continuous probability distribution that is useful in describing the time it takes to complete a task is the *exponential probability distribution.* The exponential random variable can be used to describe such things as the time between arrivals at a car wash, the time required to load a truck, the distance between major defects in a highway, and so on. The exponential probability density function follows.

FIGURE 6.10
Exponential Probability Distribution for the Schips Loading Dock Example

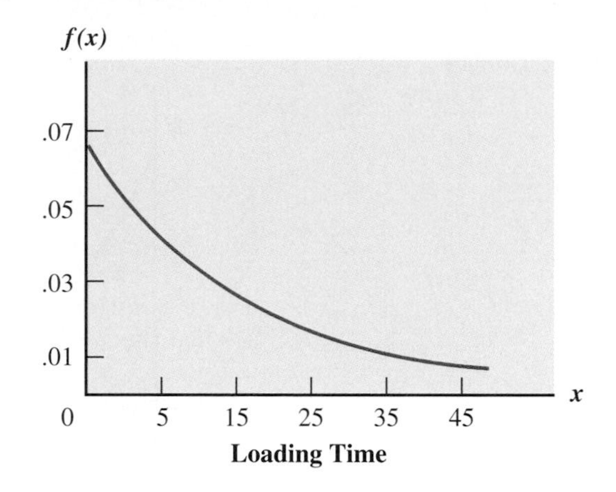

Exponential Probability Density Function

$$f(x) = \frac{1}{\mu} e^{-x/\mu} \qquad \text{for } x \geq 0, \mu > 0 \tag{6.4}$$

As an example of the exponential probability distribution, assume that the time it takes to load a truck at the Schips loading dock follows such a distribution. If the mean, or average, time to load a truck is 15 minutes ($\mu = 15$), the appropriate probability density function is

$$f(x) = \frac{1}{15} e^{-x/15}$$

Figure 6.10 is the graph of this density function.

Computing Probabilities for the Exponential Distribution

As with any continuous probability distribution, the area under the curve corresponding to some interval provides the probability that the random variable assumes a value in that interval. In the Schips loading dock example, the probability that loading a truck will take six minutes or less ($x \leq 6$) is defined to be the area under the curve from $x = 0$ to $x = 6$. Similarly, the probability that loading a truck will take 18 minutes or less ($x \leq 18$) is the area under the curve from $x = 0$ to $x = 18$. Note also that the probability that loading a truck will take between six minutes and 18 minutes ($6 \leq x \leq 18$) is given by the area under the curve from $x = 6$ to $x = 18$.

To compute exponential probabilities such as those just described, we use equation (6.5). It provides the probability of obtaining a value for the exponential random variable of less than or equal to some specific value of x, denoted by x_0.

Exponential Distribution Probabilities

$$P(x \leq x_0) = 1 - e^{-x_0/\mu} \tag{6.5}$$

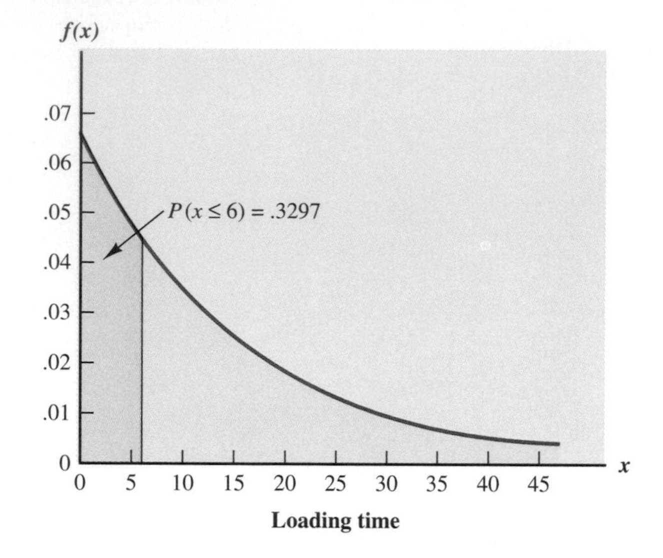

FIGURE 6.11
Probability of a Loading Time of Six Minutes or Less

For the Schips loading dock example, (6.5) can be written as

$$P(\text{loading time} \leq x_0) = 1 - e^{-x_0/15}$$

Hence, the probability that loading a truck will take six minutes or less ($x \leq 6$) is

$$P(\text{loading time} \leq 6) = 1 - e^{-6/15} = .3297$$

Figure 6.11 shows the area or probability for a loading time of six minutes or less. Note also that the probability of loading a truck in 18 minutes or less ($x \leq 18$) is

$$P(\text{loading time} \leq 18) = 1 - e^{-18/15} = .6988$$

Thus, the probability that loading a truck will take between six minutes and 18 minutes is equal to .6988 – .3297 = .3691. Probabilities for any other interval can be computed similarly.

Relationship Between the Poisson and Exponential Distributions

In Section 5.5 we introduced the Poisson distribution as a discrete probability distribution that is often useful in examining the number of occurrences of an event over a specified interval of time or space. Recall that the Poisson probability function is

$$f(x) = \frac{\mu^x e^{-\mu}}{x!}$$

where

μ = expected value or mean number of occurrences in an interval.

The continuous exponential probability distribution is related to the discrete Poisson distribution in that, if the Poisson distribution provides an appropriate description of the number of occurrences per interval, the exponential distribution provides a description of the length of the interval between occurrences.

To illustrate this relationship, suppose the number of cars that arrive at a car wash during one hour is described by a Poisson probability distribution with a mean of 10 cars per hour. The Poisson probability function that gives the probability of x arrivals per hour is

$$f(x) = \frac{10^{x}e^{-10}}{x!}$$

Since the average number of arrivals is 10 cars per hour, the average time between cars arriving is

$$\frac{1 \text{ hour}}{10 \text{ cars}} = .1 \text{ hour/car}$$

Thus, the corresponding exponential distribution that describes the time between the arrivals has a mean of $\mu = .1$ hour per car; the appropriate exponential probability density function is

$$f(x) = \frac{1}{.1}e^{-x/.1} = 10e^{-10x}$$

exercises

Methods

33. Consider the following exponential probability density function.

$$f(x) = \frac{1}{8}e^{-x/8} \qquad \text{for } x \geq 0$$

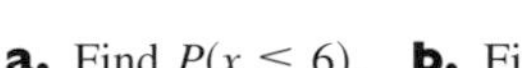

a. Find $P(x \leq 6)$. **b.** Find $P(x \leq 4)$.
c. Find $P(x \geq 6)$. **d.** Find $P(4 \leq x \leq 6)$.

Self Test

34. Consider the following exponential probability density function.

$$f(x) = \frac{1}{3}e^{-x/3} \qquad \text{for } x \geq 0$$

a. Write the formula for $P(x \leq x_0)$. **b.** Find $P(x \leq 2)$.
c. Find $P(x \geq 3)$. **d.** Find $P(x \leq 5)$.
e. Find $P(2 \leq x \leq 5)$.

Applications

35. The average life of a television set is 12 years (*Money,* April 1994). Product lifetimes often follow an exponential probability distribution. Assume that this is the case for the lifetime of a television set.

a. What is the probability that the lifetime will be six years or less?
b. What is the probability that the lifetime will be 15 years or more?
c. What is the probability that the lifetime will be between five and 10 years?

Self Test

36. The time between arrivals of vehicles at a particular intersection follows an exponential probability distribution with a mean of 12 seconds.

a. Sketch this exponential probability distribution.
b. What is the probability that the arrival time between vehicles is 12 seconds or less?
c. What is the probability that the arrival time between vehicles is six seconds or less?
d. What is the probability that there will be 30 or more seconds between vehicle arrivals?

37. The lifetime (hours) of an electronic device is a random variable with the following exponential probability density function.

$$f(x) = \frac{1}{50} e^{-x/50} \qquad \text{for } x \geq 0$$

a. What is the mean lifetime of the device?

b. What is the probability that the device will fail in the first 25 hours of operation?

c. What is the probability that the device will operate 100 or more hours before failure?

38. Waiting times are frequently assumed to follow an exponential probability distribution. A study of waiting times at fast-food restaurants conducted by *The Orlando Sentinel* in October 1993 showed that the average waiting time to get food after placing an order at McDonald's, Burger King, and Wendy's was 60 seconds. Assume that an exponential probability distribution applies to the waiting times.

a. What is the probability that a customer will wait 30 seconds or less?

b. What is the probability that a customer will wait 45 seconds or less?

c. What is the probability that a customer will wait more than two minutes?

39. For subscribers to *The Wall Street Journal,* the average number of stock transactions per year is 12 (*The Wall Street Journal* Subscriber Study, 1992). Assume that the time between transactions follows an exponential probability distribution.

a. What is the mean number of months between stock transactions?

b. What is the probability of a stock transaction within one month of a previous stock transaction?

c. What is the probability that the time between successive stock transactions is three or more months?

d. What is the probability that the time between successive stock transactions is at least one month but not more than two months?

summary

This chapter extended the discussion of probability distributions to the case of continuous random variables. The major conceptual difference between discrete and continuous probability distributions is in the method of computing probabilities. With discrete distributions, the probability function $f(x)$ provides the probability that the random variable x assumes various values. With continuous probability distributions, we associate a probability density function, denoted by $f(x)$. The probability density function does not provide probability values for a continuous random variable directly. Probabilities are given by areas under the curve or graph of the probability density function $f(x)$. Since the area under the curve above a single point is zero, we observe that the probability of any particular value is zero for a continuous random variable.

Three continuous probability distributions—the uniform, normal, and exponential distributions—were treated in detail. The normal probability distribution is used widely in statistical inference and will be used extensively in the remainder of the text.

glossary

Uniform probability distribution A continuous probability distribution where the probability that the random variable will assume a value in any interval is the same for each interval of equal length.

Probability density function The function that defines the probability distribution of a continuous random variable.

Normal probability distribution A continuous probability distribution. Its probability density function is bell shaped and determined by the mean μ and standard deviation σ.

Standard normal probability distribution A normal distribution with a mean of zero and a standard deviation of one.

Continuity correction factor A value of .5 that is added and/or subtracted from a value of x when the continuous normal probability distribution is used to approximate the discrete binomial probability distribution.

Exponential probability distribution A continuous probability distribution that is useful in computing probabilities for the time or space between occurrences of an event.

key formulas

Uniform Probability Density Function

$$f(x) = \begin{cases} \dfrac{1}{b-a} & \text{for } a \leq x \leq b \\ 0 & \text{elsewhere} \end{cases} \tag{6.1}$$

Normal Probability Density Function

$$f(x) = \frac{1}{\sqrt{2\pi}\,\sigma} e^{-(x-\mu)^2/2\sigma^2} \tag{6.2}$$

Converting to the Standard Normal Distribution

$$z = \frac{x - \mu}{\sigma} \tag{6.3}$$

Exponential Probability Density Function

$$f(x) = \frac{1}{\mu} e^{-x/\mu} \qquad \text{for } x \geq 0,\ \mu > 0 \tag{6.4}$$

Exponential Distribution Probabilities

$$P(x \leq x_0) = 1 - e^{-x_0/\mu} \tag{6.5}$$

supplementary exercises

40. The time required to complete a particular assembly operation is uniformly distributed between 30 and 40 minutes.

a. What is the mathematical expression for the probability density function?

b. Compute the probability that the assembly operation will require more than 38 minutes to complete.

c. If management wants to set a time standard for this operation, what time should be selected such that 70% of the time the operation will be completed within the time specified?

d. Find the expected value and standard deviation for the assembly time.

41. A particular make of automobile is listed as weighing 4000 pounds. Because of weight differences due to the options ordered with the car, the actual weight varies uniformly between 3900 and 4100 pounds.

a. What is the mathematical expression for the probability density function?

b. What is the probability that the car will weigh less than 3950 pounds?

42. Motorola used the normal distribution to determine the probability of defects and the number of defects expected in a production process (*APICS—The Performance Advantage,* July 1991). Assume a production process is designed to produce items with a weight of 10 ounces and that the process mean is 10. Calculate the probability of a defect and the expected number of defects for a 1000-unit production run in the following situations.

a. The process standard deviation is .15 and the process control is set at plus or minus one standard deviation. Units with weights less than 9.85 or greater than 10.15 ounces will be classified as defects.

b. Through process design improvements, the process standard deviation can be reduced to .05. Assume the process control remains the same, with weights less than 9.85 or greater than 10.15 ounces being classified as defects.

c. What is the advantage of reducing process variation and setting process control limits at a greater number of standard deviations from the mean?

43. The mean hourly operating cost of a USAir 737 airplane is $2071 (*The Tampa Tribune,* February 17, 1995). Assume that the hourly operating cost for the airplane is normally distributed.

a. If 11% of the hourly operating costs are $1800 or less, what is the standard deviation of hourly operating cost?

b. What is the probability that the hourly operating cost of a USAir 737 airplane is between $2000 and $2500?

c. What is the hourly operating cost of the 3% of the airplanes that have the lowest operating cost?

44. The sales of High-Brite Toothpaste are believed to be approximately normally distributed, with a mean of 10,000 tubes per week and a standard deviation of 1500 tubes per week.

a. What is the probability that more than 12,000 tubes will be sold in any given week?

b. To have a .95 probability that the company will have sufficient stock to cover the weekly demand, how many tubes should be produced?

45. Ward Doering Auto Sales is considering offering a special service contract that will cover the total cost of any service work required on leased vehicles. From experience, the company manager estimates that yearly service costs are approximately normally distributed, with a mean of $150 and a standard deviation of $25.

a. If the company offers the service contract to customers for a yearly charge of $200, what is the probability that any one customer's service costs will exceed the contract price of $200?

b. What is Ward's expected profit per service contract?

46. The attendance at football games at a certain stadium is normally distributed, with a mean of 45,000 and a standard deviation of 3000.

a. What percentage of the time should attendance be between 44,000 and 48,000?

b. What is the probability of the attendance exceeding 50,000?

c. For 80% of the time the attendance should be at least how many?

47. Assume that the test scores from a college admissions test are normally distributed, with a mean of 450 and a standard deviation of 100.

a. What percentage of the people taking the test score between 400 and 500?

b. Suppose someone receives a score of 630. What percentage of the people taking the test score better? What percentage score worse?

c. If a particular university will not admit anyone scoring below 480, what percentage of the persons taking the test would be acceptable to the university?

48. A survey of salaries paid to accounting graduates showed that the mean salary for public accounting managers with six to nine years of experience was $47,000 (*Student Newsbriefs,* The Ohio Society of CPAs, Fall 1992). Assume salaries are normally distributed with a standard deviation of $5500.

a. What is the probability that a manager earns between $40,000 and $50,000?

b. What is the probability that a manager earns less than $35,000?

c. What is the probability that a manager earns $55,000 or more?

d. How much do the top 1% of public accounting managers with six to nine years of experience earn?

49. A machine fills containers with a particular product. The standard deviation of filling weights is known from past data to be .6 ounce. If only 2% of the containers hold less than 18 ounces, what is the mean filling weight for the machine? That is, what must μ equal? Assume the filling weights have a normal distribution.

50. Consider a multiple-choice examination with 50 questions. Each question has four possible answers. Assume that a student who has done the homework and attended lectures has a .75 probability of answering any question correctly.

a. A student must answer 43 or more questions correctly to obtain a grade of A. What percentage of the students who have done their homework and attended lectures will obtain a grade of A on this multiple-choice examination?

b. A student who answers 35 to 39 questions correctly will receive a grade of C. What percentage of students who have done their homework and attended lectures will obtain a grade of C on this multiple-choice examination?

c. A student must answer 30 or more questions correctly to pass the examination. What percentage of the students who have done their homework and attended lectures will pass the examination?

d. Assume that a student has not attended class and has not done the homework for the course. Furthermore, assume that the student will simply guess at the answer to each question. What is the probability that this student will answer 30 or more questions correctly and pass the examination?

51. The book *100% American* by Daniel Evan Weiss reports that 64% of Americans live in the state where they were born. What is the probability that a random sample of 100 people will find between 60 and 70 people living in the state where they were born? That is, find $P(60 \leq x \leq 70)$.

52. A Labor Department survey asked working women what worries them most. Concern about low wages, job stress, and health benefits were mentioned most often, with 60% of working women concerned about low wages (*Business Week,* October 24, 1994). Consider a random sample of 500 working women from this population.

a. What is the expected number of women in this group who will express a concern about low wages?

b. What is the variance and standard deviation of the number who will express a concern about low wages?

c. What is the probability that 290 to 320 women will express a concern about low wages?

d. What is the probability that 325 or more women will express a concern about low wages?

53. The time in minutes for which a student uses a computer terminal at the computer center of a major university follows an exponential probability distribution with a mean of 36 minutes. Assume a student arrives at the terminal just as another student is beginning to work on the terminal.

a. What is the probability that the wait for the second student will be 15 minutes or less?

b. What is the probability that the wait for the second student will be between 15 and 45 minutes?

c. What is the probability that the second student will have to wait an hour or more?

54. A new automated production process has been averaging two breakdowns per day, and the number of breakdowns per day follows a Poisson probability distribution.

a. What is the mean time between breakdowns, assuming eight hours of operation per day?

b. Show the exponential probability density function that can be used for the time between breakdowns.

c. What is the probability that the process will run one hour or more before another breakdown?

d. What is the probability that the process can run a full eight-hour shift without a breakdown?

55. The time (in minutes) a checkout lane is idle between customers at a supermarket follows an exponential probability distribution with a mean of 1.2 minutes.

a. Show the probability density function for this distribution.

b. What is the probability that the next customer will arrive between .5 and 1.0 minutes after a customer is served?

c. What is the probability of the checkout lane being idle for more than a minute between customers?

56. The time (in minutes) between telephone calls at an insurance claims office has the following exponential probability distribution.

$$f(x) = .50e^{-.50x} \qquad \text{for } x \geq 0$$

a. What is the mean time between telephone calls?

b. What is the probability of having 30 seconds or less between telephone calls?

c. What is the probability of having one minute or less between telephone calls?

d. What is the probability of having five or more minutes without a telephone call?

APPENDIX 6.1 Continuous Probability Distributions with Minitab

Let us demonstrate the Minitab procedure for computing continuous probabilities by referring to the Grear Tire Company problem where tire mileage was described by a normal probability distribution with $\mu = 36{,}500$ and $\sigma = 5000$. One question asked was: What is the probability that the tire mileage will exceed 40,000 miles?

For continuous probability distributions, Minitab gives a cumulative probability; that is, Minitab gives the probability that the random variable will assume a value less than or equal to a specified constant. For the Grear tire mileage question, Minitab can be used to determine the cumulative probability that the tire mileage will be less than or equal to 40,000 miles. (The specified constant in this case is 40,000.) After obtaining the cumulative probability from Minitab, we must subtract it from one to determine the probability that the tire mileage will exceed 40,000 miles.

Prior to using Minitab to compute a probability, one must enter the specified constant into a column of the worksheet. For the Grear tire mileage question we entered the specified constant of 40,000 into column 1 of the Minitab worksheet. The steps in using Minitab to compute the cumulative probability of the normal random variable assuming a value less than or equal to 40,000 follow.

STEP 1. Select the **Calc** pull-down menu
STEP 2. Select the **Probability Distributions** pull-down menu
STEP 3. Select the **Normal** option
STEP 4. When the dialog box appears:
Select **Cumulative probability**
Enter 36500 in the **Mean** box
Enter 5000 in the **Standard deviation** box
Enter C1 in the **Input column** box (the column containing 40,000)
Select **OK** to produce the cumulative normal probability

After the user selects **OK,** Minitab will print the cumulative probability that the normal random variable assumes a value less than or equal to 40,000. Minitab will show that this probability is .7580. Since we are interested in the probability that the tire mileage will be greater than 40,000, the desired probability is 1 − .7580 = .2420.

A second question in the Grear Tire Company problem was: What mileage guarantee should Grear set to ensure that no more than 10% of the tires qualify for the guarantee? Here we are given a probability and want to find the corresponding value for the random variable. Minitab uses an inverse calculation routine to find the value of the random variable associated with a given cumulative probability. First, we must enter the cumulative probability into a column of the Minitab worksheet (say C1). In this case, the desired cumulative probability is .10. Then, the first three steps of the Minitab procedure are the same as the preceding example. In step 4, we select **Inverse cumulative probability** instead of **Cumulative probability** and complete the step as shown above. Minitab then displays the mileage guarantee of 30,100 miles.

Minitab is capable of computing probabilities for other continuous probability distributions, including the exponential probability distribution. To compute exponential probabilities, follow the procedure shown previously for the normal probability distribution and select the **Exponential** option in step 3. Step 4 is as shown, with the exception that entering the standard deviation is not required. Output for cumulative probabilities and inverse cumulative probabilities is identical to that described for the normal probability distribution.

APPENDIX 6.2 Continuous Probability Distributions with Spreadsheets

Excel has the capability of computing probabilities for several continuous probability distributions, including the normal and exponential probability distributions. In this appendix, we describe how Excel can be used to compute probabilities for any normal probability distribution. The procedures for the exponential and other continuous probability distributions are similar to the one we describe for the normal probability distribution.

Let us return to the Grear Tire Company problem where the tire mileage was described by a normal probability distribution with $\mu = 36{,}500$ and $\sigma = 5000$. Assume we are interested in the probability that tire mileage will exceed 40,000 miles. The following steps describe how to use Excel to produce the desired normal probability.

STEP 1. Select a cell in the worksheet where you want the normal probability to appear
STEP 2. Select the **Insert** pull-down menu
STEP 3. Choose the **Function** option
STEP 4. When the Function Wizard—Step 1 of 2 dialog box appears:
Choose **Statistical** from the **Function Category** box
Choose **NORMDIST** from the **Function Name** box
Select **Next>**
STEP 5. When the Function Wizard—Step 2 of 2 dialog box appears:
Enter 40000 in the **x** box
Enter 36500 in the **mean** box
Enter 5000 in the **standard deviation** box
Enter true in the **cumulative** box
Select **Finish**

At this point, .7580 will appear in the cell selected in step 1, indicating that the probability of tire mileage being less than or equal to 40,000 miles is .7580. The probability that tire mileage will exceed 40,000 miles is 1 − .7580 = .2420.

Excel uses an inverse computation to convert a given cumulative normal probability into a value for the random variable. For example, what mileage guarantee should Grear offer if the company wants no more than 10% of the tires to be eligible for the guarantee? To compute the mileage guarantee by using Excel, follow the procedure described above. However, two changes are necessary: in step 4, choose **NORMINV** from the **Function Name** box; in step 5, enter the cumulative probability of .10 in the **probability** box and then enter the mean and the standard deviation. When **Finish** is selected in step 5, the tire mileage guarantee of 30,092 or approximately 30,100 miles appears in the worksheet.

The Excel procedure for generating exponential probabilities is similar to the procedure described above. Step 4 can be used to choose the **EXPONDIST** function name. The dialog box in step 5 will guide the user through the input values required to compute the desired probability. Note that the mean is entered in the lambda box. When **Finish** is selected in step 5, the cumulative exponential probability appears in the worksheet.

CHAPTER 7

Sampling and Sampling Distributions

Mead Corporation*

Dayton, Ohio

Mead Corporation, located in Dayton, Ohio, is a diversified paper and forest products company that manufactures paper, pulp, and lumber and converts paperboard into shipping containers and beverage carriers. The company's distribution capability is used to market many of its own products, including paper, school supplies, and stationery. The company's internal consulting group uses sampling for decision analysis to provide a variety of information that enables Mead to obtain significant productivity benefits and remain competitive in its industry.

For example, Mead maintains large woodland holdings, which provide the trees that are the raw material for many of the company's products. Managers need reliable and accurate information about the timberlands and forests to evaluate the company's ability to meet its future raw material needs. What is the present volume in the forests? What is the past growth of the forests? What is the projected future growth of the forests? With answers to these important questions, Mead's managers can develop plans for the future including long-term planting and harvesting schedules for the trees.

How does Mead obtain the information it needs about its vast forest holdings? Data collected from sample plots throughout the forests are the basis for learning about the population of trees owned by the company. To identify the sample plots, the timberland holdings are first divided into three sections based on location and types of trees. Using maps and tables of random numbers, Mead analysts identify random samples of 1/5- to 1/7-acre plots in each section of the forest. The sample plots are where Mead foresters collect data and learn about the forest population.

Foresters throughout the organization participate in the field data collection process. Periodically, two-person teams gather information on each tree in every sample plot. The sample data are entered into the company's continuous forest inventory (CFI) computer system. Reports from the CFI system include a number of frequency distribution summaries containing statistics on types of trees, present forest volume, past forest growth rates, and projected future forest growth and volume. Sampling and the associated statistical summaries of the sample data provide the reports that are essential for the effective management of Mead's forests and timberland assets.

In this chapter you will learn about simple random sampling and the sample selection process. In addition, you will learn how statistics such as the sample mean and sample proportion are used to estimate the population mean and population proportion. The important concept of a sampling distribution is also introduced.

Some of Mead's well-known stationery and school supply products.

Dr. Edward P. Winkofsky, Mead Corporation, provided this Statistics in Practice.

In Chapter 1, we defined a *population* and a *sample* as two important aspects of a statistical study. The definitions are restated here.

1. A population is the set of all the elements of interest in a study.
2. A sample is a subset of the population.

The purpose of *statistical inference* is to obtain information about a population from information contained in a sample. Let us begin by citing two situations in which sampling is conducted to give a manager or decision maker information about a population.

1. A tire manufacturer has developed a new tire that may provide an increase in mileage over the firm's current line of tires. To evaluate the new tires, managers need an estimate of the mean number of miles provided by the new tires. The manufacturer selects a sample of 120 new tires for testing. The test results in a sample mean of 36,500 miles. Hence, 36,500 miles is used as an estimate of the mean tire life for the population of new tires.
2. Members of a political party are considering supporting a particular candidate for election to the United States Senate. To decide whether to enter the candidate in the upcoming primary election, party leaders need an estimate of the proportion of registered voters favoring the candidate. The time and cost associated with contacting every individual in the population of registered voters are prohibitive. Hence, a sample of 400 registered voters is selected. If 160 of the 400 voters indicate a preference for the candidate, an estimate of the proportion of the population of registered voters favoring the candidate is 160/400 = .40.

The preceding examples show how sampling and the sample results can be used to develop estimates of population characteristics. Note that in the tire mileage example, collecting the data on tire life involves wearing out each tire tested. Clearly it is not feasible to test every tire in the population; a sample is the only realistic way to obtain the desired tire mileage data. In the example involving the primary election, contacting every registered voter in the population is theoretically possible, but the time and cost in doing so are prohibitive; thus, a sample of registered voters is preferred.

The examples illustrate some of the reasons for using samples. However, it is important to realize that sample results provide only *estimates* of the values of the population characteristics. That is, we do not expect the sample mean of 36,500 miles to *exactly equal* the mean mileage for all tires in the population; neither do we expect *exactly* 40% of the population of registered voters to favor the candidate. The reason is simply that the sample contains only a portion of the population. With proper sampling methods, the sample results will provide "good" estimates of the population characteristics. But how good can we expect the sample results to be? Fortunately, statistical procedures are available for answering that question.

In this chapter we show how simple random sampling can be used to select a sample from a population. We then show how data obtained from a simple random sample can be used to compute estimates of a population mean and a population proportion. In addition, we introduce the important concept of a sampling distribution. As we show, knowledge of the appropriate sampling distribution is what enables us to make statements about the goodness of the sample results. The last section discusses some alternatives to simple random sampling that are often employed in practice.

7.1 The Electronics Associates Sampling Problem

The director of personnel for Electronics Associates, Inc. (EAI), has been assigned the task of developing a profile of the company's 2500 managers. The characteristics to be identified include the mean annual salary for the managers and the proportion of managers having completed the company's management training program.

Using the 2500 managers as the population for this study, we can find the annual salary and the training program status for each individual in the population by referring to the firm's personnel records. Let us assume that this has been done and that we have obtained the information for all 2500 managers in the population.

Using the formula for a population mean presented in Chapter 3, we can compute the mean annual salary for the population. Assume that this calculation has been performed with the following results.

Population mean: $\mu = \$51,800$

Furthermore, assume that 1500 of the 2500 managers have completed the training program. Letting p denote the proportion of the population having completed the training program, we see that $p = 1500/2500 = .60$.

A *parameter* is a numerical characteristic of a population. For example, the population mean annual salary ($\mu = \$51,800$) and the population proportion having completed the training program ($p = .60$) are parameters of the population of EAI managers.

The question we want to consider is how the firm's director of personnel can obtain estimates of these population parameters by using a sample of managers rather than all 2500 managers in the population. Assume that a sample of 30 managers will be used. Clearly, the time and the cost of developing a profile would be substantially less for 30 managers than for the entire population. If the personnel director could be assured that a sample of 30 managers would provide adequate information about the population of 2500 managers, working with a sample would be preferable to working with the entire population. Let us explore the possibility of using a sample for the EAI study by first considering how we could identify a sample of 30 managers.

7.2 Simple Random Sampling

Several methods can be used to select a sample from a population; one of the most common is *simple random sampling*. The definition of a simple random sample and the process of selecting a simple random sample depend on whether the population is *finite* or *infinite*. Since the EAI sampling problem involves a finite population of 2500 managers, we first consider sampling from finite populations.

Sampling from Finite Populations

A simple random sample of size n from a finite population of size N is defined as follows.

Simple Random Sample (Finite Population)

A simple random sample of size n from a finite population of size N is a sample selected such that each possible sample of size n has the same probability of being selected.

One procedure for identifying a simple random sample from a finite population is to select the elements for the sample *one at a time* in such a way that each of the elements remaining in the population has the *same probability* of being selected. Sampling n elements in that way will satisfy the definition of a simple random sample from a finite population.

To select a simple random sample from the finite population in the EAI problem, we first assume that the 2500 EAI managers have been numbered sequentially (i.e., 1, 2, 3, . . . , 2499, 2500) in the order that their names appear in the EAI personnel file. We could then write the numbers from 1 to 2500 on equal-size pieces of paper, place the 2500 pieces of paper in a container, and mix them thoroughly. We would begin the process of identifying managers for the sample by reaching into the container and selecting one piece of paper *randomly.* The number on the chosen piece of paper would correspond to one of the numbered managers in the file of 2500 managers; thus, that manager would be selected for the sample. The remaining 2499 pieces of paper would be thoroughly mixed again, after which another piece of paper would be selected. This second number would correspond to another EAI manager to be included in the sample. The process would continue until 30 managers have been selected from the population. The 30 managers identified in this way would form a simple random sample from the population.

In this procedure we did not place a selected (sampled) piece of paper back into the container after it was drawn. Hence, we selected a simple random sample *without replacement.* We could have followed the sampling procedure of *replacing* each sampled element before selecting subsequent elements. This form of sampling, referred to as sampling *with replacement,* would have made it possible for some elements to appear in the sample more than once. Sampling with replacement is a valid way of identifying a simple random sample, but sampling without replacement is the sampling procedure used most often. Whenever we refer to simple random sampling, we assume that the sampling is done without replacement.

Rather than labeling 2500 pieces of paper to select a simple random sample of 30 EAI managers, we can use tables of random numbers to obtain the same results much more easily. Such tables are available in a variety of handbooks* that contain page after page of random numbers. We have included one such page of random numbers as Table 8 of Appendix B. Table 7.1 is a portion of this page of random numbers. The first line of the table begins as follows.

63271 59986 71744 51102 15141 80714

Each digit shown, 6, 3, 2, . . . , is a random selection of the digits 0, 1, . . . , 9, with each digit having an equal chance of occurring. The grouping of the numbers into sets of five is simply for the convenience of making the table easy to read.

Let us see how the numbers in this random number table can be used to select a simple random sample of 30 EAI managers. Again we want to select numbers from 1 to 2500 such that every number has an equal chance of being selected. Since the largest number in the EAI population, 2500, has four digits, we select random numbers from the table in sets or groups of four digits. We could select four-digit numbers from any portion of the random number table, but suppose we start by using the first row of random numbers appearing in Table 7.1. The four-digit groupings of the first 28 random numbers in the first row follow.

6327 1599 8671 7445 1102 1514 1807

Since the numbers in the table are random, the preceding four-digit numbers are all equally probable, or equally likely.

We can now use the equally likely four-digit random numbers to give each element in the population an equal chance of being included in the sample. The first number, 6327, is greater than 2500. It does not correspond to an element in the population, and hence it is discarded. The second number, 1599, is between 1 and 2500. Thus the first individual selected for the sample is manager 1599 on the list of EAI managers. Continuing the process, we ignore 8671 and 7445 before identifying individuals 1102, 1514, and 1807 as the next managers to be included in the sample. This process of selecting managers continues until the desired simple random sample of size 30 has

*For example, The Rand Corporation, *A Million Random Digits with 100,000 Normal Deviates.* New York: The Free Press, 1983.

TABLE 7.1 Random Numbers

63271	59986	71744	51102	15141	80714	58683	93108	13554	79945
88547	09896	95436	79115	08303	01041	20030	63754	08459	28364
55957	57243	83865	09911	19761	66535	40102	26646	60147	15702
46276	87453	44790	67122	45573	84358	21625	16999	13385	22782
55363	07449	34835	15290	76616	67191	12777	21861	68689	03263
69393	92785	49902	58447	42048	30378	87618	26933	40640	16281
13186	29431	88190	04588	38733	81290	89541	70290	40113	08243
17726	28652	56836	78351	47327	18518	92222	55201	27340	10493
36520	64465	05550	30157	82242	29520	69753	72602	23756	54935
81628	36100	39254	56835	37636	02421	98063	89641	64953	99337
84649	48968	75215	75498	49539	74240	03466	49292	36401	45525
63291	11618	12613	75055	43915	26488	41116	64531	56827	30825
70502	53225	03655	05915	37140	57051	48393	91322	25653	06543
06426	24771	59935	49801	11082	66762	94477	02494	88215	27191
20711	55609	29430	70165	45406	78484	31639	52009	18873	96927
41990	70538	77191	25860	55204	73417	83920	69468	74972	38712
72452	36618	76298	26678	89334	33938	95567	29380	75906	91807
37042	40318	57099	10528	09925	89773	41335	96244	29002	46453
53766	52875	15987	46962	67342	77592	57651	95508	80033	69828
90585	58955	53122	16025	84299	53310	67380	84249	25348	04332
32001	96293	37203	64516	51530	37069	40261	61374	05815	06714
62606	64324	46354	72157	67248	20135	49804	09226	64419	29457
10078	28073	85389	50324	14500	15562	64165	06125	71353	77669
91561	46145	24177	15294	10061	98124	75732	00815	83452	97355
13091	98112	53959	79607	52244	63303	10413	63839	74762	50289

been obtained. We note that with this random number procedure for simple random sampling, a random number used previously to identify an element for the sample may reappear in the random number table. Since we want to select the simple random sample *without replacement,* previously used random numbers are ignored because the corresponding element is already included in the sample.

Random numbers can be selected from anywhere in the random number table. Although we used the first row of the table in the example, we could have started at any other point in the table and continued in any direction. Once the arbitrary starting point is selected, it is recommended that a predetermined systematic procedure, such as reading across rows or down columns, be used to pick the subsequent random numbers.

Sampling from Infinite Populations

Most sampling situations in business and economics involve finite populations, but in some situations the population is either infinite or so large that for practical purposes it must be treated as infinite. In sampling from an infinite population, we must use a new definition of a simple random sample. In addition, since the elements in an infinite population cannot be numbered, we must use a different process for selecting elements for the sample.

Suppose we want to estimate the average time between placing an order and receiving food for customers at a fast-food restaurant during the 11:30 A.M. to 1:30 P.M.

lunch period. If we consider the population as being all possible customer visits, we see that it would not be feasible to specify a finite limit on the number of possible visits. In fact, if we define the population as being all customer visits that could *conceivably* occur during the lunch period, we can consider the population as being infinite. Our task is to select a simple random sample of n customers from this population. The definition of a simple random sample from an infinite population follows.

Simple Random Sample (Infinite Population)

A simple random sample from an infinite population is a sample selected such that the following conditions are satisfied.

1. Each element selected comes from the same population.
2. Each element is selected independently.

For the problem of selecting a simple random sample of customer visits at a fast-food restaurant, we find that the first condition defined above is satisfied by any customer visit occurring during the 11:30 A.M. to 1:30 P.M. lunch period while the restaurant is operating with its regular staff under "normal" operating conditions. The second condition is satisfied by ensuring that the selection of a particular customer does not influence the selection of any other customer. That is, the customers are selected independently.

A well-known fast-food restaurant has implemented a simple random sampling procedure for just such a situation. The sampling procedure is based on the fact that some customers will present discount coupons for special prices on sandwiches, drinks, french fries, and so on. Whenever a customer presents a discount coupon, the *next* customer served is selected for the sample. Since the customers present discount coupons randomly and independently, the firm is satisfied that the sampling plan satisfies the two conditions for a simple random sample from an infinite population.

notes and comments

1. Finite populations are often defined by lists such as organization membership rosters, student enrollment records, credit-card account lists, inventory product numbers, and so on. Infinite populations are often defined by an ongoing process whereby the elements of the population consist of items generated as though the process would operate indefinitely under the same conditions; in such cases, it is impossible to obtain a list of all items in the population. For example, populations consisting of all possible parts to be manufactured, all possible customer visits, all possible bank transactions, and so on can be classified as infinite populations.
2. The number of different simple random samples of size n that can be selected from a finite population of size N is

$$\frac{N!}{n!(N-n)!}.$$

In this formula, $N!$ and $n!$ refer to the factorial computations discussed in Chapter 4. For the EAI problem with $N = 2500$ and $n = 30$, this expression can be used to show that there are approximately 2.75×10^{69} different simple random samples of 30 EAI managers.

exercises

Methods

1. Consider a finite population with five items labeled A, B, C, D, and E. Ten possible simple random samples of size two can be selected.

a. List the 10 samples beginning with AB, AC, and so on.

b. Using simple random sampling, what is the probability that each sample of size two is selected?

c. Assume random number 1 corresponds to A, random number 2 corresponds to B, and so on. List the simple random sample of two items that will be selected by using the random digits 8 0 5 7 5 3 2.

2. Assume a finite population has 350 items. Using the last three digits of each of the following five-digit random numbers, determine the first four units that will be selected for the simple random sample.

98601 73022 83448 02147 34229 27553 84147 93289 14209

Applications

Self Test

3. *Fortune* publishes data on sales, profits, assets, stockholders' equity, market value, and earnings per share for the 500 largest U.S. industrial corporations (The *Fortune* 500, 1995). Assume that you want to select a simple random sample of 10 corporations from the *Fortune* 500 list. Use column 9 of Table 7.1 beginning with 554. Read down the column and identify the numbers of the 10 corporations that would be selected.

4. The Highway Loss Data Institute reported the number of automobiles stolen in 1991 and 1992 by model (*America by the Numbers,* 1993). The 10 most frequently stolen models follow.

1. Infiniti Q45	**6.** Toyota Supra
2. Volkswagen Jetta	**7.** Ford Mustang
3. Chevrolet Camaro	**8.** Cadillac Brougham
4. Accura Legend	**9.** BMW 525/535
5. Lincoln Mark VII	**10.** BMW 318/325

a. Beginning with the first random digit in Table 7.1 (6) and reading down the column, use single-digit random numbers to select a simple random sample of five automobiles from the list.

b. According to the information in Notes & Comments (2), how many different simple random samples of size five can be selected from this list of 10 automobile models?

5. A student government organization is interested in estimating the proportion of students who favor a mandatory "pass-fail" grading policy for elective courses. A list of names and addresses of the 645 students enrolled during the current quarter is available from the registrar's office. Using row 10 of Table 7.1 and moving across the row from left to right, identify the first 10 students who would be selected by simple random sampling. The three-digit random numbers begin with 816, 283, and 610.

6. The *County and City Data Book,* published by the Bureau of the Census, lists information on 3139 counties throughout the United States. Assume that a national study will collect data from 30 randomly selected counties. Use four-digit random numbers from the last column of Table 7.1 to identify the numbers corresponding to the first five counties selected for the sample. Ignore the first digits and begin with the four-digit random numbers 9945, 8364, 5702, and so on.

7. Assume that we want to identify a simple random sample of 12 of the 372 doctors practicing in a particular city. The doctors' names are available from a local medical organization. Use the eighth column of five-digit random numbers in Table 7.1 to identify the

12 doctors for the sample. Ignore the first two random digits in each five-digit grouping of the random numbers. This process begins with random number 108 and proceeds down the column of random numbers.

8. An article published in *Psychology Today* ranked 286 cities in the United States on the basis of four psychological well-being indicators: alcoholism, suicide, divorce, and crime (*America by the Numbers,* 1993). Assume that a simple random sample of 12 of the 286 cities will be selected for a follow-up in-depth study. Use the third column of five-digit random numbers in Table 7.1, beginning with 71744, to select a simple random sample of 12 cities. Begin with city number 717 and use the first three digits in each row for your selection process. What are the numbers of the 12 cities in the sample?

9. The research group at the Paramont's King's Island theme park in Kings Mills, Ohio, uses surveys to determine what visitors like about the park.

a. Assume that the research group treats the population of visitors as an infinite population. Is this acceptable? Explain.

b. Assume that immediately after completing an interview with a visitor, the interviewer returns to the entrance gate and begins counting individuals as they enter the park. The 25th individual counted is selected as the next person to be sampled for the survey. After completing this interview, the interviewer returns to the entrance and again selects the 25th individual entering the park. Does this sampling process appear to provide a simple random sample? Explain.

10. Indicate whether the following populations should be considered finite or infinite.

a. All registered voters in the state of California.

b. All television sets that could be produced by the Allentown, Pennsylvania, plant of the TV-M Company.

c. All orders that could be processed by a mail-order firm.

d. All emergency telephone calls that could come into a local police station.

e. All components that Fibercon, Inc., produced on the second shift on May 17.

7.3 Point Estimation

Now that we have described how to select a simple random sample, let us return to the EAI problem. Assume that a simple random sample of 30 managers has been selected and that the corresponding data on annual salary and management training program participation are as shown in Table 7.2. The notation x_1, x_2, and so on is used to denote the annual salary of the first manager in the sample, the annual salary of the second manager in the sample, and so on. Participation in the management training program is indicated by Yes in the management training program column.

To estimate the value of a population parameter, we compute a corresponding characteristic of the sample, referred to as a *sample statistic.* For example, to estimate the population mean μ for the annual salary of EAI managers, we simply use the data in Table 7.2 to calculate the corresponding sample statistic: the sample mean $\bar{x}$. The sample mean is

$$\bar{x} = \frac{\Sigma x_i}{n} = \frac{1{,}554{,}420}{30} = \$51{,}814.00$$

In addition, by computing the proportion of managers in the sample who have responded Yes, we can estimate the proportion of managers in the population who have completed the management training program. Table 7.2 shows that 19 of the 30 managers in the sample have completed the training program. Thus, the sample proportion, denoted by $\bar{p}$, is given by

$$\bar{p} = \frac{19}{30} = .63$$

TABLE 7.2 Annual Salary and Training Program Status for a Simple Random Sample of 30 EAI Managers

Annual Salary ($)	Management Training Program?	Annual Salary ($)	Management Training Program?
$x_1 = 49{,}094.30$	Yes	$x_{16} = 51{,}766.00$	Yes
$x_2 = 53{,}263.90$	Yes	$x_{17} = 52{,}541.30$	No
$x_3 = 49{,}643.50$	Yes	$x_{18} = 44{,}980.00$	Yes
$x_4 = 49{,}894.90$	Yes	$x_{19} = 51{,}932.60$	Yes
$x_5 = 47{,}621.60$	No	$x_{20} = 52{,}973.00$	Yes
$x_6 = 55{,}924.00$	Yes	$x_{21} = 45{,}120.90$	Yes
$x_7 = 49{,}092.30$	Yes	$x_{22} = 51{,}753.00$	Yes
$x_8 = 51{,}404.40$	Yes	$x_{23} = 54{,}391.80$	No
$x_9 = 50{,}957.70$	Yes	$x_{24} = 50{,}164.20$	No
$x_{10} = 55{,}109.70$	Yes	$x_{25} = 52{,}973.60$	No
$x_{11} = 45{,}922.60$	Yes	$x_{26} = 50{,}241.30$	No
$x_{12} = 57{,}268.40$	No	$x_{27} = 52{,}793.90$	No
$x_{13} = 55{,}688.80$	Yes	$x_{28} = 50{,}979.40$	Yes
$x_{14} = 51{,}564.70$	No	$x_{29} = 55{,}860.90$	Yes
$x_{15} = 56{,}188.20$	No	$x_{30} = 57{,}309.10$	No

TABLE 7.3 Summary of Point Estimates Obtained from a Simple Random Sample of 30 EAI Managers

Population Parameter	Parameter Value	Point Estimator	Point Estimate
μ = Population mean annual salary	$51,800.00	$\bar{x}$ = Sample mean annual salary	$51,814.00
p = Population proportion having completed the management training program	.60	$\bar{p}$ = Sample proportion having completed the management training program	.63

Thus, the value $\bar{p} = .63$ is used as an estimate of the population proportion p.

By making the preceding computations, we have performed the statistical procedure called *point estimation.* In point estimation we use the data from the sample to compute a value of a sample statistic that serves as an estimate of a population parameter. Using the terminology of point estimation, we refer to $\bar{x}$ as the *point estimator* of the population mean μ, and $\bar{p}$ as the *point estimator* of the population proportion p. The actual numerical value obtained for $\bar{x}$ or $\bar{p}$ in a particular sample is called the *point estimate* of the parameter. Thus, for the sample of 30 EAI managers, $51,814.00 is the point estimate of μ and .63 is the point estimate of p. Table 7.3 summarizes the sample results and compares the point estimates to the actual values of the population parameters.

notes and comments

In the discussion of point estimators, we use $\bar{x}$ to denote a sample mean and $\bar{p}$ to denote a sample proportion. The use of $\bar{p}$ is based on the fact that the sample proportion is also a *sample mean.* For instance, suppose that in a sample of n items with data values $x_1, x_2, \ldots, x_n$, we let $x_i = 1$ when a characteristic of interest is present for the ith item and $x_i = 0$ when the characteristic is not present. Then the sample proportion is computed by $\Sigma x_i / n$, which is the formula for a sample mean. We also like the consistency of using the bar over the letter to remind the reader that the sample proportion $\bar{p}$ estimates the population proportion just as the sample mean $\bar{x}$ estimates the population mean. Some texts in statistics use $\hat{p}$ instead of $\bar{p}$ to denote the sample proportion.

exercises

Self Test 

Methods

11. The following data have been collected from a simple random sample.

5 8 10 7 10 14

What is the point estimate of the population mean?

12. A survey question for a sample of 150 individuals yielded 75 Yes responses, 55 No responses, and 20 No Opinions.

a. What is the point estimate of the proportion in the population who respond Yes?

b. What is the point estimate of the proportion in the population who respond No?

Applications

Self Test 

13. A simple random sample of five months of sales data provided the following information:

Month:	1	2	3	4	5
Units Sold:	94	100	85	94	92

What is the point estimate of the population mean number of units sold per month?

14. A 1992 survey conducted by the Foundation for Women and the Center for Policy Awareness asked married working women to identify the factors that would contribute most to improved family life. Suppose the data for a sample of 800 respondents are as follows:

Improve Family Life	Frequency
More flexible hours	272
Higher pay	208
More help at home	120
Better day care	56
Nothing	144

Use the sample results to obtain the following estimates.

a. The proportion of the population of married working women who believe more flexible work hours would contribute most to improved family life.

b. The proportion of the population of married working women who believe higher paying jobs would contribute most to improved family life.

15. The California Highway Patrol maintains records showing the time between a report of an accident and the arrival of an officer at the accident scene. A simple random sample of 10 records shows the following times in minutes.

12.6 3.4 4.8 5.0 6.8 2.3 3.6 8.1 2.5 10.3

What is a point estimate of the population mean time between the report of an accident and the arrival of an officer at the accident scene?

16. A report by the U.S. Department of Transportation (March 1993) indicated that one of the major complaints against the nation's air carriers is that many flights fail to arrive on schedule. A flight is considered on time if it arrives no later than 15 minutes after the scheduled time. Suppose that of a random sample of 200 flights into Los Angeles International Airport, 182 flights were classified as being on time. On the basis of these data, what is the point estimate of the proportion of all flights into Los Angeles International Airport that should be classified as *late* arrivals?

17. A 1992 Louis Harris poll called "America and the Arts" was used to study taxpayers' attitudes about federal funding of the arts. Suppose a sample of 500 taxpayers were selected and asked the following two questions.

Would you support a $5 tax to fund the arts?

Would you support a $10 tax to fund the arts?

Assume that 345 answered yes to the first question and 320 answered yes to the second question. Use these results to provide point estimates for each of the following population parameters.

a. The proportion of all taxpayers who would support a $5 tax to fund the arts.

b. The proportion of all taxpayers who would support a $10 tax to fund the arts.

7.4 Introduction to Sampling Distributions

In the preceding section we used a simple random sample of 30 EAI managers to develop point estimates of the mean annual salary for the population of all EAI managers as well as the proportion of the managers in the population who have completed the company's management training program. Suppose we select another simple random sample of 30 EAI managers, and an analysis of the data from the second sample provides the following information.

$$\text{Sample Mean } \bar{x} = \$52{,}669.70$$

$$\text{Sample Proportion } \bar{p} = .70$$

These results show that different values of $\bar{x}$ and $\bar{p}$ have been obtained with the second sample. In general, this is to be expected because the second simple random sample is not likely to contain the same elements as the first. Let us imagine carrying out the same process of selecting a new simple random sample of 30 managers over and over again, each time computing values of $\bar{x}$ and $\bar{p}$. In this way we could begin to identify the variety of values that these point estimators can have. To illustrate, we repeated the simple random sampling process for the EAI problem until we obtained 500 samples of 30 managers each and their corresponding $\bar{x}$ and $\bar{p}$ values. A portion of the results is shown in Table 7.4. Table 7.5 provides the frequency and relative frequency distributions for the 500 $\bar{x}$ values. Figure 7.1 is the relative frequency histogram for the $\bar{x}$ results.

TABLE 7.4 Values of $\bar{x}$ and $\bar{p}$ for the 500 Simple Random Samples of 30 EAI Managers

Sample Number	Sample Mean $\bar{x}$	Sample Proportion $\bar{p}$
1	\$51,814.00	.63
2	\$52,669.70	.70
3	\$51,780.30	.67
4	\$51,587.90	.53
.	.	.
.	.	.
.	.	.
500	\$51,752.00	.50

TABLE 7.5 Frequency and Relative Frequency Distributions of $\bar{x}$ for the 500 Simple Random Samples of 30 EAI Managers

Mean Annual Salary (\$)		Frequency	Relative Frequency
49,500.00–49,999.99		2	.004
50,000.00–50,499.99		16	.032
50,500.00–50,999.99		52	.104
51,000.00–51,499.99		101	.202
51,500.00–51,999.99		133	.266
52,000.00–52,499.99		110	.220
52,500.00–52,999.99		54	.108
53,000.00–53,499.99		26	.052
53,500.00–53,999.99		6	.012
	Totals	500	1.000

Recall that in Chapter 5 we defined a random variable as a numerical description of the outcome of an experiment. If we consider the process of selecting a simple random sample as an experiment, the sample mean $\bar{x}$ is the numerical description of the outcome of the experiment. Thus, the sample mean $\bar{x}$ is a random variable. Since the various possible values of $\bar{x}$ are the result of different simple random samples, the probability distribution of $\bar{x}$ is called the *sampling distribution of* $\bar{x}$. Knowledge of this sampling distribution and its properties will enable us to make probability statements about how close the sample mean $\bar{x}$ is to the population mean μ.

Let us return to Figure 7.1. We would need to enumerate every possible sample of 30 managers and compute each sample mean to completely determine the sampling distribution of $\bar{x}$. However, the histogram of 500 $\bar{x}$ values gives an approximation of this sampling distribution. From the approximation we observe the bell-shaped appearance of the distribution. We also note that the mean of the 500 $\bar{x}$ values is near the population mean $\mu = \$51,800$. We will describe the properties of the sampling distribution of $\bar{x}$ more fully in the next section.

FIGURE 7.1
Relative Frequency Histogram of $\bar{x}$ for the 500 Simple Random Samples of 30 EAI Managers

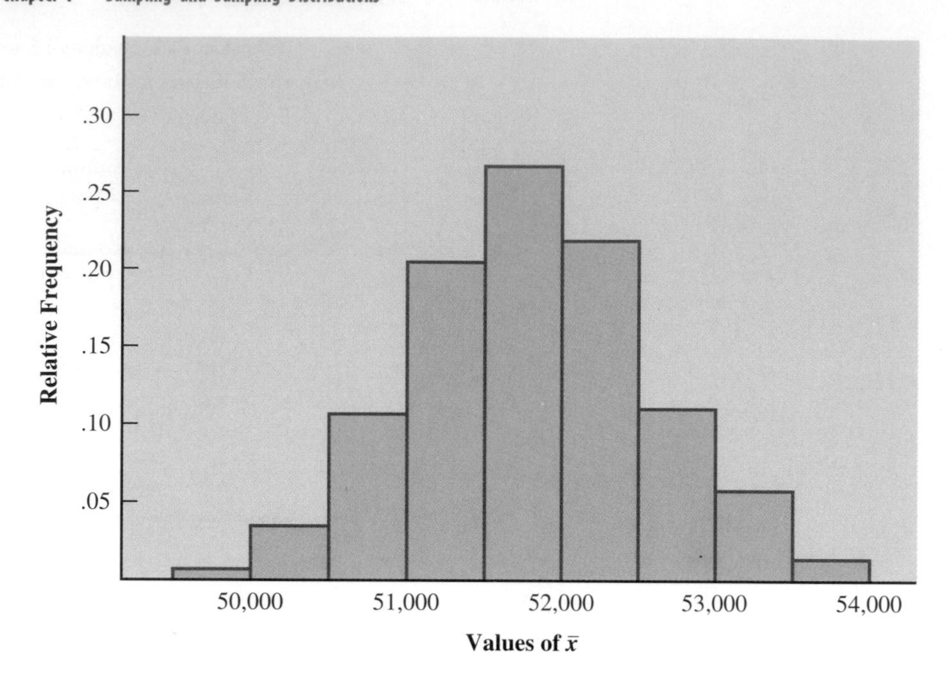

FIGURE 7.2
Relative Frequency Histogram of $\bar{p}$ for the 500 Simple Random Samples of 30 EAI Managers

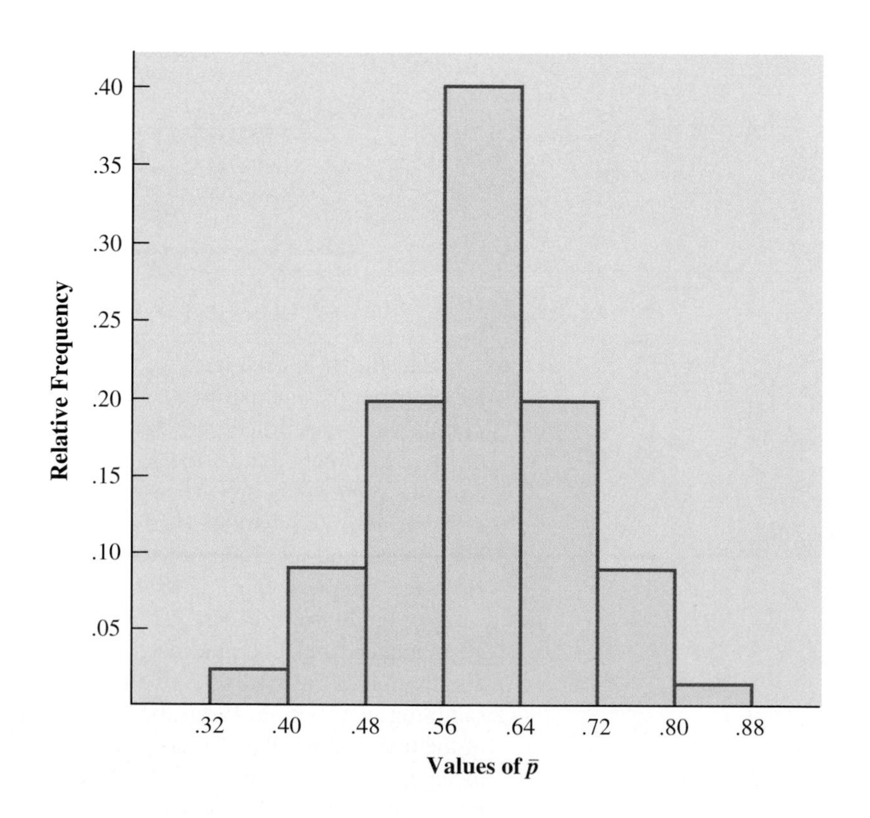

The 500 values of the sample proportion $\bar{p}$ are summarized by the relative frequency histogram in Figure 7.2. As in the case of $\bar{x}$, $\bar{p}$ is a random variable that provides a numerical description of the outcome of a simple random sample. If every possible sample of size 30 were selected from the population and a value $\bar{p}$ were computed for each sample, the resulting probability distribution would be called the *sampling distribution of* $\bar{p}$. The relative frequency histogram of the 500 sample values in Figure 7.2 provides a general idea of the appearance of this sampling distribution.

In practice, we select *only one simple random sample* from the population. We repeated the sampling process 500 times in this section simply to illustrate that many different samples are possible and that the different samples generate a variety of values for the sample statistics $\bar{x}$ and $\bar{p}$. The probability distribution of any particular sample statistic is called the sampling distribution of the statistic. In Section 7.5 we show the characteristics of the sampling distribution of $\bar{x}$. In Section 7.6 we show the characteristics of the sampling distribution of $\bar{p}$.

7.5 Sampling Distribution of $\bar{x}$

One of the most common statistical procedures is the use of a sample mean $\bar{x}$ to make inferences about a population mean μ. This process is shown in Figure 7.3. On each repetition of the process, we can anticipate obtaining a different value for the sample mean $\bar{x}$. The probability distribution for all possible values of the sample mean $\bar{x}$ is called the sampling distribution of the sample mean $\bar{x}$.

Sampling Distribution of $\bar{x}$

The sampling distribution of $\bar{x}$ is the probability distribution of all possible values of the sample mean $\bar{x}$.

The purpose of this section is to describe the properties of the sampling distribution of $\bar{x}$, including the expected value or mean of $\bar{x}$, the standard deviation of $\bar{x}$, and the shape or form of the sampling distribution itself. As we shall see, knowledge of the sampling distribution of $\bar{x}$ will enable us to make probability statements about the error involved when $\bar{x}$ is used to estimate μ. Let us begin by considering the mean of all possible $\bar{x}$ values or, simply, the expected value of $\bar{x}$.

Expected Value of $\bar{x}$

In the EAI sampling problem we saw that different simple random samples result in a variety of values for the sample mean $\bar{x}$. Since many different values of the random variable $\bar{x}$ are possible, we are often interested in the mean of all possible values of $\bar{x}$ that can be generated by the various simple random samples. The mean of the $\bar{x}$ random variable is the expected value of $\bar{x}$. Let $E(\bar{x})$ represent the expected value of $\bar{x}$ and μ represent the mean of the population from which the sample is drawn. It can be shown that with simple random sampling, these two values are equal.

FIGURE 7.3
The Statistical Process of Using a Sample Mean to Make Inferences About a Population Mean

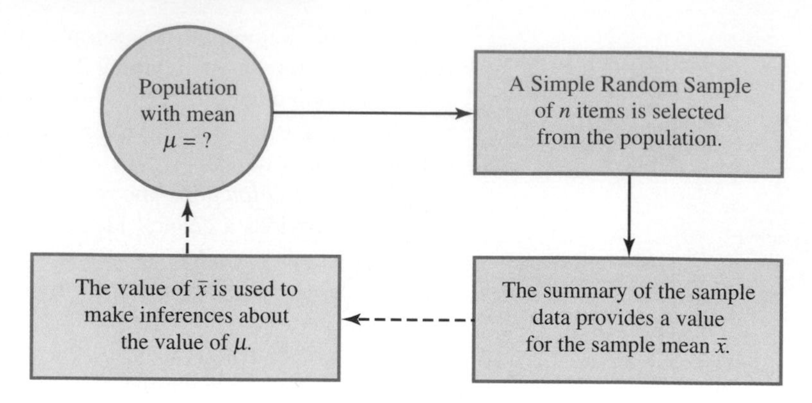

Expected Value of $\bar{x}$

$$E(\bar{x}) = \mu \tag{7.1}$$

where

$E(\bar{x})$ = the expected value of the random variable $\bar{x}$

μ = the population mean

This result shows that with simple random sampling, the expected value or mean for $\bar{x}$ is equal to the mean of the population. In statistical terminology, (7.1) allows us to state that $\bar{x}$ is an *unbiased* estimator of μ. From Section 7.1 we saw that the mean annual salary for the population of EAI managers is $\mu = \$51{,}800$. Thus, according to (7.1), the mean of all possible sample means for the EAI study is also $51,800.

Standard Deviation of $\bar{x}$

Let us define the standard deviation of the sampling distribution of $\bar{x}$. We will use the following notation.

$\sigma_{\bar{x}}$ = the standard deviation of the sampling distribution of $\bar{x}$
σ = the standard deviation of the population
n = the sample size
N = the population size

With simple random sampling, the standard deviation of $\bar{x}$ depends on whether the population is finite or infinite. The two expressions for the standard deviation of $\bar{x}$ follow.

Standard Deviation of $\bar{x}$

Finite Population *Infinite Population*

$$\sigma_{\bar{x}} = \sqrt{\frac{N-n}{N-1}}\left(\frac{\sigma}{\sqrt{n}}\right) \qquad \sigma_{\bar{x}} = \frac{\sigma}{\sqrt{n}} \tag{7.2}$$

In comparing the two expressions in (7.2), we see that the factor $\sqrt{(N - n)/(N - 1)}$ is required for the finite population but not for the infinite population. This factor is commonly referred to as the *finite population correction factor*. In many practical sampling situations, we find that the population involved, although finite, is "large," whereas the sample size is relatively "small." In such cases the finite population correction factor $\sqrt{(N - n)/(N - 1)}$ is close to 1. As a result, the difference between the values of the standard deviation of $\bar{x}$ for the finite and infinite population cases becomes negligible. When this occurs, $\sigma_{\bar{x}} = \sigma/\sqrt{n}$ becomes a very good approximation to the standard deviation of $\bar{x}$ even though the population is finite. A general guideline or rule of thumb for computing the standard deviation of $\bar{x}$ follows.

Use the Following Expression to Calculate the Standard Deviation of $\bar{x}$

$$\sigma_{\bar{x}} = \frac{\sigma}{\sqrt{n}} \tag{7.3}$$

whenever

1. The population is infinite; or
2. The population is finite, *and* the sample size is less than or equal to 5% of the population size—that is, $n/N \leq .05$.

In cases where $n / N > .05$, the finite population version of (7.2) should be used in the computation of $\sigma_{\bar{x}}$. Unless otherwise noted, throughout the text we will assume that $n/N \leq .05$ and the finite population correction factor is unnecessary. Thus (7.3) can be used to compute $\sigma_{\bar{x}}$.

Now let us return to the EAI study and determine the standard deviation of all possible sample means that can be generated with samples of 30 EAI managers. In this illustration, we will assume that the population standard deviation for the annual salary data is given with $\sigma = 4000$. The population is finite, with $N = 2500$. However, with a sample size of 30, we have $n / N = 30/2500 = .012$. Following the rule of thumb given in (7.3), we can ignore the finite population correction factor and use (7.3) to compute the standard deviation of $\bar{x}$.

$$\sigma_{\bar{x}} = \frac{\sigma}{\sqrt{n}} = \frac{4000}{\sqrt{30}} = 730.30$$

Later we will see that the value of $\sigma_{\bar{x}}$ is helpful in determining how far the sample mean may be from the population mean. Because of the role that $\sigma_{\bar{x}}$ plays in computing possible errors, $\sigma_{\bar{x}}$ is referred to as the *standard error of the mean.*

Central Limit Theorem

The final step in identifying the characteristics of the sampling distribution of $\bar{x}$ is to determine the form of the probability distribution of $\bar{x}$. We consider two cases: one in which the population distribution is unknown and one in which the population distribution is known to be normal.

When the population distribution is unknown, we rely on one of the most important theorems in statistics—the *central limit theorem.* A statement of the central limit theorem as it applies to the sampling distribution of $\bar{x}$ follows.

FIGURE 7.4
Illustration of the Central Limit Theorem for Three Populations

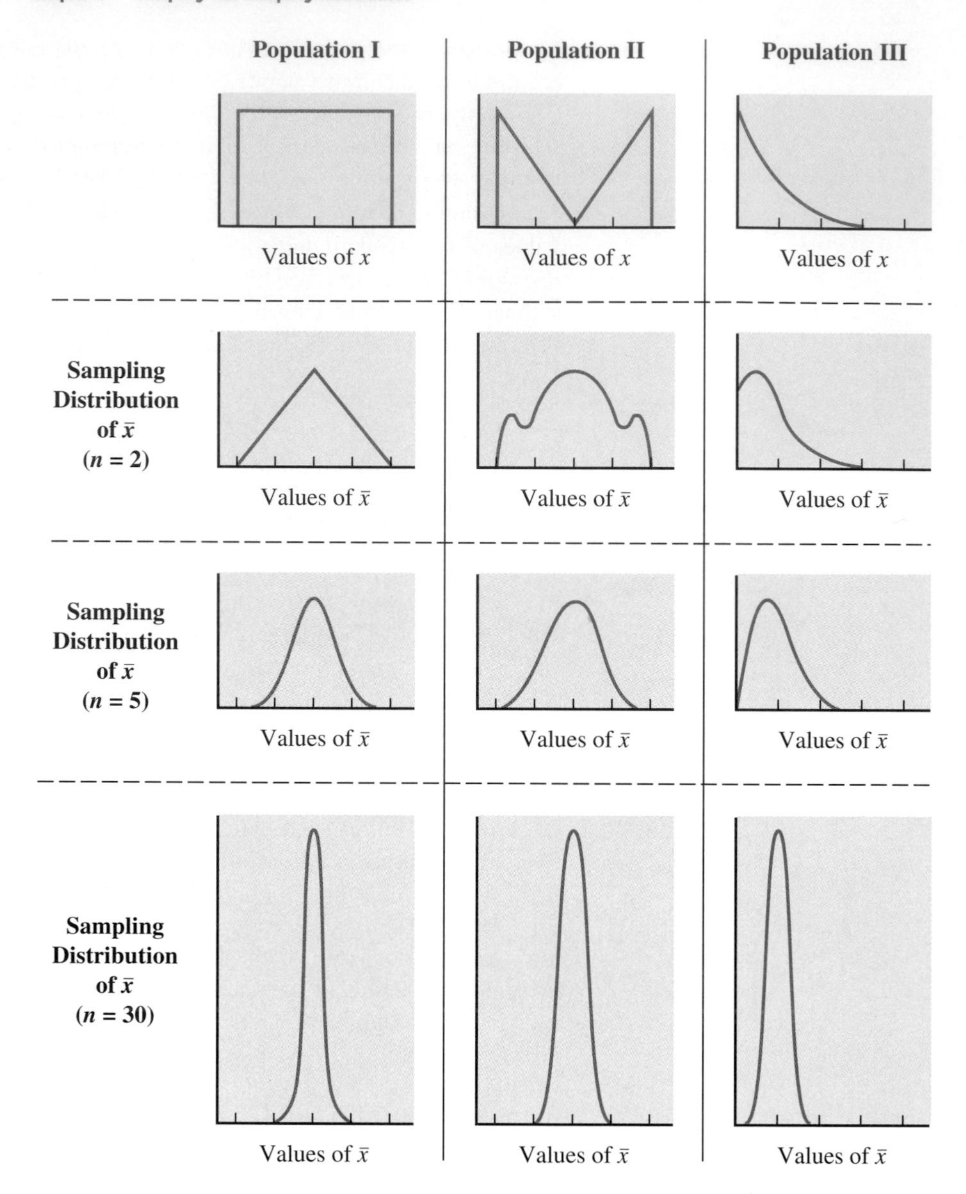

Central Limit Theorem

In selecting simple random samples of size n from a population, the sampling distribution of the sample mean $\bar{x}$ can be approximated by a *normal probability distribution* as the sample size becomes large.

Figure 7.4 shows how the central limit theorem works for three different populations; in each case the population clearly is not normal. However, note what begins to happen to the sampling distribution of $\bar{x}$ as the sample size is increased. When the samples are of size two, we see that the sampling distribution of $\bar{x}$ begins to take on an appearance different from that of the population distribution. For samples of size five, we see all three sampling distributions beginning to take on a bell-shaped appearance. Finally, the samples of size 30 show all three sampling distributions to be approximately normal probability distributions. Thus, for sufficiently large samples, the sampling distribution of $\bar{x}$ can be approximated by a normal probability distribution. However, how large must the sample size be before we can assume that the central limit theorem applies? Statistical researchers have investigated this question by studying the sampling distribution of $\bar{x}$ for a variety of populations and a variety of sample sizes. Whenever the population distribution is mound-shaped and symmetrical, sample sizes as small as five to 10 can be enough for the central limit theorem to apply. However, if the population distribution is highly skewed and clearly nonnormal, larger sample sizes are needed. General statistical practice is to assume that for most applications, the sampling distribution of $\bar{x}$ can be approximated by a normal probability distribution whenever the *sample size is 30 or more.* In effect, a sample size of 30 or more is assumed to satisfy the large-sample condition of the central limit theorem. This observation is so important that we restate it.

The sampling distribution of $\bar{x}$ can be approximated by a normal probability distribution whenever the sample size is large. The large-sample-size condition can be assumed for simple random samples of size 30 or more.

The central limit theorem is the key to identifying the form of the sampling distribution of $\bar{x}$ whenever the population distribution is unknown. However, we may encounter some sampling situations in which the population is assumed or believed to have a normal probability distribution. When this condition occurs, the following result identifies the form of the sampling distribution of $\bar{x}$.

Whenever the population has a normal probability distribution, the sampling distribution of $\bar{x}$ is a normal probability distribution for any sample size.

In summary, if we use a large simple random sample ($n \geq 30$), the central limit theorem enables us to conclude that the sampling distribution of $\bar{x}$ can be approximated by a normal probability distribution. When the simple random sample is small ($n < 30$), the sampling distribution of $\bar{x}$ can be considered normal only if we assume that the population has a normal probability distribution.

Sampling Distribution of $\bar{x}$ for the EAI Problem

For the EAI study we have shown that $E(\bar{x}) = 51{,}800$ and $\sigma_{\bar{x}} = 730.30$. Since we are using a simple random sample of 30 managers, the central limit theorem enables us to conclude that the sampling distribution of $\bar{x}$ can be approximated by a normal probability distribution as shown in Figure 7.5.

Practical Value of the Sampling Distribution of $\bar{x}$

Whenever a simple random sample is selected and the value of the sample mean $\bar{x}$ is used to estimate the value of the population mean σ, we cannot expect the sample mean

FIGURE 7.5
Sampling Distribution of $\bar{x}$, the Sample Mean Annual Salary, for a Simple Random Sample of 30 EAI Managers

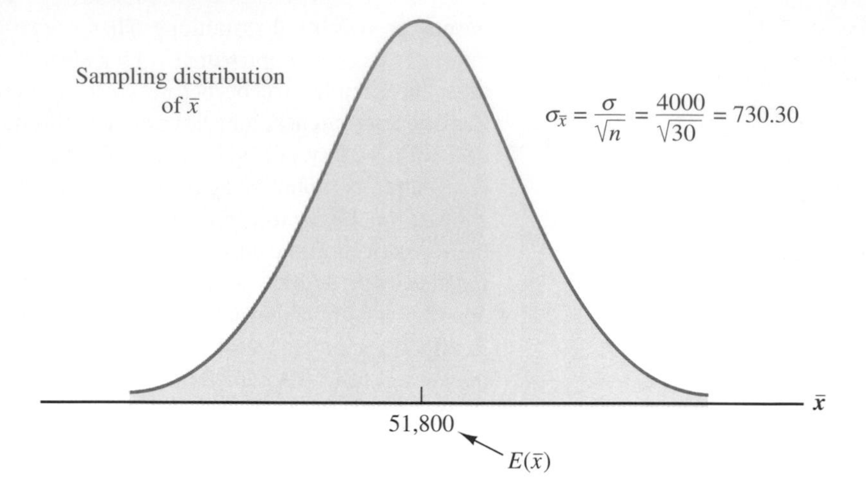

to *exactly equal* the population mean. The absolute value of the difference between the value of the sample mean $\bar{x}$ and the value of the population mean μ, $|\bar{x} - \mu|$, is called the *sampling error*. The practical reason we are interested in the sampling distribution of $\bar{x}$ is that it can be used to provide probability information about the size of the sampling error. To demonstrate this use, let us return to the EAI problem.

Suppose the personnel director believes the sample mean will be an acceptable estimate of the population mean if the sample mean is within \$500 of the population mean. In probability terms, the personnel director is really concerned with the following question: What is the probability that the sample mean we obtain from a simple random sample of 30 EAI managers will be within \$500 of the population mean?

Since we have identified the properties of the sampling distribution of $\bar{x}$ (see Figure 7.5), we will use this distribution to answer the probability question. Refer to the sampling distribution of $\bar{x}$ shown again in Figure 7.6. The personnel director is asking about the probability that the sample mean is between \$51,300 and \$52,300. If the value of the sample mean $\bar{x}$ is in this interval, the value of $\bar{x}$ will be within \$500 of the population mean. The appropriate probability is given by the area of the sampling distribution shown in Figure 7.6. Since the sampling distribution is normal, with mean 51,800 and standard deviation 730.30, we can use the standard normal probability distribution table to find the area or probability. At $\bar{x} = 51{,}300$, we have

$$z = \frac{51{,}300 - 51{,}800}{730.30} = -.68$$

Referring to the standard normal probability distribution table, we find an area between $z = 0$ and $z = -.68$ of .2518. Similar calculations for $\bar{x} = 52{,}300$ show an area between $z = 0$ and $z = +.68$ of .2518. Thus, the probability that the value of the sample mean is between 51,300 and 52,300 is $.2518 + .2518 = .5036$.

The preceding computations show that a simple random sample of 30 EAI managers has a .5036 probability of providing a sample mean $\bar{x}$ that is within \$500 of the population mean. Thus, there is a $1 - .5036 = .4964$ probability that the sample mean will miss the population mean by more than \$500. In other words, a simple random sample of 30 EAI managers has roughly a 50–50 chance of providing a sample mean within the allowable \$500. Perhaps a larger sample size should be considered. Let us explore this possibility by considering the relationship between the sample size and the sampling distribution of $\bar{x}$.

FIGURE 7.6
The Probability of a Sample Mean Being Within $500 of the Population Mean

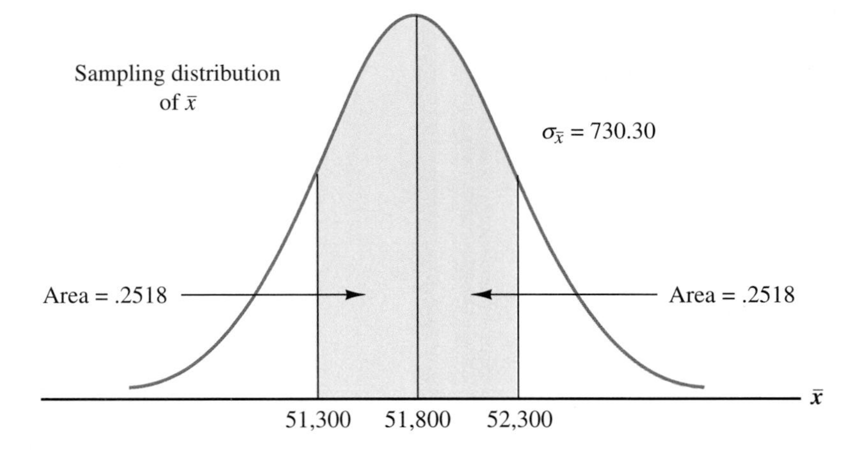

Relationship Between the Sample Size and the Sampling Distribution of $\bar{x}$

Suppose that in the EAI sampling problem we select a simple random sample of 100 EAI managers instead of the 30 originally considered. Intuitively, it would seem that with more data provided by the larger sample size, the sample mean based on $n = 100$ should provide a better estimate of the population mean than the sample mean based on $n = 30$. To see how much better, let us consider the relationship between the sample size and the sampling distribution of $\bar{x}$.

First note that $E(\bar{x}) = \mu$ regardless of the sample size. Thus, the mean of all possible values of $\bar{x}$ is equal to the population mean μ regardless of the sample size n. However, note that the standard error of the mean, $\sigma_{\bar{x}} = \sigma/\sqrt{n}$, is related to the square root of the sample size. Specifically, whenever the sample size is increased, the standard error of the mean, $\sigma_{\bar{x}}$, is decreased. With $n = 30$, the standard error of the mean for the EAI problem is 730.30. However, increasing the sample size to $n = 100$ decreases the standard error of the mean to

$$\sigma_{\bar{x}} = \frac{\sigma}{\sqrt{n}} = \frac{4000}{\sqrt{100}} = 400$$

The sampling distributions of $\bar{x}$ with $n = 30$ and $n = 100$ are shown in Figure 7.7. Since the sampling distribution with $n = 100$ has a smaller standard error, the values of $\bar{x}$ have less variation and tend to be closer to the population mean than the values of $\bar{x}$ with $n = 30$.

We can use the sampling distribution of $\bar{x}$ for the case with $n = 100$ to compute the probability that a simple random sample of 100 EAI managers will provide a sample mean that is within $500 of the population mean. Since the sampling distribution is a normal probability distribution with mean 51,800 and standard deviation 400, we can use the standard normal probability distribution table to find the area or probability. At $\bar{x} = 51{,}300$ (Figure 7.8), we have

$$z = \frac{51{,}300 - 51{,}800}{400} = -1.25$$

Referring to the standard normal probability distribution table, we find an area between $z = 0$ and $z = -1.25$ of .3944. With a similar calculation for $\bar{x} = 52{,}300$, we see that the probability of the value of the sample mean being between 51,300 and 52,300 is $.3944 + .3944 = .7888$. Thus, by increasing the sample size from 30 to 100 EAI

FIGURE 7.7
A Comparison of the Sampling Distributions of $\bar{x}$ for Simple Random Samples of $n = 30$ and $n = 100$ EAI Managers

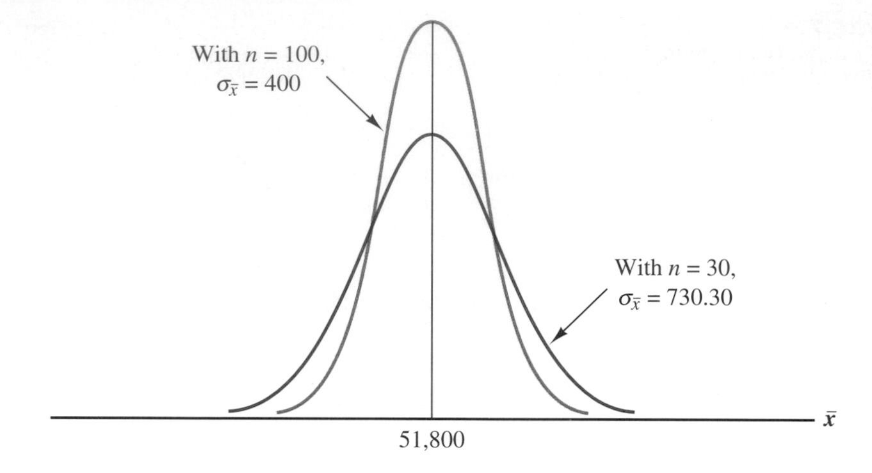

FIGURE 7.8
The Probability of a Sample Mean Being Within $500 of the Population Mean When a Simple Random Sample of 100 EAI Managers Is Used

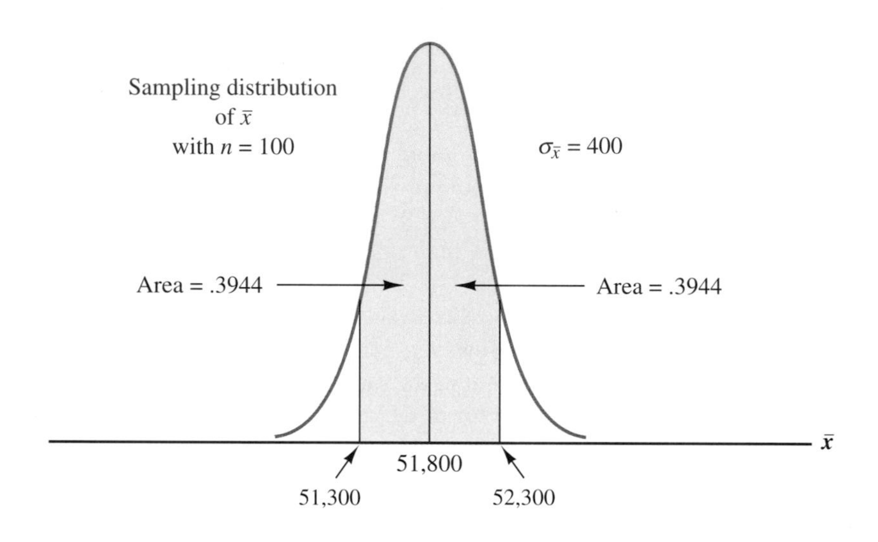

managers, we have increased the probability of obtaining a sample mean within $500 of the population mean from .5036 to .7888.

Perhaps an even larger sample size should be considered. However, the important point in this discussion is that as the sample size is increased, the standard error of the mean is decreased. As a result, the sampling distribution of $\bar{x}$ will have less variation. In effect, the larger sample size will provide a higher probability that the value of the sample mean will be within a specified distance of the population mean.

notes and comments

In presenting the sampling distribution of $\bar{x}$ for the EAI problem, we took advantage of the fact that the population mean $\mu = 51{,}800$ and the population standard deviation $\sigma = 4000$ were provided. However, in general, the values of the population mean μ and the population standard deviation σ will be unknown. In Chapter 8 we will show how the sample mean $\bar{x}$ and the sample standard deviation s are used when μ and σ are unknown.

exercises

Methods

18. A population has a mean of 200 and a standard deviation of 50. A simple random sample of size 100 will be taken and the sample mean $\bar{x}$ will be used to estimate the population mean.

a. What is the expected value of $\bar{x}$?

b. What is the standard deviation of $\bar{x}$?

c. Show the sampling distribution of $\bar{x}$.

d. What does the sampling distribution of $\bar{x}$ show?

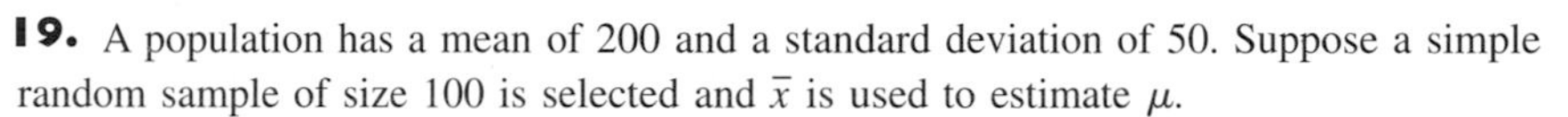

Self Test

19. A population has a mean of 200 and a standard deviation of 50. Suppose a simple random sample of size 100 is selected and $\bar{x}$ is used to estimate μ.

a. What is the probability that the sample mean will be within $\pm$ 5 of the population mean?

b. What is the probability that the sample mean will be within $\pm$ 10 of the population mean?

20. Assume the population standard deviation is $\sigma = 25$. Compute the standard error of the mean, $\sigma_{\bar{x}}$, for sample sizes of 50, 100, 150, and 200. What can you say about the size of the standard error of the mean as the sample size is increased?

21. Suppose a simple random sample of size 50 is selected from a population with $\sigma = 10$. Find the value of the standard error of the mean in each of the following cases (use the finite population correction factor if appropriate).

a. The population size is infinite.

b. The population size is $N = 50{,}000$.

c. The population size is $N = 5000$.

d. The population size is $N = 500$.

22. A population has a mean of 400 and a standard deviation of 50. The probability distribution of the population is unknown.

a. A researcher will use simple random samples of either 10, 20, 30, or 40 items to collect data about the population. With which of these sample-size alternatives will we be able to use a normal probability distribution to describe the sampling distribution of $\bar{x}$? Explain.

b. Show the sampling distribution of $\bar{x}$ for the instances in which the normal probability distribution is appropriate.

23. A population has a mean of 100 and a standard deviation of 16. What is the probability that a sample mean will be within $\pm$ 2 of the population mean for each of the following sample sizes?

a. $n = 50$ **b.** $n = 100$ **c.** $n = 200$ **d.** $n = 400$

e. What is the advantage of a larger sample size?

Applications

24. Refer to the EAI sampling problem. Suppose the simple random sample had contained 60 managers.

a. Sketch the sampling distribution of $\bar{x}$ when simple random samples of size 60 are used.

b. What happens to the sampling distribution of $\bar{x}$ if simple random samples of size 120 are used?

c. What general statement can you make about what happens to the sampling distribution of $\bar{x}$ as the sample size is increased? Does this seem logical? Explain.

Self Test

25. In the EAI sampling problem, we showed that for $n = 30$, there was .5036 probability of obtaining a sample mean within $\pm$ \$500 of the population mean.

a. What is the probability that $\bar{x}$ is within $500 of the population mean if a sample of size 60 is used?

b. Answer part (a) for a sample of size 120.

26. A 1993 survey conducted by the American Automobile Association showed that a family of four spends an average of $215.60 per day when on vacation. Assume that $215.60 is the population mean expenditure per day for a family of four and that $85.00 is the population standard deviation. Assume that a random sample of 40 families will be selected for further study.

a. Show the sampling distribution of the sample mean $\bar{x}$, where $\bar{x}$ is the mean expenditure per day for a family of four; assume a random sample of 40 families.

b. What is the probability that the simple random sample of 40 families will provide a sample mean that is within $20 of the population mean?

c. What is the probability that the simple random sample of 40 families will provide a sample mean that is within $10 of the population mean?

27. Annual surveys of starting salaries for college graduates are conducted by the College Placement Council. The mean annual starting salary for accounting majors is $26,542 (*USA Today,* September 9, 1991). Assume that the population of graduates with accounting majors has a mean annual starting salary of $26,542 and a standard deviation of $2000.

a. What is the probability that a simple random sample of accounting majors will have a sample mean within ± $250 of the population mean for each of the following sample sizes: 30, 50, 100, 200, and 400?

b. What is the advantage of a larger sample size

28. In 1993, women took an average of 8.5 weeks of unpaid leave from their jobs after the birth of a child (*U.S. News & World Report,* December 27, 1993). Assume that 8.5 weeks is the population mean and 2.2 weeks is the population standard deviation.

a. What is the probability that a simple random sample of 50 women provides a sample mean leave after the birth of a child of between 7.5 and 9.5 weeks?

b. What is the probability that a simple random sample of 50 women provides a sample mean leave after the birth of a child of between eight and nine weeks?

29. The population mean price for a new automobile is $16,012 (*U.S. News & World Report,* September 9, 1991). Assume that the population standard deviation is $4200 and that a sample of 100 new automobile purchases will be selected.

a. Show the sampling distribution of the sample mean price for new automobiles based on the sample of 100.

b. What is the probability that the sample mean for the 100 purchases will be within $1000 of the population mean?

c. Repeat part (b) for values of $500, $250, and $100.

d. To estimate the population mean price to within ± $250 or ± $100, what would you recommend?

30. In 1992, the National Fisheries Institute reported that tuna was the favorite seafood consumed in the United States, with a population mean annual consumption of 3.6 pounds per person. Assume that the population standard deviation is 1.5 pounds.

a. Show the sampling distribution of the sample mean $\bar{x}$, where $\bar{x}$ is the mean number of pounds of tuna consumed per year for a sample of 100 individuals.

b. What is the probability that the sample mean consumption is four or more pounds?

c. What is the probability that the sample mean consumption is between 3.2 and four pounds?

31. To estimate the mean age for a population of 4000 employees, a simple random sample of 40 employees is selected.

a. Would you use the finite population correction factor in calculating the standard error of the mean? Explain.

b. If the population standard deviation is $\sigma = 8.2$ years, compute the standard error both with and without using the finite population correction factor. What is the rationale for ignoring the finite population correction factor whenever $n / N \leq .05$?

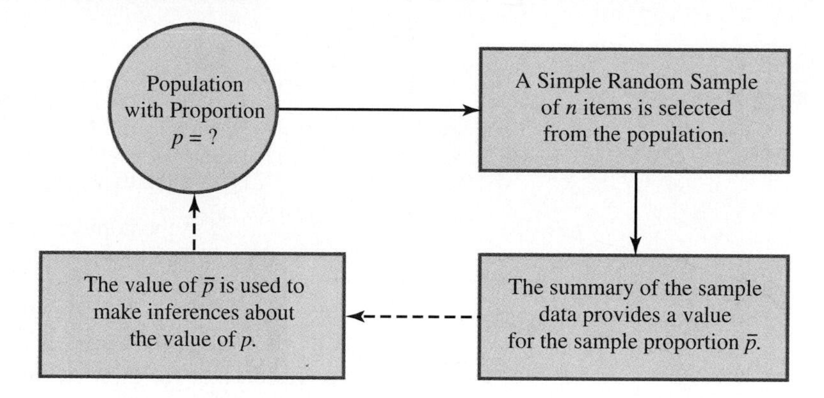

FIGURE 7.9
The Statistical Process of Using a Sample Proportion to Make Inferences About a Population Proportion

c. What is the probability that the sample mean age of the employees will be within ± 2 years of the population mean age?

32. A library checks out an average of $\mu = 320$ books per day, with a standard deviation of $\sigma = 75$ books. Consider a sample of 30 days of operation, with $\bar{x}$ being the sample mean number of books checked out per day.

a. Show the sampling distribution of $\bar{x}$.
b. What is the standard deviation of $\bar{x}$?
c. What is the probability that the sample mean for the 30 days will be between 300 and 340 books?
d. What is the probability that the sample mean will show 325 or more books checked out?

7.6 Sampling Distribution of $\bar{p}$

In many situations in business and economics, we use the sample proportion $\bar{p}$ to make statistical inferences about the population proportion p. This process is depicted in Figure 7.9. On each repitition of the process, we can anticipate obtaining a different value for the sample proportion $\bar{p}$. The probability distribution for all possible values of the sample proportion $\bar{p}$ is called the sampling distribution of the sample proportion $\bar{p}$.

Sampling Distribution of $\bar{p}$

The sampling distribution of $\bar{p}$ is the probability distribution of all possible values of the sample proportion $\bar{p}$.

To determine how close the sample proportion $\bar{p}$ is to the population proportion p, we need to understand the properties of the sampling distribution of $\bar{p}$: the expected value of $\bar{p}$, the standard deviation of $\bar{p}$, and the shape of the sampling distribution of $\bar{p}$.

Expected Value of $\bar{p}$

The expected value of $\bar{p}$ (i.e., the mean of all possible values of $\bar{p}$) can be expressed as follows.

Expected Value of $\bar{p}$

$$E(\bar{p}) = p \tag{7.4}$$

where

$E(\bar{p})$ = the expected value of the random variable $\bar{p}$

p = the population proportion

Equation (7.4) shows that the mean of all possible $\bar{p}$ values is equal to the population proportion p. Recall that for the EAI population, the proportion of managers who had participated in the company's management training program is $p = .60$. Thus, the expected value of $\bar{p}$ for the EAI sampling problem is .60.

Standard Deviation of $\bar{p}$

Different simple random samples generate a variety of values for $\bar{p}$. We now are interested in determining the standard deviation of $\bar{p}$, which is referred to as the *standard error of the proportion.* Just as we found for the sample mean $\bar{x}$, the standard deviation of $\bar{p}$ depends on whether the population is finite or infinite. The two expressions for the standard deviation of $\bar{p}$ follow.

Standard Deviation of $\bar{p}$

Finite Population *Infinite Population*

$$\sigma_{\bar{p}} = \sqrt{\frac{N-n}{N-1}}\sqrt{\frac{p(1-p)}{n}} \qquad \sigma_{\bar{p}} = \sqrt{\frac{p(1-p)}{n}} \tag{7.5}$$

Comparing the two expressions in (7.5), we see that the only difference is the use of the finite population correction factor $\sqrt{(N-n)/(N-1)}$.

As was the case with the sample mean $\bar{x}$, we find that the difference between the expressions for the finite population and the infinite population becomes negligible if the size of the finite population is large in comparison to the sample size. We follow the same rule of thumb that we recommended for the sample mean. That is, if the population is finite with $n/N \leq .05$, we will use $\sigma_{\bar{p}} = \sqrt{p(1-p)/n}$. However, if the population is finite and $n/N > .05$, the finite population correction factor should be used, as shown in (7.5). Again, unless specifically noted, throughout the text we will assume that the population size is large in relation to the sample size and that the finite population correction factor is unnecessary.

For the EAI study we know that the population proportion of managers who have participated in the management training program is $p = .60$. With $n/N = 30/2500 = .012$, we can ignore the finite population correction factor when we compute the standard deviation of $\bar{p}$. For the simple random sample of 30 managers, $\sigma_{\bar{p}}$ is

$$\sigma_{\bar{p}} = \sqrt{\frac{p(1-p)}{n}} = \sqrt{\frac{.60(1-.60)}{30}} = \sqrt{.008} = .0894$$

FIGURE 7.10
Sampling Distribution of $\bar{p}$ for the Proportion of EAI Managers Who Have Participated in the Management Training Program

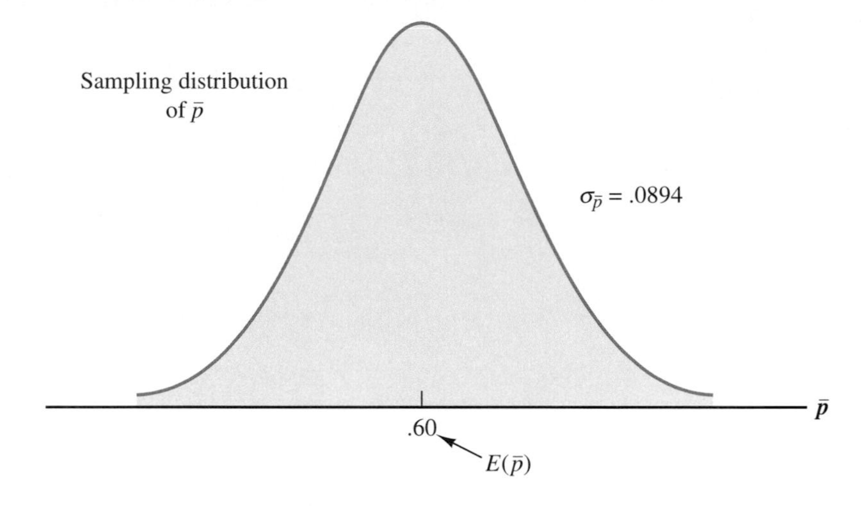

Form of the Sampling Distribution of $\bar{p}$

Now that we know the mean and standard deviation of $\bar{p}$, we want to consider the form of the sampling distribution of $\bar{p}$. Applying the central limit theorem as it relates to $\bar{p}$ produces the following result.

> The sampling distribution of $\bar{p}$ can be approximated by a normal probability distribution whenever the sample size is large.

With $\bar{p}$, the sample size can be considered large whenever the following two conditions are satisfied.

$$np \geq 5$$
$$n(1 - p) \geq 5$$

Recall that for the EAI sampling problem we know that the population proportion of managers who have participated in the training program is $p = .60$. With a simple random sample of size 30, we have $np = 30(.60) = 18$ and $n(1 - p) = 30(.40) = 12$. Thus, the sampling distribution of $\bar{p}$ can be approximated by a normal probability distribution as shown in Figure 7.10.

Practical Value of the Sampling Distribution of $\bar{p}$

Whenever a simple random sample is selected and the value of the sample proportion $\bar{p}$ is used to estimate the value of the population proportion p, we anticipate some sampling error. In this case, the sampling error is the absolute value of the difference between the value of the sample proportion $\bar{p}$ and the value of the population proportion p. The practical value of the sampling distribution of $\bar{p}$ is that it can be used to provide probability information about the sampling error.

FIGURE 7.11
Sampling Distribution of $\bar{p}$ for the EAI Sampling Problem

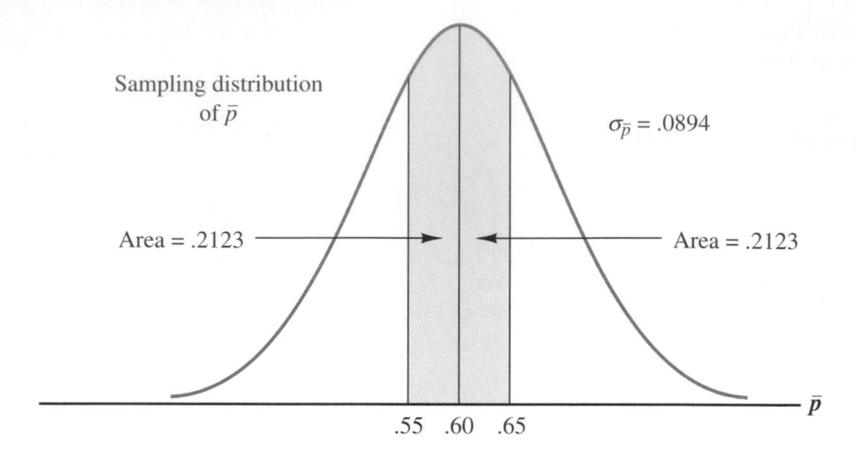

Suppose, in the EAI problem, the personnel director wants to know the probability of obtaining a value of $\bar{p}$ that is within .05 of the population proportion of EAI managers who have participated in the training program. That is, what is the probability of obtaining a sample with a sample proportion $\bar{p}$ between .55 and .65? The area in Figure 7.11 shows this probability. Using the fact that the sampling distribution of $\bar{p}$ can be approximated by a normal probability distribution with mean .60 and standard deviation $\sigma_{\bar{p}} = .0894$, we find that the standard normal random variable corresponding to $\bar{p} = .55$ has a value of $z = (.55 - .60)/.0894 = -.56$. Referring to the standard normal probability distribution table, we see that the area between $z = -.56$ and $z = 0$ is .2123. Similarly, at $\bar{p} = .65$ we find an area between $z = 0$ and $z = .56$ of .2123. Thus, the probability of selecting a sample that provides a sample proportion $\bar{p}$ within .05 of the population proportion p is $.2123 + .2123 = .4246$.

If we consider increasing the sample size to $n = 100$, the standard error of the proportion becomes

$$\sigma_{\bar{p}} = \sqrt{\frac{.60(1 - .60)}{100}} = \sqrt{.0024} = .0490$$

With a sample size of 100 EAI managers, the probability of the sample proportion having a value within .05 of the population proportion can now be computed. Since the sampling distribution can be approximated by a normal probability distribution with mean .60 and standard deviation .0490, we can use the standard normal probability distribution table to find the area or probability. At $\bar{p} = .55$, we have $z = (.55 - .60)/.0490 = -1.02$. Referring to the standard normal probability distribution table, we see that the area between $z = -1.02$ and $z = 0$ is .3461. Similarly, at .65 the area between $z = 0$ and $z = 1.02$ is .3461. Thus, if the sample size is increased from 30 to 100, the probability that the sample proportion $\bar{p}$ is within .05 of the population proportion p will increase from .4246 to $.3461 + .3461 = .6922$.

exercises

Methods

33. A simple random sample of size 100 is selected from a population with $p = .40$.

a. What is the expected value of $\bar{p}$?

b. What is the standard deviation of $\bar{p}$?

c. Show the sampling distribution of $\bar{p}$.

d. What does the sampling distribution of $\bar{p}$ show?

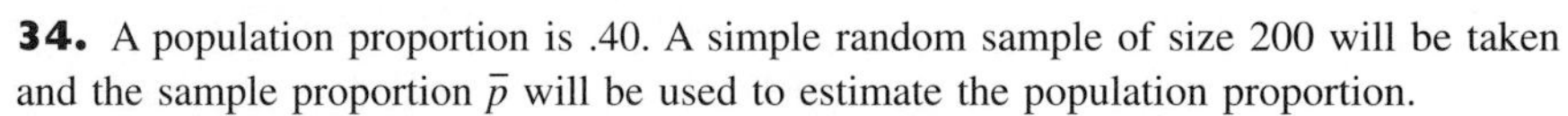

Self Test

34. A population proportion is .40. A simple random sample of size 200 will be taken and the sample proportion $\bar{p}$ will be used to estimate the population proportion.

a. What is the probability that the sample proportion will be within $\pm$ .03 of the population proportion?

b. What is the probability that the sample proportion will be within $\pm$ .05 of the population proportion?

35. Assume that the population proportion is .55. Compute the standard error of the proportion, $\sigma_{\bar{p}}$, for sample sizes of 100, 200, 500, and 1000. What can you say about the size of the standard error of the proportion as the sample size is increased?

36. The population proportion is .30. What is the probability that a sample proportion will be within $\pm$ .04 of the population proportion for each of the following sample sizes?

a. $n = 100$ **b.** $n = 200$ **c.** $n = 500$ **d.** $n = 1000$

e. What is the advantage of a larger sample size?

Applications

Self Test

37. The president of Doerman Distributors, Inc., believes that 30% of the firm's orders come from new or first-time customers. A simple random sample of 100 orders will be used to estimate the proportion of new or first-time customers. The results of the sample will be used to verify the president's claim of $p = .30$.

a. Assume that the president is correct and $p = .30$. What is the sampling distribution of $\bar{p}$ for this study?

b. What is the probability that the sample proportion $\bar{p}$ will be between .20 and .40?

c. What is the probability that the sample proportion will be within $\pm$ .05 of the population proportion $p = .30$?

38. The Grocery Manufacturers of America reported that 76% of consumers read the ingredients listed on a product's label (*America by the Numbers,* 1993). Assume the population proportion is $p = .76$ and a sample of 400 consumers is selected from the population.

a. Show the sampling distribution of the sample proportion $\bar{p}$ where $\bar{p}$ is the proportion of the sampled consumers who read the ingredients listed on a product's label.

b. What is the probability that the sample proportion will be within $\pm$ .03 of the population proportion?

c. Answer part (b) for a sample of 750 consumers.

39. Louis Harris & Associates, Inc., conducted a survey of 1253 adults to learn how individuals feel about the United States' position in the global economy (*Business Week,* April 6, 1992). One question asked how concerned the individual was about U.S. industry becoming less competitive in the global economy. Assume that for the entire population, 55% of the adults are very concerned about U.S. industry becoming less competitive. Let $\bar{p}$ be the sample proportion of the polled adults who are very concerned about this issue.

a. Show the sampling distribution of $\bar{p}$ if the population proportion is $p = .55$.

b. What is the probability that the Harris poll sample proportion will have a sampling error of $\pm$ .02 or less?

c. What is the probability that the Harris poll sample proportion will have a sampling error of $\pm$ .03 or less?

d. Comment on why the Harris poll stated, "Results should be accurate to within 3 percentage points."

40. The American Association of Individual Investors reported results from a survey on a range of investment-related topics (*AAII Journal,* April 1992). One question was: Do you favor relaxed SEC rules for financial reporting by foreign corporations to allow more foreign stocks to be traded in the United States? Assume that the population proportion is .42 and that the sample size is 300.

a. What is the probability that the sample proportion will be within $\pm$ .03 of the population proportion?

b. What is the probability that the sample proportion will be .45 or greater?

c. What is the probability that the sample proportion will show 50% or more favoring relaxed SEC rules for foreign corporations?

41. The Institute for Women's Policy Research reported that women now constitute 37% of all union members, an all-time high percentage (*The Wall Street Journal,* July 26, 1994). Suppose the population proportion of women who are union members is $p = .37$ and a simple random sample of 1000 union members is selected.

a. Show the sampling distribution of $\bar{p}$, the proportion of women in the sample.

b. What is the probability that the sample proportion will be within $\pm$ .03 of the population proportion?

c. Answer part (b) for a simple random sample of 500.

42. Assume that 15% of the items produced in an assembly line operation are defective, but that the firm's production manager is not aware of this situation. Assume further that 50 parts are tested by the quality assurance department to determine the quality of the assembly operation. Let $\bar{p}$ be the sample proportion defective found by the quality assurance test.

a. Show the sampling distribution for $\bar{p}$.

b. What is the probability that the sample proportion will be within $\pm$ .03 of the population proportion defective?

c. If the test shows $\bar{p} = .10$ or more, the assembly line operation will be shut down to check for the cause of the defects. What is the probability that the sample of 50 parts will lead to the conclusion that the assembly line should be shut down?

43. The Food Marketing Institute shows that 17% of households spend more than \$100 per week on groceries (*USA Today,* June 21, 1994). Assume the population proportion is $p = .17$ and a simple random sample of 800 households will be selected from the population.

a. Show the sampling distribution of $\bar{p}$, the sample proportion of households spending more than \$100 per week on groceries.

b. What is the probability that the sample proportion will be within $\pm$ .02 of the population proportion?

c. Answer part (b) for a sample of 1600 households.

7.7 Other Sampling Methods

We have described the simple random sampling procedure and discussed the properties of the sampling distributions of $\bar{x}$ and $\bar{p}$ when simple random sampling is used. However, simple random sampling is not the only sampling method available. Such methods as stratified random sampling, cluster sampling, and systematic sampling are alternatives that in some situations have advantages over simple random sampling. In this section we briefly describe some of these alternative sampling methods.

Stratified Random Sampling

In *stratified random sampling,* the population is first divided into groups of elements called *strata,* such that each item in the population belongs to one and only one stratum. The basis for forming the strata, such as department, location, age, industry type, and so on, is at the discretion of the designer of the sample. However, best results are obtained when the elements within each stratum are as much alike as possible. Figure 7.12 is a diagram of a population divided into H strata.

After the strata are formed, a simple random sample is taken from each stratum. Formulas are available for combining the results for the individual stratum samples into one estimate of the population parameter of interest. The value of stratified

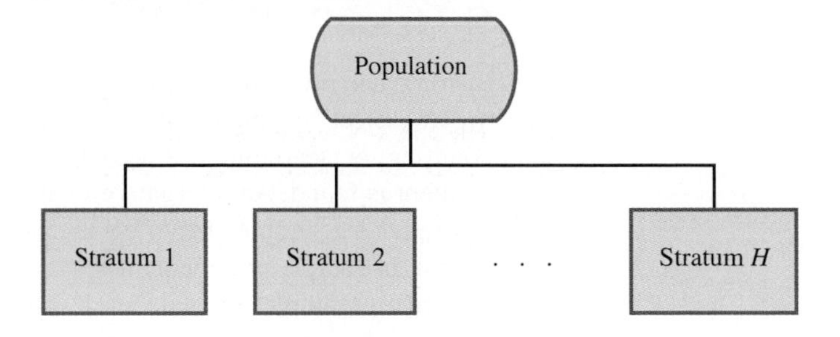

FIGURE 7.12
Diagram for Stratified Simple Random Sampling

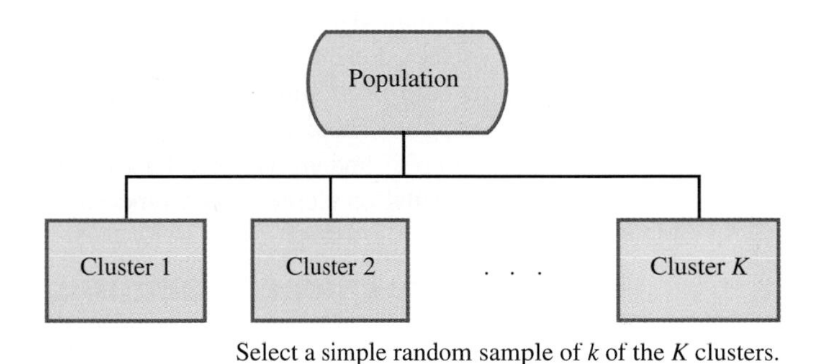

FIGURE 7.13
Diagram for Cluster Sampling

random sampling depends on how homogeneous the elements are within the strata. If units within strata are alike (homogeneity), the strata will have low variances. Thus relatively small sample sizes can be used to obtain good estimates of the strata characteristics. If strata are homogeneous, the stratified random sampling procedure will provide results just as precise as those of simple random sampling but with a smaller total sample size.

Cluster Sampling

In *cluster sampling,* the population is first divided into separate groups of elements called *clusters.* Each element of the population belongs to one and only one cluster (see Figure 7.13). A simple random sample of the clusters is then taken. All elements within each sampled cluster form the sample. Cluster sampling tends to provide best results when the elements within the clusters are heterogeneous (not alike). In the ideal case, each cluster is a representative small-scale version of the entire population. The value of cluster sampling depends on how representative each cluster is of the entire population. If all clusters are alike in this regard, sampling a small number of clusters will provide good estimates of the population parameters.

One of the primary applications of cluster sampling is area sampling, where clusters are city blocks or other well-defined areas. Cluster sampling generally requires a larger total sample size than either simple random sampling or stratified random sampling. However, it can result in cost savings because of the fact that when an interviewer is sent to a sampled cluster (say a city-block location), many sample observations can be obtained in a relatively short time. Hence, a larger sample size may be obtainable with a significantly lower cost per element and thus possibly a lower total cost.

Systematic Sampling

In some sampling situations, especially those with large populations, it is timeconsuming to select a simple random sample by first finding a random number and then counting or searching through the list of population items until the corresponding element is found. An alternative to simple random sampling is *systematic sampling*. For example, if a sample size of 50 is desired from a population containing 5000 elements, we might sample one element for every 5000/50 = 100 elements in the population. A systematic sample for this case involves selecting randomly one of the first 100 elements from the population list. Other sample elements are identified by starting with the first sampled element and then selecting every 100th element that follows in the population list. In effect, the sample of 50 is identified by moving systematically through the population and identifying every 100th element after the first randomly selected element. The sample of 50 usually will be easier to identify in this way than it would be if simple random sampling were used. Since the first element selected is a random choice, a systematic sample is usually assumed to have the properties of a simple random sample. This assumption is especially applicable when the list of the population elements is a random ordering of the elements in the population.

Convenience Sampling

The sampling methods discussed thus far are referred to as *probability sampling* techniques. Elements selected from the population have a known probability of being included in the sample. The advantage of probability sampling is that the sampling distribution of the appropriate sample statistic generally can be identified. Formulas such as the ones for simple random sampling presented in this chapter can be used to determine the properties of the sampling distribution. Then the sampling distribution can be used to make probability statements about possible sampling errors associated with the sample results.

Convenience sampling is a *nonprobability sampling* technique. As the name implies, the sample is identified primarily by convenience. Items are included in the sample without prespecified or known probabilities of being selected. For example, a professor conducting research at a university may use student volunteers to constitute a sample simply because they are readily available and will participate as subjects for little or no cost. Similarly, an inspector may sample a shipment of oranges by selecting oranges haphazardly from among several crates. Labeling each orange and using a probability method of sampling would be impractical. Samples such as wildlife captures and volunteer panels for consumer research are also convenience samples.

Convenience samples have the advantage of relatively easy sample selection and data collection; however, it is impossible to evaluate the "goodness" of the sample in terms of its representativeness of the population. A convenience sample may provide good results or it may not. There is no statistically justified procedure that will allow a probability analysis and inference about the quality of the sample results. Sometimes researchers apply statistical methods designed for probability samples to a convenience sample, arguing that the convenience sample can be treated as though it were a random sample. However, this argument cannot be supported, and we should be very cautious in interpreting the results of convenience samples that are used to make inferences about populations.

Judgment Sampling

One additional nonprobability sampling technique is *judgment sampling*. In this approach, the person most knowledgeable on the subject of the study selects individuals or other elements of the population that he or she feels are most representative of the population. Often this is a relatively easy way of selecting a sample. For example, a reporter may sample two or three senators, judging that those senators reflect the general opinion of all senators. However, the quality of the sample results depends on the judgment of the person selecting the sample. Again, great caution is warranted in drawing conclusions based on judgment samples used to make inferences about populations.

notes and comments

We recommend sampling by one of the probability sampling methods: simple random sampling, stratified simple random sampling, cluster sampling, or systematic sampling. For these methods, formulas are available for evaluating the "goodness" of the sample results in terms of the closeness of the results to the population characteristics being estimated. An evaluation of the goodness cannot be made with convenience or judgment sampling. Thus, great care should be used in interpreting the results when nonprobability sampling methods have been used to obtain statistical information.

summary

In this chapter we presented the concepts of simple random sampling and sampling distributions. We demonstrated how a simple random sample can be selected and how the data collected for the sample can be used to develop point estimates of population parameters. Since different simple random samples provided a variety of different values for the point estimators, point estimators such as $\bar{x}$ and $\bar{p}$ are random variables. The probability distribution of such a random variable is called a sampling distribution. In particular, we described the sampling distributions of the sample mean $\bar{x}$ and the sample proportion $\bar{p}$.

In considering the characteristics of the sampling distributions of $\bar{x}$ and $\bar{p}$, we stated that $E(\bar{x}) = \mu$ and $E(\bar{p}) = p$. After developing the standard deviation or standard error formulas for these estimators, we showed how the central limit theorem provided the basis for using a normal probability distribution to approximate these sampling distributions in the large-sample case. Rules of thumb were given for determining when large-sample-size conditions were satisfied. Other sampling methods such as stratified random sampling, cluster sampling, systematic sampling, convenience sampling, and judgment sampling were discussed briefly.

glossary

Parameter A numerical characteristic of a population, such as a population mean μ, a population standard deviation σ, a population proportion p, and so on.

Simple random sampling Finite population: a sample selected such that each possible sample of size n has the same probability of being selected. Infinite population:

a sample selected such that each element comes from the same population and the successive elements are selected independently.

Sampling without replacement Once an element from the population has been included in the sample, it is removed from the population and cannot be selected a second time.

Sampling with replacement As each element is selected for the sample, it is returned to the population. A previously selected element can be selected again and therefore may appear in the sample more than once.

Sample statistic A sample characteristic, such as a sample mean $\bar{x}$ or a sample proportion $\bar{p}$. The value of the sample statistic is used to estimate the value of the population parameter.

Sampling distribution A probability distribution consisting of all possible values of a sample statistic.

Point estimate A single numerical value used as an estimate of a population parameter.

Point estimator The sample statistic, such as $\bar{x}$ and $\bar{p}$, that provides the point estimate of the population parameter.

Finite population correction factor The term $\sqrt{(N-n)/(N-1)}$ that is used in the formulas for $\sigma_{\bar{x}}$ and $\sigma_{\bar{p}}$ whenever a finite population, rather than an infinite population, is being sampled. The generally accepted rule of thumb is to ignore the finite population correction factor whenever $n/N \leq .05$.

Standard error The standard deviation of a point estimator.

Central limit theorem A theorem that enables one to use the normal probability distribution to approximate the sampling distribution of $\bar{x}$ and $\bar{p}$ whenever the sample size is large.

Stratified simple random sampling A probabilistic method of selecting a sample in which the population is first divided into strata and a simple random sample is then taken from each stratum.

Cluster sampling A probabilistic method of selecting a sample in which the population is first divided into clusters and then one or more clusters is selected for sampling.

Systematic sampling A probabilistic method of selecting a sample by randomly selecting one of the first k elements and then selecting every kth element thereafter.

Convenience sampling A nonprobabilistic method of sampling whereby elements are selected for the sample on the basis of convenience.

Judgment sampling A nonprobabilistic method of sampling whereby elements are selected for the sample based on the judgment of the person doing the study.

key formulas

Expected Value of $\bar{x}$

$$E(\bar{x}) = \mu \tag{7.1}$$

Standard Deviation of $\bar{x}$

Finite Population *Infinite Population*

$$\sigma_{\bar{x}} = \sqrt{\frac{N-n}{N-1}}\left(\frac{\sigma}{\sqrt{n}}\right) \qquad \sigma_{\bar{x}} = \frac{\sigma}{\sqrt{n}} \tag{7.2}$$

Expected Value of $\bar{p}$

$$E(\bar{p}) = p \tag{7.4}$$

Standard Deviation of $\bar{p}$

Finite Population *Infinite Population*

$$\sigma_{\bar{p}} = \sqrt{\frac{N-n}{N-1}}\sqrt{\frac{p(1-p)}{n}} \qquad \sigma_{\bar{p}} = \sqrt{\frac{p(1-p)}{n}} \tag{7.5}$$

supplementary exercises

44. Nationwide Supermarkets has 4800 retail stores in 32 states. At the end of each year, a sample of 35 stores is selected for physical inventories. Results from the inventory samples are used in annual tax reports. Assume that the retail stores are listed sequentially on a computer printout. Begin at the bottom of the second column of random numbers in Table 7.1. Ignoring the first digit in each group and using four-digit random numbers beginning with 8112, read *up* the column to identify the first five stores to be included in the simple random sample.

45. A 1993 study conducted by *The Orlando Sentinel* provided data on how much time customers wait for service at Burger King, McDonald's, and Wendy's restaurants. Assume that the population mean waiting time for a drive-through order is four minutes and that the population standard deviation is 1.5 minutes.

a. Suppose a random sample of 60 drive-through customers will be selected and the sample mean waiting time computed. Show the sampling distribution of the sample mean.

b. What is the probability that the sample mean waiting time will be within $\pm$ .25 minutes of the population mean waiting time?

46. The U.S. Department of Transportation reported the number of car miles traveled per day for residents of the 75 largest metropolitan areas in the United States (*1994 Information Please Environmental Almanac*). Tulsa, Oklahoma, provided the largest value, with residents driving 30 miles per day. Assume that this is the population mean and that the population standard deviation is 12 miles per day. A random sample of 50 Tulsa residents is selected.

a. Show the sampling distribution of $\bar{x}$ where $\bar{x}$ is the sample mean number of miles Tulsa residents travel per day in a car.

b. What is the probability that the sample mean is within $\pm$ 2 miles of the population mean?

47. The population mean household income in Pittsburgh, Pennsylvania, is $49,000 (*U.S. News & World Report,* April 6, 1992). Assume that the population standard deviation is $12,000. Furthermore, assume that a simple random sample of households in the Pittsburgh area will be selected and the sample mean will be used to estimate the population mean.

a. Show the sampling distribution of the sample mean if a sample of 100 households is used.

b. What is the probability that a random sample of 100 households will provide a sampling error of $1000 or less?

c. What is the probability that a random sample of 200 households will provide a sampling error of $1000 or less?

d. What is the probability that a random sample of 400 households will provide a sampling error of $1000 or less?

e. How large a simple random sample would be required if we wanted a .95 probability of having a sampling error of $1000 or less?

48. The speed of automobiles on a section of I-75 in northern Florida has a mean of $\mu = 67$ miles per hour with a standard deviation of $\sigma = 6$ miles per hour. Assume the population has a *normal distribution* and a random sample of 16 automobiles will be selected to compute a sample mean automobile speed.

a. What is the expected value of $\bar{x}$?

b. What is the standard deviation of $\bar{x}$?

c. Show the sampling distribution of $\bar{x}$.

d. What is the probability that the sample mean will be 65 miles per hour or more?

e. What is the probability that the sample mean will be between 66 and 68 miles per hour?

49. According to *USA Today* (April 11, 1995), the mean number of days per year that business travelers are on the road for business is 115. The standard deviation is 60 days per year. Assume that these results apply to the population of business travelers and that a random sample of 50 business travelers will be selected from the population.

a. What is the value of the standard error of the mean?

b. What is the probability that the sample mean will be more than 115 days per year?

c. What is the probability that the sample mean will be within $\pm$ 5 days of the population mean?

d. How would the probability change in part (c) if the sample size were increased to 100?

50. In the EAI study the population of managers had annual salaries with $\mu = \$51,800$ and $\sigma = \$4000$. Samples of size 30 provided a .5036 probability of obtaining a value of $\bar{x}$ within $\pm$ $500 of the population mean. How large a sample should be selected if the personnel director wanted a .95 probability of a sample mean $\bar{x}$ being within $\pm$ $500 of μ?

51. Three firms have inventories that differ in size. Firm A has a population of 2000 items, firm B has a population of 5000 items, and firm C has a population of 10,000 items. For each firm, the population standard deviation for the cost of the items is $\sigma = 144$. A statistical consultant recommends that each firm take a simple random sample of 50 items from its population to estimate the population mean cost per item. Managers of the small firm state that since it has the smallest population, it should be able to obtain the data from a much smaller sample than that required by the larger firms. However, the consultant states that to obtain the same standard error and thus the same precision in the sample results, all firms should use the same sample size regardless of population size.

a. Using the finite population correction factor, compute the standard error for each of the three firms given a sample of size 50.

b. What is the probability that for each firm the sample mean $\bar{x}$ will be within $\pm$ 25 of the population mean μ?

52. The grade point average for all juniors at Strausser College has a standard deviation of .50.

a. A random sample of 20 students is to be used to estimate the population mean grade point average. What assumption is necessary to compute the probability of obtaining a sample mean within $\pm$ .2 of the population mean?

b. Provided that the assumption in (a) can be made, what is the probability of $\bar{x}$ being within $\pm$ .2 of the population mean?

c. If the assumption in (a) cannot be made, what would you recommend doing?

53. Assume that the proportion of persons having a college degree is $p = .35$.

a. Explain how the sampling distribution of $\bar{p}$ results from random samples of size 80 being used to estimate the proportion of individuals having a college degree.

b. Show the sampling distribution for $\bar{p}$ in this case.

c. If the sample size is increased to 200, what happens to the sampling distribution of $\bar{p}$? Compare the standard error for the $n = 80$ and $n = 200$ alternatives.

54. What is the most important factor for business travelers when they are staying in a hotel? According to *USA Today,* 74% of business travelers state that having a smoke-free room is the most important factor (*USA Today,* April 11, 1995). Assume that the population proportion is $p = .74$ and that a sample of 200 business travelers will be selected.

a. Show the sampling distribution of $\bar{p}$, the sample proportion of business travelers stating that a smoke-free room is the most important factor when staying in a hotel.

b. What is the probability that the sample proportion will be within $\pm$ .04 of the population proportion?

c. What is the probability that the sample proportion will be within $\pm$ .02 of the population proportion?

55. A market research firm conducts telephone surveys with a 40% historical response rate. What is the probability that in a new sample of 400 telephone numbers, at least 150 individuals will cooperate and respond to the questions? In other words, what is the probability that the sample proportion will be at least 150/400 = .375?

56. A production run is not acceptable for shipment to customers if a sample of 100 items contains 5% or more defective items. If a production run has a population proportion defective of $p = .10$, what is the probability that $\bar{p}$ will be at least .05?

CHAPTER 8

Interval Estimation

Dollar General Corporation*

Nashville, Tennessee

Dollar General Corporation was founded in 1939 as a dry goods wholesale company. After World War II, the company began opening retail locations in rural south-central Kentucky. Today Dollar General Corporation operates more than 2000 neighborhood stores in 23 states. Serving predominantly low- and middle-income customers, Dollar General markets soft goods, and health, beauty, and cleaning supplies at low everyday prices.

Being in an inventory-intense business with approximately 17,000 different products, Dollar General made the decision to adopt the LIFO (last-in-first-out) method of inventory valuation. This method matches current costs against current revenues, which minimizes the effect of radical price changes on profit and loss results. In addition, the LIFO method reduces net income and thereby income taxes during periods of inflation. This in turn brings disposable cash generated from sales in line with income and allows for the replacement of inventory at current costs.

Accounting practices require that a LIFO index be established for inventory under the LIFO method of valuation. For example, a LIFO index of 1.048 indicates that the company's inventory value at current costs reflects a 4.8% increase due to inflation over the most recent one-year period.

The establishment of a LIFO index requires that the year-end inventory count for each product be valued at the current year-end cost and at the preceding year-end cost. To avoid counting the inventory of every product in more than 2000 retail locations, a random sample of 800 products is selected from 100 retail locations and three warehouses. Physical inventories for the 800 sampled products are taken at the end of the year. Accounting personnel then provide the current-year and preceding-year costs needed to construct the LIFO index.

For a recent year, the LIFO index was 1.034. However, since this index is only a sample estimate of the population's LIFO index, a statement about the precision of the estimate was required. On the basis of the sample results the margin of error was .006. Thus, the interval from 1.028 to 1.040 provides the 95% confidence interval estimate of the population LIFO index. This precision was judged to be very good.

In this chapter you will learn how to make a probability statement about the sampling error associated with the sample mean and sample proportion. Then, you will learn how to use this information to construct and interpret confidence interval estimates of a population mean and a population proportion. You will also learn how to determine the sample size needed to ensure that the sampling error will be within acceptable limits.

This store in Bedford, Indiana, is one of more than 2000 Dollar General Stores.

Mr. Robert S. Knaul, Controller, Dollar General Corporation, provided this Statistics in Practice.

In Chapter 7 we showed that the sample mean $\bar{x}$ and the sample proportion $\bar{p}$ provide point estimates of the population mean μ and the population proportion p, respectively. Since point estimates are based on a sample of the population, they cannot be expected to equal the value of the corresponding population parameter. In this chapter we will show how interval estimates of the population mean and the population proportion provide information about the precision of an estimate. The sampling distributions of $\bar{x}$ and $\bar{p}$ presented in Chapter 7 play an important role in the development of interval estimates of μ and p.

8.1 Population Mean: Large-Sample Case

In this section we show how the sampling distribution of $\bar{x}$ can be used to develop an interval estimate of a population mean μ. We begin with the large-sample case ($n \geq 30$) in which the population standard deviation σ is *known*. Once we show how to develop an interval estimate of μ for this case, we will show how to do so for the large-sample case in which σ is unknown.

Let us begin by considering a sampling study conducted by CJW, Inc., a mail-order firm that specializes in sporting equipment and accessories. The company works hard to provide the best possible customer service. To monitor the quality of its service, CJW selects a simple random sample of mail-order customers each month; each customer sampled is contacted and asked a series of questions pertaining to the level of customer service. The answers to the questions are used to compute a satisfaction score for each customer sampled; these scores range from 0 (worst possible rating) to 100 (best possible rating). A sample mean satisfaction score is then computed and used as a point estimate of the mean satisfaction score for the population of all CJW customers.

Previous monthly surveys have shown that although the sample mean satisfaction score changes from month to month, the standard deviation of satisfaction scores has tended to stabilize at a value of 20. Hence, we will assume that the population standard deviation is $\sigma = 20$. The most recent CJW customer satisfaction survey provided data on the satisfaction scores of 100 customers ($n = 100$); the sample mean satisfaction score was $\bar{x} = 82$. In the discussion that follows, we will use the sample results to develop an interval estimate of the population mean satisfaction score μ.

Sampling Error

Anytime a sample mean is used to provide a point estimate of a population mean, someone may ask: How good is the estimate? The "how good" question is a way of asking about the error involved when the value of $\bar{x}$ is used as the point estimate of μ. In general, the absolute value of the difference between an unbiased point estimator and the population parameter it estimates is called the *sampling error.* For the case of a sample mean estimating a population mean, the sampling error is

$$\text{Sampling Error} = |\bar{x} - \mu| \tag{8.1}$$

In practice, the value of the sampling error cannot be determined because the population mean μ is unknown. However, the sampling distribution of $\bar{x}$ can be used to make probability statements about the size of the sampling error. We will illustrate how this is done for the CJW sampling study.

With a sample size of $n = 100$ and a population standard deviation of $\sigma = 20$, the central limit theorem, introduced in Chapter 7, enables us to conclude that the sampling distribution of $\bar{x}$ can be approximated by a normal probability distribution with a mean μ and a standard deviation $\sigma_{\bar{x}} = \sigma/\sqrt{n} = 20/\sqrt{100} = 2$. This sampling distribution is shown in Figure 8.1. Since the sampling distribution of $\bar{x}$ shows how values of $\bar{x}$ are distributed around μ, it provides information about the possible differences between $\bar{x}$ and μ. We can use this information to develop probability statements about the sampling error.

FIGURE 8.1
Sampling Distribution of the Sample Mean Satisfaction Score for Simple Random Samples of 100 Customers

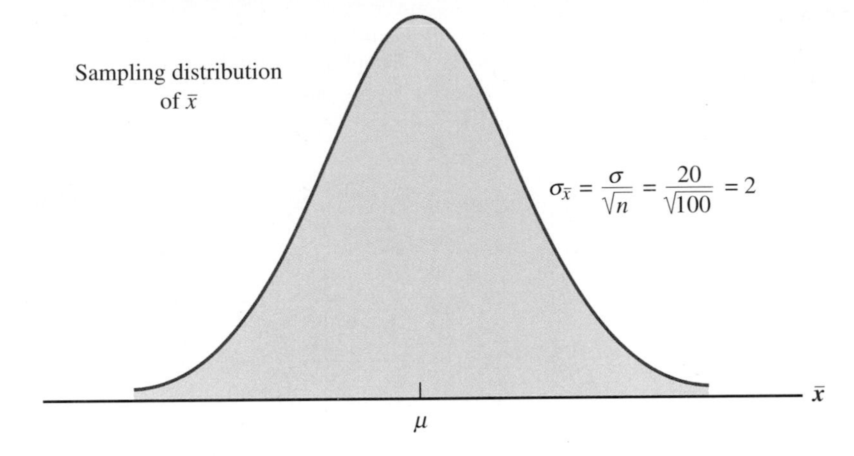

FIGURE 8.2
Sampling Distribution of $\bar{x}$ Showing the Location of Sample Means That Provide a Sampling Error of 3.92 or Less

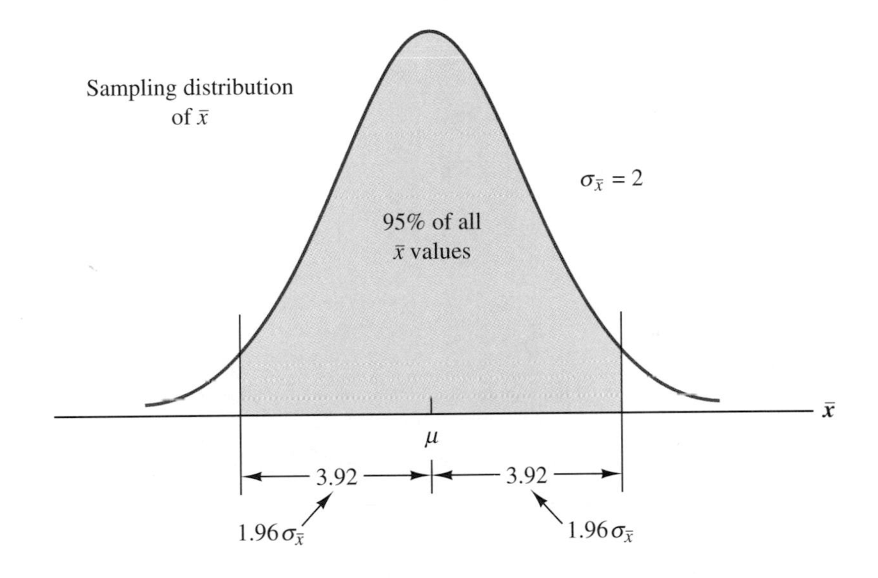

Probability Statements About the Sampling Error

Using the table of areas for the standard normal probability distribution, we find that 95% of the values of any normally distributed random variable are within $\pm$ 1.96 standard deviations of the mean. Hence, for the sampling distribution in Figure 8.1, 95% of all $\bar{x}$ values must be within $\pm$ 1.96 standard deviations of μ. Since $1.96\sigma_{\bar{x}} = 1.96(2) = 3.92$, 95% of the sample means must be within $\pm$ 3.92 of the population mean.

The location of the sample means that provide a sampling error of 3.92 or less is shown in Figure 8.2. Note that if a sample mean is in the region denoted "95% of all $\bar{x}$ values," the sampling error is 3.92 or less. However, if a sample mean is in either the lower tail or the upper tail of the distribution, the sampling error will be greater than 3.92. We therefore can make the following probability statement about the sampling error for the CJW problem.

> There is a .95 probability that the sample mean will provide a sampling error of 3.92 or less.

FIGURE 8.3
Sampling Distribution of $\bar{x}$ Showing the Location of 99% of the $\bar{x}$ Values

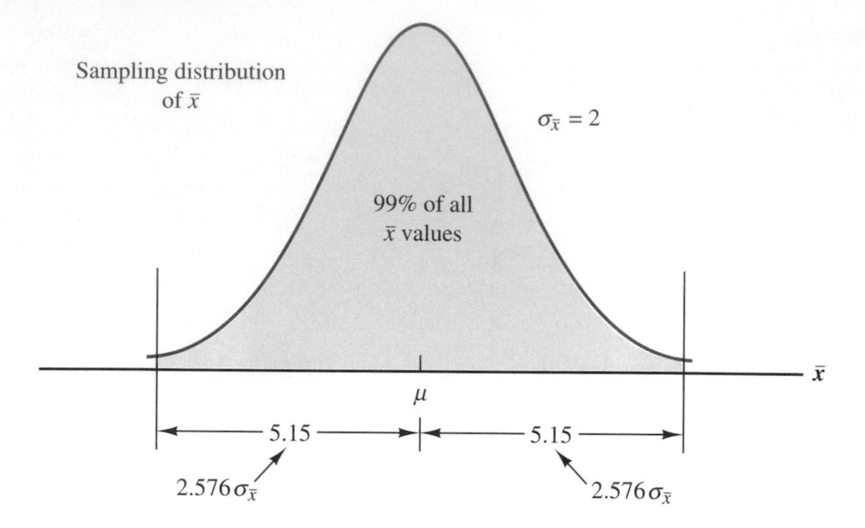

FIGURE 8.4
Area of a Sampling Distribution of $\bar{x}$ Used to Make Probability Statements About the Sampling Error

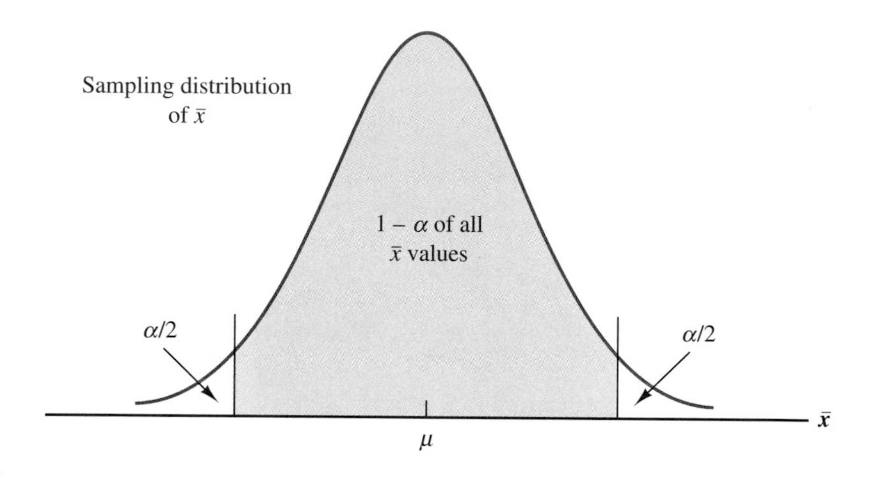

This probability statement about the sampling error is a *precision statement* telling CJW about the error that can expected if a simple random sample of 100 customers is used to estimate the population mean. Although a .95 probability is frequently used in making precision statements, other probability values such as .90 and .99 can be used. For instance, Figure 8.3 shows the location of 99% of the sample means for the CJW problem. Using the standard normal probability distribution table, we find that 99% of the $\bar{x}$ values are within ± 2.576 standard deviations of μ. Since $2.576\sigma_{\bar{x}} = 2.576(2) = 5.15$, there is a .99 probability that the sample mean will provide a sampling error of 5.15 or less. Similarly, there is a .90 probability that the sample mean will provide a sampling error of $1.645\sigma_{\bar{x}} = 1.645(2) = 3.29$ or less.

Let us generalize the procedure we use to make precision statements about the sampling error whenever the value of a sample mean is used to estimate a population mean. We will use the Greek letter α to indicate the probability that the sampling error is *larger* than the sampling error in the precision statement. In Figure 8.4, $\alpha/2$ denotes the area, or probability, in each tail of the sampling distribution and $1 - \alpha$ denotes the area, or probability, that a sample mean will provide a sampling error less than or equal to the sampling error in the precision statement. For example, the statement that there is a .95 probability that the value of a sample mean will provide a sampling error of 3.92 or less is based on $\alpha = .05$ and $1 - \alpha = .95$. The area in each tail of the sampling distribution is $\alpha/2 = .025$.

Using z to denote the value of the standard normal random variable, we will place a subscript on z to denote the *area in the upper tail* of the distribution. In general, $z_{\alpha/2}$ is the value of the standard normal random variable corresponding to an area of $\alpha/2$ in the upper tail of the distribution. With this notation, the following precision statement defines the size of the sampling error whenever $\bar{x}$ is used to estimate μ.

Precision Statement

There is a $1 - \alpha$ probability that the value of a sample mean will provide a sampling error of $z_{\alpha/2}\sigma_{\bar{x}}$ or less.

The quantity $z_{\alpha/2}\sigma_{\bar{x}}$ is referred to as the *margin of error*.

When σ Is Known

In the CJW example, we can state that there is a .95 probability that the value of the sample mean will provide a margin of error of 3.92. Hence, we can construct an interval estimate for μ by subtracting 3.92 from $\bar{x}$ and adding 3.92 to $\bar{x}$; that is, $\bar{x} \pm 3.92$. To interpret interval estimate of μ, let us consider possible values of $\bar{x}$ that could be obtained from three different simple random samples, each consisting of 100 customers.

Suppose the first sample mean turns out to have the value shown in Figure 8.5 as $\bar{x}_1$. In this case, Figure 8.5 shows that the interval formed by subtracting 3.92 from $\bar{x}_1$ and adding 3.92 to $\bar{x}_1$ includes the population mean μ. Now consider what happens if the sample mean turns out to have the value shown in Figure 8.5 as $\bar{x}_2$. Although this sample mean is different from the first sample mean, we see that the interval based on $\bar{x}_2$ also includes the population mean μ. However, the interval based on the third sample mean, denoted by $\bar{x}_3$, does not include the population mean; the reason is that $\bar{x}_3$ is in a tail of the distribution at a distance farther than 3.92 from μ. Hence, subtracting and adding 3.92 to $\bar{x}_3$ forms an interval that does not include μ.

In general, any sample mean $\bar{x}$ that is within the shaded region of Figure 8.5 will provide an interval that contains the population mean μ. Since 95% of the possible sample means will be in this region, 95% of all intervals formed by subtracting 3.92 from $\bar{x}$ and adding 3.92 to $\bar{x}$ will include μ. We therefore say that we are 95% confident that an interval constructed from $\bar{x} - 3.92$ to $\bar{x} + 3.92$ will include the population mean. Using common statistical terminology, we refer to the interval as a *confidence interval.* Since 95% of the sample means will result in a confidence interval that includes the population mean μ, we say that the confidence interval is established at the 95% *confidence level.* The value .95 is referred to as the *confidence coefficient.*

Recall that CJW's sample of 100 customers provided a sample mean satisfaction score of $\bar{x} = 82$. Using the interval $\bar{x} \pm 3.92$, we find that the 95% confidence interval estimate of the population mean is 82 ± 3.92, or 78.08 to 85.92. Hence, at the 95% confidence level, CJW can conclude that the mean satisfaction score for the population of all mail-order customers is between 78.08 and 85.92.

FIGURE 8.5
Intervals Formed from Selected Sample Means at Locations $\bar{x}_1$, $\bar{x}_2$, and $\bar{x}_3$

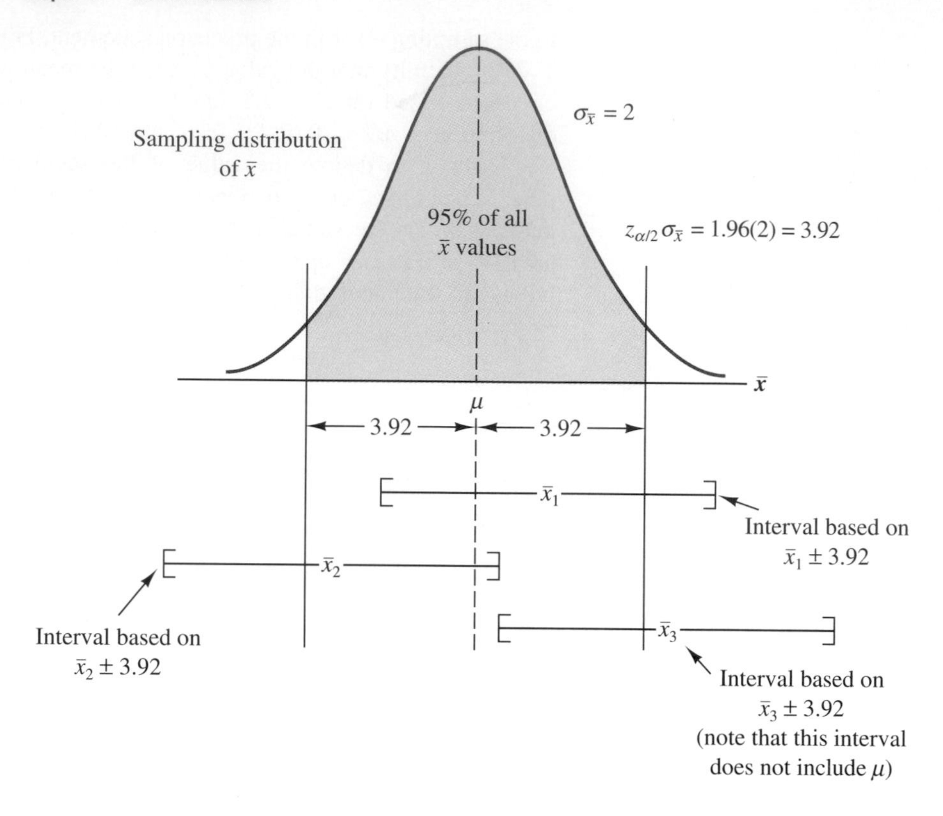

TABLE 8.1 Values of $z_{\alpha/2}$ for the Most Commonly Used Confidence Levels

Confidence Level	α	$\alpha/2$	$z_{\alpha/2}$
90%	.10	.05	1.645
95%	.05	.025	1.96
99%	.01	.005	2.576

Let us now state the general procedure for computing an interval estimate of a population mean for the large-sample case with σ known. As previously noted, there is a $1 - \alpha$ probability that the value of the sample mean will provide a margin of error of $z_{\alpha/2}\sigma_{\bar{x}}$. Using the fact that $\sigma_{\bar{x}} = \sigma/\sqrt{n}$, we can write the procedure for calculating the *interval estimate* of a population mean as follows.

Interval Estimate of a Population Mean: Large-Sample Case ($n \geq 30$) with σ Known

$$\bar{x} \pm z_{\alpha/2}\frac{\sigma}{\sqrt{n}} \tag{8.2}$$

where $1 - \alpha$ is the confidence coefficient and $z_{\alpha/2}$ is the z value providing an area of $\alpha/2$ in the upper tail of the standard normal probability distribution.

The values of $z_{\alpha/2}$ for the most commonly used confidence levels are given in Table 8.1.

When σ Is Unknown

A difficulty in using (8.2) is that in most sampling situations the value of the population standard deviation σ is unknown. In the large-sample case ($n \geq 30$), we simply use the

TABLE 8.2 Ages of Life Insurance Policyholders from a Simple Random Sample of 36 Statewide Policyholders

Policyholder	Age	Policyholder	Age	Policyholder	Age
1	32	13	39	25	23
2	50	14	46	26	36
3	40	15	45	27	42
4	24	16	39	28	34
5	33	17	38	29	39
6	44	18	45	30	34
7	45	19	27	31	35
8	48	20	43	32	42
9	44	21	54	33	53
10	47	22	36	34	28
11	31	23	34	35	49
12	36	24	48	36	39

LIFEINS

value of the sample standard deviation s as the point estimate of the population standard deviation σ to obtain the following interval estimate.

Interval Estimate of a Population Mean: Large-Sample Case ($n \geq 30$) with σ Unknown

$$\bar{x} \pm z_{\alpha/2} \frac{s}{\sqrt{n}} \tag{8.3}$$

where s is the sample standard deviation, $1 - \alpha$ is the confidence coefficient, and $z_{\alpha/2}$ is the z value providing an area of $\alpha/2$ in the upper tail of the standard normal probability distribution.

As an illustration of the interval estimation procedure when σ is unknown, let us consider a sampling study conducted by the Statewide Insurance Company. Suppose that as part of an annual review of life insurance policies, Statewide selects a simple random sample of 36 Statewide life insurance policyholders. The corresponding life insurance policies are reviewed in terms of the amount of coverage, the cash value of the policy, disability options, and so on. For the current study, a manager has asked for a 90% confidence interval estimate of the mean age for the population of life insurance policyholders.

Table 8.2 reports the age data collected from a simple random sample of 36 policyholders. The sample mean age of $\bar{x} = 39.5$ years is the point estimate of the population mean age. In addition, the sample standard deviation for the data in Table 8.2 is $s = 7.77$. At 90% confidence, $z_{.05} = 1.645$. Using (8.3), we obtain

$$39.5 \pm 1.645 \frac{7.77}{\sqrt{36}}$$

$$39.5 \pm 2.13$$

Hence, the 90% confidence interval estimate of the population mean is 37.37 to 41.63. The manager can be 90% confident that the mean age for the population of Statewide life insurance policyholders is between 37.37 years and 41.63 years.

Computer-Generated Confidence Intervals

To illustrate the use of computer software packages in developing interval estimates of a population mean, we used Minitab to develop a 90% confidence interval estimate of the population mean age for the Statewide Insurance policyholders. The Minitab output is shown in panel A of Figure 8.6. The sample standard deviation $s = 7.77$ is shown to be the assumed value for the population standard deviation σ. Additional output shows that the sample of 36 policyholders provides a sample mean of 39.50 years, a sample standard deviation of 7.77 years, and a standard error of the mean of 1.29 years. The 90% confidence interval estimate is 37.37 years to 41.63 years.

To compute an interval estimate of the population mean with a different confidence coefficient, we simply change the value for the confidence level in the Minitab procedure. Panel B of Figure 8.6 shows that the 95% confidence interval is 36.96 years to 42.04 years; note that, as is expected, the 95% confidence interval estimate is wider than the 90% confidence interval estimate.

FIGURE 8.6
Computer-Generated Confidence Intervals for the Statewide Insurance Company Sampling Problem

Panel A: 90% Confidence Interval

```
The assumed sigma =7.77

 N     Mean      StDev    SE Mean      90.0 % C.I.
36    39.50       7.77       1.29    (  37.37,    41.63)
```

Panel B: 95% Confidence Interval

```
The assumed sigma =7.77

 N     Mean      StDev    SE Mean      95.0 % C.I.
36    39.50       7.77       1.29    (  36.96,    42.04)
```

notes and comments

1. In developing an interval estimate of the population mean, we specify the desired confidence coefficient $(1 - \alpha)$ before selecting the sample. Thus, prior to selecting the sample, we conclude that there is a $1 - \alpha$ probability that the confidence interval we eventually compute will contain the population mean μ. However, once the sample is taken, the sample mean $\bar{x}$ is computed, and the particular interval estimate is determined, the resulting interval *may or may not* contain μ. If $1 - \alpha$ is reasonably large, we can be confident that the resulting interval will contain μ because we know that if we use this procedure repeatedly, $100(1 - \alpha)$ percent of all possible intervals developed in this way will contain μ.

2. Note that the sample size n appears in the denominator of the interval estimation expressions (8.2) and (8.3). Thus, if a particular sample size provides too wide an interval to be of any practical use, we may want to consider increasing the sample size. With n in the denominator, a larger sample size will provide a narrower interval and a greater precision. The procedure for determining the size of a simple random sample necessary to obtain a desired precision is discussed in Section 8.3.

exercises

Methods

1. A simple random sample of 40 items resulted in a sample mean of 25. The population standard deviation is $\sigma = 5$.

a. What is the standard error of the mean, $\sigma_{\bar{x}}$?

b. At a 95% probability, what can be said about the margin of error?

Self Test

2. A simple random sample of 50 items resulted in a sample mean of 32 and a sample standard deviation of 6.

a. Provide a 90% confidence interval for the population mean.

b. Provide a 95% confidence interval for the population mean.

c. Provide a 99% confidence interval for the population mean.

Applications

Self Test

3. In an effort to estimate the mean amount spent per customer for dinner at a major Atlanta restaurant, data were collected for a sample of 49 customers over a three-week period.

a. Assume a population standard deviation of \$2.50. What is the standard error of the mean?

b. With a .95 probability, what statement can be made about the margin of error?

c. If the sample mean is \$22.60, what is the 95% confidence interval for the population mean?

4. Data on the automotive industry and automobile purchasing characteristics were presented in *Financial World* (April 14, 1992). The mean car payment per month was reported to be \$310. Assume that this result was based on a sample of 250 car payment records and that the sample standard deviation was \$100. Compute the 95% confidence interval for the mean monthly car payment.

5. The results of an annual survey of 1404 mutual funds were reported in *Forbes* (September 2, 1991). A sample of 75 funds showed a mean return over the previous 12 months of 10.2%. The sample standard deviation was 3%. Provide a 95% confidence interval for the mean return for the population of mutual funds.

6. In a study of student loan subsidies, the Department of Education reported that four-year Stafford Loan borrowers will owe an average of $12,168 upon graduation (*USA Today,* April 5, 1995). Assume that this average or mean amount owed is based on a sample of 480 student loans and that the population standard deviation for the amount owed upon graduation is $2200.

a. Develop a 90% confidence interval estimate of the population mean amount owed.

b. Develop a 95% confidence interval estimate of the population mean amount owed.

c. Develop a 99% confidence interval estimate of the population mean amount owed.

d. Discuss what happens to the width of the confidence interval as the confidence level is increased. Does this seem reasonable? Explain.

7. A container-filling operation has a historical standard deviation of 5.5 ounces. A quality control inspector periodically selects 36 containers at random and uses the sample mean filling weight to estimate the population mean filling weight for the production process.

a. What is the standard error of the mean?

b. With .75, .90, and .99 probabilities, what statements can be made about the margin of error? What happens to the margin of error when the probability is increased? Why does this happen?

c. What is the 99% confidence interval for the population mean filling weight for the process if the sample mean is 48.6 ounces?

8. The profitability of used car sales was determined in a study conducted by the National Automobile Dealers Association (*USA Today,* April 12, 1995). Assume that a sample of 200 used car sales provided a sample mean profit of $300 per car and a sample standard deviation of $150. Use this information to develop a 95% confidence interval estimate of the mean profit for the population of used car sales.

9. J. D. Power & Associates' annual quality survey for automobiles found that the industry average number of defects per new car is 1.07 (*The Wall Street Journal,* January 27, 1994). Suppose a sample of 30 new automobiles taken by a particular manufacturer provides the following data on number of defects per car.

0	1	1	2	1	0	2	3	2	1	0	4	3	1	1
0	2	0	0	2	3	0	2	0	2	0	3	1	0	2

a. Using these data, what is the sample mean number of defects per car?

b. What is the sample standard deviation?

c. Provide a 95% confidence interval estimate of the mean number of defects per car for the population of cars produced by this manufacturer.

d. After viewing the confidence interval estimate in part (c), a statistical analyst suggested that the manufacturer test a larger number of new cars before drawing a conclusion about how the quality of its cars compares to the J. D. Powers & Associates industry average of 1.07 defects per car. Do you support this idea? Why or why not?

10. The International Air Transport Association surveys business travelers to develop quality ratings for transatlantic gateway airports. The maximum possible rating is 10. The highest rated airport is Amsterdam with an average rating of 7.93, followed by Toronto with a rating of 7.17 (*Newsweek,* June 13, 1994). Suppose a simple random sample of 50 business travelers is selected and each traveler is asked to provide a rating for the Miami International Airport. The ratings obtained from the sample of 50 follow.

MIAMI

6	4	6	8	7	7	6	3	3	8	10	4	8
7	8	7	5	9	5	8	4	3	8	5	5	4
4	4	8	4	5	6	2	5	9	9	8	4	8
9	9	5	9	7	8	3	10	8	9	6		

Develop a 95% confidence interval estimate of the population mean rating for Miami.

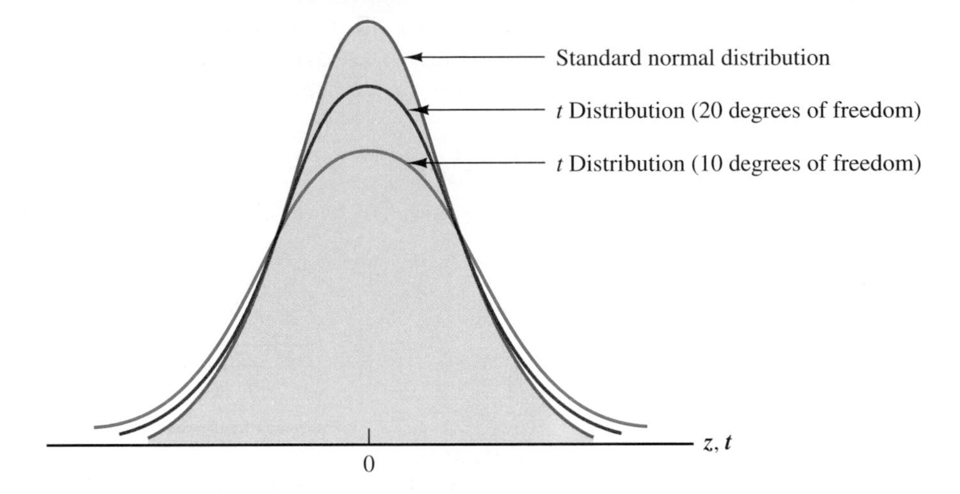

FIGURE 8.7
Comparison of the Standard Normal Distribution with t Distributions Having 10 and 20 Degrees of Freedom

8.2 Population Mean: Small-Sample Case

In the small-sample case ($n < 30$), the sampling distribution of $\bar{x}$ depends on the probability distribution of the population. *If the population has a normal probability distribution,* the methodology presented in this section can be used to develop a confidence interval for a population mean. However, if the assumption of a normal probability distribution for the population is not appropriate, the only alternative is to increase the sample size to $n \geq 30$ and rely on the large-sample interval-estimation procedures given by (8.2) and (8.3).

If the population has a normal probability distribution, the sampling distribution of $\bar{x}$ is a normal probability distribution regardless of the sample size. In this case, if the population standard deviation σ is *known,* (8.2) can be used to compute an interval estimate of a population mean even with a small sample. However, if the population standard deviation σ is *unknown,* the sample standard deviation s is used to estimate σ, and the appropriate confidence interval is based on a probability distribution known as the *t distribution.*

The t distribution is a family of similar probability distributions, with a specific t distribution depending on a parameter known as the *degrees of freedom.* That is, there is a unique t distribution with one degree of freedom, with two degrees of freedom, with three degrees of freedom, and so on. As the number of degrees of freedom increases, the difference between the t distribution and the standard normal probability distribution becomes smaller and smaller. Figure 8.7 shows t distributions with 10 and 20 degrees of freedom and their relationship to the standard normal probability distribution. Note that a t distribution with more degrees of freedom has less dispersion and more closely resembles the standard normal probability distribution. Note also that the mean of the t distribution is zero.

We will use a subscript for t to indicate the area in the upper tail of the t distribution. For example, just as we used $z_{.025}$ to indicate the z value providing a .025 area in the upper tail of a standard normal probability distribution, we will use $t_{.025}$ to indicate a .025 area in the upper tail of the t distribution. In general, we will use the notation $t_{\alpha/2}$ to represent a t value with an area of $\alpha/2$ in the upper tail of the t distribution. See Figure 8.8.

Table 8.3 is a table for the t distribution. This table is also shown inside the front cover of the text. Note, for example, that for a t distribution with 10 degrees of freedom,

TABLE 8.3 *t* Distribution Table for Areas in the Upper Tail. Example: With 10 Degrees of Freedom, $t_{.025} = 2.228$

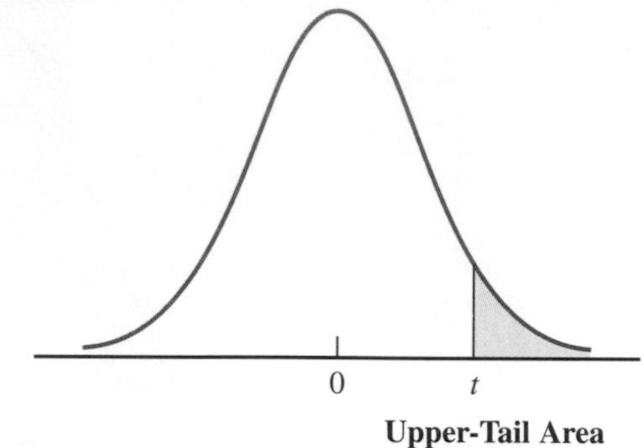

Degrees of Freedom	Upper-Tail Area .10	.05	.025	.01	.005
1	3.078	6.314	12.706	31.821	63.657
2	1.886	2.920	4.303	6.965	9.925
3	1.638	2.353	3.182	4.541	5.841
4	1.533	2.132	2.776	3.747	4.604
5	1.476	2.015	2.571	3.365	4.032
6	1.440	1.943	2.447	3.143	3.707
7	1.415	1.895	2.365	2.998	3.499
8	1.397	1.860	2.306	2.896	3.355
9	1.383	1.833	2.262	2.821	3.250
10	1.372	1.812	2.228	2.764	3.169
11	1.363	1.796	2.201	2.718	3.106
12	1.356	1.782	2.179	2.681	3.055
13	1.350	1.771	2.160	2.650	3.012
14	1.345	1.761	2.145	2.624	2.977
15	1.341	1.753	2.131	2.602	2.947
16	1.337	1.746	2.120	2.583	2.921
17	1.333	1.740	2.110	2.567	2.898
18	1.330	1.734	2.101	2.552	2.878
19	1.328	1.729	2.093	2.539	2.861
20	1.325	1.725	2.086	2.528	2.845
21	1.323	1.721	2.080	2.518	2.831
22	1.321	1.717	2.074	2.508	2.819
23	1.319	1.714	2.069	2.500	2.807
24	1.318	1.711	2.064	2.492	2.797
25	1.316	1.708	2.060	2.485	2.787
26	1.315	1.706	2.056	2.479	2.779
27	1.314	1.703	2.052	2.473	2.771
28	1.313	1.701	2.048	2.467	2.763
29	1.311	1.699	2.045	2.462	2.756
30	1.310	1.697	2.042	2.457	2.750
40	1.303	1.684	2.021	2.423	2.704
60	1.296	1.671	2.000	2.390	2.660
120	1.289	1.658	1.980	2.358	2.617
∞	1.282	1.645	1.960	2.326	2.576

FIGURE 8.8
t **Distribution with α/2 Area or Probability in the Upper Tail**

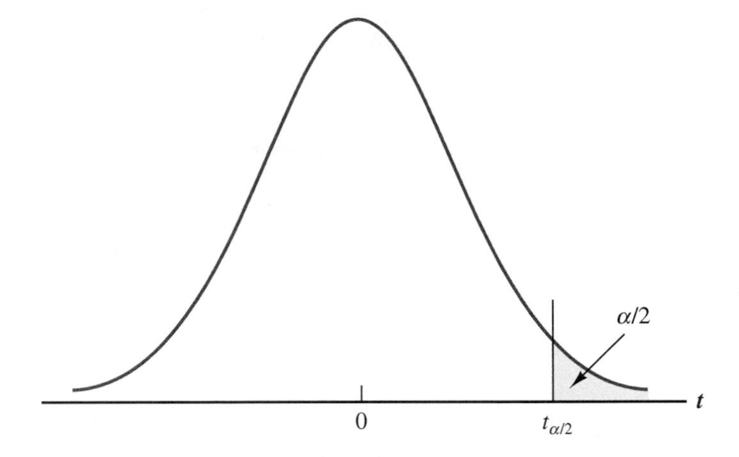

$t_{.025} = 2.228$. Similarly, for a *t* distribution with 20 degrees of freedom, $t_{.025} = 2.086$. As the degrees of freedom continue to increase, $t_{.025}$ approaches $z_{.025} = 1.96$.

Now that we have an idea of what the *t* distribution is, let us see how it is used to develop an interval estimate of a population mean. Assume that the population has a normal probability distribution and that the sample standard deviation *s* is used to estimate the population standard deviation σ. The following interval-estimation procedure is applicable.

Interval Estimate of a Population Mean: Small-Sample Case ($n < 30$) with σ Unknown

$$\bar{x} \pm t_{\alpha/2}\frac{s}{\sqrt{n}} \tag{8.4}$$

where $1 - \alpha$ is the confidence coefficient, $t_{\alpha/2}$ is the *t* value providing an area of $\alpha/2$ in the upper tail of a *t* distribution with $n - 1$ degrees of freedom, and *s* is the sample standard deviation. The population is assumed to have a normal probability distribution.

The reason the number of degrees of freedom associated with the *t* value in (8.4) is $n - 1$ has to do with the use of *s* as an estimate of the population standard deviation σ. The expression for the sample standard deviation is

$$s = \sqrt{\frac{\Sigma(x_i - \bar{x})^2}{n - 1}}$$

Degrees of freedom refers to the number of independent pieces of information that go into the computation of $\Sigma(x_i - \bar{x})^2$. The *n* pieces of information involved in computing $\Sigma(x_i - \bar{x})^2$ are as follows: $x_1 - \bar{x}, x_2 - \bar{x}, \ldots, x_n - \bar{x}$. In Section 3.2 we indicated that $\Sigma(x_i - \bar{x}) = 0$ for any data set. Thus, only $n - 1$ of the $x_i - \bar{x}$ values are independent; that is, if we know $n - 1$ of the values, the remaining value can be determined exactly by using the condition that the sum of the $x_i - \bar{x}$ values must be 0. Thus, $n - 1$ is the number of degrees of freedom associated with $\Sigma(x_i - \bar{x})^2$ and hence the number of degrees of freedom for the *t* distribution used in (8.4).

Let us demonstrate the small-sample interval-estimation procedure by considering the training program evaluation conducted by Scheer Industries. Scheer's director of manufacturing is interested in a computer-assisted program that can be used to train the

TABLE 8.4 Training Time in Days for the Computer-Assisted Training Program at Scheer Industries

Employee	Time	Employee	Time	Employee	Time
1	52	6	59	11	54
2	44	7	50	12	58
3	55	8	54	13	60
4	44	9	62	14	62
5	45	10	46	15	63

firm's maintenance employees for machine-repair operations. The expectation is that the computer-assisted method will reduce the time needed to train employees. To evaluate the training method, the director of manufacturing has requested an estimate of the mean training time required for the computer-assisted program.

Suppose management has agreed to train 15 employees with the new approach. The data on training days required for each employee in the sample are listed in Table 8.4. The sample mean and sample standard deviation for these data follow:

$$\bar{x} = \frac{\Sigma x_i}{n} = \frac{808}{15} = 53.87 \text{ days}$$

$$s = \sqrt{\frac{\Sigma(x_i - \bar{x})^2}{n - 1}} = \sqrt{\frac{651.73}{14}} = 6.82 \text{ days}$$

The point estimate of the mean training time for the population of employees is 53.87 days. We can obtain information about the precision of this estimate by developing an interval estimate of the population mean. Since the population standard deviation is unknown, we will use the sample standard deviation $s = 6.82$ days as the point estimate of σ. With the small sample size, $n = 15$, we will use (8.4) to develop an interval estimate of the population mean at 95% confidence. If we assume that the population of training times has a normal probability distribution, the t distribution with $n - 1 = 14$ degrees of freedom is the appropriate probability distribution for the interval-estimation procedure. We see from Table 8.3 that with 14 degrees of freedom, $t_{\alpha/2} = t_{.025} = 2.145$. Using (8.4), we have

$$\bar{x} \pm t_{.025} \frac{s}{\sqrt{n}}$$

$$53.87 \pm 2.145\left(\frac{6.82}{\sqrt{15}}\right)$$

$$53.87 \pm 3.78$$

Thus, the 95% confidence interval estimate of the population mean training time is 50.09 days to 57.65 days.

The preceding approach, in which the t distribution is used to develop an interval estimate of μ, is applicable whenever the population standard deviation is unknown and the population being sampled has a normal probability distribution. However, statistical research has shown that (8.4) is applicable even if the population being sampled is not quite normal. That is, confidence intervals based on the t distribution can be used as long as the population distribution does not differ extensively from a normal probability distribution.

FIGURE 8.9
Computer-Generated Confidence Intervals for the Scheer Industries Sampling Problem

N	Mean	StDev	SE Mean	95.0 % C.I.
15	53.87	6.82	1.76	(50.09, 57.65)

Computer-Generated Confidence Intervals

To illustrate the use of computer software packages in developing interval estimates of a population mean for the small-sample case, we used Minitab to develop a 95% confidence interval estimate of the population mean training time for Scheer Industries. Figure 8.9 is the Minitab output. For the sample of 15 employees, the output shows a sample mean of 53.87 days, a sample standard deviation of 6.82 days, and a standard error of the mean of 1.76 days. The 95% confidence interval estimate is 50.09 days to 57.65 days as previously computed.

notes and comments

The t distribution is not restricted to the small-sample situation. Actually, the t distribution is applicable whenever the population has a normal probability distribution and whenever the sample standard deviation is used to estimate the population standard deviation. If these conditions are met, the t distribution can be used for any sample size. However, (8.3) shows that with a large sample ($n \geq 30$), interval estimation of a population mean can be based on the standard normal probability distribution and the value $z_{\alpha/2}$. Thus, with (8.3) available for the large-sample case, we generally do not consider the use of the t distribution until we encounter a small-sample case.

exercises

Methods

11. For a t distribution with 12 degrees of freedom, find the area, or probability, that is in each region.

a. To the left of 1.782
b. To the right of −1.356
c. To the right of 2.681
d. To the left of −1.782
e. Between −2.179 and +2.179
f. Between −1.356 and +1.782

12. Find the t value(s) for each of the following examples.

a. Upper tail area of .05 with 18 degrees of freedom
b. Lower tail area of .10 with 22 degrees of freedom
c. Upper tail area of .01 with 5 degrees of freedom
d. 90% of the area is between these two t values with 14 degrees of freedom
e. 95% of the area is between these two t values with 28 degrees of freedom

Self Test

13. The following data have been collected from a sample of eight items from a normal population: 10, 8, 12, 15, 13, 11, 6, 5.

a. What is the point estimate of the population mean?

b. What is the point estimate of the population standard deviation?

c. What is the 95% confidence interval for the population mean?

14. A simple random sample of 20 items from a normal population resulted in a sample mean of 17.25 and a sample standard deviation of 3.3.

a. Develop a 90% confidence interval for the population mean.

b. Develop a 95% confidence interval for the population mean.

c. Develop a 99% confidence interval for the population mean.

Applications

15. In the testing of a new production method, 18 employees were selected randomly and asked to try the new method. The sample mean production rate for the 18 employees was 80 parts per hour and the sample standard deviation was 10 parts per hour. Provide 90% and 95% confidence intervals for the population mean production rate for the new method, assuming the population has a normal probability distribution.

16. The Money & Investing section of the *The Wall Street Journal* contains a summary of daily investment performance for the New York Stock Exchange, the American Stock Exchange, overseas markets, options, commodities, futures, and so on. In the New York Stock Exchange section, information is provided on each stock's 52-week high price per share, 52-week low price per share, dividend rate, yield, P/E ratio, daily volume, daily high price per share, daily low price per share, closing price per share, and daily net change. The P/E (price–earnings) ratio for each stock is determined by dividing the price of a share of stock by the earnings per share reported by the company for the most recent four quarters. A sample of 10 stocks taken from *The Wall Street Journal* (May 19, 1995) provided the following data on P/E ratios: 5, 7, 9, 10, 14, 23, 20, 15, 3, 26.

a. What is the point estimate of the mean P/E ratio for the population of all stocks listed on the New York Stock Exchange?

b. What is the value of the sample standard deviation?

c. With a .95 confidence coefficient, what is the interval estimate of the mean P/E ratio for the population of all stocks listed on the New York Stock Exchange? Assume the population has a normal distribution.

d. Comment on the precision of the results.

17. The American Association of Advertising Agencies records data on nonprogramming minutes per half hour of prime-time television programming (*U.S. News & World Report,* April 13, 1992). Representative data for a sample of prime-time programs on major networks at 8:30 P.M. are listed here. Provide a point estimate and a 95% confidence interval for the mean number of nonprogramming minutes on half-hour prime-time television shows at 8:30 P.M.

6.0	7.0	7.2	7.0	6.0	7.3	6.0	6.6	6.3	5.7
6.5	6.5	7.6	6.2	5.8	6.2	6.4	6.2	7.2	6.8

18. Sales personnel for Skillings Distributors are required to submit weekly reports listing the customer contacts made during the week. A sample of 61 weekly contact reports showed a mean of 22.4 customer contacts per week for the sales personnel. The sample standard deviation was five contacts.

a. Use the large-sample case (8.3) to develop a 95% confidence interval for the mean number of weekly customer contacts for the population of sales personnel.

b. Assume that the population of weekly contact data has a normal distribution. Use the t distribution with 60 degrees of freedom to develop a 95% confidence interval for the mean number of weekly customer contacts.

c. Compare your answers for parts (a) and (b). Comment on why in the large-sample case it is permissible to base interval estimates on the procedure used in part (a) even though the t distribution may also be applicable.

19. The U.S. Department of Transportation reported the number of miles that residents of metropolitan areas travel per day in a car (*1994 Information Please Environmental Almanac*). Suppose a simple random sample of 15 residents of Cleveland provided the following data on car miles per day.

20 20 28 16 11 17 23 16 22 18 10 22 29 19 32

a. Compute a 95% confidence interval estimate of the population mean number of miles residents of Cleveland travel per day in a car.
b. What assumption about the population was necessary to obtain an answer to part (a)?
c. Suppose it is desirable to estimate the population mean number of miles to within ±2 miles at 95% confidence. Do the data provide this desired level of precision? What action, if any, would you recommend be taken?

20. The duration (in minutes) for a sample of 20 flight-reservation telephone calls is as follows.

2.1	10.4	4.8	5.5	5.9	10.5	4.5	4.8	3.3	5.8
2.8	6.6	7.5	4.8	5.5	3.5	5.3	3.6	7.8	6.0

a. What is the point estimate of the population mean time for flight-reservation phone calls?
b. Assuming that the population has a normal distribution, develop a 95% confidence interval for the population mean time.

8.3 Determining the Sample Size

In Section 8.1 we were able to make the following statement about the sampling error whenever a sample mean was used to provide a point estimate of a population mean:

> There is a $1 - \alpha$ probability that the value of the sample mean will provide a sampling error of $z_{\alpha/2}\sigma_{\bar{x}}$ or less.

Since $\sigma_{\bar{x}} = \sigma/\sqrt{n}$, we can rewrite the preceeding statement as follows:

> There is a $1 - \alpha$ probability that the value of the sample mean will provide a sampling error of $z_{\alpha/2}(\sigma/\sqrt{n})$ or less.

The quantity $z_{\alpha/2}(\sigma/\sqrt{n})$ is referred to as the margin of error. From this statement we see that the values of $z_{\alpha/2}$, σ, and the sample size n combine to determine the margin of error. Once we select a confidence coefficient or probability of $1 - \alpha$, $z_{\alpha/2}$ can be determined. Given values for $z_{\alpha/2}$ and σ, we can determine the sample size n needed to provide any margin of error. Development of the formula used to compute the required sample size n follows.

Let $E =$ the desired margin of error. We have

$$E = z_{\alpha/2}\frac{\sigma}{\sqrt{n}}$$

Solving for $\sqrt{n}$, we have

$$\sqrt{n} = \frac{z_{\alpha/2}\sigma}{E}$$

Squaring both sides of this equation, we obtain the following expression for the sample size.

Sample Size for Estimation of a Population Mean

$$n = \frac{(z_{\alpha/2})^2\sigma^2}{E^2} \tag{8.5}$$

This sample size will provide a precision statement with a $1 - \alpha$ probability that the margin of error will be E.

In (8.5) the value E is the margin of error that the user is willing to accept at the given confidence level, and the value of $z_{\alpha/2}$ follows directly from the confidence level to be used in developing the interval estimate. Although user preference must be considered, 95% confidence is the most frequently chosen value ($z_{.025} = 1.96$).

Finally, use of (8.5) requires a value for the population standard deviation σ. In most cases, σ will be unknown. However, we can use (8.5) if we have a preliminary or *planning value* for σ. In practice, one of the following procedures can be chosen.

1. Use the sample standard deviation from a previous sample.
2. Use a pilot study to select a preliminary sample. The sample standard deviation from the preliminary sample can be used as the planning value for σ.
3. Use judgment or a "best guess" for the value of σ. For example, we might begin by estimating the largest and smallest data values in the population. The difference between the largest and smallest values provides an estimate of the range for the data. Finally, the range divided by four is often suggested as a rough approximation of the standard deviation and thus an acceptable planning value for σ.

Let us return to the Scheer Industries example in Section 8.2 to see how (8.5) can be used to determine the sample size for the study. Previously we showed that with a 95% level of confidence, a sample of 15 Scheer employees generated a population mean training time estimate of 53.87 ± 3.78 days. Assume that after viewing these results, Scheer's director of manufacturing is not satisfied with the degree of precision, feeling that a margin of error of 3.78 days is too large. Furthermore, suppose the director makes the following statement about the desired precision: "I would like a .95 probability that the value of the sample mean will provide a margin of error of two days." We can see that the director is specifying $E = 2$ days. In addition, the .95 probability indicates that a 95% confidence level is to be used; thus, $z_{\alpha/2} = z_{.025} = 1.96$. We need a planning value for σ to use in (8.5) to determine the sample size. Do we have a planning value for σ in the Scheer Industries example? Although σ is unknown, let us take advantage of the data provided for the 15 employees in Section 8.2. We can view these data as being from a pilot study, with the sample standard deviation $s = 6.82$ days providing the planning value for σ. Thus, using (8.5), we have

$$n = \frac{(z_{\alpha/2})^2\sigma^2}{E^2} = \frac{(1.96)^2(6.82)^2}{2^2} = 44.67$$

In cases where the computed n is a fraction, we round up to the next integer value; hence, the recommended sample size for the Scheer Industries example is 45 employees.

Finally, note that in the Scheer Industries example, $z_{.025}$ was used to determine the sample size even though the original computations for 15 employees had employed the t distribution. The reason for the use of $z_{.025}$ is that since the sample size is yet to be determined, we are anticipating that n will be larger than 30, making $z_{.025}$ the appropriate value. In addition, if n is yet to be determined, we do not know the $(n - 1)$ degrees of freedom necessary to use the t distribution. Hence, the use of (8.5) to determine the sample size will always be based on a z value rather than a t value.

exercises

Methods

21. How large a sample should one select to be 95% confident that the margin of error is 5? Assume that the population standard deviation is 25.

22. The range for a set of data is estimated to be 36.

a. What is the planning value for the population standard deviation?

b. How large a sample should one take to be 95% confident that the margin of error is 3?

c. How large a sample should one take to be 95% confident that the margin of error is 2?

Applications

Self Test

23. What sample size would have been recommended for the Scheer Industries example if the director of manufacturing had specified a .95 probability for a margin of error of 1.5 days? How large a sample would have been necessary if the precision statement had specified a .90 probability for a margin of error of two days? Use $\sigma = 6.82$ days.

24. In Section 8.1 the Statewide Insurance Company used a simple random sample of 36 policyholders to estimate the mean age of the population of policyholders. The resulting precision statement reported a .95 probability that the value of the sample mean provided a margin of error of 2.35 years. This statement was based on a sample standard deviation of 7.2 years.

a. How large a simple random sample would have been necessary to reduce the margin of error to two years? To 1.5 years? To one year?

b. Would you recommend that Statewide attempt to estimate the population mean age of the policyholders with $E = 1$ year? Explain.

25. Annual starting salaries for college graduates with business administration degrees are believed to have a standard deviation of \$2000. Assume that a 95% confidence interval estimate of the mean annual starting salary is desired. How large a sample should be taken if the margin of error is

a. \$500? **b.** \$200? **c.** \$100?

26. The mean number of days a house is on the market prior to selling was reported for 100 different cities (*U.S. News & World Report,* April 6, 1992). In a particular city the standard deviation of the number of days a house is on the market prior to selling is 20. How many house sales records would have to be collected to estimate the population mean with a margin of error of 2 days? Use a 95% level of confidence.

27. In Exercise 19, we noted that the U.S. Department of Transportation reported the number of miles that residents of metropolitan areas travel per day in a car (*1994 Information Please Environmental Almanac*). Suppose a preliminary simple random sample of residents of Cleveland is used to develop a planning value of 6.17 for the population standard deviation.

a. If we want to estimate the mean number of miles that Cleveland residents travel per day in a car with a margin of error of 2 miles, what sample size should be selected? Assume 95% confidence.

b. If we want to estimate the mean number of miles that Cleveland residents travel per day in a car with a margin of error of 1 mile, what sample size should be selected? Assume 95% confidence.

28. From Exercise 16, the sample standard deviation of P/E ratios for stocks listed on the New York Stock Exchange is $s = 7.8$ (*The Wall Street Journal,* May 19, 1995). Assume that we are interested in estimating the population mean P/E ratio for all stocks listed on the New York Stock Exchange. How many stocks should be included in the sample if we want a .95 probability that the margin of error is 2?

8.4 Population Proportion

In the Scheer Industries example presented in Section 8.2, the objective was to estimate the mean employee training time for a new machine-repair training program. To evaluate the program from a different perspective, management has requested that some measure of program quality be developed. The degree of success of the training program has previously been measured by the scores the employees obtain on a standard examination given at the end of the training program. From experience, the company has found that an individual passing the examination has an excellent chance of high performance on the job. After some discussion, management has agreed to base the quality evaluation of the new training method on the proportion of employees who pass the examination. Let us assume that Scheer has implemented the sample-size recommendation of the preceding section. Thus, we now have a sample of 45 employees that can be used to develop an interval estimate for the proportion of the population who will pass the examination.

In Chapter 7 we showed that a sample proportion $\bar{p}$ is an unbiased estimator of a population proportion p and that for large samples the sampling probability distribution of $\bar{p}$ can be approximated by a normal probability distribution, as shown in Figure 8.10. Recall that the use of the normal distribution as an approximation of the sampling distribution of $\bar{p}$ is based on the condition that both np and $n(1 - p)$ are 5 or more. We will be using the sampling distribution of $\bar{p}$ to make probability statements about the sampling error whenever a sample proportion $\bar{p}$ is used to estimate a population proportion p. In this case, the sampling error is defined as the absolute value of the difference between $\bar{p}$ and p, written $|\bar{p} - p|$.

The probability statements that can be made about the sampling error for the proportion take the following form.

> There is a $1 - \alpha$ probability that the value of the sample proportion will provide a sampling error of $z_{\alpha/2}\sigma_{\bar{p}}$ or less.

Hence, for a proportion, the quantity $z_{\alpha/2}\sigma_{\bar{p}}$ is the *margin of error.*

The rationale for the preceding statement is the same one we gave when the value of a sample mean was used as an estimate of a population mean. Namely, since we know that the sampling distribution of $\bar{p}$ can be approximated by a normal probability distribution, we can use the value of $z_{\alpha/2}$ and the value of the standard error of the proportion $\sigma_{\bar{p}}$ to determine the margin of error.

Once we see that the margin of error is $z_{\alpha/2}\sigma_{\bar{p}}$, we can subtract and add this value to $\bar{p}$ to obtain an interval estimate of the population proportion. Such an interval estimate is given by

$$\bar{p} \pm z_{\alpha/2}\sigma_{\bar{p}} \tag{8.6}$$

where $1 - \alpha$ is the confidence coefficient. Since $\sigma_{\bar{p}} = \sqrt{p(1-p)/n}$, we can rewrite (8.6)

$$\bar{p} \pm z_{\alpha/2}\sqrt{\frac{p(1-p)}{n}} \tag{8.7}$$

However, to use (8.7) to develop an estimate of a population proportion p, the value of p would have to be *known.* Since the value of p is *unknown,* we simply substitute the

FIGURE 8.10
The Sampling Distribution of $\bar{p}$ When $np \geq 5$ and $n(1 - p) \geq 5$

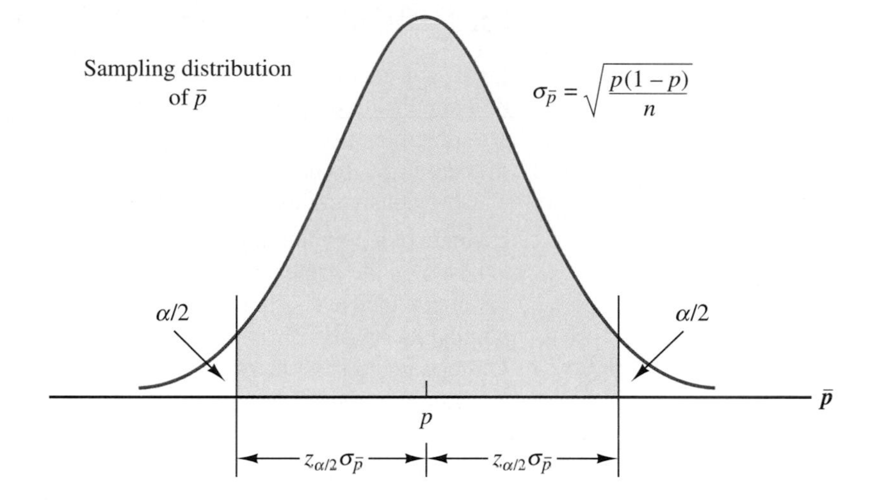

sample proportion $\bar{p}$ for p. The resulting general expression for a confidence interval estimate of a population proportion follows.*

Interval Estimate of a Population Proportion

$$\bar{p} \pm z_{\alpha/2}\sqrt{\frac{\bar{p}(1 - \bar{p})}{n}} \tag{8.8}$$

where $1 - \alpha$ is the confidence coefficient and $z_{\alpha/2}$ is the z value providing an area of $\alpha/2$ in the upper tail of the standard normal probability distribution.

Let us return to the Scheer Industries example. Assume that in the sample of 45 employees who completed the new training program, 36 passed the examination. Thus, the point estimate of the proportion in the population who pass the examination is $\bar{p} = 36/45 = .80$. Using (8.8) and a .95 confidence coefficient, we see that the interval estimate for the population proportion is given by

$$\bar{p} \pm z_{.025}\sqrt{\frac{\bar{p}(1 - \bar{p})}{n}}$$

$$.80 \pm 1.96\sqrt{\frac{.80(1 - .80)}{45}}$$

$$.80 \pm .12$$

Thus, at the 95% confidence level, the interval estimate of the population proportion is .68 to .92.

*An unbiased estimate of $\sigma^2_{\bar{p}}$ is $\bar{p}(1 - \bar{p})/(n - 1)$, which suggests that $\sqrt{\bar{p}(1-\bar{p})/(n-1)}$ should be used in place of $\sqrt{\bar{p}(1-\bar{p})/n}$ in (8.8). However, the bias introduced by using n in the denominator does not cause any difficulty because large samples are generally used in making estimates about population proportions. In such cases the numerical difference between the results obtained by using n and those obtained by using $n - 1$ is negligible.

Determining the Sample Size

Let us consider the question of how large the sample size should be to obtain an estimate of a population proportion with a specified margin of error. The rationale for the sample-size determination in developing interval estimates of p is similar to the rationale used in Section 8.3 to determine the sample size for estimating a population mean.

Previously in this section we pointed out that the margin of error associated with an estimate of a population proportion is $z_{\alpha/2}\sigma_{\bar{p}}$. With $\sigma_{\bar{p}} = \sqrt{p(1-p)/n}$, the margin of error is based on the values of $z_{\alpha/2}$, the population proportion p, and the sample size n. For a given confidence coefficient $1 - \alpha$, $z_{\alpha/2}$ can be determined. Then, since the value of the population proportion is fixed, the margin of error is determined by the sample size n. Larger sample sizes provide a smaller margin of error and better precision.

Let E = the desired margin of error; thus

$$E = z_{\alpha/2}\sqrt{\frac{p(1-p)}{n}}$$

Solving the equation for n provides the following formula for the sample size.

Sample Size for Estimation of a Population Proportion

$$n = \frac{(z_{\alpha/2})^2 p(1-p)}{E^2} \tag{8.9}$$

In (8.9), the margin of error E must be specified by the user; in most cases, E is .10 or less. User preference also specifies the confidence level and thus the corresponding value of $z_{\alpha/2}$. Finally, use of (8.9) requires a planning value for the population proportion p. In practice, this planning value can be chosen by one of the following procedures.

1. Use the sample proportion from a previous sample.
2. Use a pilot study to select a preliminary sample. The sample proportion from this sample can be used as the planning value for p.
3. Use judgment or a "best guess" for the value of p.
4. If none of the preceding alternatives apply, use $p = .50$.

Let us return to the Scheer Industries example where we were interested in estimating the proportion of employees who pass the training program examination. How large a sample of employees should be used if Scheer's director of manufacturing wants to estimate the population proportion with a margin of error of .10 at a 95% confidence level? With $E = .10$ and $z_{.025} = 1.96$, we need a planning value for p to answer the sample-size question. Previously in this section we reported that 36 of the 45 employees who took the examination passed. Therefore, $\bar{p} = 36/45 = .80$ can be used as the planning value for p. Using (8.9), we obtain

$$n = \frac{(1.96)^2 .80(1-.80)}{(.10)^2} = 61.47$$

Hence, a sample size of 62 employees is recommended.

The fourth alternative suggested for selecting a planning value for p is to use $p = .50$. This value of p is frequently used when no other information is available. To

TABLE 8.5 Some Possible Values for $p(1 - p)$

p	$p(1 - p)$	
.10	(.10)(.90) = .09	
.30	(.30)(.70) = .21	
.40	(.40)(.60) = .24	
.50	(.50)(.50) = .25	← Largest value for $p(1 - p)$
.60	(.60)(.40) = .24	
.70	(.70)(.30) = .21	
.90	(.90)(.10) = .09	

understand why, note that the numerator of (8.9) shows the sample size is proportional to the quantity $p(1 - p)$. A larger value for the quantity $p(1 - p)$ will result in a larger sample size. Table 8.5 gives some possible values of $p(1 - p)$. Note that the largest value of $p(1 - p)$ occurs when $p = .50$. Thus, if there is uncertainty about an appropriate planning value for p, we know that $p = .50$ will provide the largest sample-size recommendation. In effect, we are being on the safe or conservative side in recommending the largest possible sample size. If the proportion turns out to be different from the .50 planning value, the precision statement will be better than anticipated. In any case, in using $p = .50$, we are guaranteeing that the sample size will be sufficient to obtain the desired margin of error.

In the Scheer Industries example, a planning value of $p = .50$ would have provided the following recommended sample size.

$$n = \frac{(1.96)^2 .50(1 - .50)}{(.10)^2} = 96$$

This larger recommended sample size reflects the caution inherent in using the conservative planning value for the population proportion.

notes and comments

The margin of error for estimating a population proportion is almost always .10 or less. In national public opinion polls conducted by organizations such as Gallup and Harris, a .03 or .04 margin of error is generally reported. The use of these margins of error—E in Equation (8.9)—will generally provide a sample size that is large enough to satisfy the central limit theorem requirements of $np \geq 5$ and $n(1 - p) \geq 5$.

exercises

Methods

29. A simple random sample of 400 items provides 100 Yes responses.

a. What is the point estimate of the proportion of the population who would provide a Yes response?

b. What is the standard error of the proportion, $\sigma_{\bar{p}}$?
c. Compute the 95% confidence interval for the population proportion.

30. A simple random sample of 800 units generates a sample proportion $\bar{p} = .70$.
a. Provide a 90% confidence interval for the population proportion.
b. Provide a 95% confidence interval for the population proportion.

31. In a survey, the planning value for the population proportion p is given as .35. How large a sample should be taken to be 95% confident that the sample proportion is within $\pm$ .05 of the population proportion?

32. How large a sample should be taken to be 95% confident that the margin of error for the estimation of a population proportion is .03? Assume past data are not available for developing a planning value for p.

Applications

33. In a Louis Harris survey of 400 senior executives, 248 of the executives stated that the U.S. legal system significantly hampers the ability of U.S. companies to compete with Japanese and European companies (*Business Week,* April 13, 1992).
a. What is the point estimate of the population proportion of executives who believe the legal system hampers the ability to compete?
b. What is the 90% confidence interval for the population proportion?

34. The Bureau of National Affairs, Inc., selected a sample of 617 companies and found that 56 companies required their employees to surrender airline frequent-flier mileage awards for business-related travel (*The Wall Street Journal,* March 28, 1994).
a. What is the point estimate of the proportion of all companies that require their employees to surrender airline frequent-flier mileage awards?
b. Develop a 95% confidence interval estimate of the population proportion.

35. A *Time*/CNN survey of 600 adults was conducted to elicit public opinion about a variety of welfare-reform proposals (*Time,* May 23, 1994). When asked whether money should be withheld from the paychecks of fathers who refuse to make child-support payments, 570 of the respondents said yes. Develop a 95% confidence interval estimate of the proportion of all adults who believe money should be withheld from the paychecks of fathers who refuse to make child-support payments.

36. A *Fortune* subscriber survey conducted by Pulse On America, Inc., showed that 665 of 831 subscribers use a personal computer at work (1994 *Fortune* National Subscriber Portrait).
a. Develop a 95% confidence interval estimate of the proportion of *Fortune* subscribers who use a personal computer at work.
b. How large a sample should be taken to estimate the proportion of *Fortune* subscribers who use a personal computer with a margin of error of .02?

37. The Tourism Institute for the State of Florida plans to sample visitors at major beaches throughout the state to estimate the proportion of beach visitors who are not residents of Florida. Preliminary estimates are that 55% of the beach visitors are not Florida residents.
a. How large a sample should be taken to estimate the proportion of out-of-state visitors with a margin of error of 3%? Use a 95% confidence level.
b. How large a sample should be taken if the margin of error is increased to 6%?

38. A 1994 survey conducted by Louis Harris & Associates of 529 mutual-fund investors showed that 497 investors were confident that their investments were safe and only 127 investors planned to reduce their holdings (*Business Week,* August 15, 1994).
a. Develop a 95% confidence interval estimate of the proportion of mutual-fund investors who are confident that their investments are safe.
b. Develop a 95% confidence interval estimate of the proportion of mutual-fund investors who plan to reduce their holdings.

39. A firm provides national survey and interview services designed to estimate the proportion of the population who have certain beliefs or preferences. Typical questions seek to find the proportion favoring gun control, abortion, a particular political candidate, and so on. Assume that all interval estimates of population proportions are conducted at the 95% confidence level. How large a sample size would you recommend if the firm wants the margin of error to be

a. 3% **b.** 2% **c.** 1%

40. A survey of female executives conducted by Louis Harris & Associates showed that 33% of those surveyed rated their own company as an excellent place for women executives to work (*Working Woman,* November 1994). Suppose *Working Woman* wants to conduct an annual survey to monitor this proportion. With $p = .33$ as a planning value for the population proportion, how many female executives should be sampled for each of the following margins of error? Assume that all interval estimates are conducted at the 95% confidence level.

a. 10%
b. 5%
c. 2%
d. 1%
e. In general, what happens to the sample size as the margin of error decreases?

41. The National Automobile Dealers Association collects data on sales numbers, prices, and usage of both new and used automobiles. One statistic of interest is the percentage of automobiles that are still on the road after 10 years (*U.S. News & World Report,* September 9, 1991).

a. How large a sample should be taken if we want to be 95% confident that the sample percentage is within ± 2.5% of the actual percentage of automobiles that are still on the road after 10 years? Use $p = .25$ as a planning value for the population proportion.
b. Using your sample size in part (a), assume that 357 of the automobiles sampled in 1991 were still on the road after 10 years. Provide the point estimate and a 95% confidence interval for the population proportion.
c. In 1980, only 21% of automobiles were still on the road after 10 years. What conclusion can you make after viewing the confidence interval results for 1991 in part (b)?

summary

In this chapter we presented methods for developing a confidence interval for a population mean μ and a population proportion p. The purpose of developing a confidence interval is to give the user a better understanding of the margin of error that may be present. A wide confidence interval indicates poor precision; in such cases, the sample size can be increased to reduce the width of the confidence interval and improve the precision of the estimate.

Figure 8.11 summarizes the interval-estimation procedures for a population mean and provides a practical guide for computing the interval estimate. The figure shows that the expression used to compute an interval estimate depends on whether the sample size is large ($n \geq 30$) or small ($n < 30$), whether the population standard deviation is known, and in some cases whether or not the population has a normal or approximately normal probability distribution. If the sample size is large, no assumption is required about the distribution of the population and $z_{\alpha/2}$ is used in the computation of the interval estimate. If the sample size is small, the population must have a normal or approximately normal probability distribution in order to develop an interval estimate of μ. If this is the case, $z_{\alpha/2}$ is used in the computation of the interval estimate when σ is known, whereas $t_{\alpha/2}$ is used when σ is estimated by the sample standard deviation s. Finally, if the sample size is small and the assumption of a normally distributed population is inappropriate, we

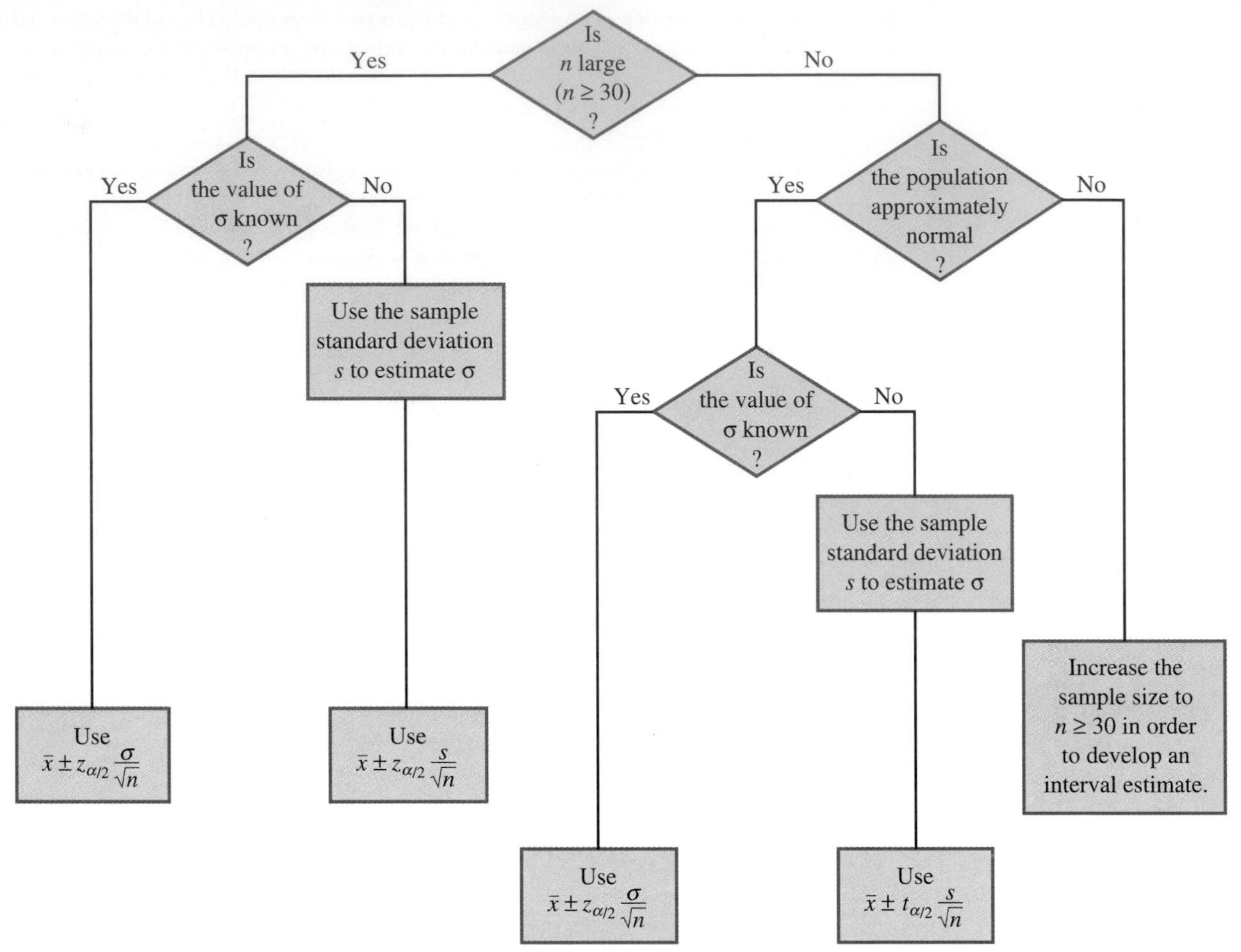

FIGURE 8.11
Summary of Interval Estimation Procedures for a Population Mean

recommend increasing the sample size to $n \geq 30$ to develop a large-sample interval estimate of the population mean.

In addition, we showed how to determine the sample size so that estimates of μ and p would have a specified margin of error. In practice, the sample sizes required for interval estimates of a population proportion are generally large. Hence, we provided the large-sample interval-estimation formulas for a population proportion where both $np \geq 5$ and $n(1 - p) \geq 5$.

glossary

Interval estimate An estimate of a population parameter that provides an interval believed to contain the value of the parameter.

Sampling error The absolute value of the difference between the value of an unbiased point estimator, such as the sample mean $\bar{x}$, and the value of the population parameter it estimates, such as the population mean μ; in this case the sampling error is $|\bar{x} - \mu|$. In the case of the population proportion, the sampling error is $|\bar{p} - p|$.

Precision A probability statement about the sampling error.

Confidence level The confidence associated with an interval estimate. For example, if an interval-estimation procedure provides intervals such that 95% of the intervals formed using the procedure will include the population parameter, the interval estimate is said to be constructed at the 95% confidence level; note that .95 is referred to as the *confidence coefficient.*

Margin of error The maximum sampling error defined in the precision statement. Generally the ± value used to construct the confidence interval.

***t* Distribution** A family of probability distributions that can be used to develop interval estimates of a population mean whenever the population standard deviation is unknown and the population has a normal or near-normal probability distribution.

Degrees of freedom A parameter of the t distribution. When the t distribution is used in the computation of an interval estimate of a population mean, the appropriate t distribution has $n - 1$ degrees of freedom, where n is the size of the simple random sample.

key formulas

Sampling Error When Estimating μ

$$|\bar{x} - \mu| \tag{8.1}$$

Interval Estimate of a Population Mean: Large-Sample Case With σ Known

$$\bar{x} \pm z_{\alpha/2}\frac{\sigma}{\sqrt{n}} \tag{8.2}$$

Interval Estimate of a Population Mean: Large-Sample Case With σ Unknown

$$\bar{x} \pm z_{\alpha/2}\frac{s}{\sqrt{n}} \tag{8.3}$$

Interval Estimate of a Population Mean: Small-Sample Case With σ Unknown

$$\bar{x} \pm t_{\alpha/2}\frac{s}{\sqrt{n}} \tag{8.4}$$

Sample Size for an Interval Estimate of a Population Mean

$$n = \frac{(z_{\alpha/2})^2\sigma^2}{E^2} \tag{8.5}$$

Interval Estimate of a Population Proportion

$$\bar{p} \pm z_{\alpha/2}\sqrt{\frac{\bar{p}(1-\bar{p})}{n}} \qquad (8.8)$$

Sample Size for an Interval Estimate of a Population Proportion

$$n = \frac{(z_{\alpha/2})^2 p(1-p)}{E^2} \qquad (8.9)$$

supplementary exercises

42. A study showed that 99% of U.S. households have at least one television set and 98% of Americans watch some television every day (*In Health,* January 1992). In addition, the study indicated a mean of 2.25 television sets per household. Assume that the mean number of sets per household was based on a sample of 300 households and that the sample standard deviation was 1.2 television sets per household. Provide a 95% confidence interval estimate of the population mean number of television sets per household.

43. A 1993 survey conducted by the American Automobile Association showed that a family of four spends an average of $215.60 per day while on vacation. Suppose a sample of 64 families of four vacationing at Niagara Falls resulted in a sample mean of $252.45 per day and a sample standard deviation of $74.50.

a. Develop a 95% confidence interval estimate of the mean amount spent per day by a family of four visiting Niagara Falls.

b. With the confidence interval from part (a), does it appear that the population mean amount spent per day by families visiting Niagara Falls is different from the mean reported by the American Automobile Association? Explain.

44. What is the mean annual compensation paid to the chief executive officers of the largest firms in the United States? A sample of eight firms provided the annual salary/bonus data (*Business Week,* April 25, 1994) is as follows.

Firm	Annual Salary/ Bonus ($1000s)
Coca-Cola	3654
General Motors	1375
Intel	2184
Motorola	1736
Readers Digest	1708
Sears	3095
Sprint	1692
Wells Fargo	2125

a. What is the point estimate of the population mean annual salary/bonus for chief executives?

b. What is the point estimate of the population standard deviation?

c. What is the 95% confidence interval estimate of the population mean annual compensation for chief executives?

45. The Atlantic Fishing and Tackle Company has developed a new synthetic fishing line. To estimate the breaking strength of this line (pounds), testers subjected six lengths of line to breakage testing. The following data were obtained from a normal probability distribution.

Line	1	2	3	4	5	6
Breaking Strength (pounds)	18	24	19	21	20	18

Develop a 95% confidence interval for the mean breaking strength of the new line.

46. Sample assembly times for a particular manufactured part were 8, 10, 10, 12, 15, and 17 minutes. If the mean of the sample is used to estimate the mean of the population of assembly times, provide a point estimate and a 90% confidence interval for the population mean. Assume that the population has a normal probability distribution.

47. Consider the Atlantic Fishing and Tackle Company problem presented in Exercise 45. How large a sample would be necessary to estimate the mean breaking strength of the new line with a .99 probability of a margin of error of one pound?

48. Mileage tests are conducted for a particular model of automobile. If the desired precision is stated such that there is a .98 probability that the margin of error is one mile per gallon, how many automobiles should be used in the test? Assume that preliminary mileage tests indicate the standard deviation to be 2.6 miles per gallon.

49. In developing patient appointment schedules, a medical center wants an estimate of the mean time that a staff member spends with each patient. How large a sample should be taken if the margin of error is to be 2 minutes at a 95% confidence level? How large a sample should be taken for a 99% confidence level? Use a planning value for the population standard deviation of eight minutes.

50. Exercise 44 provides annual salary/bonus data for chief executives of firms in the United States (*Business Week,* April 25, 1994). The sample standard deviation is $656.64, with data provided in thousands of dollars. How many chief executives should be in the sample if we want to estimate the population mean annual salary/bonus with a margin of error of $100,000? Note that the margin of error is $100 in thousands of dollars. Use a 95% confidence level.

51. A University of Michigan study (January 1992) found that drug use among college students had continued its decade-long decline. However, the survey showed that alcohol consumption remained steady. Among the 1400 college students surveyed, 602 indicated they had consumed five or more drinks within the two weeks prior to the survey. Compute a 95% confidence interval for the proportion of all college students who had five or more drinks within the two-week period.

52. Towers Perrin, a compensation consultant, asked 500 U.S. companies whether they encouraged quality by giving top performers recognition and/or rewards such as cash and stock (*Business Week,* December 1991). Results showed that 56% of the companies used recognition to encourage quality, whereas 26% of the companies used cash and stock rewards. Develop 95% confidence intervals for the population in terms of both the proportion of companies that use recognition and the proportion that use cash and stock rewards to encourage quality.

53. A *Time*/CNN telephone poll of 1400 American adults asked, "Where would you rather go in your spare time?" (*Time,* April 6, 1992). The top response by 504 adults was a shopping mall.

a. What is the point estimate of the proportion of adults who would prefer going to a shopping mall in their spare time?

b. At 95% confidence, what is the margin of error associated with this estimate?

54. A well-known bank credit-card firm is interested in estimating the proportion of credit-card holders who carry a nonzero balance at the end of the month and incur an interest charge. Assume that the desired margin of error for the proportion estimate is 3% at a 98% confidence level.

a. How large a sample should be selected if it is anticipated that roughly 70% of the firm's cardholders carry a nonzero balance at the end of the month?

b. How large a sample should be selected if no planning value for the population proportion could be specified?

55. A sample of 200 people were asked to identify their major source of news information; 110 stated that their major source was television news coverage.

a. Construct a 95% confidence interval for the proportion of people in the population who consider television their major source of news information.

b. How large a sample would be necessary to estimate the population proportion with a margin of error of .05 at a 95% confidence level?

56. Although airline schedules and cost are important factors for business travelers when choosing an airline carrier, a *USA Today* survey found that business travelers list an airline's frequent-flyer program as the most important factor (*USA Today,* April 11, 1995). From a sample of business travelers who responded to the survey, 618 of 1993 business travelers listed a frequent-flyer program as the most important factor.

a. What is the point estimate of the proportion of the population of business travelers who believe a frequent-flyer program is the most important factor when choosing an airline carrier?

b. Develop a 95% confidence interval estimate of the population proportion.

c. How large a sample would be required to report a margin of error of .01 at 95% confidence? Would you recommend that *USA Today* attempt to provide this degree of precision? Why or why not?

computer case

Bock Investment Services

Lisa Rae Bock started Bock Investment Services (BIS) in 1994 with the goal of making BIS the leading money market advisory service in South Carolina. To provide better service for her present clients and to attract new clients, she has developed a weekly newsletter. Lisa has been considering adding a new feature to the newsletter that will report the results of a weekly telephone survey of fund managers. To investigate the feasibility of offering this service, and to determine what type of information to include in the newsletter, Lisa selected a simple random sample of 45 money market funds. A portion of the data obtained is shown in Table 8.6, which reports fund assets and yields for the past seven and 30 days (*Barron's,* October 3, 1994). Before calling the money market fund managers to obtain additional data, Lisa decided to do some preliminary analysis of the data already collected (see Table 8.6). The data are available in the data set BOCK.

Managerial Report

1. Use appropriate descriptive statistics to summarize the data on assets and yields for the money market funds.

2. Develop a 95% confidence interval estimate of the mean assets, mean seven-day yield, and mean 30-day yield for the population of money market funds. Provide a managerial interpretation of each interval estimate.

3. Discuss the implication of your findings in terms of how Lisa could use this type of information in preparing her weekly newsletter.

4. What other information would you recommend that Lisa gather to provide the most useful information to her clients?

TABLE 8.6 Data for Bock Investment Services Computer Case

BOCK

Money Market Fund	Assets (Mil $)	7-day Yield (%)	30-day Yield (%)
Amcore	103.9	4.10	4.08
Alger	156.7	4.79	4.73
Arch MM/Trust	496.5	4.17	4.13
BT Instit Treas	197.8	4.37	4.32
Benchmark Div	2755.4	4.54	4.47
Bradford	707.6	3.88	3.83
Capital Cash	1.7	4.29	4.22
Cash Mgt Trust	2707.8	4.14	4.04
Composite	122.8	4.03	3.91
Cowen Standby	694.7	4.25	4.19
Cortland	217.3	3.57	3.51
Declaration	38.4	2.67	2.61
Dreyfus	4832.8	4.01	3.89
Elfun	81.7	4.51	4.41
FFB Cash	506.2	4.17	4.11
Federated Master	738.7	4.41	4.34
Fidelity Cash	13272.8	4.51	4.42
Flex-fund	172.8	4.60	4.48
Fortis	105.6	3.87	3.85
Franklin Money	996.8	3.97	3.92
Freedom Cash	1079.0	4.07	4.01
Galaxy Money	801.4	4.11	3.96
Government Cash	409.4	3.83	3.82
Hanover Cash	794.3	4.32	4.23
Heritage Cash	1008.3	4.08	4.00
Infinity/Alpha	53.6	3.99	3.91
John Hancock	226.4	3.93	3.87
Landmark Funds	481.3	4.28	4.26
Liquid Cash	388.9	4.61	4.64
MarketWatch	10.6	4.13	4.05
Merrill Lynch Money	27005.6	4.24	4.18
NCC Funds	113.4	4.22	4.20
Nationwide	517.3	4.22	4.14
Overland	291.5	4.26	4.17
Pierpont Money	1991.7	4.50	4.40
Portico Money	161.6	4.28	4.20
Prudential MoneyMart	6835.1	4.20	4.16
Reserve Primary	1408.8	3.91	3.86
Schwab Money	10531.0	4.16	4.07
Smith Barney Cash	2947.6	4.16	4.12
Stagecoach	1502.2	4.18	4.13
Strong Money	470.2	4.37	4.29
Transamerica Cash	175.5	4.20	4.19
United Cash	323.7	3.96	3.89
Woodward Money	1330.0	4.24	4.21

Source: *Barron's,* October 3, 1994.

computer case

Metropolitan Research, Inc.

Metropolitan Research, Inc., is a consumer research organization that takes surveys designed to evaluate a wide variety of products and services available to consumers. In one particular study, Metropolitan was interested in learning about consumer satisfaction with the performance of automobiles produced by a major Detroit manufacturer. A questionnaire sent to owners of one of the manufacturer's full-sized cars revealed several complaints about early transmission problems. To learn more about the transmission failures, Metropolitan used a sample of actual transmission repairs provided by a transmission repair firm in the Detroit area. The following data show the actual number of miles that 50 vehicles had been driven at the time of transmission failure. The data are available in the data set AUTO.

AUTO

85,092	32,609	59,465	77,437	32,534	64,090	32,464	59,902
39,323	89,641	94,219	116,803	92,857	63,436	65,605	85,861
64,342	61,978	67,998	59,817	101,769	95,774	121,352	69,568
74,276	66,998	40,001	72,069	25,066	77,098	69,922	35,662
74,425	67,202	118,444	53,500	79,294	64,544	86,813	116,269
37,831	89,341	73,341	85,288	138,114	53,402	85,586	82,256
77,539	88,798						

Managerial Report

1. Use appropriate descriptive statistics to summarize the transmission failure data.
2. Develop a 95% confidence interval for the mean number of miles driven until transmission failure for the population of automobiles that have had transmission failure. Provide a managerial interpretation of the interval estimate.
3. Discuss the implication of your statistical finding in terms of the belief that some owners of the automobiles have experienced early transmission failures.
4. How many repair records should be sampled if the research firm wants the population mean number of miles driven until transmission failure to be estimated with a margin of error of 5000 miles at 95% confidence?
5. What other information would you like to gather to evaluate the transmission failure problem more fully?

APPENDIX 8.1 Confidence Interval Estimation with Minitab

Large-Sample Case

In Section 8.1 we discussed the use of computer-generated confidence interval estimates by showing how Minitab can be used to obtain interval estimates for the mean age in the Statewide Insurance Company study. The data are in Table 8.2. With the sample standard deviation, $s = 7.77$, as an estimate of the population standard deviation σ, the following steps can be used to produce the 90% confidence-interval output shown in panel A of Figure 8.6. We assume the age data have been entered in column C1 of the Minitab worksheet.

STEP 1. Select the **Stat** pull-down menu
STEP 2. Select the **Basic Statistics** pull-down menu
STEP 3. Select the **1-Sample Z** option
STEP 4. When the dialog box appears:
Enter C1 in the **Variables** box
Enter 90 in the **Confidence interval Level** box
Enter 7.77 in the **Sigma** box
Select **OK**

The output in panel B of Figure 8.6 was obtained by simply changing the value of 90 to 95 in the **Confidence interval Level** box (step 4). If the user does not specify a value in the **Confidence interval Level** box, Minitab will use a default value of 95.

Small-Sample Case

In Section 8.2 we discussed the use of computer-generated confidence interval estimates for the small-sample case by showing how Minitab can be used to develop interval estimates for the Scheer Industries problem. With the data from Table 8.4 entered in column C1, the following steps can be used to produce a 95% confidence interval as shown in Figure 8.9.

STEP 1. Select the **Stat** pull-down menu
STEP 2. Select the **Basic Statistics** pull-down menu
STEP 3. Select the **1-Sample t** option
STEP 4. When the dialog box appears:
Enter C1 in the **Variables** box
Enter 95 in the **Confidence interval Level** box
Select **OK**

A 90% confidence interval can be obtained by simply changing the value of 95 to 90 in the **Confidence interval Level** box (step 4). If the user does not specify a value in the **Confidence interval Level** box, Minitab will use a default value of 95.

APPENDIX 8.2 Confidence Interval Estimation with Spreadsheets

Large-Sample Case

We will show how Excel can be used to develop confidence intervals for a population mean in the large-sample case by describing how to develop a 90% confidence interval for the population mean age in the Statewide Insurance Company study introduced in Section 8.1. We assume the user has already used the Data Analysis Tools described in Appendix 3.2 to compute the sample mean of 39.5 and the sample standard deviation of 7.77.

STEP 1. Select an empty cell in the Excel worksheet
STEP 2. Select the **Insert** pull-down menu
STEP 3. Choose the **Function** option
STEP 4. When The Function Wizard—Step 1 of 2 dialog box appears:
Choose **Statistical** in the Function Category box

Choose **Confidence** in the Function Name box
Select **Next** >

STEP 5. When The Function Wizard—Step 2 of 2 dialog box appears:
Enter .10 in the **alpha** box*
Enter 7.77 in the **standard_dev** box
Enter 36 in the **size** box
Select **Finish**

The margin of error will appear in the cell selected in step 1. With the sample mean of 39.5 and a margin of error of 2.13, the 90% confidence interval obtained by subtracting 2.13 from 39.5 and adding 2.13 to 39.5 is 37.37 to 41.63.

Small-Sample Case

To illustrate how to develop a confidence interval for the small-sample case, we will compute a 95% confidence interval for the Scheer Industries study introduced in Section 8.2. We assume that the user has entered the training-time data for the 15 employees (Table 8.4) into worksheet rows 1 to 15 of column A. The following steps can be used to produce a 95% confidence interval.

STEP 1. Select the **Tools** pull-down menu

STEP 2. Choose the **Data Analysis** option

STEP 3. When the Data Analysis dialog box appears:
Choose **Descriptive Statistics**
Select **OK**

STEP 4. When the **Descriptive Statistics** dialog box appears:
Enter A1:A15 in the **Input Range** box
Select **Confidence Level for Mean** and enter 95 in the box
Select **Output Range** and enter B1 in the box
Select **Summary Statistics**
Select **OK**

The value of the sample mean, 53.87, appears in cell C3 and the value of the margin of error, 3.78, appears in cell C16; note that the Excel label for the margin of error is Confidence Level (95.0%). With the sample mean of 53.87, the 95% confidence interval estimate obtained by subtracting 3.78 from 53.87 and adding 3.78 to 53.87 is 50.09 to 57.65.

---

*Alpha is 1 minus the confidence coefficient. For a .90 confidence coefficient the value of alpha is 1 − .90 = .10.

CHAPTER 9

Hypothesis Testing

Harris Corporation*

Melbourne, Florida

Harris Corporation's RF Communications Division, in Melbourne, Florida, is a major manufacturer of point-to-point radio communications equipment. It is a horizontally integrated manufacturing company with a multiplant facility. Most of the Harris products require medium- to high-volume production operations, including printed circuit assembly, final product assembly, and testing.

One of the company's high-volume products has an assembly called an RF deck. Each RF deck consists of 16 electronic components soldered to a machined casting that forms the plated surface of the deck. During a manufacturing run, a problem developed in the soldering process; the flow of solder onto the deck did not meet the quality criteria established for the product. After considering a variety of factors that might affect the soldering process, an engineer made the preliminary determination that the soldering problem was most likely due to defective platings.

The engineer wondered whether the proportion of defective platings in the Harris inventory exceeded that set by the supplier's design specifications. With p indicating the proportion of defective platings in the Harris inventory and p_0 indicating the proportion of defective platings set by the supplier's design specifications, the following hypotheses were formulated.

$$H_0: p \leq p_0$$

$$H_a: p > p_0$$

H_0 indicates that the Harris inventory has a defective plating proportion less than or equal to that set by the design specifications. Such a proportion would be judged acceptable, and the engineer would need to look for other causes of the soldering problem. However, H_a indicates that the Harris inventory has a defective plating proportion greater than that set by the design specifications. In that case, excessive defective platings may well be the cause of the soldering problem and action should be taken to determine why the defective proportion in inventory is so high.

Tests made on a sample of platings from the Harris inventory resulted in the rejection of H_0. The conclusion was that H_a was true and that the proportion of defective platings in inventory exceeded that set by the supplier's design specifications. Further investigation of the inventory area led to the conclusion that the underlying problem was shelf contamination during storage. By altering the storage environment, the engineer was able to solve the problem.

In this chapter you will learn how to formulate hypotheses about a population mean and a population proportion. Through the analysis of sample data, you will be able to determine whether a hypothesis should or should not be rejected. Appropriate conclusions and actions will be demonstrated for testing research hypotheses, testing the validity of assumptions, and decision making.

Maintaining quality of components is a high priority at Harris Corporation.

The authors are indebted to Richard A. Marshall of the Harris Corporation for providing this Statistics in Practice.

In Chapters 7 and 8 we showed how a sample could be used to develop point and interval estimates of population parameters. In this chapter we continue the discussion of statistical inference by showing how *hypothesis testing* can be used to determine whether a statement about the value of a population parameter should or should not be rejected.

In hypothesis testing we begin by making a tentative assumption about a population parameter. This tentative assumption is called the *null hypothesis* and is denoted by

H_0. We then define another hypothesis, called the *alternative hypothesis,* which is the opposite of what is stated in the null hypothesis. The alternative hypothesis is denoted by H_a. The hypothesis-testing procedure involves using data from a sample to test the two competing statements indicated by H_0 and H_a.

The purpose of this chapter is to show how hypothesis tests can be conducted about a population mean and a population proportion. We begin by providing examples that illustrate approaches to developing null and alternative hypotheses.

9.1 Developing Null and Alternative Hypotheses

In some applications it may not be obvious how the null and alternative hypotheses should be formulated. Care must be taken to be sure the hypotheses are structured appropriately and that the hypothesis-testing conclusion provides the information the researcher or decision maker wants. Guidelines for establishing the null and alternative hypotheses will be given for three types of situations in which hypothesis-testing procedures are commonly employed.

Testing Research Hypotheses

Consider a particular automobile model that currently attains an average fuel efficiency of 24 miles per gallon. A product-research group has developed a new carburetor specifically designed to increase the miles-per-gallon rating. To evaluate the new carburetor, several will be manufactured, installed in automobiles, and subjected to research-controlled driving tests. Note that the product-research group is looking for evidence to conclude that the new design *increases* the mean miles-per-gallon rating. In this case, the research hypothesis is that the new carburetor will provide a mean miles-per-gallon rating exceeding 24; that is, $\mu > 24$. As a general guideline, a research hypothesis such as this should be formulated as the *alternative hypothesis*. Hence, the appropriate null and alternative hypotheses for the study are:

$$H_0\colon \mu \leq 24$$
$$H_a\colon \mu > 24$$

If the sample results indicate that H_0 cannot be rejected, researchers cannot conclude that the new carburetor is better. Perhaps more research and subsequent testing should be conducted. However, if the sample results indicate that H_0 can be rejected, researchers can make the inference that H_a: $\mu > 24$ is true. With this conclusion, the researchers have the statistical support necessary to state that the new carburetor increases the mean number of miles per gallon. Action to begin production with the new carburetor may be undertaken.

In research studies such as these, the null and alternative hypotheses should be formulated so that the rejection of H_0 supports the conclusion and action being sought. In such cases, the research hypothesis should be expressed as the alternative hypothesis.

Testing the Validity of a Claim

As an illustration of testing the validity of a claim, consider the situation of a manufacturer of soft drinks who states that two-liter containers of its products have an average of at least 67.6 fluid ounces. A sample of two-liter containers will be selected, and the contents will be measured to test the manufacturer's claim. In this type of hypothesis-testing situation, we generally begin by assuming that the manufacturer's claim is true. Using this approach for the soft-drink example, we would state the null and alternative hypotheses as follows.

$$H_0: \mu \geq 67.6$$
$$H_a: \mu < 67.6$$

If the sample results indicate H_0 cannot be rejected, the manufacturer's claim cannot be challenged. However, if the sample results indicate H_0 can be rejected, the inference will be made that $H_a: \mu < 67.6$ is true. With this conclusion, statistical evidence indicates that the manufacturer's claim is incorrect and that the soft-drink containers are being filled with a mean less than the claimed 67.6 ounces. Appropriate action against the manufacturer may be considered.

In any situation that involves testing the validity of a product claim, the null hypothesis is generally based on the assumption that the claim is true. The alternative hypothesis is then formulated so that rejection of H_0 will provide statistical evidence that the stated assumption is incorrect. Action to correct the claim should be considered whenever H_0 is rejected.

Testing in Decision-Making Situations

In testing research hypotheses or testing the validity of a claim, action is taken if H_0 is rejected. In many instances, however, action must be taken both when H_0 cannot be rejected and when H_0 can be rejected. In general, this type of situation occurs when a decision maker must choose between two courses of action, one associated with the null hypothesis and another associated with the alternative hypothesis. For example, on the basis of a sample of parts from a shipment that has just been received, a quality-control inspector must decide whether to accept the entire shipment or to return the shipment to the supplier because it does not meet specifications. Assume that specifications for a particular part state that a mean length of two inches is desired. If the mean is greater or less than two inches, the parts will cause quality problems in the assembly operation. In this case, the null and alternative hypotheses would be formulated as follows.

$$H_0: \mu = 2$$
$$H_a: \mu \neq 2$$

If the sample results indicate H_0 cannot be rejected, the quality-control inspector will have no reason to doubt that the shipment meets specifications, and the shipment will be accepted. However, if the sample results indicate H_0 should be rejected, the conclusion will be that the parts do not meet specifications. In this case, the quality-control inspector will have sufficient evidence to return the shipment to the supplier. Thus, we see that for these types of situations, action is taken both when H_0 cannot be rejected and when H_0 can be rejected.

Summary of Forms for Null and Alternative Hypotheses

Let μ_0 denote the specific numerical value being considered in the null and alternative hypotheses. In general, a hypothesis test about the values of a population mean μ must take one of the following three forms.

$$H_0: \mu \geq \mu_0 \qquad H_0: \mu \leq \mu_0 \qquad H_0: \mu = \mu_0$$
$$H_a: \mu < \mu_0 \qquad H_a: \mu > \mu_0 \qquad H_a: \mu \neq \mu_0$$

In many situations, the choice of H_0 and H_a is not obvious and judgment is necessary to select the proper form. However, as the preceding forms show, the equality part of the expression (either $\geq$, $\leq$, or $=$) *always* appears in the null hypothesis. In selecting the proper form of H_0 and H_a, keep in mind that the alternative hypothesis is what the test is attempting to establish. Hence, asking whether the user is looking for evidence to support $\mu < \mu_0$, $\mu > \mu_0$, or $\mu \neq \mu_0$ will help determine H_a. The following exercises are designed to provide practice in choosing the proper form for a hypothesis test.

exercises

1. The manager of the Danvers-Hilton Resort Hotel has stated that the mean amount spent by guests for a weekend is $400 or less. A member of the hotel's accounting staff has noticed that the total charges for weekend guests have been increasing in recent months. The accountant will use a sample of weekend guest bills to test the manager's claim.

a. Which form of the hypotheses should be used to test the manager's claim? Explain.

H_0: $\mu \geq 400$	H_0: $\mu \leq 400$	H_0: $\mu = 400$
H_a: $\mu < 400$	H_a: $\mu > 400$	H_a: $\mu \neq 400$

b. What conclusion is appropriate when H_0 cannot be rejected?

c. What conclusion is appropriate when H_0 can be rejected?

2. The manager of an automobile dealership is considering a new bonus plan designed to increase sales volume. Currently, the mean sales volume is 14 automobiles per month. The manager wants to conduct a research study to see whether the new bonus plan increases sales volume. To collect data on the plan, a sample of sales personnel will be allowed to sell under the new bonus plan for a one-month period.

Self Test

a. Develop the null and alternative hypotheses that are most appropriate for this research situation.

b. Comment on the conclusion when H_0 cannot be rejected.

c. Comment on the conclusion when H_0 can be rejected.

3. A production line operation is designed to put 32 ounces of laundry detergent in each carton. A sample of cartons is periodically selected and weighed to determine whether underfilling or overfilling is occurring. If the sample data lead to a conclusion of underfilling or overfilling, the production line will be shut down and adjusted to obtain proper filling.

a. Formulate the null and alternative hypotheses that will help in deciding whether to shut down and adjust the production line.

b. Comment on the conclusion and the decision when H_0 cannot be rejected.

c. Comment on the conclusion and the decision when H_0 can be rejected.

4. Because of high production-changeover time and costs, a director of manufacturing must convince management that a proposed manufacturing method reduces costs before the new method can be implemented. The current production method operates with a mean cost of $220 per hour. A research study is to be conducted in which the cost of the new method will be measured over a sample production period.

a. Develop the null and alternative hypotheses that are most appropriate for this study.

b. Comment on the conclusion when H_0 cannot be rejected.

c. Comment on the conclusion when H_0 can be rejected.

TABLE 9.1 Errors and Correct Conclusions in Hypothesis Testing

		Population Condition	
		H_0 *True*	H_a *True*
Conclusion	*Accept* H_0	Correct Conclusion	Type II Error
	Reject H_0	Type I Error	Correct Conclusion

9.2 Type I and Type II Errors

The null and alternative hypotheses are competing statements about a population parameter. Either the null hypothesis H_0 is true or the alternative hypothesis H_a is true, but not both. Ideally the hypothesis testing procedure should lead to the acceptance of H_0 when H_0 is true and the rejection of H_0 when H_a is true. Unfortunately, this is not always possible. Since hypothesis tests are based on sample information, we must allow for the possibility of errors. Table 9.1 illustrates the two kinds of errors that can be made in hypothesis testing.

The first row of Table 9.1 shows what can happen when the conclusion is to accept H_0. Since either H_0 is true or H_a is true, if H_0 is true and the conclusion is to accept H_0, this conclusion is correct. However if H_a is true and the conclusion is to accept H_0, we have made a *Type II error;* that is, we have accepted H_0 when it is false. The second row of Table 9.1 shows what can happen when the conclusion is to reject H_0. In this case, if H_0 is true, we have made a *Type I error;* that is, we have rejected H_0 when it is true. However, if H_a is true, and the conclusion is to reject H_0, this conclusion is correct.

Although we cannot eliminate the possibility of errors in hypothesis testing, we can consider the probability of their occurrence. Using common statistical notation, we denote the probabilities of making the two errors as follows.

$$\alpha = \text{the probability of making a Type I error}$$

$$\beta = \text{the probability of making a Type II error}$$

Recall the hypothesis-testing illustration discussed in Section 9.1 in which an automobile product-research group had developed a new carburetor designed to increase the miles-per-gallon rating of a particular automobile. With the current model obtaining an average of 24 miles per gallon, the hypothesis test was formulated as follows.

$$H_0\text{: } \mu \le 24$$
$$H_a\text{: } \mu > 24$$

The alternative hypothesis, H_a: $\mu > 24$, indicates that the researchers are looking for sample evidence that will support the conclusion that the mean miles per gallon is greater than 24.

In this application, the Type I error of rejecting H_0 when it is true corresponds to the researchers claiming that the new carburetor improves the miles per gallon rating

($\mu > 24$) when in fact the new carburetor is not any better than the current carburetor. In contrast, the Type II error of accepting H_0 when it is false corresponds to the researchers concluding that the new carburetor is not any better than the current carburetor ($\mu \leq 24$) when in fact the new carburetor improves miles-per-gallon performance.

In practice, the person conducting the hypothesis test specifies the maximum allowable probability of making a Type I error, called the *level of significance* for the test. Common choices for the level of significance are .05 and .01. Referring to the second row of Table 9.1, note that the conclusion to *reject* H_0 indicates that either a Type I error or a correct conclusion has been made. Thus, if the probability of making a Type I error is controlled for by selecting a small value for the level of significance, we have a high degree of confidence that the conclusion to reject H_0 is correct. In such cases, we have statistical support for concluding that H_0 is false and H_a is true. Any action suggested by the alternative hypothesis H_a is appropriate.

Although most applications of hypothesis testing control for the probability of making a Type I error, they do not always control for the probability of making a Type II error. Hence, if we decide to accept H_0, we cannot determine how confident we can be with that decision. Because of the uncertainty associated with making a Type II error, statisticians often recommend that we use the statement "do not reject H_0" instead of "accept H_0." Using the statement "do not reject H_0" carries the recommendation to withhold both judgment and action. In effect, by never directly accepting H_0, the statistician avoids the risk of making a Type II error. Whenever the probability of making a Type II error has not been determined and controlled, we will not make the conclusion to accept H_0. In such cases, only two conclusions are possible: *do not reject* H_0 or *reject* H_0.

notes and comments

Many applications of hypothesis testing have a decision-making goal. The conclusion *reject* H_0 provides the statistical support to conclude that H_a is true and take whatever action is appropriate. The statement "do not reject H_0," although inconclusive, often forces managers to behave as though H_0 is true. In this case, managers need to be aware of the fact that such behavior may result in a Type II error.

exercises

Self Test

5. Americans spend an average of 8.6 minutes per day reading newspapers (*USA Today,* April 10, 1995). A researcher believes that individuals in management positions spend more than the national average time per day reading newspapers. A sample of individuals in management positions will be selected by the researcher. Data on newspaper-reading times will be used to test the following null and alternative hypotheses.

$$H_0: \mu \leq 8.6$$
$$H_a: \mu > 8.6$$

a. What is the Type I error in this situation? What are the consequences of making this error?

b. What is the Type II error in this situation? What are the consequences of making this error?

6. The label on a three-quart container of orange juice indicates that the orange juice contains an average of one gram of fat or less. Answer the following questions for a hypothesis test that could be used to test the claim on the label.

a. Develop the appropriate null and alternative hypotheses.

b. What is the Type I error in this situation? What are the consequences of making this error?

c. What is the Type II error in this situation? What are the consequences of making this error?

7. Carpetland salespersons have had sales averaging $8000 per week. Steve Contois, the firm's vice president, has proposed a compensation plan with new selling incentives. Steve hopes that the results of a trial selling period will enable him to conclude that the compensation plan increases the mean sales per salesperson.

a. Develop the appropriate null and alternative hypotheses.

b. What is the Type I error in this situation? What are the consequences of making this error?

c. What is the Type II error in this situation? What are the consequences of making this error?

8. Suppose a new production method will be implemented if a hypothesis test supports the conclusion that the new method reduces the mean operating cost per hour.

a. State the appropriate null and alternative hypotheses if the mean cost for the current production method is $220 per hour.

b. What is the Type I error in this situation? What are the consequences of making this error?

c. What is the Type II error in this situation? What are the consequences of making this error?

9.3 One-Tailed Tests About a Population Mean: Large-Sample Case

The Federal Trade Commission (FTC) periodically conducts studies designed to test the claims manufacturers make about their products. For example, the label on a large can of Hilltop Coffee states that the can contains at least three pounds of coffee. Suppose we want to check this claim by using hypothesis testing.

The first step is to develop the null and alternative hypotheses. We begin by tentatively assuming that the manufacturer's claim is correct. If the population of coffee cans has a mean weight of three or more pounds per can, Hilltop's claim about its product is correct. However, if the population of coffee cans has a mean weight less than three pounds per can, Hilltop's claim is incorrect.

With μ denoting the mean weight of cans for the population, the null and the alternative hypotheses are as follows.

$$H_0: \mu \geq 3$$
$$H_a: \mu < 3$$

If the sample data indicate that H_0 cannot be rejected, the statistical evidence does not support the conclusion that a label violation has occurred. Hence, no action would be taken against Hilltop. However, if sample data indicate that H_0 can be rejected, we will conclude that the alternative hypothesis, H_a: $\mu < 3$, is true. In that case, a conclusion of underfilling and a charge of a label violation would be appropriate.

FIGURE 9.1
Sampling Distribution of $\bar{x}$ for the Hilltop Coffee Study When the Null Hypothesis Is True (μ = 3)

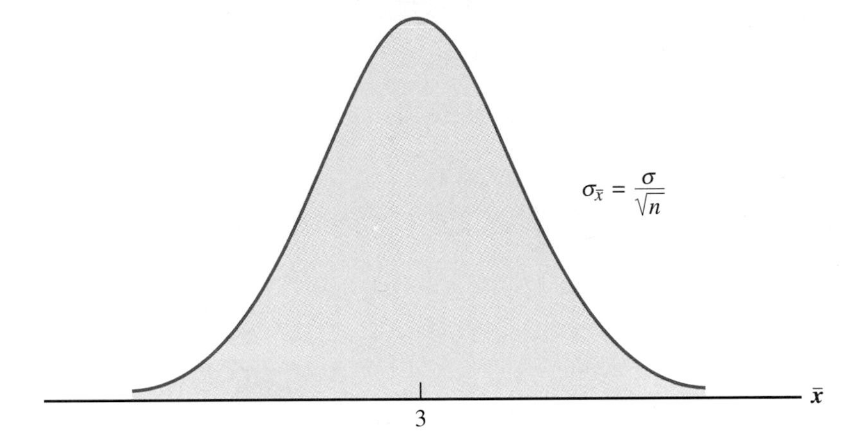

FIGURE 9.2
The Probability That $\bar{x}$ Is More Than 1.645 Standard Deviations Below the Mean of μ = 3

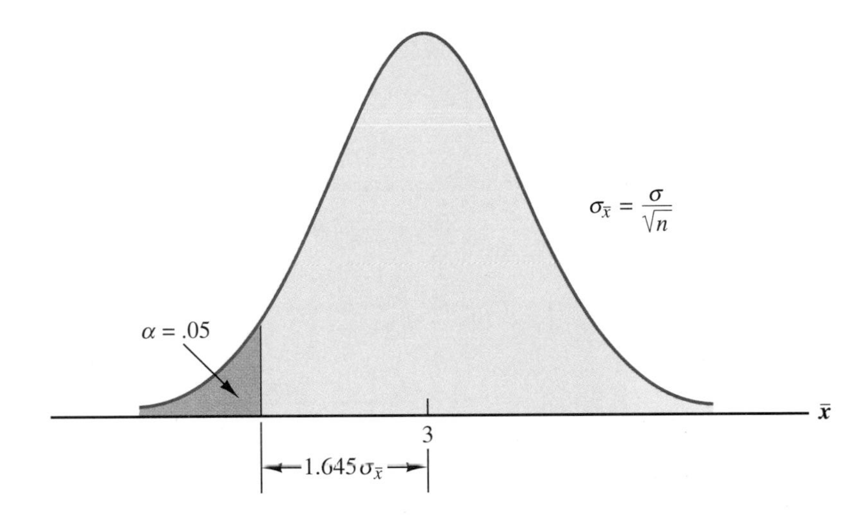

Suppose a random sample of 36 cans of coffee is selected. Note that if the mean filling weight for the sample of 36 cans is less than three pounds, the sample results will begin to cast doubt on the null hypothesis H_0: $\mu \geq 3$. But how much less than three pounds must $\bar{x}$ be before we would be willing to risk making a Type I error and falsely accuse the company of a label violation?

To answer this question, let us tentatively assume that the null hypothesis is true with $\mu = 3$. From the study of sampling distributions in Chapter 7, we know that whenever the sample size is large ($n \geq 30$), the sampling distribution of $\bar{x}$ can be approximated by a normal probability distribution. Figure 9.1 shows the sampling distribution of $\bar{x}$ when the null hypothesis is true at $\mu = 3$.

The value of $z = (\bar{x} - 3)/\sigma_{\bar{x}}$ gives the number of standard deviations $\bar{x}$ is from $\mu = 3$. For hypothesis tests about a population mean, we will use z as a *test statistic* to determine whether $\bar{x}$ deviates enough from $\mu = 3$ to justify rejecting the null hypothesis. Note that a value of $z = -1$ means that $\bar{x}$ is 1 standard deviation below $\mu = 3$, a value of $z = -2$ means that $\bar{x}$ is 2 standard deviations below $\mu = 3$, and so on. Obtaining a value of $z < -3$ is very unlikely if the null hypothesis is true. The key question is: How small must the test statistic z be before we have enough evidence to reject the null hypothesis?

Figure 9.2 shows that the probability of observing a value for $\bar{x}$ of more than 1.645 standard deviations below the mean of $\mu = 3$ is .05. Hence, if we were to

FIGURE 9.3
Hilltop Coffee Rejection Rule Has a Level of Significance of α = .01

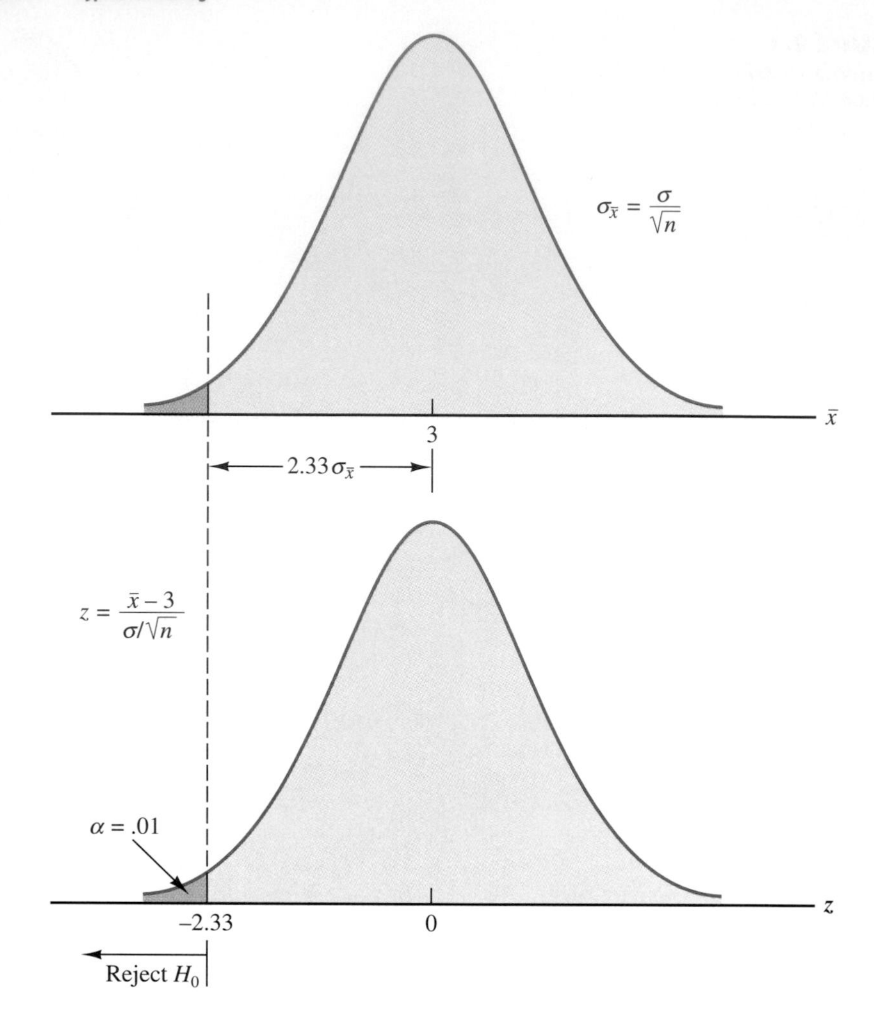

reject the null hypothesis whenever the value of the test statistic $z = (\bar{x} - 3)/\sigma_{\bar{x}}$ is less than – 1.645, the probability of making a Type I error would be .05. If the FTC considered .05 to be an acceptable probability of making a Type I error, we would reject the null hypothesis whenever the test statistic indicates that the sample mean is more than 1.645 standard deviations below $\mu = 3$. Thus, we would reject H_0 if $z < -1.645$.

The methodology of hypothesis testing requires that we specify the maximum allowable probability of a Type I error. As noted in the preceding section, this maximum probability is called the level of significance for the test; it is denoted by α, and it represents the probability of making a Type I error when the null hypothesis is true as an equality. The manager must specify the level of significance. If the cost of making a Type I error is high, a small value should be chosen for the level of significance. If the cost is not high, a larger value may be appropriate.

In the Hilltop Coffee study, the director of the FTC's testing program has made the following statement: "If the company is meeting its weight specifications exactly ($\mu = 3$), I would like a 99% chance of not taking any action against the company. While I do not want to accuse the company wrongly of underfilling its product, I am willing to risk a 1% chance of making this error."

From the director's statement, the maximum probability of a Type I error is .01. Hence, the level of significance for the hypothesis test is $\alpha = .01$. Figure 9.3 shows the sampling distributions of both $\bar{x}$ and $z = (\bar{x} - \mu)/\sigma_{\bar{x}}$ for the Hilltop Coffee example.

FIGURE 9.4
Value of the Test Statistic for $\bar{x} = 2.92$ Is in the Rejection Region

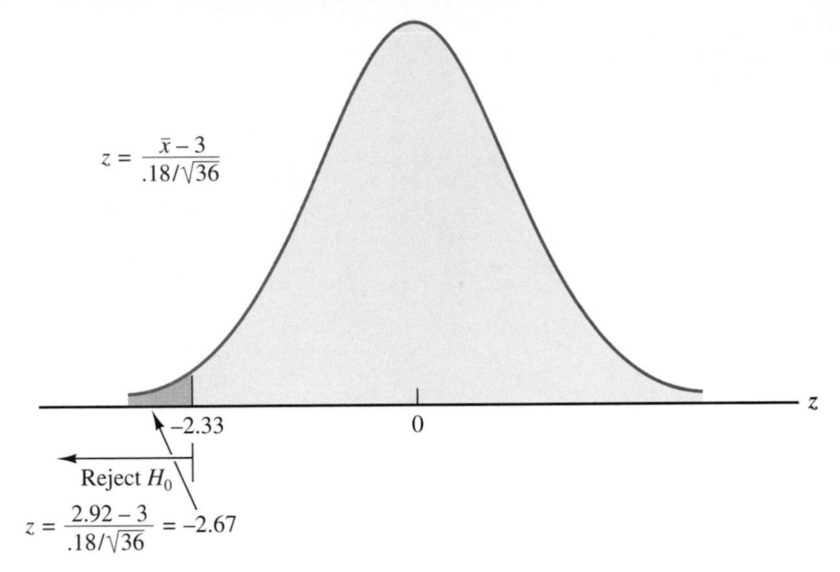

Note that when the null hypothesis is true at $\mu = 3$, the probability is .01 that $\bar{x}$ is more than 2.33 standard deviations below the mean of 3. Therefore, we establish the following rejection rule.

$$\text{Reject } H_0 \text{ if } z = \frac{\bar{x} - \mu}{\sigma_{\bar{x}}} < -2.33$$

If the value of $\bar{x}$ is such that the test statistic z is in the rejection region, we reject H_0 and conclude that H_a is true. If the value of $\bar{x}$ is such that the test statistic z is not in the rejection region, we cannot reject H_0. Note that the rejection region in Figure 9.3 is in only one tail of the sampling distribution. In such cases, we say the test is a *one-tailed* hypothesis test.

Suppose a sample of 36 cans provides a mean of $\bar{x} = 2.92$ pounds and we know from previous studies that the population standard deviation is $\sigma = .18$. With $\sigma_{\bar{x}} = \sigma/\sqrt{n}$, the value of the test statistic is given by

$$z = \frac{\bar{x} - 3}{\sigma/\sqrt{n}} = \frac{2.92 - 3}{.18/\sqrt{36}} = -2.67$$

Figure 9.4 shows that the value of the test statistic is in the rejection region. We are now justified in concluding that $\mu < 3$ at a .01 level of significance. The director has statistical justification for taking action against Hilltop Coffee for underfilling its product.

Suppose the sample of 36 cans had provided a sample mean of $\bar{x} = 2.97$. In this case, the value of the test statistic would be

$$z = \frac{\bar{x} - 3}{\sigma/\sqrt{n}} = \frac{2.97 - 3}{.18/\sqrt{36}} = -1.00$$

Since $z = -1.00$ is greater than -2.33, the value of the test statistic is not in the rejection region (see Figure 9.5). Hence we cannot reject the null hypothesis. Thus, with a sample mean of $\bar{x} = 2.97$, no statistical justification is provided for taking action against Hilltop Coffee.

The value $z = -2.33$ establishes the boundary of the rejection region and is called the *critical value.* In establishing the critical value, we tentatively assume the null hypothesis is true. For Hilltop Coffee, the null hypothesis is true whenever $\mu \geq 3$; however, we considered only the case when $\mu = 3$. What about the case when

FIGURE 9.5
Value of the Test Statistic for $\bar{x} = 2.97$ Is Not in the Rejection Region

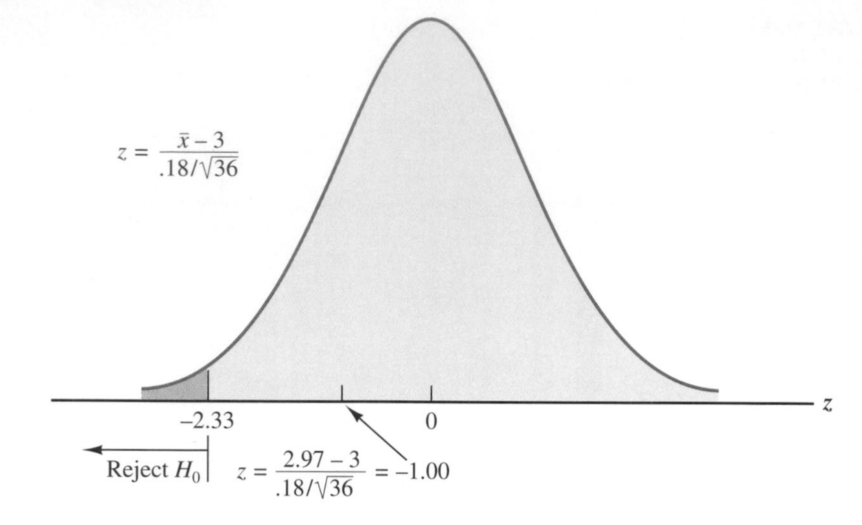

FIGURE 9.6
Rejection Region for an Upper-Tail Hypothesis Test About a Population Mean

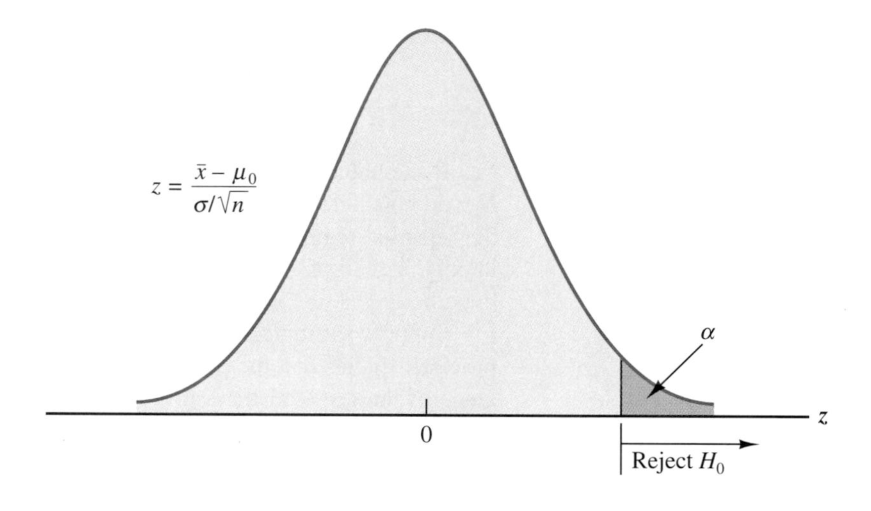

$\mu > 3$? If $\mu > 3$, the probability of making a Type I error will be less than it is when $\mu = 3$; that is, we are even less likely to find a value of the test statistic that is in the rejection region. Since the objective of the hypothesis-testing procedure is to control for the maximum probability of making a Type I error, the critical value for the test is established by assuming $\mu = 3$.

Summary: One-Tailed Tests About a Population Mean

Let us generalize the hypothesis-testing procedure for one-tailed tests about a population mean. We consider the large-sample case ($n \geq 30$) in which the central limit theorem enables us to assume that the sampling distribution $\bar{x}$ can be approximated by a normal probability distribution. In the large-sample case with σ unknown, we simply substitute the sample standard deviation s for σ in computing the test statistic. The general form of a lower-tail test, where μ_0 is a stated value for the population mean, follows.

Large-Sample ($n \geq 30$) Hypothesis Test About a Population Mean for a One-Tailed Test of the Form

$$H_0\colon \mu \geq \mu_0$$
$$H_a\colon \mu < \mu_0$$

Test Statistic: σ Known

$$z = \frac{\bar{x} - \mu_0}{\sigma/\sqrt{n}} \tag{9.1}$$

Test Statistic: σ Unknown

$$z = \frac{\bar{x} - \mu_0}{s/\sqrt{n}}$$

Rejection Rule at a Level of Significance of α

$$\text{Reject } H_0 \text{ if } z < -z_\alpha$$

A second form of the one-tailed test rejects the null hypothesis when the test statistic is in the upper tail of the sampling distribution. This one-tailed test and rejection rule are summarized next (see Figure 9.6). Again, we are considering the large-sample case; when σ is unknown, s can be substituted for σ in the computation of the test statistic z.

Large-Sample ($n \geq 30$) Hypothesis Test About a Population Mean for a One-Tailed Test of the Form

$$H_0\colon \mu \leq \mu_0$$
$$H_a\colon \mu > \mu_0$$

Test Statistic: σ Known

$$z = \frac{\bar{x} - \mu_0}{\sigma/\sqrt{n}} \tag{9.2}$$

Test Statistic: σ Unknown

$$z = \frac{\bar{x} - \mu_0}{s/\sqrt{n}}$$

Rejection Rule at a Level of Significance of α

$$\text{Reject } H_0 \text{ if } z > z_\alpha$$

FIGURE 9.7
p-Value for the Hilltop Coffee Study When $\bar{x} = 2.92$ and $z = -2.67$

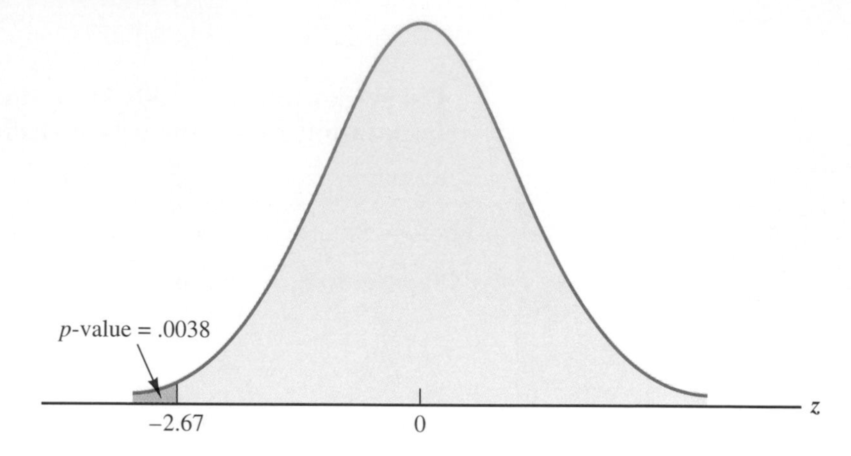

The Use of p-Values

Another approach that can be used to decide whether to reject H_0 is based on a probability called a *p-value.* If we assume that the null hypothesis is true, the p-value is the probability of obtaining a sample result that is at least as unlikely as what is observed. In the Hilltop Coffee example, the rejection region is in the lower tail; therefore the p-value is the probability of observing a sample mean less than or equal to what is observed. The p-value is often called the observed level of significance.

Let us compute the p-value associated with the sample mean $\bar{x} = 2.92$ in the Hilltop Coffee example. The p-value in this case is the probability of obtaining a value for the sample mean that is less than or equal to the observed value of $\bar{x} = 2.92$, given the hypothesized value for the population mean of $\mu = 3$. Previously we showed that the test statistic $z = -2.67$ corresponded to $\bar{x} = 2.92$. Thus, as shown in Figure 9.7, the p-value is the area in the tail of the standard normal probability distribution for $z = -2.67$. Using the standard normal probability distribution table, we find that the area between the mean and $z = -2.67$ is .4962. Hence, there is a $.5000 - .4962 = .0038$ probability of obtaining a sample mean that is less than or equal to the observed $\bar{x} = 2.92$. The p-value is therefore .0038. This p-value shows that there is a very small probability of obtaining a sample mean as small as $\bar{x} = 2.92$ when sampling from a population with $\mu = 3$.

The p-value can be used to make the decision in a hypothesis test by noting that if the p-value is *less than the level of significance* α, the value of the test statistic is in the rejection region. Similarly, if the p-value is *greater than or equal to* α, the value of the test statistic is not in the rejection region. For the Hilltop Coffee example, the fact that the p-value of .0038 is less than the level of significance, $\alpha = .01$, indicates that the null hypothesis should be rejected. Given the level of significance α, the decision of whether to reject H_0 can be made as follows.

p-Value Criterion for Hypothesis Testing

Reject H_0 if p-value $< \alpha$

The p-value and the corresponding test statistic will always provide the same hypothesis-testing conclusion. When the rejection region is in the lower tail of the sampling distribution, the p-value is the area under the curve less than or equal to the test statistic. When the rejection region is in the upper tail of the sampling distribution, the p-value is the area under the curve greater than or equal to the test statistic. A small p-value indicates a sample result that is unusual given the assumption that H_0 is true. Small p-values lead to rejection of H_0, whereas large p-values indicate that the null hypothesis cannot be rejected.

The Steps of Hypothesis Testing

In conducting the Hilltop Coffee hypothesis test, we carried out the steps that are required for any hypothesis-testing procedure. A summary of the steps that can be applied to any hypothesis test follows.

Steps of Hypothesis Testing

1. Determine the null and alternative hypotheses that are appropriate for the application.
2. Select the test statistic that will be used to decide whether to reject the null hypothesis.
3. Specify the level of significance α for the test.
4. Use the level of significance to develop the rejection rule that indicates the values of the test statistic that will lead to the rejection of H_0.
5. Collect the sample data and compute the value of the test statistic.
6. **a.** Compare the value of the test statistic to the critical value(s) specified in the rejection rule to determine whether H_0 should be rejected.
or
b. Compute the p-value based on the test statistic in step 5. Use the p-value to determine whether H_0 should be rejected.

notes and comments

The p-value, the observed level of significance, is a measure of the likelihood of the sample results when the null hypothesis is assumed to be true. The smaller the p-value, the less likely it is that the sample results came from a population where the null hypothesis is true. Most statistical software packages provide the p-value associated with a hypothesis test. The user can then compare the p-value to the level of significance α and draw a hypothesis test conclusion without referring to a statistical table.

exercises

Methods

9. Consider the following hypothesis test.

$$H_0\colon \mu \geq 10$$
$$H_a\colon \mu < 10$$

A sample with n=50 provides a sample mean of 9.46 and sample standard deviation of 2.

a. At α = .05, what is the critical value for z? What is the rejection rule?

b. Compute the value of the test statistic z. What is your conclusion?

Self Test

10. Consider the following hypothesis test.

$$H_0\colon \mu \leq 15$$
$$H_a\colon \mu > 15$$

A sample of 40 provides a sample mean of 16.5 and sample standard deviation of 7.

a. At α = .02, what is the critical value for z, and what is the rejection rule?

b. Compute the value of the test statistic z.

c. What is the p-value?

d. What is your conclusion?

11. Consider the following hypothesis test.

$$H_0\colon \mu \geq 25$$
$$H_a\colon \mu < 25$$

A sample of 100 is used and the population standard deviation is 12. Provide the value of the test statistic z and your conclusion for each of the following sample results. Use α = .05.

a. $\bar{x}$ = 22.0 **b.** $\bar{x}$ = 24.0 **c.** $\bar{x}$ = 23.5 **d.** $\bar{x}$ = 22.8

12. Consider the following hypothesis test.

$$H_0\colon \mu \leq 5$$
$$H_a\colon \mu > 5$$

Assume the following test statistics. Compute the corresponding p-values and make the appropriate conclusions based on α = .05.

a. z = 1.82 **b.** z = .45 **c.** z = 1.50 **d.** z = 3.30 **e.** z = −1.00

Applications

Self Test

13. Individuals filing 1994 federal income tax returns prior to March 31, 1995, had an average refund of $1056 (*USA Today,* April 5, 1995). Consider the population of "last-minute" filers who mail their returns during the last five days of the income tax period (typically April 10 to April 15).

a. A researcher suggests that one of the reasons individuals wait until the last five days to file their returns is that on average those individuals have a lower refund than early filers. Develop appropriate hypotheses such that rejection of H_0 will support the researcher's contention.

b. For a sample of 400 individuals who filed a return between April 10 and April 15, the sample mean refund was $910 and the sample standard deviation was $1600. At α = .05, what is your conclusion?

c. What is the p-value for the test?

14. In 1990, The Motor Vehicle Manufacturers Association, Detroit, Michigan, reported statistics on the average number of years passenger cars were being used. In 1980, the population mean number of years passenger cars were being used was reported to be

$\mu = 6.5$. Assume that 1990 data from a sample of 100 passenger cars showed a sample mean of 7.8 years and a sample standard deviation of 2.2 years.

a. Formulate the null and alternative hypotheses for a researcher who is looking for evidence to show that individuals were driving cars longer in 1990.

b. Do the data support the conclusion that individuals were driving cars longer? Use a .01 level of significance.

c. What implications does this finding have for vehicle manufacturers?

15. According to the National Automobile Dealers Association, the mean price for used cars is $10,192 (*USA Today,* April 12, 1995). A manager of a Kansas City used car dealership reviewed a sample of 100 recent used car sales at the dealership. The sample mean price was $9300 and the sample standard deviation was $4500. Letting μ denote the population mean price for used cars at the Kansas City dealership, test H_0: $\mu \geq 10{,}192$ and H_a: $\mu < 10{,}192$ at a .05 level of significance.

a. What is the hypothesis-testing conclusion?

b. What is the p-value?

c. What information does the hypothesis test result provide for the manager of the Kansas City dealership? What follow-up action might the manager want to consider?

16. New tires manufactured by a company in Findlay, Ohio, are designed to provide a mean of at least 28,000 miles. Tests with 30 randomly selected tires showed a sample mean of 27,500 miles and a sample standard deviation of 1000 miles. Using a .05 level of significance, test whether there is sufficient evidence to reject the claim of a mean of at least 28,000 miles. What is the p-value?

17. A company currently pays its production employees a mean wage of $15.00 per hour. The company is planning to build a new factory and is considering several locations. The availability of labor at a rate less than $15.00 per hour is a major factor in the location decision. For one location, a sample of 40 workers showed a current mean hourly wage of $\bar{x} = \$14.00$ and a sample standard deviation of $s = \$2.40$.

a. At a .10 level of significance, do the sample data indicate that the location has a mean wage rate significantly below the $15.00 per hour rate?

b. What is the p-value?

18. A new diet program claims that participants will lose on average at least eight pounds during the first week of the program. A random sample of 40 people participating in the program showed a sample mean weight loss of seven pounds. The sample standard deviation was 3.2 pounds.

a. What is the rejection rule with $\alpha = .05$?

b. What is your conclusion about the claim made by the diet program?

c. What is the p-value?

9.4 Two-Tailed Tests About a Population Mean: Large-Sample Case

Two-tailed hypothesis tests differ from one-tailed tests in that the rejection region is placed in both the lower and the upper tails of the sampling distribution. Let us introduce an example to show how and why two-tailed tests are conducted.

The United States Golf Association (USGA) has established rules that manufacturers of golf equipment must meet if their products are to be acceptable for use in USGA events. One of the rules for the manufacture of golf balls states: "A brand of golf ball, when tested on apparatus approved by the USGA on the outdoor range at the USGA Headquarters . . . shall not cover an average distance in carry and roll exceeding 280 yards. . . ." Suppose Superflight, Inc., has recently developed a high-technology manufacturing method that can produce golf balls having an average distance in carry and roll of 280 yards.

FIGURE 9.8
Rejection Region for the Two-Tailed Hypothesis Test for Superflight, Inc.

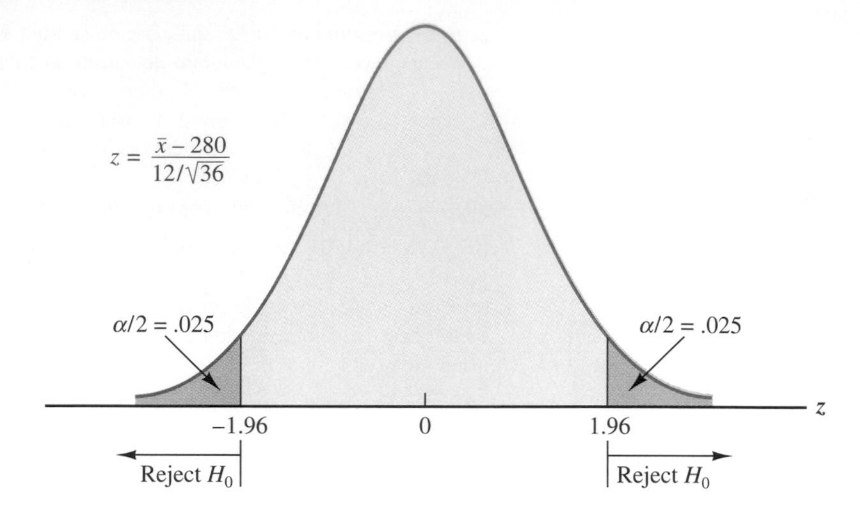

Superflight realizes, however, that if the new manufacturing process goes out of adjustment, the process may produce balls for which the average distance is less than or more than 280 yards. In the former case, sales may decline as a result of marketing an inferior product, and in the latter case the golf balls may be rejected by the USGA. Superflight's managers therefore have instituted a quality-control program to monitor the new manufacturing process.

As part of the quality-control program, an inspector periodically selects a sample of golf balls from the production line and subjects them to tests that are equivalent to those performed by the USGA. Having no reason to doubt that the manufacturing process is functioning correctly, we establish the following null and alternative hypotheses.

$$H_0: \mu = 280$$
$$H_a: \mu \neq 280$$

As usual, we make the tentative assumption that the null hypothesis is true. A rejection region must be established for the test statistic z. We want to reject the claim that $\mu = 280$ when the z value indicates that the population mean is either less than or great than 280 yards. Thus, H_0 should be rejected for values of the test statistic in either the lower tail or the upper tail of the sampling distribution. As a result, the test is called a *two-tailed* hypothesis test.

Following the hypothesis-testing procedure developed in the previous sections, we specify a level of significance by determining a maximum allowable probability of making a Type I error. Suppose we choose $\alpha = .05$. This means that there will be a .05 probability of concluding that the mean distance is not 280 yards when in fact it is. The test statistic is

$$z = \frac{\bar{x} - \mu_0}{\sigma/\sqrt{n}}$$

Figure 9.8 shows the sampling distribution of z with the two-tailed rejection region for $\alpha = .05$. With two-tailed hypothesis tests, we will always determine the rejection region by placing an area or probability of $\alpha/2$ in each tail of the distribution. The values of z that provide an area of .025 in each tail can be found from the standard normal probability distribution table. We see in Figure 9.8 that $-z_{.025} = -1.96$ identifies an area of .025 in the lower tail and $z_{.025} = +1.96$ identifies an area of .025 in the upper tail. Referring to Figure 9.8, we can establish the following rejection rule.

$$\text{Reject } H_0 \text{ if } z < -1.96 \text{ or if } z > 1.96$$

Suppose that a simple random sample of 36 golf balls provided the data in Table 9.2. For these data we obtained a sample mean of $\bar{x} = 278.5$ yards and a sample standard deviation of $s = 12$ yards. Using $\mu_0 = 280$ from the null hypothesis and the sample standard deviation of $s = 12$ as an estimate of the population standard deviation σ, the value of the test statistic is

$$z = \frac{\bar{x} - \mu_o}{\sigma/\sqrt{n}} = \frac{278.5 - 280}{12/\sqrt{36}} = -.75$$

According to the rejection rule, H_0 cannot be rejected. The sample results indicate that the quality-control manager has no reason to doubt the assumption that the manufacturing process is producing golf balls with a population mean distance of 280 yards.

Summary: Two-Tailed Tests About a Population Mean

Let μ_0 represent the value of the population mean in the hypotheses. The general form of the two-tailed hypothesis test about a population mean follows.

Large-Sample ($n \geq 30$) Hypothesis Test About a Population Mean for a Two-Tailed Test of the Form

$$H_0\colon \mu = \mu_0$$

$$H_a\colon \mu \neq \mu_0$$

Test Statistic: σ Known

$$z = \frac{\bar{x} - \mu_0}{\sigma/\sqrt{n}} \tag{9.3}$$

Test Statistic: σ Unknown

$$z = \frac{\bar{x} - \mu_0}{s/\sqrt{n}}$$

Rejection Rule at a Level of Significance of α

$$\text{Reject } H_0 \text{ if } z < -z_{\alpha/2} \text{ or if } z > z_{\alpha/2}$$

p-Values for Two-Tailed Tests

Let us compute the p-value for the Superflight golf ball example. The sample mean of $\bar{x} = 278.5$ has a corresponding z value of $-.75$. The table for the standard normal

TABLE 9.2 Distance Data for a Simple Random Sample of 36 Superflight Golf Balls

DISTANCE

Ball	Yards	Ball	Yards	Ball	Yards
1	269	13	296	25	272
2	300	14	265	26	285
3	268	15	271	27	293
4	278	16	279	28	281
5	282	17	284	29	269
6	263	18	260	30	299
7	301	19	275	31	263
8	295	20	282	32	264
9	288	21	260	33	273
10	278	22	266	34	291
11	276	23	270	35	274
12	286	24	293	36	277

probability distribution shows that the area between the mean and $z = -.75$ is .2734. Thus, the area in the lower tail is .5000 − .2734 = .2266. Looking at Figure 9.8, we see that the lower-tail portion of the rejection region has an area or probability of $\alpha/2 = .05/2 = .025$. Thus, with $.2266 > .025$, the test statistic is not in the rejection region, and the null hypothesis cannot be rejected.

One question remains: What value should we report as the *p*-value for the two-tailed test? At first glance, you may be inclined to say the *p*-value is .2266. If that is your choice, you will have to remember two different rules: one for the one-tailed test, which is to reject H_0 if the $p\text{-value} < \alpha$, and another for the two-tailed test, which is to reject H_0 if the $p\text{-value} < \alpha/2$. Alternatively, suppose we define the *p*-value for a two-tailed test as *double* the area found in the tail of the distribution. Thus, for the Superflight example, we would define the *p*-value to be $2(.2266) = .4532$. The advantage of this definition of the *p*-value for a two-tailed test is that the *p*-value can be compared directly to the level of significance α. Hence, with $.4532 > .05$, we see that the null hypothesis cannot be rejected. By remembering that the *p*-value for a two-tailed test is simply double the area found in the tail of the distribution, the previous rule to reject H_0 if the $p\text{-value} < \alpha$ can be used for all hypothesis tests.

The Role of the Computer

We illustrate how computer software packages can be applied in hypothesis testing by using Minitab to perform the analysis for the Superflight golf ball study. After entering the distance data for the sample of 36 golf balls into a Minitab worksheet, we obtained the hypothesis testing output in Figure 9.9. The first line of output shows that the hypothesis test is $\mu = 280$ versus $\mu \neq 280$. The assumed sigma of 12 indicates that the sample standard deviation $s = 12$ has been used to estimate the population standard deviation σ. The sample size 36, the sample mean 278.5, the sample standard deviation 12, and the standard error of the mean 2 are shown. The value of the test statistic $z = -.75$ and the $p\text{-value} = .45$ can be used to draw the hypothesis testing conclusion. Thus, at a .05 level of significance, the null hypothesis H_0: $\mu = 280$ cannot be rejected.

FIGURE 9.9
Minitab Output for the Superflight Golf Ball Hypothesis Test

```
Test of mu = 280.00 vs mu not = 280.00
The assumed sigma = 12.0

   N      Mean    StDev    SE Mean       Z    P value
  36    278.50    12.00       2.00   -0.75       0.45
```

The Relationship Between Interval Estimation and Hypothesis Testing

In Chapter 8 we showed how to develop a confidence interval estimate of a population mean. In the large-sample case, the confidence interval estimate of a population mean corresponding to a $1 - \alpha$ confidence coefficient is given by

$$\bar{x} \pm z_{\alpha/2}\frac{\sigma}{\sqrt{n}} \tag{9.4}$$

when σ is known and

$$\bar{x} \pm z_{\alpha/2}\frac{s}{\sqrt{n}} \tag{9.5}$$

when σ is unknown.

Conducting a hypothesis test requires us first to make an assumption about the value of a population parameter. In the case of the population mean, the two-tailed hypothesis test has the form

$$H_0\colon \mu = \mu_0$$

$$H_a\colon \mu \neq \mu_0$$

where μ_0 is the hypothesized value for the population mean. Using the rejection rule provided by (9.3), we see that the region over which we do not reject H_0 includes all values of the sample mean $\bar{x}$ that are within $-z_{\alpha/2}$ and $+z_{\alpha/2}$ standard errors of μ_0. Thus, the do-not-reject region for the sample mean $\bar{x}$ in a two-tailed hypothesis test is given by

$$\mu_0 \pm z_{\alpha/2}\frac{\sigma}{\sqrt{n}} \tag{9.6}$$

when σ is known and

$$\mu_0 \pm z_{\alpha/2}\frac{s}{\sqrt{n}} \tag{9.7}$$

when σ is unknown.

A close look at (9.4) and (9.6) provides insight about the relationship between the estimation and hypothesis-testing approaches to statistical inference. Note in particular that both procedures require the computation of the values $z_{\alpha/2}$ and $\sigma/\sqrt{n}$. Focusing on α, we see that a confidence coefficient of $(1 - \alpha)$ for interval estimation corresponds to a level of significance of α in hypothesis testing. For example, a 95% confidence interval corresponds to a .05 level of significance for hypothesis testing.

Furthermore, (9.4) and (9.6) show that since $z_{\alpha/2}\,(\sigma/\sqrt{n})$ is the plus or minus value for both expressions, if $\bar{x}$ is in the do-not-reject region defined by (9.6), the hypothesized value μ_0 will be in the confidence interval defined by (9.4). Conversely, if the hypothesized value μ_0 is in the confidence interval defined by (9.4), the sample mean $\bar{x}$ will be in the do-not-reject region for the hypothesis H_0: $\mu = \mu_0$ as defined by (9.6). These observations lead to the following procedure for using confidence interval results to draw hypothesis-testing conclusions.

A Confidence Interval Approach to Hypothesis Testing

Form of Hypothesis:

$$H_0: \mu = \mu_0$$
$$H_a: \mu \neq \mu_0$$

1. Select a simple random sample from the population and use the value of the sample mean $\bar{x}$ to develop the confidence interval for the population mean μ. If σ is known, compute the interval estimate by using

$$\bar{x} \pm z_{\alpha/2}\frac{\sigma}{\sqrt{n}}$$

If σ is unknown, compute the interval estimate by using

$$\bar{x} \pm z_{\alpha/2}\frac{s}{\sqrt{n}}$$

2. If the confidence interval contains the hypothesized value μ_0, do not reject H_0. Otherwise, reject H_0.

Let us return to the Superflight golf ball study and the following two-tailed test.

$$H_0: \mu = 280$$
$$H_a: \mu \neq 280$$

To test this hypothesis with a level of significance of $\alpha = .05$, we sampled 36 golf balls and found a sample mean of $\bar{x} = 278.5$ yards and a sample standard deviation of $s = 12$ yards. Using these results with $z_{.025} = 1.96$, we find that the 95% confidence interval estimate of the population mean becomes

$$\bar{x} \pm z_{.025}\frac{s}{\sqrt{n}}$$

$$278.5 \pm 1.96\frac{12}{\sqrt{36}}$$

$$278.5 \pm 3.92$$

or

$$274.58 \text{ to } 282.42$$

This finding enables the quality-control manager to conclude with 95% confidence that the mean distance for the population of golf balls is between 274.58 and 282.42 yards.

Since the hypothesized value for the population mean, $\mu_0 = 280$, is in this interval, the hypothesis-testing conclusion is that the null hypothesis, H_0: $\mu = 280$, cannot be rejected.

Note that this discussion and example pertain to two-tailed hypothesis tests about a population mean. However, the same confidence interval and hypothesis-testing relationship exists for other population parameters. In addition, the relationship can be extended to make one-tailed tests about population parameters. Doing so, however, requires the development of one-sided confidence intervals.

notes and comments

1. The p-value depends only on the sample outcome. However, it is necessary to know whether the hypothesis test being investigated is one-tailed or two-tailed. Given the value of $\bar{x}$ in a sample, the p-value for a two-tailed test will always be *twice* the area in the tail of the sampling distribution.
2. The interval-estimation approach to hypothesis testing helps to highlight the role of the sample size. From (9.4) we can see that larger sample sizes lead to narrower confidence intervals. Thus, for a given level of significance α, a larger sample is less likely to lead to an interval containing μ_0 when the null hypothesis is false. That is, the larger sample size will provide a higher probability of rejecting H_0 when H_0 is false.

exercises

Methods

19. Consider the following hypothesis test.

$$H_0: \mu = 10$$

$$H_a: \mu \neq 10$$

A sample of 36 provides a sample mean of 11 and sample standard deviation of 2.5.

a. At $\alpha = .05$, what is the rejection rule?
b. Compute the value of the test statistic z. What is your conclusion?

Self Test

20. Consider the following hypothesis test.

$$H_0: \mu = 15$$

$$H_a: \mu \neq 15$$

A sample of 50 gives a sample mean of 14.2 and sample standard deviation of 5.

a. At $\alpha = .02$, what is the rejection rule?
b. Compute the value of the test statistic z.
c. What is the p-value?
d. What is your conclusion?

21. Consider the following hypothesis test.

$$H_0: \mu = 25$$

$$H_a: \mu \neq 25$$

A sample of 80 is used and the population standard deviation is 10. Use $\alpha = .05$. Compute the value of the test statistic z and specify your conclusion for each of the following sample results.

a. $\bar{x} = 22.0$ **b.** $\bar{x} = 27.0$ **c.** $\bar{x} = 23.5$ **d.** $\bar{x} = 28.0$

22. Consider the following hypothesis test.

$$H_0: \mu = 5$$
$$H_a: \mu \neq 5$$

Assume the following test statistics. Compute the corresponding p-values and specify your conclusions based on $\alpha = .05$.

a. $z = 1.80$ **b.** $z = -.45$ **c.** $z = 2.05$ **d.** $z = -3.50$ **e.** $z = -1.00$

Applications

Self Test

23. The Bureau of Economic Analysis in the U.S. Department of Commerce reported that the mean annual income for a resident of North Carolina is \$18,688 (*USA Today,* August 24, 1995). A researcher for the state of South Carolina wants to test $H_0: \mu = 18{,}688$ and $H_a: \mu \neq 18{,}688$, where μ is the mean annual income for a resident of South Carolina.

a. What is the appropriate conclusion if a sample of 400 residents of South Carolina shows a sample mean annual income of \$16,860 and a sample standard deviation of \$14,624? Use a .05 level of significance.

b. What is the p-value for this test?

24. A study of the operation of a city-owned parking garage showed a historical mean parking time of 220 minutes per car. The garage area has recently been remodeled and the parking charges have been increased. The city manager would like to know whether these changes have had any effect on the mean parking time. Test $H_0: \mu = 220$ and $H_a: \mu \neq 220$ at a .05 level of significance.

a. What is your conclusion if a sample of 50 cars showed $\bar{x} = 208$ and $s = 80$?

b. What is the p-value?

25. A production line operates with a filling mean weight of 16 ounces per container. Overfilling or underfilling is a serious problem, and the production line should be shut down if either occurs. From past data, σ is known to be .8 ounces. A quality-control inspector samples 30 items every two hours and at that time makes the decision of whether to shut the line down for adjustment.

a. With a .05 level of significance, what is the rejection rule?

b. If a sample mean of $\bar{x} = 16.32$ ounces were found, what action would you recommend?

c. If $\bar{x} = 15.82$ ounces, what action would you recommend?

d. What is the p-value for parts (b) and (c)?

26. An automobile assembly-line operation has a scheduled mean completion time of 2.2 minutes. Because of the effect of completion time on both preceding and subsequent assembly operations, it is important to maintain the 2.2-minute mean completion time. A random sample of 45 assemblies shows a sample mean completion time of 2.39 minutes, with a sample standard deviation of .20 minutes. Use a .02 level of significance and test whether the operation is meeting its 2.2-minute mean completion time.

27. Historically, evening long-distance phone calls from a particular city have averaged 15.2 minutes per call. In a random sample of 35 calls, the sample mean time was 14.3 minutes per call, with a sample standard deviation of five minutes. Use this sample information to test whether there has been a change in the mean duration of long distance phone calls. Use a .05 level of significance. What is the p-value?

28. The mean salary for full professors in the United States is \$61,650 (*The American Almanac of Jobs and Salaries,* 1994–1995 Edition). A sample of 36 full professors at business colleges showed $\bar{x} = \$72{,}800$ and $s = \$5{,}000$.

a. Develop a 95% confidence interval for the population mean salary of business college professors.
b. Use the confidence interval to conduct the hypothesis test: H_0: μ = 61,650 and H_a: $\mu \neq$ 61,650. What is your conclusion?

9.5 Tests About a Population Mean: Small-Sample Case

Assume that the sample size is small ($n < 30$) and that the sample standard deviation s is used to estimate the population standard deviation σ. If it is also reasonable to assume that the population has a normal probability distribution, the t distribution can be used to make inferences about the value of the population mean.* In this case, the test statistic is

$$t = \frac{\bar{x} - \mu_0}{s/\sqrt{n}} \tag{9.8}$$

This test statistic has a t distribution with $n - 1$ degrees of freedom.

Let us consider an example of a one-tailed hypothesis test about a population mean for the small-sample case. The International Air Transport Association surveys business travelers to develop ratings of transatlantic gateway airports. The maximum possible score is 10. A magazine devoted to business travel has decided to classify airports according to the rating they receive. Airports that have a population mean rating of 7 or more will be designated as providing superior service. Suppose a simple random sample of 12 business travelers have been asked to rate London's Heathrow airport, and that the 12 ratings obtained are 7, 8, 10, 8, 6, 9, 6, 7, 7, 8, 9, and 8. The sample mean is $\bar{x} = 7.75$ and the sample standard deviation is $s = 1.215$. Assuming that the population of ratings can be approximated by a normal probability distribution, should Heathrow be designated as providing superior service?

Using a .05 level of significance, we need a test to determine whether the population mean rating for the Heathrow airport is greater than 7. The null and alternative hypotheses follows.

$$H_0: \mu \leq 7$$

$$H_a: \mu > 7$$

The Heathrow airport will be designated as providing superior service if H_0 can be rejected indicating that the population mean rating is greater than 7.

The rejection region is in the upper tail of the sampling distribution. With $n - 1 = 12 - 1 = 11$ degrees of freedom, Table 2 of Appendix B shows that $t_{.05} = 1.796$. Thus, the rejection rule is

$$\text{Reject } H_0 \text{ if } t > 1.796$$

Using (9.8) with $\bar{x} = 7.75$ and $s = 1.215$, we have the following value for the test statistic.

$$t = \frac{\bar{x} - \mu_0}{s/\sqrt{n}} = \frac{7.75 - 7}{1.215/\sqrt{12}} = 2.14$$

*If the sample size is small ($n < 30$), if the population standard deviation σ is known, and if the population has a normal probability distribution, $z = (\bar{x} - \mu_0)(\sigma/\sqrt{n})$ can be used as the test statistic for the small-sample hypothesis test.

FIGURE 9.10
Value of the Test Statistic (t = 2.14) for the Heathrow Airport Hypothesis Test

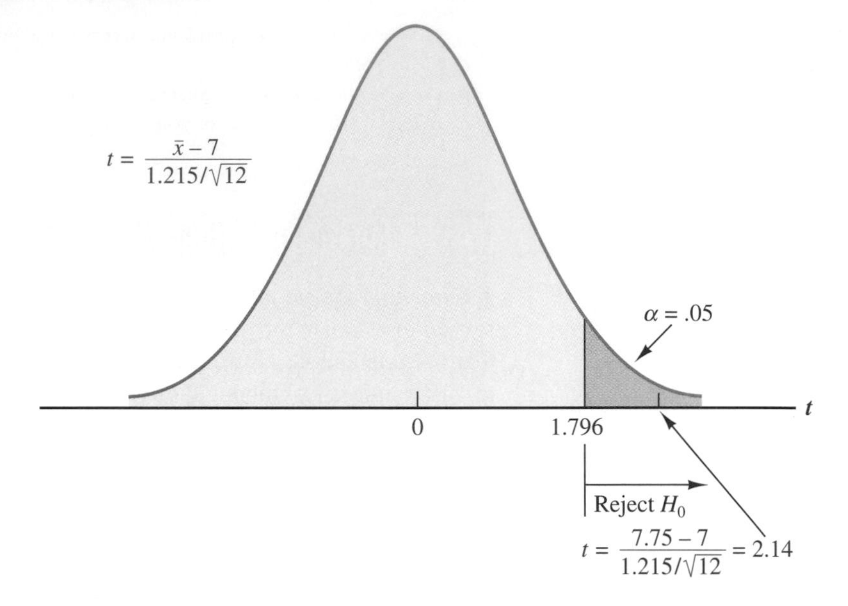

Since 2.14 is greater than 1.796, the null hypothesis is rejected. At the .05 level of significance, we can conclude that the population mean rating for the Heathrow airport is greater than 7. Thus, Heathrow can be designated as providing superior service. Figure 9.10 shows that the value of the test statistic is in the rejection region.

p-Values and the t Distribution

Let us consider the p-value for the Heathrow airport hypothesis test. The usual rule applies: If the p-value is less than the level of significance α, the null hypothesis can be rejected. Unfortunately, the format of the t distribution table provided in most statistics textbooks does not have sufficient detail to determine the exact p-value for the test. However, we can still use the t distribution table to identify a range for the p-value. For example, the t distribution used in the Heathrow airport hypothesis test has 11 degrees of freedom. Referring to Table 2 of Appendix B, we see that row 11 provides the following information about a t distribution with 11 degrees of freedom.

Area in Upper Tail	.10	.05	.025	.01	.005
***t* Value**	1.363	1.796	2.201	2.718	3.106

The computed t value for the hypothesis test was $t = 2.14$. The p-value is the area in the tail corresponding to $t = 2.14$. From the information above, we see 2.14 is between 1.796 and 2.201. Thus, although we cannot determine the exact p-value associated with $t = 2.14$, we do know that the p-value must be between .05 and .025. With a level of significance of $\alpha = .05$, we know that the p-value must be less than .05; thus, the null hypothesis is rejected.

The Role of the Computer

An advantage of computer software packages in testing hypotheses in the small-sample case is that the computer output will provide the p-value for the t distribution. To illustrate

FIGURE 9.11
Minitab Output for the Heathrow Airport Rating Hypothesis Test

```
Test of mu = 7.000 vs mu > 7.000

    N      Mean     StDev    SE Mean      T     P Value

   12     7.750     1.215      0.351   2.14       0.028
```

how computer software packages can be used to perform hypothesis testing with small samples, we used Minitab to perform the analysis for the Heathrow airport rating study. With the data for the sample of 12 business travelers entered into a Minitab worksheet, we obtained the hypothesis-testing output in Figure 9.11. The first line of output shows that the hypothesis test is being performed with the alternative hypothesis $\mu > 7$. The sample size 12, the sample mean 7.75, the sample standard deviation 1.215, and the standard error of the mean .351 are shown. The value of the test statistic $t = 2.14$ and the p-value $= .028$ can be used to draw the hypothesis-testing conclusion. Thus, at a .05 level of significance, the null hypothesis H_0: $\mu \leq 7$ is rejected. The conclusion that $\mu > 7$ and that Heathrow provides superior service is supported.

A Two-Tailed Test

As an example of a two-tailed hypothesis test about a population mean using a small sample, consider the following production problem. A production process is designed to fill containers with a mean filling weight of $\mu = 16$ ounces. If the process underfills containers, the consumer will not receive the amount of product indicated on the container label. If the process overfills containers, the firm loses money since more product is placed in a container than is required. To monitor the process, quality-assurance personnel periodically select a simple random sample of eight containers and test the following two-tailed hypotheses.

$$H_0\colon \mu = 16$$
$$H_a\colon \mu \neq 16$$

If H_0 is rejected, the production process will be stopped and the mechanism for regulating filling weights will be readjusted to ensure a mean filling weight of 16 ounces. If the sample yields data values of 16.02, 16.22, 15.82, 15.92, 16.22, 16.32, 16.12, and 15.92 ounces and the level of significance is .05, what action should be taken? Assume that the population of filling weights is normally distributed.

Since the data have not been summarized, we must first compute the sample mean and sample standard deviation. Doing so provides the following results.

$$\bar{x} = \frac{\Sigma x_i}{n} = \frac{128.56}{8} = 16.07 \text{ ounces}$$

and

$$s = \sqrt{\frac{\Sigma(x_i - \bar{x})^2}{n - 1}} = \sqrt{\frac{.22}{7}} = .18 \text{ ounces}$$

With a two-tailed test and a level of significance of $\alpha = .05$, the value of $-t_{.025}$ and the value of $t_{.025}$ are needed to determine the rejection region for the test. Using the table for the t distribution, we find that with $n - 1 = 8 - 1 = 7$ degrees of freedom, $-t_{.025} = -2.365$ and $t_{.025} = +2.365$. Thus, the rejection rule is written

$$\text{Reject } H_0 \text{ if } t < -2.365 \text{ or if } t > 2.365$$

Using $\bar{x} = 16.07$ and $s = .18$, we have

$$t = \frac{\bar{x} - \mu_0}{s/\sqrt{n}} = \frac{16.07 - 16}{.18/\sqrt{8}} = 1.10$$

Since $t = 1.10$ is not in the rejection region, the null hypothesis cannot be rejected. There is not enough evidence to stop the production process.

Using Table 2 of Appendix B and the row for seven degrees of freedom, we see that the computed t value of 1.10 has an upper tail area of *more than* .10. Although the format of the t distribution table prevents us from being more specific, we can at least conclude that the two-tailed p-value is greater than $2(.10) = .20$. Since this value is greater than the .05 level of significance, we see that the p-value leads to the same conclusion; that is, do not reject H_0. The computer solution for this problem shows $t = 1.10$ and the exact p-value $= .31$.

exercises

Methods

29. Consider the following hypothesis test.

$$H_0: \mu \leq 10$$
$$H_a: \mu > 10$$

A sample of 16 provides a sample mean of 11 and sample standard deviation of 3.

a. With $\alpha = .05$, what is the rejection rule?
b. Compute the value of the test statistic t. What is your conclusion?

Self Test

30. Consider the following hypothesis test.

$$H_0: \mu = 20$$
$$H_a: \mu \neq 20$$

Data from a sample of six items are: 18, 20, 16, 19, 17, 18.

a. Compute the sample mean.
b. Compute the sample standard deviation.
c. With $\alpha = .05$, what is the rejection rule?
d. Compute the value of the test statistic t.
e. What is your conclusion?

31. Consider the following hypothesis test.

$$H_0: \mu \geq 15$$
$$H_a: \mu < 15$$

A sample of 22 is used and the sample standard deviation is 8. Use $\alpha = .05$. Provide the value of the test statistic t and your conclusion for each of the following sample results.

a. $\bar{x} = 13.0$ **b.** $\bar{x} = 11.5$ **c.** $\bar{x} = 15.0$ **d.** $\bar{x} = 19.0$

32. Consider the following hypothesis test.

$$H_0: \mu \leq 50$$
$$H_a: \mu > 50$$

Assume a sample of 16 items provides the following test statistics. What can you say about the p-values in each case? What are your conclusions based on $\alpha = .05$?

a. $t = 2.602$ **b.** $t = 1.341$ **c.** $t = 1.960$ **d.** $t = 1.055$ **e.** $t = 3.261$

Applications

Self Test

33. In February 1995, the mean cost for an airline round trip with a discount fare was $258 (*USA Today,* March 30, 1995). A random sample of 15 round-trip discount fares during the month of March provided the following data.

310 260 265 255 300 310 230 250 265 280 290 240 285 250 260

a. What is the sample mean round trip discount fare in March?
b. What is the sample standard deviation?
c. Using $\alpha = .05$, test to see whether the mean round-trip discount fare has *increased* in March. What is your conclusion?
d. What is the p-value?

34. The average hourly wage in the United States is $10.05 (*The Tampa Tribune,* December 15, 1991). Assume that a sample of 25 individuals in Phoenix, Arizona, showed a sample mean wage of $10.83 per hour with a sample standard deviation of $3.25 per hour. Test $H_0: \mu = 10.05$ and $H_a: \mu \neq 10.05$ to see whether the population mean in Phoenix differs from the mean throughout the United States. At a .05 level of significance, what is your conclusion?

35. On the average, a housewife with a husband and two children is estimated to work 55 hours or less per week on household-related activities. The hours worked during a week for a sample of eight housewives are: 58, 52, 64, 63, 59, 62, 62, and 55.

a. Use $\alpha = .05$ to test $H_0: \mu \leq 55$, $H_a: \mu > 55$. What is your conclusion about the mean number of hours worked per week?
b. What can you say about the p-value?

36. A study of a drug designed to reduce blood pressure used a sample of 25 men between the ages of 45 and 55. With μ indicating the mean change in blood pressure for the population of men receiving the drug, the hypotheses in the study were written: $H_0: \mu \geq 0$ and $H_a: \mu < 0$. Rejection of H_0 shows that the mean change is negative, indicating that the drug is effective in lowering blood pressure.

a. At a .05 level of significance, what conclusion should be drawn if $\bar{x} = -10$ and $s = 15$?
b. What can you say about the p-value?

37. Last year the number of lunches served at an elementary-school cafeteria was normally distributed with a mean of 300 lunches per day. At the beginning of the current year, the price of a lunch was raised by 25¢. A sample of six days during the months of September, October, and November provided the following numbers of children being served lunches: 290, 275, 310, 260, 270, and 275. Do these data indicate that the mean number of lunches per day has dropped since last year? Test $H_0: \mu \geq 300$ against the alternative $H_a: \mu < 300$ at a .05 level of significance.

38. Joan's Nursery specializes in custom-designed landscaping for residential areas. The estimated labor cost associated with a particular landscaping proposal is based on the number of plantings of trees, shrubs, and so on to be used for the project. For cost-estimating purposes, managers use two hours of labor time for the planting of a medium-size tree. Actual times from a sample of 10 plantings during the past month follow (times in hours).

1.9 1.7 2.8 2.4 2.6 2.5 2.8 3.2 1.6 2.5

Using a .05 level of significance, test if the mean tree-planting time exceeds two hours. What is your conclusion, and what recommendations would you consider making to the managers?

9.6 Tests About a Population Proportion

With p denoting the population proportion and p_0 denoting a particular hypothesized value for the population proportion, the three forms for a hypothesis test about a population proportion are as follows.

$$H_0\colon p \geq p_0 \qquad H_0\colon p \leq p_0 \qquad H_0\colon p = p_0$$
$$H_a\colon p < p_0 \qquad H_a\colon p > p_0 \qquad H_a\colon p \neq p_0$$

The first two forms are one-tailed tests, whereas the third form is a two-tailed test. The specific form used depends on the application.

Hypothesis tests about a population proportion are based on the difference between the sample proportion $\bar{p}$ and the hypothesized population proportion p_0. The methods used to conduct the tests are similar to the procedures used for hypothesis tests about a population mean. The only difference is that we use the sample proportion $\bar{p}$ and its standard deviation $\sigma_{\bar{p}}$ in developing the test statistic. We begin by formulating null and alternative hypotheses about the value of the population proportion. Then, using the value of the sample proportion $\bar{p}$ and its standard deviation $\sigma_{\bar{p}}$, we compute a value for the test statistic z. Comparing the value of the test statistic to the critical value enables us to determine whether the null hypothesis should be rejected.

Let us illustrate hypothesis testing for a population proportion by considering the situation faced by Pine Creek golf course. Over the past few months, 20% of the players at Pine Creek have been women. In an effort to increase the proportion of women playing, Pine Creek used a special promotion to attract women golfers. After one week, a random sample of 400 players showed 300 men and 100 women. Course managers would like to determine whether the data support the conclusion that the proportion of women playing at Pine Creek has increased.

To determine whether the effect of the promotion has been to increase the proportion of women golfers, we state the following null and alternative hypotheses.

$$H_0\colon p \leq .20$$
$$H_a\colon p > .20$$

As usual, we begin the hypothesis-testing procedure by assuming that H_0 is true with $p = .20$. Using the sample proportion $\bar{p}$ to estimate p, we next consider the sampling distribution of $\bar{p}$. Since $\bar{p}$ is an unbiased estimator of p, we know that if $p = .20$, the mean of the sampling distribution of $\bar{p}$ is .20. In addition, we know from Chapter 7 that the standard deviation of $\bar{p}$ is given by

$$\sigma_{\bar{p}} = \sqrt{\frac{p(1-p)}{n}}$$

With the assumed value of $p = .20$ and a sample size of $n = 400$, the standard deviation of $\bar{p}$ is

$$\sigma_{\bar{p}} = \sqrt{\frac{.20(1-.20)}{400}} = .02$$

FIGURE 9.12
Sampling Distribution of $\bar{p}$ for the Proportion of Women Golfers at Pine Creek Golf Course

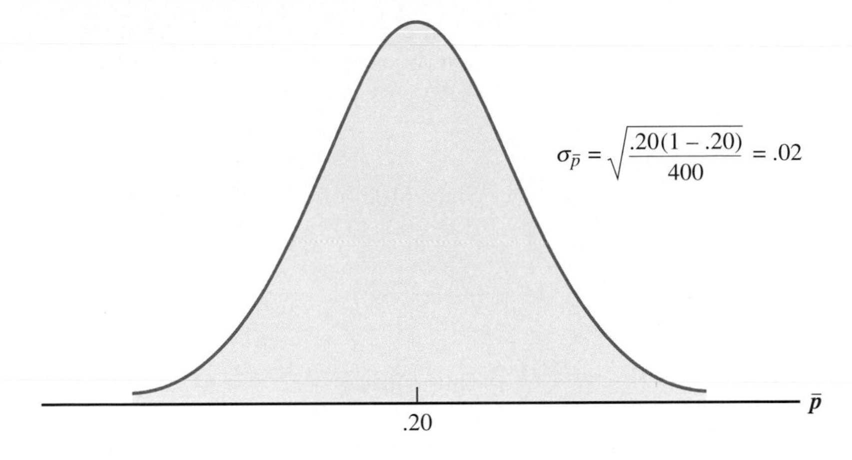

FIGURE 9.13
Rejection Region for the Pine Creek Golf Course Hypothesis Test

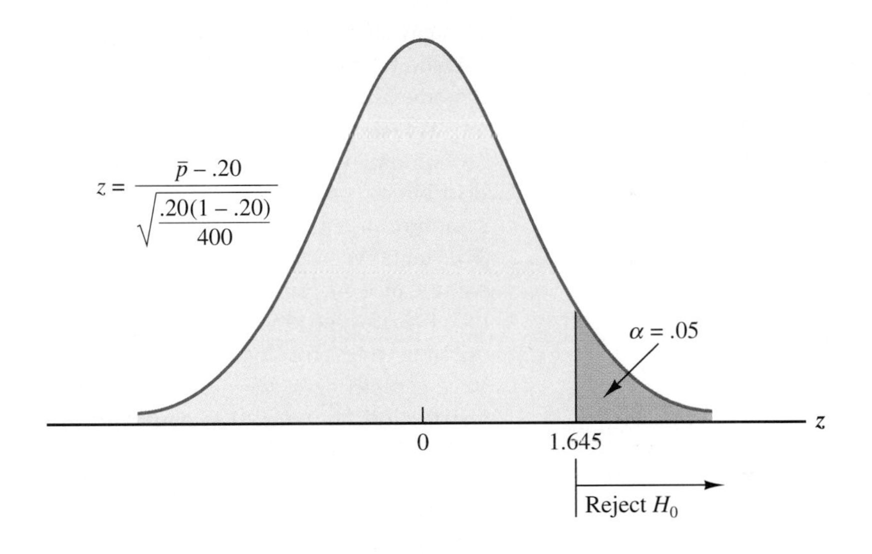

In Chapter 7 we saw that the sampling distribution of $\bar{p}$ can be approximated by a normal probability distribution if both np and $n(1 - p)$ are greater than or equal to 5. In the Pine Creek case, $np = 400(.20) = 80$ and $n(1 - p) = 400(.80) = 320$; thus, the normal probability distribution approximation is appropriate. The sampling distribution of $\bar{p}$ is shown in Figure 9.12.

Since the sampling distribution of $\bar{p}$ is approximately normal, the following test statistic can be used.

Test Statistic for Tests About a Population Proportion

$$z = \frac{\bar{p} - p_0}{\sigma_{\bar{p}}} \tag{9.9}$$

where

$$\sigma_{\bar{p}} = \sqrt{\frac{p_0(1 - p_0)}{n}} \tag{9.10}$$

Let us assume that $\alpha = .05$ has been selected as the level of significance for the test. With $z_{.05} = 1.645$, the upper-tail rejection region for the hypothesis test (see Figure 9.13) provides the following rejection rule.

$$\text{Reject } H_0 \text{ if } z > 1.645$$

Since 100 of the 400 players during the promotion were women, we obtain $\bar{p} = 100/400 = .25$. With $\sigma_{\bar{p}} = .02$, the value of the test statistic is

$$z = \frac{\bar{p} - p_0}{\sigma_{\bar{p}}} = \frac{.25 - .20}{.02} = 2.5$$

Thus, since $z = 2.5 > 1.645$, we can reject H_0. The Pine Creek managers can conclude that there has been an increase in the proportion of women players.

Using the table of areas for the standard normal probability distribution, we find that the p-value for the test can also be computed. For example, with $z = 2.50$, the table of areas shows a .4938 area or probability between the mean and $z = 2.50$. Thus, the p-value for the test is $.5000 - .4938 = .0062$. Since the p-value is less than α, the null hypothesis can be rejected.

We see that hypothesis tests about a population proportion and a population mean are similar; the primary difference is that the test statistic is based on the sampling distribution of $\bar{x}$ when the hypothesis test involves a population mean and on the sampling distribution of $\bar{p}$ when the hypothesis test involves a population proportion. The tentative assumption that the null hypothesis is true, the use of the level of significance to establish the critical value, and the comparison of the test statistic to the critical value are identical in the two testing procedures. Figure 9.14 summarizes the decision rules for hypothesis tests about a population proportion. We assume the large-sample case, $np \geq 5$ and $n(1 - p) \geq 5$, where the normal probability distribution can be used to approximate the sampling distribution of $\bar{p}$.

notes and comments

We have not shown the procedure for small-sample hypothesis tests involving population proportions. In the small-sample case, the sampling distribution of $\bar{p}$ follows the binomial distribution and hence the normal approximation is not applicable. More advanced texts show how hypothesis tests are conducted for this situation. However, in practice small-sample tests are rarely conducted for a population proportion.

FIGURE 9.14
Summary of Rejection Rules for Hypothesis Tests About a Population Proportion

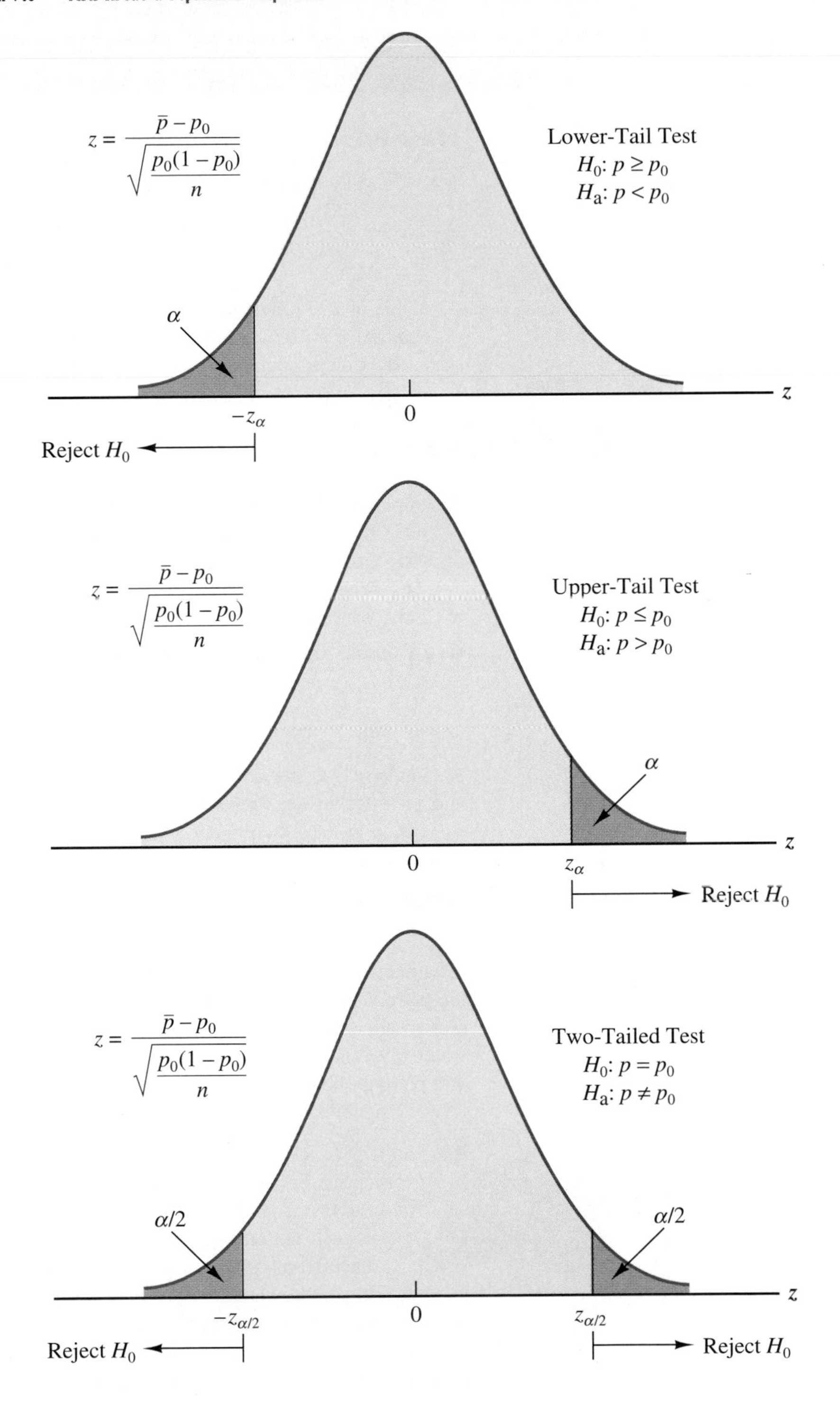

exercises

Methods

39. Consider the following hypothesis test.

$$H_0: p \leq .50$$
$$H_a: p > .50$$

A sample of 200 provided a sample proportion $\bar{p} = .57$.

a. At $\alpha = .05$, what is the rejection rule?

b. Compute the value of the test statistic z. What is your conclusion?

40. Consider the following hypothesis test.

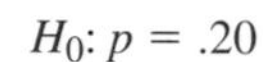

$$H_0: p = .20$$
$$H_a: p \neq .20$$

A sample of 400 provided a sample proportion of $\bar{p} = .175$.

a. At $\alpha = .05$, what is the rejection rule?

b. Compute the value of the test statistic z.

c. What is the p-value?

d. What is your conclusion?

41. Consider the following hypothesis test.

$$H_0: p \geq .75$$
$$H_a: p < .75$$

A sample of 300 is selected. Use $\alpha = .05$. Provide the value of the test statistic z, the p-value, and your conclusion for each of the following sample results.

a. $\bar{p} = .68$ **b.** $\bar{p} = .72$ **c.** $\bar{p} = .70$ **d.** $\bar{p} = .77$

Applications

42. The Honolulu Board of Water Supply suggested the water-by-request rule be adopted at restaurants on the island of Oahu to conserve water. A restaurant manager stated that 30% of the patrons do not drink water (*The Honolulu Advertiser,* December 28, 1991). Hence, the conservation of water would come not only from the unused water in each glass, but also the water saved in washing the glasses. Test H_0: p=.30 versus H_a: $p \neq .30$. Assume a sample of 480 patrons showed that 128 patrons did not drink water. Test the manager's claim at a .05 level of significance. What is the p-value and what is your conclusion?

43. A study by *Consumer Reports* (September 1993) showed that 64% of supermarket shoppers believed supermarket brands to be as good as national name brands in terms of product quality. To investigate whether this result applies to its own product, the manufacturer of a national name-brand ketchup product asked 100 supermarket shoppers whether they believed the supermarket brand of ketchup was as good as the national name-brand ketchup. Use the fact that 52 of the shoppers in the sample indicated that the supermarket brand was as good as the national name brand to test H_0: $p \geq .64$ and H_a: $p < .64$. Use a .05 level of significance. What is your conclusion?

44. The director of a college placement office claims that at least 80% of graduating seniors have made employment commitments one month prior to graduation. At a .05 level of significance, what is your conclusion if a sample of 100 seniors shows that 75 made employment commitments one month prior to graduation? Should the director's claim be rejected? What is the p-value?

45. A magazine claims that 25% of its readers are college students. A random sample of 200 readers, showed 42 were college students. Use a .10 level of significance to test H_0: $p = .25$ and H_a: $p \neq .25$. What is the p-value?

46. A new television series must prove that it has more than 25% of the viewing audience after its initial 13-week run if it is to be judged successful. Assume that in a sample of 400 households, 112 were watching the new series.

a. At a .10 level of significance, can the series be judged successful on the basis of the sample information?

b. What is the p-value? What is your hypothesis-testing conclusion?

47. An accountant believes that a company's cash-flow problems are a direct result of the slow collection of accounts receivable. The accountant claims that at least 70% of the current accounts receivable are more than two months old. A sample of 120 accounts receivable showed that 78 are more than two months old. Test the accountant's claim at the .05 level of significance.

48. In a 1992 study of the contamination of fish in the nation's rivers and lakes, the Environmental Protection Agency found that 91% of water quality test sites showed the presence of PCB, a cancer-causing agent (*America by the Numbers,* 1993). Suppose a follow-up study of 200 rivers and lakes in 1996 showed the presence of PCB in 160 cases. Does the statistical evidence support the conclusion that as of 1996 water clean-up programs have reduced the proportion of locations with PCB? Use a .05 level of significance.

49. At least 20% of all workers are believed to be willing to work fewer hours for less pay to obtain more time for personal and leisure activities. A *USA Today*/CNN/Gallup poll with a sample of 596 respondents found 83 willing to work fewer hours for less pay to obtain more personal and leisure time (*USA Today,* April 10, 1995). Test H_0: $p \geq .20$ and H_a: $p < .20$ using a .05 level of significance. What is your conclusion?

summary

Hypothesis testing is a statistical procedure that uses sample data to determine whether a statement about the value of a population parameter should be rejected. The hypotheses, which come from a variety of sources, are two competing statements about a population parameter: a null hypothesis H_0 and an alternative hypothesis H_a. In some applications it is not obvious how the null and alternative hypotheses should be formulated. We suggested guidelines for developing hypotheses in three types of situations.

1. Testing research hypotheses: The research hypothesis should be formulated as the alternative hypothesis. The null hypothesis is based on an established theory or the statement that the research treatment will have no effect. Whenever the sample data contradict the null hypothesis, the null hypothesis is rejected. In this case, the alternative or research hypothesis is supported and can be claimed true.

2. Testing the validity of a claim: Generally, the claim made is chosen as the null hypothesis; the challenge to the claim is chosen as the alternative hypothesis. Action against the claim will be taken whenever the sample data contradict the null hypothesis. When that occurs, the challenge implied by the alternative hypothesis is concluded to be true.

3. Testing in decision-making situations: Often a decision maker must choose between two courses of action, one associated with the null hypothesis and one associated with the alternative hypothesis.

Figure 9.15 summarizes the test statistics used in hypothesis tests about a population mean and provides a practical guide for selecting the hypothesis-testing procedure. The figure shows that the test statistic depends on whether the sample size is large, whether the population standard deviation is known, and in some cases whether the population has a normal or approximately normal probability distribution. If the sample size is large, ($n \geq 30$), the z test statistic is used to conduct the hypothesis test. If the sample size is small, ($n < 30$), the population must have a normal or approximately normal probability distribution to conduct a hypothesis test about the value of μ. If that is the case, the z test statistic is used if σ is known, whereas the t test statistic is used if σ is estimated by the sample standard deviation s. Finally, note that if the sample size is small and the assumption of a normal probability distribution for the population is inappropriate, we recommend increasing the sample size to $n \geq 30$ to test the hypothesis about a population mean.

Hypothesis tests about a population proportion were developed for the large-sample case, $np \geq 5$ and $n(1 - p) \geq 5$. The test statistic used is

$$z = \frac{\bar{p} - p_0}{\sqrt{\frac{p_0(1 - p_0)}{n}}}$$

The rejection rule for all the hypothesis-testing procedures involves comparing the value of the test statistic with a critical value. For lower-tail tests, the null hypothesis is rejected if the value of the test statistic is less than the critical value. For upper-tail tests,

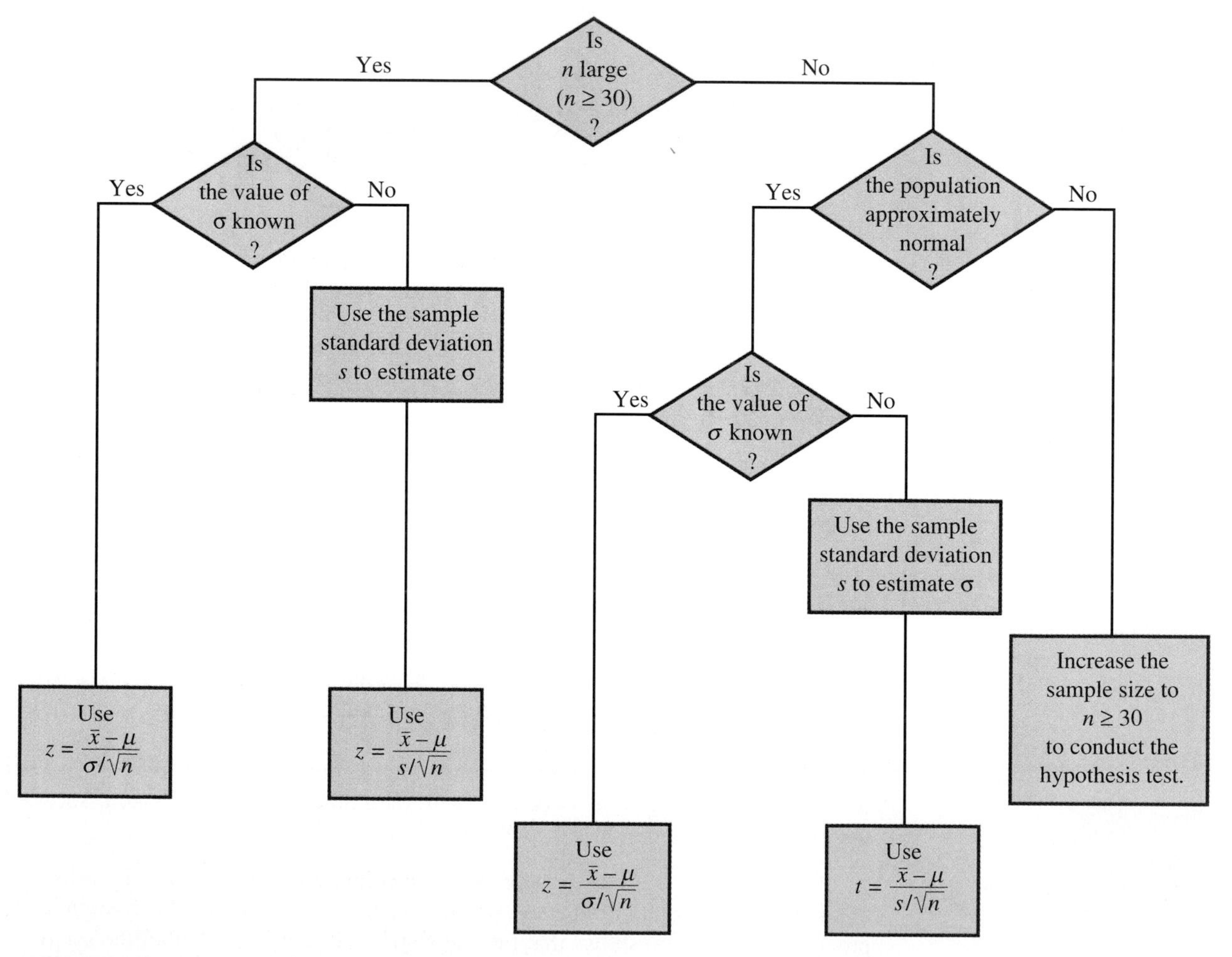

FIGURE 9.15
Summary of the Test Statistics to Be Used in a Hypothesis Test About a Population Mean

the null hypothesis is rejected if the test statistic is greater than the critical value. For two-tailed tests, the null hypothesis is rejected for values of the test statistic in either tail of the sampling distribution.

We also saw that p-values could be used for hypothesis testing. The p-value yields the probability, when the null hypothesis is true, of obtaining a sample result that is at least as unlikely as what is observed. When p-values are used to conduct a hypothesis test, the rejection rule calls for rejecting the null hypothesis whenever the p-value is less than α. The p-value is often called the observed level of significance.

glossary

Null hypothesis The hypothesis tentatively assumed true in the hypothesis-testing procedure.

Alternative hypothesis The hypothesis concluded to be true if the null hypothesis is rejected.

Type I error The error of rejecting H_0 when it is true.

Type II error The error of accepting H_0 when it is false.

Critical value A value that is compared with the test statistic to determine whether H_0 should be rejected.

Level of significance The maximum probability of a Type I error.

One-tailed test A hypothesis test in which rejection of the null hypothesis occurs for values of the test statistic in one tail of the sampling distribution.

Two-tailed test A hypothesis test in which rejection of the null hypothesis occurs for values of the test statistic in either tail of the sampling distribution.

***p*-Value** The probability, when the null hypothesis is true, of obtaining a sample result that is at least as unlikely as what is observed. It is often called the observed level of significance.

key formulas

Test Statistics for a Large-Sample ($n \geq 30$) Hypothesis Test About a Population Mean

σ Known *σ Unknown*

$$z = \frac{\bar{x} - \mu_0}{\sigma/\sqrt{n}} \qquad z = \frac{\bar{x} - \mu_0}{s/\sqrt{n}} \tag{9.1}$$

Test Statistic for a Small-Sample ($n < 30$) Hypothesis Test About a Population Mean

$$t = \frac{\bar{x} - \mu_0}{s/\sqrt{n}} \tag{9.8}$$

Test Statistic for Hypothesis Tests About a Population Proportion

$$z = \frac{\bar{p} - p_0}{\sigma_{\bar{p}}} \tag{9.9}$$

where

$$\sigma_{\bar{p}} = \sqrt{\frac{p_0(1 - p_0)}{n}} \tag{9.10}$$

supplementary exercises

50. The Ford Taurus is listed as having a highway fuel efficiency average of 30 miles per gallon (1995 *Motor Trend* New Car Buyer's Guide). A consumer interest group conducts automobile mileage tests seeking statistical evidence to show that automobile manufacturers overstate the miles per gallon ratings for particular models. In the case of the Ford Taurus, hypotheses for the test would be stated H_0: $\mu \geq 30$ and H_a: $\mu < 30$. In a sample of 50 mileage tests with Ford Taurus, the consumer interest group finds a sample mean highway milage rating of 29.5 miles per gallon and a sample standard deviation of 1.8 miles per gallon. What conclusion should be drawn from the sample results? Use a .01 level of significance.

51. The manager of the Keeton Department Store has assumed that the mean annual income of the store's customers is at least $28,000 per year. A sample of 58 customers shows a sample mean of $27,200 and a sample standard deviation of $3000. At the .05 level of significance, should this assumption be rejected? What is the *p*-value?

52. The monthly rent for a two-bedroom apartment in a particular city is reported to average $550. Suppose we want to test H_0: $\mu = 550$ versus H_a: $\mu \neq 550$. A sample of 36 two-bedroom apartments is selected. The sample mean is $\bar{x} = \$562$ and the sample standard deviation is $s = \$40$.

- **a.** Conduct this hypothesis test with a .05 level of significance.
- **b.** Compute the *p*-value.
- **c.** Use the sample results to construct a 95% confidence interval for the population mean. What hypothesis-testing conclusion would you draw from the confidence interval result?

53. In making bids on building projects, Sonneborn Builders, Inc. assumes construction workers are idle no more than 15% of the time. Hence, for a normal eight-hour shift, the mean idle time per worker should be 72 minutes or less per day. A sample of 30 construction workers had a mean idle time of 80 minutes per day. The sample standard deviation was 20 minutes. Suppose a hypothesis test is to be designed to test the validity of the company's assumption.

- **a.** What is the *p*-value associated with the sample result?
- **b.** Using a .05 level of significance and the *p*-value, test H_0: $\mu \leq 72$. What is your conclusion?

54. The Immigration and Naturalization service reported that 79% of foreign travelers visiting the United States in 1992 stated that the primary purpose of their visit was to enjoy a vacation (*America by the Numbers,* 1993). As a follow-up study conducted in 1996, suppose a sample of 500 foreign visitors is selected and that 360 say that their primary reason for visiting the United States is to enjoy a vacation. Is the proportion of foreign travelers vacationing in the United States in 1996 less than the proportion reported in 1992? Support your conclusion with a statistical test using a .05 level of significance.

55. In 1991, the proportion of individuals trying to follow expert guidelines on eating right was .44 (*Time,* October 11, 1993). In a 1993 survey of 1000 individuals conducted by the

American Dietetic Association, 390 individuals were trying to follow expert guidelines on eating right. Conduct a statistical test to determine whether the proportion of individuals trying to follow expert guidelines on eating right decreased over the two-year period. Use a .05 level of significance.

56. The Gallup Organization conducted a survey of 1350 people for the National Occupational Information Coordinating Committee, a panel Congress created to improve the use of job information (*The Arizona Republic,* January 12, 1990). A research question related to the study was: Do individuals hold jobs that they planned to hold or do they hold jobs for such reasons as chance or lack of choice? Let p indicate the population proportion of individuals who hold jobs that they planned to hold.

a. If the hypotheses are stated H_0: $p \geq .50$ and H_a: $p < .50$, discuss the research hypothesis H_a in terms of what the researcher is investigating.

b. The Gallup poll found that 41% of the respondents hold jobs they planned to hold. What is your conclusion at a .01 level of significance? Discuss.

57. A well-known doctor hypothesized that 75% of women wear shoes that are too small. A 1991 study of 356 women by the American Orthopedic Foot and Ankle Society found 313 women who wore shoes that were at least one size too small (*New York Times,* March 10, 1991). Test H_0: $p = .70$ and H_a: $p \neq .70$ at $\alpha = .01$. What is your conclusion?

58. In 1984 the percentage of individuals who reported having reduced their consumption of alcoholic beverages during the previous five years was 29% (*The Gallup Poll Monthly,* June 1994). In a 1994 Gallup study of trends in alcoholic beverage consumption, a sample of 663 individuals found that 272 individuals reported having reduced their consumption of alcoholic beverages during the previous five years. Do the results for the 1994 study indicate an increase in the proportion of individuals who are reducing their consumption of alcoholic beverages? Support your conclusion with a statistical testing using a .05 level of significance.

computer case

Quality Associates, Inc.

Quality Associates, Inc., is a consulting firm that advises its clients about sampling and statistical procedures that can be used to control manufacturing processes. In one particular application, a client provided Quality Associates with a sample of 800 observations that were taken during a time in which the client's process was operating satisfactorily. The sample standard deviation for these data was .21; hence, the population standard deviation was assumed to be .21. Quality Associates then suggested that random samples of size 30 be taken periodically to monitor the process on an ongoing basis. By analyzing the new samples, the client could quickly learn whether the process was operating satisfactorily. When the process was not operating satisfactorily, corrective action could be taken to eliminate the problem. The design specification indicated the mean for the process should be 12. The hypotheses test suggested by Quality Associates follows.

$$H_0: \mu = 12$$
$$H_a: \mu \neq 12$$

Corrective action will be taken any time H_0 is rejected.

The following four samples were collected during the first day of operation of the new statistical process-control procedure. These data are available in the QUALITY data set.

Sample 1	Sample 2	Sample 3	Sample 4
11.55	11.62	11.91	12.02
11.62	11.69	11.36	12.02
11.52	11.59	11.75	12.05
11.75	11.82	11.95	12.18
11.90	11.97	12.14	12.11
11.64	11.71	11.72	12.07
11.80	11.87	11.61	12.05
12.03	12.10	11.85	11.64
11.94	12.01	12.16	12.39
11.92	11.99	11.91	11.65
12.13	12.20	12.12	12.11
12.09	12.16	11.61	11.90
11.93	12.00	12.21	12.22
12.21	12.28	11.56	11.88
12.32	12.39	11.95	12.03
11.93	12.00	12.01	12.35
11.85	11.92	12.06	12.09
11.76	11.83	11.76	11.77
12.16	12.23	11.82	12.20
11.77	11.84	12.12	11.79
12.00	12.07	11.60	12.30
12.04	12.11	11.95	12.27
11.98	12.05	11.96	12.29
12.30	12.37	12.22	12.47
12.18	12.25	11.75	12.03
11.97	12.04	11.96	12.17
12.17	12.24	11.95	11.94
11.85	11.92	11.89	11.97
12.30	12.37	11.88	12.23
12.15	12.22	11.93	12.25

Managerial Report

1. Conduct the hypothesis test for each sample at the .01 level of significance and determine what action, if any, should be taken. Provide the test statistic and *p*-value for each test.

2. Consider the standard deviation for each of the four samples. Does the assumption of .21 for the population standard deviation appear reasonable?

3. Compute limits for the sample mean $\bar{x}$ around $\mu = 12$ such that, as long as a new sample mean is within those limits, the process will be considered to be operating satisfactorily. If $\bar{x}$ exceeds the upper limit or if $\bar{x}$ is below the lower limit, corrective action will be taken. These limits are referred to as upper and lower control limits for quality-control purposes.

4. Discuss the implications of changing the level of significance to a larger value. What mistake or error could increase if that were done?

APPENDIX 9.1 Hypothesis Testing with Minitab

Large-Sample Case

In Section 9.4 we showed the results obtained using Minitab to perform a hypothesis test for the Superflight golf ball study. Assume that the user has entered the distance data for the 36 golf balls from Table 9.2 into column 1 of a Minitab worksheet. With the sample standard deviation $s = 12$ as an estimate of the population standard deviation σ, the following steps can be used to produce the computer output in Figure 9.9.

STEP 1. Select the **Stat** pull-down menu
STEP 2. Select the **Basic Statistics** pull-down menu
STEP 3. Select the **1-Sample Z** option
STEP 4. When the dialog box appears,
Enter C1 in the **Variables** box
Select the **Test mean** option and enter 280 in the box
Select **not equal** in the **Alternative** box
Enter 7.77 in the **Sigma** box
Select **OK**

Although the Superflight golf ball study involved a two-tailed hypothesis test, the procedure described can be easily modified for one-tailed hypothesis tests. In Step 4 we can conduct a one-tailed test by simply selecting the less than or the greater than option in the **Alternative** box.

Small-Sample Case

In Section 9.5 we showed the results obtained using Minitab to perform a hypothesis test for the Heathrow airport rating study. Assume that the user has entered the ratings for the 12 business travelers into column 1 of a Minitab worksheet. The following steps can be used to produce the computer output in Figure 9.11.

STEP 1. Select the **Stat** pull-down menu
STEP 2. Select the **Basic Statistics** pull-down menu
STEP 3. Select the **1-Sample t** option
STEP 4. When the dialog box appears,
Enter C1 in the **Variables** box
Select the **Test mean** option and enter 7 in the box
Select the **greater than** option in the **Alternative** box
Select **OK**

CHAPTER 10

Comparisons Involving Means

Fisons Corporation*

Rochester, New York

statistics in practice

Fisons Corporation, Rochester, New York, is a unit of Fisons Plc., UK. Fisons opened its United States operations in 1966.

Fisons' Pharmaceutical Division uses extensive statistical procedures to test and develop new drugs. The testing process in the pharmaceutical industry usually consists of three stages: (1) preclinical testing, (2) testing for long-term usage and safety, and (3) clinical efficacy testing. At each successive stage, the chance that a drug will pass the rigorous tests decreases; however, the cost of further testing increases dramatically. Industry surveys indicate that on average the research and development for one new drug costs $250 million and takes 12 years. Hence, it is important to eliminate unsuccessful new drugs in the early stages of the testing process, as well as identify promising ones for further testing.

Statistics plays a major role in pharmaceutical research, where government regulations are stringent and rigorously enforced. In preclinical testing, a two- or three-population statistical study typically is used to determine whether a new drug should continue to be studied in the long-term usage and safety program. The populations may consist of the new drug, a control, and a standard drug. The preclinical testing process begins when a new drug is sent to the pharmacology group for evaluation of efficacy—the capacity of the drug to produce the desired effects. As part of the process, a statistician is asked to design an experiment that can be used to test the new drug. The design must specify the sample size and the statistical methods of analysis. In a two-population study, one sample is used to obtain data on the efficacy of the new drug (population 1) and a second sample is used to obtain data on the efficacy of a standard drug (population 2). Depending on the intended use, the new and standard drugs are tested in such disciplines as neurology, cardiology, and immunology. In most studies, the statistical method involves hypothesis testing for the difference between the means of the new drug population and the standard drug population. If a new drug lacks efficacy or produces undesirable effects in comparison with the standard drug, the new drug is rejected and withdrawn from further testing. Only new drugs that show promising comparisons with the standard drugs are forwarded to the long-term usage and safety testing program.

Further data collection and multipopulation studies are conducted in the long-term usage and safety testing program and in the clinical testing programs. The Food and Drug Administration (FDA) requires that statistical methods be defined prior to such testing to avoid data-related biases. In addition, to avoid human biases, some of the clinical trials are double or triple blind. That is, neither the subject nor the investigator knows what drug is administered. If the new drug meets all requirements in relation to the standard drug, a new drug application (NDA) is filed with the FDA. The application is rigorously scrutinized by statisticians and scientists at the agency.

In this chapter you will learn how to construct interval estimates and make hypothesis tests about means with two or more populations. Techniques will be presented for analyzing independent random samples as well as matched samples.

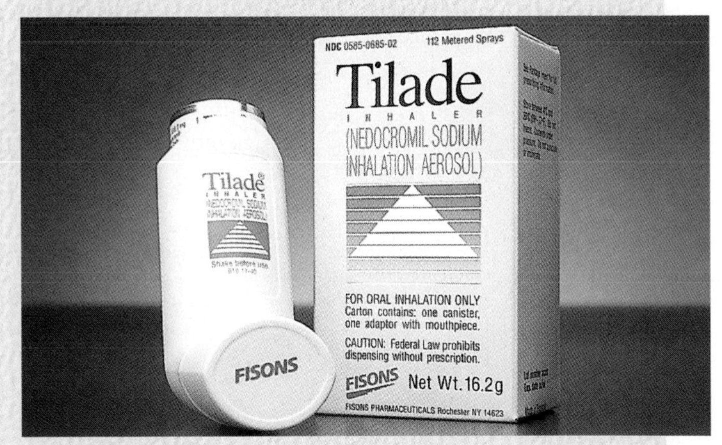

Some of the drug products manufactured by the pharmaceutical division of Fisons Corporation.

The authors are indebted to Dr. M. C. Trivedi, Fisons Corporation, for providing this Statistics in Practice.

In Chapters 8 and 9 we showed how to develop interval estimates and conduct hypothesis tests for situations involving one population mean. In this chapter we continue our discussion of statistical inference by showing how interval estimates and hypothesis tests can be developed for situations involving two or more populations. For example, we may want to develop an interval estimate of the difference between the mean starting salary for a population of men and the mean starting salary for a population of women or test a hypothesis that the mean number of hours between breakdowns is the same for four machines. We will begin our discussion by showing how an interval estimate of the difference between the means of two populations can be developed for a sampling study conducted by Greystone Department Stores, Inc.

10.1 Estimation of the Difference Between the Means of Two Populations: Independent Samples

Greystone Department Stores, Inc., operates two stores in Buffalo, New York; one is in the inner city and the other is in a suburban shopping center. The regional manager has noticed that products that sell well in one store do not always sell well in the other. She believes this situation may be attributable to differences in customer demographics at the two locations. Customers differ in age, education, income, and so on. Suppose the regional manager has asked us to investigate the difference between the mean ages of the customers who shop at the two stores.

Let us define population 1 as all customers who shop at the inner-city store and population 2 as all customers who shop at the suburban store. Let

μ_1 = mean of population 1 (i.e., the mean age of all customers who shop at the inner-city store) and

μ_2 = mean of population 2 (i.e., the mean age of all customers who shop at the suburban store).

The difference between the two population means is $\mu_1 - \mu_2$.

To estimate $\mu_1 - \mu_2$, we will select a simple random sample of n_1 customers from population 1 and a simple random sample of n_2 customers from population 2. Since the simple random sample of n_1 customers is selected independently of the simple random sample of n_2 customers, we have the case of *independent simple random samples.* Let

$\bar{x}_1$ = sample mean age for the simple random sample of n_1 inner-city customers and

$\bar{x}_2$ = sample mean age for the simple random sample of n_2 suburban customers.

Since $\bar{x}_1$ is a point estimator of μ_1 and $\bar{x}_2$ is a point estimator of μ_2, the point estimator of the difference in the two population means is expressed as follows.

Point Estimator of the Difference Between the Means of Two Populations

$$\bar{x}_1 - \bar{x}_2 \qquad (10.1)$$

Thus, we see that the point estimator of the difference between the two population means is the difference between the sample means of the two independent simple random

FIGURE 10.1
Estimating the Difference Between the Means of Two Populations

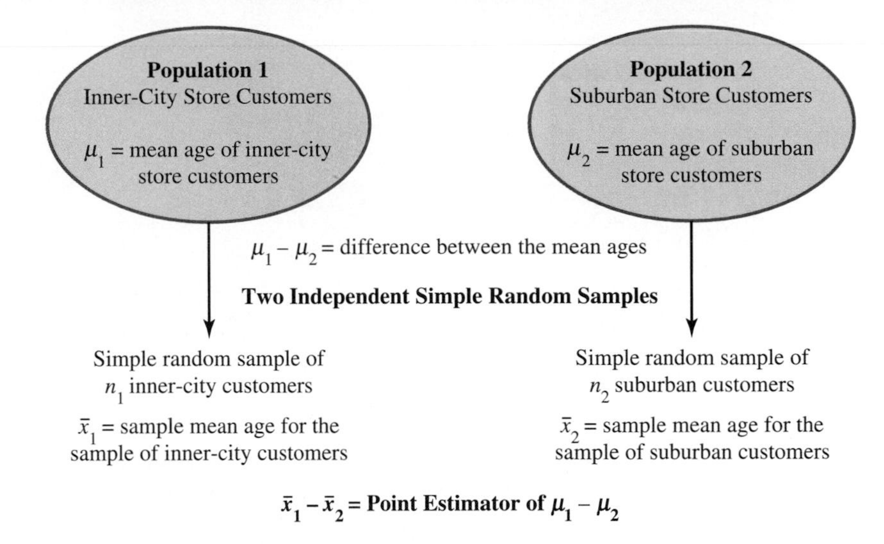

samples. Figure 10.1 provides an overview of the process used to estimate the difference between two sample means based on two independent simple random samples.

Assume the customer age data collected from the two independent simple random samples of Greystone customers provide the following results.

Store	Number of Customers Sampled	Sample Mean Age	Sample Standard Deviation
Inner City	36	$\bar{x}_1 = 40$ years	$s_1 = 9$ years
Suburban	49	$\bar{x}_2 = 35$ years	$s_2 = 10$ years

Using (10.1), we find that a point estimate of the difference between the mean ages of the two populations is $\bar{x}_1 - \bar{x}_2 = 40 - 35 = 5$ years. Thus, we are led to believe that the customers at the inner-city store have a mean age five years greater than the mean age of the suburban store customers. However, as with all point estimates, we know that five years is only one of many possible estimates of the difference between the mean ages of the two populations. If Greystone selected another simple random sample of 36 inner-city customers and another simple random sample of 49 suburban customers, the difference between the two new sample means would probably not equal five years. The sampling distribution of $\bar{x}_1 - \bar{x}_2$ is the probability distribution of the difference in sample means for all possible sets of two samples.

Sampling Distribution of $\bar{x}_1 - \bar{x}_2$

We can use the sampling distribution of $\bar{x}_1 - \bar{x}_2$ to develop an interval estimate of the difference between the two population means in much the same way as we used the sampling distribution of $\bar{x}$ for interval estimation with a single population mean. The sampling distribution of $\bar{x}_1 - \bar{x}_2$ has the following properties.

Sampling Distribution of $\bar{x}_1 - \bar{x}_2$

$$\text{Expected Value: } E(\bar{x}_1 - \bar{x}_2) = \mu_1 - \mu_2 \quad (10.2)$$

$$\text{Standard Deviation: } \sigma_{\bar{x}_1 - \bar{x}_2} = \sqrt{\frac{\sigma_1^2}{n_1} + \frac{\sigma_2^2}{n_2}} \quad (10.3)$$

where

σ_1 = standard deviation of population 1

σ_2 = standard deviation of population 2

n_1 = sample size for the simple random sample from population 1

n_2 = sample size for the simple random sample from population 2

Distribution form: If the sample sizes are both *large* ($n_1 \geq 30$ and $n_2 \geq 30$), the sampling distribution of $\bar{x}_1 - \bar{x}_2$ can be approximated by a normal probability distribution.

Figure 10.2 shows the sampling distribution of $\bar{x}_1 - \bar{x}_2$ and its relationship to the individual sampling distributions of $\bar{x}_1$ and $\bar{x}_2$.

Let us now develop an interval estimate of the difference between the means of two populations. We consider two cases, one in which the sample sizes are large ($n_1 \geq 30$ and $n_2 \geq 30$) and another in which one or both sample sizes are small ($n_1 < 30$ and/or $n_2 < 30$). We consider the large-sample case first.

Interval Estimate of $\mu_1 - \mu_2$: Large-Sample Case

In the large-sample case, the sampling distribution of $\bar{x}_1 - \bar{x}_2$ can be approximated by a normal probability distribution. With this approximation we can use the following expression to develop an interval estimate of the difference between the means of the two populations.

Interval Estimate of the Difference between the Means of Two Populations: Large-Sample Case ($n_1 \geq 30$ and $n_2 \geq 30$) with σ_1 and σ_2 Known

$$\bar{x}_1 - \bar{x}_2 \pm z_{\alpha/2}\sigma_{\bar{x}_1 - \bar{x}_2} \quad (10.4)$$

where $1 - \alpha$ is the confidence coefficient.

FIGURE 10.2
Sampling Distribution of $\bar{x}_1 - \bar{x}_2$ and Its Relationship to the Individual Sampling Distributions of $\bar{x}_1$ and $\bar{x}_2$

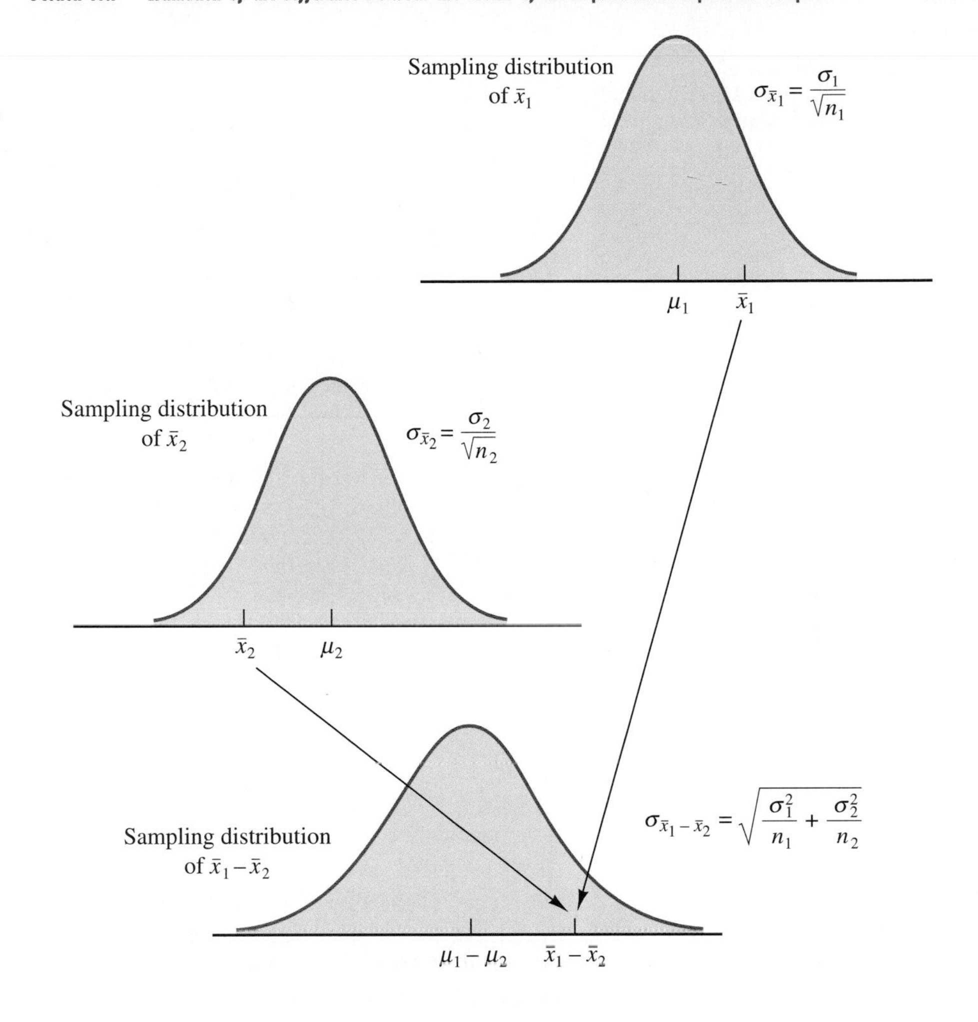

The quantity $z_{\alpha/2}\,\sigma_{\bar{x}_1 - \bar{x}_2}$ is the margin of error for the estimate. To compute the margin of error, we must know the value of $\sigma_{\bar{x}_1 - \bar{x}_2}$, the standard deviation of the sampling distribution of $\bar{x}_1 - \bar{x}_2$. However, (10.3) shows that the value of $\sigma_{\bar{x}_1 - \bar{x}_2}$ depends on the values of σ_1 and σ_2, the standard deviations of each of the populations. When the population standard deviations are unknown, we can use the sample standard deviations as estimates of the population standard deviations and estimate $\sigma_{\bar{x}_1 - \bar{x}_2}$ as follows.

Point Estimator of $\sigma_{\bar{x}_1 - \bar{x}_2}$

$$s_{\bar{x}_1 - \bar{x}_2} = \sqrt{\frac{s_1^2}{n_1} + \frac{s_2^2}{n_2}} \qquad (10.5)$$

In the large-sample case, we can use this point estimator of $\sigma_{\bar{x}_1 - \bar{x}_2}$ to develop an approximate confidence interval estimate of the difference between the two population means.

FIGURE 10.3
Sampling Distribution of $\bar{x}_1 - \bar{x}_2$ When the Populations Have Normal Distributions With Equal Variances

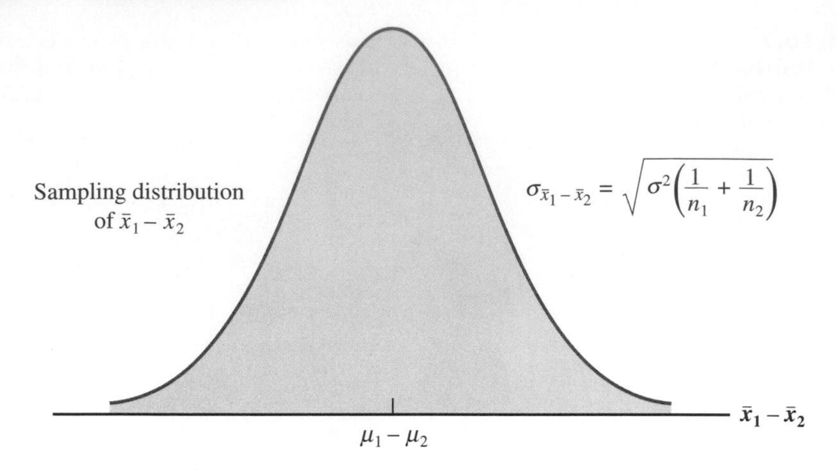

Interval Estimate of the Difference between the Means of Two Populations: Large-Sample Case ($n_1 \geq 30$ and $n_2 \geq 30$) with σ_1 and σ_2 Unknown

$$\bar{x}_1 - \bar{x}_2 \pm z_{\alpha/2} s_{\bar{x}_1 - \bar{x}_2} \tag{10.6}$$

where $1 - \alpha$ is the confidence coefficient.

Let us use (10.6) to develop a confidence interval estimate of the difference between the mean ages of the two customer populations in the Greystone Department Store study. Recall that the sample mean age and sample standard deviation for the simple random sample of 36 inner-city customers are $\bar{x}_1 = 40$ years and $s_1 = 9$ years, respectively; the sample mean and sample standard deviation for the simple random sample of 49 suburban customers are $\bar{x}_2 = 35$ years and $s_2 = 10$ years, respectively. Using (10.5) to estimate $\sigma_{\bar{x}_1 - \bar{x}_2}$, we have

$$s_{\bar{x}_1 - \bar{x}_2} = \sqrt{\frac{(9)^2}{36} + \frac{(10)^2}{49}} = \sqrt{4.29} = 2.07$$

With $z_{\alpha/2} = z_{.025} = 1.96$, (10.6) provides the following 95% confidence interval.

$$5 \pm (1.96)(2.07)$$

or

$$5 \pm 4.06$$

With a margin of error of 4.06, the 95% confidence interval for the difference between the mean ages of the two Greystone populations is .94 years to 9.06 years.

Interval Estimate of $\mu_1 - \mu_2$: Small-Sample Case

Let us now consider the interval-estimation procedure for the difference between the means of two populations whenever one or both sample sizes are less than 30—that is, $n_1 < 30$ and/or $n_2 < 30$. This will be referred to as the small-sample case.

In Chapter 8 we presented a procedure for interval estimation of the mean for a single population with a small sample. Recall that the procedure required the assumption that the population had a normal probability distribution. With the sample standard deviation s used as an estimate of the population standard deviation σ, the t distribution was used to develop an interval estimate of the population mean.

To develop interval estimates for the two-population small-sample case, we will make two assumptions about the two populations and the samples selected from the two populations.

1. Both populations have normal probability distributions.
2. The variances of the populations are equal ($\sigma_1^2 = \sigma_2^2 = \sigma^2$).

Given these assumptions, the sampling distribution of $\bar{x}_1 - \bar{x}_2$ is normally distributed regardless of the sample sizes. The expected value of $\bar{x}_1 - \bar{x}_2$ is $\mu_1 - \mu_2$. Because of the equal variances assumption, (10.3) can be written

$$\sigma_{\bar{x}_1 - \bar{x}_2} = \sqrt{\frac{\sigma^2}{n_1} + \frac{\sigma^2}{n_2}} = \sqrt{\sigma^2\left(\frac{1}{n_1} + \frac{1}{n_2}\right)} \tag{10.7}$$

The sampling distribution of $\bar{x}_1 - \bar{x}_2$ is shown in Figure 10.3.

If the variance σ^2 of the populations is known, (10.4) can be used to develop the interval estimate of the difference between the two population means. However, in most cases, σ^2 is unknown; thus, the two sample variances s_1^2 and s_2^2 must be used to develop the estimate of σ^2 in (10.7). Since (10.7) is based on the assumption that $\sigma_1^2 = \sigma_2^2 = \sigma^2$, we do not need separate estimates of σ_1^2 and σ_2^2. In fact, we can combine the data from the two samples to provide the best single estimate of σ^2. The process of combining the results of two independent simple random samples to provide one estimate of σ^2 is referred to as *pooling*. The *pooled estimator* of σ^2, denoted by s^2, is a weighted average of the two sample variances s_1^2 and s_2^2. The formula for the pooled estimator of σ^2 follows.

Pooled Estimator of σ^2

$$s^2 = \frac{(n_1 - 1)s_1^2 + (n_2 - 1)s_2^2}{n_1 + n_2 - 2} \tag{10.8}$$

With s^2 as the pooled estimator of σ^2 and using (10.7), we can obtain the following estimator of the standard deviation of $\bar{x}_1 - \bar{x}_2$.

Point Estimator of $\sigma_{\bar{x}_1-\bar{x}_2}$ when $\sigma_1^2 = \sigma_2^2 = \sigma^2$

$$s_{\bar{x}_1 - \bar{x}_2} = \sqrt{s^2\left(\frac{1}{n_1} + \frac{1}{n_2}\right)} \tag{10.9}$$

The t distribution can now be used to compute an interval estimate of the difference between the means of the two populations. There are $n_1 - 1$ degrees of freedom associated with the random sample from population 1 and $n_2 - 1$ degrees of freedom associated with the random sample from population 2. Thus, the t distribution will have $n_1 + n_2 - 2$ degrees of freedom. The interval-estimation procedure follows.

Interval Estimate of the Difference between the Means of Two Populations: Small-Sample Case ($n_1 < 30$ and/or $n_2 < 30$)

$$\bar{x}_1 - \bar{x}_2 \pm t_{\alpha/2} s_{\bar{x}_1 - \bar{x}_2} \tag{10.10}$$

where the t value is based on a t distribution with $n_1 + n_2 - 2$ degrees of freedom and where $1-\alpha$ is the confidence coefficient.

Let us demonstrate the interval-estimation procedure for a sampling study conducted by the Clearview National Bank. Independent random samples of checking account balances for customers at two Clearview branch banks yielded the following results.

Branch Bank	Number of Checking Accounts	Sample Mean Balance	Sample Standard Deviation
Cherry Grove	12	$\bar{x}_1 = \$1000$	$s_1 = \$150$
Beechmont	10	$\bar{x}_2 = \$\ 920$	$s_2 = \$120$

Let us use these data to develop a 90% confidence interval for the difference between the mean checking account balances at the two branch banks. Suppose that checking account balances are normally distributed at both branches and that the variances of checking account balances at both branches are equal. Using (10.8), we see that the pooled estimate of the population variance becomes

$$s^2 = \frac{(n_1 - 1)s_1^2 + (n_2 - 1)s_2^2}{n_1 + n_2 - 2} = \frac{(11)(150)^2 + (9)(120)^2}{12 + 10 - 2} = 18{,}855$$

The corresponding estimate of the standard deviation of $\bar{x}_1 - \bar{x}_2$ is

$$s_{\bar{x}_1 - \bar{x}_2} = \sqrt{s^2\left(\frac{1}{n_1} + \frac{1}{n_2}\right)} = \sqrt{18{,}855\left(\frac{1}{12} + \frac{1}{10}\right)} = 58.79$$

The appropriate t distribution for the interval-estimation procedure has $n_1 + n_2 - 2 = 12 + 10 - 2 = 20$ degrees of freedom. With $\alpha = .10$, $t_{\alpha/2} = t_{.05} = 1.725$. Thus, using (10.10), we see that the interval estimate becomes

$$\bar{x}_1 - \bar{x}_2 \pm t_{.05}\, s_{\bar{x}_1 - \bar{x}_2}$$

$$1000 - 920 \pm (1.725)(58.79)$$

$$80 \pm 101.41$$

With a margin of error of 101.41, the 90% confidence interval for the difference between the mean account balances at the two branch banks is –$21.41 to $181.41. The fact that the interval includes a negative range of values indicates that the actual difference between the two means, $\mu_1 - \mu_2$, may be negative. Thus, μ_2 could actually be larger than μ_1, indicating that the population mean balance could be greater for the Beechmont branch even though the results show a greater sample mean balance at the Cherry Grove branch. The fact that the confidence interval contains the value 0 can be interpreted as indicating that we do not have sufficient evidence to conclude that the population mean account balances differ between the two branches.

notes and comments

1. The use of the t distribution in the small-sample procedure presented in this section is based on the assumptions that both populations have a normal probability distribution and that $\sigma_1^2 = \sigma_2^2$. Fortunately, this procedure is a *robust* statistical procedure, meaning that it is relatively insensitive to these assumptions. For instance, if $\sigma_1^2 \neq \sigma_2^2$, the procedure provides acceptable results if n_1 and n_2 are approximately equal.
2. The t distribution is not restricted to the small-sample situation; it is applicable whenever both populations are normally distributed and the variances of the populations are equal. However, (10.4) and (10.6) show how to determine an interval estimate of the difference between the means of two populations when the sample sizes are large. Thus, in the large-sample case, use of the t distribution and its corresponding assumptions is not required. We therefore do not need to refer to the t distribution until we have a small-sample case.

exercises

Methods

1. Consider the results shown here for two independent random samples taken from two populations.

Sample 1	$n_1 = 50$	$\bar{x}_1 = 13.6$	$s_1 = 2.2$
Sample 2	$n_2 = 35$	$\bar{x}_2 = 11.6$	$s_2 = 3.0$

a. What is the point estimate of the difference between the two population means?
b. Provide a 90% confidence interval for the difference between the two population means.
c. Provide a 95% confidence interval for the difference between the two population means.

2. Consider the following results for two independent random samples taken from two populations.

Sample 1	Sample 2
$n_1 = 10$	$n_2 = 8$
$\bar{x}_1 = 22.5$	$\bar{x}_2 = 20.1$
$s_1 = 2.5$	$s_2 = 2.0$

a. What is the point estimate of the difference between the two population means?
b. What is the pooled estimate of the population variance?
c. Develop a 95% confidence interval for the difference between the two population means.

3. Consider the following data for two independent random samples taken from two populations.

Sample 1	10	12	9	7	7	9
Sample 2	8	8	6	7	4	9

a. Compute the two sample means.
b. Compute the two sample standard deviations.
c. What is the point estimate of the difference between the two population means?
d. What is the pooled estimate of the population variance?
e. Develop a 95% confidence interval for the difference between the two population means.

Applications

4. Data gathered by the U.S. Department of Transportation (*1994 Information Please Environmental Almanac*) show the number of miles that residents of the 75 largest metropolitan areas travel per day in a car. Suppose that for a simple random sample of 50 Buffalo residents the mean is 22.5 miles a day and the standard deviation is 8.4 miles a day, and for an independent simple random sample of 100 Boston residents the mean is 18.6 miles a day and the standard deviation is 7.4 miles a day.
a. What is the point estimate of the difference between the mean number of miles that Buffalo residents travel per day and the mean number of miles that Boston residents travel per day?
b. What is the 95% confidence interval for the difference between the two population means?

5. The International Air Transport Association surveyed business travelers to determine ratings of transatlantic gateway airports. The maximum possible score was 10, the highest rated airport was Amsterdam with an average rating of 7.93, followed by Toronto with a rating of 7.17 (*Newsweek,* June 13, 1994). Suppose a simple random sample of 50 business travelers were asked to rate the Miami airport and an independent simple random sample of 50 business travelers were asked to rate the Los Angeles airport. The rating scores follow.

AIRPORT

Miami

6	4	6	8	7	7	6	3	3	8	10	4	8
7	8	7	5	9	5	8	4	3	8	5	5	4
4	4	8	4	5	6	2	5	9	9	8	4	8
9	9	5	9	7	8	3	10	8	9	6		

Los Angeles

10	9	6	7	8	7	9	8	10	7	6	5	7
3	5	6	8	7	10	8	4	7	8	6	9	9
5	3	1	8	9	6	8	5	4	6	10	9	8
3	2	7	9	5	3	10	3	5	10	8		

Develop a 95% confidence interval estimate of the difference between the mean ratings of the Miami and Los Angeles airports.

6. The Butler County Bank and Trust Company wants to estimate the difference between the mean credit-card balances at two of its branch banks. Independent random samples of credit-card customers generated the following results.

Branch 1	Branch 2
$n_1 = 32$	$n_2 = 36$
$\bar{x}_1 = \$500$	$\bar{x}_2 = \$375$
$s_1 = \$150$	$s_2 = \$130$

a. Develop a point estimate of the difference between the mean balances at the two branches.

b. Develop a 99% confidence interval for the difference between the mean balances.

7. An urban-planning group is interested in estimating the difference between the mean household incomes for two neighborhoods in a large metropolitan area. Independent random samples of households in the neighborhoods provided the following results.

Neighborhood 1	Neighborhood 2
$n_1 = 8$	$n_2 = 12$
$\bar{x}_1 = \$15{,}700$	$\bar{x}_2 = \$14{,}500$
$s_1 = \$700$	$s_2 = \$850$

a. Develop a point estimate of the difference between the mean incomes in the two neighborhoods.

b. Develop a 95% confidence interval for the difference between the mean incomes in the two neighborhoods.

c. What assumptions were made to compute the interval estimates in part (b)?

8. Women who are union members earn \$2.50 per hour more than women who are not union members (*The Wall Street Journal,* July 26, 1994). Suppose independent random samples of 15 unionized women and 20 nonunionized women in manufacturing have been selected and the following hourly wage rates are found.

Union Workers

22.40	18.90	16.70	14.05	16.20	20.00	16.10	16.30	19.10
16.50	18.50	19.80	17.00	14.30	17.20			

Nonunion Workers

17.60	14.40	16.60	15.00	17.65	15.00	17.55	13.30	11.20
15.90	19.20	11.85	16.65	15.20	15.30	17.00	15.10	14.30
13.90	14.50							

a. Suppose we want to develop an interval estimate of the difference in the mean wage rates between unionized and nonunionized women working in manufacturing. What assumptions must be made about the two populations?
b. What is the pooled estimate of the population variance?
c. Develop a 95% confidence interval estimate of the difference between the two population means.
d. Does there appear to be any difference in the mean wage rate between these two groups? Explain.

10.2 Hypothesis Tests About the Difference Between the Means of Two Populations: Independent Samples

In this section we present procedures that can be used to test hypotheses about the difference between the means of two populations. The methodology is again divided into large-sample ($n_1 \geq 30$, $n_2 \geq 30$) and small-sample ($n_1 < 30$ and / or $n_2 < 30$) cases.

Large-Sample Case

As part of a study to evaluate differences in educational quality between two training centers, a standardized examination is given to individuals who are trained at the two centers. The examination scores are a major factor in assessing any quality differences between the centers.

Let

μ_1 = the mean examination score for the population of individuals trained at center A and

μ_2 = the mean examination score for the population of individuals trained at center B.

We begin with the tentative assumption that there is no difference in the training quality provided at the two centers. Hence, in terms of the mean examination scores, the null hypothesis is that $\mu_1 - \mu_2 = 0$. If sample evidence leads to the rejection of this hypothesis, we will conclude that the mean examination scores differ for the two populations. This conclusion indicates a quality differential between the two centers and that a follow-up study investigating the reasons for the differential may be warranted. The null and alternative hypotheses are written as follows.

$$H_0: \mu_1 - \mu_2 = 0$$

$$H_a: \mu_1 - \mu_2 \neq 0$$

FIGURE 10.4
Sampling Distribution of $\bar{x}_1 - \bar{x}_2$ With H_0: $\mu_1 - \mu_2 = 0$

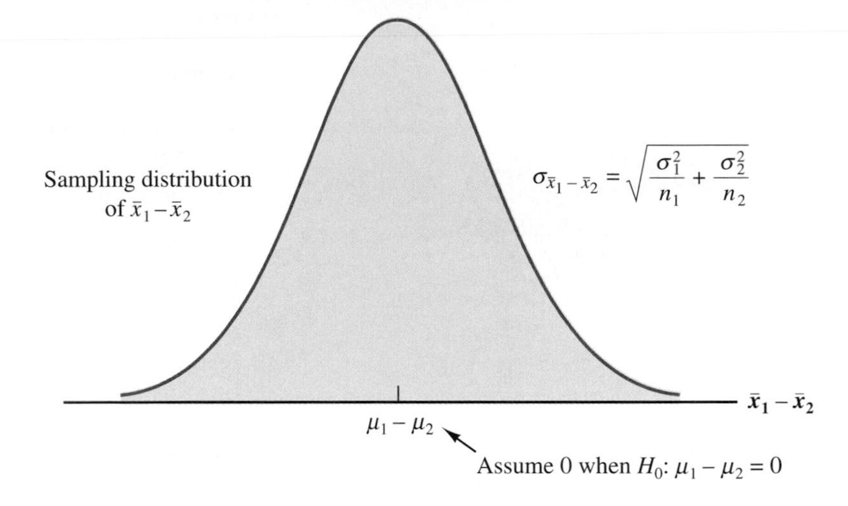

FIGURE 10.5
Rejection Region for the Two-Tailed Hypothesis Test with $\alpha = .05$

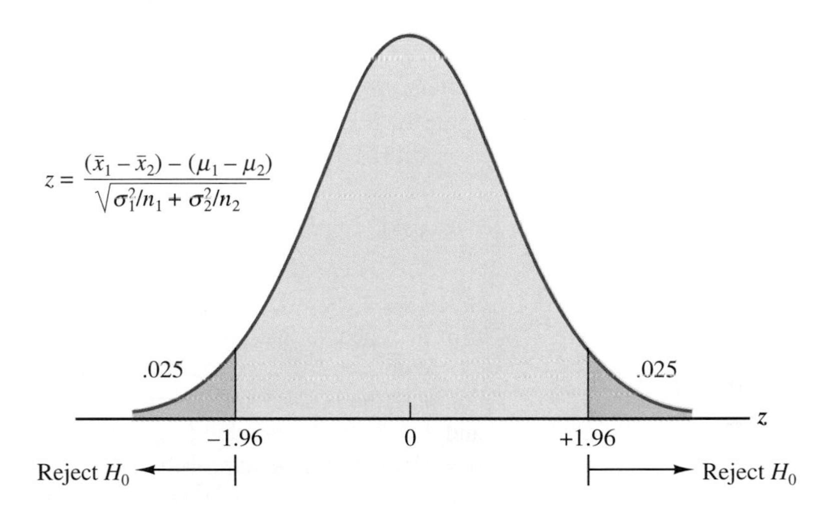

Following the hypothesis-testing procedure from Chapter 9, we make the tentative assumption that H_0 is true. Using the difference between the sample means as the point estimator of the difference between the population means, we consider the sampling distribution of $\bar{x}_1 - \bar{x}_2$ when H_0 is true. For the large-sample case, this distribution is as shown in Figure 10.4. Since the sampling distribution of $\bar{x}_1 - \bar{x}_2$ can be approximated by a normal probability distribution, the following test statistic is used.

$$z = \frac{(\bar{x}_1 - \bar{x}_2) - (\mu_1 - \mu_2)}{\sqrt{\sigma_1^2/n_1 + \sigma_2^2/n_2}} \tag{10.11}$$

Whenever $n_1 \geq 30$ and $n_2 \geq 30$, we will use s_1^2 and s_2^2 as estimates of σ_1^2 and σ_2^2 to compute the test statistic.

The value of z given by (10.11) can be interpreted as the number of standard deviations $\bar{x}_1 - \bar{x}_2$ is from the value of $\mu_1 - \mu_2$ specified in H_0. For $\alpha = .05$ and thus $z_{\alpha/2} = z_{.025} = 1.96$, the rejection region for the two-tailed hypothesis test is shown in Figure 10.5. The rejection rule is

TABLE 10.1 Examination Score Data

Training Center A			Training Center B			
97	83	91	64	66	91	84
90	84	87	85	83	78	85
94	76	73	72	74	87	85
79	82	92	64	70	93	84
78	85	64	74	82	89	59
87	85	74	93	82	79	62
83	91	88	70	75	84	91
89	72	88	79	78	65	83
76	86	74	79	99	78	80
84	70	73	75	57	66	76

Reject H_0 if $z < -1.96$ or if $z > +1.96$.

TABLE 10.2 Examination Score Results

Training Center A	Training Center B
$n_1 = 30$	$n_2 = 40$
$\bar{x}_1 = 82.5$	$\bar{x}_2 = 78$
$s_1 = 8$	$s_2 = 10$

Let us assume that independent random samples of individuals trained at the two centers provide the examination scores in Table 10.1; summary statistics are given in Table 10.2. Using s_1^2 and s_2^2 to estimate σ_1^2 and σ_2^2, we find that the test statistic z given by (10.11) for the null hypothesis H_0: $\mu_1 - \mu_2 = 0$ becomes

$$z = \frac{(82.5 - 78) - 0}{\sqrt{(8)^2/30 + (10)^2/40}} = 2.09$$

Since $z = 2.09 > 1.96$, the conclusion is to reject H_0. Thus, the sample scores lead the firm to conclude that the two centers differ in educational quality.

With $z = 2.09$, the standard normal probability distribution table can be used to compute the p-value for this two-tailed test. With an area of .4817 between the mean and $z = 2.09$, the p-value is $2(.5000 - .4817) = .0366$; since the p-value is less than $\alpha = .05$, the p-value approach also results in the rejection of H_0.

In this hypothesis test, we were interested in determining whether the means of the two populations differ. Since we did not have a prior belief that one mean might be greater than or less than the other, the hypotheses H_0: $\mu_1 - \mu_2 = 0$ and H_a: $\mu_1 - \mu_2 \neq 0$ were appropriate. In other hypothesis tests about the difference between the means of two populations, we may want to find out whether one of the means is greater than or perhaps less than the other mean. In these cases, a one-tailed hypothesis test would be appropriate. The two forms of a one-tailed test about the difference between two population means follow.

$$H_0: \mu_1 - \mu_2 \leq 0 \qquad H_0: \mu_1 - \mu_2 \geq 0$$
$$H_a: \mu_1 - \mu_2 > 0 \qquad H_a: \mu_1 - \mu_2 < 0$$

These hypotheses are tested by using the test statistic z given by (10.11). The rejection region is determined in the same way as in the one-tailed approach presented in Chapter 9.

Small-Sample Case

Let us now consider hypothesis tests about the difference between the means of two populations for the small-sample case; that is, where $n_1 < 30$ and/or $n_2 < 30$. The procedure we will use is based on the t distribution with $n_1 + n_2 - 2$ degrees of

TABLE 10.3 Data for the Current and New Software Technology Study

Current	New
299	315
360	200
276	214
310	263
340	334
388	344
278	282
365	307
281	290
315	288
378	318
310	301

freedom. As discussed in Section 10.1, assumptions are made that both populations have normal probability distributions and that the variances of the populations are equal.

The problem situation that we will use to illustrate the small-sample case involves a new computer software package that has been developed to help systems analysts reduce the time required to design, develop, and implement an information system. To evaluate the benefits of the new software package, a random sample of 24 systems analysts is selected. Each analyst is given specifications for a hypothetical information system, and 12 of the analysts are instructed to produce the information system by using current technology. The other 12 analysts are first trained in the use of the new software package and then instructed to use it to produce the information system.

In this study, there are two populations: a population of systems analysts using the current technology and a population of systems analysts using the new software package. In terms of the time required to complete the information-system design project, the population means are

μ_1 = the mean project-completion time for systems analysts using the current technology

and

μ_2 = the mean project-completion time for systems analysts using the new software package.

The researcher in charge of the new software-evaluation project hopes to show that the new software package will provide a shorter mean project-completion time. Thus, the researcher is looking for evidence to conclude that μ_2 is less than μ_1; in this case, the difference between the two population means, $\mu_1 - \mu_2$, will be greater than zero. The research hypothesis $\mu_1 - \mu_2 > 0$ is stated as the alternative hypothesis.

$$H_0\colon \mu_1 - \mu_2 \leq 0$$
$$H_a\colon \mu_1 - \mu_2 > 0$$

TABLE 10.4 Results of Study

Current Technology	New Software Package
$n_1 = 12$	$n_2 = 12$
$\bar{x}_1 = 325$ hours	$\bar{x}_2 = 288$ hours
$s_1 = 40$ hours	$s_2 = 44$ hours

In tentatively assuming H_0 is true, we are taking the position that using the new software package takes as much time or perhaps even more than the current technology. The researcher is looking for evidence to reject H_0 and conclude that the new software package ensures a shorter mean completion time.

Suppose that the 24 analysts complete the study with the results shown in Table 10.3; summary statistics are given in Table 10.4. Under the assumption that the variances of the populations are equal, (10.8) is used to compute the pooled estimate of σ^2.

$$s^2 = \frac{(n_1 - 1)s_1^2 + (n_2 - 1)s_2^2}{n_1 + n_2 - 2} = \frac{11(40)^2 + 11(44)^2}{12 + 12 - 2} = 1768$$

The test statistic for the small-sample case is

$$t = \frac{(\bar{x}_1 - \bar{x}_2) - (\mu_1 - \mu_2)}{\sqrt{s^2\left(\frac{1}{n_1} + \frac{1}{n_2}\right)}} \tag{10.12}$$

In the case of two independent random samples of sizes n_1 and n_2, the t distribution will have $n_1 + n_2 - 2$ degrees of freedom. For $\alpha = .05$, the t distribution table shows that with $12 + 12 - 2 = 22$ degrees of freedom, $t_{.05} = 1.717$. Thus, the rejection region for the one-tailed test is

```
Twosample T for CURRENT vs NEW
           N       Mean      StDev    SE Mean
CURRENT   12      325.0       40.0         12
NEW       12      288.0       44.0         13

95% C.I. for mu CURRENT - mu NEW: (1, 73)
T-Test mu CURRENT = mu NEW (vs >): T= 2.16  P=0.021  DF=  22
```

FIGURE 10.6
Minitab Output for the Hypothesis Test About the Current and New Software Technology

Reject H_0 if $t > 1.717$.

The sample data and (10.12) provide the following value for the test statistic.

$$t = \frac{(325 - 288) - 0}{\sqrt{1768\left(\frac{1}{12} + \frac{1}{12}\right)}} = 2.16$$

Checking the rejection region, we see that $t = 2.16 > 1.717$. Thus, we reject H_0 at the .05 level of significance. The sample results enable the researcher to conclude that the new software package provides a shorter mean completion time.

The Role of the Computer

Computer software packages can be used for testing hypotheses about the difference between the means of two populations. We will illustrate the use of Minitab to perform the t-test for the case involving the use of the new software package to produce an information system (see Table 10.4). Assume the user has entered the completion times for the 12 systems analysts who used the current technology into column 1 of a Minitab worksheet and those of the 12 systems analysts who used the new software package into column 2; column 1 is labeled CURRENT and column 2 is labeled NEW. Using the Minitab 2-Sample t option, we obtained the hypothesis-testing output shown in Figure 10.6. The first part of the output gives the mean completion time, the standard deviation, and the standard error of the mean for the two samples. The row beginning with TTEST provides the hypothesis-testing results. The output TTEST MU CURRENT = MU NEW (VS GT) indicates that we have elected to perform a one-tailed test with a greater-than alternative hypothesis. With a p-value of .021, we can reject the null hypothesis at the .05 level of significance; thus, the conclusion is that the new software package provides a shorter mean completion time.

Note that the computer solution also provides the 95% confidence interval for the difference between the two population means. Although the hypothesis test enables us to conclude that the new software is better, the 95% confidence interval shows that the improvement that can be expected with the new software may be as little as one hour or as much as 73 hours. The wide confidence interval suggests that further study may be desirable to obtain a more precise estimate of how much improvement can be anticipated with the new software package.

notes and comments

In hypothesis tests about the difference between the means of two populations, the null hypothesis almost always contains the condition that there is no difference between the means. Hence, the following null hypotheses are possible choices.

$$H_0: \mu_1 - \mu_2 = 0 \qquad H_0: \mu_1 - \mu_2 \leq 0 \qquad H_0: \mu_1 - \mu_2 \geq 0$$

In some instances, we may want to determine whether there is a nonzero difference D_0 between the population means. The specific value chosen for D_0 depends on the application under study. However, in this case, the null hypothesis may be one of the following forms.

$$H_0: \mu_1 - \mu_2 = D_0 \qquad H_0: \mu_1 - \mu_2 \leq D_0 \qquad H_0: \mu_1 - \mu_2 \geq D_0$$

The hypothesis-testing computations remain the same with the exception that D_0 is used for the value of $\mu_1 - \mu_2$ in (10.11) and (10.12).

exercises

Self Test

Methods

9. Consider the following hypothesis test.

$$H_0: \mu_1 - \mu_2 \leq 0$$
$$H_a: \mu_1 - \mu_2 > 0$$

The results shown here are for two independent samples taken from the two populations.

Sample 1	Sample 2
$n_1 = 40$	$n_2 = 50$
$\bar{x}_1 = 25.2$	$\bar{x}_2 = 22.8$
$s_1 = 5.2$	$s_2 = 6.0$

a. With $\alpha = .05$, what is your hypothesis-testing conclusion?
b. What is the p-value?

10. Consider the following hypothesis test.

$$H_0: \mu_1 - \mu_2 = 0$$
$$H_a: \mu_1 - \mu_2 \neq 0$$

The following results are for two independent samples taken from the two populations.

Sample 1	Sample 2
$n_1 = 80$	$n_2 = 70$
$\bar{x}_1 = 104$	$\bar{x}_2 = 106$
$s_1 = 8.4$	$s_2 = 7.6$

a. With $\alpha = .05$, what is your hypothesis-testing conclusion?
b. What is the p-value?

11. Consider the following hypothesis test.

$$H_0: \mu_1 - \mu_2 = 0$$
$$H_a: \mu_1 - \mu_2 \neq 0$$

The following results are for two independent samples taken from the two populations. With $\alpha = .05$, what is your hypothesis-testing conclusion?

Sample 1	Sample 2
$n_1 = 8$	$n_2 = 7$
$\bar{x}_1 = 1.4$	$\bar{x}_2 = 1.0$
$s_1 = 0.4$	$s_2 = 0.6$

Applications

12. Refer to Exercise 5, in which two independent random samples of business travelers rated the Miami and Los Angeles airports. Summary statistics follow.

Airport	Sample Size	Sample Mean	Sample Standard Deviation
Miami	50	6.34	2.163
Los Angeles	50	6.72	2.374

Is the mean rating for the Los Angeles airport greater than the mean rating for the Miami airport? Support your conclusion with a statistical test using a .05 level of significance.

Self Test 

13. The Greystone Department Store study in Section 10.1 supplied the following data on customer ages from independent random samples taken at two store locations.

Inner-City Store	Suburban Store
$n_1 = 36$	$n_2 = 49$
$\bar{x}_1 = 40$ years	$\bar{x}_2 = 35$ years
$s_1 = 9$ years	$s_2 = 10$ years

For $\alpha = .05$, test $H_0: \mu_1 - \mu_2 = 0$ against the alternative $H_a: \mu_1 - \mu_2 \neq 0$. What is your conclusion about the mean ages of the populations of customers at the two stores?

14. A firm is studying the delivery times of two raw material suppliers. The firm is basically satisfied with supplier A and is prepared to stay with that supplier if the mean delivery time is the same as or less than that of supplier B. However, if the firm finds that the mean delivery time of supplier B is less than that of supplier A, it will begin making raw material purchases from supplier B.

a. What are the null and alternative hypotheses for this situation?

b. Assume that independent samples show the following delivery time characteristics for the two suppliers.

Supplier A	Supplier B
$n_1 = 50$	$n_2 = 30$
$\bar{x}_1 = 14$ days	$\bar{x}_2 = 12.5$ days
$s_1 = 3$ days	$s_2 = 2$ days

With $\alpha = .05$, what is your conclusion for the hypotheses from part (a)? What action do you recommend in terms of supplier selection?

15. In a wage discrimination case involving male and female employees, independent samples of male and female employees with five years' experience or more provided the following hourly wage results.

Male Employees	$n_1 = 44$	$\bar{x}_1 = \$9.25$	$s_1 = \$1.00$
Female Employees	$n_2 = 32$	$\bar{x}_2 = \$8.70$	$s_2 = \$.80$

The null hypothesis is that male employees have a mean hourly wage less than or equal to that of the female employees. Rejection of H_0 leads to the conclusion that male employees have a mean hourly wage exceeding that of the female employees. Test the hypothesis with $\alpha = .01$. Does wage discrimination appear to be present in this case?

16. Starting salary data for college graduates is reported by the College Placement Council (*USA Today,* April 6, 1992). Annual salaries in thousands of dollars for a sample of accounting majors and a sample of finance majors are listed in the following table.

Accounting	Finance
28.8	26.3
25.3	23.6
26.2	25.0
27.9	23.0
27.0	27.9
26.2	24.5
28.1	29.0
24.7	27.4
25.2	23.5
29.2	26.9
29.7	26.2
29.3	24.0

a. Use a .05 level of significance to test the hypothesis that there is no difference between the mean annual starting salary of accounting majors and the mean annual starting salary of finance majors. What is your conclusion?
b. Provide the point estimate and the 95% confidence interval for the difference between the mean starting salaries for the two majors.

10.3 Inferences About the Difference Between the Means of Two Populations: Matched Samples

Suppose a manufacturing company has two methods by which employees can perform a production task. To maximize production output, the company wants to identify the method with the shortest mean completion time per unit. Let μ_1 denote the mean completion time for production method 1 and μ_2 denote the mean completion time for production method 2. With no preliminary indication of the preferred production method, we begin by tentatively assuming that the two production methods have the same mean completion time. Thus, the null hypothesis is H_0: $\mu_1 - \mu_2 = 0$. If this hypothesis is rejected, we can conclude that the mean completion times differ. In this

TABLE 10.5 Task-Completion Times for a Matched-Sample Design

Worker	Completion Time for Method 1 (minutes)	Completion Time for Method 2 (minutes)	Difference in Completion Times (d_i)
1	6.0	5.4	.6
2	5.0	5.2	−.2
3	7.0	6.5	.5
4	6.2	5.9	.3
5	6.0	6.0	.0
6	6.4	5.8	.6

case, the method providing the shorter mean completion time would be recommended. The null and alternative hypotheses are written as follows.

$$H_0: \mu_1 - \mu_2 = 0$$
$$H_a: \mu_1 - \mu_2 \neq 0$$

In choosing the sampling procedure that will be used to collect production time data and test the hypotheses, we consider two alternative designs. One is based on *independent samples* and the other is based on *matched samples.*

1. *Independent-sample design:* A simple random sample of workers is selected and each worker uses method 1. A second independent simple random sample of workers is selected and each worker uses method 2. The test of the difference between means is based on the procedures in Section 10.2.

2. *Matched-sample design:* One simple random sample of workers is selected. Each worker first uses one method and then uses the other method. The order of the two methods is assigned randomly to the workers, with some workers performing method 1 first and others performing method 2 first. Each worker provides a pair of data values, one value for method 1 and another value for method 2.

In the matched-sample design the two production methods are tested under similar conditions (i.e., with the same workers); hence this design often leads to a smaller sampling error than the independent sample design. The primary reason is that in a matched-sample design, variation between workers is eliminated as a source of sampling error.

Let us demonstrate the analysis of a matched-sample design by assuming that it is the method used to test the difference between the two production methods. A random sample of six workers is used. The data on completion times for the six workers are given in Table 10.5. Note that each worker provides a pair of data values, one for each production method. Also note that the last column contains the difference in completion times d_i for each worker in the sample.

The key to the analysis of the matched-sample design is to realize that we consider only the column of differences. We therefore have six data values (.6, −.2, .5, .3, .0, and .6) that will be used to analyze the difference between the means of the two production methods.

Let $\mu_d =$ the mean of the *difference* values for the population of workers. With this notation, the null and alternative hypotheses are rewritten as follows.

$$H_0\colon \mu_d = 0$$
$$H_a\colon \mu_d \neq 0$$

If H_0 can be rejected, we can conclude that the mean completion times differ.

The d notation is a reminder that the matched sample provides *difference* data. The sample mean and sample standard deviation for the six difference values in Table 10.5 follow.

$$\bar{d} = \frac{\Sigma d_i}{n} = \frac{1.8}{6} = .30$$

$$s_d = \sqrt{\frac{\Sigma(d_i - \bar{d})^2}{n-1}} = \sqrt{\frac{.56}{5}} = .335$$

In Chapter 9 we stated that if the population can be assumed to be normally distributed, the t distribution with $n - 1$ degrees of freedom can be used to test the null hypothesis about a population mean. With difference data, the test statistic becomes

$$t = \frac{\bar{d} - \mu_d}{s_d/\sqrt{n}} \tag{10.3}$$

With $\alpha = .05$ and $n - 1 = 5$ degrees of freedom ($t_{.025} = 2.571$), the rejection rule for the two-tailed test becomes

Reject H_0 if $t < -2.571$ or if $t > 2.571$.

With $\bar{d} = .30$, $s_d = .335$, and $n = 6$, the value of the test statistic for the null hypothesis $H_0\colon \mu_d = 0$ is

$$t = \frac{\bar{d} - \mu_d}{s_d/\sqrt{n}} = \frac{.30 - 0}{.335/\sqrt{6}} = 2.19$$

Since $t = 2.19$ is not in the rejection region, the sample data do not provide sufficient evidence to reject H_0.

Using the sample results, we can obtain an interval estimate of the difference between the two population means by using the single-population methodology of Chapter 8. The calculation follows.

$$0.3 \pm t_{\alpha/2}\frac{s_d}{\sqrt{n}}$$
$$0.3 \pm 2.571\frac{.335}{\sqrt{6}}$$
$$0.3 \pm .35$$

Thus, the 95% confidence interval for the difference between the means of the two production methods is $-.05$ minutes to .65 minutes. Note that since the confidence interval includes the value of zero, we see again that the sample data do not provide sufficient evidence to reject H_0.

notes and comments

1. In the example presented in this section, workers performed the production task with first one method and then the other method. This example illustrates a matched-sample design in which each sampled item (worker) provides a pair of data values. It is also possible to use different but "similar" items to provide the pair of data values. For example, a worker at one location could be matched with a similar worker at another location (similarity based on age, education, sex, experience, etc.). The pairs of workers would provide the difference data that could be used in the matched-sample analysis.

2. Since a matched-sample procedure for inferences about two population means generally provides better precision than the independent-sample approach, it is the recommended design. However, in some applications the matching cannot be achieved, or perhaps the time and cost associated with matching are excessive. In such cases, the independent-sample design should be used.

3. The example presented in this section had a sample size of six workers and thus was a small-sample case. The t distribution was used in both the hypothesis-test and interval-estimation computations. If the sample size is large ($n \geq 30$), use of the t distribution is unnecessary; in such cases, statistical inferences can be based on the z values of the standard normal probability distribution.

exercises

Self Test

Methods

17. Consider the following hypothesis test.

$$H_0: \mu_d \leq 0$$
$$H_a: \mu_d > 0$$

The following data are from matched samples taken from two populations.

	Population	
Element	**1**	**2**
1	21	20
2	28	26
3	18	18
4	20	20
5	26	24

a. Compute the difference value for each element.

b. Compute $\bar{d}$.

c. Compute the standard deviation s_d.
d. Test the hypothesis using $\alpha = .05$. What is your conclusion?

18. The following data are from matched samples taken from two populations.

	Population	
Element	1	2
1	11	8
2	7	8
3	9	6
4	12	7
5	13	10
6	15	15
7	15	14

a. Compute the difference value for each element.
b. Compute $\bar{d}$.
c. Compute the standard deviation s_d.
d. What is the point estimate of the difference between the two population means?
e. Provide a 95% confidence interval for the difference between the two population means.

Applications

Self Test

19. A market research firm used a sample of individuals to rate the purchase potential of a particular product before and after the individuals saw a new television commercial about the product. The purchase-potential ratings were based on a 0 to 10 scale, with higher values indicating a higher purchase potential. The null hypothesis stated that the mean rating "after" would be less than or equal to the mean rating "before." Rejection of this hypothesis would show that the commercial improved the mean purchase-potential rating. Use $\alpha = .05$ and the following data to test the hypothesis and comment on the value of the commercial.

	Purchase Rating			Purchase Rating	
Individual	After	Before	Individual	After	Before
1	6	5	5	3	5
2	6	4	6	9	8
3	7	7	7	7	5
4	4	3	8	6	6

20. To investigate the amount of savings due to purchasing store brand versus name brand products, *Consumer Reports* shopped for a list of items at an A&P grocery store. One cart was filled with national brand products and the other cart was filled with the store brands of the same products (*Consumer Reports,* September 1993). A portion of the data obtained follows.

Product, Size	Name Brand Price ($)	Store Brand Price ($)
Ketchup, 2 lb.	1.69	0.79
Coffee, 12 oz.	2.79	1.59
Soda, 6-pack	2.79	1.64
Paper towels, 90 sheets	1.39	0.50
Ice cream, 1/2 gal.	3.99	2.39
American cheese, 1 lb.	3.99	2.99
Thin spaghetti, 1 lb.	0.89	0.53
Butter, 1 lb.	2.39	1.69
White rice, 5 lb.	3.99	1.59
Vegetable oil, qt.	2.19	1.69

a. Does the sampling procedure used by *Consumer Reports* represent an independent-sample design or a matched-sample design? Explain.
b. Develop a 95% confidence interval for the difference in price between name brand products and store brand products. What assumption was necessary to develop your interval estimate?

21. The cost of transportation from the airport to the downtown area depends on the method of transportation. One-way costs ($) for taxi and shuttle bus transportation for a sample of 10 major cities follow (*USA Today,* February 13, 1992). Provide a 95% confidence interval for the mean cost increase associated with taxi transportation.

City	Taxi	Shuttle Bus
Atlanta	15	7
Chicago	22	12.5
Denver	11	5
Houston	15	4.5
Los Angeles	26	11
Minneapolis	16.5	7.5
New Orleans	18	7
New York (LaGuardia)	16	8.5
Philadelphia	20	8
Washington, D.C.	10	5

22. A survey was made of Book-of-the-Month-Club members to ascertain whether members spend more time watching television than they do reading (*The Cincinnati Enquirer,* November 21, 1991). Assume a small sample of respondents in this survey provided the weekly hours of television watching and weekly hours of reading listed in the following table.

Respondent	Television	Reading
1	10	6
2	14	16
3	16	8
4	18	10
5	15	10
6	14	8
7	10	14
8	12	14
9	4	7
10	8	8
11	16	5
12	5	10
13	8	3
14	19	10
15	11	6

Using a .05 level of significance, can you conclude that Book-of-the-Month-Club members spend more time per week, on average, watching television than reading?

23. A manufacturer produces both a deluxe and a standard model of an automatic sander designed for home use. Selling prices obtained from a sample of retail outlets follow.

	Model Price ($)			Model Price ($)	
Retail Outlet	Deluxe	Standard	Retail Outlet	Deluxe	Standard
1	39	27	5	40	30
2	39	28	6	39	34
3	45	35	7	35	29
4	38	30			

a. The manufacturer's suggested retail prices for the two models show a $10 price differential. Using a .05 level of significance, test that the mean difference between the prices of the two models is $10.

b. What is the 95% confidence interval for the difference between the mean prices of the two models?

24. A company attempts to evaluate the potential for a new bonus plan by selecting a random sample of five salespersons to use the bonus plan for a trial period. The weekly sales volumes before and after implementation of the bonus plan are shown in the following table.

	Weekly Sales	
Salesperson	Before	After
1	15	18
2	12	14
3	18	19
4	15	18
5	16	18

a. Use $\alpha = .05$ and test to see whether the bonus plan will result in an increase in the mean weekly sales.

b. Provide a 90% confidence interval for the mean increase in weekly sales that can be expected if a new bonus plan is implemented.

10.4 An Introduction to Analysis of Variance

In Section 10.2 we showed how a hypothesis test can be conducted to determine whether the means of two populations are equal based upon two independent simple random samples. In this section we introduce a statistical technique called *analysis of variance* (ANOVA), which can be used to test the hypothesis that the means of three or more populations are equal. We begin our discussion of analysis of variance by considering a problem facing National Computer Products, Inc.

National Computer Products, Inc. (NCP), manufactures printers and fax machines at plants in Charlotte, Houston, and San Diego. To measure how much employees at these plants know about total quality management, a random sample of six employees

TABLE 10.6 Quality-Awareness Examination Scores for 18 Employees

	Plant 1	Plant 2	Plant 3
Observation	**Charlotte**	**Houston**	**San Diego**
1	85	71	59
2	75	75	64
3	82	73	62
4	76	74	69
5	71	69	75
6	85	82	67
Sample mean	79	74	66
Sample variance	34	20	32
Sample standard deviation	5.83	4.47	5.66

was selected from each plant and given a quality-awareness examination. The examination scores obtained for these 18 employees are listed in Table 10.6. The sample means, sample variances, and sample standard deviations for each group are also provided. Managers want to use these data to test the hypothesis that the mean examination score is the same for all three plants.

We will define population 1 as all employees at the Charlotte plant, population 2 as all employees at the Houston plant, and population 3 as all employees at the San Diego plant. Let

$$\mu_1 = \text{mean examination score for population 1,}$$

$$\mu_2 = \text{mean examination score for population 2, and}$$

$$\mu_3 = \text{mean examination score for population 3.}$$

Although we will never know the actual values of μ_1, μ_2, and μ_3, we want to use the sample results to test the following hypotheses.

$$H_0\colon \mu_1 = \mu_2 = \mu_3$$

$$H_a\colon \text{Not all population means are equal}$$

As we will demonstrate shortly, analysis of variance is a statistical procedure for testing whether the observed differences are significant.

Assumptions for Analysis of Variance

To test the preceding hypotheses by using analysis of variance, we must make the following three assumptions.

1. The examination scores must be normally distributed at each plant.
2. The variance of examination scores, denoted σ^2, must be the same for all three plants.
3. The examination score for each employee must be independent of the examination score for any other employee.

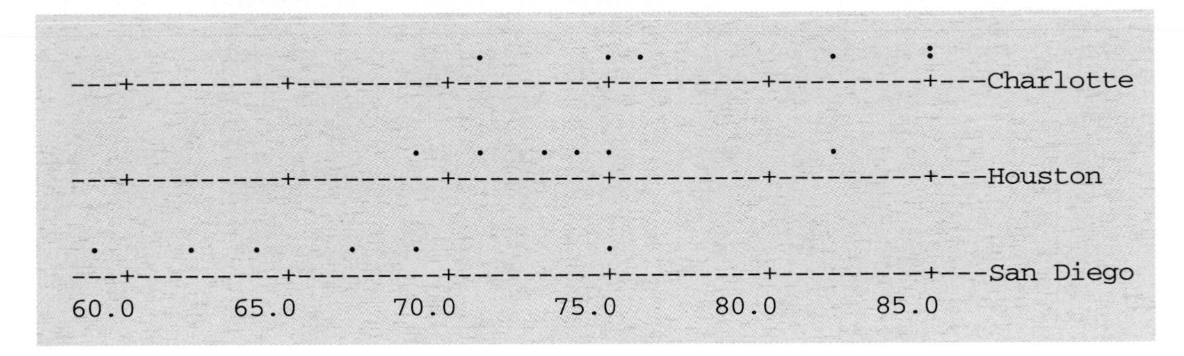

FIGURE 10.7
Minitab Dotplot for Examination Scores

A Conceptual Overview

Suppose the assumptions for analysis of variance are satisfied in the NCP example. If the null hypothesis is true ($\mu_1 = \mu_2 = \mu_3 = \mu$), each sample observation would have been drawn from the same normal probability distribution with mean μ and variance σ^2. For a visual perspective, consider the Minitab dotplots for the NCP data shown in Figure 10.7. Does it appear that the observations in each sample have been drawn from populations with the same mean? Although this is purely a subjective observation, you might agree that the employees at the Charlotte plant appear to have higher examination scores, whereas the employees at the San Diego plant appear to have lower examination scores.

If the means for the three populations are equal, we would expect the three sample means to be close together. In fact, the closer the three sample means are to one another, the more evidence we have for the conclusion that the population means are equal. Alternatively, the more the sample means differ, the more evidence we have for the conclusion that the population means are not equal. In other words, if the variability among the sample means is "small," it supports H_0; if the variability among the sample means is "large," it supports H_a.

If the null hypothesis, H_0: $\mu_1 = \mu_2 = \mu_3$, is true, we can use the variability among the sample means to develop an estimate of σ^2. First, note that, if the assumptions for analysis of variance are satisfied, each sample will have come from the same normal probability distribution with mean μ and variance σ^2. Recall from Chapter 7 that the sampling distribution of the sample mean $\bar{x}$ for a simple random sample of size n from a normal population will be normally distributed with mean μ and variance σ^2/n. Figure 10.8 illustrates such a sampling distribution.

If the null hypothesis is true, we can think of each of the three sample means, $\bar{x}_1 = 79$, $\bar{x}_2 = 74$, and $\bar{x}_3 = 66$, from Table 10.6 as values drawn at random from the sampling distribution shown in Figure 10.8. In this case, the mean and variance of the three $\bar{x}$ values can be used to estimate the mean and variance of the sampling distribution. In the NCP example, the best estimate of the mean of the sampling distribution of $\bar{x}$ is the mean or average of the three sample means. That is, $(79 + 74 + 66)/3 = 73$. We refer to this estimate as the *overall sample mean.* To estimate the variance of the sampling distribution of $\bar{x}$, we compute the variance by using the three sample means.

$$s_{\bar{x}}^2 = \frac{(79-73)^2 + (74-73)^2 + (66-73)^2}{3-1} = \frac{86}{2} = 43$$

FIGURE 10.8
Sampling Distribution of $\bar{x}$ Given H_0 Is True

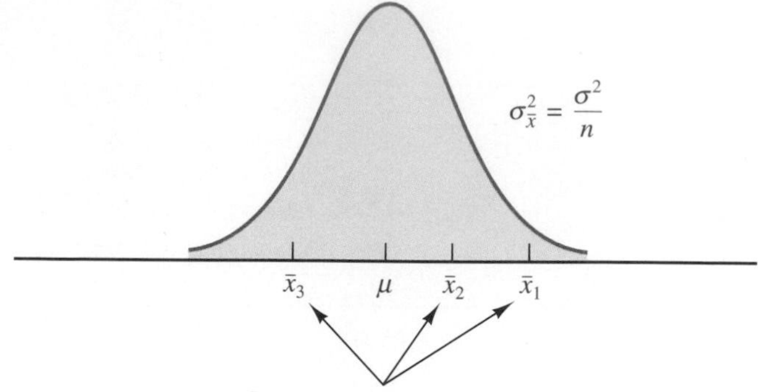

FIGURE 10.9
Sampling Distribution for $\bar{x}$ Given H_0 Is False

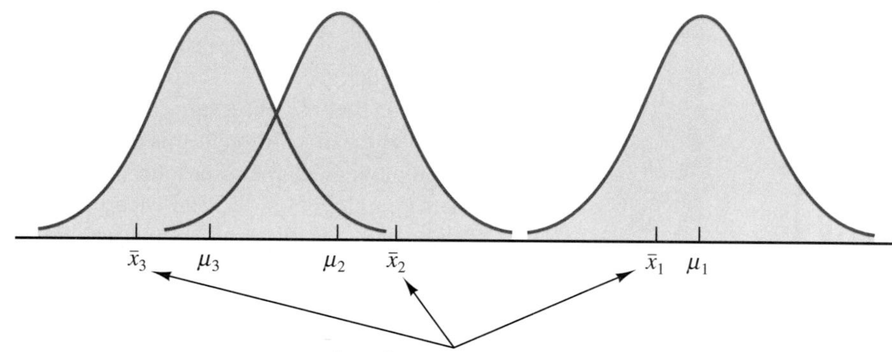

Since $\sigma_{\bar{x}}^2 = \sigma^2/n$, solving for σ^2 gives

$$\sigma^2 = n\sigma_{\bar{x}}^2$$

Hence,

$$\text{Estimate of } \sigma^2 = n\,(\text{Estimate of } \sigma_{\bar{x}}^2) = ns_{\bar{x}}^2 = 6(43) = 258$$

The result, $ns_{\bar{x}}^2 = 258$, is referred to as the *between-samples* estimate of σ^2.

The between-samples estimate of σ^2 is based on the assumption that the null hypothesis is true. In this case, each sample comes from the same population, and there is only one sampling distribution of $\bar{x}$. To illustrate what happens when H_0 is false, suppose the population means *all differ.* Note that since the three samples are from normal populations with different means, there will be three different sampling distributions. Figure 10.9 shows that in this case, the sample means are not as close together as they were when H_0 was true. Thus, $s_{\bar{x}}^2$ will be larger, causing the between-samples estimate of σ^2 to be larger. In general, when the population means are not equal, the between-samples estimate will overestimate the population variance σ^2.

The variation within each of the samples also has an effect on the conclusion we reach in analysis of variance. When a simple random sample is selected from each population, each of the sample variances provides an unbiased estimate of σ^2. Hence, we can combine or pool the individual estimates of σ^2 into one overall estimate. The estimate of σ^2 obtained in this way is called the *pooled* or *within-samples* estimate of σ^2. Because each sample variance provides an estimate of σ^2 based only on the variation within each sample, the within-samples estimate of σ^2 is not affected by whether or not the population means are equal. When the sample sizes are equal, the within-samples estimate of σ^2 can be obtained by computing the average of the individual sample variances. For the NCP example we obtain

$$\text{Within-Samples Estimate of } \sigma^2 = \frac{34 + 20 + 32}{3} = \frac{86}{3} = 28.67$$

In the NCP example, the between-samples estimate of σ^2 (258) is much larger than the within-samples estimate of σ^2 (28.67). In fact, the ratio of these two estimates is $258/28.67 = 9.00$. Recall, however, that the between-samples approach provides a good estimate of σ^2 only if the null hypothesis is true; if the null hypothesis is false, the between-samples approach *overestimates* σ^2. The within-samples approach provides a good estimate of σ^2 in either case. Thus, if the null hypothesis is true, the two estimates will be similar and their ratio will be close to 1. If the null hypothesis is false, the between-samples estimate will be larger than the within-samples estimate, and their ratio will be large. In the next section we will show how large this ratio must be to reject H_0.

In summary, the logic behind ANOVA is based on the development of two independent estimates of the common population variance σ^2. One estimate of σ^2 is based on the variability among the sample means themselves, and the other estimate of σ^2 is based on the variability of the data within each sample. By comparing these two estimates of σ^2, we will be able to determine whether the population means are equal. Since the methodology involves a comparison of variances, it is referred to as analysis of variance.

notes and comments

1. In Sections 10.2–10.3 we presented statistical methods for testing the hypothesis that the means of two populations are equal. ANOVA can also be used to test the hypothesis that the means of two populations are equal. In practice, however, analysis of variance is usually not used unless there are three or more population means.

2. In Section 10.2 we discussed how to test for the equality of two population means whenever one or both sample sizes are less than 30. As part of that discussion we illustrated the process of combining the results of two independent random samples to provide one estimate of σ^2; that process was referred to as pooling, and the resulting sample variance was referred to as the pooled estimator of σ^2. In analysis of variance, the within-samples estimate of σ^2 is simply the generalization of that concept to the case of more than two samples; that is why we also referred to the within-samples estimator as the pooled estimator of σ^2.

10.5 Analysis of Variance: Testing for the Equality of k Population Means

Analysis of variance can be used to test for the equality of k population means. The general form of the hypotheses tested is

$$H_0\colon \mu_1 = \mu_2 = \cdots = \mu_k$$
$$H_a\colon \text{Not all population means are equal}$$

where

$$\mu_j = \text{mean of the } j\text{th population.}$$

We assume that a simple random sample of size n_j has been selected from each of the k populations. Let

$$\begin{aligned} x_{ij} &= \text{the } i\text{th observation in the } j\text{th sample,} \\ n_j &= \text{the number of observations in the } j\text{th sample,} \\ \bar{x}_j &= \text{the mean of the } j\text{th sample,} \\ s_j^2 &= \text{the variance of the } j\text{th sample, and} \\ s_j &= \text{the standard deviation of the } j\text{th sample.} \end{aligned}$$

The formulas for the jth sample mean and variance follow.

$$\bar{x}_j = \frac{\sum_{i=1}^{n_j} x_{ij}}{n_j} \tag{10.14}$$

$$s_j^2 = \frac{\sum_{i=1}^{n_j} (x_{ij} - \bar{x}_j)^2}{n_j - 1} \tag{10.15}$$

The overall sample mean, denoted $\bar{\bar{x}}$, is the sum of all the observations divided by the total number of observations. That is,

$$\bar{\bar{x}} = \frac{\sum_{j=1}^{k} \sum_{i=1}^{n_j} x_{ij}}{n_T} \tag{10.16}$$

where

$$n_T = n_1 + n_2 + \cdots + n_k \tag{10.17}$$

If the size of each sample is n, $n_T = kn$; in this case (10.16) reduces to

$$\bar{\bar{x}} = \frac{\sum_{j=1}^{k} \sum_{i=1}^{n_j} x_{ij}}{nk} = \frac{\sum_{j=1}^{k} \sum_{i=1}^{n_j} x_{ij}/n}{k} = \frac{\sum_{j=1}^{k} \bar{x}_j}{k} \tag{10.18}$$

In other words, whenever the sample sizes are the same, the overall sample mean is just the average of the k sample means.

Since each sample in the NCP example consists of $n = 6$ observations, the overall sample mean can be computed by using (10.18). For the data in Table 10.6 we obtain the following result.

$$\bar{\bar{x}} = \frac{79 + 74 + 66}{3} = 73$$

Thus, if the null hypothesis is true, the overall sample mean of 73 is the best estimate of the population mean μ.

Between-Samples Estimate of Population Variance

In the preceding section we introduced the concept of a between-samples estimate of σ^2. This estimate of σ^2 is called the *mean square between* and is denoted MSB. The formula for computing MSB is

$$\text{MSB} = \frac{\sum_{j=1}^{k} n_j(\bar{x}_j - \bar{\bar{x}})^2}{k - 1} \tag{10.19}$$

The numerator in (10.19) is called the *sum of squares between* and is denoted SSB. The denominator, $k - 1$, represents the degrees of freedom associated with SSB. Hence, the mean square between can be computed by the following formula.

Mean Square Between

$$\text{MSB} = \frac{\text{SSB}}{k - 1} \tag{10.20}$$

where

$$\text{SSB} = \sum_{j=1}^{k} n_j(\bar{x}_j - \bar{\bar{x}})^2 \tag{10.21}$$

If H_0 is true, MSB provides an unbiased estimate of σ^2. However, if the means of the k populations are not equal, MSB is not an unbiased estimate of σ^2; in fact, in that case, MSB should overestimate σ^2.

For the NCP data in Table 10.6, we obtain the following results.

$$\text{SSB} = \sum_{j=1}^{k} n_j(\bar{x}_j - \bar{\bar{x}})^2 = 6(79 - 73)^2 + 6(74 - 73)^2 + 6(66 - 73)^2 = 516$$

$$\text{MSB} = \frac{\text{SSB}}{k - 1} = \frac{516}{2} = 258$$

Within-Samples Estimate of Population Variance

The second estimate of σ^2 is based on the variation of the sample observations within each sample. This estimate of σ^2 is called the *mean square within* and is denoted MSW. The formula for computing MSW is

$$\text{MSW} = \frac{\sum_{j=1}^{k}(n_j - 1)s_j^2}{n_T - k} \tag{10.22}$$

The numerator in (10.22) is called the *sum of squares within* and is denoted SSW. The denominator of MSW is referred to as the degrees of freedom associated with SSW. Hence, the formula for MSW can also be stated as follows.

Mean Square Within

$$\text{MSW} = \frac{\text{SSW}}{n_T - k} \tag{10.23}$$

where

$$\text{SSW} = \sum_{j=1}^{k}(n_j - 1)s_j^2 \tag{10.24}$$

Note that MSW is based on the variation within each of the samples; it is not influenced by whether or not the null hypothesis is true. Thus, MSW always provides an unbiased estimate of σ^2.

For the NCP data in Table 10.6 we obtain the following results.

$$\text{SSW} = \sum_{j=1}^{k}(n_j - 1)s_j^2 = (6 - 1)34 + (6 - 1)20 + (6 - 1)32 = 430$$

$$\text{MSW} = \frac{\text{SSW}}{n_T - k} = \frac{430}{18 - 3} = \frac{430}{15} = 28.67$$

Comparing the Variance Estimates: the F Test

Let us assume that the null hypothesis is true. In that case, MSB and MSW provide two independent, unbiased estimates of σ^2. If the null hypothesis is true and the ANOVA assumptions are valid, the sampling distribution of MSB/MSW is an F distribution with numerator degrees of freedom equal to $k - 1$ and denominator degrees of freedom equal to $n_T - k$.

Each specific F distribution depends on the number of degrees of freedom associated with the numerator and the number of degrees of freedom associated with the denominator. We use F_α to denote the F value that provides an area or probability of α in the upper tail of the F distribution. For instance, $F_{.05}$ provides an area in the upper tail of .05. Table 4 in Appendix B provides F values with areas of .05, .025, and .01 in the upper tail of the distribution. Let us now see how the F distribution is used in the analysis of variance procedure.

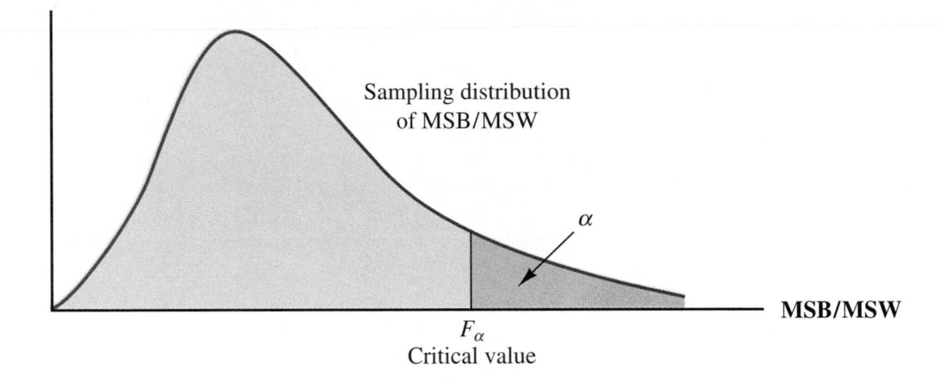

FIGURE 10.10
Sampling Distribution of MSB/MSW; the Critical Value for Rejecting the Null Hypothesis of Equality of Means Is F_α

If the means of the k populations are not equal, the value of MSB/MSW will be inflated because MSB overestimates σ^2. Hence, we will reject H_0 if the resulting value of MSB/MSW appears to be too large to have been selected at random from an F distribution with degrees of freedom $k - 1$ in the numerator and $n_T - k$ in the denominator. The value of MSB/MSW that will cause us to reject H_0 depends on α, the level of significance. Once α is selected, a critical value can be determined. Figure 10.10 shows the sampling distribution of MSB/MSW and the rejection region associated with a level of significance equal to α where F_α denotes the critical value. A summary of the overall procedure follows.

Test for the Equality of k Population Means

$$H_0: \mu_1 = \mu_2 = \cdots = \mu_k$$

$$H_a: \text{Not all population means are equal}$$

Test Statistic

$$F = \frac{\text{MSB}}{\text{MSW}} \tag{10.25}$$

Rejection Rule at a Level of Significance α

$$\text{Reject } H_0 \text{ if } F > F_\alpha$$

where the value of F_α is based on an F distribution with $k - 1$ numerator degrees of freedom and $n_T - k$ denominator degrees of freedom.

Suppose the manager responsible for making the decision in the National Computer Products example was willing to accept a probability of a Type I error of at most $\alpha = .05$. From Table 4 of Appendix B we can determine the critical F value by locating the value corresponding to numerator degrees of freedom equal to $k - 1 = 3 - 1 = 2$ and denominator degrees of freedom equal to $n_T - k = 18 - 3 = 15$. Thus, we obtain the value $F_{.05} = 3.68$. Note that this tells us that if we were to select a value at random from an F distribution with two numerator degrees of freedom and 15

TABLE 10.7 Analysis of Variance Table for the NCP Example

Source of Variation	Sum of Squares	Degrees of Freedom	Mean Square	F
Between	516	2	258.00	9.00
Within	430	15	28.67	
Total	946	17		

denominator degrees of freedom, only 5% of the time would we observe a value greater than 3.68. Moreover, the theory behind the analysis of variance tells us that if the null hypothesis is true, the ratio of MSB/MSW would be a value from such an F distribution. Hence, the appropriate rejection rule for the NCP example is written

$$\text{Reject } H_0 \text{ if MSB/MSW} > 3.68.$$

Recall that MSB = 258 and MSW = 28.67. Since MSB/MSW = 258/28.67 = 9.00 is greater than the critical value, $F_{.05} = 3.68$, we have sufficient evidence to reject the null hypothesis that the means of the three populations are equal. In other words, analysis of variance supports the conclusion that the population mean examination scores at the three NCP plants are not equal.

The ANOVA Table

The results of the preceding calculations can be displayed conveniently in a table referred to as the *analysis of variance (ANOVA) table.* Table 10.7 is the analysis of variance table for the National Computer Products example. The sum of squares associated with the source of variation referred to as "Total" is called the total sum of squares (SST). Note that the results for the NCP example suggest that SST = SSB + SSW, and that the degrees of freedom associated with this total sum of squares is the sum of the degrees of freedom associated with the between-samples estimate of σ^2 and the within-samples estimate of σ^2.

We point out that SST divided by its degrees of freedom $n_T - 1$ is nothing more than the overall sample variance that would be obtained if we treated the entire set of 18 observations as one data set. With the entire data set as one sample, the formula for computing the total sum of squares, SST, is

$$\text{SST} = \sum_{j=1}^{k} \sum_{i=1}^{n_j} (x_{ij} - \bar{\bar{x}})^2 \tag{10.26}$$

It can be shown that the results we observed for the analysis of variance table for the NCP example also apply to other problems. That is,

$$\text{SST} = \text{SSB} + \text{SSW} \tag{10.27}$$

In other words, SST can be partitioned into two sums of squares: the sum of squares between and the sum of squares within. Note also that the degrees of freedom corre-

```
Analysis of Variance
Source      DF        SS        MS         F        p
Factor       2     516.0     258.0      9.00    0.003
Error       15     430.0      28.7
Total       17     946.0
                                  Individual 95% CIs For Mean
                                  Based on Pooled StDev
 Level       N      Mean     StDev  --- --------- --------- --------- ---
PLANT 1      6    79.000     5.831                    (------*------)
PLANT 2      6    74.000     4.472              (------*-----)
PLANT 3      6    66.000     5.657   (-----*------)
                                    --- --------- --------- --------- ---
Pooled StDev =     5.354            63.0      70.0      77.0      84.0
```

FIGURE 10.11
Minitab Output for the NCP Analysis of Variance

sponding to SST, $n_T - 1$, can be partitioned into the degrees of freedom corresponding to SSB, $k - 1$, and the degrees of freedom corresponding to SSW, $n_T - k$. The analysis of variance can be viewed as the process of partitioning the total sum of squares and the degrees of freedom into their corresponding sources: between and within. Dividing the sum of squares by the appropriate degrees of freedom provides the variance estimates and the F value used to test the hypothesis of equal population means.

Computer Results for Analysis of Variance

Because of the widespread availability of statistical computer packages, analysis of variance computations with large sample sizes and/or a large number of populations can be performed easily. In Figure 10.11 we show output for the NCP example obtained from the Minitab computer package. The first part of the computer output contains the familiar ANOVA table format. Comparing Figure 10.11 with Table 10.7, we see that the same information is available, although some of the headings are slightly different. The heading SOURCE is used for the source of variation column, FACTOR identifies the between-samples row, and ERROR identifies the within-samples row. The sum of squares and degrees of freedom columns are interchanged, and a p-value is provided for the F test.

Note that below the ANOVA table the computer output contains the respective sample sizes, the sample means, and the standard deviations. In addition, Minitab provides a figure that shows individual 95% confidence interval estimates of each population mean. In developing these confidence interval estimates, Minitab uses MSW as the estimate of σ^2. Thus, the square root of MSW provides the best estimate of the population standard deviation σ. This estimate of σ on the computer output is Pooled StDev; it is equal to 5.354. To provide an illustration of how these interval estimates are developed, we will compute a 95% confidence interval estimate of the population mean for the Charlotte plant, identified as PLANT 1 in the computer output.

From our study of interval estimation in Chapter 8, we know that the general form of an interval estimate of a population mean is

$$\bar{x} \pm t_{\alpha/2}\frac{s}{\sqrt{n}} \tag{10.28}$$

where s is the estimate of the population standard deviation σ. Since in the analysis of variance the best estimate of σ is provided by the square root of MSW or the Pooled StDev, we use a value of 5.354 for s in (10.28). The degrees of freedom for the t value is 15, the degrees of freedom associated with the within-samples estimate of σ^2. Hence, with $t_{.025} = 2.131$ we obtain

$$79 \pm 2.131\frac{5.354}{\sqrt{6}} = 79 \pm 4.66$$

From this calculation, we see that the interval shown on the Minitab output for Plant 1 depicts an interval from 74.34 to 83.66. Since the sample sizes are equal for the NCP example, the confidence intervals for Plants 2 and 3 are also constructed by adding and subtracting 4.66 from each sample mean. Thus, in the figure provided by Minitab we see that the widths of the confidence intervals are the same.

notes and comments

1. The overall sample mean can also be computed as a weighted average of the k sample means.

$$\bar{\bar{x}} = \frac{n_1\bar{x}_1 + n_2\bar{x}_2 + \cdots + n_k\bar{x}_k}{n_T}$$

In problems where the sample means are provided, this formula is simpler than (10.16) for computing the overall mean.

2. If each sample consists of n observations, (10.19) can be written as

$$\text{MSB} = \frac{n\sum_{j=1}^{k}(\bar{x}_j - \bar{\bar{x}})^2}{k-1} = n\left[\frac{\sum_{j=1}^{k}(\bar{x}_j - \bar{\bar{x}})^2}{k-1}\right] = ns_{\bar{x}}^2$$

Note that this is the same result we presented in Section 10.4 when we introduced the concept of the between-samples estimate of σ^2. Equation (10.19) is simply a generalization of this result to the unequal sample-size case.

3. If each sample has n observations, $n_T = kn$; thus, $n_T - k = k(n-1)$, and (10.22) can be rewritten as

$$\text{MSW} = \frac{\sum_{j=1}^{k}(n-1)s_j^2}{k(n-1)} = \frac{(n-1)\sum_{j=1}^{k}s_j^2}{k(n-1)} = \frac{\sum_{j=1}^{k}s_j^2}{k}$$

In other words, if the sample sizes are the same, the within-samples estimate of σ^2 is just the average of the k sample variances. Note that this is the result we used in Section 10.4 when we introduced the concept of the within-samples estimate of σ^2.

exercises

Methods

25. Samples of five observations were selected from each of three populations. The data obtained follow.

Observation	Sample 1	Sample 2	Sample 3
1	32	44	33
2	30	43	36
3	30	44	35
4	26	46	36
5	32	48	40
Sample mean	30	45	36
Sample variance	6.00	4.00	6.50

a. Develop the dotplots for these data. From your subjective evaluation of the dotplots, do the observations in each sample appear to have been drawn from the same population?
b. Compute the between-samples estimate of σ^2.
c. Compute the within-samples estimate of σ^2.
d. At the $\alpha = .05$ level of significance, can we reject the null hypothesis that the means of the three populations are equal?
e. Set up the ANOVA table for this problem.

26. Four observations were selected from each of three populations. The data obtained follow.

Observation	Sample 1	Sample 2	Sample 3
1	165	174	169
2	149	164	154
3	156	180	161
4	142	158	148
Sample mean	153	169	158
Sample variance	96.67	97.33	82.00

a. Compute the between-samples estimate of σ^2.
b. Compute the within-samples estimate of σ^2.
c. At the $\alpha = .05$ level of significance, can we reject the null hypothesis that the three population means are equal? Explain.
d. Set up the ANOVA table for this problem.

27. A random sample of 16 observations was selected from each of four populations. A portion of the ANOVA table follows.

Source of Variation	Sum of Squares	Degrees of Freedom	Mean Square	F
Between			400	
Within				
Total	1500			

a. Provide the missing entries for the ANOVA table.

b. At the $\alpha = .05$ level of significance, can we reject the null hypothesis that the means of the four populations are equal?

Applications

Self Test

28. To test whether the mean time needed to mix a batch of material is the same for machines produced by three manufacturers, the Jacobs Chemical Company obtained the data shown in the following table on the time (in minutes) needed to mix the material.

	Manufacturer		
	1	2	3
	20	28	20
	26	26	19
	24	31	23
	22	27	22
$\bar{x}_j$	23	28	21
s_j^2	6.67	4.67	3.33

Use these data to test whether the population mean times for mixing a batch of material differ for the three manufacturers. Use $\alpha = .05$.

29. Managers at all levels of an organization need adequate information to perform their respective tasks. A recent study investigated the effect the source has on the dissemination of the information (*Journal of Management Information Systems,* Fall 1988). In this particular study the sources of information were a superior, a peer, and a subordinate. In each case, a measure of dissemination was obtained, with higher values indicating greater dissemination of information. Using $\alpha = .05$ and the following data, test whether the source of information significantly affects dissemination. What is your conclusion, and what does it suggest about the use and dissemination of information?

	Superior	Peer	Subordinate
	8	6	6
	5	6	5
	4	7	7
	6	5	4
	6	3	3
	7	4	5
	5	7	7
	5	6	5
$\bar{x}_j$	5.75	5.5	5.25
s_j^2	1.64	2.00	1.93

30. A study investigated the perception of corporate ethical values among individuals specializing in marketing (*Journal of Marketing Research,* July 1989). Suppose the following data were obtained in a similar study (higher scores indicate higher ethical values). Using $\alpha = .05$, test for significant differences in perception among the three groups of specialists.

	Marketing Managers	Marketing Research	Advertising
	6	5	6
	5	5	7
	4	4	6
	5	4	5
	6	5	6
	4	4	6
$\bar{x}_j$	5	4.5	6
s_j^2	.8	.3	.4

31. To test for any significant difference in the number of hours between breakdowns for four machines, the following data were obtained.

	Machine			
	1	2	3	4
	6.4	8.7	11.1	9.9
	7.8	7.4	10.3	12.8
	5.3	9.4	9.7	12.1
	7.4	10.1	10.3	10.8
	8.4	9.2	9.2	11.3
	7.3	9.8	8.8	11.5
$\bar{x}_j$	7.1	9.1	9.9	11.4
s_j^2	1.21	.93	.70	1.02

At the $\alpha = .05$ level of significance, is there any difference in the population mean times among the four machines?

32. The *Business Week* Global 1000 ranks companies on the basis of their market value (*Business Week,* July 11, 1994). The following table shows the P/E ratios for 30 companies classified as being in the finance economic sector. An industry code of 1 indicates a banking firm, a code of 2 a financial services firm, and a code of 3 an insurance firm. At the .05 level of significance, test whether the mean price/earnings ratio is the same for these three groups of financial firms.

Company	Industry Code	P/E	Company	Industry Code	P/E
Citicorp	1	8.0	Dean Witter, Discover	2	10.0
NationsBank	1	10.0	MBNA	2	18.0
Wells Fargo	1	13.0	Cincinnati Financial	2	14.0
First Union	1	3.4	Franklin Resources	2	14.0
KeyCorp	1	11.0	Fannie Mae	2	11.0
Chase Manhattan	1	6.0	American International Group	3	15.0
Fifth Third Bancorp	1	16.0			
Bank of New York	1	9.0	Allstate	3	16.0
First Chicago	1	6.0	Marsh & McLennan	3	18.0
Mellon Bank	1	10.0	American General	3	10.0
Fleet Financial Group	1	12.0	Cigna	3	11.0
			Lincoln National	3	9.0
First Bank System	1	15.0	AFLAC	3	14.0
American Express	2	12.0	Equitable	3	15.0
Travelers	2	22.0	Chubb	3	11.0
Merrill Lynch	2	17.0	General Re	3	16.0

summary

In this chapter we discussed procedures for developing interval estimates and conducting hypothesis tests involving two or more population means. First, we showed how to make inferences about the difference between the means of two populations when independent simple random samples are selected. We considered both the large- and small-sample cases. The z values from the standard normal probability distribution are used for inferences about the difference between two population means when the sample sizes are large. In the small-sample case, if the populations are normally distributed with equal variances, the t distribution is used for inferences.

Inferences about the difference between the means of two populations were then discussed for the matched-sample design. In the matched-sample design each element provides a pair of data values, one from each population. The difference between the paired data values is then used in the statistical analysis. The matched-sample design is generally preferred to the independent-sample design because the matched-sample procedure often reduces the sampling error and improves the precision of the estimate.

Finally, we showed how analysis of variance can be used to test for differences among means of several populations. We showed that the basis for the statistical tests used in analysis of variance is the development of two independent estimates of the population variance, σ^2. By computing the ratio of these two estimator (the F statistic), we developed a rejection rule for determining whether to reject the null hypothesis that the population means are equal.

glossary

Independent samples Samples selected from two (or more) populations in such a way that the elements making up one sample are chosen independently of the elements making up the other sample(s).

Pooled variance An estimate of the variance of a population based on the combination of two (or more) sample results. The pooled variance estimate is appropriate whenever the variances of two (or more) populations are assumed equal.

Matched samples Samples in which each data value of one sample is matched with a corresponding data value of the other sample.

Analysis of variance A statistical technique that can be used to test the hypothesis that the means of three or more populations are equal.

ANOVA table A table used to summarize the analysis of variance computations and results.

key formulas

Point Estimator of the Difference Between the Means of Two Populations

$$\bar{x}_1 - \bar{x}_2 \tag{10.1}$$

Expected Value of $\bar{x}_1 - \bar{x}_2$

$$E(\bar{x}_1 - \bar{x}_2) = \mu_1 - \mu_2 \tag{10.2}$$

Standard Deviation of $\bar{x}_1 - \bar{x}_2$

$$\sigma_{\bar{x}_1 - \bar{x}_2} = \sqrt{\frac{\sigma_1^2}{n_1} + \frac{\sigma_2^2}{n_2}} \tag{10.3}$$

Interval Estimate of the Difference Between the Means of Two Populations: Large-Sample Case ($n_1 \geq 30$ and $n_2 \geq 30$) with σ_1 and σ_2 Known

$$\bar{x}_1 - \bar{x}_2 \pm z_{\alpha/2}\sigma_{\bar{x}_1 - \bar{x}_2} \tag{10.4}$$

Point Estimator of $\sigma_{\bar{x}_1 - \bar{x}_2}$

$$s_{\bar{x}_1 - \bar{x}_2} = \sqrt{\frac{s_1^2}{n_1} + \frac{s_2^2}{n_2}} \tag{10.5}$$

Interval Estimate of the Difference Between the Means of Two Populations: Large-Sample Case ($n_1 \geq 30$ and $n_2 \geq 30$) with σ_1 and σ_2 Unknown

$$\bar{x}_1 - \bar{x}_2 \pm z_{\alpha/2}\, s_{\bar{x}_1 - \bar{x}_2} \tag{10.6}$$

Standard Deviation of $\bar{x}_1 - \bar{x}_2$ When $\sigma_1^2 = \sigma_2^2 = \sigma^2$

$$\sigma_{\bar{x}_1 - \bar{x}_2} = \sqrt{\frac{\sigma^2}{n_1} + \frac{\sigma^2}{n_2}} = \sqrt{\sigma^2\left(\frac{1}{n_1} + \frac{1}{n_2}\right)} \tag{10.7}$$

Pooled Estimator of σ^2

$$s^2 = \frac{(n_1 - 1)s_1^2 + (n_2 - 1)s_2^2}{n_1 + n_2 - 2} \tag{10.8}$$

Point Estimator of $\sigma_{\bar{x}_1 - \bar{x}_2}$ When $\sigma_1^2 = \sigma_2^2 = \sigma^2$

$$s_{\bar{x}_1 - \bar{x}_2} = \sqrt{s^2\left(\frac{1}{n_1} + \frac{1}{n_2}\right)} \tag{10.9}$$

Interval Estimate of the Difference Between the Means of Two Populations: Small-Sample Case ($n_1 < 30$ and/or $n_2 < 30$)

$$\bar{x}_1 - \bar{x}_2 \pm t_{\alpha/2} s_{\bar{x}_1 - \bar{x}_2} \tag{10.10}$$

Test Statistic for Hypothesis Tests About the Difference Between the Means of Two Populations (Large-Sample Case)

$$z = \frac{(\bar{x}_1 - \bar{x}_2) - (\mu_1 - \mu_2)}{\sqrt{\sigma_1^2/n_1 + \sigma_2^2/n_2}} \tag{10.11}$$

Test Statistic for Hypothesis Tests About the Difference Between the Means of Two Populations (Small-Sample Case)

$$t = \frac{(\bar{x}_1 - \bar{x}_2) - (\mu_1 - \mu_2)}{\sqrt{s^2\left(\frac{1}{n_1} + \frac{1}{n_2}\right)}} \tag{10.12}$$

Sample Mean for Matched Samples

$$\bar{d} = \frac{\Sigma d_i}{n}$$

Sample Standard Deviation for Matched Samples

$$s_d = \sqrt{\frac{\Sigma(d_i - \bar{d})^2}{n - 1}}$$

Test Statistic for Matched Samples

$$t = \frac{\bar{d} - \mu_d}{s_d/\sqrt{n}} \tag{10.13}$$

***j*th Sample Mean**

$$\bar{x}_j = \frac{\sum_{i=1}^{n_j} x_{ij}}{n_j} \tag{10.14}$$

jth Sample Variance

$$s_j^2 = \frac{\sum_{i=1}^{n_j} (x_{ij} - \bar{x}_j)^2}{n_j - 1} \tag{10.15}$$

Overall Sample Mean

$$\bar{\bar{x}} = \frac{\sum_{j=1}^{k} \sum_{i=1}^{n_j} x_{ij}}{n_T} \tag{10.16}$$

where

$$n_T = n_1 + n_2 + \cdots + n_k \tag{10.17}$$

Mean Square Between

$$\text{MSB} = \frac{\text{SSB}}{k - 1} \tag{10.20}$$

Sum of Squares Between

$$\text{SSB} = \sum_{j=1}^{k} n_j (\bar{x}_j - \bar{\bar{x}})^2 \tag{10.21}$$

Mean Square Within

$$\text{MSW} = \frac{\text{SSW}}{n_T - k} \tag{10.23}$$

Sum of Squares Within

$$\text{SSW} = \sum_{j=1}^{k} (n_j - 1) s_j^2 \tag{10.24}$$

Test for the Equality of k Population Means

$$F = \frac{\text{MSB}}{\text{MSW}} \tag{10.25}$$

Total Sum of Squares

$$\text{SST} = \sum_{j=1}^{k} \sum_{i=1}^{n_j} (x_{ij} - \bar{\bar{x}})^2 \tag{10.26}$$

Partition of Sum of Squares

$$\text{SST} = \text{SSB} + \text{SSW} \tag{10.27}$$

supplementary exercises

33. Starting annual salaries for individuals with master's and bachelor's degrees in business were collected in two independent random samples. Use the following data to develop a 90% confidence interval estimate of the increase in starting salary that can be expected upon completion of a master's program.

Master's Degree	Bachelor's Degree
$n_1 = 60$	$n_2 = 80$
$\bar{x}_1 = \$30,000$	$\bar{x}_2 = \$26,000$
$s_1 = \$2,500$	$s_2 = \$2,000$

34. Safegate Foods, Inc., is redesigning the checkout lanes in its supermarkets throughout the country. Two designs have been suggested. Tests on customer checkout times have been conducted at two stores where the two new systems have been installed. A summary of the sample data follows.

System A	System B
$n_1 = 120$	$n_2 = 100$
$\bar{x}_1 = 4.1$ minutes	$\bar{x}_2 = 3.3$ minutes
$s_1 = 2.2$ minutes	$s_2 = 1.5$ minutes

Test at the .05 level of significance to determine whether there is a difference between the mean checkout times of the two systems. Which system is preferred?

35. Samples of final examination scores for two statistics classes with different instructors provided the following results. With $\alpha = .05$, test whether these data are sufficient to conclude that the mean grades for the two classes differ.

Instructor A	Instructor B
$n_1 = 12$	$n_2 = 15$
$\bar{x}_1 = 72$	$\bar{x}_2 = 78$
$s_1 = 8$	$s_2 = 10$

36. In a study of job attitudes and job satisfaction, a sample of 50 men and 50 women were asked to rate their overall job satisfaction on a 1 to 10 scale. High ratings indicate a high degree of job satisfaction. From the following sample results, do you find a significant difference between the levels of job satisfaction for men and women? Use $\alpha = .05$.

Men	Women
$\bar{x}_1 = 7.2$	$\bar{x}_2 = 6.4$
$s_1 = 1.7$	$s_2 = 1.4$

37. Figure Perfect, Inc., is a women's figure salon that specializes in weight-reduction programs. Weights for a sample of clients before and after a six-week introductory program are listed in the following table. Using $\alpha = .05$, test to determine whether the introductory program provides a statistically significant weight loss.

	Weight	
Client	Before	After
1	140	132
2	160	158
3	210	195
4	148	152
5	190	180
6	170	164

38. A simple random sample of the asking prices ($1000s) of four houses currently for sale in each of two residential areas resulted in the following data.

	Area 1	Area 2
	92	90
	89	102
	98	96
	105	88
$\bar{x}_j$	96	94
s_j^2	50	40

a. Use the procedure developed in Section 10.2 to test whether the mean asking price is the same in both areas. Use $\alpha = .05$.

b. Use the ANOVA procedure to test whether the mean asking price is the same. Compare your analysis with part (a). Use $\alpha = .05$.

39. Suppose that in exercise 38 data were collected for another residential area. The asking prices for the simple random sample from the third area were $81,000, $86,000, $75,000, and $90,000; the sample mean and sample variance for these data are 83 and 42, respectively. Is the mean asking price the same for all three areas? Use $\alpha = .05$.

40. Executives rated service quality in each of several American industries (*Journal of Accountancy,* February 1992). Assume the following sample of ratings for the airline, retail, hotel, and automotive industries was obtained; higher scores indicate a higher service quality rating. At the $\alpha = .05$ level of significance, test for a significant difference in the population mean quality ratings for the four industries. What is your conclusion?

	Airlines	Retail	Hotel	Automotive
	59	63	70	49
	56	49	68	55
	47	60	62	48
	46	54	69	49
	55	56	59	50
	54	55		
	48			
$\bar{x}_j$	52.14	56.17	65.6	50.20
s_j^2	25.81	23.77	23.20	7.70

41. Three different assembly methods have been proposed for a new product. To determine which assembly method results in the greatest number of parts produced per hour, 30 workers were randomly selected and assigned to use one of the proposed methods. The number of units produced by each worker follows.

ASSEMBLY

	Method		
	A	B	C
	97	93	99
	73	100	94
	93	93	87
	100	55	66
	73	77	59
	91	91	75
	100	85	84
	86	73	72
	92	90	88
	95	83	86
$\bar{x}_j$	90	84	81
s_j^2	98.00	168.44	159.78

Use these data and test to see whether the mean number of parts produced is the same with each method. Use $\alpha = .05$.

MKTCAP

42. The following data show stocks with high earnings growth divided into groups based on market capitalization: large cap (above $1.5 billion), medium cap ($250 million to $1.5 billion), and shadow stocks (small firms with low institutional interest). Since the price–earnings ratios for stocks with high growth potential tend to be above the market average, many analysts look at the ratio of the market price per share to the forecasted earnings per share; such ratios are provided in the columns labeled P/E Est. (*AAII Journal,* June 1994). At the .05 level of significance, test whether the mean ratio of the market price per share to the forecasted earnings per share is the same for the three groups.

Large Cap	P/E Est.	Medium Cap	P/E Est.	Shadow Stocks	P/E Est.
U.S. Healthcare	16.7	Vencor Inc.	20.4	Ashworth Inc.	23.4
Cisco Systems	24.8	Westcott Communications	22.3	Homecare Management Inc.	30.8
Parametric Technology	25.4	CML Group	10.4	Methode Electric B	20.6
United Healthcare	25.2	Snyder Oil	33.1	Marten Transport	10.5
EMC Corp.	18.6	Owens & Minor	22.1	Gates/F.A. Distribution	15.6
American Power Conversion	29.7	Xilinx Inc.	24.1	Rotech Medical Corp.	20.2
Cabletron Systems	19.4	Invacare Corp.	15.3	Cosmetic Center B	16.5
CUC International	30.0	Tech Data Corp.	17.0	Volunteer Capital	26.1
Intel Corp.	10.3	Briggs & Stratton	13.5	BGS Systems	8.1
BMC Software	15.0	KCS Energy Inc.	10.1		
Microsoft Corp.	23.6	Applebee's International	35.8		
Blockbuster Entertainment	20.1	Bowne & Co.	10.6		
Linear Technology	31.9	Horizon Healthcare	26.0		
Sysco Corp	21.5	Oakwood Homes	14.7		
Home Depot	31.8	Progress Software	18.5		

computer case

Par, Inc.

Par, Inc., is a major manufacturer of golf equipment. Management believes that Par's market share could be increased with the introduction of a cut-resistant, longer-lasting golf ball. Therefore, the research group at Par has been investigating a new golf ball coating that is designed to resist cuts and provide a more durable ball. The tests with the coating have been very promising.

One of the researchers voiced concern about the effect of the new coating on driving distances. Par would like the new cut-resistant ball to offer driving distances comparable to those of the current-model golf ball. To compare the driving distances for the two balls, 40 balls of both the new and current models were subjected to distance tests. The testing was performed with a mechanical hitting machine so that any difference between the mean distances for the two models could be attributed to a difference in the design. The results of the tests, with distances measured to the nearest yard, follow. These data are available on the data disk in the file named GOLF.

GOLF

Model		Model		Model		Model	
Current	New	Current	New	Current	New	Current	New
264	277	270	272	263	274	281	283
261	269	287	259	264	266	274	250
267	263	289	264	284	262	273	253
272	266	280	280	263	271	263	260
258	262	272	274	260	260	275	270
283	251	275	281	283	281	267	263
258	262	265	276	255	250	279	261
266	289	260	269	272	263	274	255
259	286	278	268	266	278	276	263
270	264	275	262	268	264	262	279

Managerial Report

1. Formulate and present the rationale for a hypothesis test that Par could use to compare the driving distances of the current and new golf balls.
2. Analyze the data to provide the hypothesis-testing conclusion. What is the p-value for your test? What is your recommendation for Par, Inc.?
3. Provide descriptive statistical summaries of the data for each model.
4. What is the 95% confidence interval for the population mean of each model and what is the 95% confidence interval for the difference between the means of the two populations?
5. Do you see a need for larger sample sizes and more testing with the golf balls? Discuss.

computer case

Wentworth Medical Center

As part of a long-term study of individuals 65 years of age or older, sociologists and physicians at the Wentworth Medical Center in upstate New York conducted a study to investigate the relationship between geographic location and depression. A sample of 60 individuals, all in reasonably good health, was selected; 20 individuals were residents of Florida, 20 were residents of New York, and 20 were residents of North Carolina. Each of the individuals sampled was given a standardized test to measure depression. The data collected follow; higher test scores indicate higher levels of depression. These data are available on the data disk in the file MEDICAL1.

A second part of the study considered the relationship between geographic location and depression for individuals 65 years of age or older who had a chronic health condition such as arthritis, hypertension, and/or heart ailment. A sample of 60 individuals with such conditions was identified. Again, 20 were residents of Florida, 20 were residents of New York, and 20 were residents of North Carolina. The levels of depression recorded for this study follow. These data are available on the data disk in the file MEDICAL2.

MEDICAL1

MEDICAL2

Data from MEDICAL1			Data from MEDICAL2		
Florida	New York	North Carolina	Florida	New York	North Carolina
3	8	10	13	14	10
7	11	7	12	9	12
7	9	3	17	15	15
3	7	5	17	12	18
8	8	11	20	16	12
8	7	8	21	24	14
8	8	4	16	18	17
5	4	3	14	14	8
5	13	7	13	15	14
2	10	8	17	17	16
6	6	8	12	20	18
2	8	7	9	11	17
6	12	3	12	23	19
6	8	9	15	19	15
9	6	8	16	17	13
7	8	12	15	14	14
5	5	6	13	9	11
4	7	3	10	14	12
7	7	8	11	13	13
3	8	11	17	11	11

Managerial Report

1. Use descriptive statistics to summarize the data from the two studies. What are your preliminary observations about the depression scores?
2. Use analysis of variance on both data sets. State the hypotheses being tested in each case. What are your conclusions?

	A	B	C	D	E	F	G
1	Charlotte	Houston	San Diego				
2	85	71	59				
3	75	75	64				
4	82	73	62				
5	76	74	69				
6	71	69	75				
7	85	82	67				
8							
9							
10	Anova: Single Factor						
11							
12	SUMMARY						
13	*Groups*	*Count*	*Sum*	*Average*	*Variance*		
14	Charlotte	6	474	79	34		
15	Houston	6	444	74	20		
16	San Diego	6	396	66	32		
17							
18							
19	ANOVA						
20	*Source of Variation*	*SS*	*df*	*MS*	*F*	*P-value*	*F crit*
21	Between Groups	516	2	258	9	0.003	3.6823
22	Within Groups	430	15	28.67			
23							
24	Total	946	17				

FIGURE 10.12
Excel Solution for the NCP Analysis of Variance Example

3. Use inferences about individual treatment means where appropriate. What are your conclusions?
4. Discuss extensions of this study or other analyses that you feel might be helpful.

APPENDIX 10.1 Analysis of Variance with Minitab

To illustrate how Minitab can be used to test for the equality of k population means, we show how to test whether the mean examination score is the same at each plant in the National Computer Products example introduced in Section 10.4. We assume that the user has entered the examination score data into the first three columns of a Minitab worksheet, and that column 1 is labeled Plant 1, column 2 is labeled Plant 2, and column 3 is labeled Plant 3. The following steps produce the Minitab output in Figure 10.11.

STEP 1. Select the **Stat** pull-down menu
STEP 2. Select the **ANOVA** pull-down menu
STEP 3. Select the **Oneway (Unstacked)** option
STEP 4. When the dialog box appears:
Enter C1–C3 in the **Responses (in separate columns) box**
Select **OK**

APPENDIX 10.2 Analysis of Variance with Spreadsheets

To illustrate how Excel can be used to test for the equality of k population means, we show how to test whether the mean examination score is the same at each plant in the National Computer Products example introduced in Section 10.4. We assume that the user has entered the examination score data into worksheet rows 2 to 7 of columns A, B, and C as shown in Figure 10.12; note that the cells in row 1 for columns A, B, and C are labeled Plant 1, Plant 2, and Plant 3. The following steps are used to test whether the mean examination scores are equal.

STEP 1. Select the **Tools** pull-down menu
STEP 2. Select the **Data Analysis** option
STEP 3. When the Data Analysis dialog box appears:
Choose **ANOVA: Single Factor**
Select **OK**
STEP 4. When the Anova: Single Factor dialog box appears:
Enter A1:C7 in the **Input Range** box
Select **Columns**
Select **Labels in First Row**
Select **Output Range** and enter C10 in the box
Select **OK**

CHAPTER 11

Comparisons Involving Proportions

United Way*

Rochester, New York

The United Way of Greater Rochester is a nonprofit fund-raising and social planning organization dedicated to improving the quality of life of residents in the six counties it serves. The annual United Way/Red Cross campaign, conducted each spring, helps support more than 140 human service agencies. These agencies meet a wide variety of human needs—physical, mental, and social—and serve people of all ages, backgrounds, and economic means. Because of widespread volunteer involvement, United Way is able to hold its operating costs to less than nine cents of every dollar raised.

statistics in practice

The United Way of Greater Rochester decided to conduct a survey to learn more about community perceptions of charities. Focus-group interviews were held with professional, service, and general worker groups to get preliminary information on perceptions. The information obtained was then used to help develop the questionnaire for the survey. The questionnaire was pretested, modified, and distributed to 440 individuals; 323 completed questionnaires were obtained.

A variety of descriptive statistics, including frequency distributions and crosstabulations, were provided from the data collected. An important part of the analysis involved the use of contingency tables and chi-square tests of independence. One use of such statistical tests was to determine whether perceptions of administrative expenses were independent of occupation.

The hypotheses for the test of independence were:

H_0: Perception of United Way administrative expenses is independent of the occupation of the respondent.

H_a: Perception of United Way administrative expenses is not independent of the occupation of the respondent.

Two questions in the survey provided the data for the statistical test. One question obtained data on perceptions of the percentage of funds going to administrative expenses (up to 10%, 11–20%, and 21% or more). The other question asked for the occupation of the respondent.

The chi-square test at a 5% level of significance led to rejection of the null hypothesis of independence and to the conclusion that perceptions of United Way's administrative expenses did vary by occupation. Actual administrative expenses were less than 9%, but 35% of the respondents perceived that administrative expenses were 21% or more. Hence, many had inaccurate perceptions of administrative costs. In this group, production-line, clerical, sales, and professional-technical employees had more inaccurate perceptions than other groups.

In this chapter, you will learn how a statistical test of independence, such as that described here, is conducted. The community perceptions study helped United Way of Rochester to develop adjustments to its program and fund-raising activities.

This mural represents the United Way spirit of volunteerism.

The authors are indebted to Dr. Philip R. Tyler, Marketing Consultant to the United Way, for providing this Statistics in Practice.

In many statistical applications it is of interest to compare population proportions. In Section 11.1 we will describe statistical inferences that can be made concerning differences in the proportions for two populations. Two samples are required, one from each population. The second section is concerned with hypothesis testing comparing the proportions of a single multinomial population with the values stated in a null hypothesis. One sample from the multinomial population is used. In the last section of the chapter, we show how contingency tables are used to test for the independence of two variables. One sample is used for the test of independence, but measures on two variables are required for each sampled element. Both Sections 11.2 and 11.3 rely on the use of a chi-square statistical test.

11.1 Inferences About the Difference Between the Proportions of Two Populations

A tax preparation firm is interested in comparing the quality of work at two of its regional offices. By randomly selecting samples of tax returns prepared at each office and having the sample returns verified for accuracy, the firm will be able to estimate the proportion of erroneous returns prepared at each office. Of particular interest is the difference between these proportions. Let

p_1 = proportion of erroneous returns for population 1(office 1),
p_2 = proportion of erroneous returns for population 2 (office 2),
$\bar{p}_1$ = sample proportion for a simple random sample from population 1, and
$\bar{p}_2$ = sample proportion for a simple random sample from population 2.

The difference between the two population proportions is given by $p_1 - p_2$. The point estimator of $p_1 - p_2$ follows.

Point Estimator of the Difference Between the Proportions of Two Populations

$$\bar{p}_1 - \bar{p}_2 \tag{11.1}$$

Thus, the point estimator of the difference between two population proportions is the difference between the sample proportions of two independent simple random samples.

Sampling Distribution of $\bar{p}_1 - \bar{p}_2$

In the study of the difference between two population proportions, $\bar{p}_1 - \bar{p}_2$ is the point estimator of interest. As we have seen in several preceding cases, the sampling distribution of a point estimator is a key factor in developing interval estimates and in testing hypotheses about parameters of interest. The properties of the sampling distribution of $\bar{p}_1 - \bar{p}_2$ follow.

FIGURE 11.1
Sampling Distribution of $\bar{p}_1 - \bar{p}_2$

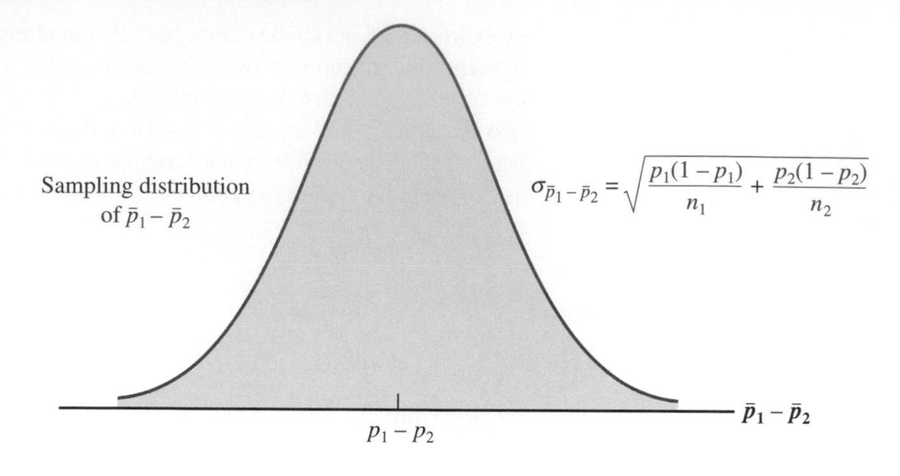

Sampling Distribution of $\bar{p}_1 - \bar{p}_2$

$$\text{Expected Value}: E(\bar{p}_1 - \bar{p}_2) = p_1 - p_2 \tag{11.2}$$

$$\text{Standard Deviation}: \sigma_{\bar{p}_1 - \bar{p}_2} = \sqrt{\frac{p_1(1-p_1)}{n_1} + \frac{p_2(1-p_2)}{n_2}} \tag{11.3}$$

where

n_1 = sample size for the simple random sample from population 1

n_2 = sample size for the simple random sample from population 2

Distribution form: If the sample sizes are large—i.e., n_1p_1, $n_1(1 - p_1)$, n_2p_2, and $n_2(1 - p_2)$ are all greater than or equal to 5—the sampling distribution of $\bar{p}_1 - \bar{p}_2$ can be approximated by a normal probability distribution.

Figure 11.1 shows the sampling distribution of $\bar{p}_1 - \bar{p}_2$.

Interval Estimation of $p_1 - p_2$

Let us assume that independent simple random samples of tax returns from the two offices provide the following information.

Office 1	Office 2
$n_1 = 250$	$n_2 = 300$
Number of returns with errors = 35	Number of returns with errors = 27

The sample proportions for the two offices are as follows:

$$\bar{p}_1 = \frac{35}{250} = .14$$

$$\bar{p}_2 = \frac{27}{300} = .09$$

The point estimate of the difference between the proportions of erroneous tax returns for the two populations is $\bar{p}_1 - \bar{p}_2 = .14 - .09 = .05$. Thus, we estimate that office 1 has a .05 greater error rate than office 2.

Before developing a confidence interval for the difference between population proportions, we must know the value of $\sigma_{\bar{p}_1 - \bar{p}_2}$. However (11.3) cannot be used directly because p_1 and p_2 will not be known in practice. The approach taken is to substitute $\bar{p}_1$ as an estimator of p_1 and $\bar{p}_2$ as an estimator of p_2 in (11.3). Doing so provides the point estimator of $\sigma_{\bar{p}_1 - \bar{p}_2}$ given by (11.4).

Point Estimator of $\sigma_{\bar{p}_1 - \bar{p}_2}$

$$s_{\bar{p}_1 - \bar{p}_2} = \sqrt{\frac{\bar{p}_1(1 - \bar{p}_1)}{n_1} + \frac{\bar{p}_2(1 - \bar{p}_2)}{n_2}} \tag{11.4}$$

The following expression provides an interval estimate of the difference between the proportions of the two populations in the large sample case. The use of $z_{\alpha/2}$ in (11.5) is due to the fact that we can assume a normal sampling distribution for $\bar{p}_1 - \bar{p}_2$ in the large sample case.

Interval Estimate of the Difference Between the Proportions of Two Populations: Large-Sample Case with $n_1p_1, n_1(1 - p_1), n_2p_2,$ and $n_2(1 - p_2) \geq 5$

$$\bar{p}_1 - \bar{p}_2 \pm z_{\alpha/2} s_{\bar{p}_1 - \bar{p}_2} \tag{11.5}$$

where $1 - \alpha$ is the confidence coefficient.

Let us now use (11.5) to develop an interval estimate of the difference in proportions for the two tax preparation offices in our example. Using (11.4), we find

$$s_{\bar{p}_1 - \bar{p}_2} = \sqrt{\frac{.14(.86)}{250} + \frac{.09(.91)}{300}} = .0275$$

With a 90% confidence interval, we have $z_{\alpha/2} = z_{.05} = 1.645$, and a margin of error of $1.645(.0275) = .045$. The interval estimate is

$$(.14 - .09) \pm 1.645(.0275)$$
$$.05 \pm .045$$

Hence, the 90% confidence interval for the difference in error rates at the two offices is .005 to .095.

Hypothesis Tests About $p_1 - p_2$

As an example of hypothesis testing concerning the difference between two population proportions, let us reconsider the example concerning the error rates at two offices of the

FIGURE 11.2
Sampling Distribution of $\bar{p}_1 - \bar{p}_2$ With $H_0: p_1 - p_2 = 0$

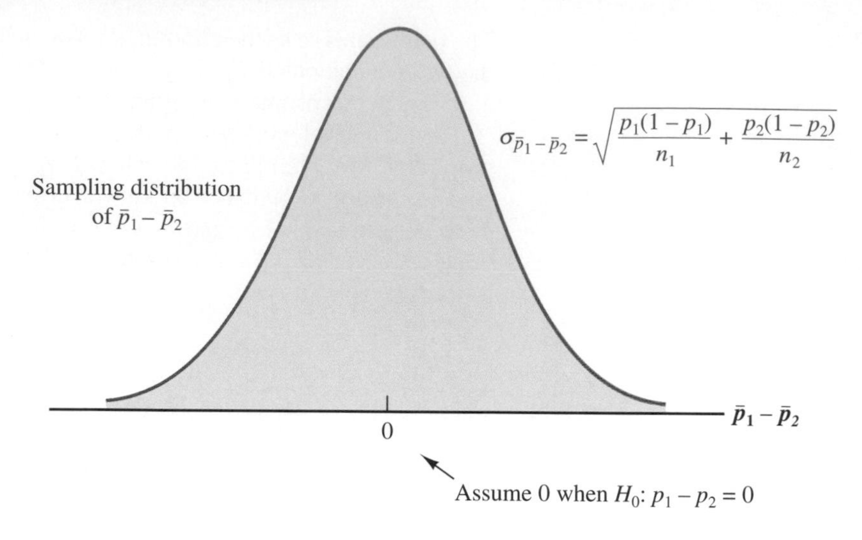

tax preparation firm. Suppose the firm had simply been interested in whether there was a significant difference in error rates at the two offices. The appropriate null and alternative hypotheses are

$$H_0: p_1 - p_2 = 0$$

$$H_a: p_1 - p_2 \neq 0$$

Figure 11.2 shows the sampling distribution of $\bar{p}_1 - \bar{p}_2$ based on the assumption that there is no difference between the two population proportions. That is, $p_1 - p_2 = 0$. In the large sample case, the sampling distribution is approximately normal and the test statistic for the difference between two population proportions can be written

$$z = \frac{(\bar{p}_1 - \bar{p}_2) - (p_1 - p_2)}{\sigma_{\bar{p}_1 - \bar{p}_2}} \tag{11.6}$$

Using $\alpha = .10$ and $z_{\alpha/2} = z_{.05} = 1.645$, the rejection rule is

$$\text{Reject } H_0 \text{ if } z < -1.645 \text{ or if } z > 1.645.$$

The computation of z in (11.6) requires a value for the standard error of the difference between proportions $\sigma_{\bar{p}_1 - \bar{p}_2}$. As in the case of interval estimation, it will be necessary to estimate $\sigma_{\bar{p}_1 - \bar{p}_2}$ since p_1 and p_2 will not be known in practice. But for the special case of a null hypothesis in which there is no difference in the proportions ($H_0: p_1 = p_2$), we do not use (11.4). In this case, there is only one population proportion; that is, $p_1 = p_2 = p$.

In this case, we can combine or *pool* the two sample proportions to provide one estimate of p given by $\bar{p}$.

$$\bar{p} = \frac{n_1\bar{p}_1 + n_2\bar{p}_2}{n_1 + n_2} \tag{11.7}$$

With $\bar{p}$ used in place of both $\bar{p}_1$ and $\bar{p}_2$, (11.4) is revised to

$$s_{\bar{p}_1 - \bar{p}_2} = \sqrt{\bar{p}(1 - \bar{p})\left(\frac{1}{n_1} + \frac{1}{n_2}\right)} \tag{11.8}$$

Using (11.7) and (11.8), we can now proceed with an hypothesis test for the tax preparation firm example.

$$\bar{p} = \frac{250(.14) + 300(.09)}{550} = \frac{62}{550} = .113$$

$$s_{\bar{p}_1 - \bar{p}_2} = \sqrt{(.113)(.887)\left(\frac{1}{250} + \frac{1}{300}\right)} = .0271$$

The value of the test statistic becomes

$$z = \frac{(\bar{p}_1 - \bar{p}_2) - (p_1 - p_2)}{s_{\bar{p}_1 - \bar{p}_2}} = \frac{(.14 - .09) - 0}{.0271} = 1.85$$

Since 1.85 > 1.645, at the .10 level of significance the null hypothesis is rejected. The sample evidence indicates that there is a difference between the error proportions at the two offices.

As we saw with the hypothesis tests about differences between two population means (Chapter 10), one-tailed tests can also be developed for the difference between two population proportions. The one-tailed rejection regions are established in a manner similar to the one-tailed hypothesis-testing procedures for a single-population proportion.

exercises

Methods

1. Consider the following results for two independent samples taken from two populations.

Sample 1	Sample 2
$n_1 = 400$	$n_2 = 300$
$\bar{p}_1 = .48$	$\bar{p}_2 = .36$

a. What is the point estimate of the difference between the two population proportions?
b. Develop a 90% confidence interval for the difference between the two population proportions.
c. Develop a 95% confidence interval for the difference between the two population proportions.

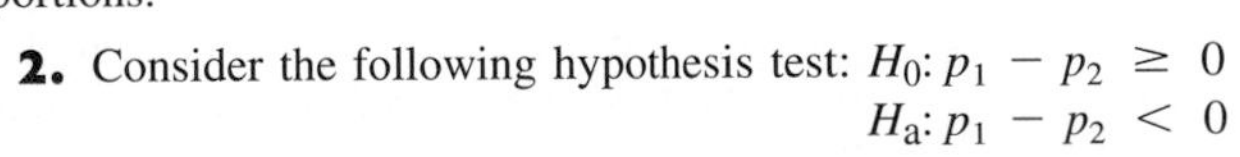

2. Consider the following hypothesis test: H_0: $p_1 - p_2 \geq 0$
H_a: $p_1 - p_2 < 0$

The following results are for two independent samples taken from the two populations.

Sample 1	Sample 2
$n_1 = 200$	$n_2 = 300$
$\bar{p}_1 = .22$	$\bar{p}_2 = .16$

a. Using $\alpha = .05$, what is your hypothesis-testing conclusion?
b. What is the p-value?

Applications

3. *Business Week*/ Harris polls compared the views adults had about their children's future in 1989 with the views adults held in 1992 (*Business Week,* April 6, 1992). In a 1989 poll, 59% of the adults sampled felt their children would have a better life than they had. In a 1992 poll, 34% of the adults sampled felt their children would have a better life. Assume that 1250 adults were used in both polls. Provide a 95% confidence interval estimate of the difference between the proportions in 1989 and 1992. What is your interpretation of the interval estimate and the difference shown?

4. A 1994 Gallup poll found that 16% of 505 men and 25% of 496 women surveyed favored a law fobidding the sale of all beer, wine, and liquor throughout the nation (*The Gallup Poll Monthly,* June 1994). Develop a 95% confidence interval for the difference between the proportion of women who favor such a ban and the proportion of men who favor such a ban.

5. In December 1993, the women and family issues committee of the American Institute of Certified Public Accountants (AICPA) mailed surveys asking about family-friendly policies and women's upward mobility to 5300 firms of all sizes (excluding sole practitioners). Of the 1710 responses, 57% were from firms with five or fewer AICPA members, 26% were from firms with six to 10 members, 9% were from firms with 11 to 20 members, and 8% were from firms with more than 20 members. For the firms with fewer than five AICPA members, 58% of the hires within the previous three years were women and 42% were men. In contrast, firms with more than 20 AICPA members hired 43% women and 57% men (*Journal of Accountancy,* October 1994). Develop a 95% confidence interval estimate of the difference between the proportion of women hired by firms with fewer than five AICPA members and the proportion of women hired by firms with more than 20 AICPA members.

6. Two loan officers at the North Ridge National Bank show the following data for defaults on loans that they have approved (the data are based on samples of loans granted over the past five years).

Loan Officer	Loans Reviewed in the Sample	Defaulted Loans
A	60	9
B	80	6

Using $\alpha = .05$, test the hypothesis that the default rates are the same for the two loan officers.

7. A sample of 1545 men and an independent sample of 1691 women were used to compare the amount of housework done by women and men in dual-earner marriages. The study showed that 67.5% of the men felt the division of housework was fair and 60.8% of the women felt the division of housework was fair (*American Journal of Sociology,* September 1994). Is the proportion of men who felt the division of housework was fair greater than the proportion of women who felt the division of housework was fair? Support your conclusion with a statistical test using a .05 level of significance.

8. A survey firm conducts door-to-door surveys on a variety of issues. Some individuals cooperate with the interviewer and complete the interview questionnaire and others do not. The sample data are shown.

Respondents	Sample Size	Number Cooperating
Men	200	110
Women	300	210

a. Using $\alpha = .05$, test the hypothesis that the response rate is the same for both men and women.
b. Compute the 95% confidence interval for the difference between the proportions of men and women who cooperate with the survey.

9. In a test of the quality of two television commercials, each commercial was shown in a separate test area six times over a one-week period. The following week a telephone survey was conducted to identify individuals who had seen the commercials. The individuals who had seen the commercials were asked to state the primary message in the commercials. The following results were recorded.

Commercial	Number Who Saw Commercial	Number Who Recalled Primary Message
A	150	63
B	200	60

a. Using $\alpha = .05$, test the hypothesis that there is no difference in the recall proportions for the two commercials.
b. Compute a 95% confidence interval for the difference between the recall proportions for the two populations.

10. In a study designed to test the effectiveness of a new drug for treating rheumatoid arthritis, investigators divided 73 rheumatoid arthritis patients between the ages of 18 and 75 into three groups. Patients in one group were given a high dose of the drug, patients in another group were given a low dose, and patients in the third group were given a placebo. After four weeks, 19 of 24 patients in the high-dosage group indicated that they felt better, whereas 11 of the 25 patients in the low-dosage group and two of the 24 patients in the placebo group felt better (*The Lancet,* October 1994). Is the proportion of patients who felt better in the high-dosage group greater than the proportion of patients who felt better in the low-dosage group? What is your conclusion? Use $\alpha = .05$ for any statistical test.

11.2 A Hypothesis Test for Proportions of a Multinomial Population

In this section, we consider hypothesis tests concerning the proportion of elements in a population belonging to each of several classes or categories. In contrast to the preceding section, we will be dealing with a single population: a *multinomial population.* The parameters of the multinomial population are the proportion of elements belonging to each category; the hypothesis tests we describe concern the values of those parameters.

The multinomial probability distribution can be thought of as an extension of the binomial distribution to the case of three or more categories of outcomes. On each trial of a multinomial experiment, such as sampling an item from a multinomial population, one and only one of the outcomes occurs. Each trial of a multinomial experiment is assumed to be independent, and the probabilities must stay the same for each trial.

As an example, consider the market share evaluation being conducted by Scott Marketing Research. Over the past year market shares have stabilized at 30% for company A, 50% for company B, and 20% for company C. Recently company C has developed a "new and improved" product that has replaced its current entry in the market. Scott Marketing Research has been retained by company C to determine whether the new product will alter market shares.

In this example, the population of interest is a multinomial population; each customer is classified as buying from company A, company B, or company C. Thus, we have a multinomial population with three classifications or categories. Let us use the following notation for the proportions.

$$p_A = \text{market share for company A}$$
$$p_B = \text{market share for company B}$$
$$p_C = \text{market share for company C}$$

Scott Marketing Research will conduct a sample survey and compute the proportion preferring each company's product. A hypothesis test will then be conducted to see whether the new product has caused a change in market shares. Assuming that company C's new product will not alter the market shares, the null and alternative hypotheses would be stated as follows.

$$H_0\text{: } p_A = .30,\ p_B = .50, \text{ and } p_C = .20$$
$$H_a\text{: The population proportions are not } p_A = .30,\ p_B = .50, \text{ and } p_C = .20$$

If the sample results lead to the rejection of H_0, Scott will have evidence that the introduction of the new product has had an impact on the market shares.

Let us assume that the market research firm has used a consumer panel of 200 customers for the study. Each individual has been asked to specify a purchase preference among the three alternatives: company A's product, company B's product, and company C's new product. The 200 responses are summarized here.

Observed Results

Company A's Product	Company B's Product	Company C's New Product
48	98	54

We now can perform a *goodness of fit test* that will determine whether the sample of 200 customer purchase preferences is consistent with the null hypothesis. The goodness of fit test is based on a comparison of the sample of *observed* results, such as those shown, with the *expected* results under the assumption that the null hypothesis is true. Hence, the next step in our example is to compute expected purchase preferences for the 200 customers under the assumption that $p_A=.30$, $p_B=.50$, and $p_C=.20$. Doing so provides the expected results.

Expected Results

Company A's Product	Company B's Product	Company C's New Product
200(.30) = 60	200(.50) = 100	200(.20) = 40

Thus, we see that the expected frequency for each category is found by multiplying the sample size of 200 by the hypothesized proportion for the category.

The goodness of fit test now focuses on the differences between the observed frequencies and the expected frequencies. Large differences between observed and expected frequencies cast doubt on the assumption that the hypothesized proportions or market shares are correct. Whether the differences between the observed and expected frequencies are "large" or "small" is a question answered with the aid of the following test statistic.

Test Statistic for Goodness of Fit

$$\chi^2 = \sum_{i=1}^{k} \frac{(f_i - e_i)^2}{e_i} \tag{11.9}$$

where

f_i = observed frequency for category i

e_i = expected frequency for category i based on the assumption that H_0 is true

k = the number of categories

The test statistic has a chi square probability distribution with $k - 1$ degrees of freedom provided that the expected frequencies are 5 *or more* for all categories.

Let us return to the market-share data for the three companies. Since the expected frequencies are all 5 or more, we can proceed with the computation of the chi-square test statistic.

$$\chi^2 = \frac{(48-60)^2}{60} + \frac{(98-100)^2}{100} + \frac{(54-40)^2}{40} = 2.40 + .04 + 4.90 = 7.34$$

Suppose we test the null hypothesis that the multinomial population has the proportions of p_A=.30, p_B=.50, and p_C=.20 at the $\alpha = .05$ level of significance. Since we will reject the null hypothesis if the differences between the observed and expected frequencies are *large,* we will place a rejection area of .05 in the upper tail of the chi-square distribution. Checking the chi-square distribution table (Table 3 of Appendix B), we find that with $k - 1 = 3 - 1 = 2$ degrees of freedom, $\chi^2_{.05} = 5.99$. Since $7.34 > 5.99$, we reject H_0. In rejecting H_0 we are concluding that the introduction of the new product by company C will alter the current market-share structure. The goodness of fit test itself allows no further conclusions, but we can compare the observed and expected frequencies informally to obtain an idea of how the market-share structure has changed.

Considering company C, we find that the observed frequency of 54 is larger than the expected frequency of 40. Since the expected frequency was based on current market shares, the larger observed frequency suggests that the new product will have a positive effect on company C's market share. Comparisons of the observed and expected frequencies for the other two companies indicate that company C's gain in market share will hurt company A more than company B.

As illustrated in the example, the goodness of fit test uses the chi-square distribution to determine whether a hypothesized multinomial probability distribution for a population provides a good fit. The hypothesis test is based on differences between the observed frequencies in a sample and the expected frequencies based on the assumed population distribution. Let us outline the general steps that can be used to conduct a goodness of fit test for any hypothesized multinomial population distribution.

Multinomial Distribution Goodness of Fit Test: a Summary

1. Set up the null and alternative hypotheses.

H_0: The population follows a multinomial probability distribution with specified probabilities for each of k categories.

H_a: The population does not follow a multinomial probability distribution with the specified probabilities for each of the k categories.

2. Select a random sample and record the observed frequencies, f_i, for each category.
3. Assuming the null hypothesis is true, determine the expected frequency, e_i, in each category by multiplying the category probability by the sample size.
4. Compute the value of the test statistic.

$$\chi^2 = \sum_{i=1}^{k} \frac{(f_i - e_i)^2}{e_i}$$

5. Rejection rule:

$$\text{Reject } H_0 \text{ if } \chi^2 > \chi^2_\alpha$$

where α is the level of significance for the test and there are $k - 1$ degrees of freedom.

notes and comments

For χ^2 goodness of fit tests the rejection region is always in the upper tail. The differences between observed and expected frequencies are squared, and larger differences lead to larger values for χ^2.

exercises

Methods

11. Test the following hypotheses using the χ^2 goodness of fit test.

H_0: $p_A = .40$, $p_B = .40$, and $p_C = .20$

H_a: The population proportions are not $p_A = .40$, $p_B = .40$, and $p_C = .20$

A sample of size 200 yielded 60 in category A, 120 in category B, and 20 in category C. Use $\alpha = .01$.

12. Suppose we have a multinomial population with four categories: A, B, C, and D. The null hypothesis is that the proportion of items is the same in every category. The null hypothesis is

$$H_0: p_A = p_B = p_C = p_D = .25$$

A sample of size 300 yielded the following numbers in each category.

A: 85 B: 95 C: 50 D: 70

Use $\alpha = .05$ and the χ^2 goodness of fit test to determine whether H_0 should be rejected.

Applications

13. During the first 13 weeks of the television season, the Saturday evening 8:00 P.M. to 9:00 P.M. audience proportions were recorded as ABC 29%, CBS 28%, NBC 25%, and independents 18%. A sample of 300 homes two weeks after a Saturday night schedule revision yielded the following viewing audience data: ABC 95 homes, CBS 70 homes, NBC 89 homes, and independents 46 homes. Test with $\alpha = .05$ to determine whether the viewing audience proportions have changed.

Self Test

14. In November 1993, 17% of American manufacturers felt we were "well on our way to a national recovery," 29% felt we had "just entered a recovery from a recession," 46% were "uncertain whether we were heading back into a recession or into a recovery," 7% felt we were still in a recession, and 1% were not sure. A survey of 500 manufacturers in May 1994 elicited the following responses (Grant Thornton Survey of American Manufacturers, 1994).

Opinion Category	Number of Respondents
Well on our way to recovery	165
Just entering a recovery	105
Uncertain	210
Still in a recession	15
Not sure	5

Perform a goodness of fit test to see whether opinions have changed. Use $\alpha = .01$.

15. The four major competitors in the computer-workstation market were reported to be Sun Microsystems (29%), Hewlett-Packard (18.8%), IBM (16%), and Digital Equipment (11.6%), with other manufacturers holding 24.6% of the market (*USA Today,* February 13, 1992). Assume that one year later a survey of 400 computer workstations finds 106 Sun, 72 Hewlett-Packard, 80 IBM, 48 Digital, and 94 other systems in use. Do the data suggest any changes have occurred during the one-year period? Test at a .05 level of significance.

16. A new container design has been adopted by a manufacturer. Color preferences indicated in a sample of 150 individuals follow.

Red	Blue	Green
40	64	46

Use $\alpha = .10$ and test for a difference in preference among the three colors. Hint: Formulate the null hypothesis as $H_0: p_1 = p_2 = p_3 = 1/3$.

17. Consumer panel preferences for three proposed store displays follow.

Display A	Display B	Display C
43	53	39

Use $\alpha = .05$ and test to see whether there is a difference in preference among the three display designs.

18. Grade-distribution guidelines for a statistics course at a major university are: 10% A, 30% B, 40% C, 15% D, and 5% F. A sample of 120 statistics grades at the end of a semester showed 18 As, 30 Bs, 40 Cs, 22 Ds, and 10 Fs. Use $\alpha = .05$ and test whether the actual grades deviate significantly from the grade-distribution guidelines.

11.3 Test of Independence: Contingency Tables

Another important application of the chi-square probability distribution involves using sample data to test for the independence of two variables. Let us illustrate the test of independence by considering the study conducted by the Alber's Brewery of Tucson, Arizona. Alber's manufactures and distributes three types of beer: light, regular, and dark. In an analysis of the market segments for the three beers, the firm's market research group has raised the question of whether preferences for the three beers differ among male and female beer drinkers. If beer preference is independent of the sex of the beer drinker, one advertising campaign will be initiated for all of Alber's beers. However, if beer preference depends on the sex of the beer drinker, the firm will tailor its promotions to different target markets.

A test of independence addresses the question of whether the beer preference (light, regular, or dark) is independent of the sex of the beer drinker (male, female). The hypotheses for this test of independence are:

H_0: Beer preference is independent of the sex of the beer drinker.

H_a: Beer preference is not independent of the sex of the beer drinker.

Table 11.1 can be used to describe the situation being studied. After identification of the population as all male and female beer drinkers, a sample can be selected and each individual asked to state his or her preference for the three Alber's beers. Every individual in the sample will be classified in one of the six cells in the table. For example, an individual may be a male preferring regular beer (cell (1,2)), a female preferring light beer (cell (2,1)), a female preferring dark beer (cell (2,3)), and so on. Since we have listed all possible combinations of beer preference and sex or, in other words, listed all possible contingencies, Table 11.1 is called a *contingency table.* The test of independence takes the contingency table format and for that reason is sometimes referred to as a *contingency table test.*

Suppose a simple random sample of 150 beer drinkers has been selected. After tasting each beer, the individuals in the sample are asked to state their preference or first choice. The crosstabulation in Table 11.2 summarizes the responses for the study. As we see, the data for the test of independence are collected in terms of counts or frequencies for each cell or category. Of the 150 individuals in the sample, 20 were men who favored light beer, 40 were men who favored regular beer, 20 were men who favored dark beer, and so on.

TABLE 11.1 Contingency Table for Beer Preference and Sex of Beer Drinker

		Beer Preference		
		Light	*Regular*	*Dark*
Sex	*Male*	cell(1,1)	cell(1,2)	cell(1,3)
	Female	cell(2,1)	cell(2,2)	cell(2,3)

TABLE 11.2 Sample Results for Beer Preferences of Male and Female Beer Drinkers (Observed Frequencies)

		Beer Preference			
		Light	*Regular*	*Dark*	**Total**
Sex	*Male*	20	40	20	80
	Female	30	30	10	70
	Total	50	70	30	150

The data in Table 11.2 are the sample or observed frequencies for each of six classes or categories. If we can determine the expected frequencies under the assumption of independence between beer preference and sex of the beer drinker, we can use the chi-square distribution, just as we did in the preceding section, to determine whether there is a significant difference between observed and expected frequencies.

Expected frequencies for the cells of the contingency table are based on the following rationale. First we assume that the null hypothesis of independence between beer preference and sex of the beer drinker is true. Then we note that in the entire sample of 150 beer drinkers, a total of 50 prefer light beer, 70 prefer regular beer, and 30 prefer dark beer. In terms of fractions we conclude that $50/150 = 1/3$ of the beer drinkers prefer light beer, $70/150 = 7/15$ prefer regular beer, and $30/150 = 1/5$ prefer dark beer. If the *independence* assumption is valid, we argue that these fractions must be applicable to both male and female beer drinkers. Thus, under the assumption of independence, we would expect the sample of 80 male beer drinkers to show that $(1/3)80 = 26.67$ prefer light beer, $(7/15)80 = 37.33$ prefer regular beer, and $(1/5)80 = 16$ prefer dark beer. Application of the same fractions to the 70 female beer drinkers provides the expected frequencies shown in Table 11.3.

Let e_{ij} denote the expected frequency for the contingency table category in row i and column j. With this notation, let us reconsider the expected frequency calculation for males (row $i = 1$) who prefer regular beer (column $j = 2$)—that is, expected frequency e_{12}. Following the preceding argument for the computation of expected frequencies, we see that

TABLE 11.3 Expected Frequencies If Beer Preference Is Independent of the Sex of the Beer Drinker

		Beer Preference			Total
		Light	*Regular*	*Dark*	
Sex	*Male*	26.67	37.33	16.00	80
	Female	23.33	32.67	14.00	70
Total		50.00	70.00	30.00	150

$$e_{12} = (7/15)80 = 37.33$$

This expression can be rewritten slightly differently as

$$e_{12} = (7/15)80 = (70/150)80 = \frac{(80)(70)}{150} = 37.33$$

Note that 80 in the expression is the total number of males (row 1 total), 70 is the total number of individuals preferring regular beer (column 2 total), and 150 is the total sample size. Hence, we see that

$$e_{12} = \frac{\text{(Row 1 Total)(Column 2 Total)}}{\text{Sample Size}}$$

Generalization of the expression shows that the following formula provides the expected frequencies for a contingency table in the test of independence.

Expected Frequencies for Contingency Tables Under the Assumption of Independence

$$e_{ij} = \frac{(\text{Row } i \text{ Total})(\text{Column } j \text{ Total})}{\text{Sample Size}} \tag{11.10}$$

Using the formula for male beer drinkers who prefer dark beer, we find an expected frequency of $e_{13} = (80)(30)/150 = 16.00$, as shown in Table 11.3. Use (11.10) to verify the other expected frequencies shown in Table 11.3.

The test procedure for comparing the observed frequencies of Table 11.2 with the expected frequencies of Table 11.3 is similar to the goodness of fit calculations made in the preceding section. Specifically, the χ^2 value based on the observed and expected frequencies is computed as follows.

Test Statistic for Independence

$$\chi^2 = \sum_i \sum_j \frac{(f_{ij} - e_{ij})^2}{e_{ij}} \tag{11.11}$$

where

f_{ij} = observed frequency for contingency table category in row i and column j

e_{ij} = expected frequency for contingency table category in row i and column j based on the assumption of independence

With n rows and m columns in the contingency table, the test statistic has a chi-square distribution with $(n - 1)(m - 1)$ degrees of freedom provided that the expected frequencies are 5 *or more* for all categories.

The double summation in (11.11) is used to indicate that the calculation must be made for all the cells in the contingency table.

By reviewing the expected frequencies in Table 11.3, we see that the expected frequencies are 5 or more for each category. We therefore proceed with the computation of the chi-square test statistic. Using Tables 11.2 and 11.3, we obtain

$$\chi^2 = \frac{(20 - 26.67)^2}{26.67} + \frac{(40 - 37.33)^2}{37.33} + \cdots + \frac{(10 - 14.00)^2}{14.00}$$
$$= 1.67 + .19 + \cdots + 1.14 = 6.13$$

The number of degrees of freedom for the appropriate chi-square distribution is computed by multiplying the number of rows minus 1 by the number of columns minus 1. With two rows and three columns, we have $(2 - 1)(3 - 1) = (1)(2) = 2$ degrees of freedom for the test of independence of beer preference and sex of the beer drinker. With $\alpha = .05$ for the level of significance of the test, Table 3 of Appendix B shows an upper-tail χ^2 value of $\chi^2_{.05} = 5.99$. Note that we are again using the upper-tail value because we will reject the null hypothesis only if the differences between observed and expected frequencies provide a large χ^2 value. In our example, $\chi^2 = 6.13$ is greater than the critical value of $\chi^2_{.05} = 5.99$. Thus, we reject the null hypothesis of independence and conclude that beer preference is not independent of the sex of the beer drinker.

Although the test for independence allows no further conclusions, again we can compare the observed and expected frequencies informally to obtain an idea of how the dependence between beer preference and sex of the beer drinker comes about. Refer to Tables 11.2 and 11.3. We see that male beer drinkers have higher observed than expected frequencies for both regular and dark beers, whereas female beer drinkers have a higher observed than expected frequency only for light beer. These observations give us insight about the beer preference differences between male and female beer drinkers.

Let us summarize the steps in a contingency table test of independence.

Contingency Table Test: a Summary

1. Set up the null and alternative hypotheses.

    H_0: The column variable is independent of the row variable

    H_a: The column variable is not independent of the row variable

2. Select a random sample and record the observed frequencies for each cell of the contingency table.
3. Use (11.10) to compute the expected frequency for each cell.
4. Use (11.11) to compute a χ^2 value as a test statistic.
5. Rejection rule:

$$\text{Reject } H_0 \text{ if } \chi^2 > \chi^2_\alpha$$

where α is the level of significance for the test and with n rows and m columns there are $(n - 1)(m - 1)$ degrees of freedom.

notes and comments

The test statistic for the chi-square tests in this chapter requires an expected frequency of 5 for each category. When a category has fewer than 5 items, it is often appropriate to combine two adjacent categories to obtain an expected frequency of 5 or more. Exercise 30 provides an example in which adjacent categories must be combined.

exercises

Methods

Self Test

19. Shown is a 2 × 3 contingency table with observed frequencies for a sample of 200. Test for independence of the row and column factors by using the χ^2 test with $\alpha = .025$.

Row Factor	Column Factor A	B	C
P	20	44	50
Q	30	26	30

20. Below is a 3 × 3 contingency table with observed frequencies for a sample of 240. Test for independence of the row and column factors by using the χ^2 test with $\alpha = .05$.

Row Factor	Column Factor A	B	C
P	20	30	20
Q	30	60	25
R	10	15	30

Applications

21. The 1992 NCAA basketball championship final four teams were Duke, Michigan, Indiana, and Cincinnati. The following data are the season three-point shooting records (*NCAA Final Four Program,* April 1992) for the four teams. At the .05 level of significance, is there a difference in three-point shooting abilities among the four teams? What is your conclusion?

Self Test

3-Point Shooting	Duke	Michigan	Indiana	Cincinnati
Made	160	113	154	202
Missed	214	228	215	331

22. The numbers of units sold by three salespersons over a three-month period are shown. Use $\alpha = .05$ and test for the independence of salesperson and type of product. What is your conclusion?

Salesperson	Product A	B	C
Troutman	14	12	4
Kempton	21	16	8
McChristian	15	5	10

23. Starting positions for business and engineering graduates are classified by industry as shown in the following table.

Degree Major	Industry: Oil	Chemical	Electrical	Computer
Business	30	15	15	40
Engineering	30	30	20	20

Use $\alpha = .01$ and test for independence of degree major and industry type.

24. A CBS News/*New York Times* poll (February 24, 1991) asked a sample of individuals a series of questions about the involvement of the United States in the Persian Gulf war. One question asked men and women: "Do you think the United States did the right thing in starting the ground war against Iraq, or should the United States have waited longer to see if bombing from the air worked?" Assume the responses for men and women were summarized in the contingency table shown. Use the chi-square test of independence to analyze the data. What is your conclusion at a .05 level of significance?

Response	Men	Women
Right thing	243	207
Waited longer	48	66
Not sure	9	27

25. Medical researchers at Harvard and Boston University randomly assigned 227 General Electric employees with alcohol problems to one of three alcohol-treatment groups: a 28-day hospitalization followed by Alcoholics Anonymous (AA) meetings, AA meetings only with no hospitalization, or a choice of the two programs (*USA Today,* September 12, 1991). Two years later, the researchers identified the patients who had remained sober after completing a program. Assume the data are as shown in the following contingency table.

	Program		
Status	Hospitalization	AA Only	Choice
Remained sober	28	13	12
Did not remain sober	48	63	63

Use the chi-square test of independence with a .01 level of significance to analyze the data. What is your conclusion and recommendation?

26. A research study (*GMAC Occasional Papers,* March 1988) provided data on the primary reason for application to an MBA program by full-time and part-time students. Do the data suggest full-time and part-time students differ in their reasons for applying to MBA programs? Explain. Use $\alpha = .01$.

	Primary Reason for Application		
Student Status	Program Quality	Convenience/Cost	Other
Full-time	421	393	76
Part-time	400	593	46

27. A sport preference poll yielded the following data for men and women:

	Favorite Sport		
Sex	Baseball	Basketball	Football
Men	19	15	24
Women	16	18	16

Use $\alpha = .05$ and test for the same sport preferences by men and women. What is your conclusion?

28. Three suppliers provide the following data on defective parts.

	Part Quality		
Supplier	Good	Minor Defect	Major Defect
A	90	3	7
B	170	18	7
C	135	6	9

Use $\alpha = .05$ and test for independence between supplier and part quality. What does the result of your analysis tell the purchasing department?

29. A study of educational levels of voters and their political party affiliations yielded the following results.

	Party Affiliation		
Educational Level	Democratic	Republican	Independent
Did not complete high school	40	20	10
High school degree	30	35	15
College degree	30	45	25

Use $\alpha = .01$ and determine whether party affiliation is independent of the educational level of the voters.

30. In a *Business Week*/Harris executive poll, senior executives were asked: "Compared with the last 12 months, do you think the rate of growth of the gross domestic product will go up, go down, or stay the same over the next 12 months?" The poll was repeated at three successive points in time. The results are summarized here (*Business Week,* January 9, 1995).

		Date of Survey			
		12/94	*6/94*	*12/93*	**Total**
Outlook	*Go Up*	152	177	101	430
	Go Down	104	72	36	212
	Stay the Same	144	152	261	557
	Not Sure	0	0	4	4
	Total	400	401	402	1203

Have the executives changed their outlook over time? Use $\alpha = .01$ to test. (Hint: Combine the "not sure" category with the "stay the same" category to obtain 5 or more for each expected frequency.)

summary

In this chapter we introduced statistical procedures appropriate for comparisons involving proportions and the contingency table test of independence. In the first section we described how inferences concerning differences between proportions from two different populations could be made. In the second section the focus was on a single multinomial population. There we saw how to conduct hypothesis tests comparing the proportions for the categories of the multinomial population with hypothesized values. The chi-square goodness of fit test was used to make the comparison. The final section was concerned with tests of independence of two variables. Again a chi-square goodness of fit test provided the statistical support for the conclusion.

glossary

Pooled variance An estimate of the variance of a population based on the combination of two (or more) sample results. The pooled variance estimate is appropriate whenever the variances of two (or more) populations are assumed equal.

Goodness of fit test A statistical test conducted to determine whether to reject a hypothesized probability distribution for a population.

Contingency table A table used to summarize observed and expected frequencies for a test of independence.

key formulas

Point Estimator of the Difference Between the Proportions of Two Populations

$$\bar{p}_1 - \bar{p}_2 \tag{11.1}$$

Expected Value of $\bar{p}_1 - \bar{p}_2$

$$E(\bar{p}_1 - \bar{p}_2) = p_1 - p_2 \tag{11.2}$$

Standard Deviation of $\bar{p}_1 - \bar{p}_2$

$$\sigma_{\bar{p}_1 - \bar{p}_2} = \sqrt{\frac{p_1(1 - p_1)}{n_1} + \frac{p_2(1 - p_2)}{n_2}} \tag{11.3}$$

Point Estimator of $\sigma_{\bar{p}_1 - \bar{p}_2}$

$$s_{\bar{p}_1 - \bar{p}_2} = \sqrt{\frac{\bar{p}_1(1 - \bar{p}_1)}{n_1} + \frac{\bar{p}_2(1 - \bar{p}_2)}{n_2}} \tag{11.4}$$

Interval Estimate of the Difference Between the Proportions of Two Populations: Large-Sample Case With n_1p_1, $n_1(1 - p_1)$, n_2p_2, and $n_2(1 - p_2) \geq 5$

$$\bar{p}_1 - \bar{p}_2 \pm z_{\alpha/2} s_{\bar{p}_1 - \bar{p}_2} \tag{11.5}$$

Test Statistic for Hypothesis Tests About the Difference Between Proportions of Two Populations

$$z = \frac{(\bar{p}_1 - \bar{p}_2) - (p_1 - p_2)}{\sigma_{\bar{p}_1 - \bar{p}_2}} \tag{11.6}$$

Pooled Estimator of the Population Proportion

$$\bar{p} = \frac{n_1\bar{p}_1 + n_2\bar{p}_2}{n_1 + n_2} \tag{11.7}$$

Point Estimator of $\sigma_{\bar{p}_1 - \bar{p}_2}$ when $p_1 = p_2$

$$s_{\bar{p}_1 - \bar{p}_2} = \sqrt{\bar{p}(1 - \bar{p})\left(\frac{1}{n_1} + \frac{1}{n_2}\right)} \tag{11.8}$$

Test Statistic for Goodness of Fit

$$\chi^2 = \sum_{i=1}^{k} \frac{(f_i - e_i)^2}{e_i} \tag{11.9}$$

Expected Frequencies for Contingency Tables Under the Assumption of Independence

$$e_{ij} - \frac{(\text{Row } i \text{ Total})(\text{Column } j \text{ Total})}{\text{Sample Size}} \tag{11.10}$$

Test Statistic for Independence

$$\chi^2 = \sum_i \sum_j \frac{(f_{ij} - e_{ij})^2}{e_{ij}} \tag{11.11}$$

supplementary exercises

31. A cable television firm is considering submitting bids for rights to operate in two regions of the state of Florida. Surverys of the two regions provided the following data on customer acceptance of the cable television service.

Region I	Region II
$n_1 = 500$ Number indicating an intent to purchase = 175	$n_2 = 800$ Number indicating an intent to purchase = 360

Develop a 99% confidence interval for the difference between population proportions of customer acceptance in the two regions.

32. A group of physicians in Denmark conducted a year-long study on the effectiveness of nicotine chewing gum in helping people to stop smoking (*New England Journal of Medicine,* 1988). The 113 people who participated in the study were all smokers. Sixty were given chewing gum with 2 milligrams of nicotine and 53 were given a placebo chewing gum with no nicotine content. No one in the study knew which type of gum he or she had been given. All were told to use the gum and refrain from smoking.

a. State the null and alternative hypotheses that would be appropriate if the researchers hoped to show that the group given nicotine chewing gum had a higher proportion of non-smokers one year after the study began.

b. Results showed that 23 of the smokers given nicotine chewing gum had remained non-smokers for the one-year period and 12 of the smokers given the placebo had remained nonsmokers during the same period. Do these results support the conclusion that nicotine gum can help a person to stop smoking? Test using $\alpha = .05$. What is the p-value?

33. A large automobile-insurance company selected samples of single and married male policyholders and recorded the number who had made an insurance claim over the preceding three-year period.

Single Policyholders	Married Policyholders
$n_1 = 400$	$n_2 = 900$
Number making claims = 76	Number making claims = 90

a. Using $\alpha = .05$, test to determine whether the claim rates differ between single and married male policyholders.

b. Provide a 95% confidence interval for the difference between the proportions for the two populations.

34. Medical tests were conducted to learn about drug-resistant tuberculosis (*The New York Times,* January 24,1992). Of 142 cases tested in New Jersey, nine were found to be drug-resistant. Of 268 cases tested in Texas, five were found to be drug-resistant. Do these data suggest a statistically significant difference between the proportions of drug-resistant cases in the two states? Test H_0: $p_1 - p_2 = 0$ at the .02 level of significance. What is the p-value and what is your conclusion?

35. In setting sales quotas, the marketing manager makes the assumption that order potentials are the same for each of four sales territories. A sample of 200 sales shows the numbers of orders from the territories.

Sales Territories			
I	II	III	IV
60	45	59	36

Should the manager's assumption be rejected? Use $\alpha = .05$.

36. Seven percent of mutual fund investors rate corporate stocks "very safe," 58% rate them "somewhat safe," 24% rate them "not very safe," 4% rate them "not at all safe," and 7% are "not sure." A *Business Week*/Harris poll asked 529 mutual fund investors how they would rate corporate bonds on safety (*Business Week,* August 15, 1994). The responses are shown.

Safety Rating	Frequency
Very safe	48
Somewhat safe	323
Not very safe	79
Not at all safe	16
Not sure	63
Total	529

Do mutual fund investors' attitudes toward corporate bonds differ from their attitudes toward corporate stock? Support your conclusion with a statistical test using $\alpha = .01$.

37. A community park is to be opened soon. A sample of 140 individuals has been asked to state their preference for when they would most like to visit the park. The sample results follow.

Week Day	Saturday	Sunday	Holiday
20	20	40	60

In developing a staffing plan, should the park manager plan on the same number visiting the park each day? Support your conclusion with a statistical test using $\alpha = .05$.

38. A regional transit authority was concerned about the number of riders on one of its bus routes. In setting up the route, the assumption was that the number of riders was the same on every day from Monday through Friday. Using the data shown, test with $\alpha = .05$ to determine whether the transit authority's assumption is correct.

Day	Number of Riders
Monday	13
Tuesday	16
Wednesday	28
Thursday	17
Friday	16

39. A sample of parts provided the following contingency table data on part quality by production shift.

Shift	Number Good	Number Defective
First	368	32
Second	285	15
Third	176	24

Use $\alpha = .05$ and test the hypothesis that part quality is independent of the production shift. What is your conclusion?

40. The Graduate Management Admission Council (GMAC) sponsored a survey of MBA students to learn about characteristics of the population of students interested in graduate education in business administration. The following table was published in the *GMAC Occasional Papers* (March 1988). Do the data suggest male and female students differ in their reasons for application to MBA programs? Explain. Use $\alpha = .05$.

	Primary Reason for Application		
MBA Students	Program Quality	Convenience/Cost	Other
Male	519	599	86
Female	298	390	36

41. A lending institution supplied data on loan approvals by four loan officers.

Loan Officer	Loan Approval Decision: Approved	Rejected
Miller	24	16
McMahon	17	13
Games	35	15
Runk	11	9

Use α = .05 and test to determine whether the loan approval decision is independent of the loan officer reviewing the loan application.

42. A survey of commercial buildings served by the Cincinnati Gas & Electric Company was concluded in 1992 (CG&E Commercial Building Characteristics Survey, November 25, 1992). One question asked the main type of heating fuel used and another asked the year of construction. A crosstabulation of the findings follows.

		Fuel Type					
		Electricity	*Natural Gas*	*Oil*	*Propane*	*Other*	**Total**
Year Constructed	*1973 or before*	40	183	12	5	7	247
	1974–1979	24	26	2	2	0	54
	1980–1986	37	38	1	0	6	82
	1987–1991	48	70	2	0	1	121
	Total	149	317	17	7	14	504

Do fuel type and year constructed appear to be related? Conduct a statistical test with α = .025 to support your conclusion.

43. Below is a crosstabulation of industry type and P/E ratio for 20 companies in the consumer products and banking industries (*Business Week,* August 15, 1994).

		P/E Ratio					
		5–9	*10–14*	*15–19*	*20–24*	*25–29*	**Total**
Industry	*Consumer*	0	3	5	1	1	10
	Banking	4	4	2	0	0	10
	Total	4	7	7	1	1	20

Does there appear to be a relationship between industry type and P/E ratio? Support your conclusion with a statistical test using $\alpha = .05$.

44. The following data were collected on the number of emergency ambulance calls for an urban county and a rural county in Virginia (*Journal of the Operational Research Society,* November 1986).

		Day of Week							Total
		Sun	*Mon*	*Tue*	*Wed*	*Thur*	*Fri*	*Sat*	
County	*Urban*	61	48	50	55	63	73	43	393
	Rural	7	9	16	13	9	14	10	78
	Total	68	57	66	68	72	87	53	471

Conduct a test for independence using $\alpha = .05$. What is your conclusion?

45. The 1991 office occupancy rates were reported for four California metropolitan areas (*Business Week,* 1991). Do the following data suggest that the office vacancies were independent of metropolitan area? Use a .05 level of significance. What is your conclusion?

Occupancy Status	Los Angeles	San Diego	San Francisco	San Jose
Occupied	160	116	192	174
Vacant	40	34	33	26

CHAPTER 12

Regression Analysis

Polaroid Corporation*

Cambridge, Massachusetts

Polaroid's consumer photography business began in 1947 when the company's founder, Dr. Edwin Land, announced a one-step dry process for producing a finished photograph within one minute after taking the picture. The first Polaroid Land camera and Polaroid Land film went on sale in 1949. Since then, Polaroid's continuous experimentation and development in chemistry, optics, and electronics have produced photographic systems of ever higher quality, reliability, and convenience.

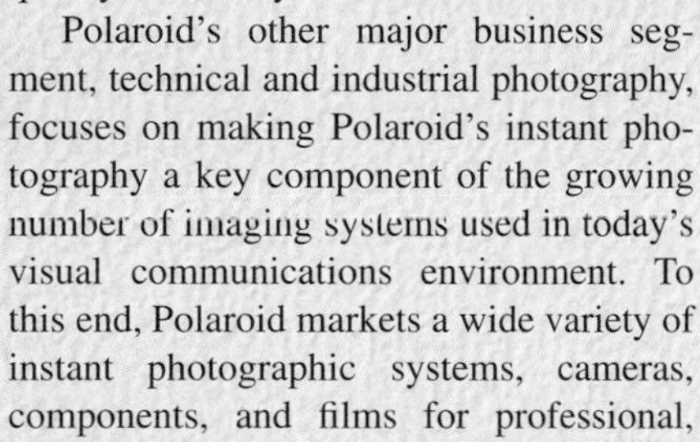

Polaroid's other major business segment, technical and industrial photography, focuses on making Polaroid's instant photography a key component of the growing number of imaging systems used in today's visual communications environment. To this end, Polaroid markets a wide variety of instant photographic systems, cameras, components, and films for professional, industrial, scientific, and medical uses. Other businesses include magnetics, sunglasses, industrial polarizers, chemicals, custom coating, and holography.

Sensitometry, the measurement of the sensitivity of photographic materials, provides information on many characteristics of film, such as its useful exposure range. Within Polaroid's central sensitometry laboratory, scientists systematically sample and analyze instant films that have been stored at temperature and humidity levels approximating those to which the films will be subjected after they have been purchased by consumers. To investigate the relationship between film speed and the age of a Polaroid extended range, color professional print film, Polaroid's central sensitometry lab selected film samples ranging in age (time since manufacture) from one to 13 months. The data showed that film speed decreases with age, and that a straight-line or linear relationship could be used to approximate the relationship between change in film speed and age of the film.

Using regression analysis, Polaroid was able to develop the following equation relating the change in the film speed to the film's age.

$$\hat{y} = -19.8 - 7.6x$$

where

$$\hat{y} = \text{change in film speed}$$
$$x = \text{film age in months}$$

This equation shows that the average decrease in film speed is 7.6 units per month. The information provided by this analysis, when coupled with consumer purchase and use patterns, enables Polaroid to make manufacturing adjustments that help the company produce films with the performance levels its customers require.

In this chapter you will learn how regression analysis can be used to develop an equation relating two or more variables, such as the change in film speed to the age of the film in the Polaroid example.

The authors are indebted to Mr. Lawrence Friedman, Manager, Photographic Quality, for providing this Statistics in Practice.

Polaroid is the leader in quality instant photography products.

Managerial decisions often are based on the relationship between two or more variables. For example, after considering the relationship between advertising expenditures and sales, a marketing manager might attempt to predict sales for a given level of advertising expenditures. In another case, a public utility might use the relationship between the daily high temperature and the demand for electricity to predict electricity usage on the basis of next month's anticipated daily high temperatures. Sometimes a manager will rely on intuition to judge how two variables are related. However, if data can be obtained, a statistical procedure called *regression analysis* can be used to develop an equation showing how the variables are related.

In regression terminology, the variable that is being predicted is called the *dependent* variable. The variable or variables being used to predict the value of the dependent variable are called the *independent* variables. For example, in analyzing the effect of advertising expenditures on sales, a marketing manager's desire to predict sales would suggest making sales the dependent variable. Advertising expenditure would be the independent variable used to help predict sales. In statistical notation, y denotes the dependent variable and x denotes the independent variable.

The simplest type of regression analysis involves one independent variable and one dependent variable in which the relationship between the variables is approximated by a straight line. This is called *simple linear regression.* Regression analysis involving two or more independent variables is called multiple regression analysis.

12.1 The Simple Linear Regression Model

As an illustration of regression analysis, let us consider the situation of Armand's Pizza Parlors, a chain of Italian-food restaurants located in a five-state area. The most successful locations for Armand's have been near college campuses. The managers believe that quarterly sales for these restaurants (denoted by y) are related positively to the size of the student population (denoted by x); that is, restaurants near campuses with a large population tend to generate more sales than those located near campuses with a small population. Using regression analysis, we can develop an equation showing how the dependent variable y is related to the independent variable x.

The Regression Model and the Regression Equation

In the Armand's Pizza Parlors example, every restaurant has associated with it a value of x (student population) and a corresponding value of y (quarterly sales). The equation that describes how y is related to x and an error term (denoted by ϵ) is called the *regression model.* The regression model used in simple linear regression follows.

Simple Linear Regression Model

$$y = \beta_0 + \beta_1 x + \epsilon \tag{12.1}$$

In the simple linear regression model, β_0 and β_1 are the parameters and ϵ (the Greek letter epsilon) is a random variable. A close examination of the simple linear regression model reveals that y is a linear function of x (the $\beta_0 + \beta_1 x$ part) plus ϵ. The random variable ϵ is an error term that accounts for the variability in y that cannot be explained by the linear relationship between x and y.

In Section 12.4 we will discuss all of the assumptions for the simple linear regression model and ϵ. One of the assumptions is that the mean or expected value of ϵ is zero. A consequence of this assumption is that the mean or expected value of y, denoted $E(y)$, is equal to $\beta_0 + \beta_1 x$; in other words, the mean value of y is a linear function of x. The equation that describes how the mean value of y is related to x is called the *regression equation.* The regression equation for simple linear regression follows.

Simple Linear Regression Equation

$$E(y) = \beta_0 + \beta_1 x \tag{12.2}$$

In simple linear regression, the graph of the regression equation is a straight line; β_0 is the y intercept of the regression line, β_1 is the slope, and $E(y)$ is the mean or expected value of y for a given value of x. Examples of possible regression lines for simple linear regression are shown in Figure 12.1. The regression line in panel A shows that the mean value of y is related positively to x, with larger values of $E(y)$ associated with larger values of x. The regression line in panel B shows that the mean value of y is related negatively to x, with smaller values of $E(y)$ associated with larger values of x. The regression line in panel C shows the case in which y is not related to x; that is, the mean value of y is the same for every value of x.

The Estimated Regression Equation

If the values of the parameters β_0 and β_1 were known, we could use (12.2) to compute the mean value of y for a known value of x. Unfortunately, the parameter values are not known in practice and must be estimated by using sample data. Sample statistics (denoted b_0 and b_1) are computed as estimates of the parameters β_0 and β_1. Substituting the values of the sample statistics b_0 and b_1 for β_0 and β_1 in the regression equation, we obtain the *estimated regression equation.* In simple linear regression, the estimated regression equation is written in the following form.

Estimated Simple Linear Regression Equation

$$\hat{y} = b_0 + b_1 x \tag{12.3}$$

FIGURE 12.1
Possible Regression Lines in Simple Linear Regression

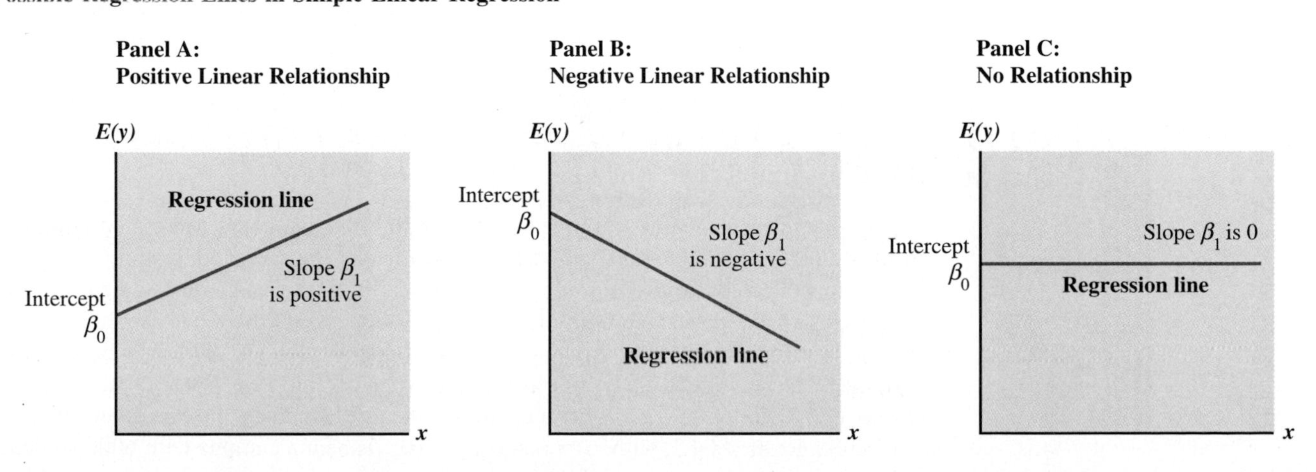

FIGURE 12.2
The Estimation Process in Simple Linear Regression

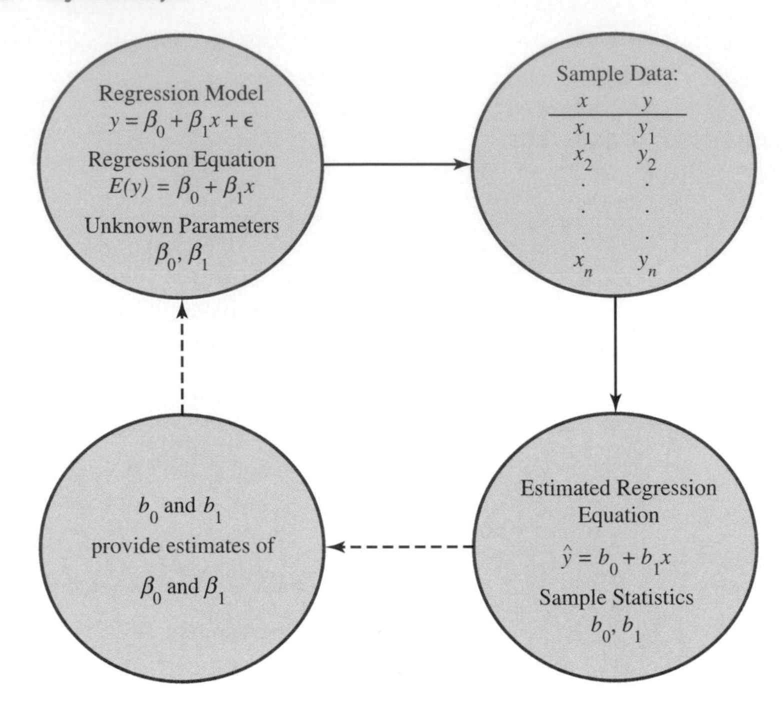

In simple linear regression, the graph of the estimated regression equation is called the *estimated regression line;* b_0 is the y intercept, b_1 is the slope, and $\hat{y}$ is the estimated value of y for a given value of x. In the next section, we show how the least squares method can be used to compute the values of b_0 and b_1 in the estimated regression equation. Figure 12.2 is a summary of the estimation process for simple linear regression.

notes and comments

We caution the reader that regression analysis *cannot* be interpreted as a procedure for establishing a *cause-and-effect* relationship between variables. It can only indicate how or to what extent variables are *associated* with each other. Any conclusions about cause and effect must be based on the *judgment* of the individual or individuals most knowledgeable about the application.

12.2 The Least Squares Method

The *least squares method* is a procedure for finding the estimated regression equation. To illustrate the least squares method for the Armand's Pizza Parlors example, suppose data were collected from a sample of 10 Armand's restaurants located near college campuses. For the ith restaurant in the sample, x_i is the size of the student population (in thousands) and y_i is the quarterly sales (in thousands of dollars). The values of x_i and y_i for the 10 restaurants in the sample are summarized in Table 12.1. We see that restaurant 1, with $x_1 = 2$ and $y_1 = 58$, is near a campus with 2000 students and has quarterly sales of \$58,000. Restaurant 2, with $x_2 = 6$ and $y_2 = 105$, is near a campus with 6000 students

TABLE 12.1 Student Population and Quarterly Sales Data for 10 Armand's Pizza Parlors

Restaurant i	Student Population (1000s) x_i	Quarterly Sales ($1000s) y_i
1	2	58
2	6	105
3	8	88
4	8	118
5	12	117
6	16	137
7	20	157
8	20	169
9	22	149
10	26	202

and has quarterly sales of $105,000. The largest sales value is for restaurant 10, which is near a campus with 26,000 students and has quarterly sales of $202,000.

Figure 12.3 is a scatter diagram of the data in Table 12.1. The size of the student population is shown on the horizontal axis and the value of quarterly sales is shown on the vertical axis. Scatter diagrams for regression analysis are constructed with values of the independent variable x on the horizontal axis and values of the

FIGURE 12.3
Scatter Diagram of Student Population and Quarterly Sales for Armand's Pizza Parlors

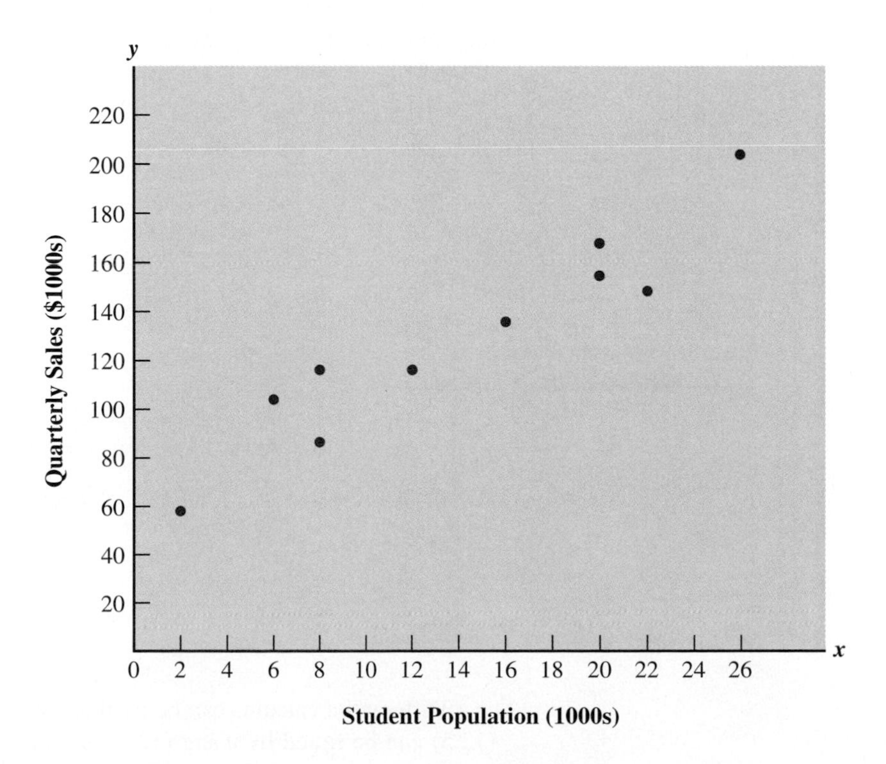

dependent variable y on the vertical axis. The scatter diagram enables us to observe the data graphically and to draw preliminary conclusions about the possible relationship between the variables.

What preliminary conclusions can be drawn from Figure 12.3? Sales appear to be higher at campuses with larger student populations. In addition, for these data the relationship between the size of the student population and sales appears to be approximated by a straight line; indeed, there seems to be a positive linear relationship between x and y. We therefore choose the simple linear regression model to represent the relationship between quarterly sales and student population. Given that choice, our next task is to use the sample data in Table 12.1 to determine the values of b_0 and b_1 in the estimated simple linear regression equation. For the ith restaurant, the estimated regression equation provides

$$\hat{y}_i = b_0 + b_1 x_i \tag{12.4}$$

where

x_i = size of the student population (1000s) for the ith restaurant,

b_0 = the y intercept of the estimated regression line,

b_1 = the slope of the estimated regression line, and

$\hat{y}_i$ = estimated value of quarterly sales ($1000s) for the ith restaurant

With y_i denoting the observed (actual) sales for restaurant i and $\hat{y}_i$ in (12.4) representing the estimated value of sales for restaurant i, every restaurant in the sample will have an observed value of sales y_i and an estimated value of sales $\hat{y}_i$. For the estimated regression line to provide a good fit to the data, we want the differences between the observed sales values and the estimated sales values to be small.

The least squares method uses the sample data to provide the values of b_0 and b_1 that minimize the *sum of the squares of the deviations* between the observed values of the dependent variable y_i and the estimated values of the dependent variable $\hat{y}_i$. The criterion for the least squares method is given by (12.5).

Least Squares Criterion

$$\min \Sigma(y_i - \hat{y}_i)^2 \tag{12.5}$$

where

y_i = observed value of the dependent variable for the ith observation

$\hat{y}_i$ = estimated value of the dependent variable for the ith observation

Differential calculus can be used to show that the values of b_0 and b_1 that minimize (12.5) can be found by using (12.6) and (12.7).

Slope and y-Intercept for the Estimated Regression Equation

$$b_1 = \frac{\Sigma x_i y_i - (\Sigma x_i \, \Sigma y_i)/n}{\Sigma x_i^2 - (\Sigma x_i)^2/n} \tag{12.6}$$

$$b_0 = \bar{y} - b_1 \bar{x} \tag{12.7}$$

where

x_i = value of the independent variable for the ith observation

y_i = observed value of the dependent variable for the ith observation

$\bar{x}$ = sample mean for the independent variable

$\bar{y}$ = sample mean for the dependent variable

n = total number of observations

In computing b_1 with a calculator, it is best to carry as many significant digits as possible in the intermediate calculations; we recommend carrying at least four significant digits.

Some of the calculations necessary to develop the least squares estimated regression equation for Armand's Pizza Parlors are shown in Table 12.2. In this example there are 10 restaurants, or observations; hence $n = 10$. Using (12.6), (12.7), and the information in Table 12.2, we can compute the slope and intercept of the estimated regression equation for Armand's Pizza Parlors. The calculation of the slope (b_1) proceeds as follows.

$$\begin{aligned} b_1 &= \frac{\Sigma x_i y_i - (\Sigma x_i \Sigma y_i)/n}{\Sigma x_i^2 - (\Sigma x_i)^2/n} \\ &= \frac{21{,}040 - (140)(1300)/10}{2528 - (140)^2/10} \\ &= \frac{2840}{568} \\ &= 5 \end{aligned}$$

TABLE 12.2 Calculations for the Least Squares Estimated Regression Equation for Armand's Pizza Parlors

Restaurant i		x_i	y_i	$x_i y_i$	x_i^2
1		2	58	116	4
2		6	105	630	36
3		8	88	704	64
4		8	118	944	64
5		12	117	1,404	144
6		16	137	2,192	256
7		20	157	3,140	400
8		20	169	3,380	400
9		22	149	3,278	484
10		26	202	5,252	676
	Totals	140	1300	21,040	2528
		Σx_i	Σy_i	$\Sigma x_i y_i$	Σx_i^2

The calculation of the y intercept (b_0) follows.

$$\bar{x} = \frac{\Sigma x_i}{n} = \frac{140}{10} = 14$$

$$\bar{y} = \frac{\Sigma y_i}{n} = \frac{1300}{10} = 130$$

$$\begin{aligned} b_0 &= \bar{y} - b_1\bar{x} \\ &= 130 - 5(14) \\ &= 60 \end{aligned}$$

Thus, the estimated regression equation found by using the least squares method is

$$\hat{y} = 60 + 5x$$

Figure 12.4 shows the graph of this equation on the scatter diagram.

The slope of the estimated regression equation ($b_1 = 5$) is positive, implying that as student population increases, sales increase. In fact, we can conclude (since sales are measured in \$1000s and student population in 1000s) that an increase in the student population of 1000 is associated with an increase of \$5000 in expected sales; that is, sales are expected to increase by \$5.00 per student.

If we believe the least squares estimated regression equation adequately describes the relationship between x and y, it would seem reasonable to use the estimated regression equation to predict the value of y for a given value of x. For example, if we wanted to predict sales for a restaurant to be located near a campus with 16,000 students, we would compute

$$\begin{aligned} \hat{y} &= 60 + 5(16) \\ &= 140 \end{aligned}$$

FIGURE 12.4
Graph of the Estimated Regression Equation for Armand's Pizza Parlors: $\hat{y} = 60 + 5x$

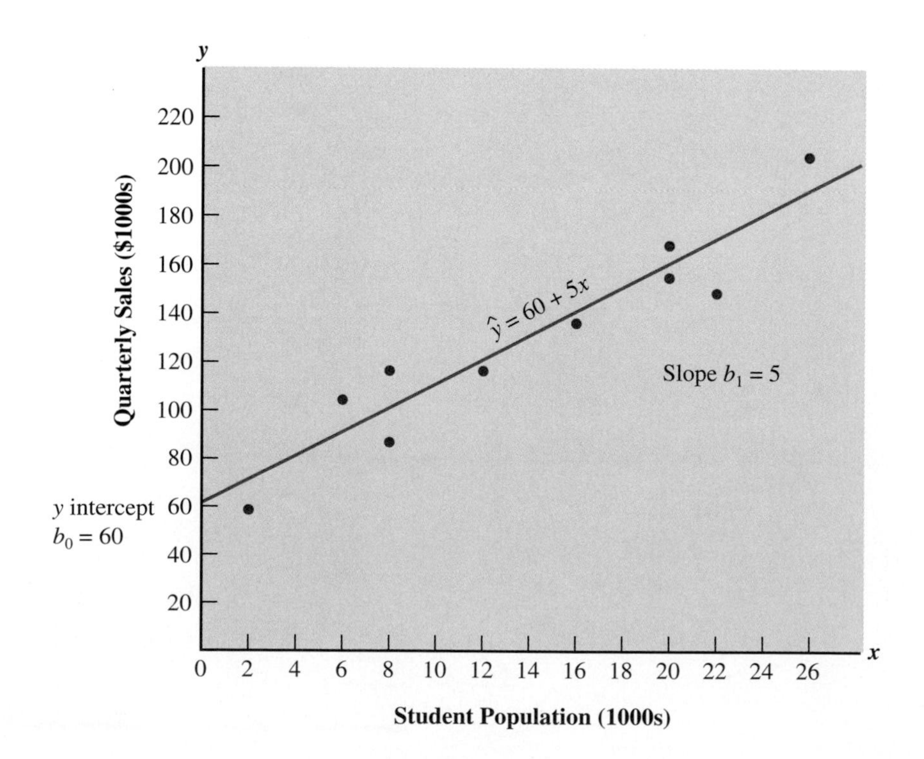

Hence, we would predict quarterly sales of $140,000 for this restaurant. In the following sections we will discuss methods for assessing the appropriateness of using the estimated regression equation for estimation and prediction.

notes and comments

The least squares method provides an estimated regression equation that minimizes the sum of squared deviations between the observed values of the dependent variable y_i and the estimated values of the dependent variable $\hat{y}_i$. This is the least squares criterion for choosing the equation that provides the best fit. If some other criterion were used, such as minimizing the sum of the absolute deviations between y_i and $\hat{y}_i$, a different equation would be obtained. In practice, the least squares method is the most widely used.

exercises

Methods

1. Given are five observations for two variables, x and y.

x_i	1	2	3	4	5
y_i	3	7	5	11	14

a. Develop a scatter diagram for these data.
b. What does the scatter diagram developed in (a) indicate about the relationship between the two variables?
c. Try to approximate the relationship between x and y by drawing a straight line through the data.
d. Develop the estimated regression equation by computing the values of b_0 and b_1 using (12.6) and (12.7).
e. Use the estimated regression equation to predict the value of y when $x = 4$.

2. Given are five observations for two variables, x and y.

x_i	2	3	5	1	8
y_i	25	25	20	30	16

a. Develop a scatter diagram for these data.
b. What does the scatter diagram developed in (a) indicate about the relationship between the two variables?
c. Try to approximate the relationship between x and y by drawing a straight line through the data.
d. Develop the estimated regression equation by computing the values of b_0 and b_1 using (12.6) and (12.7).
e. Use the estimated regression equation to predict the value of y when $x = 6$.

3. Given are five observations collected in a regression study on two variables.

x_i	2	4	5	7	8
y_i	2	3	2	6	4

a. Develop a scatter diagram for these data.
b. Develop the estimated regression equation for these data.
c. Use the estimated regression equation to predict the value of y when $x = 4$.

Applications

Self Test

4. The following data are the monthly starting salaries and the grade point averages (GPA) of students who have obtained a bachelor's degree in business administration.

GPA	2.6	3.4	3.6	3.2	3.5	2.9
Monthly Salary ($)	1800	2100	2500	2000	2400	2100

a. Develop a scatter diagram for these data with GPA as the independent variable.
b. What does the scatter diagram developed in (a) indicate about the relationship between the two variables?
c. Draw a straight line through the data to approximate a linear relationship between GPA and salary.
d. Use the least squares method to develop the estimated regression equation.
e. Predict the monthly starting salary for a student with a 3.0 GPA and for a student with a 3.5 GPA.

5. Performance data for a Century Coronado 21 with a 310-hp MerCruiser V-8 gasoline inboard engine was reported in *Boating* (September 1991). Data on how the boat speed in miles per hour (mph) affected fuel consumption in gallons per hour (gph) follow.

Speed (mph)	Fuel Consumption (gph)
6.1	2.3
10.7	4.8
20.9	7.5
27.5	9.2
31.5	12.4

a. Develop a scatter diagram for these data.
b. What does the scatter diagram developed in (a) indicate about the relationship between the two variables?
c. Develop the estimated regression equation showing how fuel consumption is related to the boat speed.
d. What is the estimated fuel consumption if the boat speed is 25 mph?

6. Major hotels frequently provide special rates for business travelers. The lowest rates are charged when reservations are made 14 days in advance. The following table reports the business rates and the 14-day-advance super-saver rates for one night at a sample of six ITT Sheraton Hotels (*Sky Magazine,* January 1995).

Hotel Location	Business Rate	14-Day Advance Rate
Birmingham	$ 89	$ 81
Miami	130	115
Atlanta	98	89
Chicago	149	138
New Orleans	199	149
Nashville	114	94

a. Develop a scatter diagram for these data with business rates as the independent variable.
b. Develop the least squares estimated regression equation.
c. The ITT Sheraton Hotel in Tampa offers a business rate of $135 per night. Estimate the 14-day-advance super-saver rate at this hotel.

7. *Consumer Reports* uses a survey to collect data on the annual cost of repairs for more than 300 makes and models of automobiles (*Consumer Reports* 1992 Buying Guide). The following data are the average annual repair cost ($) and the age of the automobile (years).

Age	1	2	3	4	5
Repair	135	175	320	300	450

a. Develop the estimated regression equation showing how annual repair cost is related to the age of an automobile.
b. What is the estimated annual repair cost for a three-year-old automobile?

8. The following table reports the data a sales manager has collected on annual sales and years of experience.

Salesperson	Years of Experience	Annual Sales ($1000s)
1	1	80
2	3	97
3	4	92
4	4	102
5	6	103
6	8	111
7	10	119
8	10	123
9	11	117
10	13	136

a. Develop a scatter diagram for these data with years of experience as the independent variable.
b. Develop an estimated regression equation that can be used to predict annual sales given the years of experience.
c. Use the estimated regression equation to predict annual sales for a salesperson with nine years of experience.

9. Tire rating and load-carrying capacity for a sample of automobile tires follow (*Road & Track,* October 1994).

Tire Rating	Load-Carrying Capacity
75	853
82	1047
85	1135
87	1201
88	1235
91	1356
92	1389
93	1433
105	2039

a. Develop a scatter diagram for these data with tire rating as the independent variable.
b. Develop the least squares estimated regression equation.
c. Estimate the load-carrying capacity for a tire that has a rating of 90.

10. The following table gives the percentage of women working in each company (x) and the percentage of management jobs held by women in that company (y); the data represent companies in retailing and trade (*Louis Rukeyser's Business Almanac*).

Company	x_i	y_i
Federated Department Stores	72	61
Kroger	47	16
Marriott	51	32
McDonald's	57	46
Sears	55	36

a. Develop a scatter diagram for these data.
b. What does the scatter diagram developed in (a) indicate about the relationship between x and y?
c. Develop the estimated regression equation for these data.
d. Predict the percentage of management jobs held by women in a company that has 60% women employees.
e. Use the estimated regression equation to predict the percentage of management jobs held by women in a company where 55% of the jobs are held by women. How does this predicted value compare to the 36% value observed for Sears, a company in which 55% of the employees are women?

11. The following table reports the median income and the median home price for a sample of six cities (*Who's Buying Homes in America,* Chicago Title and Trust Company, 1994). Data are in thousands of dollars.

City	Median Income	Median Home Price
Atlanta	$65.2	$120.2
Cleveland	49.8	92.7
Denver	53.8	111.7
Dallas	62.7	104.7
Orlando	50.9	98.5
Minneapolis	53.1	105.8

a. Develop a scatter diagram for these data with median income as the independent variable.
b. Develop the least squares estimated regression equation.
c. Estimate the median home price in a city with a median income of $60,000.

12. To the Internal Revenue Service, the reasonableness of total itemized deductions depends on the taxpayer's adjusted gross income. Large deductions, which include charity and medical deductions, are more reasonable for taxpayers with large adjusted gross incomes. If a taxpayer claims greater than average itemized deductions for a given level of income, the chances of an IRS audit are increased. Data on adjusted gross income and the average or reasonable amount of itemized deductions follow (*Money,* October 1994). The data are in thousands of dollars.

Adjusted Gross Income	Total Itemized Deductions
$ 22	$ 9.6
27	9.6
32	10.1
48	11.1
65	13.5
85	17.7
120	25.5

a. Develop a scatter diagram for these data with adjusted gross income as the independent variable.
b. Use the least squares method to develop the estimated regression equation.
c. Estimate a reasonable level of total itemized deductions for a taxpayer with an adjusted gross income of $52,500. If this taxpayer has claimed total itemized deductions of $20,400, would the IRS agent's request for an audit appear justified? Explain.

12.3 The Coefficient of Determination

For the Armand's Pizza Parlors example, we developed the estimated regression equation $\hat{y} = 60 + 5x$ to approximate the linear relationship between the size of the student population x and quarterly sales y. A question now is: How well does the estimated regression equation fit the data? In this section, we show that the *coefficient of determination* provides a measure of the goodness of fit for the estimated regression equation.

For the ith observation in the sample that we used to estimate b_0 and b_1, the deviation between the observed value of the dependent variable, y_i, and the estimated value of the dependent variable, $\hat{y}_i$, is called the *ith residual.* The ith residual represents the error in using $\hat{y}_i$ to estimate y_i. Thus, for the ith observation, the residual is $y_i - \hat{y}_i$. The sum of squares of these residuals or errors is the quantity that is minimized by the least squares method. This quantity, also known as the *sum of squares due to error,* is denoted by SSE.

Sum of Squares Due to Error

$$\text{SSE} = \Sigma(y_i - \hat{y}_i)^2 \qquad (12.8)$$

TABLE 12.3 Calculation of SSE for Armand's Pizza Parlors

Restaurant i	x_i = Student Population (1000s)	y_i = Quarterly Sales ($1000s)	$\hat{y}_i = 60 + 5x_i$	$y_i - \hat{y}_i$	$(y_i - \hat{y}_i)^2$
1	2	58	70	−12	144
2	6	105	90	15	225
3	8	88	100	−12	144
4	8	118	100	18	324
5	12	117	120	−3	9
6	16	137	140	−3	9
7	20	157	160	−3	9
8	20	169	160	9	81
9	22	149	170	−21	441
10	26	202	190	12	144
					SSE = 1530

The value of SSE is a measure of the error in using the estimated regression equation to estimate the values of the dependent variable in the sample.

In Table 12.3 we show the calculations required to compute the sum of squares due to error for the Armand's Pizza Parlors example. For instance, for restaurant 1 the values of the independent and dependent variables are $x_1 = 2$ and $y_1 = 58$. Using the estimated regression equation, we find that the estimated value of sales for restaurant 1 is $\hat{y}_1 = 60 + 5(2) = 70$. Thus, the error in using $\hat{y}_1$ to estimate y_1 for restaurant 1 is $y_1 - \hat{y}_1 = 58 - 70 = -12$. The squared error, $(-12)^2 = 144$, is shown in the last column of Table 12.3. After computing and squaring the residuals for each restaurant in the sample, we sum them to obtain SSE = 1530. Thus, SSE = 1530 measures the error in using the estimated regression equation $\hat{y} = 60 + 5x$ to predict sales.

Now suppose we are asked to develop an estimate of sales without knowledge of the size of the student population. Without knowledge of any related variables, we would use the sample mean as an estimate of sales at any given restaurant. Table 12.2 shows that for the sales data, $\Sigma y_i = 1300$. Hence, the mean value of sales for the sample of 10 Armand's restaurants is $\bar{y} = \Sigma y_i / n = 1300/10 = 130$. In Table 12.4 we show the sum of squared deviations obtained by using the sample mean $\bar{y} = 130$ to estimate the value of sales for each restaurant in the sample. For the ith restaurant in the sample, the difference $y_i - \bar{y}$ provides a measure of the error involved in using $\bar{y}$ to estimate sales. The corresponding sum of squares, called the *total sum of squares,* is denoted SST.

Total Sum of Squares

$$\text{SST} = \Sigma(y_i - \bar{y})^2 \tag{12.9}$$

The sum at the bottom of the last column in Table 12.4 is the total sum of squares for Armand's Pizza Parlors; it is SST = 15,730.

TABLE 12.4 Computation of the Total Sum of Squares for Armand's Pizza Parlors

Restaurant i	x_i = Student Population (1000s)	y_i = Quarterly Sales ($1000s)	$y_i - \bar{y}$	$(y_i - \bar{y})^2$
1	2	58	−72	5,184
2	6	105	−25	625
3	8	88	−42	1,764
4	8	118	−12	144
5	12	117	−13	169
6	16	137	7	49
7	20	157	27	729
8	20	169	39	1,521
9	22	149	19	361
10	26	202	72	5,184
				SST = 15,730

In Figure 12.5 we show the estimated regression line $\hat{y} = 60 + 5x$ and the line corresponding to $\bar{y} = 130$. Note that the points cluster more closely around the estimated regression line than they do about the line $\bar{y} = 130$. For example, for the 10th restaurant in the sample we see that the error is much larger when $\bar{y} = 130$ is used as an estimate of y_{10} than when $\hat{y}_{10} = 60 + 5(26) = 190$ is used. We can think of SST as a measure of how well the observations cluster about the $\bar{y}$ line and SSE as a measure of how well the observations cluster about the $\hat{y}$ line.

To measure how much the $\hat{y}$ values on the estimated regression line deviate from $\bar{y}$, another sum of squares is computed. This sum of squares, called the *sum of squares due to regression,* is denoted SSR.

FIGURE 12.5
Deviations About the Estimated Regression Line and the Line $y = \bar{y}$ for Armand's Pizza Parlors

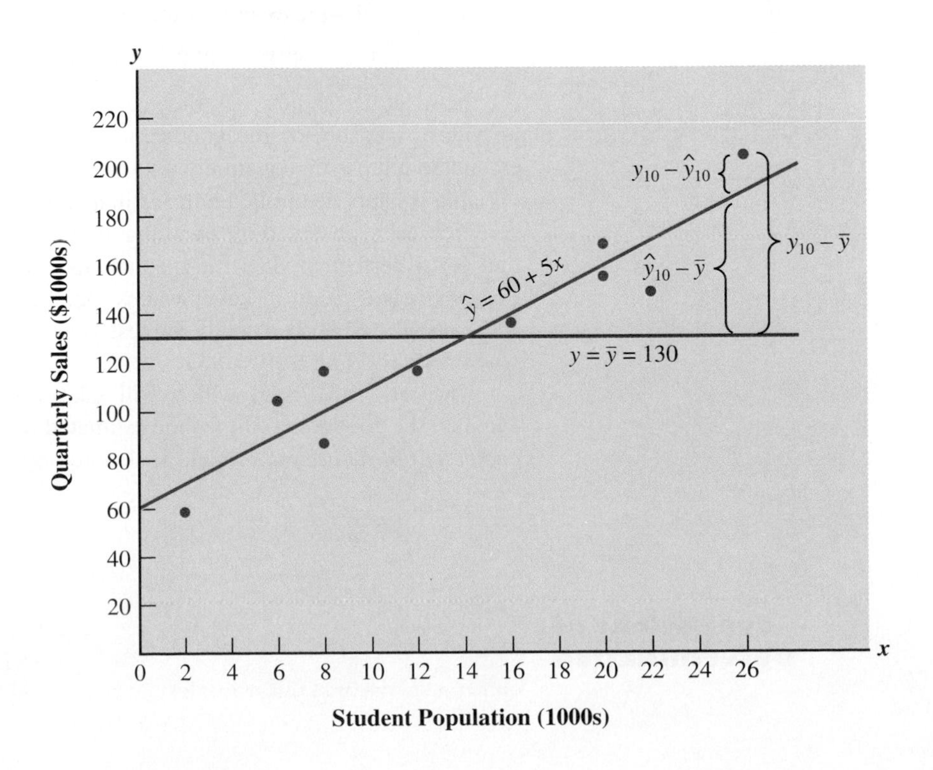

Sum of Squares Due to Regression

$$SSR = \Sigma(\hat{y}_i - \bar{y})^2 \qquad (12.10)$$

From the preceding discussion, we should expect that SST, SSR, and SSE are related. Indeed, the relationship among these three sums of squares provides one of the most important results in statistics.

Relationship Among SST, SSR, and SSE

$$SST = SSR + SSE \qquad (12.11)$$

where

SST = total sum of squares

SSR = sum of squares due to regression

SSE = sum of squares due to error

Equation (12.11) shows that the total sum of squares can be partitioned into two components, the regression sum of squares and the sum of squares due to error. Hence, if the values of any two of these sum of squares are known, the third sum of squares can be computed easily. For instance, in the Armand's Pizza Parlors example, we already know that SSE = 1530 and SST = 15,730; therefore, solving for SSR in (12.11), we find that the sum of squares due to regression is

$$SSR = SST - SSE = 15{,}730 - 1530 = 14{,}200$$

Now let us see how the three sums of squares, SST, SSR, and SSE, can be used to provide a measure of the goodness of fit for the estimated regression equation. The estimated regression equation would provide a perfect fit if every value of the dependent variable y_i happened to lie on the estimated regression line. In this case, $y_i - \hat{y}_i$ would be zero for each observation, resulting in SSE = 0. Since SST = SSR + SSE, we see that for a perfect fit SSR must equal SST, and the ratio (SSR/SST) must equal one. Poorer fits will result in larger values for SSE. Solving for SSE in (12.11), we see that SSE = SST − SSR. Hence, the largest value for SSE (and hence the poorest fit) occurs when SSR = 0 and SSE = SST.

The ratio SSR/SST, which will take values between zero and one, is used to evaluate the goodness of fit for the estimated regression equation. This ratio is called the *coefficient of determination* and is denoted by r^2.

Coefficient of Determination

$$r^2 = \frac{SSR}{SST} \qquad (12.12)$$

For the Armand's Pizza Parlors example, the value of the coefficient of determination is

$$r^2 = \frac{\text{SSR}}{\text{SST}} = \frac{14{,}200}{15{,}730} = .9027$$

When we express the coefficient of determination as a percentage, r^2 can be interpreted as the percentage of the total sum of squares that can be explained by using the estimated regression equation. For Armand's Pizza Parlors, we can conclude that 90.27% of the total sum of squares can be explained by using the estimated regression equation $\hat{y} = 60 + 5x$ to predict sales. In other words, 90.27% of the variation in sales can be explained by the linear relationship between the size of the student population and sales. We should be pleased to find such a good fit for the estimated regression equation.

Computational Efficiencies

In modern applications of regression analysis, a computer software package is almost always used to perform the calculations required to determine the estimated regression equation and the value of the coefficient of determination. However, when solving a small problem with a calculator, we can realize computational efficiencies by using alternative formulas for SST and SSR. We illustrate with the Armand's Pizza Parlors example.

The total sum of squares can be computed by the following alternate formula.

Computational Formula for SST

$$\text{SST} = \Sigma y_i^2 - (\Sigma y_i)^2/n \tag{12.13}$$

TABLE 12.5 Computing SST and SSR for Armand's Pizza Parlors Using the Computational Formulas

Restaurant i		x_i	y_i	$x_i y_i$	x_i^2	y_i^2
1		2	58	116	4	3,364
2		6	105	630	36	11,025
3		8	88	704	64	7,744
4		8	118	944	64	13,924
5		12	117	1,404	144	13,689
6		16	137	2,192	256	18,769
7		20	157	3,140	400	24,649
8		20	169	3,380	400	28,561
9		22	149	3,278	484	22,201
10		26	202	5,252	676	40,804
	Totals	140	1300	21,040	2528	184,730
		Σx_i	Σy_i	$\Sigma x_i y_i$	Σx_i^2	Σy_i^2

Using the information in Table 12.5 and (12.13), we obtain SST = 184,730 – $(1300)^2/10$ = 15,730 (the same value as shown previously). Next, the sum of squares due to regression, SSR, can be calculated directly by the following alternate formula.

Computational Formula for SSR

$$\text{SSR} = \frac{[\Sigma x_i y_i - (\Sigma x_i \Sigma y_i)/n]^2}{\Sigma x_i^2 - (\Sigma x_i)^2/n} \tag{12.14}$$

Using the data in Table 12.5, we have

$$\text{SSR} = \frac{[21{,}040 - (140)(1300)/10]^2}{2528 - (140)^2/10} = \frac{8{,}065{,}600}{568} = 14{,}200$$

With SST and SSR known, we can compute the coefficient of determination by using (12.12).

$$r^2 = \frac{\text{SSR}}{\text{SST}} = \frac{14{,}200}{15{,}730} = .9027$$

Finally, if SSE is also desired, we can use the relationship among SST, SSR, and SSE as follows.

$$\text{SSE} = \text{SST} - \text{SSR} = 15{,}730 - 14{,}200 = 1530$$

The Correlation Coefficient

In Chapter 3 we introduced the *correlation coefficient* as a descriptive measure of the strength of linear association between two variables, x and y. Values of the correlation coefficient are always between –1 and +1. A value of +1 indicates that the two variables x and y are perfectly related in a positive linear sense. That is, all data points are on a straight line that has a positive slope. A value of –1 indicates that x and y are perfectly related in a negative linear sense, with all data points on a straight line that has a negative slope. Values of the correlation coefficient close to zero indicate that x and y are not linearly related.

In Section 3.5 we presented the equation for computing the sample correlation coefficient. If a regression analysis has already been performed and the coefficient of determination r^2 has been computed, the sample correlation coefficient can be computed as follows.

Sample Correlation Coefficient

$$\begin{aligned} r_{xy} &= (\text{the sign of } b_1)\sqrt{\text{Coefficient of Determination}} \\ &= \pm\sqrt{r^2} \end{aligned} \tag{12.15}$$

where

b_1 = the slope of the estimated regression equation $\hat{y} = b_0 + b_1 x$

That is, the sample correlation coefficient is plus or minus the square root of the coefficient of determination. The sign for the sample correlation coefficient is positive if the estimated regression equation has a positive slope ($b_1 > 0$) and negative if the estimated regression equation has a negative slope ($b_1 < 0$).

For the Armand's Pizza Parlor example, the value of the coefficient of determination corresponding to the estimated regression equation $\hat{y} = 60 + 5x$ is .9027. Since the slope of the estimated regression equation is positive, (12.15) shows that the sample correlation coefficient is $+\sqrt{.9027} = +.9501$. With a sample correlation coefficient of $r_{xy} = +.9501$, we would conclude that there is a strong positive linear association between x and y.

In the case of a linear relationship between two variables, both the coefficient of determination and the sample correlation coefficient provide measures of the strength of the relationship. The coefficient of determination provides a measure between zero and one whereas the sample correlation coefficient provides a measure between −1 and +1. Although the sample correlation coefficient is restricted to a linear relationship between two variables, the coefficient of determination can be used for nonlinear relationships and for relationships that have two or more independent variables. In that sense, the coefficient of determination has a wider range of applicability.

notes and comments

1. In developing the least squares estimated regression equation and computing the coefficient of determination, we made no probabilistic assumptions and no statistical tests for significance of the relationship between x and y. Larger values of r^2 simply imply that the least squares line provides a better fit to the data; that is, the observations are more closely grouped about the least squares line. But, using only r^2, we can draw no conclusion about whether the relationship between x and y is statistically significant. Such a conclusion must be based on considerations that involve the sample size and the properties of the appropriate sampling distributions of the least squares estimators.

2. As a practical matter, for typical data found in the social sciences, values of r^2 as low as .25 are often considered useful. For data in the physical and life sciences, r^2 values of .60 or greater are often found; in fact, in some cases, r^2 values greater than .90 can be found. In business applications, r^2 values vary greatly, depending on the unique characteristics of each application.

exercises

Methods

13. The data from Exercise 1 follow.

x_i	1	2	3	4	5
y_i	3	7	5	11	14

The estimated regression equation for these data is $\hat{y} = .20 + 2.60x$.

a. Compute SSE, SST, and SSR using (12.8), (12.9), and (12.11).

b. Compute the coefficient of determination r^2. Comment on the goodness of fit.

c. Recompute SST and SSR using (12.13) and (12.14). Do you get the same results as in (a)?

d. Compute the sample correlation coefficient.

14. The data from Exercise 2 follow.

x_i	2	3	5	1	8
y_i	25	25	20	30	16

The estimated regression equation for these data is $\hat{y} = 30.33 - 1.88x$.

a. Compute SSE, SST, and SSR.

b. Compute the coefficient of determination r^2. Comment on the goodness of fit.

c. Compute the sample correlation coefficient.

15. The data from Exercise 3 follow.

x_i	2	4	5	7	8
y_i	2	3	2	6	4

The estimated regression equation for these data is $\hat{y} = .75 + .51x$. What percentage of the total sum of squares can be accounted for by the estimated regression equation? What is the value of the sample correlation coefficient?

Applications

Self Test

16. In Exercise 4, data were collected on the monthly salaries y and the grade point averages x for students who had obtained a bachelor's degree in business administration. The data from Exercise 4 are shown here. The estimated regression equation for these data is $\hat{y} = -109.46 + 581.08x$.

GPA	2.6	3.4	3.6	3.2	3.5	2.9
Monthly Salary ($)	1800	2100	2500	2000	2400	2100

a. Compute SST, SSR, and SSE.

b. Compute the coefficient of determination r^2. Comment on the goodness of fit.

c. What is the value of the sample correlation coefficient?

17. The data from Exercise 6 follow.

Hotel Location	Business Rate	14-Day Advance Rate
Birmingham	$ 89	$ 81
Miami	130	115
Atlanta	98	89
Chicago	149	138
New Orleans	199	149
Nashville	114	94

The estimated regression equation for these data is $\hat{y} = 25.21 + .6608x$. What percentage of the total sum of squares can be accounted for by the estimated regression equation? Comment on the goodness of fit. What is the sample correlation coefficient?

18. A medical laboratory at Duke University estimates the amount of protein in liver samples through the use of a regression model. A spectrometer emitting light shines through a substance containing the sample, and the amount of light absorbed is used to estimate the amount of protein in the sample. A new estimated regression equation is developed daily because of differing amounts of dye in the solution. On one day, six samples with known protein concentrations gave the following absorbence readings.

Absorbence Reading (x_i)	.509	.756	1.020	1.400	1.570	1.790
Milligrams of Protein (y_i)	0	20	40	80	100	127

a. Use these data to develop an estimated regression equation relating the light absorbence reading to milligrams of protein present in the sample.
b. Compute r^2. Would you feel comfortable using this regression model to estimate the amount of protein in a sample?
c. In a sample just received, the light absorbence reading was .941. Estimate the amount of protein in the sample.

19. An important application of regression analysis in accounting is in the estimation of cost. By collecting data on volume and cost and using the least squares method to develop an estimated regression equation relating volume and cost, an accountant can estimate the cost associated with a particular manufacturing operation (*Managerial Accounting,* D. Ricketts, 1991). Consider the following sample of production volumes and total cost data for a manufacturing operation.

Production Volume (Units)	Total Cost ($)
400	4000
450	5000
550	5400
600	5900
700	6400
750	7000

a. Use these data to develop an estimated regression equation that could be used to predict the total cost for a given production volume.
b. What is the variable, or additional, cost per unit produced?
c. Compute the coefficient of determination. What percentage of the variation in total cost can be explained by volume?
d. The company's production schedule shows 500 units must be produced next month. What is the estimated total cost for this operation?

20. Are company presidents and chief executive officers paid according to the profit performance of the company? The following table lists corporate data on percentage change in return on equity over a two-year period and percentage change in the pay of presidents and chief executive officers immediately after the two-year period (*Business Week,* April 25, 1994).

Company	Two-Year Change in Return on Equity (%)	Change in Executive Compensation (%)
Dupont	−64.4	0
Monsanto	29.6	55
Morton International	−9.3	17
Union Carbide	−16.4	58
W. R. Grace	3.4	−11
International Flavors	18.8	24
Great Lakes Chemical	17.0	10

a. Develop the estimated regression equation with the two-year percentage change in return on equity as the independent variable.
b. Compute r^2. Would you feel comfortable using the percentage change in return on equity over a two-year period to predict the percentage change in the pay of presidents and chief executive officers? Discuss.
c. What is the sample correlation coefficient? Does it reflect a strong or weak relationship between return on equity and executive compensation?

12.4 Model Assumptions

In conducting a regression analysis, we begin by making an assumption about the appropriate model for the relationship between the dependent and independent variable(s). For the case of simple linear regression, the assumed regression model is

$$y = \beta_0 + \beta_1 x + \epsilon$$

An important step in determining whether the assumed model is appropriate involves testing for the significance of the relationship. The tests of significance in regression analysis are based on the following assumptions about the error term ϵ.

Assumptions About the Error Term ϵ in the Regression Model

1. The error term ϵ is a random variable with a mean or expected value of zero; that is, $E(\epsilon) = 0$.
Implication: Since β_0 and β_1 are constants, $E(\beta_0) = \beta_0$ and $E(\beta_1) = \beta_1$; thus, for a given value of x, the expected value of y is

$$E(y) = \beta_0 + \beta_1 x. \tag{12.16}$$

As we indicated previously, equation (12.16) is referred to as the regression equation.

2. The variance of ϵ, denoted by σ^2, is the same for all values of x.
Implication: The variance of y equals σ^2 and is the same for all values of x.

3. The values of ϵ are independent.
Implication: The value of ϵ for a particular value of x is not related to the value of ϵ for any other value of x; thus, the value of y for a particular value of x is not related to the value of y for any other value of x.

4. The error term ϵ is a normally distributed random variable.
Implication: Since y is a linear function of ϵ, y is also a normally distributed random variable.

FIGURE 12.6
Assumptions for the Regression Model

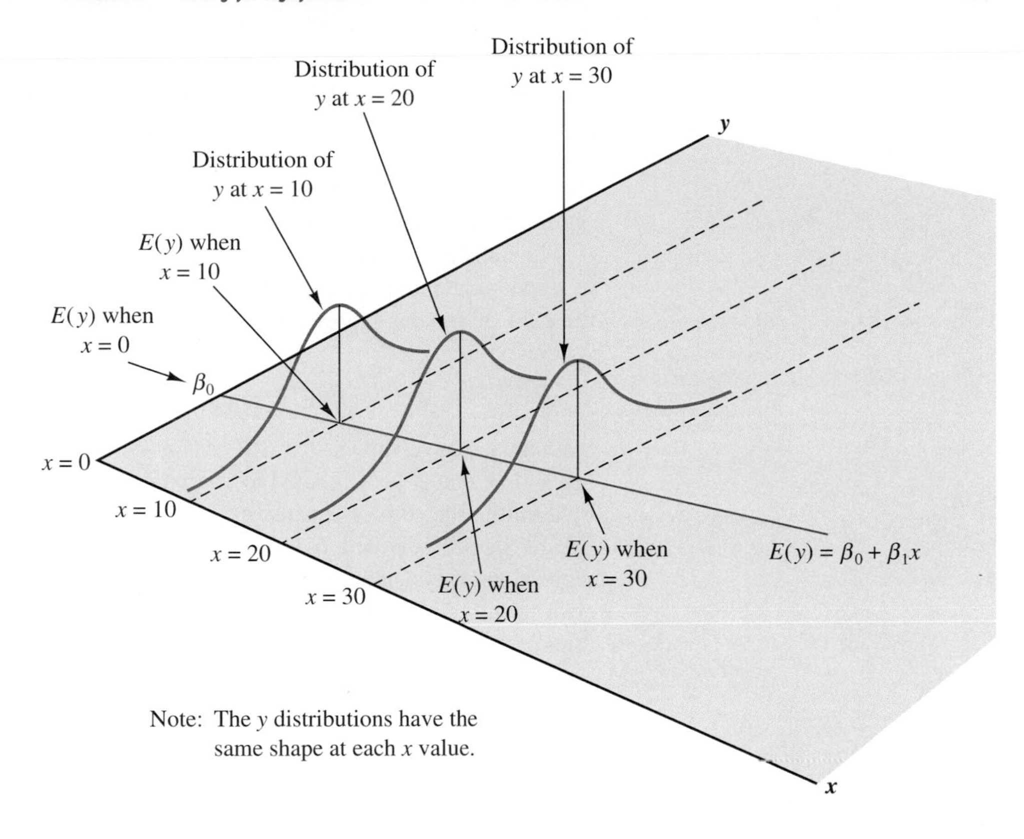

Figure 12.6 is an illustration of the model assumptions and their implications; note that in this graphical interpretation, the value of $E(y)$ changes according to the specific value of x considered. However, regardless of the x value, the probability distribution of ϵ and hence the probability distributions of y are normally distributed, each with the same variance. The specific value of the error ϵ at any particular point depends on whether the actual value of y is greater than or less than $E(y)$.

At this point, we must keep in mind that we are also making an assumption or hypothesis about the form of the relationship between x and y. That is, we have assumed that a straight line represented by $\beta_0 + \beta_1 x$ is the basis for the relationship between the variables. We must not lose sight of the fact that some other model, for instance $y = \beta_0 + \beta_1 x^2 + \epsilon$, may turn out to be a better model for the underlying relationship.

12.5 Testing for Significance

The simple linear regression equation shows that the mean or expected value of y is a linear function of x: $E(y) = \beta_0 + \beta_1 x$. If the value of β_1 is zero, $E(y) = \beta_0 + (0)x = \beta_0$. In this case, the mean value of y does not depend on the value of x and hence we would conclude that x and y are not linearly related. Alternatively, if the value of β_1 is not equal to zero, we would conclude that the two variables are related. Thus, to test for a significant regression relationship, we must conduct a hypothesis test to determine whether the value of β_1 is zero. Two tests are commonly used. Both require an estimate of σ^2, the variance of ϵ in the regression model.

An Estimate of σ^2

From the regression model and its assumptions we can conclude that σ^2, the variance of ϵ, also represents the variance of the y values about the regression line. Recall that the deviations of the y values about the estimated regression line are called residuals. Thus, SSE, the sum of squared residuals, is a measure of the variability of the actual observations about the estimated regression line.

In statistics, every sum of squares has associated with it a number called its degrees of freedom and a number called its mean square. The degrees of freedom depends upon the sum of squares being considered. For instance, with $\hat{y}_i = b_0 + b_1 x_i$, SSE can be written as

$$\text{SSE} = \Sigma(y_i - \hat{y}_i)^2 = \Sigma(y_i - b_0 - b_1 x_i)^2$$

Statisticians have shown that SSE has $n - 2$ degrees of freedom since two parameters (β_0 and β_1) must be estimated to compute SSE. The value of mean square is computed by dividing the sum of squares by its degrees of freedom. Thus, the mean square error (MSE) is SSE divided by $n - 2$. MSE is one of the most important statistics in regression analysis because it provides an unbiased estimator of σ^2, the variance of the error term ϵ. Since the value of MSE provides an estimate of σ^2, the notation s^2 is also used.

Mean Square Error (Estimate of σ^2)

$$s^2 = \text{MSE} = \frac{\text{SSE}}{n - 2} \tag{12.17}$$

In Section 12.3 we showed that for the Armand's Pizza Parlors example, SSE = 1530; hence,

$$s^2 = \text{MSE} = \frac{1530}{8} = 191.25$$

provides an unbiased estimate of σ^2.

To estimate σ we take the square root of s^2. The resulting value, s, is referred to as the *standard error of the estimate.*

Standard Error of the Estimate

$$s = \sqrt{\text{MSE}} = \sqrt{\frac{\text{SSE}}{n - 2}} \tag{12.18}$$

For the Armand's Pizza Parlors example, $s = \sqrt{\text{MSE}} = \sqrt{191.25} = 13.829$. In the following discussion, we use the standard error of the estimate in the tests for a significant relationship between x and y.

t Test

The simple linear regression model is $y = \beta_0 + \beta_1 x + \epsilon$. If x and y are linearly related, we must have $\beta_1 \neq 0$. The purpose of the t test is to see whether we can conclude that $\beta_1 \neq 0$. We will use the sample data to test the following hypotheses about the parameter β_1.

$$H_0\colon \beta_1 = 0$$
$$H_a\colon \beta_1 \neq 0$$

If H_0 is rejected, we will conclude that $\beta_1 \neq 0$ and that there is a statistically significant relationship between the two variables. However, if H_0 cannot be rejected, we will have insufficient evidence to conclude that a significant relationship exists. As usual, the properties of the sampling distribution of b_1, the least squares estimator of β_1, provide the basis for the hypothesis test.

First, let us consider what would have happened if we had used a different random sample for the same regression study. For example, suppose that Armand's Pizza Parlors had used the sales records of a different sample of 10 restaurants. A regression analysis of this new sample might result in an estimated regression equation similar to our previous estimated regression equation $\hat{y} = 60 + 5x$. However, it is doubtful that we would obtain exactly the same equation (with an intercept of exactly 60 and a slope of exactly 5). Indeed, b_0 and b_1, the least squares estimators, are sample statistics that have their own sampling distributions. The properties of the sampling distribution of b_1 follow.

Sampling Distribution of b_1

Expected Value

$$E(b_1) = \beta_1$$

Standard Deviation

$$\sigma_{b_1} = \frac{\sigma}{\sqrt{\Sigma x_i^2 - (\Sigma x_i)^2/n}} \tag{12.19}$$

Distribution Form

Normal

Note that the expected value of b_1 is equal to β_1, so b_1 is an unbiased estimator of β_1.

Since we do not know the value of σ, we develop an estimate of σ_{b_1}, denoted s_{b_1}, by estimating σ with s in (12.19). Thus, we obtain the following estimate of σ_{b_1}.

Estimated Standard Deviation of b_1

$$s_{b_1} = \frac{s}{\sqrt{\Sigma x_i^2 - (\Sigma x_i)^2/n}} \tag{12.20}$$

For Armand's Pizza Parlors, $s = 13.829$. Hence, using $\Sigma x_i^2 = 2528$ and $\Sigma x_i = 140$ as shown in Table 12.5, we have

$$s_{b_1} = \frac{13.829}{\sqrt{2528 - (140)^2/10}} = .5803$$

as the estimated standard deviation of b_1.

The t test for a significant relationship is based on the fact that the test statistic

$$\frac{b_1 - \beta_1}{s_{b_1}}$$

follows a t distribution with $n - 2$ degrees of freedom. If the null hypothesis is true, then $\beta_1 = 0$ and $t = b_1/s_{b_1}$. With b_1/s_{b_1} as the test statistic, the steps of the t test for a significant relationship are as follows.

t Test for Significance in Regression

$$H_0\colon \beta_1 = 0$$

$$H_a\colon \beta_1 \neq 0$$

Test Statistic

$$t = \frac{b_1}{s_{b_1}} \tag{12.21}$$

Rejection Rule

Reject H_0 if $t < -t_{\alpha/2}$ or if $t > t_{\alpha/2}$

where $t_{\alpha/2}$ is based on a t distribution with $n - 2$ degrees of freedom.

Let us conduct this test of significance for Armand's Pizza Parlors. The test statistic (12.21) is

$$t = \frac{b_1}{s_{b_1}} = \frac{5}{.5803} = 8.62$$

From Table 2 of Appendix B we find that the two-tailed t value corresponding to $\alpha = .01$ and $n - 2 = 10 - 2 = 8$ degrees of freedom is $t_{.005} = 3.355$. With $8.62 > 3.355$, we reject H_0 and conclude at the .01 level of significance that β_1 is not equal to zero. The statistical evidence is sufficient to conclude that we have a significant relationship between student population and sales.

F Test

An F test, based on the F probability distribution, can also be used to test for significance in regression. With only one independent variable, the F test will provide the same conclusion as the t test; that is, if the t test indicates $\beta_1 \neq 0$ and hence a significant relationship, the F test will also indicate a significant relationship. But with more than one independent variable, only the F test can be used to test for an overall significant relationship.

The logic behind the use of the F test for determining whether the regression relationship is statistically significant is based on the development of two independent estimates of σ^2. We have just seen that MSE provides an estimate of σ^2. If the null hypothesis H_0: $\beta_1 = 0$ is true, the sum of squares due to regression, SSR, divided by its degrees of freedom provides another independent estimate of σ^2. This estimate is called the *mean square due to regression,* or simply the *mean square regression,* and is denoted MSR. In general,

$$\text{MSR} = \frac{\text{SSR}}{\text{regression degrees of freedom}}$$

For the models we consider in this text, the regression degrees of freedom is always equal to the number of independent variables; thus,

$$\text{MSR} = \frac{\text{SSR}}{\text{number of independent variables}} \tag{12.22}$$

Since there is only one independent variable in the Armand's Pizza Parlors example, we have MSR = SSR/1 = SSR. Hence, for Armand's Pizza Parlors, MSR = SSR = 14,200.

If the null hypothesis is true ($\beta_1 = 0$), MSR and MSE are two independent estimates of σ^2 and the sampling distribution of MSR/MSE follows an F distribution with numerator degrees of freedom equal to one and denominator degrees of freedom equal to $n - 2$. Therefore, when $\beta_1 = 0$, the value of MSR/MSE should be close to one. However, if the null hypothesis is false ($\beta_1 \neq 0$), MSR will overestimate σ^2 and the value of MSR/MSE will be inflated; thus, large values of MSR/MSE lead to the rejection of H_0 and the conclusion that the relationship between x and y is statistically significant. A summary of how the F test is used to test for a significant relationship follows.

F Test for Significance in Simple Linear Regression

$$H_0: \beta_1 = 0$$

$$H_a: \beta_1 \neq 0$$

Test Statistic

$$F = \frac{\text{MSR}}{\text{MSE}} \tag{12.23}$$

Rejection Rule

$$\text{Reject } H_0 \text{ if } F > F_\alpha$$

where F_α is based on an F distribution with 1 degree of freedom in the numerator and $n - 2$ degrees of freedom in the denominator.

Let us conduct the F test for the Armand's Pizza Parlors example. The test statistic is

$$F = \frac{\text{MSR}}{\text{MSE}} = \frac{14{,}200}{191.25} = 74.25$$

From Table 4 of Appendix B we find that the F value corresponding to $\alpha = .01$ with one degree of freedom in the numerator and $n - 2 = 10 - 2 = 8$ degrees of freedom in the denominator is $F_{.01} = 11.26$. With $74.25 > 11.26$, we reject H_0 and conclude at the .01 level of significance that β_1 is not equal to zero. The F test has provided the statistical evidence necessary to conclude that we have a significant relationship between student population and sales.

In Chapter 10 we covered analysis of variance (ANOVA) and showed how an ANOVA table could be used to provide a convenient summary of the computational aspects of analysis of variance. A similar ANOVA table can be used to summarize the results of the F test for significance in regression. Table 12.6 is the general form of the ANOVA table for simple linear regression. Table 12.7 is the ANOVA table with the F test computations we have just performed for Armand's Pizza Parlors. Regression, error, and total are listed as the three sources of variation, with SSR, SSE, and SST appearing as the corresponding sum of squares in column two. The degrees of freedom, 1 for regression, $n - 2$ for error, and $n - 1$ for total are shown in column three. Column 4 contains the values of MSR and MSE and column 5 contains the value of F = MSR/MSE. Almost all computer printouts of regression analysis include an ANOVA table summary of the test for significance.

Some Cautions Regarding the Interpretation of Significance Tests

Rejecting the null hypothesis H_0: $\beta_1 = 0$ and concluding that the relationship between x and y is significant does not enable us to conclude that a *cause-and-effect* relationship is present between x and y. Concluding a cause-and-effect relationship is warranted only if the analyst has some type of theoretical justification that the relationship is in fact causal. In the Armand's Pizza Parlors example, we can conclude that there is a significant relationship between the size of the student population x and sales y; moreover, the estimated regression equation $\hat{y} = 60 + 5x$ provides the least squares estimate of the relationship. We cannot, however, conclude that changes in student population x *cause* changes in sales y just because we have identified a statistically significant relationship. The appropriateness of such a cause-and-effect conclusion is left to supporting theoretical justification and to good judgment on the part of the

TABLE 12.6 General Form of the ANOVA Table for Simple Linear Regression

Source of Variation	Sum of Squares	Degrees of Freedom	Mean Square	F
Regression	SSR	1	$\text{MSR} = \frac{\text{SSR}}{1}$	$F = \frac{\text{MSR}}{\text{MSE}}$
Error	SSE	$n - 2$	$\text{MSE} = \frac{\text{SSE}}{n-2}$	
Total	SST	$n - 1$		

TABLE 12.7 ANOVA Table for the Armand's Pizza Parlors Problem

Source of Variation	Sum of Squares	Degrees of Freedom	Mean Square	F
Regression	14,200	1	$\frac{14,200}{1} = 14,200$	$\frac{14,200}{191.25} = 74.25$
Error	1,530	8	$\frac{1530}{8} = 191.25$	
Total	15,730	9		

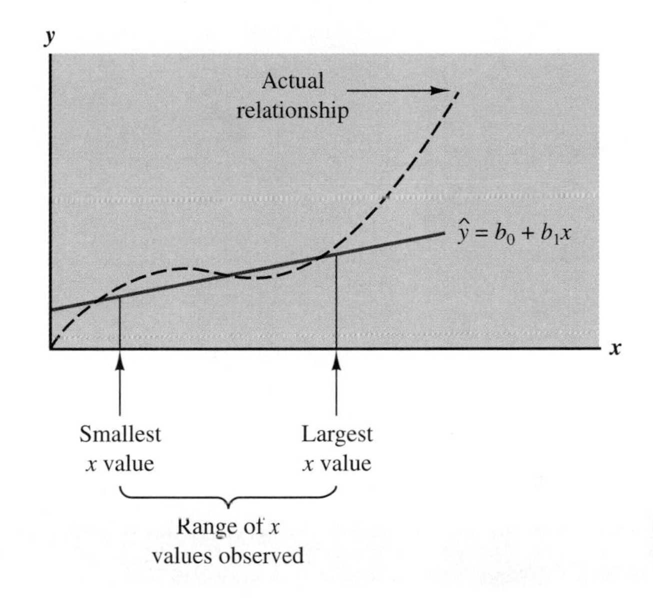

FIGURE 12.7
Example of a Linear Approximation of a Nonlinear Relationship

analyst. Armand's managers felt that increases in the student population were a likely cause of increased sales. Thus, the result of the significance test enabled them to conclude that a cause-and-effect relationship was present.

In addition, just because we are able to reject H_0: $\beta_1 = 0$ and demonstrate statistical significance does not enable us to conclude that the relationship between x and y is linear. We can state only that x and y are related and that a linear relationship explains a significant portion of the variability in y over the range of values for x observed in the sample. Figure 12.7 illustrates this situation. The test for significance has rejected the null hypothesis H_0: $\beta_1 = 0$ and has led to the conclusion that x and y are significantly related, but the figure shows that the actual relationship between x and y is not linear. Although the linear approximation provided by $\hat{y} = b_0 + b_1x$ is very good over the range of x values observed in the sample, it becomes very poor for x values outside that range.

Given a significant relationship, we should feel confident in using the estimated regression equation for predictions corresponding to x values within the range of the x values observed in the sample. For Armand's Pizza Parlors, since the regression relationship has been found significant at the .01 level, we should feel confident using it to predict sales for restaurants where the associated student population is between 2000 and 26,000. But unless there are reasons to believe the model is valid beyond this range, predictions outside the range of the independent variable should bemade with caution.

notes and comments

1. The assumptions made about the error term (Section 12.4) are what allow the tests of statistical significance in this section. The properties of the sampling distribution of b_1 and the subsequent t and F tests follow directly from these assumptions.

2. Do not confuse statistical significance with practical significance. With very large sample sizes, statistically significant results can be obtained for small values of b_1; in such cases, one must exercise care in concluding that the relationship has practical significance.

3. A test of significance for a linear relationship between x and y can also be performed by using the sample correlation coefficient r_{xy}. With ρ_{xy} denoting the population correlation coefficient, the hypotheses are as follows.

$$H_0: \rho_{xy} = 0$$

$$H_a: \rho_{xy} \neq 0$$

A significant relationship can be concluded if H_0 is rejected. However, the t and F tests presented previously in this section give the *same result* as the test for significance using the correlation coefficient. Conducting a test for significance using the correlation coefficient therefore is not necessary if a t or F test has already been conducted.

exercises

Methods

21. The data from Exercise 1 follow.

x_i	1	2	3	4	5
y_i	3	7	5	11	14

a. Compute the mean square error using (12.17).
b. Compute the standard error of the estimate using (12.18).
c. Compute the estimated standard deviation of b_1 using (12.20).
d. Use the t test to test the following hypotheses ($\alpha = .05$):

$$H_0: \beta_1 = 0$$

$$H_a: \beta_1 \neq 0$$

e. Use the F test to test the hypotheses in (d) at a .05 level of significance. Present the results in the analysis of variance table format (see Table 12.6).

22. The data from Exercise 2 follow.

x_i	2	3	5	1	8
y_i	25	25	20	30	16

a. Compute the mean square error using (12.17).
b. Compute the standard error of the estimate using (12.18).
c. Compute the estimated standard deviation of b_1 using (12.20).
d. Use the t test to test the following hypotheses ($\alpha = .05$):

$$H_0: \beta_1 = 0$$
$$H_a: \beta_1 \neq 0$$

e. Use the F test to test the hypotheses in (d) at a .05 level of significance. Present the results in the analysis of variance table format.

23. The data from Exercise 3 follow.

x_i	2	4	5	7	8
y_i	2	3	2	6	4

a. What is the value of the standard error of the estimate?
b. Test for a significant relationship by using the t test. Use $\alpha = .05$.
c. Use the F test to test for a significant relationship. Use $\alpha = .05$. What is your conclusion?

24. In Exercise 4 the data on grade point average and monthly salary were as shown in the following table.

GPA	2.6	3.4	3.6	3.2	3.5	2.9
Monthly Salary ($)	1800	2100	2500	2000	2400	2100

a. Does the t test indicate a significant relationship between grade point average and monthly salary? What is your conclusion? Use $\alpha = .05$.
b. Test for a significant relationship using the F test. What is your conclusion? Use $\alpha = .05$.
c. Show the ANOVA table.

25. Refer to Exercise 8, where an estimated regression equation relating years of experience and annual sales was developed. At a .05 level of significance, determine whether years of experience and annual sales are related.

26. Refer to Exercise 9, where an estimated regression equation relating tire rating to load carrying capacity was developed (*Road and Track,* October 1994). At a .01 level of significance, test whether these two variables are related. Show the ANOVA table. What is your conclusion?

27. Refer to Exercise 19, where data on production volume and cost were used to develop an estimated regression equation relating production volume and cost for a particular manufacturing operation. Using $\alpha = .05$, test whether the production volume is significantly related to the total cost. Show the ANOVA table. What is your conclusion?

28. Refer to Exercise 20, where the following data were used to determine whether company presidents and chief executive officers are paid on the basis of company profit performance (*Business Week,* April 25, 1994).

Company	Two-Year Change in Return on Equity (%)	Change in Executive Compensation (%)
Dupont	−64.4	0
Monsanto	29.6	55
Morton International	−9.3	17
Union Carbide	−16.4	58
W. R. Grace	3.4	−11
International Flavors	18.8	24
Great Lakes Chemical	17.0	10

Is there evidence of a significant relationship between the two variables? Conduct the appropriate statistical test and state your conclusion. Use $\alpha = .05$.

12.6 Using the Estimated Regression Equation for Estimation and Prediction

The simple linear regression model is an assumption about the relationship between x and y. Using the least squares method, we obtained the estimated simple linear regression equation. If the results show a statistically significant relationship between x and y, and the fit provided by the estimated regression equation appears to be good, the estimated regression equation should be useful for estimation and prediction.

Point Estimation

In the Armand's Pizza Parlors example, the estimated regression equation $\hat{y} = 60 + 5x$ provides an estimate of the relationship between the size of the student population x and quarterly sales y. We can use the estimated regression equation to develop a point estimate of the mean value of y for a particular value of x or to predict an individual value of y corresponding to a given value of x. For instance, suppose Armand's managers want a point estimate of the mean sales for all restaurants located near college campuses with 10,000 students. Using the estimated regression equation $\hat{y} = 60 + 5x$, we see that for $x = 10$ (10,000 students), $\hat{y} = 60 + 5(10) = 110$. Thus, a point estimate of the mean sales for all restaurants located near campuses with 10,000 students is \$110,000.

Now suppose Armand's managers want to predict sales for an individual restaurant located near Talbot College, a school with 10,000 students. In this case we are not interested in the mean value for all restaurants located near campuses with 10,000 students; we are just interested in predicting sales for one individual restaurant. As it turns out, the point estimate is the same as the point estimate for the mean value of y. Hence, we would predict sales of $\hat{y} = 60 + 5(10) = 110$ or \$110,000 for this one restaurant.

Interval Estimation

Point estimates do not provide any idea of the precision associated with the estimate. For that we must develop interval estimates much like those in Chapters 8, 10, and 11.

The first type of interval estimate, a *confidence interval estimate,* is an interval estimate of the *mean value of* y for a given value of x. The second type of interval estimate, a *prediction interval estimate,* is used whenever we want an interval estimate of an *individual value of* y corresponding to a given value of x. With point estimation we obtain the same value whether we are estimating the mean value of y or predicting an individual value of y, but with interval estimates we obtain different values.

Confidence Interval Estimate of the Mean Value of y

The estimated regression equation provides a point estimate of the mean value of y for a given value of x. In describing the confidence interval estimation procedure, we will use the following notation.

$$x_p = \text{the particular or given value of the independent variable } x$$

$$E(y_p) = \text{the mean or expected value of the dependent variable } y \text{ corresponding to the given } x_p$$

$$\hat{y}_p = b_0 + b_1 x_p = \text{the estimate of } E(y_p) \text{ when } x = x_p$$

Using this notation to estimate the mean sales for all Armand's restaurants located near a campus with 10,000 students, we have $x_p = 10$, and $E(y_p)$ denotes the unknown mean value of sales for all restaurants where $x_p = 10$. The point estimate of $E(y_p)$ is provided by $\hat{y}_p = 60 + 5(10) = 110$.

In general, we cannot expect $\hat{y}_p$ to equal $E(y_p)$ exactly. If we want to make an inference about how close $\hat{y}_p$ is to the true mean value $E(y_p)$, we will have to consider the variance of the estimates based on the estimated regression equation. Statisticians have developed a formula for estimating the variance of $\hat{y}_p$ given x_p . This estimate, denoted by $s^2_{\hat{y}_p}$, is

$$s^2_{\hat{y}_p} = s^2\left[\frac{1}{n} + \frac{(x_p - \bar{x})^2}{\Sigma x_i^2 - (\Sigma x_i)^2/n}\right] \tag{12.24}$$

The estimate of the standard deviation of $\hat{y}_p$ is given by the square root of (12.24).

$$s_{\hat{y}_p} = s\sqrt{\frac{1}{n} + \frac{(x_p - \bar{x})^2}{\Sigma x_i^2 - (\Sigma x_i)^2/n}} \tag{12.25}$$

The computational results for Armand's Pizza Parlors in Section 12.5 provided $s = 13.829$. With $\Sigma x_i^2 = 2528$, $\Sigma x_i = 140$, $\bar{x} = \Sigma x_i/n = 140/10 = 14$, and $x_p = 10$, we can use (12.25) to obtain

$$s_{\hat{y}_p} = 13.829\sqrt{\frac{1}{10} + \frac{(10 - 14)^2}{2528 - (140)^2/10}}$$

$$= 13.829\sqrt{.1282} = 4.95$$

The general expression for a confidence interval estimate of $E(y_p)$ at a given x_p follows.

Confidence Interval Estimate of $E(y_p)$

$$\hat{y}_p \pm t_{\alpha/2} s_{\hat{y}_p} \tag{12.26}$$

where the confidence coefficient is $1 - \alpha$ and $t_{\alpha/2}$ is based on a t distribution with $n - 2$ degrees of freedom.

Using (12.26) to develop a 95% confidence interval estimate of the mean sales for all Armand's restaurants located near campuses with 10,000 students, we need the value of t for $\alpha/2 = .025$ and $n - 2 = 10 - 2 = 8$ degrees of freedom. Using Table 2 of Appendix B, we have $t_{.025} = 2.306$. Thus, with $\hat{y}_p = 110$ and $s_{\hat{y}_p} = 4.95$, we have

$$110 \pm 2.306(4.95)$$

$$110 \pm 11.415$$

In dollars, the 95% confidence interval for the mean sales of all restaurants near campuses with 10,000 students is \$110,000 ± \$11,415. Therefore, the confidence interval estimate for the mean sales when the student population is 10,000 is \$98,585 to \$121,415.

Note that the estimated standard deviation of $\hat{y}_p$ given by (12.25) is smallest when $x_p = \bar{x}$ and the quantity $x_p - \bar{x} = 0$. In this case, the estimated standard deviation of $\hat{y}_p$ becomes

$$s_{\hat{y}_p} = s\sqrt{\frac{1}{n} + \frac{(x_p - \bar{x})^2}{\Sigma x_i^2 - (\Sigma x_i)^2/n}} = s\sqrt{\frac{1}{n}}$$

This result implies that we can make the best or most precise estimate of the mean value of y whenever we are using the mean value of the independent variable; that is, whenever $x_p = \bar{x}$. In fact, the further x_p is from $\bar{x}$, the larger $x_p - \bar{x}$ becomes. As a result, confidence intervals for the mean value of y will become wider as x_p deviates more from $\bar{x}$. This pattern is shown graphically in Figure 12.8.

Prediction Interval Estimate of an Individual Value of *y*

Suppose that instead of estimating the mean value of sales for all Armand's restaurants located near campuses with 10,000 students, we want to estimate the sales for an individual restaurant located near Talbot College, a school with 10,000 students. As noted previously, the point estimate of an individual value of y given $x = x_p$ is provided by the estimated regression equation with $\hat{y}_p = b_0 + b_1 x_p$. For the restaurant at Talbot College, we have $x_p = 10$ and a corresponding estimated sales of $\hat{y}_p = 60 + 5(10) = 110$, or \$110,000. Note that this value is the same as the point estimate of the mean sales for all restaurants located near campuses with 10,000 students.

To develop a prediction interval estimate, we must first determine the variance associated with using $\hat{y}_p$ as an estimate of an individual value of y when $x = x_p$. This variance is made up of the sum of the following two components.

1. The variance of individual y values about the mean $E(y_p)$, an estimate of which is given by s^2.
2. The variance associated with using $\hat{y}_p$ to estimate $E(y_p)$, an estimate of which is given by $s^2_{\hat{y}_p}$.

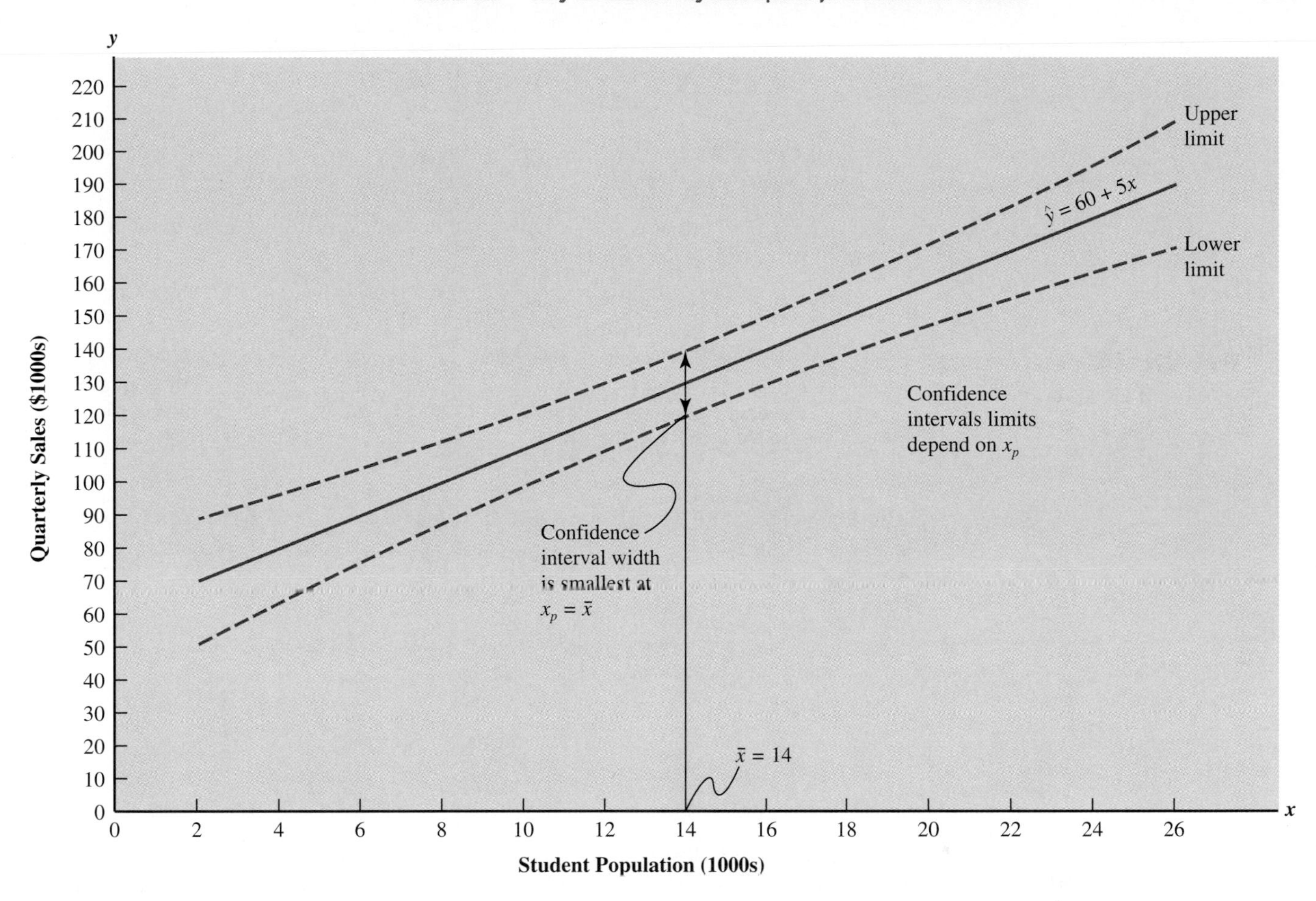

FIGURE 12.8
Confidence Intervals for the Mean Sales *y* at Given Values of Student Population *x*

Statisticians have shown that an estimate of the variance of an individual value of y_p, which we denote s^2_{ind}, is given by

$$s^2_{ind} = s^2 + s^2_{\hat{y}_p}$$

$$= s^2 + s^2\left[\frac{1}{n} + \frac{(x_p - \bar{x})^2}{\Sigma x_i^2 - (\Sigma x_i)^2/n}\right]$$

$$= s^2\left[1 + \frac{1}{n} + \frac{(x_p - \bar{x})^2}{\Sigma x_i^2 - (\Sigma x_i)^2/n}\right] \quad (12.27)$$

Hence, an estimate of the standard deviation of an individual value of y_p is given by

$$s_{ind} = s\sqrt{1 + \frac{1}{n} + \frac{(x_p - \bar{x})^2}{\Sigma x_i^2 - (\Sigma x_i)^2/n}} \quad (12.28)$$

For Armand's Pizza Parlors, the estimated standard deviation corresponding to the prediction of sales for one specific restaurant located near a campus with 10,000 students is computed as follows.

$$s_{\text{ind}} = 13.829 \sqrt{1 + \frac{1}{10} + \frac{(10 - 14)^2}{2528 - (140)^2/10}}$$
$$= 13.829\sqrt{1.1282}$$
$$= 14.69$$

The general expression for a prediction interval estimate for an individual value of y at a given x_p follows.

Prediction Interval Estimate of y_p

$$\hat{y}_p \pm t_{\alpha/2} s_{\text{ind}} \tag{12.29}$$

where the confidence coefficient is $1 - \alpha$ and $t_{\alpha/2}$ is based on a t distribution with $n - 2$ degrees of freedom.

The 95% prediction interval for sales at Armand's Talbot College restaurant can be found by using $t_{.025} = 2.306$ and $s_{\text{ind}} = 14.69$. Expression (12.29) provides

$$110 \pm 2.306(14.69)$$
$$110 \pm 33.875$$

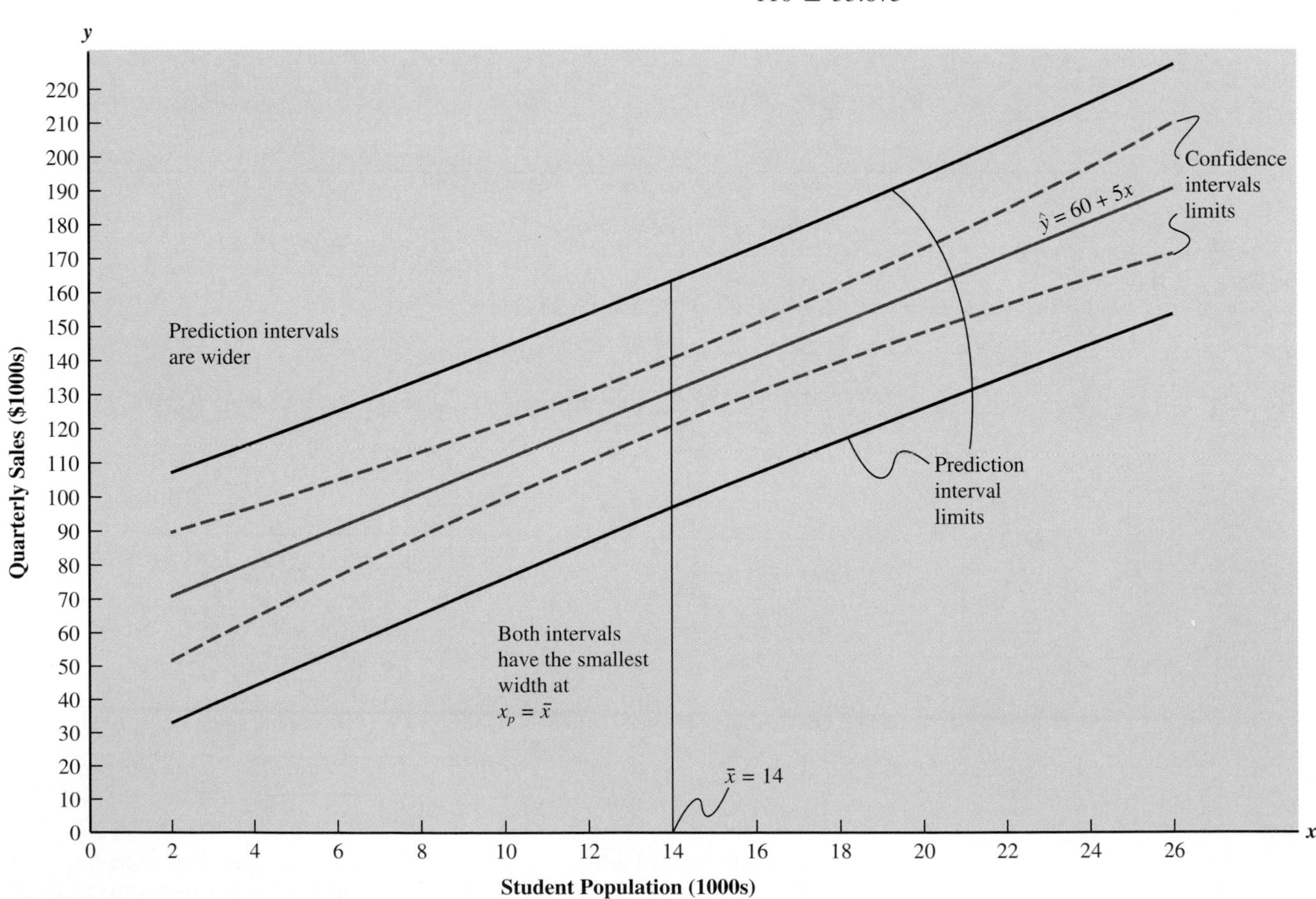

FIGURE 12.9
Confidence and Prediction Intervals for Sales y at Given Values of Student Population x

In dollars, this prediction interval is \$110,000 ± \$33,875 or \$76,125 to \$143,875. Note that this prediction interval for the individual restaurant is wider than the confidence interval for the mean sales of all restaurants located near campuses with 10,000 students (\$98,585 to \$121,415). The difference reflects the fact that we are able to estimate the mean value of y more precisely than we can predict any one particular or individual value of y.

Both confidence interval estimates and prediction interval estimates are most precise when the value of the independent variable is $x_p = \bar{x}$. The general shapes of confidence intervals and the wider prediction intervals are shown together in Figure 12.9.

exercises

Methods

29. The data from Exercise 1 follow.

x_i	1	2	3	4	5
y_i	3	7	5	11	14

a. Use (12.25) to estimate the standard deviation of $\hat{y}_p$ when $x = 4$.
b. Use (12.26) to develop a 95% confidence interval estimate of the expected value of y when $x = 4$.
c. Use (12.28) to estimate the standard deviation of an individual value when $x = 4$.
d. Use (12.29) to develop a 95% prediction interval for $x = 4$.

30. The data from Exercise 2 follow.

x_i	2	3	5	1	8
y_i	25	25	20	30	16

a. Estimate the standard deviation of $\hat{y}_p$ when $x = 3$.
b. Develop a 95% confidence interval estimate of the expected value of y when $x = 3$.
c. Estimate the standard deviation of an individual value when $x = 3$.
d. Develop a 95% prediction interval when $x = 3$.

31. The data from Exercise 3 follow.

x_i	2	4	5	7	8
y_i	2	3	2	6	4

Develop the 95% confidence and prediction intervals when $x = 3$. Explain why these two intervals are different.

Applications

Self Test

32. In Exercise 4, the data on grade point average x and monthly salary y provided the estimated regression equation $\hat{y} = 290.54 + 581.08x$.

a. Develop a 95% confidence interval estimate of the mean starting salary for all students with a 3.0 GPA.

b. Develop a 95% prediction interval estimate of the starting salary for Joe Heller, a student with a GPA of 3.0.

33. In Exercise 9, data on tire ratings x and load-carrying capacities y of automobile tires provided the estimated regression equation $\hat{y} = -2196.89 + 39.42x$ (*Road & Track,* October 1994).

a. Verify that the point estimate of the load-carrying capacity of a tire rated 90 is 1351 pounds.

b. Develop a 95% confidence interval estimate of the mean load-carrying capacity for all tires that have a rating of 90.

c. Develop a 95% prediction interval estimate of the load-carrying capacity for one tire that has a rating of 90.

d. Discuss the differences in your answers to parts (b) and (c).

34. In Exercise 11, the following data on median income x and the median home price y for a sample of six cities were provided (*Who's Buying Homes in America,* Chicago Title and Trust Company, 1994). Data are in thousands of dollars.

City	Median Income	Median Home Price
Atlanta	$65.2	$120.2
Cleveland	49.8	92.7
Denver	53.8	111.7
Dallas	62.7	104.7
Orlando	50.9	98.5
Minneapolis	53.1	105.8

a. The estimated regression equation for these data is $\hat{y} = 43.4 + 1.11x$. Phoenix has a median income of $51,100 or $51.1 thousand dollars. What is the estimate of the median home price in Phoenix?

b. Develop an interval estimate of the median home price in Phoenix. Use $\alpha = .05$. Is this a confidence interval estimate or a prediction interval estimate? Explain.

c. According to *Who's Buying Homes in America* (Chicago Title and Trust Company, 1994), the median home price in Phoenix was $105,700. What was the error involved in using the estimated regression equation to predict the median home price in Phoenix?

35. In Exercise 12, data were given on the adjusted gross income x and the amount of itemized deductions taken by taxpayers (*Money,* October 1994). Data were reported in thousands of dollars. With the estimated regression equation $\hat{y} = 4.68 + .16x$, the point estimate of a reasonable level of total itemized deductions for a taxpayer with an adjusted gross income of $52,500 is $13,080.

a. Develop a 95% confidence interval estimate of the mean amount of total itemized deductions for all taxpayers with an adjusted gross income of $52,500.

b. Develop a 95% prediction interval estimate for the amount of total itemized deductions for a particular taxpayer with an adjusted gross income of $52,500.

c. If the particular taxpayer referred to in part (b) has claimed total itemized deductions of $20,400, would the IRS agent's request for an audit appear to be justified?

d. Using your answer to part (b), give the IRS agent a guideline as to the amount of total itemized deductions a taxpayer with an adjusted gross income of $52,500 should have before an audit is recommended.

36. Refer to Exercise 19, where data on the production volume x and total cost y for a particular manufacturing operation were used to develop the estimated regression equation $\hat{y} = 1246.67 + 7.6x$.

a. The company's production schedule shows that 500 units must be produced next month. What is the point estimate of the total cost for next month?

b. Develop a 99% prediction interval estimate of the total cost for next month.

c. If an accounting cost report at the end of next month shows that the actual production cost during the month was $6000, should managers be concerned about incurring such a high total cost for the month? Discuss.

12.7 Computer Solution

Performing regression analysis computations without the help of a computer can be quite time consuming. In this section we discuss how the computational burden can be minimized by using a computer software package such as Minitab.

We entered Armand's student population data into a Minitab worksheet. The independent variable was named POP and the dependent variable was named SALES to assist with interpretation of the computer output. Using Minitab, we obtained the printout for Armand's Pizza Parlors shown in Figure 12.10.* The interpretation of this printout follows.

1. Minitab prints the estimated regression equation as SALES = 60.0 + 5.00 POP.

2. A table is printed that shows the values of the coefficients b_0 and b_1, the standard deviation of each coefficient, the t value obtained by dividing each coefficient value by its standard deviation, and the p-value associated with the t test. Thus, to test H_0: $\beta_1 = 0$ versus H_a: $\beta_1 \neq 0$, we could compare 8.62 (located in the t-ratio column) to the appropriate critical value. This is the procedure described for the t test in Section 12.5. Alternatively, we could use the p-value provided by Minitab to perform the same test. Recall from Chapter 9 that the p-value is the probability of obtaining a sample result more unlikely than what is observed. Since the p-value in this case is zero (to three decimal places), the sample results indicate that the null hypothesis (H_0: $\beta_1 = 0$) should be rejected.

3. Minitab prints the standard error of the estimate, $s = 13.83$, as well as information about the goodness of fit. Note that "R-sq = 90.3%" is the coefficient of determination expressed as a percentage. The output "R-Sq (adj) = 89.1%" is discussed in Section 12.9.

4. The ANOVA table is printed below the heading Analysis of Variance. Note that DF is an abbreviation for degrees of freedom and that MSR is given as 14,200 and MSE as 191. The ratio of these two values provides the F value of 74.25; in Section 12.5 we showed how the F value can be used to determine whether there is a significant relationship between SALES and POP. Minitab also prints the p-value associated with this F test. Since the p-value is zero (to three decimal places), the relationship is judged statistically significant.

5. The 95% confidence interval estimate of the expected sales and the 95% prediction interval estimate of sales for an individual restaurant located near a campus with 10,000 students are printed below the ANOVA table. The confidence interval is (98.58, 121.42) and the prediction interval is (76.12, 143.88) as we showed in Section 12.6.

---

* The Minitab steps necessary to generate the output are given in Appendix 12.1.

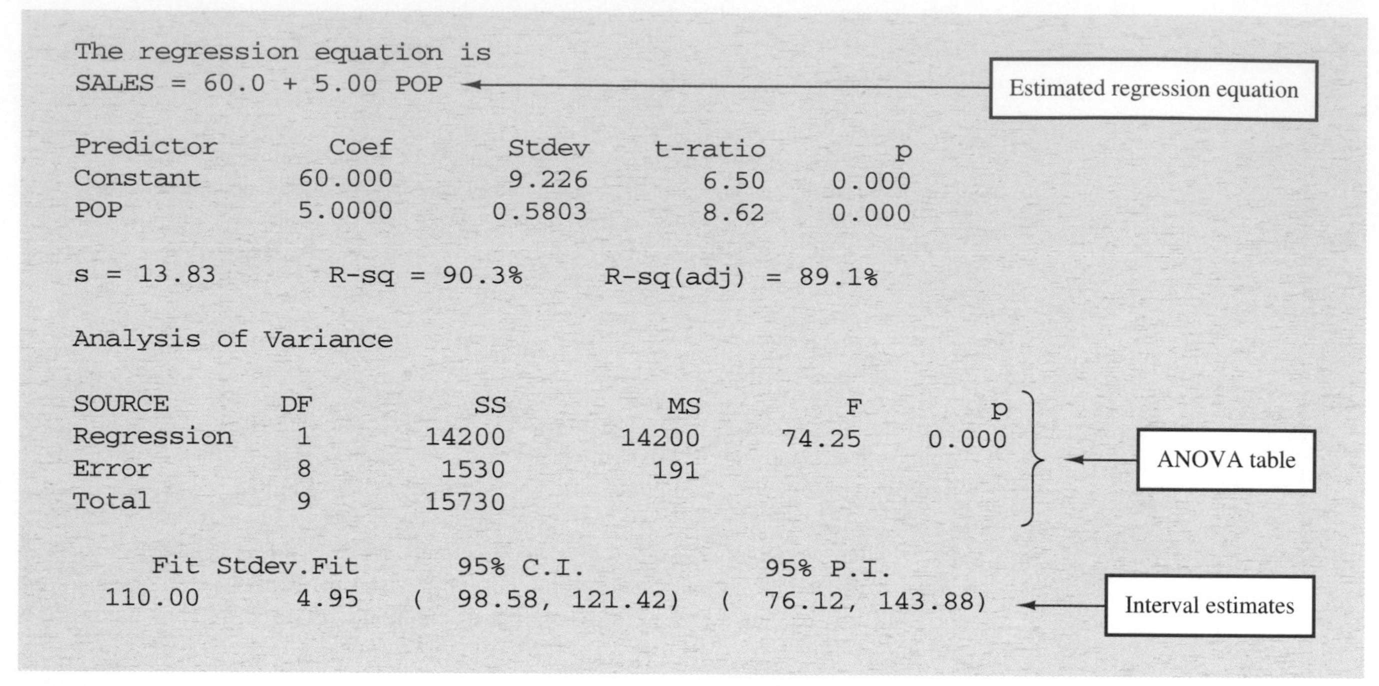

```
The regression equation is
SALES = 60.0 + 5.00 POP

Predictor       Coef       Stdev    t-ratio       p
Constant      60.000       9.226       6.50   0.000
POP           5.0000      0.5803       8.62   0.000

s = 13.83       R-sq = 90.3%      R-sq(adj) = 89.1%

Analysis of Variance

SOURCE        DF        SS          MS         F        p
Regression     1     14200       14200     74.25    0.000
Error          8      1530         191
Total          9     15730

    Fit Stdev.Fit       95% C.I.             95% P.I.
 110.00      4.95   (  98.58, 121.42)   (  76.12, 143.88)
```


FIGURE 12.10
Minitab Output for the Armand's Pizza Parlors Problem

exercises

Applications

37. The commercial division of a real estate firm is conducting a regression analysis of the relationship between x, annual gross rents ($1000s), and y, selling price ($1000s) for apartment buildings. Data have been collected on several properties recently sold, and the following output has been obtained in a computer run.

```
The regression equation is
Y = 20.0 + 7.21 X

Predictor         Coef       Stdev  t-ratio
Constant        20.000      3.2213     6.21
X                7.210      1.3626     5.29

Analysis of Variance

SOURCE          DF          SS
Regression       1     41587.3
Error            7
Total            8      51984.1
```

a. How many apartment buildings were in the sample?
b. Write the estimated regression equation.
c. What is the value of s_{b_1}?
d. Use the F statistic to test the significance of the relationship at a .05 level of significance.
e. Estimate the selling price of an apartment building with gross annual rents of $50,000.

38. Following is a portion of the computer output for a regression analysis relating y = maintenance expense (dollars per month) to x = usage (hours per week) of a particular brand of computer terminal.

```
The regression equation is

Y = 6.1092 + .8951 X

Predictor          Coef       Stdev
Constant         6.1092      0.9361
X                0.8951      0.1490

Analysis of Variance

SOURCE         DF           SS          MS

Regression      1      1575.76     1575.76
Error           8       349.14       43.64
Total           9      1924.90
```

a. Write the estimated regression equation.
b. Use a t test to determine whether monthly maintenance expense is related to usage at the .05 level of significance.
c. Use the estimated regression equation to predict monthly maintenance expense for any terminal that is used 25 hours per week.

39. A regression model relating x, number of salespersons at a branch office, to y, annual sales at the office ($1000s), has been developed. The computer output from a regression analysis of the data follows.

```
The regression equation is
Y = 80.0 + 50.00 X

Predictor         Coef      Stdev   t-ratio
Constant          80.0     11.333      7.06
X                 50.0      5.482      9.12

Analysis of Variance

SOURCE          DF          SS              MS
Regression       1      6828.6          6828.6
Error           28      2298.8            82.1
Total           29      9127.4
```

a. Write the estimated regression equation.
b. How many branch offices were involved in the study?
c. Compute the F statistic and test the significance of the relationship at a .05 level of significance.
d. Predict the annual sales at the Memphis branch office. This branch has 12 salespersons.

40. The following data show the dollar value of prescriptions for 13 pharmacies in Iowa and the population of the city served by the given pharmacy ("The Use of Categorical Variables in Data Envelopment Analysis," R. Banker and R. Morey, *Management Science,* December 1986).

PRESCRIP

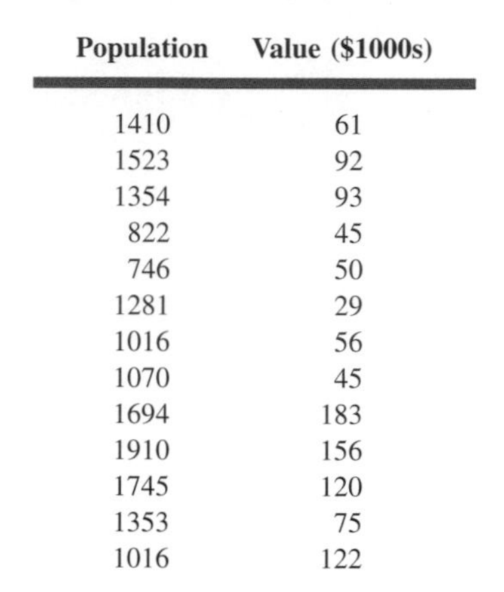

Population	Value ($1000s)
1410	61
1523	92
1354	93
822	45
746	50
1281	29
1016	56
1070	45
1694	183
1910	156
1745	120
1353	75
1016	122

a. Use a computer package to develop a scatter diagram for these data; plot population on the horizontal axis.
b. Does there appear to be any relationship between these two variables?
c. Use the computer package to develop the estimated regression line that could be used to predict the dollar value of prescriptions given the population of the city.
d. Test for the significance of the relationship at a .05 level of significance.
e. Predict the dollar value for a particular city with a population of 1500 people. Use $\alpha = .05$.

41. The National Association of Home Builders compared the median home prices with the median household incomes in cities throughout the United States (*USA Today,* September 10, 1991). Twenty-three of the most affordable cities are listed in the following table. Both home prices and household incomes are shown in thousands of dollars.

City	Median Income	Median Home Price
Amarillo, Texas	$36.7	$69.0
Brazoria, Texas	42.4	80.0
Canton, Ohio	34.1	66.0
Davenport, Iowa	38.4	59.0
Daytona Beach, Florida	31.0	63.0
Detroit, Michigan	44.6	77.0
Fort Walton Beach, Florida	34.2	65.0
Grand Rapids, Michigan	40.3	73.0
Jackson, Michigan	36.8	60.0
Kansas City, Missouri	41.1	77.0
Lansing, Michigan	40.0	70.0
Lorain, Ohio	$38.8	$72.5
Mansfield, Ohio	35.6	58.5
Milwaukee, Wisconsin	41.8	72.0
Oklahoma City, Oklahoma	34.5	63.0
Omaha, Nebraska	38.8	65.0
Rockford, Illinois	41.6	73.5
Saginaw, Michigan	39.7	61.0
Shreveport, Louisiana	34.4	66.0
Toledo, Ohio	39.4	65.0
Tulsa, Oklahoma	36.2	68.0
Winter Haven, Florida	30.2	56.0
Youngstown, Ohio	34.9	59.0

a. Use a computer package to develop a scatter diagram for these data; plot median income on the horizontal axis.
b. Does there appear to be any relationship between these two variables?
c. Use a computer package to develop the estimated regression equation that could be used to predict the median home price given the median income.
d. Test the significance of the relationship at the .05 level of significance.
e. Did the estimated regression equation provide a good fit? Explain.
f. Predict the expected median home price for cities with a median income of $35,000.
g. Predict the median home price for Elmira, New York, a city with a median income of $35,000.

12.8 Residual Analysis: Validating Model Assumptions

As we have previously noted, the *ith residual* is the difference between the observed value of the dependent variable (y_i) and the estimated value of the dependent variable ($\hat{y}_i$).

Residual for Observation *i*

$$y_i - \hat{y}_i \tag{12.30}$$

where

y_i is the observed value of the dependent variable
$\hat{y}_i$ is the estimated value of the dependent variable

In other words, the *i*th residual is the error resulting from using the estimated regression equation to predict the value of y_i. The residuals for the Armand's Pizza Parlors example are computed in Table 12.8. The observed values of the dependent variable are in the second column and the estimated values of the dependent variable, obtained using the estimated regression equation $\hat{y} = 60 + 5x$, are in the third column. The corresponding residuals are in the fourth column. An analysis of these residuals will help determine whether the assumptions that have been made about the regression model are appropriate.

Let us now review the regression assumptions for the Armand's Pizza Parlors example. A simple linear regression model was assumed.

$$y = \beta_0 + \beta_1 x + \epsilon \tag{12.31}$$

This model indicates that we assumed that sales (y) are a linear function of the size of the student population (x) plus an error term ϵ. In Section 12.4 we made the following assumptions about the error term ϵ.

1. $E(\epsilon) = 0$.
2. The variance of ϵ, denoted by σ^2, is the same for all values of x.
3. The values of ϵ are independent.
4. The error term ϵ has a normal probability distribution.

These assumptions provide the theoretical basis for the t test and the F test used to determine whether the relationship between x and y is significant, and for the confidence and prediction interval estimates presented in Section 12.6. If the assumptions about the

TABLE 12.8 Residuals for Armand's Pizza Parlors

Student Population x_i	Sales y_i	Estimated Sales $\hat{y}_i = 60 + 5x_i$	Residuals $y_i - \hat{y}_i$
2	58	70	−12
6	105	90	15
8	88	100	−12
8	118	100	18
12	117	120	−3
16	137	140	−3
20	157	160	−3
20	169	160	9
22	149	170	−21
26	202	190	12

error term ϵ appear questionable, the hypothesis tests about significance of the regression relationship and the interval estimation results may not be valid.

The residuals provide the best information about ϵ; hence an analysis of the residuals is an important step in determining whether the assumptions for ϵ are appropriate. Much of residual analysis is based on an examination of graphical plots. In this section, we show how to develop and interpret a plot of the residuals against values of the independent variable x.

Residual Plot Against x

A residual plot against the independent variable x is a graph in which the values of the independent variable are represented by the horizontal axis and the corresponding residual values are represented by the vertical axis. A point is plotted for each residual. The first coordinate for each point is given by the value of x_i and the second coordinate is given by the corresponding value of the residual $y_i - \hat{y}_i$. For a residual plot against x

FIGURE 12.11
Plot of the Residuals Against the Independent Variable x for Armand's Pizza Parlors

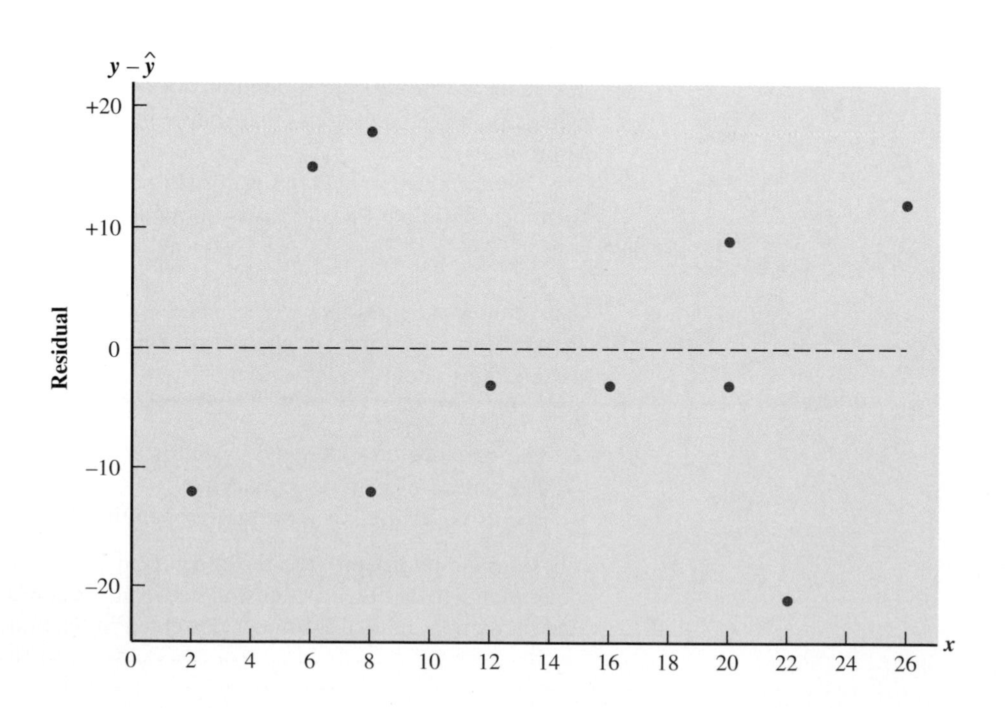

with the Armand's Pizza Parlors data from Table 12.8, the coordinates of the first point are (2, −12), corresponding to $x_1 = 2$ and $y_1 - \hat{y}_1 = -12$; the coordinates of the second point are (6, 15), corresponding to $x_2 = 6$ and $y_2 - \hat{y}_2 = 15$, and so on. Figure 12.11 is the resulting residual plot.

Before interpreting the results for this residual plot, let us consider some general patterns that might be observed in any residual plot. Three examples are shown in Figure 12.12. If the assumption that the variance of ϵ is the same for all values of x and

FIGURE 12.12
Residual Plots from Three Regression Studies

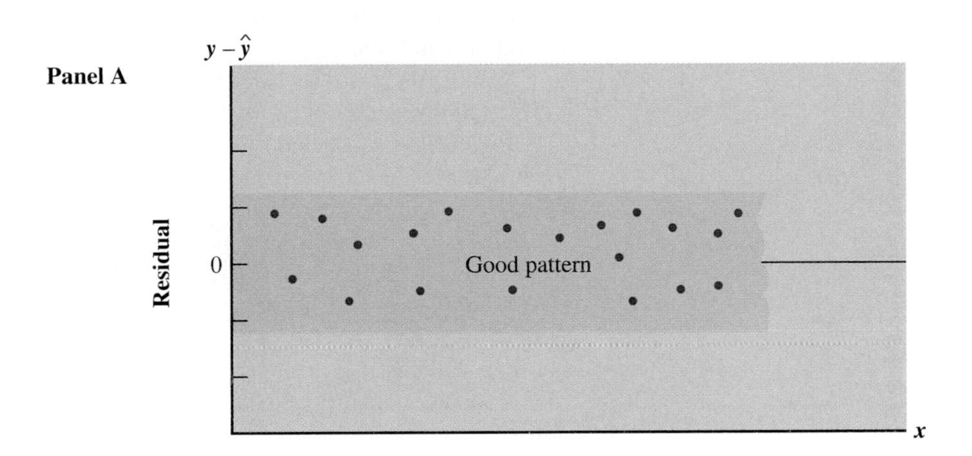

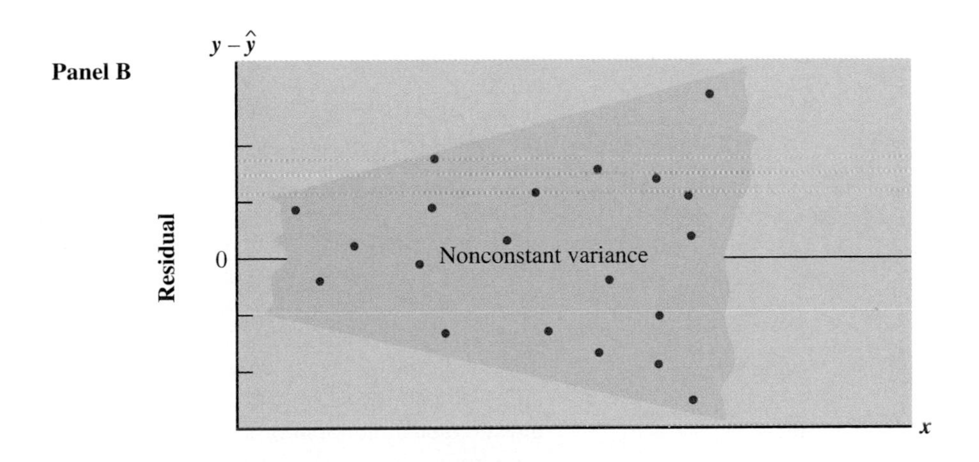

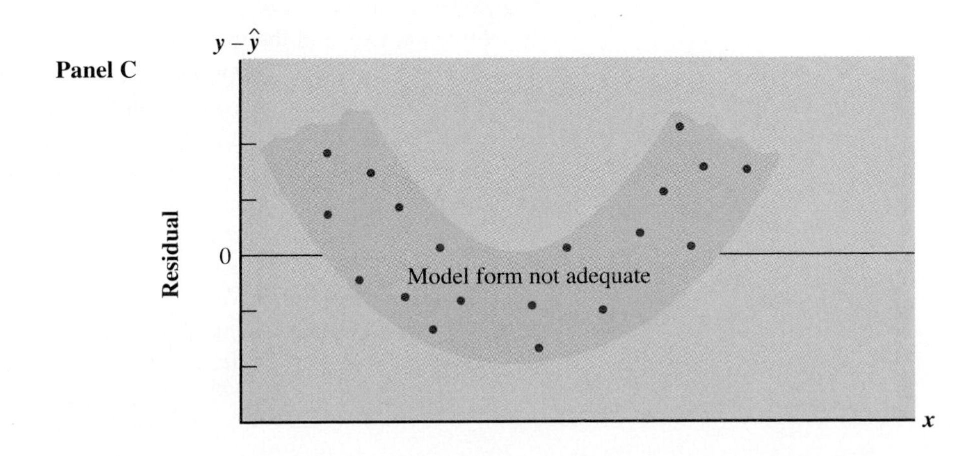

the assumed regression model is an adequate representation of the relationship between the variables, the residual plot should give an overall impression of a horizontal band of points such as the one in panel A of Figure 12.12. However, if the variance of ϵ is not the same for all values of x—for example, if variability about the regression line is greater for larger values of x—a pattern such as the one in panel B of Figure 12.12 could be observed. In this case, the assumption of a constant variance of ϵ is violated. Another possible residual plot is shown in panel C. In this case, we would conclude that the simple linear regression model is not an adequate representation of the relationship between the variables. A curvilinear regression model or multiple regression model should be considered.

Now let us return to the residual plot for Armand's Pizza Parlors shown in Figure 12.11. The residuals appear to approximate the horizontal pattern in panel A of Figure 12.12. Hence, we conclude that the residual plot does not provide evidence that the assumptions made for Armand's regression model should be challenged. At this point, we are confident in the conclusion that Armand's simple linear regression model is valid.

Experience and good judgment are always factors in the effective interpretation of residual plots. Seldom does a residual plot conform precisely to one of the patterns shown in Figure 12.12. Yet analysts who frequently conduct regression studies and frequently review residual plots become very good at understanding the differences between patterns that are reasonable and patterns that indicate the assumptions of the model should be questioned. A residual plot as shown here is one of the techniques that is used to assess the validity of the assumptions for a regression model.

notes and comments

1. The analysis of residuals is the primary method statisticians use to verify that the assumptions associated with a regression model are valid. Even if no violations are found, it does not necessarily folllow that the model will yield good predictions. However, if in addition the statistical tests support the conclusion of significance and the coefficient of determination is large, we should be able to develop good estimates using the estimated regression equation.

2. We use residual plots to validate the assumptions of a regression model. The appropriate corrective action when assumptions are violated must be based on good judgment; recommendations from an experienced statistician can be valuable. If the residual analysis indicates that the one or more assumptions are questionable, but not necessarily invalid, the user should take caution in using and interpreting the regression results.

3. Another residual plot places the predicted value of the dependent variable $\hat{y}$ on the horizontal axis and the residual values on the vertical axis. A point is plotted for each residual; the first coordinate for each point is given by $\hat{y}_i$ and the second coordinate is given by the corresponding value of the ith residual $y_i - \hat{y}_i$. For simple linear regression, both the residual plot against x and the residual plot against $\hat{y}$ provide the same information. In multiple regression analysis, the residual plot against $\hat{y}$ is more widely used.

4. Many of the residual plots provided by computer software packages use a standardized version of the residuals, which are obtained by dividing each residual by its standard deviation. We should expect to see approximately 95% of the standardized residuals between −2 and +2 if the normal probability assumption for the error term ϵ is appropriate.

exercises

Methods

42. Given are data for two variables, x and y.

x_i	6	11	15	18	20
y_i	6	8	12	20	30

a. Develop an estimated regression equation for these data.
b. Compute the residuals.
c. Develop a plot of the residuals against the independent variable x. Do the assumptions about the error terms seem to be satisfied?

43. The data that follow were used in a regression study.

Observation	x_i	y_i
1	2	4
2	3	5
3	4	4
4	5	6
5	7	4
6	7	6
7	7	9
8	8	5
9	9	11

a. Develop an estimated regression equation for these data.
b. Construct a plot of the residuals. Do the assumptions about the error terms seem to be satisfied?

Applications

44. Data on advertising expenditures ($) and revenue (in thousands of dollars) for the Four Seasons Restaurant follow.

Advertising Expenditures	Revenue
1	19
2	32
4	44
6	40
10	52
14	53
20	54

a. Let x equal advertising expenditures ($1000s) and y equal sales ($1000s). Use the method of least squares to develop a straight line approximation of the relationship between the two variables.
b. Test whether sales and advertising expenditures are related at a .05 level of significance.

c. Develop a plot of the residuals against the independent variable x.

d. What conclusions can you draw from residual analysis? Should this model be used, or should we look for a better one?

45. Refer to Exercise 8, where an estimated regression equation relating years of experience and annual sales was developed.

a. Compute the residuals and construct a residual plot for this problem.

b. Do the assumptions about the error terms seem reasonable in light of the residual plot?

46. The following table lists the number of employees and the yearly revenue for the 10 largest wholesale bakers (*Louis Rukeyser's Business Almanac*).

Company	Employees	Revenues ($Millions)
Nabisco Brands USA	9,500	1,734
Continental Baking Co.	22,400	1,600
Campbell Taggart, Inc.	19,000	1,044
Keebler Company	8,943	988
Interstate Bakeries Corp.	11,200	704
Flowers Industries, Inc.	10,200	557
Sunshine Biscuits, Inc.	5,000	490
American Bakeries Co.	6,600	461
Entenmann's Inc.	3,734	450
Kitchens of Sara Lee	1,550	405

a. Use a computer package to develop an estimated regression equation relating yearly revenues y to the number of employees x.

b. Develop a plot of the residuals against the independent variable.

c. Do the assumptions about the error terms and model form seem reasonable in light of the residual plot?

12.9 Multiple Regression

Multiple regression analysis is the study of how a dependent variable y is related to two or more independent variables. In the general case, we will use p to denote the number of independent variables. The concepts of a regression model and a regression equation introduced in Section 12.1 are applicable in the multiple regression case. The equation that describes how the dependent variable y is related to the independent variables $x_1, x_2, \ldots, x_p$ and an error term is called the *regression model.* We begin with the assumption that the multiple regression model has the following form.

Multiple Regression Model

$$y = \beta_0 + \beta_1 x_1 + \beta_2 x_2 + \ldots + \beta_p x_p + \epsilon \quad (12.32)$$

The assumptions about the error term ϵ in the multiple regression model are the same as the assumptions for ϵ in the simple linear regression model (see Section 12.4). Recall that one of the assumptions is that the mean or expected value of ϵ is zero. A consequence of this assumption is that the mean or expected value of y, denoted $E(y)$, is equal to $\beta_0 + \beta_1 x_1 + \beta_2 x_2 + \ldots + \beta_p x_p$. The equation that describes how the mean value of y is related to $x_1, x_2, \ldots, x_p$ is called the *multiple regression equation.*

Multiple Regression Equation

$$E(y) = \beta_0 + \beta_1 x_1 + \beta_2 x_2 + \ldots + \beta_p x_p \quad (12.33)$$

If the values of $\beta_0, \beta_1, \beta_2, \ldots, \beta_p$ were known, (12.33) could be used to compute the mean value of y at given values of $x_1, x_2, \ldots, x_p$. Unfortunately, these parameter values will not, in general, be known and must be estimated from sample data. A simple random sample is used to compute sample statistics $b_0, b_1, b_2, \ldots, b_p$ that are used as the point estimators of the parameters $\beta_0, \beta_1, \beta_2, \ldots, \beta_p$. These sample statistics provide the following estimated multiple regression equation.

Estimated Multiple Regression Equation

$$\hat{y} = b_0 + b_1 x_1 + b_2 x_2 + \ldots + b_p x_p \quad (12.34)$$

where

$b_0, b_1, b_2, \ldots, b_p$ are the estimates of $\beta_0, \beta_1, \beta_2, \ldots, \beta_p$

$\hat{y}$ = estimated value of the dependent variable

The estimation process for multiple regression is shown in Figure 12.13.

In Section 12.2, we used the least squares method to develop the estimated regression equation that best approximated the straight-line relationship between the dependent and independent variables. This same approach is used to develop the estimated multiple regression equation. The least squares criterion is restated as follows.

Least Squares Criterion

$$\min \Sigma(y_i - \hat{y}_i)^2 \quad (12.35)$$

where

y_i = observed value of the dependent variable for the ith observation

$\hat{y}_i$ = estimated value of the dependent variable for the ith observation

The estimated values of the dependent variable are computed by using the estimated multiple regression equation,

$$\hat{y} = b_0 + b_1 x_1 + b_2 x_2 + \ldots + b_p x_p$$

FIGURE 12.13
The Estimation Process for Multiple Regression

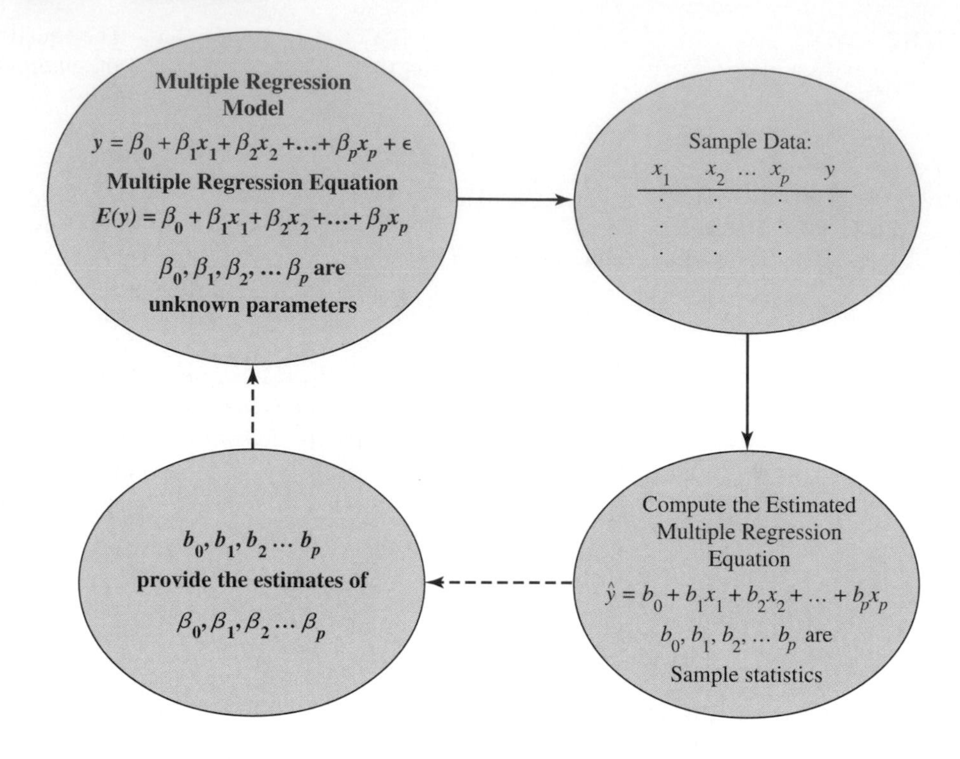

As (12.35) shows, the least squares method uses sample data to provide the values of b_0, b_1, b_2, . . ., b_p that make the sum of squared residuals [the deviations between the observed values of the dependent variable (y_i) and the estimated values of the dependent variable ($\hat{y}_i$)] a minimum.

In Section 12.2 we presented formulas for computing the least squares estimators b_0 and b_1 for the estimated simple linear regression equation $\hat{y} = b_0 + b_1 x$. In multiple regression, the presentation of the formulas for the regression coefficients b_0, b_1, b_2, . . ., b_p involves the use of matrix algebra and is beyond the scope of this text. Therefore, in presenting multiple regression, we will focus on how computer software packages can be used to obtain the estimated regression equation and other information. The emphasis will be on how to interpret the computer output rather than on how to make the multiple regression computations.

TABLE 12.9 Preliminary Data for Butler Trucking

Driving Assignment	x_1 = Miles Traveled	y = Travel Time (hours)
1	100	9.3
2	50	4.8
3	100	8.9
4	100	6.5
5	50	4.2
6	80	6.2
7	75	7.4
8	65	6.0
9	90	7.6
10	90	6.1

An Example: Butler Trucking Company

As an illustration of multiple regression analysis, we will consider a problem faced by the Butler Trucking Company, an independent trucking company in southern California. A major portion of Butler's business involves deliveries throughout its local area. To develop better work schedules, the managers want to estimate the total daily travel time for their drivers.

Initially the managers believed that the total daily travel time would be closely related to the number of miles traveled in making the daily deliveries. A simple random sample of 10 driving assignments provided the data shown in Table 12.9 and the scatter diagram shown in Figure 12.14. After reviewing this scatter diagram, the managers hypothesized that the simple linear regression model $y = \beta_0 + \beta_1 x_1 + \epsilon$ could be used to describe the relationship between the total travel time (y) and the number of miles traveled (x_1). To estimate the parameters β_0 and β_1, the least squares method was used to develop the estimated regression equation,

$$\hat{y} = b_0 + b_1 x_1 \qquad (12.36)$$

FIGURE 12.14
Scatter Diagram of Preliminary Data for Butler Trucking

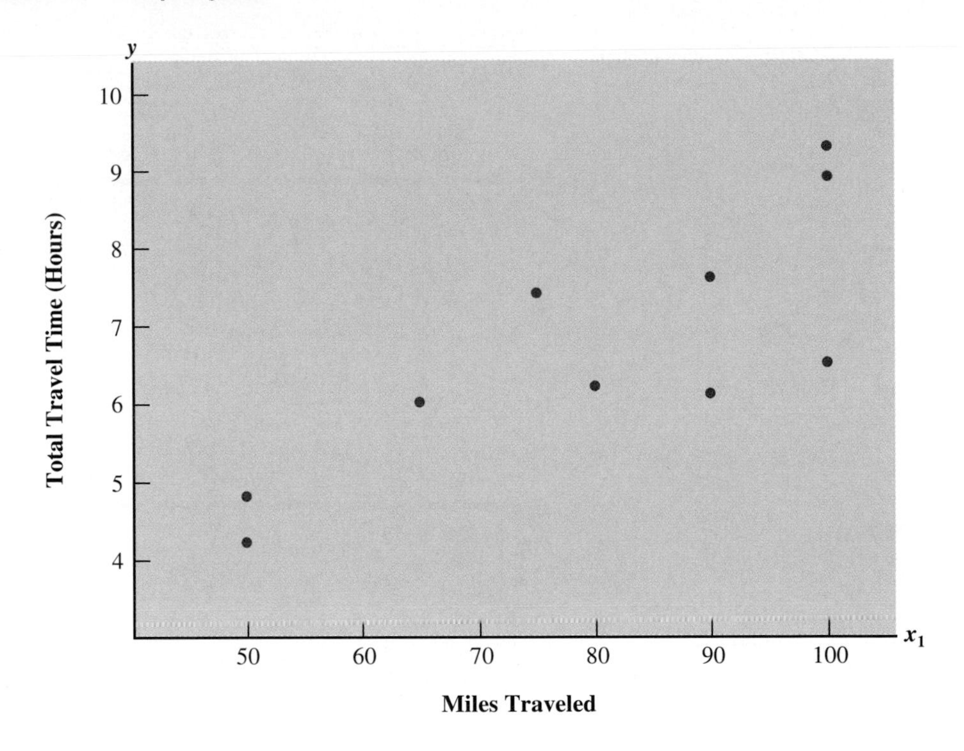

FIGURE 12.15
Minitab Output for Butler Trucking With One Independent Variable

```
The regression equation is
Y = 1.27 + 0.0678 X1

Predictor        Coef       Stdev    t-ratio        p
Constant        1.274       1.401       0.91    0.390
X1            0.06783     0.01706       3.98    0.004

s = 1.002       R-sq = 66.4%      R-sq(adj) = 62.2%

Analysis of Variance

SOURCE        DF         SS          MS         F        p
Regression     1     15.871      15.871     15.81    0.004
Error          8      8.029       1.004
Total          9     23.900
```

In Figure 12.15, we show the Minitab computer output from applying simple linear regression to the data in Table 12.9. The estimated regression equation is

$$\hat{y} = 1.27 + .0678x_1$$

At the .05 level of significance, the F value of 15.81 and its corresponding p-value of .004 indicate that the relationship is significant; that is, we can reject H_0: $\beta_1 = 0$ since the p-value is less than $\alpha = .05$. Note that the same conclusion is obtained from the t value of 3.98 and its associated p-value of .004. Thus, we can conclude that the relationship between the total travel time and the number of miles traveled is significant;

TABLE 12.10 Data for Butler Trucking with Miles Traveled (x_1) and Number of Deliveries (x_2) as the Independent Variables

Driving Assignment	x_1 = Miles Traveled	x_2 = Number of Deliveries	y = Travel Time (hours)
1	100	4	9.3
2	50	3	4.8
3	100	4	8.9
4	100	2	6.5
5	50	2	4.2
6	80	2	6.2
7	75	3	7.4
8	65	4	6.0
9	90	3	7.6
10	90	2	6.1

FIGURE 12.16
Minitab Output for Butler Trucking With Two Independent Variables

```
The regression equation is
Y = - 0.869 + 0.0611 X1 + 0.923 X2

Predictor        Coef       Stdev    t-ratio       p
Constant      -0.8687      0.9515      -0.91   0.392
X1           0.061135    0.009888       6.18   0.000
X2             0.9234      0.2211       4.18   0.004

s = 0.5731       R-sq = 90.4%     R-sq(adj) = 87.6%

Analysis of Variance

SOURCE        DF         SS          MS         F         p
Regression     2     21.601      10.800     32.88     0.000
Error          7      2.299       0.328
Total          9     23.900
```

longer travel times are associated with more miles traveled. With a coefficient of determination (expressed as a percentage) of R-sq = 66.4%, we see that 66.4% of the variability in travel time can be explained by the linear effect of the number of miles traveled. This finding is fairly good, but the managers might want to consider adding a second independent variable to explain some of the remaining variability in the dependent variable.

In attempting to identify another independent variable, the managers felt that the number of deliveries could also contribute to the total travel time. The Butler Trucking data, with the number of deliveries added, are shown in Table 12.10. The Minitab computer solution with both miles traveled (x_1) and number of deliveries (x_2) as independent variables is shown in Figure 12.16. The estimated regression equation is

$$\hat{y} = -.869 + .0611x_1 + .923x_2 \tag{12.37}$$

A Note on Interpretation of Coefficients

One observation can be made at this point about the relationship between the estimated regression equation with only the miles traveled as an independent variable and the equation that includes the number of deliveries as a second independent variable. The value of b_1 is not the same in both cases. In simple linear regression, we interpret b_1 as an estimate of the change in y for a one-unit change in the independent variable. For example, in the Butler Trucking Company problem involving only one independent variable, number of miles traveled, $b_1 = .0678$. Thus, .0678 is an estimate of the expected increase in travel time corresponding to an increase of one mile in the distance traveled. In multiple regression analysis, this interpretation must be modified somewhat. That is, in multiple regression analysis, we interpret each regression coefficient as follows: b_i represents an estimate of the change in y corresponding to a one-unit change in x_i when all other independent variables are held constant. In the Butler Trucking example involving two independent variables, $b_1 = .0611$. Thus, .0611 hours is an estimate of the expected increase in travel time corresponding to an increase of one mile in the distance traveled when the number of deliveries is held constant. Similarly, since $b_2 = .923$, an estimate of the expected increase in travel time corresponding to an increase of one delivery when the number of miles traveled is held constant is .923 hours.

Testing for Significance

The significance tests used in simple linear regression were a t and an F test. In simple linear regression, both tests provided the same conclusion; that is, if the null hypothesis was rejected, we concluded $\beta_1 \neq 0$. In multiple regression, the t and the F test have different purposes.

1. The F test is used to determine if there is a significant relationship between the dependent variable and the set of all the independent variables; we will refer to the F test as the test for *overall significance*.
2. If the F test shows an overall significance, the t test is used to determine whether each of the individual independent variables is significant. A separate t test is conducted for each of the independent variables in the model; we refer to each of these t tests as a test for *individual significance*.

In the following discussion, we will show how the F test and the t test apply to the Butler Trucking Company multiple regression problem.

With two independent variables, the hypotheses used to determine if there is a significant relationship between travel time and the two independent variables, miles traveled and number of deliveries, are written as follows:

$$H_0\colon \beta_1 = \beta_2 = 0$$

$$H_a\colon \beta_1 \text{ and/or } \beta_2 \neq 0$$

Recall that in simple linear regression the F test statistic is MSR/MSE. In multiple regression the F test statistic is also MSR/MSE; however, the degrees of freedom associated with MSR and MSE change because of the additional independent variables. In the analysis of variance portion of Figure 12.16, we see that MSR = 10.8, MSE = .328, and the value of the F test statistic is 32.88. With a level of significance $\alpha = .01$, Table 4 of Appendix B shows that with 2 degrees of freedom in the numerator (the degrees of freedom associated with MSR) and 7 degrees of freedom in the denominator (the degrees of freedom associated with MSE), $F_{.01} = 9.55$. Since $32.88 > 9.55$, we reject $H_0\colon \beta_1 = \beta_2 = 0$ and conclude that a significant relationship exists between travel time y and the two independent variables, miles traveled and number of deliveries. The p-value = 0.000 in the last column of the analysis of variance table (Figure 12.16) also indicates that we can reject H_0: $\beta_1 = \beta_2 = 0$ since the p-value $< \alpha$.

If the F test first shows that the multiple regression relationship is significant, a t test can be conducted to determine the significance of each of the individual parameters. The t test for individual significance is $t = b_i / s_{b_i}$. In this test statistic, s_{b_i} is the estimate of the standard deviation of b_i. The value of s_{b_i} will be provided by the computer software package. Let us conduct the t test for the Butler Trucking multiple regression problem. The Minitab output shown in Figure 12.16 shows that $b_1 = .061135$, $b_2 = .9234$, $s_{b_1} = .009888$ and $s_{b_2} = .2211$. Thus, the test statistics for the hypotheses involving parameters β_1 and β_2 are as follows:

$$t = .061135 / .009888 = 6.18$$

$$t = .9234 / .2211 = 4.18$$

Note that both of these t-ratios are provided by the Minitab output of Figure 12.16. Using $\alpha = .01$ and $n - p - 1 = 10 - 2 - 1 = 7$ degrees of freedom, Table 2 of Appendix B can be used to show $t_{.005} = 3.499$. With $6.18 > 3.499$, we reject the hypothesis H_0: $\beta_1 = 0$. Similarly, with $4.18 > 3.499$, we also reject the hypothesis H_0: $\beta_2 = 0$. Note also that the p-values of .000 and .004 on the Minitab output indicate rejection of these hypotheses. Thus, both parameters tested individually are statistically significant.

The Multiple Coefficient of Determination

The multiple coefficient of determination, denoted by R^2, is computed using the same formula as for the coefficient of determination in simple linear regression; that is $R^2 = \text{SSR/SST}$. As was the case for the coefficient of determination in simple linear regression, the multiple coefficient of determination can be interpreted as the proportion of the variability in the dependent variable that can be explained by the estimated multiple regression equation. Thus, when it is multiplied by 100, it can be interpreted as the percentage of variation in y that can be explained by the estimated regression equation. In the Minitab output shown in Figure 12.16 the multiple coefficient of determination is denoted by R-sq = 90.4%. Thus, the estimated multiple regression equation has explained 90.4% of the variability in travel time.

Figure 12.15 shows that the R-sq value for the estimated regression equation with only one independent variable, the number of miles traveled (x_1), is 66.4%. Thus, the percentage of the variability in travel times that is explained by the estimated regression equation increased from 66.4% to 90.4% by adding the number of deliveries as a second independent variable. In general, R^2 always increases as more independent variables are added to the model.

Estimation and Prediction

Estimating the mean value of y and predicting an individual value of y in multiple regression is similar to that for the case of simple linear regression. In multiple regression analysis we substitute the given values of the independent variables into the estimated regression equation and use the corresponding value of $\hat{y}$ as the point estimate. Although the methodology behind the development of confidence interval estimates and prediction interval estimates in multiple regression analysis is beyond the scope of the text, computer software packages such as Minitab can be used to develop the interval estimates.

notes and comments

1. As pointed out earlier, the mean square error is an unbiased estimator of σ^2, the variance of the error term ϵ. Thus, referring to Figure 12.16, we see that the estimate of σ^2 is MSE = .328. The square root of MSE is the estimate of the standard deviation of the error term. As defined in Section 12.5, this standard deviaiton is called the *standard error of the estimate* and is denoted by s. Thus we have $s = \sqrt{\text{MSE}} = \sqrt{.328} = .5731$. Note that the value of the standard error of the estimate also appears in the Minitab output shown in Figure 12.16.

2. Many analysts prefer adjusting R^2 for the number of independent variables to avoid overestimating the impact of adding an independent variable on the amount of variability explained by the estimated regression equation; the resulting measure is referred to as the *adjusted multiple coefficient of determination.* Although the computation of the adjusted multiple coefficient of determination is beyond the scope of the text, its value is routinely provided by computer packages. On the Minitab output in Figure 12.16 the adjusted multiple coefficient of determination is denoted by R-sq(adj) = 87.6%.

exercises

Note to student: The exercises involving data in this and subsequent sections were designed to be solved by means of a computer software package.

Methods

47. The estimated regression equation for a model involving two independent variables and 10 observations follows.

$$\hat{y} = 29.1270 + .5906x_1 + .4980x_2$$

a. Interpret b_1 and b_2 in this estimated regression equation.
b. Estimate y when $x_1 = 180$ and $x_2 = 310$.

48. Consider the following data for a dependent variable y and two independent variables, x_1 and x_2.

x_1	x_2	y
30	12	94
47	10	108
25	17	112
51	16	178
40	5	94
51	19	175
74	7	170
36	12	117
59	13	142
76	16	211

The estimated regression equation for these data is

$$\hat{y} = -18.4 + 2.01x_1 + 4.74x_2$$

Here SST = 15,182.9, SSR = 14052.2, s_{b_1}, = .2471, and s_{b_2} = .9484.

a. Test for a significant relationship among x_1, x_2, and y. Use $\alpha = .05$.
b. Is β_1 significant? Use $\alpha = .05$.
c. Is β_2 significant? Use $\alpha = .05$.
d. Compute R^2. Comment on the goodness of fit.

Applications

49. The owner of Showtime Movie Theaters, Inc., would like to estimate weekly gross revenue as a function of advertising expenditures. Historical data for a sample of eight weeks follow.

Weekly Gross Revenue ($1000s)	Television Advertising ($1000s)	Newspaper Advertising ($1000s)
96	5.0	1.5
90	2.0	2.0
95	4.0	1.5
92	2.5	2.5
95	3.0	3.3
94	3.5	2.3
94	2.5	4.2
94	3.0	2.5

a. Develop an estimated regression equation with the amount of television advertising as the independent variable.
b. Develop an estimated regression equation with both television advertising and newspaper advertising as the independent variables.
c. Is the estimated regression equation coefficient for television advertising expenditures the same in part (a) and in part (b)? Interpret the coefficient in each case.
Note: in parts (d) through (g), assume that both independent variables are used to estimate weekly gross revenue.
d. Use the F test to determine the overall significance of the relationship. What is your conclusion at the .05 level of significance?
e. Use the t test to determine the significance of β_1 and β_2. What is your conclusion at the .05 level of significance?
f. Compute R^2. Comment on the goodness of fit.
g. What is the estimate of the gross revenue for a week when $3500 is spent on television advertising and $1800 is spent on newspaper advertising?

50. The following table reports the horsepower, time required to go from zero to 60 miles per hour, and price in thousands of dollars for 10 popular sports cars (*Road & Track,* October 1994).

Sports Car	Horsepower	Zero to 60 mph (Seconds)	Price ($1000s)
BMW M3	240	6.0	38.4
Corvette	300	5.7	41.4
Dodge Viper	400	4.8	54.8
Ford Mustang	240	6.9	25.8
Honda Prelude	190	7.1	25.6
Mitsubishi 3000GT	320	5.7	43.7
Toyota Supra	320	5.3	48.2
Nissan 300ZX	300	6.0	40.8
Alfa Romeo	320	7.6	38.1
Mazda RX-7	255	5.5	35.0

a. Use horsepower as the independent variable and price as the dependent variable. What is the estimated regression equation?

b. Use horsepower and zero to 60 as two independent variables and price as the dependent variable. What is the estimated regression equation?

c. If your goal is to estimate the price of a sports car, do you prefer simple linear regression in part (a) or the multiple regression in part (b)? Discuss.

d. A new sports car has been advertised as having a horsepower of 236 and being able to go from zero to 60 in 5.9 seconds. Use your regression analysis to estimate the price of this new sports car.

51. Heller Company manufactures lawnmowers and related lawn equipment. The managers believe the quantity of lawnmowers sold depends on the price of the mower and the price of a competitor's mower. Let

$$y = \text{quantity sold (1000s)},$$
$$x_1 = \text{price of competitor's mower (\$), and}$$
$$x_2 = \text{price of Heller's mower (\$)}.$$

The managers want an estimated regression equation that relates quantity sold to the prices of the Heller mower and the competitor's mower. The following table lists prices in 10 cities.

Competitor's Price (x_1)	Heller's Price (x_2)	Quantity Sold (y)
120	100	102
140	110	100
190	90	120
130	150	77
155	210	46
175	150	93
125	250	26
145	270	69
180	300	65
150	250	85

a. Determine the estimated regression equation that can be used to predict the quantity sold given the competitor's price and Heller's price.

b. Interpret b_1 and b_2.

c. Test for a significant relationship among x_1, x_2, and y. Use $\alpha = .05$.

d. Is β_1 significant? Use $\alpha = .05$.

e. Is β_2 significant? Use $\alpha = .05$.

f. Predict the quantity sold in a city where Heller prices its mower at \$160 and the competitor prices its mower at \$170.

52. Data on housing markets were provided for 100 cities in the United States (*U.S. News & World Report,* April 6, 1992). The following table reports the median cost of a new home, the number of new housing starts during 1991–1992, and the average household income for sample of 16 cities. All data are in 1000s.

HOUSING

City	Median Cost	Housing Starts	Household Income
Chicago	181.8	12.9	\$61.0
Dayton	107.8	3.8	48.4
Atlanta	100.6	24.2	54.7
Oklahoma City	68.9	3.3	53.2
Columbia	90.3	3.1	57.4
Tacoma	96.1	4.1	51.1
Mobile	68.5	1.0	41.0
Baltimore	121.8	11.1	62.8

City	Median Cost	Housing Starts	Household Income
West Palm Beach	130.4	8.9	\$58.1
San Antonio	72.5	1.5	57.0
Pittsburgh	79.5	4.9	49.2
Jacksonville	82.1	8.0	47.5
Cleveland	122.9	5.4	54.0
Gary	98.2	3.2	45.7
Scranton	81.6	2.5	44.8
Richmond	102.8	6.0	64.5

a. Using these data, develop the estimated regression equation relating cost to the number of housing starts and the household income.

b. Estimate the median cost for a city with 8000 housing starts and an average household income of \$50,000.

summary

In this chapter we first showed how regression analysis can be used to determine how a dependent variable y is related to an independent variable x. In simple linear regression, the regression model is $y = \beta_0 + \beta_1 x + \epsilon$. The simple linear regression equation $E(y) = \beta_0 + \beta_1 x$ describes how the mean or expected value of y is related to x. We used sample data and the least squares method to develop the estimated regression equation $\hat{y} = b_0 + b_1 x$. In effect, b_0 and b_1 are the sample statistics used to estimate the unknown model parameters β_0 and β_1.

The coefficient of determination was presented as a measure of the goodness of fit for the estimated regression equation; it can be interpreted as the proportion of the variation in the dependent variable y that can be explained by the estimated regression equation. We reviewed correlation as a descriptive measure of the strength of a linear relationship between two variables.

The assumptions about the regression model and its associated error term ϵ were discussed, and t and F tests, based on those assumptions, were presented as a means for determining whether the relationship between two variables is statistically significant. We showed how to use the estimated regression equation to develop confidence interval estimates of the mean value of y and prediction interval estimates of individual values of y. We concluded the discussion of simple linear regression by showing how computer software can ease the computational burden of regression analysis; we then showed how residual analysis can be used to validate the model assumptions.

Multiple regression analysis was introduced as an extension of simple linear regression analysis. Multiple regression analysis enables us to understand how a dependent variable is related to two or more independent variables. Computer printouts were used throughout the discussion of multiple regression analysis to emphasize the fact that statistical software packages are the only realistic means of performing the numerous computations required in multiple regression analysis.

glossary

Dependent variable The variable that is being predicted or explained. It is denoted by y.

Independent variable The variable that is doing the predicting or explaining. It is denoted by x.

Simple linear regression Regression analysis involving one independent variable and one dependent variable in which the relationship between the variables is approximated by a straight line.

Simple linear regression model The probability model describing how y is related to x; $y = \beta_0 + \beta_1 x + \epsilon$.

Simple linear regression equation The equation that describes how the mean or expected value of the dependent variable is related to the independent variable; $E(y) = \beta_0 + \beta_1 x$.

Estimated simple linear regression equation The estimate of the regression equation developed from sample data by using the least squares method; $\hat{y} = b_0 + b_1 x$.

Scatter diagram A graph of bivariate data in which the independent variable is on the horizontal axis and the dependent variable is on the vertical axis.

Least squares method The procedure used to develop the estimated regression equation. The objective is to minimize $\Sigma(y_i - \hat{y}_i)^2$.

Coefficient of determination A measure of the goodness of fit of the estimated regression equation. It can be interpreted as the proportion of the variation in the dependent variable y that is explained by the estimated regression equation.

Residual The difference between the observed value of the dependent variable and the value predicted by using the estimated regression equation; that is, for the ith observation the residual is $y_i - \hat{y}_i$.

Correlation coefficient A measure of the strength of the linear relationship between two variables (previously discussed in Chapter 3).

Mean square error The unbiased estimate of the variance of the error term, σ^2. It is denoted by MSE or s^2.

Standard error of the estimate The square root of the mean square error, denoted by s. It is the estimate of σ, the standard deviation of the error term ϵ.

ANOVA table The analysis of variance table used to summarize the computations associated with the F test for significance.

Confidence interval estimate The interval estimate of the mean value of y for a given value of x.

Prediction interval estimate The interval estimate of an individual value of y for a given value of x.

Residual analysis The analysis of the residuals used to determine whether the assumptions made about the regression model appear to be valid.

Residual plots Graphical representations of the residuals that can be used to determine whether the assumptions made about the regression model appear to be valid.

Multiple regression Regression analysis involving two or more independent variables.

Multiple regression model The mathematical equation that describes how the dependent variable y is related to the independent variables $x_1, x_2, \ldots, x_p$ and an error term ϵ; $y = \beta_0 + \beta_1 x_1 + \beta_2 x_2 + \ldots + \beta_p x_p + \epsilon$.

Multiple regression equation The mathematical equation relating the expected value or mean value of the dependent variable to the values of the independent variables; $E(y) = \beta_0 + \beta_1 x_1 + \beta_2 x_2 + \ldots + \beta_p x_p$.

Estimated multiple regression equation The estimate of the multiple regression equation based on sample data and the least squares method; $\hat{y} = b_0 + b_1 x_1 + b_2 x_2 + \ldots + b_p x_p$.

Multiple coefficient of determination A measure of the goodness of fit of the estimated multiple regression equation. It can be interpreted as the proportion of the variation in the dependent variable that is explained by the estimated regression equation.

Adjusted multiple coefficient of determination A measure of the goodness of fit of the estimated multiple regression equation that adjusts for the number of independent variables in the model and thus avoids overestimating the impact of adding more independent variables.

key formulas

Simple Linear Regression Model

$$y = \beta_0 + \beta_1 x + \epsilon \tag{12.1}$$

Simple Linear Regression Equation

$$E(y) = \beta_0 + \beta_1 x \tag{12.2}$$

Estimated Simple Linear Regression Equation

$$\hat{y} = b_0 + b_1 x \tag{12.3}$$

Least Squares Criterion

$$\text{Min } \Sigma(y_i - \hat{y}_i)^2 \tag{12.5}$$

Slope and y-Intercept for the Estimated Regression Equation

$$b_1 = \frac{\Sigma x_i y_i - (\Sigma x_i \, \Sigma y_i)/n}{\Sigma x_i^2 - (\Sigma x_i)^2/n} \tag{12.6}$$

$$b_0 = \bar{y} - b_1\bar{x} \tag{12.7}$$

Sum of Squares Due to Error

$$\text{SSE} = \Sigma(y_i - \hat{y}_i)^2 \tag{12.8}$$

Total Sum of Squares

$$\text{SST} = \Sigma(y_i - \bar{y})^2 \tag{12.9}$$

Sum of Squares Due to Regression

$$\text{SSR} = \Sigma(\hat{y}_i - \bar{y})^2 \tag{12.10}$$

Relationship among SST, SSR, and SSE

$$\text{SST} = \text{SSR} + \text{SSE} \tag{12.11}$$

Coefficient of Determination

$$r^2 = \frac{\text{SSR}}{\text{SST}} \tag{12.12}$$

Computational Formula for SST

$$\text{SST} = \Sigma y_i^2 - (\Sigma y_i)^2/n \tag{12.13}$$

Computational Formula for SSR

$$\text{SSR} = \frac{[\Sigma x_i y_i - (\Sigma x_i \, \Sigma y_i)/n]^2}{\Sigma x_i^2 - (\Sigma x_i)^2/n} \tag{12.14}$$

Sample Correlation Coefficient

$$r_{xy} = (\text{sign of } b_1)\sqrt{\text{Coefficient of Determination}} = \pm\sqrt{r^2} \tag{12.15}$$

Mean Square Error (Estimate of σ^2)

$$s^2 = \text{MSE} = \frac{\text{SSE}}{n-2} \tag{12.17}$$

Standard Error of the Estimate

$$s = \sqrt{\text{MSE}} = \sqrt{\frac{\text{SSE}}{n-2}} \tag{12.18}$$

Standard Deviation of b_1

$$\sigma_{b_1} = \frac{\sigma}{\sqrt{\Sigma x_i^2 - (\Sigma x_i)^2/n}} \tag{12.19}$$

Estimated Standard Deviation of b_1

$$s_{b_1} = \frac{s}{\sqrt{\Sigma x_i^2 - (\Sigma x_i)^2/n}} \tag{12.20}$$

t Test Statistic

$$t = \frac{b_1}{s_{b_1}} \tag{12.21}$$

Mean Square Due to Regression

$$\text{MSR} = \frac{\text{SSR}}{\text{Number of independent variables}} \tag{12.22}$$

The F Test Statistic

$$F = \frac{\text{MSR}}{\text{MSE}} \tag{12.23}$$

Estimated Standard Deviation of $\hat{y}_\text{p}$

$$s_{\hat{y}_\text{p}} = s\sqrt{\frac{1}{n} + \frac{(x_\text{p} - \bar{x})^2}{\Sigma x_i^2 - (\Sigma x_i)^2/n}} \tag{12.25}$$

Confidence Interval Estimate of $E(\hat{y}_\text{p})$

$$\hat{y}_\text{p} \pm t_{\alpha/2} s_{\hat{y}_\text{p}} \tag{12.26}$$

Estimated Standard Deviation when Predicting an Individual Value

$$s_\text{ind} = s\sqrt{1 + \frac{1}{n} + \frac{(x_\text{p} - \bar{x})^2}{\Sigma x_i^2 - (\Sigma x_i)^2/n}} \tag{12.28}$$

Prediction Interval Estimate of y_p

$$\hat{y}_\text{p} \pm t_{\alpha/2} s_\text{ind} \tag{12.29}$$

Residual for Observation i

$$y_i - \hat{y}_i \tag{12.30}$$

Multiple Regression Model

$$y = \beta_0 + \beta_1 x_1 + \beta_2 x_2 + \cdots + \beta_p x_p + \epsilon \tag{12.32}$$

Multiple Regression Equation

$$E(y) = \beta_0 + \beta_1 x_1 + \beta_2 x_2 + \cdots + \beta_p x_p \tag{12.33}$$

Estimated Multiple Regression Equation

$$\hat{y} = b_0 + b_1 x_1 + b_2 x_2 + \cdots + b_p x_p \tag{12.34}$$

supplementary exercises

53. A study of how much supermarket shoppers could save by purchasing store brand products rather than name brand products was reported in *Consumer Reports* (September 1993). The following data are from a sample of commonly purchased items with brand product prices between $1.00 and $4.00.

Product and Size	Name Brand Price	Store Brand Price
Kraft mayonnaise, qt.	$2.79	$1.19
Heinz ketchup, 2 lb.	1.69	.79
Lipton tea, 100 bags	2.79	1.39
Folgers coffee, 12 oz.	2.79	1.59
Coca-Cola Classic, 6 pack	2.79	1.64
Planters peanuts, 12 oz.	3.79	2.39
Del Monte peaches, 1 lb.	1.19	.83
Kleenex, 250 count	1.59	1.29
Breyers ice cream, 1/2 gal.	3.99	2.39
Oscar Mayer bacon, 1 lb.	3.49	2.09

a. Develop an estimated regression equation with name brand price as the independent variable and store brand price as the dependent variable.
b. What is the coefficient of determination? Did the estimated regression equation provide a good fit?
c. A common price for name brand items is $1.99. On average, what would you expect to pay for an equivalent store brand item? What percentage savings would be realized by purchasing the store brand item?
d. Johnson & Johnson shampoo (20 oz.) has a price of $3.29. What would you expect to pay for an equivalent store brand shampoo product?

54. The law in Hamilton County, Ohio, requires the publication of delinquent property tax information. The county publication lists the name of the property owner, the property valuation, and the amount of taxes, assessments, interest, and penalties due. The property valuation and the amount of taxes due for a sample of 10 delinquent properties are shown in the following table (*Delinquent Land Tax, Hamilton County,* November 17, 1994). The property valuation is in thousands of dollars.

Property Valuation	Amount Due
$18.8	$ 445
24.4	539
20.4	1212
35.8	2237
14.8	479
40.4	1181
49.0	4187
14.5	409
37.3	1002
54.7	2062

a. Develop the estimated regression equation that could be used to estimate the amount of taxes due given the property valuation.

b. Use the estimated regression equation to estimate the taxes due for a property located on Red Bank Road. The valuation of the property is $42,400.

c. Do you believe the estimated regression equation would provide a good prediction of the amount of tax due? Use r^2 to support your answer.

55. *Value Line* (February 24, 1995) reported that the market beta for Woolworth Corporation was 1.25. Market betas for individual stocks are determined by simple linear regression. For each stock, the dependent variable is its quarterly percentage return (capital appreciation plus dividends) minus the percentage return that could be obtained from a risk-free investment (the Treasury Bill rate is used as the risk-free rate). The independent variable is the quarterly percentage return (capital appreciation plus dividends) for the stock market (S&P 500) minus the percentage return from a risk-free investment. An estimated regression equation is developed with quarterly data; the market beta for the stock is the slope of the estimated regression equation (b_1). The value of the market beta is often interpreted as a measure of the risk associated with the stock. Market betas greater than 1 indicate that the stock is more volatile than the market average; market betas less than 1 indicate that the stock is less volatile than the market average. Shown here are the differences between the percentage return and the risk-free return for 10 quarters for the S&P 500 and IBM.

S&P 500	IBM
1.2	−0.7
−2.5	−2.0
−3.0	−5.5
2.0	4.7
5.0	1.8
1.2	4.1
3.0	2.6
−1.0	2.0
.5	−1.3
2.5	5.5

a. Develop an estimated regression equation that can be used to determine the market beta for IBM. What is IBM's market beta?

b. Use the market betas of Woolworth and IBM to compare the risk associated with the two stocks.

c. Did the estimated regression equation provide a good fit? Explain.

56. In a manufacturing process the assembly line speed (feet per minute) was thought to affect the number of defective parts found during the inspection process. To test this theory, managers devised a situation in which the same batch of parts was inspected visually at a variety of line speeds. The following table lists the collected data.

Line Speed	Number of Defective Parts Found
20	21
20	19
40	15
30	16
60	14
40	17

a. Develop the estimated regression equation that relates line speed to the number of defective parts found.

b. At a .05 level of significance, determine whether line speed and number of defective parts found are related.
c. Did the estimated regression equation provide a good fit to the data?
d. Develop a 95% confidence interval to predict the mean number of defective parts for a line speed of 50 feet per minute.

57. Exercise 53 gives data on name brand price and store brand price for products frequently purchased in a supermarket (*Consumer Reports,* September 1993).
a. Use the t test to determine whether there is a significant relationship between name brand price and store brand price. Use $\alpha = .05$. What is your conclusion?
b. Use the F test to determine whether there is a significant relationship between name brand price and store brand price. Use $\alpha = .05$. Show the ANOVA table. What is your conclusion?
c. A common price for name brand items is $1.99. Develop a 95% confidence interval estimate of the mean store brand price for items with a name brand price of $1.99.
d. S & W olives (6 oz.) has a price of $1.99. Develop a 95% prediction interval estimate of the store brand price for olives.

58. A sociologist was hired by a large city hospital to investigate the relationship between the number of unauthorized days that employees are absent per year and the distance (miles) between home and work for the employees. A sample of 10 employees was chosen, and the following data were collected.

Distance to Work	Number of Days Absent
1	8
3	5
4	8
6	7
8	6
10	3
12	5
14	2
14	4
18	2

a. Develop a scatter diagram for these data. Does a linear relationship appear reasonable? Explain.
b. Develop the least squares estimated regression equation.
c. Is there a significant relationship between the two variables? Use $\alpha = .05$.
d. Did the estimated regression equation provide a good fit? Explain.
e. Use the estimated regression equation developed in (b) to develop a 95% confidence interval estimate of the expected number of days absent for employees living five miles from the company.

59. Exercise 55 gives data on quarterly percentage returns that can be used to compute the market beta for IBM.
a. Use the t test to determine whether there is a significant relationship. Use $\alpha = .05$. What is your conclusion?
b. Use the F test to determine whether there is a significant relationship. Use $\alpha = .05$. What is your conclusion?
c. Present the results of the F test in the analysis of variance table format.

60. Exercise 53 gives data on name brand price and store brand price for items frequently purchased in a supermarket (*Consumer Reports,* September 1993). The estimated regression equation is $\hat{y} = .042 + .564x$.

a. Compute the residuals for this data set.
b. What product has the largest residual? What is your interpretation of the value of that residual?
c. What product has the smallest residual? What is your interpretation of the value of that residual?
d. Prepare a residual plot against the name brand price x.
e. Does the residual plot provide any indication that the assumptions of the regression model should be questioned?

61. Do big-budget motion pictures bring in big money at the box office? Data for 1994 motion pictures follow (*Entertainment Weekly,* September 1994).

Motion Picture	Budget ($ Millions)	Gross Sales ($ Millions)
Forrest Gump	50	275
The Flintstones	46	130
Speed	30	120
Maverick	60	100
Color of Night	40	20
The Mask	23	105
City Slickers II	45	44
Clear and Present Danger	62	120
Beverly Hills Cop III	50	42
The Crow	23	51

A regression analysis was performed with budget as the independent variable and gross sales as the dependent variable. Using the following partial computer printout, answer the following questions.

```
Predictor        Coef      Stdev    t-ratio
Constant        50.43      80.95
Budget          1.172      1.804    _______

s = 75.21   R-sq = _______%

Analysis of Variance

SOURCE           DF          SS        MS         F
Regression        1        2388     _______    _______
Error             8       45258     _______
Total             9     _______
```

a. What is the estimated regression equation?
b. What is the coefficient of determination?
c. What is the value of the standard error of the estimate?
d. Conduct a t test at a .05 level of significance.

62. The admissions officer for Clearwater College developed the following estimated regression equation relating the final college GPA to the student's SAT mathematics score and high-school GPA.

$$\hat{y} = -1.41 + .0235x_1 + .00486x_2$$

where

x_1 = high-school grade point average
x_2 = SAT mathematics score
y = final college grade point average

a. Interpret the coefficients in this estimated regression equation.

b. Estimate the final college GPA for a student who has a high-school average of 84 and a score of 540 on the SAT mathematics test.

63. The personnel director for Electronics Associates developed the following estimated regression equation relating an employee's score on a job satisfaction test to his or her length of service and wage rate.

$$\hat{y} = 14.4 - 8.69x_1 + 13.5x_2$$

where

x_1 = length of service (years)

x_2 = wage rate (dollars)

y = job satisfaction test score (higher scores indicate more job satisfaction)

a. Interpret the coefficients in this estimated regression equation.

b. Develop an estimate of the job satisfaction test score for an employee who has four years of service and makes $6.50 per hour.

64. In a regression analysis involving 18 observations and four independent variables, it was determined that SSR = 18,051.63 and SSE = 1014.3.

a. Determine R^2.

b. Test for the significance of the relationship at $\alpha = .01$.

65. The following estimated regression equation involving three independent variables has been developed.

$$\hat{y} = 18.31 + 8.12x_1 + 17.9x_2 - 3.6x_3$$

Computer output indicates that $s_{b_1} = 2.1$, $s_{b_2} = 9.72$, and $s_{b_3} = .71$. There were 15 observations in the study.

a. Test H_0: $\beta_1 = 0$ at $\alpha = .05$.

b. Test H_0: $\beta_2 = 0$ at $\alpha = .05$.

c. Test H_0: $\beta_3 = 0$ at $\alpha = .05$.

d. Would you recommend dropping any of the independent variables from the model?

66. A partial computer output from a regression analysis follows.

```
The regression equation is
Y = 8.103 + 7.602 X1 + 3.111 X2

Predictor              Coef          Stdev        t-ratio
Constant             ______          2.667         _____
X1                   ______          2.105         _____
X2                   ______          0.613         _____

s = 3.35          R-sq = 92.3%

Analysis of Variance

SOURCE            DF           SS            MS           F
Regression      _____        1612         _____        _____
Error              12      _______        _____
Total           _____      _______
```

a. Compute the appropriate t-ratios.
b. Test for the significance of β_1 and β_2 at $\alpha = .05$.
c. Compute the entries in the DF, SS, and MS = SS/DF columns.

67. Recall that in Exercise 62, the admissions officer for Clearwater College developed the following estimated regression equation relating final college GPA to the student's SAT mathematics score and high-school GPA.

$$\hat{y} = -1.41 + .0235x_1 + .00486x_2$$

where

x_1 = high-school grade point average
x_2 = SAT mathematics score
y = final college grade point average

A portion of the Minitab computer output follows.

```
The regression equation is
Y = -1.41 + .0235 X1 + .00486 X2

Predictor          Coef         Stdev      t-ratio
Constant        -1.4053        0.4848      _____
X1             0.023467      0.008666      _____
X2              _______      0.001077      _____

s = 0.1298      R-sq = _____

Analysis of Variance

SOURCE        DF          SS          MS        F
Regression   _____     1.76209      _____    _____
Error        _____     _______      _____
Total          9       1.88000
```

a. Complete the missing entries in this output.
b. Compute F and test at a .05 level of significance to see whether a significant relationship is present.
c. Did the estimated regression equation provide a good fit to the data? Explain.
d. Use the t test and $\alpha = .05$ to test H_0: $\beta_1 = 0$ and H_0: $\beta_2 = 0$.

68. Recall that in Exercise 63 the personnel director for Electronics Associates developed the following estimated regression equation relating an employee's score on a job satisfaction test to length of service and wage rate.

$$\hat{y} = 14.4 - 8.69x_1 + 13.5x_2$$

where

x_1 = length of service (years)
x_2 = wage rate (dollars)
y = job satisfaction test score (higher scores indicate more job satisfaction)

A portion of the Minitab computer output follows.

```
The regression equation is
Y = 14.4 - 8.69 X1 + 13.52 X2

Predictor            Coef          Stdev     t-ratio
Constant           14.448          8.191        1.76
X1                 ______          1.555      ______
X2                 13.517          2.085      ______

s = 3.773       R-sq = ______

Analysis of Variance

SOURCE          DF          SS          MS          F
Regression       2      ______      ______     ______
Error       ______       71.17      ______
Total            7       720.0
```

a. Complete the missing entries in this output.
b. Compute F and test using $\alpha = .05$ to see whether a significant relationship is present.
c. Did the estimated regression equation provide a good fit to the data? Explain.
d. Use the t test and $\alpha = .05$ to test H_0: $\beta_1 = 0$ and H_0: $\beta_2 = 0$.

69. Bauman Construction Company makes bids on a variety of projects. In an effort to estimate the bid to be made by one of its competitors, Bauman obtained data on 15 previous bids and developed the following estimated regression equation.

$$\hat{y} = 80 + 45x_1 - 3x_2$$

where

$\hat{y}$ = competitor's bid (\$1000s)

x_1 = square feet (1000s)

x_2 = local index of construction activity

a. Estimate the competitor's bid on a project involving 50,000 square feet and an index of construction activity of 70.
b. If SSR = 19,780 and SST = 21,533, test at $\alpha = .01$ for the significance of the relationship.

computer case

U.S. Department of Transportation

As part of a study on transportation safety, the U.S. Department of Transportation collected data on the number of fatal accidents per 1000 licenses and the percentage of licensed drivers under the age of 21 in a sample of 42 cities. Data collected over a one-year period follow. These data are available on the data disk in the file named SAFETY.

SAFETY

Percent Under 21	Fatal Accidents per 1000 Licenses	Percent Under 21	Fatal Accidents per 1000 Licenses
13	2.962	17	4.100
12	0.708	8	2.190
8	0.885	16	3.623
12	1.652	15	2.623
11	2.091	9	0.835
17	2.627	8	0.820
18	3.830	14	2.890
8	0.368	8	1.267
13	1.142	15	3.224
8	0.645	10	1.014
9	1.028	10	0.493
16	2.801	14	1.443
12	1.405	18	3.614
9	1.433	10	1.926
10	0.039	14	1.643
9	0.338	16	2.943
11	1.849	12	1.913
12	2.246	15	2.814
14	2.855	13	2.634
14	2.352	9	0.926
11	1.294	17	3.256

Managerial Report

1. Develop numerical and graphical summaries of the data.
2. Use regression analysis to investigate the relationship between the number of fatal accidents and the percentage of drivers under the age of 21. Discuss your findings.
3. What conclusion and/or recommendations can you derive from your analysis?

computer case

Consumer Research, Inc.

Consumer Research, Inc., is an independent agency that conducts research on consumer attitudes and behaviors for a variety of firms. In one study, a client asked for an investigation of consumer characteristics that can be used to predict the amount charged by credit-card users. Data were collected on annual income, household size, and annual credit-card charges for a sample of 50 consumers. The data follow and are on the data disk in the file named CONSUMER.

Income ($1000s)	Household Size	Amount Charged ($)	Income ($1000s)	Household Size	Amount Charged ($)
54	3	4016	54	6	5573
30	2	3159	30	1	2583
32	4	5100	48	2	3866
50	5	4742	34	5	3586
31	2	1864	67	4	5037
55	2	4070	50	2	3605
37	1	2731	67	5	5345
40	2	3348	55	6	5370
66	4	4764	52	2	3890
51	3	4110	62	3	4705
25	3	4208	64	2	4157
48	4	4219	22	3	3579
27	1	2477	29	4	3890
33	2	2514	39	2	2972
65	3	4214	35	1	3121
63	4	4965	39	4	4183
42	6	4412	54	3	3730
21	2	2448	23	6	4127
44	1	2995	27	2	2921
37	5	4171	26	7	4603
62	6	5678	61	2	4273
21	3	3623	30	2	3067
55	7	5301	22	4	3074
42	2	3020	46	5	4820
41	7	4828	66	4	5149

Managerial Report

1. Use methods of descriptive statistics to summarize the data. Comment on the findings.
2. Develop estimated regression equations, first using annual income as the independent variable and then using household size as the independent variable. Which variable is the better predictor of annual credit-card charges? Discuss your findings.
3. Develop an estimated regression equation with annual income and household size as the independent variables. Discuss your findings.
4. What is the predicted annual credit-card charge for a three-person household with an annual income of $40,000?
5. Discuss the need for other independent variables that could be added to the model. What additional variables might be helpful?

APPENDIX 12.1 Regression Analysis with Minitab

In Section 12.7 we discussed the computer solution of regression problems by showing Minitab's output for the Armand's Pizza Parlors problem. In this appendix, we describe the steps required to generate the Minitab computer solution. First, the data must be entered in a Minitab worksheet. Student population data were entered in column C1 and sales data were entered in column C2. The variable names POP and SALES were entered as the column headings on the worksheet. In subsequent steps, we could refer to the data by using the variable names POP and SALES or the column indicators C1 and C2. The following steps describe how to use Minitab to produce the regression results.

STEP 1. Select the **Stat** pull-down menu
STEP 2. Select the **Regression** pull-down menu
STEP 3. Choose the **Regression** option
STEP 4. When the Regression dialog box appears:
Enter SALES in the **Response** box
Enter POP in the **Predictors** box
Select **OK** to obtain the regression analysis

The summary output is shown in Figure 12.10. The Minitab regression dialog box provides additional output information that can be obtained by selecting the desired options. For example, residuals, can be obtained in that way.

APPENDIX 12.2 Regression Analysis with Spreadsheets

Let us demonstrate regression analysis with spreadsheets by using Excel to provide the computations for the Armand's Pizza Parlors problem. Figure 12.17 shows that we have entered Armand's data in columns A and B of the spreadsheet. Student population data are in cells 2 to 11 of column A and sales data are in cells 2 to 11 of column B. The following steps describe how to use Excel to produce the regression results.

STEP 1. Select the **Tools** pull-down menu
STEP 2. Choose the **Data Analysis** option
STEP 3. When the Analysis Tools dialog box appears,
Choose **Regression**
STEP 4. When the Regression dialog box appears:
Enter B2:B11 in the **Input Y Range** box
Enter A2:A11 in the **Input X Range** box
Enter A14 in the **Output Range** box
(The upper-left corner cell indicating where the output is to begin should be entered here.)
Select **OK** to begin the regression analysis

The summary output begins with row 14 in Figure 12.17. The Multiple R cell gives the sample correlation coefficient of .9501 as before. The coefficient of determination

	A	B	C	D	E	F	G
1	Population	Sales					
2	2	58					
3	6	105					
4	8	88					
5	8	118					
6	12	117					
7	16	137					
8	20	157					
9	20	169					
10	22	149					
11	26	202					
12							
13							
14	SUMMARY OUTPUT						
15							
16	*Regression Statistics*						
17	Multiple R	0.9501					
18	R Square	0.9027					
19	Adjusted R Square	0.8906					
20	Standard Error	13.829					
21	Observations	10					
22							
23	ANOVA						
24		*df*	*SS*	*MS*	*F*	*Significance F*	
25	Regression	1	14200	14200	74.25	0.000	
26	Residual	8	1530	191.25			
27	Total	9	15730				
28							
29		*Coefficients*	*Standard Error*	*t Stat*	*P-value*	*Lower 95%*	*Upper 95%*
30	Intercept	60	9.2260	6.50	0.000	38.72	81.28
31	X Variable 1	5	0.5803	8.62	0.000	3.66	6.34

FIGURE 12.17
Excel Spreadsheet Solution for the Armand's Pizza Parlors Problem

(90.27%) and the standard error of the estimate (13.829) are also shown as before. The analysis of variance table is next. The *p*-value under the heading Significance *F* shows that the null hypothesis H_0: $\beta_1 = 0$ can be rejected, indicating that the relationship between student population and sales is significant. The regression coefficients of 60 for the intercept and 5 for the slope of the estimated regression line are significant. They show that the estimated regression equation is $\hat{y} = 60 + 5x$. The *t* statistic and the *p*-value for the *t* test also show that the relationship is significant. Finally, the Lower 95% and Upper 95% cells provide the 95% confidence interval estimates for the regression model parameters β_0 and β_1. For example, we can be 95% confident that β_0 is between 38.72 and 81.28 and 95% confident that β_1 is between 3.66 and 6.34.

CHAPTER 13

Statistical Methods for Quality Control

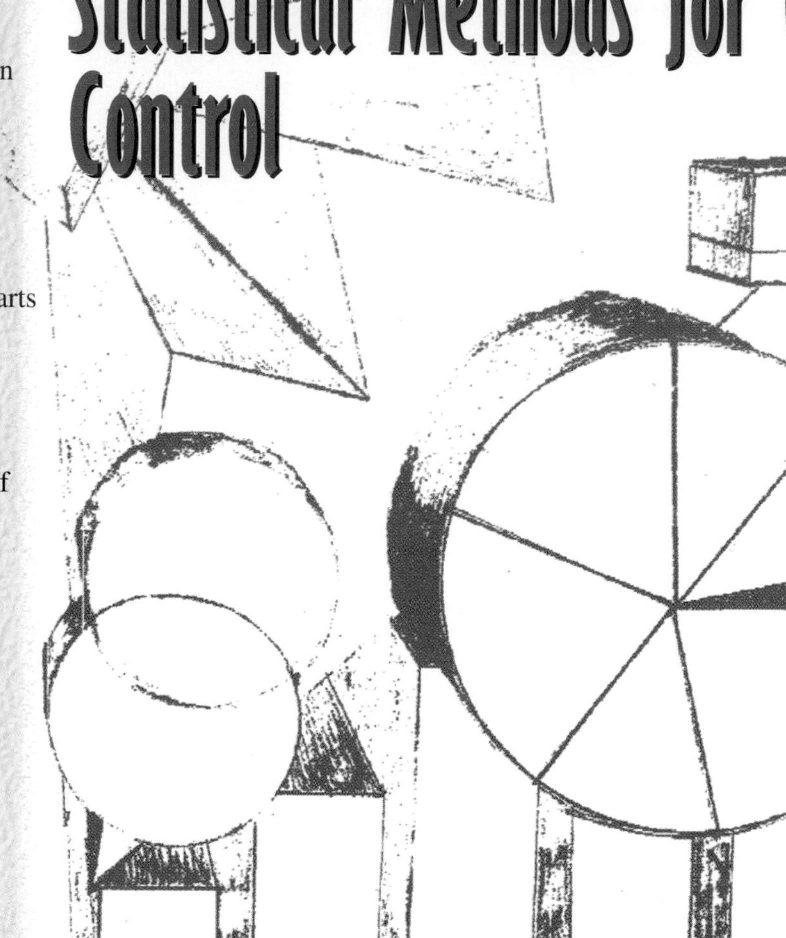

Dow Chemical U.S.A.*

Freeport, Texas

Dow Chemical U.S.A., Texas Operations, began in 1940 when The Dow Chemical Company purchased 800 acres of Texas land on the Gulf Coast to build a magnesium production facility. That original site has expanded to cover more than 5000 acres and is now one of the largest petrochemical complexes in the world. Among the products from Texas Operations are magnesium, styrene, plastics, adhesives, solvent, glycol, and chlorine. Some products are made solely for use in other processes, but many end up as essential ingredients in products such as pharmaceuticals, toothpastes, dog food, water hoses, ice chests, milk cartons, garbage bags, shampoos, and furniture.

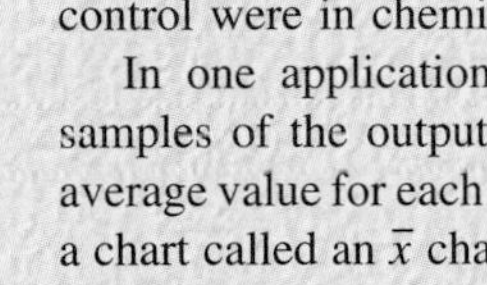

Dow's Texas Operations produces more than 30% of the world's magnesium, an extremely lightweight metal used in products ranging from tennis racquets to suitcases to "mag" wheels. The Magnesium Department was the first group in Texas Operations to train its technical people and managers in the use of statistical quality control. Some of the earliest successful applications of statistical quality control were in chemical processing.

In one application involving the operation of a drier, samples of the output were taken at periodic intervals; the average value for each sample was computed and recorded on a chart called an $\bar{x}$ chart. Such a chart enabled Dow analysts to monitor trends in the output that might indicate the process was not operating correctly. In one instance, analysts began to observe values for the sample mean that were not indicative of a process operating within its design limits. On further examination of the control chart and the operation itself, the analysts found that the variation could be traced to problems involving one operator. The $\bar{x}$ chart recorded after that operator was retrained showed a significant improvement in the process quality.

Dow Chemical has achieved quality improvements everywhere statistical quality control has been used. Documented savings of several hundred thousand dollars per year have been realized, and new applications are continually being discovered.

In this chapter we will show how an $\bar{x}$ chart such as the one used by Dow Chemical can be developed. Such charts are a part of statistical quality control known as statistical process control. We will also discuss methods of quality control for situations in which a decision to accept or reject a group of items is based on a sample.

Magnesium is the primary product at Dow's Texas operations.

The authors are indebted to Clifford B. Wilson, Magnesium Technical Manager, The Dow Chemical Company, for providing this Statistics in Practice.

The American Society for Quality Control (ASQC) defines *quality* as "the totality of features and characteristics of a product or service that bears on its ability to satisfy given needs." In other words, quality measures how well a product or service meets customer needs. Organizations recognize that to be competitive in today's global economy, they must strive for high levels of quality. As a result, there has been an increased emphasis on methods for monitoring and maintaining quality.

Quality assurance refers to the entire system of policies, procedures, and guidelines established by an organization to achieve and maintain quality. Quality assurance consists of two principal functions: quality engineering and quality control. The objective of *quality engineering* is to include quality in the design of products and processes and to identify potential quality problems prior to production. *Quality control* consists of making a series of inspections and measurements to determine whether quality standards are being met. If quality standards are not being met, corrective and/or preventive action can be taken to achieve and maintain conformance. As we will show in this chapter, statistical techniques are extremely useful in quality control.

Traditional manufacturing approaches to quality control have been found to be less than satisfactory and are being replaced by improved managerial tools and techniques. Ironically, it was two U.S. consultants, Dr. W. Edwards Deming and Dr. Joseph Juran, who helped educate the Japanese in quality management. In recent years, U.S. firms have relearned those lessons from Japan.

Although quality is everybody's job, Deming stressed that quality must be led by managers. He developed a list of 14 points that he believed are the key responsibilities of managers. For instance, Deming stated that managers must cease dependence on mass inspection; must end the practice of awarding business solely on the basis of price; must seek continual improvement in all production processes and services; must foster a team-oriented environment; and must eliminate numerical goals, slogans, and work standards that prescribe numerical quotas. Perhaps most important, managers must create a work environment in which a commitment to quality and productivity is maintained at all times.

In 1987, the U.S. Congress enacted Public Law 107, the Malcolm Baldrige National Quality Improvement Act. The Baldrige Award is given annually to U.S. firms that excel in quality. This award, along with the perspectives of individuals like Dr. Deming and Dr. Juran, has helped top managers recognize that improving service quality and product quality is the most critical challenge facing their companies. Winners of the Malcolm Baldrige Award include Motorola, IBM, Xerox, and Federal Express. In this chapter we present two statistical methods used in quality control. The first method, *statistical process control,* uses graphical displays known as *control charts* to monitor a production process; the goal is to determine whether the process can be continued or whether it should be adjusted to achieve a desired quality level. The second method, *acceptance sampling,* is used in situations where a decision to accept or reject a group of items must be based on the quality found in a sample.

13.1 Statistical Process Control

In this section we consider quality-control procedures for a production process whereby goods are manufactured continuously. On the basis of sampling and inspection of production output, a decision will be made to either continue the production process or adjust it to bring the items or goods being produced up to acceptable quality standards.

Despite high standards of quality in manufacturing and production operations, machine tools will invariably wear out, vibrations will throw machine settings out of adjustment, purchased materials will be defective, and human operators will make mistakes. Any or all of these factors can result in poor-quality output. Fortunately, procedures are available for monitoring production output so that poor quality can be detected early and the production process can be adjusted or corrected.

If the variation in the quality of the production output is due to *assignable causes* such as tools wearing out, incorrect machine settings, poor-quality raw materials, or operator error, the process should be adjusted or corrected as soon as possible. Alternatively, if the variation is due to what are called common causes—that is, randomly occurring variations in materials, temperature, humidity, and so on, which the manufacturer cannot possibly control—the process does not need to be adjusted. The main objective of statistical process control is to determine whether variations in output are due to assignable causes or common causes.

		State of Production Process	
		H_0 *True* *Process in Control*	H_0 *False* *Process Out of Control*
Decision	*Continue Process*	Correct decision	Type II error (allowing an out-of-control process to continue)
	Adjust Process	Type I error (adjusting an in-control process)	Correct decision

TABLE 13.1 The Outcomes of Statistical Process Control

Whenever assignable causes are detected, we conclude that the process is *out of control.* In that case, corrective action will be taken to bring the process back to an acceptable level of quality. However, if the variation in the output of a production process is due only to common causes, we conclude that the process is *in statistical control,* or simply *in control;* in such cases, no changes or adjustments are necessary.

The statistical procedures for process control are based on the hypothesis-testing methodology presented in Chapter 9. The null hypothesis H_0 is formulated in terms of the production process being in control. The alternative hypothesis H_a is formulated in terms of the production process being out of control. Table 13.1 shows that correct decisions to continue an in-control process and adjust an out-of-control process are possible. However, as with other hypothesis-testing procedures, both a Type I error (adjusting an in-control process) and a Type II error (allowing an out-of-control process to continue) are also possible.

Control Charts

A *control chart* provides a basis for deciding whether the variation in the output is due to common causes (in control) or assignable causes (out of control). Whenever an out-of-control situation is detected, adjustments and/or other corrective action will be taken to bring the process back into control.

Control charts can be classified by the type of data they contain. An $\bar{x}$ chart is used if the quality of the output is measured in terms of a variable such as length, weight, temperature, and so on. In that case, the decision to continue or to adjust the production process will be based on the mean value found in a sample of the output. To introduce some of the concepts common to all control charts, let us consider some specific features of an $\bar{x}$ chart.

Figure 13.1 shows the general structure of an $\bar{x}$ chart. The center line of the chart corresponds to the mean of the process when the process is *in control.* The vertical line identifies the scale of measurement for the variable of interest. Each time a sample is taken from the production process, a value of the sample mean $\bar{x}$ is computed and a data point showing the value of $\bar{x}$ is plotted on the control chart.

The two lines labeled UCL and LCL are important in determining whether the process is in control or out of control. The lines are called the *upper control limit* and the *lower control limit,* respectively. They are chosen so that when the process is in control, there will be a high probability that the value of $\bar{x}$ will be between the two control limits.

FIGURE 13.1
$\bar{x}$ Chart Structure

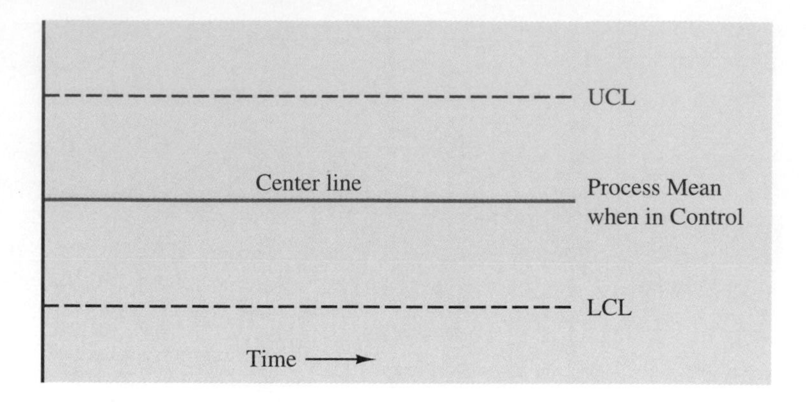

Values outside the control limits provide strong statistical evidence that the process is out of control and corrective action should be taken.

Over time, more and more data points will be added to the control chart. The order of the data points will be from left to right as the process is sampled. In essence, every time a point is plotted on the control chart, we are carrying out a hypothesis test to determine whether the process is in control.

In addition to the $\bar{x}$ chart, other control charts can be used to monitor the range of the measurements in the sample (R chart), the proportion defective in the sample (p chart), and the number of defective items in the sample (np chart). In each case, the general structure of the control chart follows the format of the $\bar{x}$ chart in Figure 13.1. The major difference among the charts is the measurement scale used; for instance, in a p chart the measurement scale denotes the proportion of defective items in the sample instead of the sample mean. In the following discussion, we will illustrate the construction and use of the $\bar{x}$ chart, R chart, p chart, and np chart.

$\bar{x}$ Chart: Process Mean and Standard Deviation Known

To illustrate the construction of an $\bar{x}$ chart, let us consider the situation at KJW Packaging. This company operates a production line where cartons of cereal are filled. Suppose KJW knows that when the process is operating correctly—and hence the system is in control—the mean filling weight is $\mu = 16.05$ ounces and the process standard deviation is $\sigma = .10$ ounces. In addition, assume the filling weights have a normal probability distribution. This distribution is shown in Figure 13.2.

The sampling distribution of $\bar{x}$, as presented in Chapter 7, can be used to determine the variation that can be expected in $\bar{x}$ values for a process that is in control. To show how this is done, let us first briefly review the properties of the sampling distribution of $\bar{x}$. First, recall that the expected value or mean of $\bar{x}$ is equal to μ, the mean filling weight when the production line is in control. For samples of size n, the formula for the standard deviation of $\bar{x}$, called the standard error of the mean, is

$$\sigma_{\bar{x}} = \frac{\sigma}{\sqrt{n}} \tag{13.1}$$

In addition, since the filling weights have a normal probability distribution, the sampling distribution of $\bar{x}$ has a normal probability distribution for any sample size. Thus, the sampling distribution of $\bar{x}$ is a normal probability distribution with mean μ and standard deviation $\sigma_{\bar{x}}$. This distribution is shown in Figure 13.3.

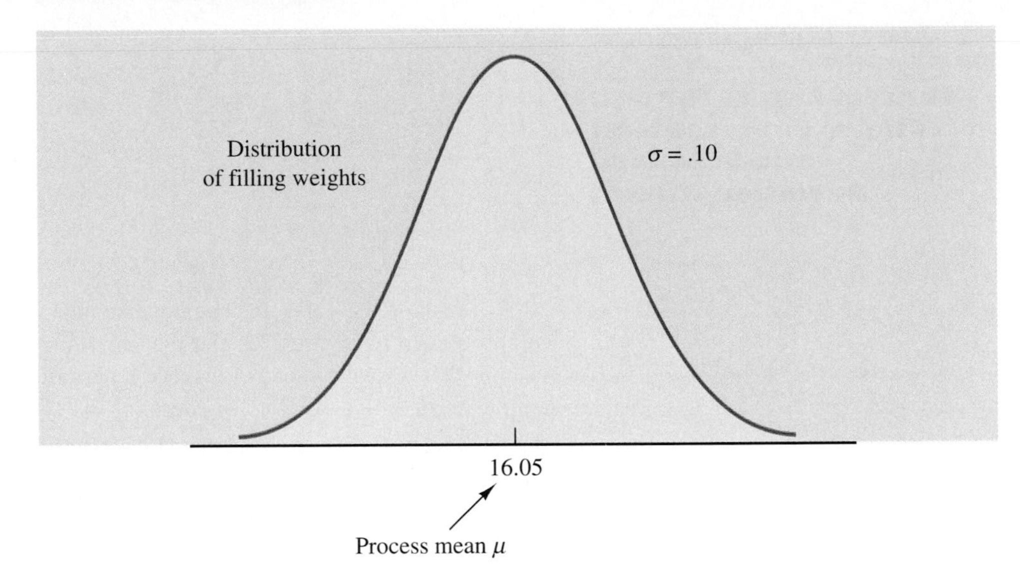

FIGURE 13.2
Distribution of Cereal-Carton Filling Weights

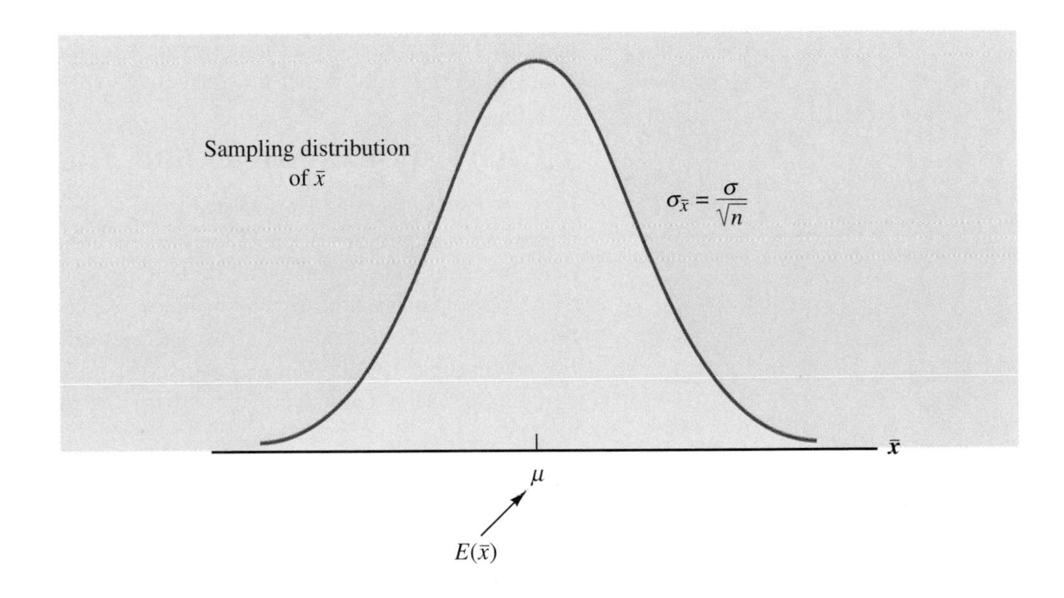

FIGURE 13.3
Sampling Distribution of $\bar{x}$

The sampling distribution of $\bar{x}$ is used to determine what values of $\bar{x}$ are reasonable if the process is in control. The general practice in quality control is to define as reasonable any value of $\bar{x}$ that is within 3 standard deviations above or below the mean value. Recall from the study of the normal probability distribution that approximately 99.7% of the values of a normally distributed random variable are within ± 3 standard deviations of its mean value. Thus, if a value of $\bar{x}$ is within the interval $\mu - 3\sigma_{\bar{x}}$ to $\mu + 3\sigma_{\bar{x}}$, we will assume that the process is in control. In summary, then, the control limits for an $\bar{x}$ chart are as follows.

Control Limits for an $\bar{x}$ Chart: Process Mean and Standard Deviation Known

$$\text{UCL} = \mu + 3\sigma_{\bar{x}} \tag{13.2}$$

$$\text{LCL} = \mu - 3\sigma_{\bar{x}} \tag{13.3}$$

Reconsider the KJW Packaging example with the process distribution of filling weights shown in Figure 13.2 and the sampling distribution of $\bar{x}$ shown in Figure 13.3. Assume that a quality-control inspector periodically samples six cartons and uses the sample mean filling weight to determine whether the process is in control or out of control. Using (13.1), we find that the standard error of the mean is $\sigma_{\bar{x}} = \sigma/\sqrt{n} = .10/\sqrt{6} = .04$. Thus, with the process mean at 16.05, the control limits are UCL $= 16.05 + 3(.04) = 16.17$ and LCL $= 16.05 - 3(.04) = 15.93$. Figure 13.4 is the control chart with the results of 10 samples taken over a 10-hour period. For ease of reading, the sample numbers 1 through 10 are listed below the chart.

Note that the mean for the fifth sample in Figure 13.4 shows that the process was out of control. In other words, the sample mean $\bar{x} = 15.89$ provides an indication that assignable causes of output variation were present and that underfilling was occurring. As a result, corrective action was taken at this point to bring the process back into control. The fact that the remaining points on the $\bar{x}$ chart are within the upper and lower control limits indicates that the corrective action was successful.

$\bar{x}$ Chart: Process Mean and Standard Deviation Unknown

In the KJW Packaging example, we showed how an $\bar{x}$ chart can be developed when the mean and standard deviation of the process are known. In most situations, those values must be estimated by using samples that are selected from the process when it is known to be operating in control. For instance, KJW might select a random sample of five boxes each morning and five boxes each afternoon for 10 days of in-control operation. For each subgroup, or sample, the mean and standard deviation of the sample are computed. The overall averages of both the sample means and the sample standard

FIGURE 13.4
The $\bar{x}$ Chart for the Cereal-Carton Filling Process

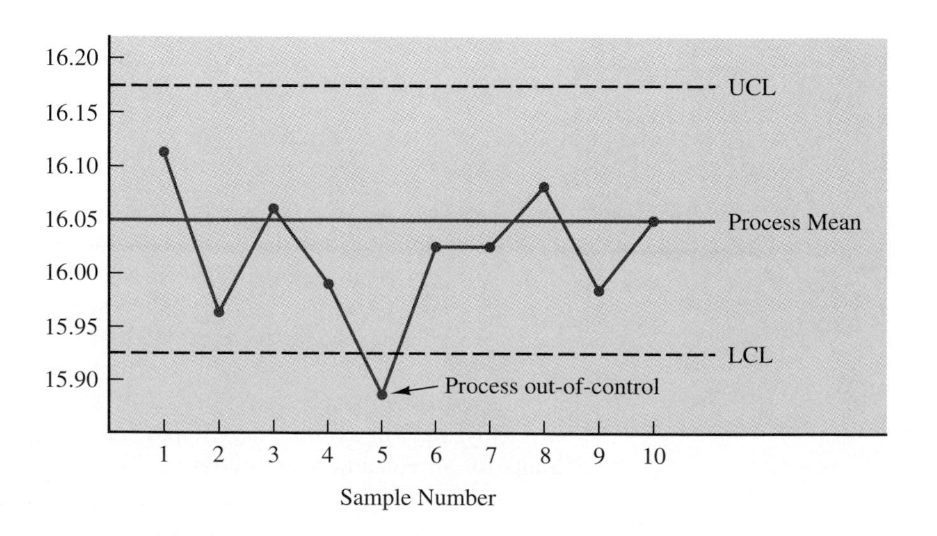

deviations are used to construct control charts for both the process mean and the process standard deviation.

In practice, it is more common to monitor the variability of the process by using the range instead of the standard deviation because the range is easier to compute. In addition to providing good estimates of the process standard deviation when the sample size is small, the range can be used to construct upper and lower control limits for the $\bar{x}$ chart with little computational effort. To illustrate, let us consider the problem facing Jensen Computer Supplies, Inc.

Jensen Computer Supplies (JCS) manufactures floppy disks that are used in personal computers. Suppose that random samples of five disks are taken during the first hour of operation, during the second hour of operation, and so on, until 20 samples have been selected. Table 13.2 lists the data, including the mean $\bar{x}_j$ and range R_j for each of the samples.

Assume that the diameter of disks produced when the process is in control is a normally distributed random variable with mean μ and standard deviation σ, and that k samples, each of size n, have been selected. The estimate of the process mean μ is given by the overall sample mean.

Overall Sample Mean

$$\bar{\bar{x}} = \frac{\bar{x}_1 + \bar{x}_2 + \cdots + \bar{x}_k}{k} \tag{13.4}$$

where

$\bar{x}_j$ = mean of the jth sample, $j = 1, 2, \ldots, k$

k = number of samples

TABLE 13.2 Data for the Jensen Computer Supplies Problem

Sample Number	Observations					Sample Mean $\bar{x}_j$	Sample Range R_j
1	3.5056	3.5086	3.5144	3.5009	3.5030	3.5065	.0135
2	3.4882	3.5085	3.4884	3.5250	3.5031	3.5026	.0368
3	3.4897	3.4898	3.4995	3.5130	3.4969	3.4978	.0233
4	3.5153	3.5120	3.4989	3.4900	3.4837	3.5000	.0316
5	3.5059	3.5113	3.5011	3.4773	3.4801	3.4951	.0340
6	3.4977	3.4961	3.5050	3.5014	3.5060	3.5012	.0099
7	3.4910	3.4913	3.4976	3.4831	3.5044	3.4935	.0213
8	3.4991	3.4853	3.4830	3.5083	3.5094	3.4970	.0264
9	3.5099	3.5162	3.5228	3.4958	3.5004	3.5090	.0270
10	3.4880	3.5015	3.5094	3.5102	3.5146	3.5047	.0266
11	3.4881	3.4887	3.5141	3.5175	3.4863	3.4989	.0312
12	3.5043	3.4867	3.4946	3.5018	3.4784	3.4932	.0259
13	3.5043	3.4769	3.4944	3.5014	3.4904	3.4935	.0274
14	3.5004	3.5030	3.5082	3.5045	3.5234	3.5079	.0230
15	3.4846	3.4938	3.5065	3.5089	3.5011	3.4990	.0243
16	3.5145	3.4832	3.5188	3.4935	3.4989	3.5018	.0356
17	3.5004	3.5042	3.4954	3.5020	3.4889	3.4982	.0153
18	3.4959	3.4823	3.4964	3.5082	3.4871	3.4940	.0259
19	3.4878	3.4864	3.4960	3.5070	3.4984	3.4951	.0206
20	3.4969	3.5144	3.5053	3.4985	3.4885	3.5007	.0259

For the JCS data in Table 13.2, the overall sample mean is $\bar{\bar{x}} = 3.4995$. This value will be the center line for the $\bar{x}$ chart. The range of each sample, denoted R_j, is simply the difference between the largest and smallest values in each sample. The average range follows.

Average Range

$$\bar{R} = \frac{R_1 + R_2 + \cdots + R_k}{k} \tag{13.5}$$

where

R_j = range of the jth sample, $j = 1, 2, \ldots, k$

k = number of samples

For the JCS data in Table 13.2, the average range is $\bar{R} = .0253$.

In the preceding section we showed that the upper and lower control limits for the $\bar{x}$ chart are

$$\bar{x} \pm 3\frac{\sigma}{\sqrt{n}} \tag{13.6}$$

Hence, to construct the control limits for the $\bar{x}$ chart, we need to estimate σ, the standard deviation of the process. An estimate of σ can be developed by using the range of each sample.

It can be shown that an estimator of the process standard deviation σ is the average range divided by d_2, a constant that depends on the sample size n. That is,

$$\text{Estimator of } \sigma = \frac{\bar{R}}{d_2} \tag{13.7}$$

The American Society for Testing and Materials Manual on Presentation of Data and Control Chart Analysis provides values for d_2 as shown in Table 9 of Appendix B. For instance, when $n = 5$, $d_2 = 2.326$ and the estimate of σ is the average range divided by 2.326. If we substitute $\bar{R}/d_2$ for σ in (13.6), we can write the control limits for the $\bar{x}$ chart as

$$\bar{\bar{x}} \pm 3\frac{\bar{R}/d_2}{\sqrt{n}} = \bar{\bar{x}} \pm \frac{3}{d_2\sqrt{n}}\bar{R} = \bar{\bar{x}} \pm A_2\bar{R} \tag{13.8}$$

Note that $A_2 = 3/(d_2\sqrt{n})$ is a constant that depends only on the sample size. Values for A_2 are provided in Table 9 of Appendix B. For $n = 5$, $A_2 = .577$; thus, the control limits for the $\bar{x}$ chart are

$$3.4995 \pm (.577)(.0253) = 3.4995 \pm .0146$$

Hence, LCL = 3.4849 and UCL = 3.5141.

R Chart

Let us now consider the use of a range chart (R chart) which can be used to control the variability of a process. To develop the R chart, we need to think of the range of a sample as a random variable with its own mean and standard deviation. The average range $\bar{R}$ provides an estimate of the mean of this random variable. Moreover, it can be shown that an estimate of the standard deviation of the range is

$$\hat{\sigma}_R = d_3 \frac{\bar{R}}{d_2} \tag{13.9}$$

where d_2 and d_3 are constants that depend on the sample size; values of d_2 and d_3 are also provided in Table 9 of Appendix B. Thus, the UCL for the R chart is given by

$$\bar{R} + 3\hat{\sigma}_R = \bar{R} + 3\, d_3 \frac{\bar{R}}{d_2} \tag{13.10}$$

and the LCL is

$$\bar{R} - 3\hat{\sigma}_R = \bar{R} - 3\, d_3 \frac{\bar{R}}{d_2} \tag{13.11}$$

If we let

$$D_4 = 1 + 3\frac{d_3}{d_2} \tag{13.12}$$

$$D_3 = 1 - 3\frac{d_3}{d_2} \tag{13.13}$$

we can write the control limits for the R chart as

$$\text{UCL} = \bar{R}D_4 \tag{13.14}$$

$$\text{LCL} = \bar{R}D_3 \tag{13.15}$$

Values for D_3 and D_4 are also provided in Table 9 of Appendix B. Note that for $n = 5$, $D_3 = 0$ and $D_4 = 2.115$. Thus, with $\bar{R} = .0253$, the control limits are

$$\text{UCL} = .0253(2.115) = .0535$$

$$\text{LCL} = .0253(0) = 0$$

Figure 13.5 is the R chart. The 20 ranges plotted on the chart do not indicate that the process is out of control. Figure 13.6 is the $\bar{x}$ chart with the 20 sample means for the JCS data; we note that these points show no indication of an out-of-control condition. These data confirm the assumption that during the time period in which the data were collected, the process was in control both in terms of its mean and its variation.

p Chart

Let us consider the case in which the decision to continue or to adjust the production process will be based on $\bar{p}$, the proportion of defective items found in a sample of the output. The control chart used for proportion-defective data is called a *p chart.*

To illustrate the construction of a p chart, consider the use of automated mail-sorting machines in a post office. These automated machines scan the zip codes on letters and divert each letter to its proper carrier route. Even when a machine is operating properly, some letters are diverted to incorrect routes. Assume that when a machine is operating correctly, or in a state of control, 3% of the letters are incorrectly diverted. Thus p, the proportion of letters incorrectly diverted when the process is in control, is .03.

FIGURE 13.5
R **Chart for the Jensen Computer Supplies Problem**

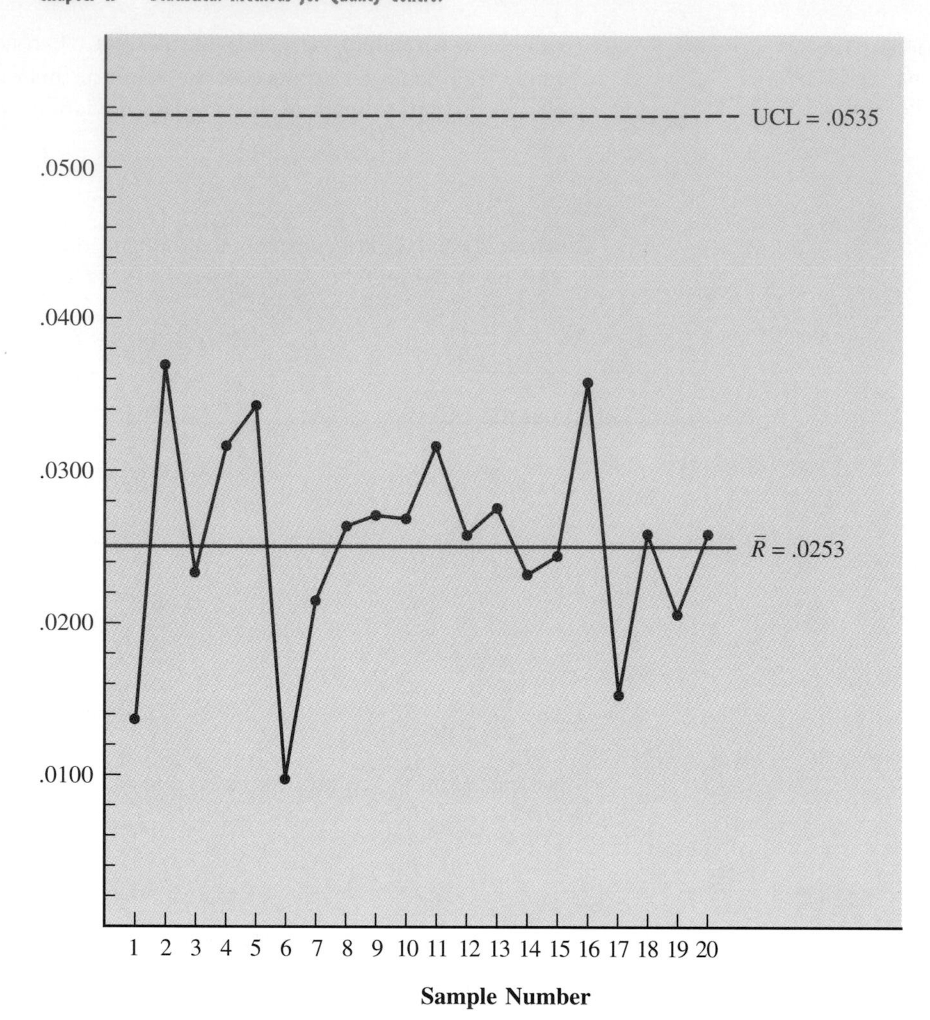

The sampling distribution of $\bar{p}$, as presented in Chapter 7, can be used to determine the variation that can be expected in $\bar{p}$ values for a process that is in control. Recall that the expected value or mean of $\bar{p}$ is p, the proportion defective when the process is in control. With samples of size n, the formula for the standard deviation of $\bar{p}$, called the standard error of the proportion, is

$$\sigma_{\bar{p}} = \sqrt{\frac{p(1-p)}{n}} \tag{13.16}$$

We also learned in Chapter 7 that the sampling distribution of $\bar{p}$ can be approximated by a normal probability distribution whenever the sample size is large. With $\bar{p}$, the sample size can be considered large whenever the following two conditions are satisfied.

$$np \geq 5$$
$$n(1-p) \geq 5$$

In summary, whenever the sample size is large, the sampling distribution of $\bar{p}$ can be approximated by a normal probability distribution with mean p and standard deviation $\sigma_{\bar{p}}$. This distribution is shown in Figure 13.7.

FIGURE 13.6
$\bar{x}$ **Chart for the Jensen Computer Supplies Problem**

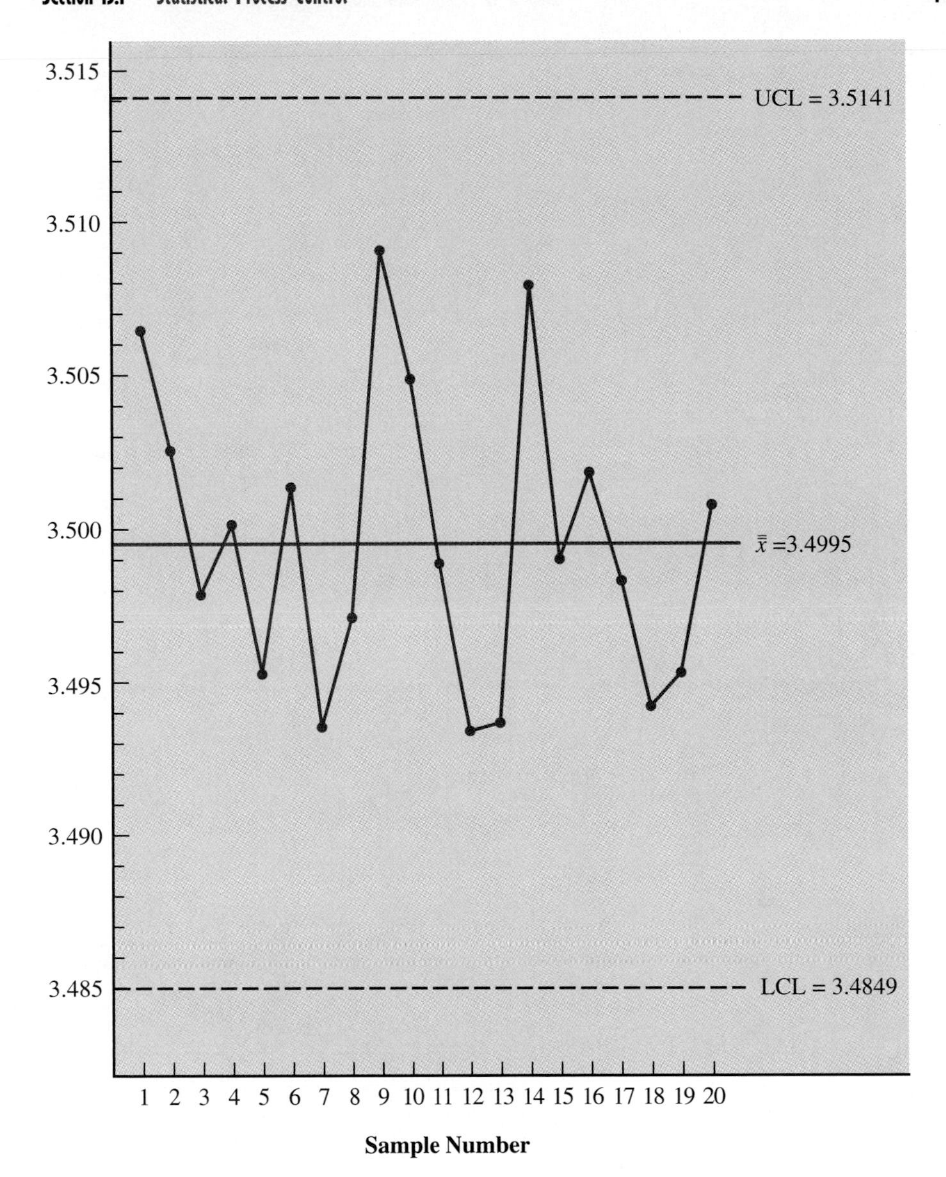

To establish control limits for a p chart, we follow the same procedure we used to establish control limits for an $\bar{x}$ chart. That is, the limits for the control chart are set at 3 standard deviations above and below the proportion defective when the process is in control. Thus, we have the following control limits.

Control Limits for a p Chart

$$\text{UCL} = p + 3\sigma_{\bar{p}} \tag{13.17}$$

$$\text{LCL} = p - 3\sigma_{\bar{p}} \tag{13.18}$$

FIGURE 13.7
Sampling Distribution of $\bar{p}$

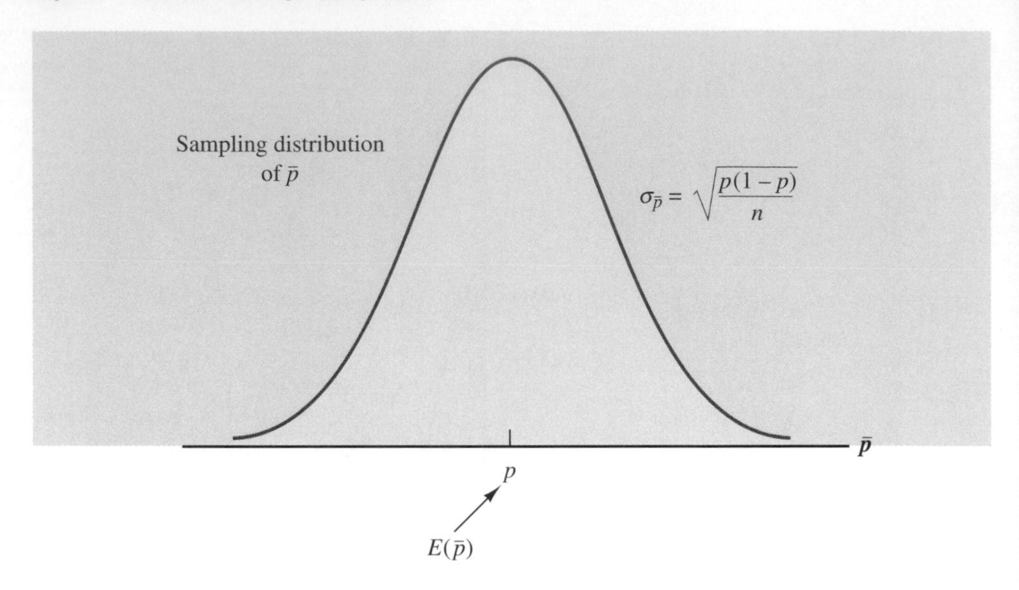

FIGURE 13.8
***p* Chart for the Proportion Defective in a Mail-Sorting Process**

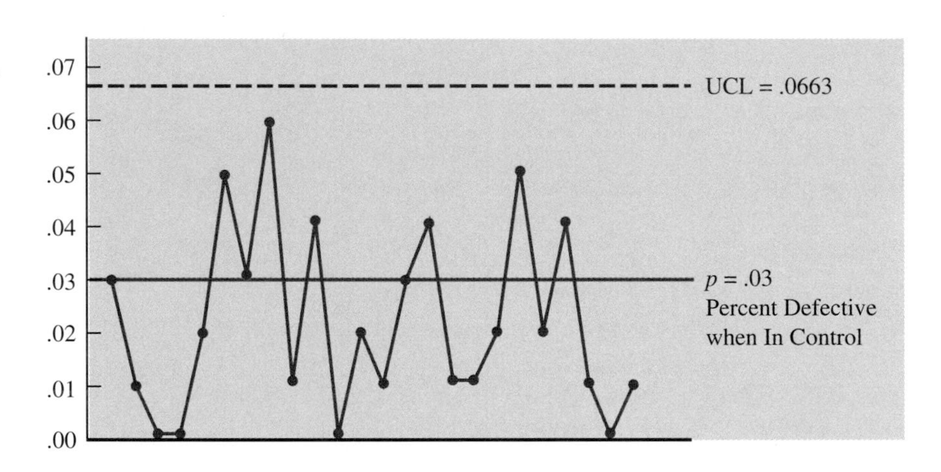

With $p = .03$ and samples of size $n = 200$, (13.16) shows that the standard error is

$$\sigma_{\bar{p}} = \sqrt{\frac{.03(1 - .03)}{200}} = .0121$$

Hence, the control limits are UCL $= .03 + 3(.0121) = .0663$ and LCL $= .03 - 3(.0121) = -.0063$. Since LCL is negative, LCL is set equal to zero in the control chart.

Figure 13.8 is the control chart for the mail-sorting process. The points plotted show the proportions defective found in samples of 200 letters taken from the process. Since all points are within the control limits, there is no evidence to conclude that the sorting process is out of control. In fact, the p chart indicates that the process should continue to operate.

If the proportion of defective items for a process that is in control is not known, that value is first estimated by using sample data. Suppose, for example, that M different samples, each of size n, are selected from a process that is in control. The fraction or proportion of defective items in each sample is then determined. Treating all the data collected as one large sample, we can determine the average number of defective items for all the data; that value can then be used to provide an estimate of p, the proportion of defective items observed when the process is in control. Note that this estimate of p also enables us to estimate the standard error of the proportion; upper and lower control limits can then be established.

np Chart

An np chart is a control chart developed for the number of defective items observed in a sample. With p denoting the probability of observing a defective item when the process is in control, the binomial probability distribution, as presented in Chapter 5, can be used to determine the probability of observing x defective items in a sample of size n. The expected value or mean of a binomial distribution is np and the standard deviation is $\sqrt{np(1-p)}$.

In Chapter 7 we learned that the normal probability distribution can be used to approximate the binomial probability distribution whenever the sample size is large. The sample size can be considered large whenever the following two conditions are satisfied.

$$np \geq 5$$

$$n(1-p) \geq 5$$

In summary, whenever the sample size is large, the distribution of the number of defective items observed in a sample of size n can be approximated by a normal probability distribution with mean np and standard deviation $\sqrt{np(1-p)}$. Thus, for the mail-sorting example, with $n = 200$ and $p = .03$, the number of defective items observed in a sample of 200 letters can be approximated by a normal probability distribution with a mean of $200(.03) = 6$ and a standard deviation of $\sqrt{200(.03)(.97)} = 2.4125$.

The control limits for an np chart are set at 3 standard deviations above and below the expected number of defective items observed when the process is in control. Thus, we have the following control limits.

Control Limits for an *np* Chart

$$\text{UCL} = np + 3\sqrt{np(1-p)} \quad (13.19)$$

$$\text{LCL} = np - 3\sqrt{np(1-p)} \quad (13.20)$$

For the mail-sorting process example, with $p = .03$ and $n = 200$, the control limits are UCL $= 6 + 3(2.4125) = 13.2375$ and LCL $= 6 - 3(2.4125) = -1.2375$. Since LCL is negative, LCL is set equal to zero in the control chart. Hence, if the number of letters diverted to incorrect routes is greater than 13, the process is concluded to be out of control.

The information provided by an np chart is equivalent to the information provided by the p chart; the only difference is that the np chart is a plot of the number of defective items observed whereas the p chart is a plot of the proportion of defective items observed. Thus, if we were to conclude that a particular process is out of control on the basis of a p chart, the process would also be concluded to be out of control on the basis of an np chart.

Interpretation of Control Charts

The location and pattern of points in a control chart enable us to determine, with a small probability of error, whether a process is in statistical control. A primary indication that a process may be out of control is a data point outside the control limits, such as point 5 in Figure 13.4. Finding such a point is statistical evidence that the process is out of control; in such cases, corrective action should be taken as soon as possible.

In addition to points outside the control limits, certain patterns of the points within the control limits can be warning signals of quality-control problems. For example, assume that all the data points are within the control limits but that a large number of points are on one side of the center line. This pattern may indicate that an equipment problem, a change in materials, or some other assignable cause of a shift in quality has occurred. Careful investigation of the production process should be undertaken to determine whether quality has changed.

Another pattern to watch for in control charts is a gradual shift, or trend, over time. For example, as tools wear out, the dimensions of machined parts will gradually deviate from their designed levels. Gradual changes in temperature or humidity, general equipment deterioration, dirt buildup, or operator fatigue may also result in a trend pattern in control charts. Six or seven points in a row that indicate either an increasing or decreasing trend should be cause for concern, even if the data points are all within the control limits. When such a pattern occurs, the process should be reviewed for possible changes or shifts in quality. Corrective action to bring the process back into control may be necessary.

notes and comments

1. Since the control limits for the $\bar{x}$ chart depend on the value of the average range, these limits will not have much meaning unless the process variability is in control. In practice, the *R* chart is usually constructed before the $\bar{x}$ chart; if the *R* chart indicates that the process variability is in control, then the $\bar{x}$ chart is constructed.

2. An *np* chart is used to monitor a process in terms of the number of defects. The Motorola Six Sigma (6σ) Quality Level sets a goal of producing no more than 3.4 defects per million operations (*American Production and Inventory Control Society,* July 1991); this goal implies $p = .0000034$.

exercises

Methods

1. A process that is in control has a mean of $\mu = 12.5$ and a standard deviation of $\sigma = .8$.
   a. Construct an $\bar{x}$ chart if samples of size four are to be used.
   b. Repeat (a) for samples of size eight and 16.
   c. What happens to the limits of the control chart as the sample size is increased? Discuss why this is reasonable.

2. Twenty-five samples, each of size five, were selected from a process that was in control. The sum of all the data collected was 677.5.

a. What is an estimate of the process mean when the process is in control?

b. Develop the control chart for this process if samples of size five will be used. Assume that the process standard deviation is .5 when the process is in control, and that the mean of the process is the estimate developed in (a).

3. Twenty-five samples of 100 items each were inspected when a process was considered to be operating satisfactorily. In the 25 samples, a total of 135 items were found to be defective.

a. What is an estimate of the proportion defective when the process is in control?

b. What is the standard error of the proportion if samples of size 100 will be used for statistical process control?

c. Compute the upper and lower control limits for the control chart.

Self Test

4. A process sampled 20 times with a sample of size eight resulted in $\bar{\bar{x}} = 28.5$ and $\bar{R} = 1.6$.

Compute the upper and lower control limits for the $\bar{x}$ and R charts for this process.

Applications

5. Temperature is used to measure the output of a production process. When the process is in control, the mean of the process is $\mu = 128.5$ and the standard deviation is $\sigma = .4$.

a. Construct an $\bar{x}$ chart if samples of size six are to be used.

b. Is the process in control for a sample providing the following data?

128.8 128.2 129.1 128.7 128.4 129.2

c. Is the process in control for a sample providing the following data?

129.3 128.7 128.6 129.2 129.5 129.0

6. A quality control process monitors the weight per carton of laundry detergent. Control limits are set at UCL=20.12 ounces and LCL=19.90 ounces. Samples of size five are used for the sampling and inspection process. What are the process mean and process standard deviation for the manufacturing operation?

7. The Goodman Tire and Rubber Company periodically tests its tires for tread wear under simulated road conditions. To study and control the manufacturing process, 20 samples, each containing three radial tires, were chosen from different shifts over several days of operation. The results are reported in the following table. Assuming that these data were collected when the manufacturing process was believed to be operating in control, develop the R and $\bar{x}$ charts.

Sample	Tread Wear*			Sample	Tread Wear*		
1	31	42	28	11	26	31	40
2	26	18	35	12	23	19	25
3	25	30	34	13	17	24	32
4	17	25	21	14	43	35	17
5	38	29	35	15	18	25	29
6	41	42	36	16	30	42	31
7	21	17	29	17	28	36	32
8	32	26	28	18	40	29	31
9	41	34	33	19	18	29	28
10	29	17	30	20	22	34	26

*Hundredths of an inch

8. Over several weeks of normal, or in-control, operation, 20 samples of 150 packages each of synthetic-gut tennis strings were tested for breaking strength. A total of 141 packages of the 3000 tested failed to conform to the manufacturer's specifications.

a. What is an estimate of the process proportion defective when the system is in control?

b. Compute the upper and lower control limits for a p chart.

c. With the results of part (b), what conclusion should be made about the process if tests on a new sample of 150 packages find 12 defective? Do there appear to be assignable causes in this situation?
d. Compute the upper and lower control limits for an *np* chart.
e. Answer part (c) using the results of part (d).
f. Which control chart would be preferred in this situation? Explain.

9. An automotive industry supplier produces pistons for several models of automobiles. Twenty samples, each consisting of 200 pistons, were selected when the process was known to be operating correctly. The numbers of defective pistons found in the samples follow.

8	10	6	4	5	7	8	12	8	15
14	10	10	7	5	8	6	10	4	8

a. What is an estimate of the proportion defective for the piston-manufacturing process when it is in control?
b. Construct a *p* chart for the manufacturing process, assuming each sample has 200 pistons.
c. With the results of part (b), what conclusion should be made if a sample of 200 has 20 defective pistons?
d. Compute the upper and lower control limits for an *np* chart.
e. Answer part (c) using the results of part (d).

13.2 Acceptance Sampling

In acceptance sampling, the items of interest can be incoming shipments of raw materials or purchased parts as well as finished goods from final assembly. Suppose we want to decide whether to accept or reject a group of items on the basis of specified quality characteristics. In quality-control terminology, the group of items is a *lot,* and *acceptance sampling* is a statistical method that enables us to base the accept-reject decision on the inspection of a sample of items from the lot.

The general steps of acceptance sampling are shown in Figure 13.9. After a lot is received, a sample of items is selected for inspection. The results of the inspection are compared to specified quality characteristics. If the quality characteristics are satisfied, the lot is accepted and sent to production or shipped to customers. If the lot is rejected, managers must decide on its disposition. In some cases, the decision may be to keep the lot and remove the unacceptable or nonconforming items during production. In other cases, the lot may be returned to the supplier at the supplier's expense; the extra work and cost placed on the supplier can motivate the supplier to provide high-quality lots. Finally, if the rejected lot consists of finished goods, the goods must be scrapped or reworked to meet acceptable quality standards.

The statistical procedure of acceptance sampling is based on the hypothesis-testing methodology presented in Chapter 9. The null and alternative hypotheses are stated as follows.

H_0: Good-quality lot

H_a: Poor-quality lot

Table 13.3, like the one in Chapter 9, shows the results of the hypothesis-testing procedure. Note that correct decisions correspond to accepting a good-quality lot and rejecting a poor-quality lot. However, as with other hypothesis-testing procedures, we

FIGURE 13.9
Acceptance Sampling Procedure

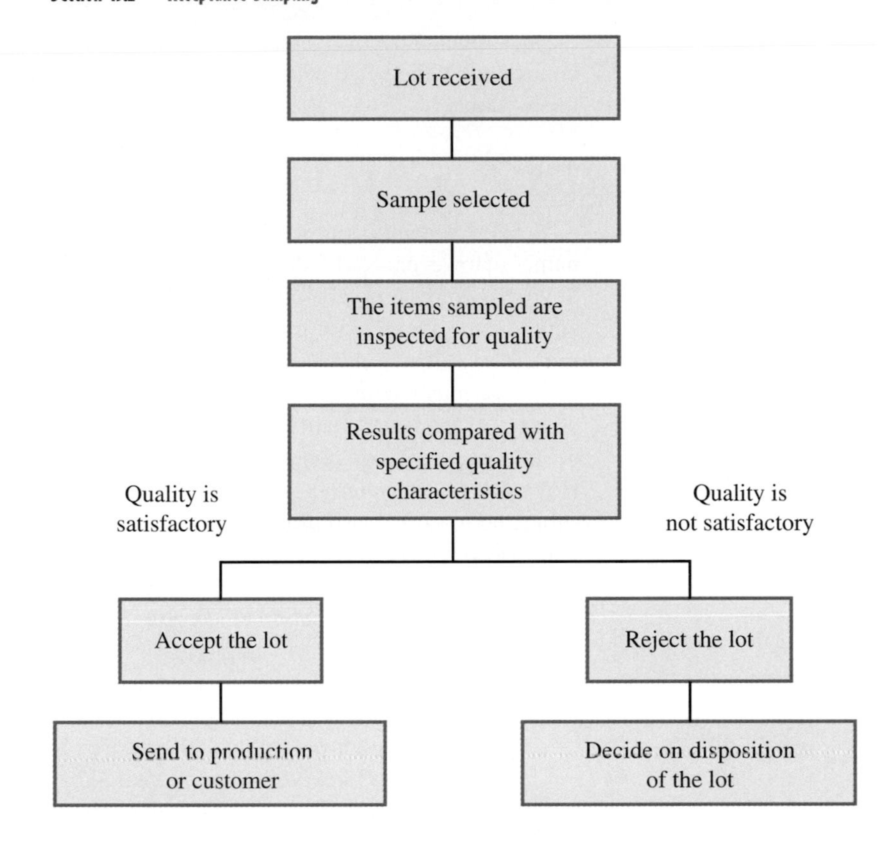

need to be aware of the possibilities of making a Type I error (rejecting a good-quality lot) or a Type II error (accepting a poor-quality lot).

Since the probability of a Type I error creates a risk for the producer of the lot, it is known as the *producer's risk.* For example, a producer's risk of .05 indicates that there is a 5% chance that a good-quality lot will be erroneously rejected. Since the probability of a Type II error creates a risk for the consumer of the lot, it is known as the *consumer's risk.* For example, a consumer's risk of .10 means that there is a 10% chance that a poor-quality lot will be erroneously accepted and thus used in production or shipped to

TABLE 13.3 The Outcomes of Acceptance Sampling

		State of the Lot	
		H_0 True *Good-Quality Lot*	*H_0 False* *Poor-Quality Lot*
Decision	*Accept the Lot*	Correct decision	Type II error (accepting a poor-quality lot)
	Reject the Lot	Type I error (rejecting a good-quality lot)	Correct decision

the customer. Specific values for the producer's risk and the consumer's risk can be controlled by the person designing the acceptance sampling procedure. To illustrate how this is done, let us consider the problem faced by KALI, Inc.

KALI, Inc.: An Example of Acceptance Sampling

KALI, Inc., manufactures home appliances that are marketed under a variety of trade names. However, KALI does not manufacture every component used in its products. Several components are purchased directly from suppliers. For example, one of the components that KALI purchases for use in home air conditioners is an overload protector, a device that turns off the compressor if it overheats. Since the compressor can be seriously damaged if the overload protector does not function properly, KALI is concerned about the quality of the overload protectors. One way to ensure quality would be to test every component received; that approach is known as 100% inspection. However, to determine proper functioning of an overload protector, the device must be subjected to time-consuming and expensive tests, and KALI cannot justify testing every overload protector it receives.

Instead, KALI uses an acceptance sampling plan to monitor the quality of the overload protectors. The acceptance sampling plan requires that KALI's quality-control inspectors select and test a sample of overload protectors from each shipment. If few defective units are found in the sample, the lot is probably of good quality and should be accepted. However, if a large number of defective units are found in the sample, the lot is probably of poor quality and should be rejected.

An *acceptance sampling plan* consists of a sample size n and an acceptance criterion c. The *acceptance criterion* is the maximum number of defective items that can be found in the sample and still indicate an acceptable lot. For example, for the KALI problem let us assume that a sample of 15 items will be selected from each incoming shipment or lot. Furthermore, assume that the manager of quality control states that the lot can be accepted only if no defective items are found. In this case, the acceptance sampling plan established by the quality-control manager is $n = 15$ and $c = 0$.

This acceptance sampling plan is easy for the quality-control inspector to implement. The inspector simply selects a sample of 15 items, performs the tests, and reaches a conclusion based on the following decision rule.

- *Accept the lot* if zero defective items are found.
- *Reject the lot* if one or more defective items are found.

Before implementing this acceptance sampling plan, the quality-control manager wants to evaluate the risks or errors possible under the plan. The plan will be implemented only if both the producer's risk (Type I error) and the consumer's risk (Type II error) are controlled at reasonable levels.

Computing the Probability of Accepting a Lot

The key to analyzing both the producer's risk and the consumer's risk is a "What-if?" type of analysis. That is, we will assume that a lot has some known percentage of defective items and compute the probability of accepting the lot for a given sampling plan. By varying the assumed percentage of defective items, we can examine the effect of the sampling plan on both types of risks.

Let us begin by assuming that a large shipment of overload protectors has been received and that 5% of the overload protectors in the shipment are defective. For a shipment or lot with 5% of the items defective, what is the probability that the $n = 15$, $c = 0$ sampling plan will lead us to accept the lot? Since each overload protector tested will be either defective or nondefective and since the lot size is large, the number of defective items in a sample of 15 has a binomial probability distribution. The binomial probability function, which was presented in Chapter 5, follows.

Binomial Probability Function for Acceptance Sampling

$$f(x) = \frac{n!}{x!(n-x)!} p^x (1-p)^{(n-x)} \tag{13.21}$$

where

n = the sample size

p = the proportion of defective items in the lot

x = the number of defective items in the sample

$f(x)$ = the probability of x defective items in the sample

For the KALI acceptance sampling plan, $n = 15$; thus, for a lot with 5% defective ($p = .05$), we have

$$f(x) = \frac{15!}{x!(15-x)!} (.05)^x (1 - .05)^{(15-x)} \tag{13.22}$$

Using (13.22), $f(0)$ will provide the probability that zero overload protectors will be defective and the lot will be accepted. In using (13.22), recall that $0! = 1$. Thus, the probability computation for $f(0)$ is

$$f(0) = \frac{15!}{0!(15-0)!} (.05)^0 (1 - .05)^{(15-0)}$$

$$= \frac{15!}{0!(15)!} (.05)^0 (.95)^{15} = (.95)^{15} = .4633$$

We now know that the $n = 15$, $c = 0$ sampling plan has a .4633 probability of accepting a lot with 5% defective items. Hence, there must be a corresponding $1 - .4633 = .5367$ probability of rejecting a lot with 5% defective items.

In Table 13.4 we show the probability that the $n = 15$, $c = 0$ sampling plan will lead to the acceptance of lots with 1%, 2%, 3%, . . . defective items. The probabilities in the table were computed by using $p = .01$, $p = .02$, $p = .03$, . . . in the binomial probability function (13.21).

TABLE 13.4 Probability of Accepting the Lot for the KALI Problem With $n = 15$ and $c = 0$

Percent Defective in the Lot	Probability of Accepting the Lot
1	.8601
2	.7386
3	.6333
4	.5421
5	.4633
10	.2059
15	.0874
20	.0352
25	.0134

Tables of binomial probabilities (see Table 5, Appendix B) can help reduce the computational effort in determining the probabilities of accepting lots. Selected binomial probabilities for $n = 15$ and $n = 20$ are listed in Table 13.5. These probabilities show that if the lot contains 10% defective items, there is a .2059 probability that the $n = 15$, $c = 0$ sampling plan will indicate an acceptable lot.

With the data in Table 13.4, a graph of the probability of accepting the lot versus the percent defective in the lot can be drawn as shown in Figure 13.10. This graph, or curve, is called the *operating characteristic* (OC) *curve* for the $n = 15$, $c = 0$ acceptance sampling plan.

TABLE 13.5 Selected Binomial Probabilities for Samples of Sizes 15 and 20

		p									
n	x	.05	.10	.15	.20	.25	.30	.35	.40	.45	.50
15	0	.4633	.2059	.0874	.0352	.0134	.0047	.0016	.0005	.0001	.0000
	1	.3658	.3432	.2312	.1319	.0668	.0305	.0126	.0047	.0016	.0005
	2	.1348	.2669	.2856	.2309	.1559	.0916	.0476	.0219	.0090	.0032
	3	.0307	.1285	.2184	.2501	.2252	.1700	.1110	.0634	.0318	.0139
	4	.0049	.0428	.1156	.1876	.2252	.2186	.1792	.1268	.0780	.0417
	5	.0006	.0105	.0449	.1032	.1651	.2061	.2123	.1859	.1404	.0916
	6	.0000	.0019	.0132	.0430	.0917	.1472	.1906	.2066	.1914	.1527
	7	.0000	.0003	.0030	.0138	.0393	.0811	.1319	.1771	.2013	.1964
	8	.0000	.0000	.0005	.0035	.0131	.0348	.0710	.1181	.1647	.1964
	9	.0000	.0000	.0001	.0007	.0034	.0116	.0298	.0612	.1048	.1527
	10	.0000	.0000	.0000	.0001	.0007	.0030	.0096	.0245	.0515	.0916
	11	.0000	.0000	.0000	.0000	.0001	.0006	.0024	.0074	.0191	.0417
	12	.0000	.0000	.0000	.0000	.0000	.0001	.0004	.0016	.0052	.0139
	13	.0000	.0000	.0000	.0000	.0000	.0000	.0001	.0003	.0010	.0032
	14	.0000	.0000	.0000	.0000	.0000	.0000	.0000	.0000	.0001	.0005
	15	.0000	.0000	.0000	.0000	.0000	.0000	.0000	.0000	.0000	.0000
20	0	.3585	.1216	.0388	.0115	.0032	.0008	.0002	.0000	.0000	.0000
	1	.3774	.2702	.1368	.0576	.0211	.0068	.0020	.0005	.0001	.0000
	2	.1887	.2852	.2293	.1369	.0669	.0278	.0100	.0031	.0008	.0002
	3	.0596	.1901	.2428	.2054	.1339	.0716	.0323	.0123	.0040	.0011
	4	.0133	.0898	.1821	.2182	.1897	.1304	.0738	.0350	.0139	.0046
	5	.0022	.0319	.1028	.1746	.2023	.1789	.1272	.0746	.0365	.0148
	6	.0003	.0089	.0454	.1091	.1686	.1916	.1712	.1244	.0746	.0370
	7	.0000	.0020	.0160	.0545	.1124	.1643	.1844	.1659	.1221	.0739
	8	.0000	.0004	.0046	.0222	.0609	.1144	.1614	.1797	.1623	.1201
	9	.0000	.0001	.0011	.0074	.0271	.0654	.1158	.1597	.1771	.1602
	10	.0000	.0000	.0002	.0020	.0099	.0308	.0686	.1171	.1593	.1762
	11	.0000	.0000	.0000	.0005	.0030	.0120	.0336	.0710	.1185	.1602
	12	.0000	.0000	.0000	.0001	.0008	.0039	.0136	.0355	.0727	.1201
	13	.0000	.0000	.0000	.0000	.0002	.0010	.0045	.0146	.0366	.0739
	14	.0000	.0000	.0000	.0000	.0000	.0002	.0012	.0049	.0150	.0370
	15	.0000	.0000	.0000	.0000	.0000	.0000	.0003	.0013	.0049	.0148
	16	.0000	.0000	.0000	.0000	.0000	.0000	.0000	.0003	.0013	.0046
	17	.0000	.0000	.0000	.0000	.0000	.0000	.0000	.0000	.0002	.0011
	18	.0000	.0000	.0000	.0000	.0000	.0000	.0000	.0000	.0000	.0002
	19	.0000	.0000	.0000	.0000	.0000	.0000	.0000	.0000	.0000	.0000
	20	.0000	.0000	.0000	.0000	.0000	.0000	.0000	.0000	.0000	.0000

FIGURE 13.10
Operating Characteristic Curve for the $n = 15$, $c = 0$ Acceptance Sampling Plan

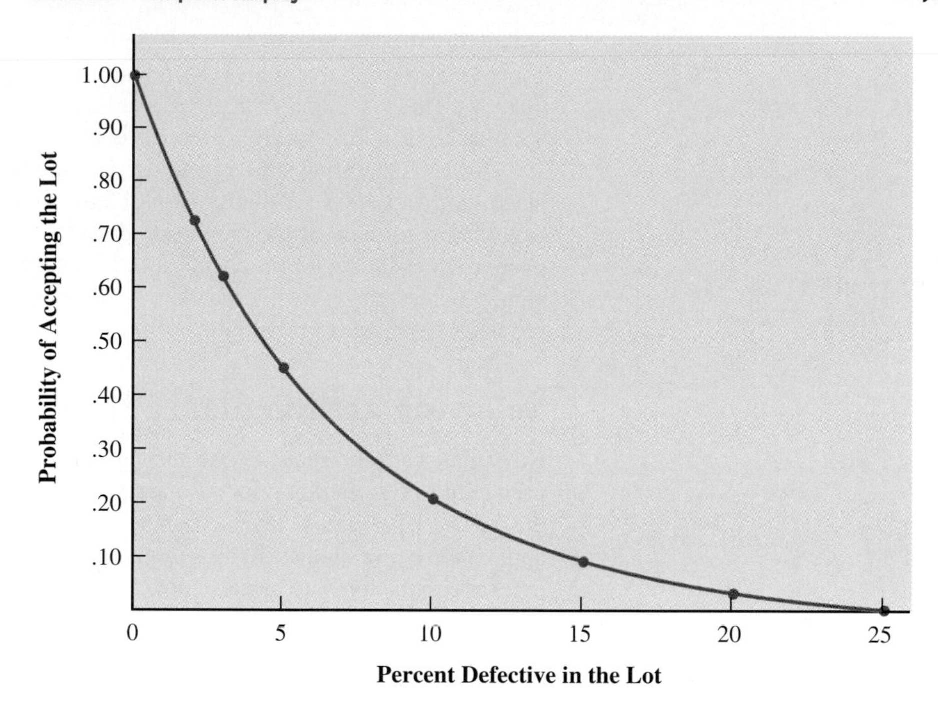

Perhaps we should consider other sampling plans, ones with different sample sizes n and/or different acceptance criteria c. First consider the case in which the sample size remains $n = 15$ but the acceptance criterion increases from $c = 0$ to $c = 1$. That is, we will now accept the lot if zero or one defective component is found in the sample. For a lot with 5% defective items ($p = .05$), the binomial probability function in (13.21) can be used to compute $f(0)$ and $f(1)$. Summing these two probabilities provides the

FIGURE 13.11
Operating Characteristic Curves for Four Acceptance Sampling Plans

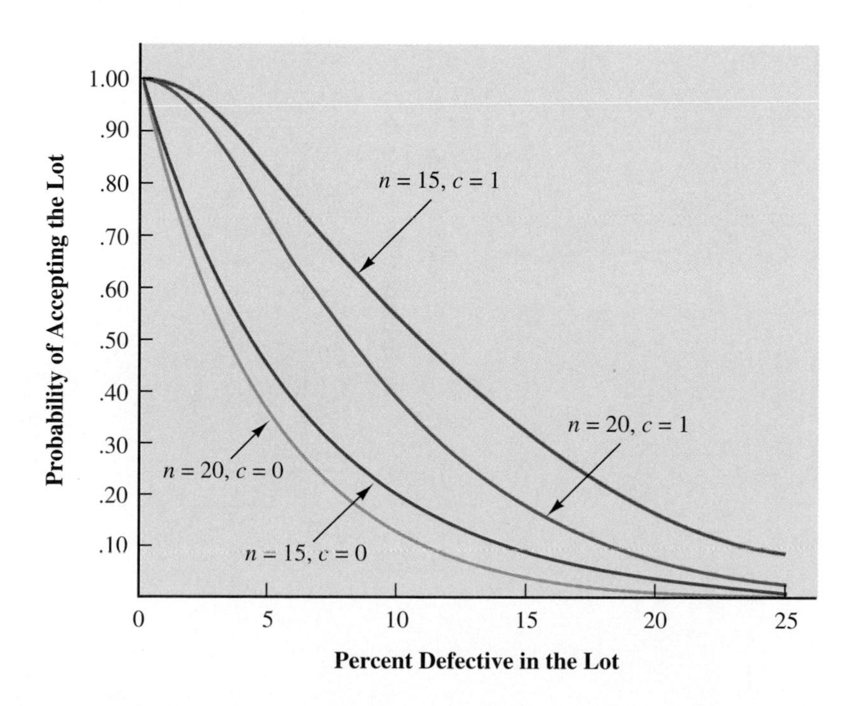

probability that the $n = 15$, $c = 1$ sampling plan will accept the lot. Alternatively, using Table 13.5, we find that with $n = 15$ and $p = .05$, $f(0) = .4633$ and $f(1) = .3658$. Thus, there is a $.4633 + .3658 = .8291$ probability that the $n = 15$, $c = 1$ plan will lead to the acceptance of a lot with 5% defective items.

Figure 13.11 shows the operating characteristic curves for four alternative acceptance sampling plans for the KALI problem. Samples of size 15 and 20 are considered. Note that regardless of the proportion defective in the lot, the $n = 15$, $c = 1$ sampling plan provides the highest probabilities of accepting the lot. The $n = 20$, $c = 0$ sampling plan provides the lowest probabilities of accepting the lot; however, that plan also provides the highest probabilities of rejecting the lot.

Selecting an Acceptance Sampling Plan

Now that we know how to use the binomial probability distribution to compute the probability of accepting a lot with a given proportion defective, we are ready to select the values of n and c that determine the desired acceptance sampling plan for the application being studied. To do this, managers must specify two values for the proportion defective in the lot. One value, denoted p_0, will be used to control for the producer's risk, and the other value, denoted p_1, will be used to control for the consumer's risk.

In showing how this can be done, we will use the following notation.

α = the producer's risk; the probability that a lot with p_0 defective will be rejected

β = the consumer's risk; the probability that a lot with p_1 defective will be accepted

Suppose that for the KALI problem, the managers specify that $p_0 = .03$ and $p_1 = .15$. From the OC curve for $n = 15$, $c = 0$ in Figure 13.12, we see that $p_0 = .03$

FIGURE 13.12
Operating Characteristic Curve for $n = 15$, $c = 0$ with $p_0 = .03$ and $p_1 = .15$

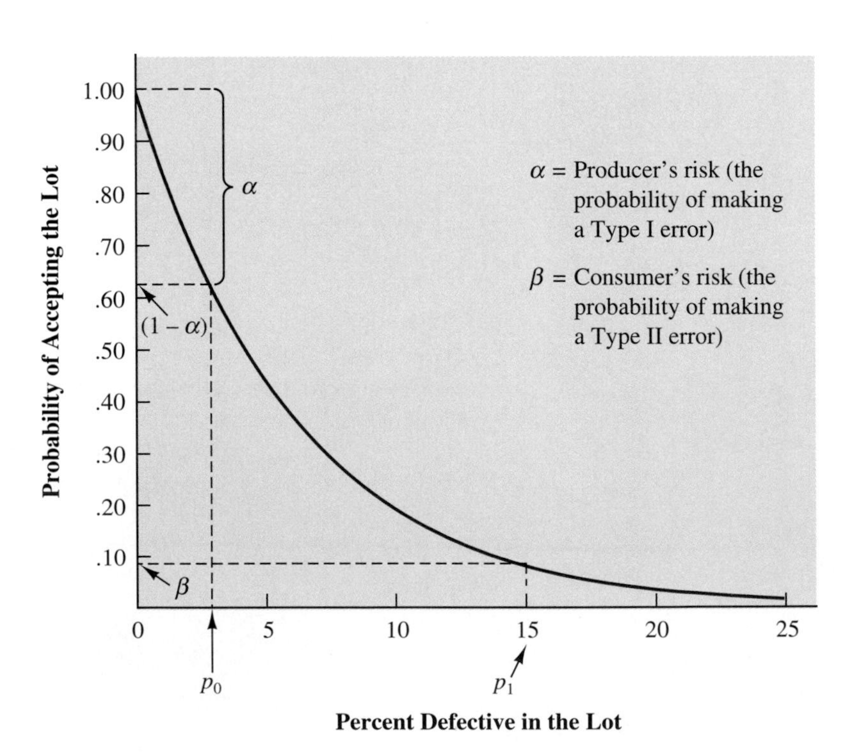

provides a producer's risk of approximately $1 - .63 = .37$, and $p_1 = .15$ provides a consumer's risk of approximately .09. Thus, if the managers are willing to tolerate both a .37 probability of rejecting a lot with 3% defective items (producer's risk) and a .09 probability of accepting a lot with 15% defective items (consumer's risk), the $n = 15$, $c = 0$ acceptance sampling plan would be acceptable.

Suppose, however, that the managers request a producer's risk of $\alpha = .10$ and a consumer's risk of $\beta = .20$. We see that now the $n = 15$, $c = 0$ sampling plan has a better-than-desired consumer's risk but an unacceptably large producer's risk. The fact that $\alpha = .37$ indicates that 37% of the lots will be erroneously rejected when only 3% of the items in them are defective. The producer's risk is too high, and a different acceptance sampling plan should be considered.

Using $p_0 = .03$, $\alpha = .10$, $p_1 = .15$, and $\beta = .20$ in Figure 13.11 shows that the acceptance sampling plan with $n = 20$ and $c = 1$ comes closest to meeting both the producer's and the consumer's risk requirements. Exercise 13 at the end of this section will ask you to compute the producer's risk and the consumer's risk for the $n = 20$, $c = 1$ sampling plan.

As shown in this section, several computations and several operating characteristic curves may need to be considered to determine the sampling plan with the desired producer's and consumer's risk. Fortunately, tables of sampling plans are published. For example, the American Military Standard Table, MIL-STD-105D, provides information helpful in designing acceptance sampling plans. More advanced texts on quality control, such as those listed in the bibliography, describe the use of such tables. The advanced texts also discuss the role of sampling costs in determining the optimal sampling plan.

Multiple Sampling Plans

The acceptance sampling procedure that we have presented for the KALI problem is a *single-sample* plan. It is called a single-sample plan because only one sample or sampling stage is used. After the number of defective components in the sample is determined, a decision must be made to accept or reject the lot. An alternative to the single-sample plan is a multiple sampling plan, in which two or more stages of sampling are used. At each stage a decision is made among three possibilities: stop sampling and accept the lot, stop sampling and reject the lot, or continue sampling. Although more complex, multiple sampling plans often result in a smaller total sample size than single-sample plans with the same α and β probabilities.

The logic of a two-stage, or double-sample, plan is shown in Figure 13.13. Initially a sample of n_1 items is selected. If the number of defective components x_1 is less than or equal to c_1, accept the lot. If x_1 is greater than or equal to c_2, reject the lot. If x_1 is between c_1 and c_2 ($c_1 < x_1 < c_2$), select a second sample of n_2 items. Determine the combined, or total, number of defective components from the first sample (x_1) and the second sample (x_2). If $x_1 + x_2 \leq c_3$, accept the lot; otherwise reject the lot. The development of the double-sample plan is more difficult because the sample sizes n_1 and n_2 and the acceptance numbers c_1, c_2, and c_3 must meet both the producer's and consumer's risks desired.

FIGURE 13.13
A Two-Stage Acceptance Sampling Plan

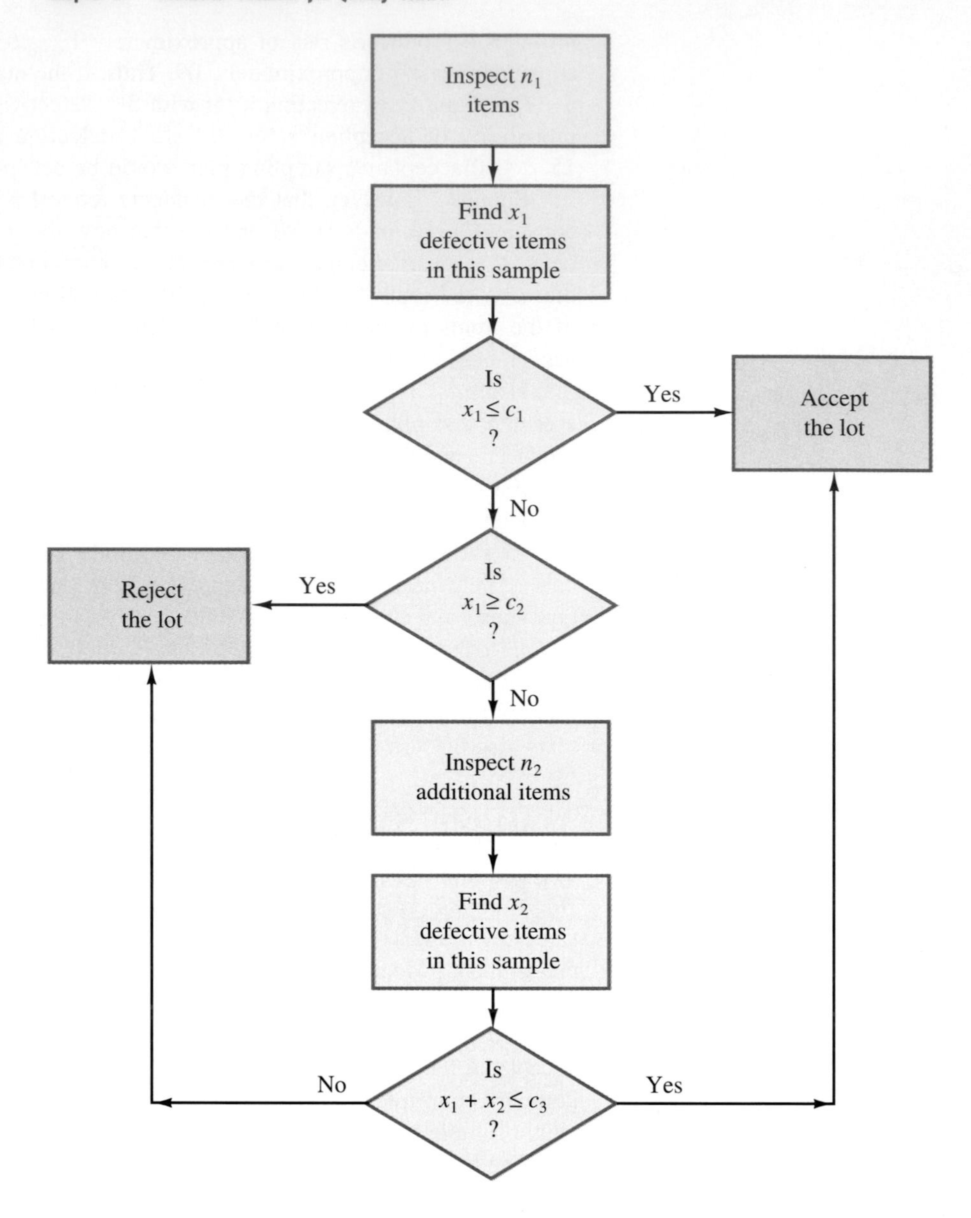

notes and comments

1. The use of the binomial probability distribution for acceptance sampling is based on the assumption of large lots. If the lot size is small, the hypergeometric probability distribution is the appropriate distribution. Experts in the field of quality control indicate that the Poisson distribution provides a good approximation for acceptance sampling when the sample size is at least 16, the lot size is at least 10 times the sample size, and p is less than .1.* For larger sample sizes, the normal approximation to the binomial probability distribution can be used.
2. In the MIL-ST-105D sampling tables, p_0 is called the acceptable quality level (AQL). In some sampling tables, p_1 is called the lot tolerance percent defective (LTPD) or the rejectable quality level (RQL). Many of the published sampling plans also use quality indexes such as the indifference quality level (IQL) and the average outgoing quality limit (AOQL). The more advanced texts listed in the bibliography provide a complete discussion of these other indexes.
3. In this section we provided an introduction to *attributes sampling plans.* In these plans each item sampled is classified as nondefective or defective. In *variables sampling plans,* a sample is taken and a measurement of the quality characteristic is taken. For example, for gold jewelry a measurement of quality may be the amount of gold it contains. A simple statistic such as the average amount of gold in the sample jewelry is computed and compared with an allowable value to determine whether to accept or reject the lot.

exercises

Methods

10. For an acceptance sampling plan with $n = 25$ and $c = 0$, find the probability of accepting a lot that has a defect rate of 2%. What is the probability of accepting the lot if the defect rate is 6%?

11. Consider an acceptance sampling plan with $n = 20$ and $c = 0$. Compute the producer's risk for each of the following cases.
- **a.** The lot has a defect rate of 2%.
- **b.** The lot has a defect rate of 6%.

12. Repeat Exercise 11 for the acceptance sampling plan with $n = 20$ and $c = 1$. What happens to the producer's risk as the acceptance number c is increased? Explain.

Applications

13. Refer to the KALI problem presented in this section. The quality-control manager requested a producer's risk of .10 when p_0 was .03 and a consumer's risk of .20 when p_1 was .15. Consider the acceptance sampling plan based on a sample size of 20 and an acceptance number of 1. Answer the following questions.
- **a.** What is the producer's risk for the $n = 20$, $c = 1$ sampling plan?
- **b.** What is the consumer's risk for the $n = 20$, $c = 1$ sampling plan?
- **c.** Does the $n = 20$, $c = 1$ sampling plan satisfy the risks requested by the quality-control manager? Discuss.

*J. M. Juran and Frank M. Gryna, Jr., *Quality Planning and Analysis,* McGraw-Hill, New York, 1980, p. 412.

14. To inspect incoming shipments of raw materials, a manufacturer is considering samples of sizes 10, 15, and 20. Use the binomial probabilities from Table 5 of Appendix B to select a sampling plan that provides a producer's risk of $\alpha = .03$ when p_0 is .05 and a consumer's risk of $\beta = .12$ when p_1 is .30.

15. A domestic manufacturer of watches purchases quartz crystals from a Swiss firm. The crystals are shipped in lots of 1000. The acceptance sampling procedure uses 20 randomly selected crystals.

a. Construct operating characteristic curves for acceptance numbers of 0, 1, and 2.

b. If $p_0 = .01$ and $p_1 = .08$, what are the producer's and consumer's risks for each sampling plan in (a)?

summary

In this chapter we discussed how statistical methods can be used to assist in the control of quality. We first presented the $\bar{x}$, *R*, *p* and *np* control charts as graphical aids in monitoring process quality. Control limits are established for each chart; samples are selected periodically and the data points plotted on the control chart. Data points outside the control limits indicate that the process is out of control and that corrective action should be taken. Patterns of data points within the control limits can also indicate potential quality-control problems and suggest that corrective action may be warranted.

We also considered the technique known as acceptance sampling. With this procedure, a sample is selected. The number of defective items in the sample provides the basis for accepting or rejecting the lot. The sample size and the acceptance criterion can be adjusted to control both the producer's risk (Type I error) and the consumer's risk (Type II error).

glossary

Quality control A series of inspections and measurements that determine whether quality standards are being met.

Common causes Normal or natural variations in process outputs that are due purely to chance. No corrective action is necessary when output variations are due to common causes.

Assignable causes Variations in process outputs that are due to factors such as machine tools wearing out, incorrect machine settings, poor-quality raw materials, operator error, and so on. Corrective action should be taken when assignable causes of output variation are detected.

Control chart A graphical tool used to help determine whether a process is in control or out of control.

$\bar{x}$ chart A control chart used when the output of a process is measured in terms of the mean value of a variable such as a length, weight, temperature, and so on.

R chart A control chart used when the output of a process is measured in terms of the range of a variable.

p chart A control chart used when the output of a process is measured in terms of the proportion defective.

np chart A control chart used to monitor the output of a process in terms of the number of defective items.

Lot A group of items such as incoming shipments of raw materials or purchased parts as well as finished goods from final assembly.

Acceptance sampling A statistical procedure in which the number of defective items found in a sample is used to determine whether a lot should be accepted or rejected.

Producer's risk The risk of rejecting a good-quality lot. This is the Type I error.

Consumer's risk The risk of accepting a poor-quality lot. This is the Type II error.

Acceptance criterion The maximum number of defective items that can be found in the sample and still allow acceptance of the lot.

Operating characteristic curve A graph showing the probability of accepting the lot as a function of the percentage defective in the lot. This curve can be used to help determine whether a particular acceptance sampling plan meets both the producer's and the consumer's risk requirements.

Multiple sampling plan A form of acceptance sampling in which more than one sample or stage is used. On the basis of the number of defective items found in a sample, a decision will be made to accept the lot, reject the lot, or continue sampling.

key formulas

Standard Error of the Mean

$$\sigma_{\bar{x}} = \frac{\sigma}{\sqrt{n}} \tag{13.1}$$

Control Limits for an $\bar{x}$ Chart: Process Mean and Standard Deviation Known

$$\text{UCL} = \mu + 3\sigma_{\bar{x}} \tag{13.2}$$

$$\text{LCL} = \mu - 3\sigma_{\bar{x}} \tag{13.3}$$

Overall Sample Mean

$$\bar{\bar{x}} = \frac{\bar{x}_1 + \bar{x}_2 + \cdots + \bar{x}_k}{k} \tag{13.4}$$

Average Range

$$\bar{R} = \frac{R_1 + R_2 + \cdots + R_k}{k} \tag{13.5}$$

Control Limits for an $\bar{x}$ Chart: Process Mean and Standard Deviation Unknown

$$\bar{\bar{x}} \pm A_2\bar{R} \tag{13.8}$$

Control Limits for an *R* Chart

$$\text{UCL} = \bar{R}D_4 \tag{13.14}$$

$$\text{LCL} = \bar{R}D_3 \tag{13.15}$$

Standard Error of the Proportion

$$\sigma_{\bar{p}} = \sqrt{\frac{p(1-p)}{n}} \tag{13.16}$$

Control Limits for a *p* Chart

$$\text{UCL} = p + 3\sigma_{\bar{p}} \tag{13.17}$$

$$\text{LCL} = p - 3\sigma_{\bar{p}} \tag{13.18}$$

Control Limits for an *np* Chart

$$\text{UCL} = np + 3\sqrt{np\,(1-p)} \tag{13.19}$$

$$\text{LCL} = np - 3\sqrt{np\,(1-p)} \tag{13.20}$$

Binomial Probability Function for Acceptance Sampling

$$f(x) = \frac{n!}{x!(n-x)!}p^x(1-p)^{(n-x)} \tag{13.21}$$

supplementary exercises

16. Samples of size five provided the 20 sample means listed in the following table for a production process that is believed to be in control.

95.72	95.44	95.40	95.50	95.56	95.72	95.60
95.24	95.46	95.44	95.80	95.22	94.82	95.78
95.18	95.32	95.08	95.22	95.04	95.46	

a. Based on these data, what is an estimate of the mean when the process is in control?
b. Assuming that the process standard deviation is $\sigma = .50$, develop a control chart for this production process. Assume that the mean of the process is the estimate developed in (a).
c. Do any of the 20 sample means indicate that the process is out of control?

17. Product filling weights have a normal probability distribution with a mean of 350 grams and a standard deviation of 15 grams.
a. Develop the control limits for samples of size 10, 20, and 30.
b. What happens to the control limits as the sample size is increased?
c. What happens when a Type I error is made?

d. What happens when a Type II error is made?
e. What is the probability of a Type I error for samples of size 10, 20, and 30?

18. Twenty-five samples of size five resulted in $\bar{\bar{x}} = 5.42$ and $\bar{R} = 2.0$. Compute control limits for the $\bar{x}$ and R charts, and estimate the standard deviation of the process.

19. Construct $\bar{x}$ and R charts for the following sample data. Assume that a sample of size five was used.

Sample	$\bar{x}$	R	Sample	$\bar{x}$	R
1	95.72	1.0	11	95.80	.6
2	95.24	.9	12	95.22	.2
3	95.18	.8	13	95.56	1.3
4	95.44	.4	14	95.22	.5
5	95.46	.5	15	95.04	.8
6	95.32	1.1	16	95.72	1.1
7	95.40	.9	17	94.82	.6
8	95.44	.3	18	95.46	.5
9	95.08	.2	19	95.60	.4
10	95.50	.6	20	95.74	.6

20. Develop $\bar{x}$ and R charts for the following data.

	Observations				
Sample	1	2	3	4	5
1	3.05	3.08	3.07	3.11	3.11
2	3.13	3.07	3.05	3.10	3.10
3	3.06	3.04	3.12	3.11	3.10
4	3.09	3.08	3.09	3.09	3.07
5	3.10	3.06	3.06	3.07	3.08
6	3.08	3.10	3.13	3.03	3.06
7	3.06	3.06	3.08	3.10	3.08
8	3.11	3.08	3.07	3.07	3.07
9	3.09	3.09	3.08	3.07	3.09
10	3.06	3.11	3.07	3.09	3.07

21. Consider the following situations. For each, comment on whether there is reason for concern about the quality of the process.
a. A p chart has LCL=0 and UCL=.068. When the process is in control, the proportion defective is .033. Plot the following seven sample results: .035, .062, .055, .049, .058, .066, and .055. Discuss.
b. An $\bar{x}$ chart has LCL=22.2 and UCL=24.5. The mean is $\mu = 23.35$ when the process is in control. Plot the following seven sample results: 22.4, 22.6, 22.65, 23.2, 23.4, 23.85, and 24.1. Discuss.

22. Managers of 1200 different retail outlets make twice-a-month restocking orders from a central warehouse. Past experience has shown that 4% of the orders have one or more errors such as wrong item shipped, wrong quantity shipped, and item requested but not shipped. Random samples of 200 orders are selected monthly and checked for accuracy.

a. Construct a control chart for this situation.

b. Six months of data show the following numbers of orders with one or more errors: 10, 15, 6, 13, 8, and 17. Plot the data on the control chart. What does your plot indicate about the order process?

23. An $n = 10$, $c = 2$ acceptance sampling plan is being considered; assume that $p_0 = .05$ and $p_1 = .20$.

a. Compute both the producer's and the consumer's risk for this acceptance sampling plan.

b. Would either the producer, the consumer, or both be unhappy with the proposed sampling plan?

c. What change in the sampling plan, if any, would you recommend?

24. An acceptance sampling plan with $n = 15$ and $c = 1$ has been designed with a producer's risk of .075.

a. Was the value of p_0 .01, .02, .03, .04, or .05? What does this value mean?

b. What is the consumer's risk associated with this plan if p_1 is .25?

25. A manufacturer produces lots of a canned food product. Let p denote the proportion of the lots that do not meet the product quality specifications. An $n = 25$, $c = 0$ acceptance sampling plan will be used.

a. Compute points on the operating characteristic curve when $p = .01$, .03, .10, and .20.

b. Plot the operating characteristic curve.

c. What is the probability that the acceptance sampling plan will reject a lot that has .01 defective?

26. Sometimes an acceptance sampling plan will be based on a large sample. In this case, the normal approximation to the binomial probability distribution can be used to compute the producer's and the consumer's risk associated with the plan. Referring to Chapter 6, we know that the normal probability distribution used to approximate binomial probabilities has a mean of np and a standard deviation of $\sqrt{np(1-p)}$. Assume that an acceptance sampling plan is $n = 250$, $c = 10$.

a. What is the producer's risk if p_0 is .02? As discussed in Chapter 6, a continuity correction factor should be used in this case. Thus, the probability of acceptance is based on the normal probability of the random variable being less than or equal to 10.5.

b. What is the consumer's risk if p_1 is .08?

c. What is an advantage of a large sample size for acceptance sampling? What is a disadvantage?

Appendixes

Appendix A References and Bibliography

General

Anderson, D. R., D. J. Sweeney, and T. A. Williams, *Statistics for Business and Economics*, 6th ed., Minneapolis/St. Paul, West, 1996.

DuToit, S. H.C., *Graphical Exploratory Data Analysis,* New York, Springer-Verlag, 1986.

Freedman, D., R. Pisani, and R. Purves, *Statistics,* 2nd ed., New York, W. W. Norton, 1991.

Freund, J. E., and R. E. Walpole, *Mathematical Statistics,* 4th ed., Englewood Cliffs, NJ, Prentice-Hall, 1987.

Hoaglin, D. C., F. Mosteller, and J. W. Tukey, *Understanding Robust and Exploratory Data Analysis,* New York, Wiley, 1983.

Hogg, R. V., and A. T. Craig, *Introduction to Mathematical Statistics,* 4th ed., New York, Macmillian, 1978.

McClave, J. T., and G. B. Benson, *Statistics for Business and Economics,* 6th ed., New York, Dellen, 1994.

Mood, A. M., F. A. Graybill, and D. C. Boes, *Introduction to the Theory of Statistics,* 3rd ed., New York, McGraw-Hill, 1974.

Moore, D. S., and G. P. McCabe, *Introduction to the Practice of Statistics,* 2nd ed., New York, Freeman, 1992.

Neter, J., W. Wasserman, and G. A. Whitmore, *Applied Statistics,* 4th ed., Boston, Allyn & Bacon, 1993.

Roberts, H., *Data Analysis for Managers,* 2nd ed., Redwood City, CA., Scientific Press, 1991.

Ryan, T. A., B. L. Joiner, and B. F. Ryan, *Minitab Handbook,* 2nd ed., Boston, PWS-Kent, 1992.

Tanur, J. M., et al., *Statistics: A Guide to the Unknown,* 3rd ed., Pacific Grove, CA, Wadsworth, 1989.

Tukey, J. W., *Exploratory Data Analysis,* Reading, MA, Addison-Wesley, 1977.

Winkler, R. L., and W. L. Hays, *Statistics: Probability, Inference, and Decision,* 2nd ed., New York, Holt, Rinehart & Winston, 1975.

Probability

Barr, D. R., and P. W. Zehna, *Probability: Modeling Uncertainty,* Reading, MA., Addison-Wesley, 1983.

Feller, W., *An Introduction to Probability Theory and Its Applications,* Vol. I, 3rd ed., New York, Wiley, 1968.

Feller, W., *An Introduction to Probability Theory and Its Applications,* Vol. II, 2nd ed., New York, Wiley, 1971.

Hogg, R. V., and Elliott A. Tanis, *Probability and Statistical Inference,* 4th ed., New York, Macmillan, 1992.

Mendenhall, W., R. L. Scheaffer, and D. Wackerly, *Mathematical Statistics with Applications,* 4th ed., Boston, PWS-Kent, 1990.

Ross, S. M., *Introduction to Probability Models* 5 ed., San Diego, Academic Press, 1993.

Regression Analysis

Belsley, D. A., E. Kuh, and R. Welsch, *Regression Diagnostics: Identifying Influential Data and Sources of Collinearity,* New York, Wiley, 1980.

Chatterjee, S., and B. Price, *Regression Analysis by Example,* New York, Wiley, 1977.

Cook, R. D., and S. Weisberg, *Residuals and Influence in Regression,* New York, Chapman and Hall, 1982.

Daniel, C., and F. Wood, *Fitting Equations to Data,* 2nd ed., New York, Wiley, 1980.

Draper, N. R., and H. Smith, *Applied Regression Analysis,* 2nd ed., New York, Wiley, 1981.

Graybill, F. A., and H. K. Iyer, *Regression Analysis: Concepts and Applications,* Belmont, CA, Duxbury, 1994.

Kleinbaum, D. G., and L. L. Kupper, *Applied Regression Analysis and Other Multivariable Methods,* 2nd ed., Boston, PWS-Kent, 1988.

Mosteller, F., and J. W. Tukey, *Data Analysis and Regression: A Second Course in Statistics,* Reading, MA, Addison-Wesley, 1977.

Myers, R. H., *Classical and Modern Regression with Applications,* 2nd ed., Boston, PWS-Kent, 1990.

Neter, J., W. Wasserman, and M. H. Kutner, *Applied Linear Statistical Models,* 2nd ed., Homewood, IL, Richard D. Irwin, 1985.

Weisberg, S., *Applied Linear Regression,* 2nd ed., New York, Wiley, 1985.

Wesolowsky, G. O., *Multiple Regression and Analysis of Variance,* New York, Wiley, 1976.

Wonnacott, T. H., and R. J. Wonnacott, *Regression: A Second Course in Statistics,* New York, Wiley, 1981.

Quality Control

Deming, W. E., *Quality, Productivity, and Competitive Position,* Cambridge, MA, MIT Center for Advanced Engineering Study, 1982.

Duncan, A. J., *Quality Control and Industrial Statistics,* 5th ed. Homewood, IL, Irwin, 1986.

Evans, J. R., and W. M. Lindsay, *The Management and Control of Quality,* 2nd ed., St. Paul, MN, West, 1992.

Gitlow, H., S. Gitlow, A. Oppenheim, and R. Oppenheim, *Tools and Methods for the Improvement of Quality,* Homewood, IL, Irwin, 1989.

Ishikawa, Kaoru, *Guide to Quality Control,* 2nd rev. ed., New York, Quality Resources, 1986.

Juran, J. M., and F. M. Gryna, Jr., *Quality Planning and Analysis,* 2nd ed., New York, McGraw-Hill, 1980.

Montgomery, D. C., Introduction to Statistical Quality Control, 2nd ed., New York, Wiley, 1991.

Appendix B Tables

TABLE I Standard Normal Distribution

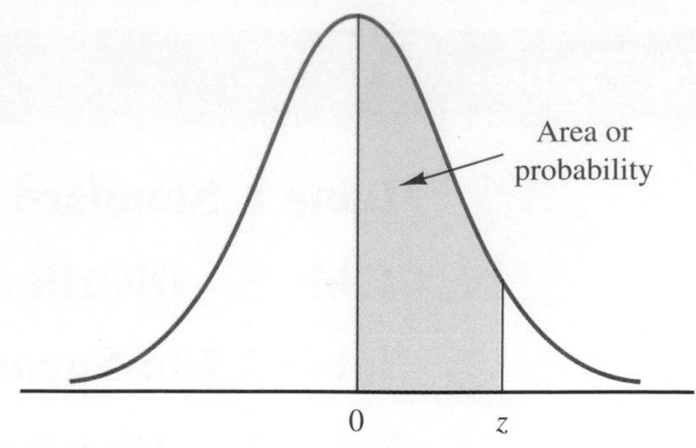

Entries in the table give the area under the curve between the mean and z standard deviations above the mean. For example, for $z = 1.25$ the area under the curve between the mean and z is .3944.

z	.00	.01	.02	.03	.04	.05	.06	.07	.08	.09
.0	.0000	.0040	.0080	.0120	.0160	.0199	.0239	.0279	.0319	.0359
.1	.0398	.0438	.0478	.0517	.0557	.0596	.0636	.0675	.0714	.0753
.2	.0793	.0832	.0871	.0910	.0948	.0987	.1026	.1064	.1103	.1141
.3	.1179	.1217	.1255	.1293	.1331	.1368	.1406	.1443	.1480	.1517
.4	.1554	.1591	.1628	.1664	.1700	.1736	.1772	.1808	.1844	.1879
.5	.1915	.1950	.1985	.2019	.2054	.2088	.2123	.2157	.2190	.2224
.6	.2257	.2291	.2324	.2357	.2389	.2422	.2454	.2486	.2518	.2549
.7	.2580	.2612	.2642	.2673	.2704	.2734	.2764	.2794	.2823	.2852
.8	.2881	.2910	.2939	.2967	.2995	.3023	.3051	.3078	.3106	.3133
.9	.3159	.3186	.3212	.3238	.3264	.3289	.3315	.3340	.3365	.3389
1.0	.3413	.3438	.3461	.3485	.3508	.3531	.3554	.3577	.3599	.3621
1.1	.3643	.3665	.3686	.3708	.3729	.3749	.3770	.3790	.3810	.3830
1.2	.3849	.3869	.3888	.3907	.3925	.3944	.3962	.3980	.3997	.4015
1.3	.4032	.4049	.4066	.4082	.4099	.4115	.4131	.4147	.4162	.4177
1.4	.4192	.4207	.4222	.4236	.4251	.4265	.4279	.4292	.4306	.4319
1.5	.4332	.4345	.4357	.4370	.4382	.4394	.4406	.4418	.4429	.4441
1.6	.4452	.4463	.4474	.4484	.4495	.4505	.4515	.4525	.4535	.4545
1.7	.4554	.4564	.4573	.4582	.4591	.4599	.4608	.4616	.4625	.4633
1.8	.4641	.4649	.4656	.4664	.4671	.4678	.4686	.4693	.4699	.4706
1.9	.4713	.4719	.4726	.4732	.4738	.4744	.4750	.4756	.4761	.4767
2.0	.4772	.4778	.4783	.4788	.4793	.4798	.4803	.4808	.4812	.4817
2.1	.4821	.4826	.4830	.4834	.4838	.4842	.4846	.4850	.4854	.4857
2.2	.4861	.4864	.4868	.4871	.4875	.4878	.4881	.4884	.4887	.4890
2.3	.4893	.4896	.4898	.4901	.4904	.4906	.4909	.4911	.4913	.4916
2.4	.4918	.4920	.4922	.4925	.4927	.4929	.4931	.4932	.4934	.4936
2.5	.4938	.4940	.4941	.4943	.4945	.4946	.4948	.4949	.4951	.4952
2.6	.4953	.4955	.4956	.4957	.4959	.4960	.4961	.4962	.4963	.4964
2.7	.4965	.4966	.4967	.4968	.4969	.4970	.4971	.4972	.4973	.4974
2.8	.4974	.4975	.4976	.4977	.4977	.4978	.4979	.4979	.4980	.4981
2.9	.4981	.4982	.4982	.4983	.4984	.4984	.4985	.4985	.4986	.4986
3.0	.4986	.4987	.4987	.4988	.4988	.4989	.4989	.4989	.4990	.4990

TABLE 2 *t* Distribution

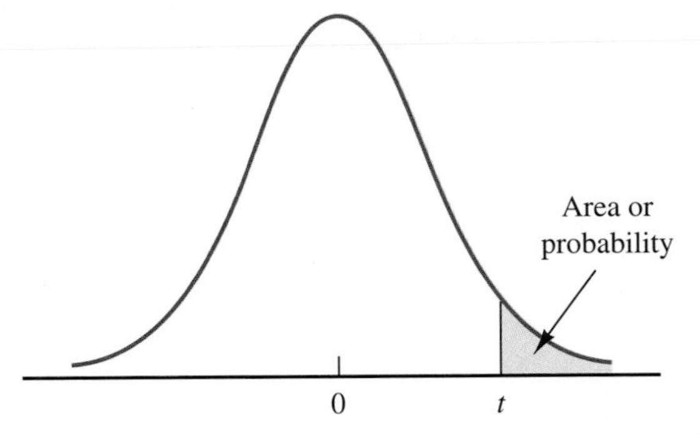

Entries in the table give t values for an area or probability in the upper tail of the t distribution. For example, with 10 degrees of freedom and a .05 area in the upper tail, $t_{.05} = 1.812$.

Degrees of Freedom	Area in Upper Tail				
	.10	.05	.025	.01	.005
1	3.078	6.314	12.706	31.821	63.657
2	1.886	2.920	4.303	6.965	9.925
3	1.638	2.353	3.182	4.541	5.841
4	1.533	2.132	2.776	3.747	4.604
5	1.476	2.015	2.571	3.365	4.032
6	1.440	1.943	2.447	3.143	3.707
7	1.415	1.895	2.365	2.998	3.499
8	1.397	1.860	2.306	2.896	3.355
9	1.383	1.833	2.262	2.821	3.250
10	1.372	1.812	2.228	2.764	3.169
11	1.363	1.796	2.201	2.718	3.106
12	1.356	1.782	2.179	2.681	3.055
13	1.350	1.771	2.160	2.650	3.012
14	1.345	1.761	2.145	2.624	2.977
15	1.341	1.753	2.131	2.602	2.947
16	1.337	1.746	2.120	2.583	2.921
17	1.333	1.740	2.110	2.567	2.898
18	1.330	1.734	2.101	2.552	2.878
19	1.328	1.729	2.093	2.539	2.861
20	1.325	1.725	2.086	2.528	2.845
21	1.323	1.721	2.080	2.518	2.831
22	1.321	1.717	2.074	2.508	2.819
23	1.319	1.714	2.069	2.500	2.807
24	1.318	1.711	2.064	2.492	2.797
25	1.316	1.708	2.060	2.485	2.787
26	1.315	1.706	2.056	2.479	2.779
27	1.314	1.703	2.052	2.473	2.771
28	1.313	1.701	2.048	2.467	2.763
29	1.311	1.699	2.045	2.462	2.756
30	1.310	1.697	2.042	2.457	2.750
40	1.303	1.684	2.021	2.423	2.704
60	1.296	1.671	2.000	2.390	2.660
120	1.289	1.658	1.980	2.358	2.617
∞	1.282	1.645	1.960	2.326	2.576

TABLE 3 Chi-Square Distribution

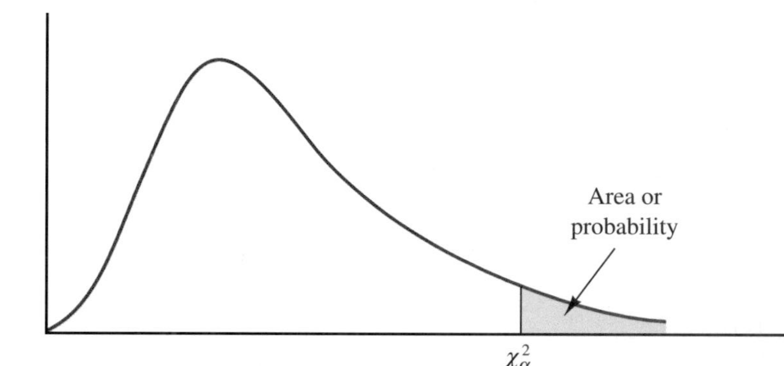

Entries in the table give χ^2_α values, where α is the area or probability in the upper tail of the chi-square distribution. For example, with 10 degrees of freedom and a .01 area in the upper tail, $\chi^2_{.01}=23.2093$.

Degrees of Freedom	Area in Upper Tail									
	.995	.99	.975	.95	.90	.10	.05	.025	.01	.005
1	$392{,}704 \times 10^{-10}$	$157{,}088 \times 10^{-9}$	$982{,}069 \times 10^{-9}$	$393{,}214 \times 10^{-8}$	.0157908	2.70554	3.84146	5.02389	6.63490	7.87944
2	.0100251	.0201007	.0506356	.102587	.210720	4.60517	5.99147	7.37776	9.21034	10.5966
3	.0717212	.114832	.215795	.351846	.584375	6.25139	7.81473	9.34840	11.3449	12.8381
4	.206990	.297110	.484419	.710721	1.063623	7.77944	9.48773	11.1433	13.2767	14.8602
5	.411740	.554300	.831211	1.145476	1.61031	9.23635	11.0705	12.8325	15.0863	16.7496
6	.675727	.872085	1.237347	1.63539	2.20413	10.6446	12.5916	14.4494	16.8119	18.5476
7	.989265	1.239043	1.68987	2.16735	2.83311	12.0170	14.0671	16.0128	18.4753	20.2777
8	1.344419	1.646482	2.17973	2.73264	3.48954	13.3616	15.5073	17.5346	20.0902	21.9550
9	1.734926	2.087912	2.70039	3.32511	4.16816	14.6837	16.9190	19.0228	21.6660	23.5893

(Table Continues)

Table 3 (Continued)

10	2.15585	2.55821	3.24697	3.94030	4.86518	15.9871	18.3070	20.4831	23.2093	25.1882
11	2.60321	3.05347	3.81575	4.57481	5.57779	17.2750	19.6751	21.9200	24.7250	26.7569
12	3.07382	3.57056	4.40379	5.22603	6.30380	18.5494	21.0261	23.3367	26.2170	28.2995
13	3.56503	4.10691	5.00874	5.89186	7.04150	19.8119	22.3621	24.7356	27.6883	29.8194
14	4.07468	4.66043	5.62872	6.57063	7.78953	21.0642	23.6848	26.1190	29.1413	31.3193
15	4.60094	5.22935	6.26214	7.26094	8.54675	22.3072	24.9958	27.4884	30.5779	32.8013
16	5.14224	5.81221	6.90766	7.96164	9.31223	23.5418	26.2962	28.8454	31.9999	34.2672
17	5.69724	6.40776	7.56418	8.67176	10.0852	24.7690	27.5871	30.1910	33.4087	35.7185
18	6.26481	7.01491	8.23075	9.39046	10.8649	25.9894	28.8693	31.5264	34.8053	37.1564
19	6.84398	7.63273	8.90655	10.1170	11.6509	27.2036	30.1435	32.8523	36.1908	38.5822
20	7.43386	8.26040	9.59083	10.8508	12.4426	28.4120	31.4104	34.1696	37.5662	39.9968
21	8.03366	8.89720	10.28293	11.5913	13.2396	29.6151	32.6705	35.4789	38.9321	41.4010
22	8.64272	9.54249	10.9823	12.3380	14.0415	30.8133	33.9244	36.7807	40.2894	42.7958
23	9.26042	10.19567	11.6885	13.0905	14.8479	32.0069	35.1725	38.0757	41.6384	44.1813
24	9.88623	10.8564	12.4011	13.8484	15.6587	33.1963	36.4151	39.3641	42.9798	45.5585
25	10.5197	11.5240	13.1197	14.6114	16.4734	34.3816	37.6525	40.6465	44.3141	46.9278
26	11.1603	12.1981	13.8439	15.3791	17.2919	35.5631	38.8852	41.9232	45.6417	48.2899
27	11.8076	12.8786	14.5733	16.1513	18.1138	36.7412	40.1133	43.1944	46.9630	49.6449
28	12.4613	13.5648	15.3079	16.9279	18.9392	37.9159	41.3372	44.4607	48.2782	50.9933
29	13.1211	14.2565	16.0471	17.7083	19.7677	39.0875	42.5569	45.7222	49.5879	52.3356
30	13.7867	14.9535	16.7908	18.4926	20.5992	40.2560	43.7729	46.9792	50.8922	53.6720
40	20.7065	22.1643	24.4331	26.5093	29.0505	51.8050	55.7585	59.3417	63.6907	66.7659
50	27.9907	29.7067	32.3574	34.7642	37.6886	63.1671	67.5048	71.4202	76.1539	79.4900
60	35.5346	37.4848	40.4817	43.1879	46.4589	74.3970	79.0819	83.2976	88.3794	91.9517
70	43.2752	45.4418	48.7576	51.7393	55.3290	85.5271	90.5312	95.0231	100.425	104.215
80	51.1720	53.5400	57.1532	60.3915	64.2778	96.5782	101.879	106.629	112.329	116.321
90	59.1963	61.7541	65.6466	69.1260	73.2912	107.565	113.145	118.136	124.116	128.299
100	67.3276	70.0648	74.2219	77.9295	82.3581	118.498	124.342	129.561	135.807	140.169

This table is reprinted by permission of Biometrika Trustees from Table 8, Percentage Points of the χ^2 Distribution, by E. S. Pearson and H. O. Hartley, *Biometrika Tables for Statisticians,* Vol. I, 3rd Edition, 1966.

TABLE 4 *F* Distribution

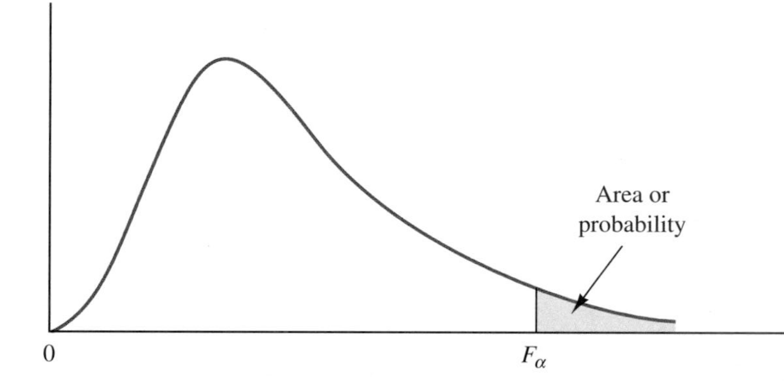

Entries in the table give F_α values, where α is the area or probability in the upper tail of the F distribution. For example, with 12 numerator degrees of freedom, 15 denominator degrees of freedom, and a .05 area in the upper tail, $F_{.05} = 2.48$.

Table of $F_{.05}$ Values

Denominator Degrees of Freedom	Numerator Degrees of Freedom																		
	1	2	3	4	5	6	7	8	9	10	12	15	20	24	30	40	60	120	∞
1	161.4	199.5	215.7	224.6	230.2	234.0	236.8	238.9	240.5	241.9	243.9	245.9	248.0	249.1	250.1	251.1	252.2	253.3	254.3
2	18.51	19.00	19.16	19.25	19.30	19.33	19.35	19.37	19.38	19.40	19.41	19.43	19.45	19.45	19.46	19.47	19.48	19.49	19.50
3	10.13	9.55	9.28	9.12	9.01	8.94	8.89	8.85	8.81	8.79	8.74	8.70	8.66	8.64	8.62	8.59	8.57	8.55	8.53
4	7.71	6.94	6.59	6.39	6.26	6.16	6.09	6.04	6.00	5.96	5.91	5.86	5.80	5.77	5.75	5.72	5.69	5.66	5.63
5	6.61	5.79	5.41	5.19	5.05	4.95	4.88	4.82	4.77	4.74	4.68	4.62	4.56	4.53	4.50	4.46	4.43	4.40	4.36

(Table Continues)

TABLE 4 (Continued)

6	5.99	5.14	4.76	4.53	4.39	4.28	4.21	4.15	4.10	4.06	4.00	3.94	3.87	3.84	3.81	3.77	3.74	3.70	3.67
7	5.59	4.74	4.35	4.12	3.97	3.87	3.79	3.73	3.68	3.64	3.57	3.51	3.44	3.41	3.38	3.34	3.30	3.27	3.23
8	5.32	4.46	4.07	3.84	3.69	3.58	3.50	3.44	3.39	3.35	3.28	3.22	3.15	3.12	3.08	3.04	3.01	2.97	2.93
9	5.12	4.26	3.86	3.63	3.48	3.37	3.29	3.23	3.18	3.14	3.07	3.01	2.94	2.90	2.86	2.83	2.79	2.75	2.71
10	4.96	4.10	3.71	3.48	3.33	3.22	3.14	3.07	3.02	2.98	2.91	2.85	2.77	2.74	2.70	2.66	2.62	2.58	2.54
11	4.84	3.98	3.59	3.36	3.20	3.09	3.01	2.95	2.90	2.85	2.79	2.72	2.65	2.61	2.57	2.53	2.49	2.45	2.40
12	4.75	3.89	3.49	3.26	3.11	3.00	2.91	2.85	2.80	2.75	2.69	2.62	2.54	2.51	2.47	2.43	2.38	2.34	2.30
13	4.67	3.81	3.41	3.18	3.03	2.92	2.83	2.77	2.71	2.67	2.60	2.53	2.46	2.42	2.38	2.34	2.30	2.25	2.21
14	4.60	3.74	3.34	3.11	2.96	2.85	2.76	2.70	2.65	2.60	2.53	2.46	2.39	2.35	2.31	2.27	2.22	2.18	2.13
15	4.54	3.68	3.29	3.06	2.90	2.79	2.71	2.64	2.59	2.54	2.48	2.40	2.33	2.29	2.25	2.20	2.16	2.11	2.07
16	4.49	3.63	3.24	3.01	2.85	2.74	2.66	2.59	2.54	2.49	2.42	2.35	2.28	2.24	2.19	2.15	2.11	2.06	2.01
17	4.45	3.59	3.20	2.96	2.81	2.70	2.61	2.55	2.49	2.45	2.38	2.31	2.23	2.19	2.15	2.10	2.06	2.01	1.96
18	4.41	3.55	3.16	2.93	2.77	2.66	2.58	2.51	2.46	2.41	2.34	2.27	2.19	2.15	2.11	2.06	2.02	1.97	1.92
19	4.38	3.52	3.13	2.90	2.74	2.63	2.54	2.48	2.42	2.38	2.31	2.23	2.16	2.11	2.07	2.03	1.98	1.93	1.88
20	4.35	3.49	3.10	2.87	2.71	2.60	2.51	2.45	2.39	2.35	2.28	2.20	2.12	2.08	2.04	1.99	1.95	1.90	1.84
21	4.32	3.47	3.07	2.84	2.68	2.57	2.49	2.42	2.37	2.32	2.25	2.18	2.10	2.05	2.01	1.96	1.92	1.87	1.81
22	4.30	3.44	3.05	2.82	2.66	2.55	2.46	2.40	2.34	2.30	2.23	2.15	2.07	2.03	1.98	1.94	1.89	1.84	1.78
23	4.28	3.42	3.03	2.80	2.64	2.53	2.44	2.37	2.32	2.27	2.20	2.13	2.05	2.01	1.96	1.91	1.86	1.81	1.76
24	4.26	3.40	3.01	2.78	2.62	2.51	2.42	2.36	2.30	2.25	2.18	2.11	2.03	1.98	1.94	1.89	1.84	1.79	1.73
25	4.24	3.39	2.99	2.76	2.60	2.49	2.40	2.34	2.28	2.24	2.16	2.09	2.01	1.96	1.92	1.87	1.82	1.77	1.71
26	4.23	3.37	2.98	2.74	2.59	2.47	2.39	2.32	2.27	2.22	2.15	2.07	1.99	1.95	1.90	1.85	1.80	1.75	1.69
27	4.21	3.35	2.96	2.73	2.57	2.46	2.37	2.31	2.25	2.20	2.13	2.06	1.97	1.93	1.88	1.84	1.79	1.73	1.67
28	4.20	3.34	2.95	2.71	2.56	2.45	2.36	2.29	2.24	2.19	2.12	2.04	1.96	1.91	1.87	1.82	1.77	1.71	1.65
29	4.18	3.33	2.93	2.70	2.55	2.43	2.35	2.28	2.22	2.18	2.10	2.03	1.94	1.90	1.85	1.81	1.75	1.70	1.64
30	4.17	3.32	2.92	2.69	2.53	2.42	2.33	2.27	2.21	2.16	2.09	2.01	1.93	1.89	1.84	1.79	1.74	1.68	1.62
40	4.08	3.23	2.84	2.61	2.45	2.34	2.25	2.18	2.12	2.08	2.00	1.92	1.84	1.79	1.74	1.69	1.64	1.58	1.51
60	4.00	3.15	2.76	2.53	2.37	2.25	2.17	2.10	2.04	1.99	1.92	1.84	1.75	1.70	1.65	1.59	1.53	1.47	1.39
120	3.92	3.07	2.68	2.45	2.29	2.17	2.09	2.02	1.96	1.91	1.83	1.75	1.66	1.61	1.55	1.50	1.43	1.35	1.25
∞	3.84	3.00	2.60	2.37	2.21	2.10	2.01	1.94	1.88	1.83	1.75	1.67	1.57	1.52	1.46	1.39	1.32	1.22	1.00

This table is reprinted by permission of the Biometrika Trustees from Table 18, Percentage Points of the *F Distribution, by E. S. Pearson and H. O. Hartley, Biometrika Tables for Statisticians,* Vol. I, 3rd Edition, 1966.

(Table Continues)

TABLE 4 (Continued)

Table of $F_{.01}$ Values

Denominator Degrees of Freedom	Numerator Degrees of Freedom 1	2	3	4	5	6	7	8	9	10	12	15	20	24	30	40	60	120	∞
1	4,052	4,999.5	5,403	5,625	5,764	5,859	5,928	5,982	6,022	6,056	6,106	6,157	6,209	6,235	6,261	6,287	6,313	6,339	6,366
2	98.50	99.00	99.17	99.25	99.30	99.33	99.36	99.37	99.39	99.40	99.42	99.43	99.45	99.46	99.47	99.47	99.48	99.49	99.50
3	34.12	30.82	29.46	28.71	28.24	27.91	27.67	27.49	27.35	27.23	27.05	26.87	26.69	26.60	26.50	26.41	26.32	26.22	26.13
4	21.20	18.00	16.69	15.98	15.52	15.21	14.98	14.80	14.66	14.55	14.37	14.20	14.02	13.93	13.84	13.75	13.65	13.56	13.46
5	16.26	13.27	12.06	11.39	10.97	10.67	10.46	10.29	10.16	10.05	9.89	9.72	9.55	9.47	9.38	9.29	9.20	9.11	9.06
6	13.75	10.92	9.78	9.15	8.75	8.47	8.26	8.10	7.98	7.87	7.72	7.56	7.40	7.31	7.23	7.14	7.06	6.97	6.88
7	12.25	9.55	8.45	7.85	7.46	7.19	6.99	6.84	6.72	6.62	6.47	6.31	6.16	6.07	5.99	5.91	5.82	5.74	5.65
8	11.26	8.65	7.59	7.01	6.63	6.37	6.18	6.03	5.91	5.81	5.67	5.52	5.36	5.28	5.20	5.12	5.03	4.95	4.86
9	10.56	8.02	6.99	6.42	6.06	5.80	5.61	5.47	5.35	5.26	5.11	4.96	4.81	4.73	4.65	4.57	4.48	4.40	4.31
10	10.04	7.56	6.55	5.99	5.64	5.39	5.20	5.06	4.94	4.85	4.71	4.56	4.41	4.33	4.25	4.17	4.08	4.00	3.91
11	9.65	7.21	6.22	5.67	5.32	5.07	4.89	4.74	4.63	4.54	4.40	4.25	4.10	4.02	3.94	3.86	3.78	3.69	3.60
12	9.33	6.93	5.95	5.41	5.06	4.82	4.64	4.50	4.39	4.30	4.16	4.01	3.86	3.78	3.70	3.62	3.54	3.45	3.36
13	9.07	6.70	5.74	5.21	4.86	4.62	4.44	4.30	4.19	4.10	3.96	3.82	3.66	3.59	3.51	3.43	3.34	3.25	3.17
14	8.86	6.51	5.56	5.04	4.69	4.46	4.28	4.14	4.03	3.94	3.80	3.66	3.51	3.43	3.35	3.27	3.18	3.09	3.00
15	8.68	6.36	5.42	4.89	4.56	4.32	4.14	4.00	3.89	3.80	3.67	3.52	3.37	3.29	3.21	3.13	3.05	2.96	2.87
16	8.53	6.23	5.29	4.77	4.44	4.20	4.03	3.89	3.78	3.69	3.55	3.41	3.26	3.18	3.10	3.02	2.93	2.84	2.75
17	8.40	6.11	5.18	4.67	4.34	4.10	3.93	3.79	3.68	3.59	3.46	3.31	3.16	3.08	3.00	2.92	2.83	2.75	2.65
18	8.29	6.01	5.09	4.58	4.25	4.01	3.84	3.71	3.60	3.51	3.37	3.23	3.08	3.00	2.92	2.84	2.75	2.66	2.57
19	8.18	5.93	5.01	4.50	4.17	3.94	3.77	3.63	3.52	3.43	3.30	3.15	3.00	2.92	2.84	2.76	2.67	2.58	2.49
20	8.10	5.85	4.94	4.43	4.10	3.87	3.70	3.56	3.46	3.37	3.23	3.09	2.94	2.86	2.78	2.69	2.61	2.52	2.42
21	8.02	5.78	4.87	4.37	4.04	3.81	3.64	3.51	3.40	3.31	3.17	3.03	2.88	2.80	2.72	2.64	2.55	2.46	2.36
22	7.95	5.72	4.82	4.31	3.99	3.76	3.59	3.45	3.35	3.26	3.12	2.98	2.83	2.75	2.67	2.58	2.50	2.40	2.31
23	7.88	5.66	4.76	4.26	3.94	3.71	3.54	3.41	3.30	3.21	3.07	2.93	2.78	2.70	2.62	2.54	2.45	2.35	2.26
24	7.82	5.61	4.72	4.22	3.90	3.67	3.50	3.36	3.26	3.17	3.03	2.89	2.74	2.66	2.58	2.49	2.40	2.31	2.21
25	7.77	5.57	4.68	4.18	3.85	3.63	3.46	3.32	3.22	3.13	2.99	2.85	2.70	2.62	2.54	2.45	2.36	2.27	2.17
26	7.72	5.53	4.64	4.14	3.82	3.59	3.42	3.29	3.18	3.09	2.96	2.81	2.66	2.58	2.50	2.42	2.33	2.23	2.13
27	7.68	5.49	4.60	4.11	3.78	3.56	3.39	3.26	3.15	3.06	2.93	2.78	2.63	2.55	2.47	2.38	2.29	2.20	2.10
28	7.64	5.45	4.57	4.07	3.75	3.53	3.36	3.23	3.12	3.03	2.90	2.75	2.60	2.52	2.44	2.35	2.26	2.17	2.06
29	7.60	5.42	4.54	4.04	3.73	3.50	3.33	3.20	3.09	3.00	2.87	2.73	2.57	2.49	2.41	2.33	2.23	2.14	2.03
30	7.56	5.39	4.51	4.02	3.70	3.47	3.30	3.17	3.07	2.98	2.84	2.70	2.55	2.47	2.39	2.30	2.21	2.11	2.01
40	7.31	5.18	4.31	3.83	3.51	3.29	3.12	2.99	2.89	2.80	2.66	2.52	2.37	2.29	2.20	2.11	2.02	1.92	1.80
60	7.08	4.98	4.13	3.65	3.34	3.12	2.95	2.82	2.72	2.63	2.50	2.35	2.20	2.12	2.03	1.94	1.84	1.73	1.60
120	6.85	4.79	3.95	3.48	3.17	2.96	2.79	2.66	2.56	2.47	2.34	2.19	2.03	1.95	1.86	1.76	1.66	1.53	1.38
∞	6.63	4.61	3.78	3.32	3.02	2.80	2.64	2.51	2.41	2.32	2.18	2.04	1.88	1.79	1.70	1.59	1.47	1.32	1.00

(Table Continues)

TABLE 4 (Continued)

Table of $F_{.025}$ Values

Denominator Degrees of Freedom	Numerator Degrees of Freedom 1	2	3	4	5	6	7	8	9	10	12	15	20	24	30	40	60	120	∞
1	647.8	799.5	864.2	899.6	921.8	937.1	948.2	956.7	963.3	968.6	976.7	984.9	993.1	997.2	1,001	1,006	1,010	1,014	1,018
2	38.51	39.00	39.17	39.25	39.30	39.33	39.36	39.37	39.39	39.40	39.41	39.43	39.45	39.46	39.46	39.47	39.48	39.49	39.50
3	17.44	16.04	15.44	15.10	14.88	14.73	14.62	14.54	14.47	14.42	14.34	14.25	14.17	14.12	14.08	14.04	13.99	13.95	13.90
4	12.22	10.65	9.98	9.60	9.36	9.20	9.07	8.98	8.90	8.84	8.75	8.66	8.56	8.51	8.46	8.41	8.36	8.31	8.26
5	10.01	8.43	7.76	7.39	7.15	6.98	6.85	6.76	6.68	6.62	6.52	6.43	6.33	6.28	6.23	6.18	6.12	6.07	6.02
6	8.81	7.26	6.60	6.23	5.99	5.82	5.70	5.60	5.52	5.46	5.37	5.27	5.17	5.12	5.07	5.01	4.96	4.90	4.85
7	8.07	6.54	5.89	5.52	5.29	5.21	4.99	4.90	4.82	4.76	4.67	4.57	4.47	4.42	4.36	4.31	4.25	4.20	4.14
8	7.57	6.06	5.42	5.05	4.82	4.65	4.53	4.43	4.36	4.30	4.20	4.10	4.00	3.95	3.89	3.84	3.78	3.73	3.67
9	7.21	5.71	5.08	4.72	4.48	4.32	4.20	4.10	4.03	3.96	3.87	3.77	3.67	3.61	3.56	3.51	3.45	3.39	3.33
10	6.94	5.46	4.83	4.47	4.24	4.07	3.95	3.85	3.78	3.72	3.62	3.52	3.42	3.37	3.31	3.26	3.20	3.14	3.08
11	6.72	5.26	4.63	4.28	4.04	3.88	3.76	3.66	3.59	3.53	3.43	3.33	3.23	3.17	3.12	3.06	3.00	2.94	2.88
12	6.55	5.10	4.47	4.12	3.89	3.73	3.61	3.51	3.44	3.37	3.28	3.18	3.07	3.02	2.96	2.91	2.85	2.79	2.72
13	6.41	4.97	4.35	4.00	3.77	3.60	3.48	3.39	3.31	3.25	3.15	3.05	2.95	2.89	2.84	2.78	2.72	2.66	2.60
14	6.30	4.86	4.24	3.89	3.66	3.50	3.38	3.29	3.21	3.15	3.05	2.95	2.84	2.79	2.73	2.67	2.61	2.55	2.49
15	6.20	4.77	4.15	3.80	3.58	3.41	3.29	3.20	3.12	3.06	2.96	2.86	2.76	2.70	2.64	2.59	2.52	2.46	2.40
16	6.12	4.69	4.08	3.73	3.50	3.34	3.22	3.12	3.05	2.99	2.89	2.79	2.68	2.63	2.57	2.51	2.45	2.38	2.32
17	6.04	4.62	4.01	3.66	3.44	3.28	3.16	3.06	2.98	2.92	2.82	2.72	2.62	2.56	2.50	2.44	2.38	2.32	2.25
18	5.98	4.56	3.95	3.61	3.38	3.22	3.10	3.01	2.93	2.87	2.77	2.67	2.56	2.50	2.44	2.38	2.32	2.26	2.19
19	5.92	4.51	3.90	3.56	3.33	3.17	3.05	2.96	2.88	2.82	2.72	2.62	2.51	2.45	2.39	2.33	2.27	2.20	2.13
20	5.87	4.46	3.86	3.51	3.29	3.13	3.01	2.91	2.84	2.77	2.68	2.57	2.46	2.41	2.35	2.29	2.22	2.16	2.09
21	5.83	4.42	3.82	3.48	3.25	3.09	2.97	2.87	2.80	2.73	2.64	2.53	2.42	2.37	2.31	2.25	2.18	2.11	2.04
22	5.79	4.38	3.78	3.44	3.22	3.05	2.93	2.84	2.76	2.70	2.60	2.50	2.39	2.33	2.27	2.21	2.14	2.08	2.00
23	5.75	4.35	3.75	3.41	3.18	3.02	2.90	2.81	2.73	2.67	2.57	2.47	2.36	2.30	2.24	2.18	2.11	2.04	1.97
24	5.72	4.32	3.72	3.38	3.15	2.99	2.87	2.78	2.70	2.64	2.54	2.44	2.33	2.27	2.21	2.15	2.08	2.01	1.94
25	5.69	4.29	3.69	3.35	3.13	2.97	2.85	2.75	2.68	2.61	2.51	2.41	2.30	2.24	2.18	2.12	2.05	1.98	1.91
26	5.66	4.27	3.67	3.33	3.10	2.94	2.82	2.73	2.65	2.59	2.49	2.39	2.28	2.22	2.16	2.09	2.03	1.95	1.88
27	5.63	4.24	3.65	3.31	3.08	2.92	2.80	2.71	2.63	2.57	2.47	2.36	2.25	2.19	2.13	2.07	2.00	1.93	1.85
28	5.61	4.22	3.63	3.29	3.06	2.90	2.78	2.69	2.61	2.55	2.45	2.34	2.23	2.17	2.11	2.05	1.98	1.91	1.83
29	5.59	4.20	3.61	3.27	3.04	2.88	2.76	2.67	2.59	2.53	2.43	2.32	2.21	2.15	2.09	2.03	1.96	1.89	1.81
30	5.57	4.18	3.59	3.25	3.03	2.87	2.75	2.65	2.57	2.51	2.41	2.31	2.20	2.14	2.07	2.01	1.94	1.87	1.79
40	5.42	4.05	3.46	3.13	2.90	2.74	2.62	2.53	2.45	2.39	2.29	2.18	2.07	2.01	1.94	1.88	1.80	1.72	1.64
60	5.29	3.93	3.34	3.01	2.79	2.63	2.51	2.41	2.33	2.27	2.17	2.06	1.94	1.88	1.82	1.74	1.67	1.58	1.48
120	5.15	3.80	3.23	2.89	2.67	2.52	2.39	2.30	2.22	2.16	2.05	1.94	1.82	1.76	1.69	1.61	1.53	1.43	1.31
∞	5.02	3.69	3.12	2.79	2.57	2.41	2.29	2.19	2.11	2.05	1.94	1.83	1.71	1.64	1.57	1.48	1.39	1.27	1.00

TABLE 5 Binomial Probabilities

Entries in the table give the probability of x successes in n trials of a binomial experiment, where p is the probability of a success on one trial. For example, with six trials and $p = .05$, the probability of two successes is .0305.

		p								
n	x	.01	.02	.03	.04	.05	.06	.07	.08	.09
2	0	.9801	.9604	.9409	.9216	.9025	.8836	.8649	.8464	.8281
	1	.0198	.0392	.0582	.0768	.0950	.1128	.1302	.1472	.1638
	2	.0001	.0004	.0009	.0016	.0025	.0036	.0049	.0064	.0081
3	0	.9703	.9412	.9127	.8847	.8574	.8306	.8044	.7787	.7536
	1	.0294	.0576	.0847	.1106	.1354	.1590	.1816	.2031	.2236
	2	.0003	.0012	.0026	.0046	.0071	.0102	.0137	.0177	.0221
	3	.0000	.0000	.0000	.0001	.0001	.0002	.0003	.0005	.0007
4	0	.9606	.9224	.8853	.8493	.8145	.7807	.7481	.7164	.6857
	1	.0388	.0753	.1095	.1416	.1715	.1993	.2252	.2492	.2713
	2	.0006	.0023	.0051	.0088	.0135	.0191	.0254	.0325	.0402
	3	.0000	.0000	.0001	.0002	.0005	.0008	.0013	.0019	.0027
	4	.0000	.0000	.0000	.0000	.0000	.0000	.0000	.0000	.0001
5	0	.9510	.9039	.8587	.8154	.7738	.7339	.6957	.6591	.6240
	1	.0480	.0922	.1328	.1699	.2036	.2342	.2618	.2866	.3086
	2	.0010	.0038	.0082	.0142	.0214	.0299	.0394	.0498	.0610
	3	.0000	.0001	.0003	.0006	.0011	.0019	.0030	.0043	.0060
	4	.0000	.0000	.0000	.0000	.0000	.0001	.0001	.0002	.0003
	5	.0000	.0000	.0000	.0000	.0000	.0000	.0000	.0000	.0000
6	0	.9415	.8858	.8330	.7828	.7351	.6899	.6470	.6064	.5679
	1	.0571	.1085	.1546	.1957	.2321	.2642	.2922	.3164	.3370
	2	.0014	.0055	.0120	.0204	.0305	.0422	.0550	.0688	.0833
	3	.0000	.0002	.0005	.0011	.0021	.0036	.0055	.0080	.0110
	4	.0000	.0000	.0000	.0000	.0001	.0002	.0003	.0005	.0008
	5	.0000	.0000	.0000	.0000	.0000	.0000	.0000	.0000	.0000
	6	.0000	.0000	.0000	.0000	.0000	.0000	.0000	.0000	.0000
7	0	.9321	.8681	.8080	.7514	.6983	.6485	.6017	.5578	.5168
	1	.0659	.1240	.1749	.2192	.2573	.2897	.3170	.3396	.3578
	2	.0020	.0076	.0162	.0274	.0406	.0555	.0716	.0886	.1061
	3	.0000	.0003	.0008	.0019	.0036	.0059	.0090	.0128	.0175
	4	.0000	.0000	.0000	.0001	.0002	.0004	.0007	.0011	.0017
	5	.0000	.0000	.0000	.0000	.0000	.0000	.0000	.0001	.0001
	6	.0000	.0000	.0000	.0000	.0000	.0000	.0000	.0000	.0000
	7	.0000	.0000	.0000	.0000	.0000	.0000	.0000	.0000	.0000
8	0	.9227	.8508	.7837	.7214	.6634	.6096	.5596	.5132	.4703
	1	.0746	.1389	.1939	.2405	.2793	.3113	.3370	.3570	.3721
	2	.0026	.0099	.0210	.0351	.0515	.0695	.0888	.1087	.1288
	3	.0001	.0004	.0013	.0029	.0054	.0089	.0134	.0189	.0255
	4	.0000	.0000	.0001	.0002	.0004	.0007	.0013	.0021	.0031
	5	.0000	.0000	.0000	.0000	.0000	.0000	.0001	.0001	.0002
	6	.0000	.0000	.0000	.0000	.0000	.0000	.0000	.0000	.0000
	7	.0000	.0000	.0000	.0000	.0000	.0000	.0000	.0000	.0000
	8	.0000	.0000	.0000	.0000	.0000	.0000	.0000	.0000	.0000

(Table Continues)

TABLE 5 (Continued)

		p								
n	x	.01	.02	.03	.04	.05	.06	.07	.08	.09
9	0	.9135	.8337	.7602	.6925	.6302	.5730	.5204	.4722	.4279
	1	.0830	.1531	.2116	.2597	.2985	.3292	.3525	.3695	.3809
	2	.0034	.0125	.0262	.0433	.0629	.0840	.1061	.1285	.1507
	3	.0001	.0006	.0019	.0042	.0077	.0125	.0186	.0261	.0348
	4	.0000	.0000	.0001	.0003	.0006	.0012	.0021	.0034	.0052
	5	.0000	.0000	.0000	.0000	.0000	.0001	.0002	.0003	.0005
	6	.0000	.0000	.0000	.0000	.0000	.0000	.0000	.0000	.0000
	7	.0000	.0000	.0000	.0000	.0000	.0000	.0000	.0000	.0000
	8	.0000	.0000	.0000	.0000	.0000	.0000	.0000	.0000	.0000
	9	.0000	.0000	.0000	.0000	.0000	.0000	.0000	.0000	.0000
10	0	.9044	.8171	.7374	.6648	.5987	.5386	.4840	.4344	.3894
	1	.0914	.1667	.2281	.2770	.3151	.3438	.3643	.3777	.3851
	2	.0042	.0153	.0317	.0519	.0746	.0988	.1234	.1478	.1714
	3	.0001	.0008	.0026	.0058	.0105	.0168	.0248	.0343	.0452
	4	.0000	.0000	.0001	.0004	.0010	.0019	.0033	.0052	.0078
	5	.0000	.0000	.0000	.0000	.0001	.0001	.0003	.0005	.0009
	6	.0000	.0000	.0000	.0000	.0000	.0000	.0000	.0000	.0001
	7	.0000	.0000	.0000	.0000	.0000	.0000	.0000	.0000	.0000
	8	.0000	.0000	.0000	.0000	.0000	.0000	.0000	.0000	.0001
	9	.0000	.0000	.0000	.0000	.0000	.0000	.0000	.0000	.0001
	10	.0000	.0000	.0000	.0000	.0000	.0000	.0000	.0000	.0001
12	0	.8864	.7847	.6938	.6127	.5404	.4759	.4186	.3677	.3225
	1	.1074	.1922	.2575	.3064	.3413	.3645	.3781	.3837	.3827
	2	.0060	.0216	.0438	.0702	.0988	.1280	.1565	.1835	.2082
	3	.0002	.0015	.0045	.0098	.0173	.0272	.0393	.0532	.0686
	4	.0000	.0001	.0003	.0009	.0021	.0039	.0067	.0104	.0153
	5	.0000	.0000	.0000	.0001	.0002	.0004	.0008	.0014	.0024
	6	.0000	.0000	.0000	.0000	.0000	.0000	.0001	.0001	.0003
	7	.0000	.0000	.0000	.0000	.0000	.0000	.0000	.0000	.0000
	8	.0000	.0000	.0000	.0000	.0000	.0000	.0000	.0000	.0000
	9	.0000	.0000	.0000	.0000	.0000	.0000	.0000	.0000	.0000
	10	.0000	.0000	.0000	.0000	.0000	.0000	.0000	.0000	.0000
	11	.0000	.0000	.0000	.0000	.0000	.0000	.0000	.0000	.0000
	12	.0000	.0000	.0000	.0000	.0000	.0000	.0000	.0000	.0000
15	0	.8601	.7386	.6333	.5421	.4633	.3953	.3367	.2863	.2430
	1	.1303	.2261	.2938	.3388	.3658	.3785	.3801	.3734	.3605
	2	.0092	.0323	.0636	.0988	.1348	.1691	.2003	.2273	.2496
	3	.0004	.0029	.0085	.0178	.0307	.0468	.0653	.0857	.1070
	4	.0000	.0002	.0008	.0022	.0049	.0090	.0148	.0223	.0317
	5	.0000	.0000	.0001	.0002	.0006	.0013	.0024	.0043	.0069
	6	.0000	.0000	.0000	.0000	.0000	.0001	.0003	.0006	.0011
	7	.0000	.0000	.0000	.0000	.0000	.0000	.0000	.0001	.0001
	8	.0000	.0000	.0000	.0000	.0000	.0000	.0000	.0000	.0000
	9	.0000	.0000	.0000	.0000	.0000	.0000	.0000	.0000	.0000
	10	.0000	.0000	.0000	.0000	.0000	.0000	.0000	.0000	.0000
	11	.0000	.0000	.0000	.0000	.0000	.0000	.0000	.0000	.0000
	12	.0000	.0000	.0000	.0000	.0000	.0000	.0000	.0000	.0000
	13	.0000	.0000	.0000	.0000	.0000	.0000	.0000	.0000	.0000
	14	.0000	.0000	.0000	.0000	.0000	.0000	.0000	.0000	.0000
	15	.0000	.0000	.0000	.0000	.0000	.0000	.0000	.0000	.0000

(Table Continues)

TABLE 5 (Continued)

n	x	p = .01	.02	.03	.04	.05	.06	.07	.08	.09
18	0	.8345	.6951	.5780	.4796	.3972	.3283	.2708	.2229	.1831
	1	.1517	.2554	.3217	.3597	.3763	.3772	.3669	.3489	.3260
	2	.0130	.0443	.0846	.1274	.1683	.2047	.2348	.2579	.2741
	3	.0007	.0048	.0140	.0283	.0473	.0697	.0942	.1196	.1446
	4	.0000	.0004	.0016	.0044	.0093	.0167	.0266	.0390	.0536
	5	.0000	.0000	.0001	.0005	.0014	.0030	.0056	.0095	.0148
	6	.0000	.0000	.0000	.0000	.0002	.0004	.0009	.0018	.0032
	7	.0000	.0000	.0000	.0000	.0000	.0000	.0001	.0003	.0005
	8	.0000	.0000	.0000	.0000	.0000	.0000	.0000	.0000	.0001
	9	.0000	.0000	.0000	.0000	.0000	.0000	.0000	.0000	.0000
	10	.0000	.0000	.0000	.0000	.0000	.0000	.0000	.0000	.0000
	11	.0000	.0000	.0000	.0000	.0000	.0000	.0000	.0000	.0000
	12	.0000	.0000	.0000	.0000	.0000	.0000	.0000	.0000	.0000
	13	.0000	.0000	.0000	.0000	.0000	.0000	.0000	.0000	.0000
	14	.0000	.0000	.0000	.0000	.0000	.0000	.0000	.0000	.0000
	15	.0000	.0000	.0000	.0000	.0000	.0000	.0000	.0000	.0000
	16	.0000	.0000	.0000	.0000	.0000	.0000	.0000	.0000	.0000
	17	.0000	.0000	.0000	.0000	.0000	.0000	.0000	.0000	.0000
	18	.0000	.0000	.0000	.0000	.0000	.0000	.0000	.0000	.0000
20	0	.8179	.6676	.5438	.4420	.3585	.2901	.2342	.1887	.1516
	1	.1652	.2725	.3364	.3683	.3774	.3703	.3526	.3282	.3000
	2	.0159	.0528	.0988	.1458	.1887	.2246	.2521	.2711	.2818
	3	.0010	.0065	.0183	.0364	.0596	.0860	.1139	.1414	.1672
	4	.0000	.0006	.0024	.0065	.0133	.0233	.0364	.0523	.0703
	5	.0000	.0000	.0002	.0009	.0022	.0048	.0088	.0145	.0222
	6	.0000	.0000	.0000	.0001	.0003	.0008	.0017	.0032	.0055
	7	.0000	.0000	.0000	.0000	.0000	.0001	.0002	.0005	.0011
	8	.0000	.0000	.0000	.0000	.0000	.0000	.0000	.0001	.0002
	9	.0000	.0000	.0000	.0000	.0000	.0000	.0000	.0000	.0000
	10	.0000	.0000	.0000	.0000	.0000	.0000	.0000	.0000	.0000
	11	.0000	.0000	.0000	.0000	.0000	.0000	.0000	.0000	.0000
	12	.0000	.0000	.0000	.0000	.0000	.0000	.0000	.0000	.0000
	13	.0000	.0000	.0000	.0000	.0000	.0000	.0000	.0000	.0000
	14	.0000	.0000	.0000	.0000	.0000	.0000	.0000	.0000	.0000
	15	.0000	.0000	.0000	.0000	.0000	.0000	.0000	.0000	.0000
	16	.0000	.0000	.0000	.0000	.0000	.0000	.0000	.0000	.0000
	17	.0000	.0000	.0000	.0000	.0000	.0000	.0000	.0000	.0000
	18	.0000	.0000	.0000	.0000	.0000	.0000	.0000	.0000	.0000
	19	.0000	.0000	.0000	.0000	.0000	.0000	.0000	.0000	.0000
	20	.0000	.0000	.0000	.0000	.0000	.0000	.0000	.0000	.0000

(Table Continues)

TABLE 5 (Continued)

		p								
n	*x*	.10	.15	.20	.25	.30	.35	.40	.45	.50
2	0	.8100	.7225	.6400	.5625	.4900	.4225	.3600	.3025	.2500
	1	.1800	.2550	.3200	.3750	.4200	.4550	.4800	.4950	.5000
	2	.0100	.0225	.0400	.0625	.0900	.1225	.1600	.2025	.2500
3	0	.7290	.6141	.5120	.4219	.3430	.2746	.2160	.1664	.1250
	1	.2430	.3251	.3840	.4219	.4410	.4436	.4320	.4084	.3750
	2	.0270	.0574	.0960	.1406	.1890	.2389	.2880	.3341	.3750
	3	.0010	.0034	.0080	.0156	.0270	.0429	.0640	.0911	.1250
4	0	.6561	.5220	.4096	.3164	.2401	.1785	.1296	.0915	.0625
	1	.2916	.3685	.4096	.4219	.4116	.3845	.3456	.2995	.2500
	2	.0486	.0975	.1536	.2109	.2646	.3105	.3456	.3675	.3750
	3	.0036	.0115	.0256	.0469	.0756	.1115	.1536	.2005	.2500
	4	.0001	.0005	.0016	.0039	.0081	.0150	.0256	.0410	.0625
5	0	.5905	.4437	.3277	.2373	.1681	.1160	.0778	.0503	.0312
	1	.3280	.3915	.4096	.3955	.3602	.3124	.2592	.2059	.1562
	2	.0729	.1382	.2048	.2637	.3087	.3364	.3456	.3369	.3125
	3	.0081	.0244	.0512	.0879	.1323	.1811	.2304	.2757	.3125
	4	.0004	.0022	.0064	.0146	.0284	.0488	.0768	.1128	.1562
	5	.0000	.0001	.0003	.0010	.0024	.0053	.0102	.0185	.0312
6	0	.5314	.3771	.2621	.1780	.1176	.0754	.0467	.0277	.0156
	1	.3543	.3993	.3932	.3560	.3025	.2437	.1866	.1359	.0938
	2	.0984	.1762	.2458	.2966	.3241	.3280	.3110	.2780	.2344
	3	.0146	.0415	.0819	.1318	.1852	.2355	.2765	.3032	.3125
	4	.0012	.0055	.0154	.0330	.0595	.0951	.1382	.1861	.2344
	5	.0001	.0004	.0015	.0044	.0102	.0205	.0369	.0609	.0938
	6	.0000	.0000	.0001	.0002	.0007	.0018	.0041	.0083	.0156
7	0	.4783	.3206	.2097	.1335	.0824	.0490	.0280	.0152	.0078
	1	.3720	.3960	.3670	.3115	.2471	.1848	.1306	.0872	.0547
	2	.1240	.2097	.2753	.3115	.3177	.2985	.2613	.2140	.1641
	3	.0230	.0617	.1147	.1730	.2269	.2679	.2903	.2918	.2734
	4	.0026	.0109	.0287	.0577	.0972	.1442	.1935	.2388	.2734
	5	.0002	.0012	.0043	.0115	.0250	.0466	.0774	.1172	.1641
	6	.0000	.0001	.0004	.0013	.0036	.0084	.0172	.0320	.0547
	7	.0000	.0000	.0000	.0001	.0002	.0006	.0016	.0037	.0078
8	0	.4305	.2725	.1678	.1001	.0576	.0319	.0168	.0084	.0039
	1	.3826	.3847	.3355	.2670	.1977	.1373	.0896	.0548	.0312
	2	.1488	.2376	.2936	.3115	.2965	.2587	.2090	.1569	.1094
	3	.0331	.0839	.1468	.2076	.2541	.2786	.2787	.2568	.2188
	4	.0046	.0185	.0459	.0865	.1361	.1875	.2322	.2627	.2734
	5	.0004	.0026	.0092	.0231	.0467	.0808	.1239	.1719	.2188
	6	.0000	.0002	.0011	.0038	.0100	.0217	.0413	.0703	.1094
	7	.0000	.0000	.0001	.0004	.0012	.0033	.0079	.0164	.0312
	8	.0000	.0000	.0000	.0000	.0001	.0002	.0007	.0017	.0039

(Table Continues)

TABLE 5 (Continued)

		p								
n	x	.10	.15	.20	.25	.30	.35	.40	.45	.50
9	0	.3874	.2316	.1342	.0751	.0404	.0207	.0101	.0046	.0020
	1	.3874	.3679	.3020	.2253	.1556	.1004	.0605	.0339	.0176
	2	.1722	.2597	.3020	.3003	.2668	.2162	.1612	.1110	.0703
	3	.0446	.1069	.1762	.2336	.2668	.2716	.2508	.2119	.1641
	4	.0074	.0283	.0661	.1168	.1715	.2194	.2508	.2600	.2461
	5	.0008	.0050	.0165	.0389	.0735	.1181	.1672	.2128	.2461
	6	.0001	.0006	.0028	.0087	.0210	.0424	.0743	.1160	.1641
	7	.0000	.0000	.0003	.0012	.0039	.0098	.0212	.0407	.0703
	8	.0000	.0000	.0000	.0001	.0004	.0013	.0035	.0083	.0176
	9	.0000	.0000	.0000	.0000	.0000	.0001	.0003	.0008	.0020
10	0	.3487	.1969	.1074	.0563	.0282	.0135	.0060	.0025	.0010
	1	.3874	.3474	.2684	.1877	.1211	.0725	.0403	.0207	.0098
	2	.1937	.2759	.3020	.2816	.2335	.1757	.1209	.0763	.0439
	3	.0574	.1298	.2013	.2503	.2668	.2522	.2150	.1665	.1172
	4	.0112	.0401	.0881	.1460	.2001	.2377	.2508	.2384	.2051
	5	.0015	.0085	.0264	.0584	.1029	.1536	.2007	.2340	.2461
	6	.0001	.0012	.0055	.0162	.0368	.0689	.1115	.1596	.2051
	7	.0000	.0001	.0008	.0031	.0090	.0212	.0425	.0746	.1172
	8	.0000	.0000	.0001	.0004	.0014	.0043	.0106	.0229	.0439
	9	.0000	.0000	.0000	.0000	.0001	.0005	.0016	.0042	.0098
	10	.0000	.0000	.0000	.0000	.0000	.0000	.0001	.0003	.0010
12	0	.2824	.1422	.0687	.0317	.0138	.0057	.0022	.0008	.0002
	1	.3766	.3012	.2062	.1267	.0712	.0368	.0174	.0075	.0029
	2	.2301	.2924	.2835	.2323	.1678	.1088	.0639	.0339	.0161
	3	.0853	.1720	.2362	.2581	.2397	.1954	.1419	.0923	.0537
	4	.0213	.0683	.1329	.1936	.2311	.2367	.2128	.1700	.1208
	5	.0038	.0193	.0532	.1032	.1585	.2039	.2270	.2225	.1934
	6	.0005	.0040	.0155	.0401	.0792	.1281	.1766	.2124	.2256
	7	.0000	.0006	.0033	.0115	.0291	.0591	.1009	.1489	.1934
	8	.0000	.0001	.0005	.0024	.0078	.0199	.0420	.0762	.1208
	9	.0000	.0000	.0001	.0004	.0015	.0048	.0125	.0277	.0537
	10	.0000	.0000	.0000	.0000	.0002	.0008	.0025	.0068	.0161
	11	.0000	.0000	.0000	.0000	.0000	.0001	.0003	.0010	.0029
	12	.0000	.0000	.0000	.0000	.0000	.0000	.0000	.0001	.0002
15	0	.2059	.0874	.0352	.0134	.0047	.0016	.0005	.0001	.0000
	1	.3432	.2312	.1319	.0668	.0305	.0126	.0047	.0016	.0005
	2	.2669	.2856	.2309	.1559	.0916	.0476	.0219	.0090	.0032
	3	.1285	.2184	.2501	.2252	.1700	.1110	.0634	.0318	.0139
	4	.0428	.1156	.1876	.2252	.2186	.1792	.1268	.0780	.0417
	5	.0105	.0449	.1032	.1651	.2061	.2123	.1859	.1404	.0916
	6	.0019	.0132	.0430	.0917	.1472	.1906	.2066	.1914	.1527
	7	.0003	.0030	.0138	.0393	.0811	.1319	.1771	.2013	.1964
	8	.0000	.0005	.0035	.0131	.0348	.0710	.1181	.1647	.1964
	9	.0000	.0001	.0007	.0034	.0116	.0298	.0612	.1048	.1527
	10	.0000	.0000	.0001	.0007	.0030	.0096	.0245	.0515	.0916
	11	.0000	.0000	.0000	.0001	.0006	.0024	.0074	.0191	.0417
	12	.0000	.0000	.0000	.0000	.0001	.0004	.0016	.0052	.0139
	13	.0000	.0000	.0000	.0000	.0000	.0001	.0003	.0010	.0032
	14	.0000	.0000	.0000	.0000	.0000	.0000	.0000	.0001	.0005
	15	.0000	.0000	.0000	.0000	.0000	.0000	.0000	.0000	.0000

(Table Continues)

TABLE 5 (Continued)

n	x	p = .10	.15	.20	.25	.30	.35	.40	.45	.50
18	0	.1501	.0536	.0180	.0056	.0016	.0004	.0001	.0000	.0000
	1	.3002	.1704	.0811	.0338	.0126	.0042	.0012	.0003	.0001
	2	.2835	.2556	.1723	.0958	.0458	.0190	.0069	.0022	.0006
	3	.1680	.2406	.2297	.1704	.1046	.0547	.0246	.0095	.0031
	4	.0700	.1592	.2153	.2130	.1681	.1104	.0614	.0291	.0117
	5	.0218	.0787	.1507	.1988	.2017	.1664	.1146	.0666	.0327
	6	.0052	.0301	.0816	.1436	.1873	.1941	.1655	.1181	.0708
	7	.0010	.0091	.0350	.0820	.1376	.1792	.1892	.1657	.1214
	8	.0002	.0022	.0120	.0376	.0811	.1327	.1734	.1864	.1669
	9	.0000	.0004	.0033	.0139	.0386	.0794	.1284	.1694	.1855
	10	.0000	.0001	.0008	.0042	.0149	.0385	.0771	.1248	.1669
	11	.0000	.0000	.0001	.0010	.0046	.0151	.0374	.0742	.1214
	12	.0000	.0000	.0000	.0002	.0012	.0047	.0145	.0354	.0708
	13	.0000	.0000	.0000	.0000	.0002	.0012	.0045	.0134	.0327
	14	.0000	.0000	.0000	.0000	.0000	.0002	.0011	.0039	.0117
	15	.0000	.0000	.0000	.0000	.0000	.0000	.0002	.0009	.0031
	16	.0000	.0000	.0000	.0000	.0000	.0000	.0000	.0001	.0006
	17	.0000	.0000	.0000	.0000	.0000	.0000	.0000	.0000	.0001
	18	.0000	.0000	.0000	.0000	.0000	.0000	.0000	.0000	.0000
20	0	.1216	.0388	.0115	.0032	.0008	.0002	.0000	.0000	.0000
	1	.2702	.1368	.0576	.0211	.0068	.0020	.0005	.0001	.0000
	2	.2852	.2293	.1369	.0669	.0278	.0100	.0031	.0008	.0002
	3	.1901	.2428	.2054	.1339	.0716	.0323	.0123	.0040	.0011
	4	.0898	.1821	.2182	.1897	.1304	.0738	.0350	.0139	.0046
	5	.0319	.1028	.1746	.2023	.1789	.1272	.0746	.0365	.0148
	6	.0089	.0454	.1091	.1686	.1916	.1712	.1244	.0746	.0370
	7	.0020	.0160	.0545	.1124	.1643	.1844	.1659	.1221	.0739
	8	.0004	.0046	.0222	.0609	.1144	.1614	.1797	.1623	.1201
	9	.0001	.0011	.0074	.0271	.0654	.1158	.1597	.1771	.1602
	10	.0000	.0002	.0020	.0099	.0308	.0686	.1171	.1593	.1762
	11	.0000	.0000	.0005	.0030	.0120	.0336	.0710	.1185	.1602
	12	.0000	.0000	.0001	.0008	.0039	.0136	.0355	.0727	.1201
	13	.0000	.0000	.0000	.0002	.0010	.0045	.0146	.0366	.0739
	14	.0000	.0000	.0000	.0000	.0002	.0012	.0049	.0150	.0370
	15	.0000	.0000	.0000	.0000	.0000	.0003	.0013	.0049	.0148
	16	.0000	.0000	.0000	.0000	.0000	.0000	.0003	.0013	.0046
	17	.0000	.0000	.0000	.0000	.0000	.0000	.0000	.0002	.0011
	18	.0000	.0000	.0000	.0000	.0000	.0000	.0000	.0000	.0002
	19	.0000	.0000	.0000	.0000	.0000	.0000	.0000	.0000	.0000
	20	.0000	.0000	.0000	.0000	.0000	.0000	.0000	.0000	.0000

(Table Continues)

TABLE 5 (Continued)

n	x	p = .55	.60	.65	.70	.75	.80	.85	.90	.95
2	0	.2025	.1600	.1225	.0900	.0625	.0400	.0225	.0100	.0025
	1	.4950	.4800	.4550	.4200	.3750	.3200	.2550	.1800	.0950
	2	.3025	.3600	.4225	.4900	.5625	.6400	.7225	.8100	.9025
3	0	.0911	.0640	.0429	.0270	.0156	.0080	.0034	.0010	.0001
	1	.3341	.2880	.2389	.1890	.1406	.0960	.0574	.0270	.0071
	2	.4084	.4320	.4436	.4410	.4219	.3840	.3251	.2430	.1354
	3	.1664	.2160	.2746	.3430	.4219	.5120	.6141	.7290	.8574
4	0	.0410	.0256	.0150	.0081	.0039	.0016	.0005	.0001	.0000
	1	.2005	.1536	.1115	.0756	.0469	.0256	.0115	.0036	.0005
	2	.3675	.3456	.3105	.2646	.2109	.1536	.0975	.0486	.0135
	3	.2995	.3456	.3845	.4116	.4219	.4096	.3685	.2916	.1715
	4	.0915	.1296	.1785	.2401	.3164	.4096	.5220	.6561	.8145
5	0	.0185	.0102	.0053	.0024	.0010	.0003	.0001	.0000	.0000
	1	.1128	.0768	.0488	.0284	.0146	.0064	.0022	.0005	.0000
	2	.2757	.2304	.1811	.1323	.0879	.0512	.0244	.0081	.0011
	3	.3369	.3456	.3364	.3087	.2637	.2048	.1382	.0729	.0214
	4	.2059	.2592	.3124	.3601	.3955	.4096	.3915	.3281	.2036
	5	.0503	.0778	.1160	.1681	.2373	.3277	.4437	.5905	.7738
6	0	.0083	.0041	.0018	.0007	.0002	.0001	.0000	.0000	.0000
	1	.0609	.0369	.0205	.0102	.0044	.0015	.0004	.0001	.0000
	2	.1861	.1382	.0951	.0595	.0330	.0154	.0055	.0012	.0001
	3	.3032	.2765	.2355	.1852	.1318	.0819	.0415	.0146	.0021
	4	.2780	.3110	.3280	.3241	.2966	.2458	.1762	.0984	.0305
	5	.1359	.1866	.2437	.3025	.3560	.3932	.3993	.3543	.2321
	6	.0277	.0467	.0754	.1176	.1780	.2621	.3771	.5314	.7351
7	0	.0037	.0016	.0006	.0002	.0001	.0000	.0000	.0000	.0000
	1	.0320	.0172	.0084	.0036	.0013	.0004	.0001	.0000	.0000
	2	.1172	.0774	.0466	.0250	.0115	.0043	.0012	.0002	.0000
	3	.2388	.1935	.1442	.0972	.0577	.0287	.0109	.0026	.0002
	4	.2918	.2903	.2679	.2269	.1730	.1147	.0617	.0230	.0036
	5	.2140	.2613	.2985	.3177	.3115	.2753	.2097	.1240	.0406
	6	.0872	.1306	.1848	.2471	.3115	.3670	.3960	.3720	.2573
	7	.0152	.0280	.0490	.0824	.1335	.2097	.3206	.4783	.6983
8	0	.0017	.0007	.0002	.0001	.0000	.0000	.0000	.0000	.0000
	1	.0164	.0079	.0033	.0012	.0004	.0001	.0000	.0000	.0000
	2	.0703	.0413	.0217	.0100	.0038	.0011	.0002	.0000	.0000
	3	.1719	.1239	.0808	.0467	.0231	.0092	.0026	.0004	.0000
	4	.2627	.2322	.1875	.1361	.0865	.0459	.0185	.0046	.0004
	5	.2568	.2787	.2786	.2541	.2076	.1468	.0839	.0331	.0054
	6	.1569	.2090	.2587	.2965	.3115	.2936	.2376	.1488	.0515
	7	.0548	.0896	.1373	.1977	.2670	.3355	.3847	.3826	.2793
	8	.0084	.0168	.0319	.0576	.1001	.1678	.2725	.4305	.6634

(Table Continues)

TABLE 5 (Continued)

		p								
n	*x*	.55	.60	.65	.70	.75	.80	.85	.90	.95
9	0	.0008	.0003	.0001	.0000	.0000	.0000	.0000	.0000	.0000
	1	.0083	.0035	.0013	.0004	.0001	.0000	.0000	.0000	.0000
	2	.0407	.0212	.0098	.0039	.0012	.0003	.0000	.0000	.0000
	3	.1160	.0743	.0424	.0210	.0087	.0028	.0006	.0001	.0000
	4	.2128	.1672	.1181	.0735	.0389	.0165	.0050	.0008	.0000
	5	.2600	.2508	.2194	.1715	.1168	.0661	.0283	.0074	.0006
	6	.2119	.2508	.2716	.2668	.2336	.1762	.1069	.0446	.0077
	7	.1110	.1612	.2162	.2668	.3003	.3020	.2597	.1722	.0629
	8	.0339	.0605	.1004	.1556	.2253	.3020	.3679	.3874	.2985
	9	.0046	.0101	.0207	.0404	.0751	.1342	.2316	.3874	.6302
10	0	.0003	.0001	.0000	.0000	.0000	.0000	.0000	.0000	.0000
	1	.0042	.0016	.0005	.0001	.0000	.0000	.0000	.0000	.0000
	2	.0229	.0106	.0043	.0014	.0004	.0001	.0000	.0000	.0000
	3	.0746	.0425	.0212	.0090	.0031	.0008	.0001	.0000	.0000
	4	.1596	.1115	.0689	.0368	.0162	.0055	.0012	.0001	.0000
	5	.2340	.2007	.1536	.1029	.0584	.0264	.0085	.0015	.0001
	6	.2384	.2508	.2377	.2001	.1460	.0881	.0401	.0112	.0010
	7	.1665	.2150	.2522	.2668	.2503	.2013	.1298	.0574	.0105
	8	.0763	.1209	.1757	.2335	.2816	.3020	.2759	.1937	.0746
	9	.0207	.0403	.0725	.1211	.1877	.2684	.3474	.3874	.3151
	10	.0025	.0060	.0135	.0282	.0563	.1074	.1969	.3487	.5987
12	0	.0001	.0000	.0000	.0000	.0000	.0000	.0000	.0000	.0000
	1	.0010	.0003	.0001	.0000	.0000	.0000	.0000	.0000	.0000
	2	.0068	.0025	.0008	.0002	.0000	.0000	.0000	.0000	.0000
	3	.0277	.0125	.0048	.0015	.0004	.0001	.0000	.0000	.0000
	4	.0762	.0420	.0199	.0078	.0024	.0005	.0001	.0000	.0000
	5	.1489	.1009	.0591	.0291	.0115	.0033	.0006	.0000	.0000
	6	.2124	.1766	.1281	.0792	.0401	.0155	.0040	.0005	.0000
	7	.2225	.2270	.2039	.1585	.1032	.0532	.0193	.0038	.0002
	8	.1700	.2128	.2367	.2311	.1936	.1329	.0683	.0213	.0021
	9	.0923	.1419	.1954	.2397	.2581	.2362	.1720	.0852	.0173
	10	.0339	.0639	.1088	.1678	.2323	.2835	.2924	.2301	.0988
	11	.0075	.0174	.0368	.0712	.1267	.2062	.3012	.3766	.3413
	12	.0008	.0022	.0057	.0138	.0317	.0687	.1422	.2824	.5404
15	0	.0000	.0000	.0000	.0000	.0000	.0000	.0000	.0000	.0000
	1	.0001	.0000	.0000	.0000	.0000	.0000	.0000	.0000	.0000
	2	.0010	.0003	.0001	.0000	.0000	.0000	.0000	.0000	.0000
	3	.0052	.0016	.0004	.0001	.0000	.0000	.0000	.0000	.0000
	4	.0191	.0074	.0024	.0006	.0001	.0000	.0000	.0000	.0000
	5	.0515	.0245	.0096	.0030	.0007	.0001	.0000	.0000	.0000
	6	.1048	.0612	.0298	.0116	.0034	.0007	.0001	.0000	.0000
	7	.1647	.1181	.0710	.0348	.0131	.0035	.0005	.0000	.0000
	8	.2013	.1771	.1319	.0811	.0393	.0138	.0030	.0003	.0000
	9	.1914	.2066	.1906	.1472	.0917	.0430	.0132	.0019	.0000
	10	.1404	.1859	.2123	.2061	.1651	.1032	.0449	.0105	.0006
	11	.0780	.1268	.1792	.2186	.2252	.1876	.1156	.0428	.0049
	12	.0318	.0634	.1110	.1700	.2252	.2501	.2184	.1285	.0307
	13	.0090	.0219	.0476	.0916	.1559	.2309	.2856	.2669	.1348
	14	.0016	.0047	.0126	.0305	.0668	.1319	.2312	.3432	.3658
	15	.0001	.0005	.0016	.0047	.0134	.0352	.0874	.2059	.4633

(Table Continues)

TABLE 5 (Continued)

		p								
n	*x*	.55	.60	.65	.70	.75	.80	.85	.90	.95
18	0	.0000	.0000	.0000	.0000	.0000	.0000	.0000	.0000	.0000
	1	.0000	.0000	.0000	.0000	.0000	.0000	.0000	.0000	.0000
	2	.0001	.0000	.0000	.0000	.0000	.0000	.0000	.0000	.0000
	3	.0009	.0002	.0000	.0000	.0000	.0000	.0000	.0000	.0000
	4	.0039	.0011	.0002	.0000	.0000	.0000	.0000	.0000	.0000
	5	.0134	.0045	.0012	.0002	.0000	.0000	.0000	.0000	.0000
	6	.0354	.0145	.0047	.0012	.0002	.0000	.0000	.0000	.0000
	7	.0742	.0374	.0151	.0046	.0010	.0001	.0000	.0000	.0000
	8	.1248	.0771	.0385	.0149	.0042	.0008	.0001	.0000	.0000
	9	.1694	.1284	.0794	.0386	.0139	.0033	.0004	.0000	.0000
	10	.1864	.1734	.1327	.0811	.0376	.0120	.0022	.0002	.0000
	11	.1657	.1892	.1792	.1376	.0820	.0350	.0091	.0010	.0000
	12	.1181	.1655	.1941	.1873	.1436	.0816	.0301	.0052	.0002
	13	.0666	.1146	.1664	.2017	.1988	.1507	.0787	.0218	.0014
	14	.0291	.0614	.1104	.1681	.2130	.2153	.1592	.0700	.0093
	15	.0095	.0246	.0547	.1046	.1704	.2297	.2406	.1680	.0473
	16	.0022	.0069	.0190	.0458	.0958	.1723	.2556	.2835	.1683
	17	.0003	.0012	.0042	.0126	.0338	.0811	.1704	.3002	.3763
	18	.0000	.0001	.0004	.0016	.0056	.0180	.0536	.1501	.3972
20	0	.0000	.0000	.0000	.0000	.0000	.0000	.0000	.0000	.0000
	1	.0000	.0000	.0000	.0000	.0000	.0000	.0000	.0000	.0000
	2	.0000	.0000	.0000	.0000	.0000	.0000	.0000	.0000	.0000
	3	.0002	.0000	.0000	.0000	.0000	.0000	.0000	.0000	.0000
	4	.0013	.0003	.0000	.0000	.0000	.0000	.0000	.0000	.0000
	5	.0049	.0013	.0003	.0000	.0000	.0000	.0000	.0000	.0000
	6	.0150	.0049	.0012	.0002	.0000	.0000	.0000	.0000	.0000
	7	.0366	.0146	.0045	.0010	.0002	.0000	.0000	.0000	.0000
	8	.0727	.0355	.0136	.0039	.0008	.0001	.0000	.0000	.0000
	9	.1185	.0710	.0336	.0120	.0030	.0005	.0000	.0000	.0000
	10	.1593	.1171	.0686	.0308	.0099	.0020	.0002	.0000	.0000
	11	.1771	.1597	.1158	.0654	.0271	.0074	.0011	.0001	.0000
	12	.1623	.1797	.1614	.1144	.0609	.0222	.0046	.0004	.0000
	13	.1221	.1659	.1844	.1643	.1124	.0545	.0160	.0020	.0000
	14	.0746	.1244	.1712	.1916	.1686	.1091	.0454	.0089	.0003
	15	.0365	.0746	.1272	.1789	.2023	.1746	.1028	.0319	.0022
	16	.0139	.0350	.0738	.1304	.1897	.2182	.1821	.0898	.0133
	17	.0040	.0123	.0323	.0716	.1339	.2054	.2428	.1901	.0596
	18	.0008	.0031	.0100	.0278	.0669	.1369	.2293	.2852	.1887
	19	.0001	.0005	.0020	.0068	.0211	.0576	.1368	.2702	.3774
	20	.0000	.0000	.0002	.0008	.0032	.0115	.0388	.1216	.3585

TABLE 6 Values of $e^{-\mu}$

μ	$e^{-\mu}$	μ	$e^{-\mu}$	μ	$e^{-\mu}$
.00	1.0000				
.05	.9512	2.05	.1287	4.05	.0174
.10	.9048	2.10	.1225	4.10	.0166
.15	.8607	2.15	.1165	4.15	.0158
.20	.8187	2.20	.1108	4.20	.0150
.25	.7788	2.25	.1054	4.25	.0143
.30	.7408	2.30	.1003	4.30	.0136
.35	.7047	2.35	.0954	4.35	.0129
.40	.6703	2.40	.0907	4.40	.0123
.45	.6376	2.45	.0863	4.45	.0117
.50	.6065	2.50	.0821	4.50	.0111
.55	.5769	2.55	.0781	4.55	.0106
.60	.5488	2.60	.0743	4.60	.0101
.65	.5220	2.65	.0707	4.65	.0096
.70	.4966	2.70	.0672	4.70	.0091
.75	.4724	2.75	.0639	4.75	.0087
.80	.4493	2.80	.0608	4.80	.0082
.85	.4274	2.85	.0578	4.85	.0078
.90	.4066	2.90	.0550	4.90	.0074
.95	.3867	2.95	.0523	4.95	.0071
1.00	.3679	3.00	.0498	5.00	.0067
1.05	.3499	3.05	.0474	5.05	.0064
1.10	.3329	3.10	.0450	5.10	.0061
1.15	.3166	3.15	.0429	5.15	.0058
1.20	.3012	3.20	.0408	5.20	.0055
1.25	.2865	3.25	.0388	5.25	.0052
1.30	.2725	3.30	.0369	5.30	.0050
1.35	.2592	3.35	.0351	5.35	.0047
1.40	.2466	3.40	.0334	5.40	.0045
1.45	.2346	3.45	.0317	5.45	.0043
1.50	.2231	3.50	.0302	5.50	.0041
1.55	.2122	3.55	.0287	5.55	.0039
1.60	.2019	3.60	.0273	5.60	.0037
1.65	.1920	3.65	.0260	5.65	.0035
1.70	.1827	3.70	.0247	5.70	.0033
1.75	.1738	3.75	.0235	5.75	.0032
1.80	.1653	3.80	.0224	5.80	.0030
1.85	.1572	3.85	.0213	5.85	.0029
1.90	.1496	3.90	.0202	5.90	.0027
1.95	.1423	3.95	.0193	5.95	.0026
2.00	.1353	4.00	.0183	6.00	.0025
				7.00	.0009
				8.00	.000335
				9.00	.000123
				10.00	.000045

TABLE 7 Poisson Probabilities

Entries in the table give the probability of x occurrences for a Poisson process with a mean μ. For example, when $\mu = 2.5$, the probability of four occurrences is .1336.

	μ									
x	0.1	0.2	0.3	0.4	0.5	0.6	0.7	0.8	0.9	1.0
0	.9048	.8187	.7408	.6703	.6065	.5488	.4966	.4493	.4066	.3679
1	.0905	.1637	.2222	.2681	.3033	.3293	.3476	.3595	.3659	.3679
2	.0045	.0164	.0333	.0536	.0758	.0988	.1217	.1438	.1647	.1839
3	.0002	.0011	.0033	.0072	.0126	.0198	.0284	.0383	.0494	.0613
4	.0000	.0001	.0002	.0007	.0016	.0030	.0050	.0077	.0111	.0153
5	.0000	.0000	.0000	.0001	.0002	.0004	.0007	.0012	.0020	.0031
6	.0000	.0000	.0000	.0000	.0000	.0000	.0001	.0002	.0003	.0005
7	.0000	.0000	.0000	.0000	.0000	.0000	.0000	.0000	.0000	.0001

	μ									
x	1.1	1.2	1.3	1.4	1.5	1.6	1.7	1.8	1.9	2.0
0	.3329	.3012	.2725	.2466	.2231	.2019	.1827	.1653	.1496	.1353
1	.3662	.3614	.3543	.3452	.3347	.3230	.3106	.2975	.2842	.2707
2	.2014	.2169	.2303	.2417	.2510	.2584	.2640	.2678	.2700	.2707
3	.0738	.0867	.0998	.1128	.1255	.1378	.1496	.1607	.1710	.1804
4	.0203	.0260	.0324	.0395	.0471	.0551	.0636	.0723	.0812	.0902
5	.0045	.0062	.0084	.0111	.0141	.0176	.0216	.0260	.0309	.0361
6	.0008	.0012	.0018	.0026	.0035	.0047	.0061	.0078	.0098	.0120
7	.0001	.0002	.0003	.0005	.0008	.0011	.0015	.0020	.0027	.0034
8	.0000	.0000	.0001	.0001	.0001	.0002	.0003	.0005	.0006	.0009
9	.0000	.0000	.0000	.0000	.0000	.0000	.0001	.0001	.0001	.0002

	μ									
x	2.1	2.2	2.3	2.4	2.5	2.6	2.7	2.8	2.9	3.0
0	.1225	.1108	.1003	.0907	.0821	.0743	.0672	.0608	.0550	.0498
1	.2572	.2438	.2306	.2177	.2052	.1931	.1815	.1703	.1596	.1494
2	.2700	.2681	.2652	.2613	.2565	.2510	.2450	.2384	.2314	.2240
3	.1890	.1966	.2033	.2090	.2138	.2176	.2205	.2225	.2237	.2240
4	.0992	.1082	.1169	.1254	.1336	.1414	.1488	.1557	.1622	.1680

(Table Continues)

TABLE 7 (Continued)

	μ									
x	2.1	2.2	2.3	2.4	2.5	2.6	2.7	2.8	2.9	3.0
5	.0417	.0476	.0538	.0602	.0668	.0735	.0804	.0872	.0940	.1008
6	.0146	.0174	.0206	.0241	.0278	.0319	.0362	.0407	.0455	.0504
7	.0044	.0055	.0068	.0083	.0099	.0118	.0139	.0163	.0188	.0216
8	.0011	.0015	.0019	.0025	.0031	.0038	.0047	.0057	.0068	.0081
9	.0003	.0004	.0005	.0007	.0009	.0011	.0014	.0018	.0022	.0027
10	.0001	.0001	.0001	.0002	.0002	.0003	.0004	.0005	.0006	.0008
11	.0000	.0000	.0000	.0000	.0000	.0001	.0001	.0001	.0002	.0002
12	.0000	.0000	.0000	.0000	.0000	.0000	.0000	.0000	.0000	.0001

	μ									
x	3.1	3.2	3.3	3.4	3.5	3.6	3.7	3.8	3.9	4.0
0	.0450	.0408	.0369	.0344	.0302	.0273	.0247	.0224	.0202	.0183
1	.1397	.1304	.1217	.1135	.1057	.0984	.0915	.0850	.0789	.0733
2	.2165	.2087	.2008	.1929	.1850	.1771	.1692	.1615	.1539	.1465
3	.2237	.2226	.2209	.2186	.2158	.2125	.2087	.2046	.2001	.1954
4	.1734	.1781	.1823	.1858	.1888	.1912	.1931	.1944	.1951	.1954
5	.1075	.1140	.1203	.1264	.1322	.1377	.1429	.1477	.1522	.1563
6	.0555	.0608	.0662	.0716	.0771	.0826	.0881	.0936	.0989	.1042
7	.0246	.0278	.0312	.0348	.0385	.0425	.0466	.0508	.0551	.0595
8	.0095	.0111	.0129	.0148	.0169	.0191	.0215	.0241	.0269	.0298
9	.0033	.0040	.0047	.0056	.0066	.0076	.0089	.0102	.0116	.0132
10	.0010	.0013	.0016	.0019	.0023	.0028	.0033	.0039	.0045	.0053
11	.0003	.0004	.0005	.0006	.0007	.0009	.0011	.0013	.0016	.0019
12	.0001	.0001	.0001	.0002	.0002	.0003	.0003	.0004	.0005	.0006
13	.0000	.0000	.0000	.0000	.0001	.0001	.0001	.0001	.0002	.0002
14	.0000	.0000	.0000	.0000	.0000	.0000	.0000	.0000	.0000	.0001

	μ									
x	4.1	4.2	4.3	4.4	4.5	4.6	4.7	4.8	4.9	5.0
0	.0166	.0150	.0136	.0123	.0111	.0101	.0091	.0082	.0074	.0067
1	.0679	.0630	.0583	.0540	.0500	.0462	.0427	.0395	.0365	.0337
2	.1393	.1323	.1254	.1188	.1125	.1063	.1005	.0948	.0894	.0842
3	.1904	.1852	.1798	.1743	.1687	.1631	.1574	.1517	.1460	.1404
4	.1951	.1944	.1933	.1917	.1898	.1875	.1849	.1820	.1789	.1755
5	.1600	.1633	.1662	.1687	.1708	.1725	.1738	.1747	.1753	.1755
6	.1093	.1143	.1191	.1237	.1281	.1323	.1362	.1398	.1432	.1462
7	.0640	.0686	.0732	.0778	.0824	.0869	.0914	.0959	.1002	.1044
8	.0328	.0360	.0393	.0428	.0463	.0500	.0537	.0575	.0614	.0653
9	.0150	.0168	.0188	.0209	.0232	.0255	.0280	.0307	.0334	.0363

(Table Continues)

TABLE 7 (Continued)

	μ									
x	4.1	4.2	4.3	4.4	4.5	4.6	4.7	4.8	4.9	5.0
10	.0061	.0071	.0081	.0092	.0104	.0118	.0132	.0147	.0164	.0181
11	.0023	.0027	.0032	.0037	.0043	.0049	.0056	.0064	.0073	.0082
12	.0008	.0009	.0011	.0014	.0016	.0019	.0022	.0026	.0030	.0034
13	.0002	.0003	.0004	.0005	.0006	.0007	.0008	.0009	.0011	.0013
14	.0001	.0001	.0001	.0001	.0002	.0002	.0003	.0003	.0004	.0005
15	.0000	.0000	.0000	.0000	.0001	.0001	.0001	.0001	.0001	.0002

	μ									
x	5.1	5.2	5.3	5.4	5.5	5.6	5.7	5.8	5.9	6.0
0	.0061	.0055	.0050	.0045	.0041	.0037	.0033	.0030	.0027	.0025
1	.0311	.0287	.0265	.0244	.0225	.0207	.0191	.0176	.0162	.0149
2	.0793	.0746	.0701	.0659	.0618	.0580	.0544	.0509	.0477	.0446
3	.1348	.1293	.1239	.1185	.1133	.1082	.1033	.0985	.0938	.0892
4	.1719	.1681	.1641	.1600	.1558	.1515	.1472	.1428	.1383	.1339
5	.1753	.1748	.1740	.1728	.1714	.1697	.1678	.1656	.1632	.1606
6	.1490	.1515	.1537	.1555	.1571	.1584	.1594	.1601	.1605	.1606
7	.1086	.1125	.1163	.1200	.1234	.1267	.1298	.1326	.1353	.1377
8	.0692	.0731	.0771	.0810	.0849	.0887	.0925	.0962	.0998	.1033
9	.0392	.0423	.0454	.0486	.0519	.0552	.0586	.0620	.0654	.0688
10	.0200	.0220	.0241	.0262	.0285	.0309	.0334	.0359	.0386	.0413
11	.0093	.0104	.0116	.0129	.0143	.0157	.0173	.0190	.0207	.0225
12	.0039	.0045	.0051	.0058	.0065	.0073	.0082	.0092	.0102	.0113
13	.0015	.0018	.0021	.0024	.0028	.0032	.0036	.0041	.0046	.0052
14	.0006	.0007	.0008	.0009	.0011	.0013	.0015	.0017	.0019	.0022
15	.0002	.0002	.0003	.0003	.0004	.0005	.0006	.0007	.0008	.0009
16	.0001	.0001	.0001	.0001	.0001	.0002	.0002	.0002	.0003	.0003
17	.0000	.0000	.0000	.0000	.0000	.0001	.0001	.0001	.0001	.0001

	μ									
x	6.1	6.2	6.3	6.4	6.5	6.6	6.7	6.8	6.9	7.0
0	.0022	.0020	.0018	.0017	.0015	.0014	.0012	.0011	.0010	.0009
1	.0137	.0126	.0116	.0106	.0098	.0090	.0082	.0076	.0070	.0064
2	.0417	.0390	.0364	.0340	.0318	.0296	.0276	.0258	.0240	.0223
3	.0848	.0806	.0765	.0726	.0688	.0652	.0617	.0584	.0552	.0521
4	.1294	.1249	.1205	.1162	.1118	.1076	.1034	.0992	.0952	.0912
5	.1579	.1549	.1519	.1487	.1454	.1420	.1385	.1349	.1314	.1277
6	.1605	.1601	.1595	.1586	.1575	.1562	.1546	.1529	.1511	.1490
7	.1399	.1418	.1435	.1450	.1462	.1472	.1480	.1486	.1489	.1490
8	.1066	.1099	.1130	.1160	.1188	.1215	.1240	.1263	.1284	.1304
9	.0723	.0757	.0791	.0825	.0858	.0891	.0923	.0954	.0985	.1014

(Table Continues)

TABLE 7 (Continued)

	μ									
x	6.1	6.2	6.3	6.4	6.5	6.6	6.7	6.8	6.9	7.0
10	.0441	.0469	.0498	.0528	.0558	.0588	.0618	.0649	.0679	.0710
11	.0245	.0265	.0285	.0307	.0330	.0353	.0377	.0401	.0426	.0452
12	.0124	.0137	.0150	.0164	.0179	.0194	.0210	.0227	.0245	.0264
13	.0058	.0065	.0073	.0081	.0089	.0098	.0108	.0119	.0130	.0142
14	.0025	.0029	.0033	.0037	.0041	.0046	.0052	.0058	.0064	.0071
15	.0010	.0012	.0014	.0016	.0018	.0020	.0023	.0026	.0029	.0033
16	.0004	.0005	.0005	.0006	.0007	.0008	.0010	.0011	.0013	.0014
17	.0001	.0002	.0002	.0002	.0003	.0003	.0004	.0004	.0005	.0006
18	.0000	.0001	.0001	.0001	.0001	.0001	.0001	.0002	.0002	.0002
19	.0000	.0000	.0000	.0000	.0000	.0000	.0000	.0001	.0001	.0001

	μ									
x	7.1	7.2	7.3	7.4	7.5	7.6	7.7	7.8	7.9	8.0
0	.0008	.0007	.0007	.0006	.0006	.0005	.0005	.0004	.0004	.0003
1	.0059	.0054	.0049	.0045	.0041	.0038	.0035	.0032	.0029	.0027
2	.0208	.0194	.0180	.0167	.0156	.0145	.0134	.0125	.0116	.0107
3	.0492	.0464	.0438	.0413	.0389	.0366	.0345	.0324	.0305	.0286
4	.0874	.0836	.0799	.0764	.0729	.0696	.0663	.0632	.0602	.0573
5	.1241	.1204	.1167	.1130	.1094	.1057	.1021	.0986	.0951	.0916
6	.1468	.1445	.1420	.1394	.1367	.1339	.1311	.1282	.1252	.1221
7	.1489	.1486	.1481	.1474	.1465	.1454	.1442	.1428	.1413	.1396
8	.1321	.1337	.1351	.1363	.1373	.1382	.1388	.1392	.1395	.1396
9	.1042	.1070	.1096	.1121	.1144	.1167	.1187	.1207	.1224	.1241
10	.0740	.0770	.0800	.0829	.0858	.0887	.0914	.0941	.0967	.0993
11	.0478	.0504	.0531	.0558	.0585	.0613	.0640	.0667	.0695	.0722
12	.0283	.0303	.0323	.0344	.0366	.0388	.0411	.0434	.0457	.0481
13	.0154	.0168	.0181	.0196	.0211	.0227	.0243	.0260	.0278	.0296
14	.0078	.0086	.0095	.0104	.0113	.0123	.0134	.0145	.0157	.0169
15	.0037	.0041	.0046	.0051	.0057	.0062	.0069	.0075	.0083	.0090
16	.0016	.0019	.0021	.0024	.0026	.0030	.0033	.0037	.0041	.0045
17	.0007	.0008	.0009	.0010	.0012	.0013	.0015	.0017	.0019	.0021
18	.0003	.0003	.0004	.0004	.0005	.0006	.0006	.0007	.0008	.0009
19	.0001	.0001	.0001	.0002	.0002	.0002	.0003	.0003	.0003	.0004
20	.0000	.0000	.0001	.0001	.0001	.0001	.0001	.0001	.0001	.0002
21	.0000	.0000	.0000	.0000	.0000	.0000	.0000	.0000	.0001	.0001

(Table Continues)

TABLE 7 (Continued)

	μ									
x	8.1	8.2	8.3	8.4	8.5	8.6	8.7	8.8	8.9	9.0
0	.0003	.0003	.0002	.0002	.0002	.0002	.0002	.0002	.0001	.0001
1	.0025	.0023	.0021	.0019	.0017	.0016	.0014	.0013	.0012	.0011
2	.0100	.0092	.0086	.0079	.0074	.0068	.0063	.0058	.0054	.0050
3	.0269	.0252	.0237	.0222	.0208	.0195	.0183	.0171	.0160	.0150
4	.0544	.0517	.0491	.0466	.0443	.0420	.0398	.0377	.0357	.0337
5	.0882	.0849	.0816	.0784	.0752	.0722	.0692	.0663	.0635	.0607
6	.1191	.1160	.1128	.1097	.1066	.1034	.1003	.0972	.0941	.0911
7	.1378	.1358	.1338	.1317	.1294	.1271	.1247	.1222	.1197	.1171
8	.1395	.1392	.1388	.1382	.1375	.1366	.1356	.1344	.1332	.1318
9	.1256	.1269	.1280	.1290	.1299	.1306	.1311	.1315	.1317	.1318
10	.1017	.1040	.1063	.1084	.1104	.1123	.1140	.1157	.1172	.1186
11	.0749	.0776	.0802	.0828	.0853	.0878	.0902	.0925	.0948	.0970
12	.0505	.0530	.0555	.0579	.0604	.0629	.0654	.0679	.0703	.0728
13	.0315	.0334	.0354	.0374	.0395	.0416	.0438	.0459	.0481	.0504
14	.0182	.0196	.0210	.0225	.0240	.0256	.0272	.0289	.0306	.0324
15	.0098	.0107	.0116	.0126	.0136	.0147	.0158	.0169	.0182	.0194
16	.0050	.0055	.0060	.0066	.0072	.0079	.0086	.0093	.0101	.0109
17	.0024	.0026	.0029	.0033	.0036	.0040	.0044	.0048	.0053	.0058
18	.0011	.0012	.0014	.0015	.0017	.0019	.0021	.0024	.0026	.0029
19	.0005	.0005	.0006	.0007	.0008	.0009	.0010	.0011	.0012	.0014
20	.0002	.0002	.0002	.0003	.0003	.0004	.0004	.0005	.0005	.0006
21	.0001	.0001	.0001	.0001	.0001	.0002	.0002	.0002	.0002	.0003
22	.0000	.0000	.0000	.0000	.0001	.0001	.0001	.0001	.0001	.0001

	μ									
x	9.1	9.2	9.3	9.4	9.5	9.6	9.7	9.8	9.9	10
0	.0001	.0001	.0001	.0001	.0001	.0001	.0001	.0001	.0001	.0000
1	.0010	.0009	.0009	.0008	.0007	.0007	.0006	.0005	.0005	.0005
2	.0046	.0043	.0040	.0037	.0034	.0031	.0029	.0027	.0025	.0023
3	.0140	.0131	.0123	.0115	.0107	.0100	.0093	.0087	.0081	.0076
4	.0319	.0302	.0285	.0269	.0254	.0240	.0226	.0213	.0201	.0189
5	.0581	.0555	.0530	.0506	.0483	.0460	.0439	.0418	.0398	.0378
6	.0881	.0851	.0822	.0793	.0764	.0736	.0709	.0682	.0656	.0631
7	.1145	.1118	.1091	.1064	.1037	.1010	.0982	.0955	.0928	.0901
8	.1302	.1286	.1269	.1251	.1232	.1212	.1191	.1170	.1148	.1126
9	.1317	.1315	.1311	.1306	.1300	.1293	.1284	.1274	.1263	.1251

(Table Continues)

TABLE 7 (Continued)

	μ									
x	9.1	9.2	9.3	9.4	9.5	9.6	9.7	9.8	9.9	10.0
10	.1198	.1210	.1219	.1228	.1235	.1241	.1245	.1249	.1250	.1251
11	.0991	.1012	.1031	.1049	.1067	.1083	.1098	.1112	.1125	.1137
12	.0752	.0776	.0799	.0822	.0844	.0866	.0888	.0908	.0928	.0948
13	.0526	.0549	.0572	.0594	.0617	.0640	.0662	.0685	.0707	.0729
14	.0342	.0361	.0380	.0399	.0419	.0439	.0459	.0479	.0500	.0521
15	.0208	.0221	.0235	.0250	.0265	.0281	.0297	.0313	.0330	.0347
16	.0118	.0127	.0137	.0147	.0157	.0168	.0180	.0192	.0204	.0217
17	.0063	.0069	.0075	.0081	.0088	.0095	.0103	.0111	.0119	.0128
18	.0032	.0035	.0039	.0042	.0046	.0051	.0055	.0060	.0065	.0071
19	.0015	.0017	.0019	.0021	.0023	.0026	.0028	.0031	.0034	.0037
20	.0007	.0008	.0009	.0010	.0011	.0012	.0014	.0015	.0017	.0019
21	.0003	.0003	.0004	.0004	.0005	.0006	.0006	.0007	.0008	.0009
22	.0001	.0001	.0002	.0002	.0002	.0002	.0003	.0003	.0004	.0004
23	.0000	.0001	.0001	.0001	.0001	.0001	.0001	.0001	.0002	.0002
24	.0000	.0000	.0000	.0000	.0000	.0000	.0000	.0001	.0001	.0001

	μ									
x	11	12	13	14	15	16	17	18	19	20
0	.0000	.0000	.0000	.0000	.0000	.0000	.0000	.0000	.0000	.0000
1	.0002	.0001	.0000	.0000	.0000	.0000	.0000	.0000	.0000	.0000
2	.0010	.0004	.0002	.0001	.0000	.0000	.0000	.0000	.0000	.0000
3	.0037	.0018	.0008	.0004	.0002	.0001	.0000	.0000	.0000	.0000
4	.0102	.0053	.0027	.0013	.0006	.0003	.0001	.0001	.0000	.0000
5	.0224	.0127	.0070	.0037	.0019	.0010	.0005	.0002	.0001	.0001
6	.0411	.0255	.0152	.0087	.0048	.0026	.0014	.0007	.0004	.0002
7	.0646	.0437	.0281	.0174	.0104	.0060	.0034	.0018	.0010	.0005
8	.0888	.0655	.0457	.0304	.0194	.0120	.0072	.0042	.0024	.0013
9	.1085	.0874	.0661	.0473	.0324	.0213	.0135	.0083	.0050	.0029
10	.1194	.1048	.0859	.0663	.0486	.0341	.0230	.0150	.0095	.0058
11	.1194	.1144	.1015	.0844	.0663	.0496	.0355	.0245	.0164	.0106
12	.1094	.1144	.1099	.0984	.0829	.0661	.0504	.0368	.0259	.0176
13	.0926	.1056	.1099	.1060	.0956	.0814	.0658	.0509	.0378	.0271
14	.0728	.0905	.1021	.1060	.1024	.0930	.0800	.0655	.0514	.0387
15	.0534	.0724	.0885	.0989	.1024	.0992	.0906	.0786	.0650	.0516
16	.0367	.0543	.0719	.0866	.0960	.0992	.0963	.0884	.0772	.0646
17	.0237	.0383	.0550	.0713	.0847	.0934	.0963	.0936	.0863	.0760
18	.0145	.0256	.0397	.0554	.0706	.0830	.0909	.0936	.0911	.0844
19	.0084	.0161	.0272	.0409	.0557	.0699	.0814	.0887	.0911	.0888

(Table Continues)

TABLE 7 (Continued)

	μ									
x	11	12	13	14	15	16	17	18	19	20
20	.0046	.0097	.0177	.0286	.0418	.0559	.0692	.0798	.0866	.0888
21	.0024	.0055	.0109	.0191	.0299	.0426	.0560	.0684	.0783	.0846
22	.0012	.0030	.0065	.0121	.0204	.0310	.0433	.0560	.0676	.0769
23	.0006	.0016	.0037	.0074	.0133	.0216	.0320	.0438	.0559	.0669
24	.0003	.0008	.0020	.0043	.0083	.0144	.0226	.0328	.0442	.0557
25	.0001	.0004	.0010	.0024	.0050	.0092	.0154	.0237	.0336	.0446
26	.0000	.0002	.0005	.0013	.0029	.0057	.0101	.0164	.0246	.0343
27	.0000	.0001	.0002	.0007	.0016	.0034	.0063	.0109	.0173	.0254
28	.0000	.0000	.0001	.0003	.0009	.0019	.0038	.0070	.0117	.0181
29	.0000	.0000	.0001	.0002	.0004	.0011	.0023	.0044	.0077	.0125
30	.0000	.0000	.0000	.0001	.0002	.0006	.0013	.0026	.0049	.0083
31	.0000	.0000	.0000	.0000	.0001	.0003	.0007	.0015	.0030	.0054
32	.0000	.0000	.0000	.0000	.0001	.0001	.0004	.0009	.0018	.0034
33	.0000	.0000	.0000	.0000	.0000	.0001	.0002	.0005	.0010	.0020
34	.0000	.0000	.0000	.0000	.0000	.0000	.0001	.0002	.0006	.0012
35	.0000	.0000	.0000	.0000	.0000	.0000	.0000	.0001	.0003	.0007
36	.0000	.0000	.0000	.0000	.0000	.0000	.0000	.0001	.0002	.0004
37	.0000	.0000	.0000	.0000	.0000	.0000	.0000	.0000	.0001	.0002
38	.0000	.0000	.0000	.0000	.0000	.0000	.0000	.0000	.0000	.0001
39	.0000	.0000	.0000	.0000	.0000	.0000	.0000	.0000	.0000	.0001

TABLE 8 Random Numbers

63271	59986	71744	51102	15141	80714	58683	93108	13554	79945
88547	09896	95436	79115	08303	01041	20030	63754	08459	28364
55957	57243	83865	09911	19761	66535	40102	26646	60147	15702
46276	87453	44790	67122	45573	84358	21625	16999	13385	22782
55363	07449	34835	15290	76616	67191	12777	21861	68689	03263
69393	92785	49902	58447	42048	30378	87618	26933	40640	16281
13186	29431	88190	04588	38733	81290	89541	70290	40113	08243
17726	28652	56836	78351	47327	18518	92222	55201	27340	10493
36520	64465	05550	30157	82242	29520	69753	72602	23756	54935
81628	36100	39254	56835	37636	02421	98063	89641	64953	99337
84649	48968	75215	75498	49539	74240	03466	49292	36401	45525
63291	11618	12613	75055	43915	26488	41116	64531	56827	30825
70502	53225	03655	05915	37140	57051	48393	91322	25653	06543
06426	24771	59935	49801	11082	66762	94477	02494	88215	27191
20711	55609	29430	70165	45406	78484	31639	52009	18873	96927
41990	70538	77191	25860	55204	73417	83920	69468	74972	38712
72452	36618	76298	26678	89334	33938	95567	29380	75906	91807
37042	40318	57099	10528	09925	89773	41335	96244	29002	46453
53766	52875	15987	46962	67342	77592	57651	95508	80033	69828
90585	58955	53122	16025	84299	53310	67380	84249	25348	04332
32001	96293	37203	64516	51530	37069	40261	61374	05815	06714
62606	64324	46354	72157	67248	20135	49804	09226	64419	29457
10078	28073	85389	50324	14500	15562	64165	06125	71353	77669
91561	46145	24177	15294	10061	98124	75732	00815	83452	97355
13091	98112	53959	79607	52244	63303	10413	63839	74762	50289
73864	83014	72457	22682	03033	61714	88173	90835	00634	85169
66668	25467	48894	51043	02365	91726	09365	63167	95264	45643
84745	41042	29493	01836	09044	51926	43630	63470	76508	14194
48068	26805	94595	47907	13357	38412	33318	26098	82782	42851
54310	96175	97594	88616	42035	38093	36745	56702	40644	83514
14877	33095	10924	58013	61439	21882	42059	24177	58739	60170
78295	23179	02771	43464	59061	71411	05697	67194	30495	21157
67524	02865	39593	54278	04237	92441	26602	63835	38032	94770
58268	57219	68124	73455	83236	08710	04284	55005	84171	42596
97158	28672	50685	01181	24262	19427	52106	34308	73685	74246
04230	16831	69085	30802	65559	09205	71829	06489	85650	38707
94879	56606	30401	02602	57658	70091	54986	41394	60437	03195
71446	15232	66715	26385	91518	70566	02888	79941	39684	54315
32886	05644	79316	09819	00813	88407	17461	73925	53037	91904
62048	33711	25290	21526	02223	75947	66466	06232	10913	75336
84534	42351	21628	53669	81352	95152	08107	98814	72743	12849
84707	15885	84710	35866	06446	86311	32648	88141	73902	69981
19409	40868	64220	80861	13860	68493	52908	26374	63297	45052
57978	48015	25973	66777	45924	56144	24742	96702	88200	66162
57295	98298	11199	96510	75228	41600	47192	43267	35973	23152
94044	83785	93388	07833	38216	31413	70555	03023	54147	06647
30014	25879	71763	96679	90603	99396	74557	74224	18211	91637
07265	69563	64268	88802	72264	66540	01782	08396	19251	83613
84404	88642	30263	80310	11522	57810	27627	78376	36240	48952
21778	02085	27762	46097	43324	34354	09369	14966	10158	76089

TABLE 9 Factors for $\bar{x}$ and R Control Charts

Observations in Sample, n	d_2	A_2	d_3	D_3	D_4
2	1.128	1.880	0.853	0	3.267
3	1.693	1.023	0.888	0	2.574
4	2.059	0.729	0.880	0	2.282
5	2.326	0.577	0.864	0	2.114
6	2.534	0.483	0.848	0	2.004
7	2.704	0.419	0.833	0.076	1.924
8	2.847	0.373	0.820	0.136	1.864
9	2.970	0.337	0.808	0.184	1.816
10	3.078	0.308	0.797	0.223	1.777
11	3.173	0.285	0.787	0.256	1.744
12	3.258	0.266	0.778	0.283	1.717
13	3.336	0.249	0.770	0.307	1.693
14	3.407	0.235	0.763	0.328	1.672
15	3.472	0.223	0.756	0.347	1.653
16	3.532	0.212	0.750	0.363	1.637
17	3.588	0.203	0.744	0.378	1.622
18	3.640	0.194	0.739	0.391	1.608
19	3.689	0.187	0.734	0.403	1.597
20	3.735	0.180	0.729	0.415	1.585
21	3.778	0.173	0.724	0.425	1.575
22	3.819	0.167	0.720	0.434	1.566
23	3.858	0.162	0.716	0.443	1.557
24	3.895	0.157	0.712	0.451	1.548
25	3.931	0.153	0.708	0.459	1.541

Appendix C Summation Notation

Summations

Definition

$$\sum_{i=1}^{n} x_i = x_1 + x_2 + \cdots + x_n \tag{C.1}$$

Example for $x_1 = 5, x_2 = 8, x_3 = 14$:

$$\begin{aligned}\sum_{i=1}^{3} x_i &= x_1 + x_2 + x_3 \\ &= 5 + 8 + 14 \\ &= 27\end{aligned}$$

Result 1

For a constant c:

$$\sum_{i=1}^{n} c = \underbrace{(c + c + \ldots + c)}_{n \text{ times}} = nc \tag{C.2}$$

Example for $c = 5, n = 10$:

$$\sum_{i=1}^{10} 5 = 10(5) = 50$$

Example for $c = \bar{x}$:

$$\sum_{i=1}^{n} \bar{x} = n\bar{x}$$

Result 2

$$\sum_{i=1}^{n} cx_i = cx_1 + cx_2 + \cdots + cx_n$$

$$= c(x_1 + x_2 + \cdots + x_n) = c \sum_{i=1}^{n} x_i \tag{C.3}$$

Example for $x_1 = 5, x_2 = 8, x_3 = 14, c = 2$:

$$\sum_{i=1}^{3} 2x_i = 2 \sum_{i=1}^{3} x_i = 2(27) = 54$$

Result 3

$$\sum_{i=1}^{n}(ax_i + by_i) = a\sum_{i=1}^{n}x_i + b\sum_{i=1}^{n}y_i \tag{C.4}$$

Example for $x_1 = 5, x_2 = 8, x_3 = 14, a = 2, y_1 = 7, y_2 = 3, y_3 = 8, b = 4$:

$$\sum_{i=1}^{3}(2x_i + 4y_i) = 2\sum_{i=1}^{3}x_i + 4\sum_{i=1}^{3}y_i$$

$$= 2(27) + 4(18)$$

$$= 54 + 72$$

$$= 126$$

Double Summations

Consider the following data involving the variable x_{ij}, where i is the subscript denoting the row position and j is the subscript denoting the column position:

		Column		
		1	2	3
Row	1	$x_{11} = 10$	$x_{12} = 8$	$x_{13} = 6$
	2	$x_{21} = 7$	$x_{22} = 4$	$x_{23} = 12$

Definition

$$\sum_{i=1}^{n}\sum_{j=1}^{m}x_{ij} = (x_{11} + x_{12} + \cdots + x_{1m}) + (x_{21} + x_{22} + \cdots + x_{2m}) + (x_{31} + x_{32} + \cdots + x_{3m}) + \cdots + (x_{n1} + x_{n2} + \cdots + x_{nm}) \tag{C.5}$$

Example:

$$\sum_{i=1}^{2}\sum_{j=1}^{3}x_{ij} = x_{11} + x_{12} + x_{13} + x_{21} + x_{22} + x_{23}$$

$$= 10 + 8 + 6 + 7 + 4 + 12$$

$$= 47$$

Definition

$$\sum_{i=1}^{n}x_{ij} = x_{1j} + x_{2j} + \cdots + x_{nj} \tag{C.6}$$

Example:

$$\sum_{i=1}^{2} x_{i2} = x_{12} + x_{22}$$
$$= 8 + 4$$
$$= 12$$

Shorthand Notation

Sometimes when a summation is for all values of the subscript, we use the following shorthand notations:

$$\sum_{i=1}^{n} x_i = \sum x_i \tag{C.7}$$

$$\sum_{i=1}^{n} \sum_{j=1}^{m} x_{ij} = \sum \sum x_{ij} \tag{C.8}$$

$$\sum_{i=1}^{n} x_{ij} = \sum_{i} x_{ij} \tag{C.9}$$

Appendix D The Data Disk

The Data Disk contains most of the larger data sets presented in the text. The Data Disk is available in formats that will enable you to retrieve any data set listed using Minitab, Excel, or The Data Analyst software package. A list of the data set filenames that appear in each chapter is shown below. In addition, each data set on The Data Disk is also identified in the text with a logo that appears in the margin.

Chapter 2

RETURN	Exercise 16
RETAIL	Exercise 18
COMPUTER	Exercise 21
APTEST	Table 2.13
JOBSAT	Exercise 27
PEFORCST	Exercise 28
SCATTER	Exercise 30
FINANCE	Table 2.18
LAWAGES	Exercise 38
COMSTOCK	Exercise 40
SHADOW	Exercise 41
CITIES	Exercise 42
BWDATA	Exercise 44
CONSOLID	Computer Case

Chapter 3

ENTRYSAL	Exercise 3
LAWAGES	Exercise 26
CHAINSAL	Exercise 27
GROWTH	Exercise 32
INJURY	Exercise 35
UTILITY	Exercise 42
CITIES	Exercise 43
PRICES	Exercise 50
DUKE	Exercise 60
CONSOLID	Computer Case
HEALTH1	Computer Case
HEALTH2	Computer Case

Chapter 8

LIFEINS	Table 8.2
MIAMI	Exercise 10
BOCK	Computer Case
AUTO	Computer Case

Chapter 9

DISTANCE	Table 9.2
QUALITY	Computer Case

Chapter 10

AIRPORT	Exercise 5
UNION	Exercise 8
EXAMDATA	Table 10.1
GOLF	Computer Case

Chapter 11

BAGS	Exercise 19
DOWJONES	Exercise 21
TRAINING	Computer Case
MACHINES	Exercise 31
MKTVALUE	Exercise 32
ASSEMBLY	Exercise 41
MKTCAP	Exercise 42
MEDICAL1	Computer Case
MEDICAL2	Computer Case

Chapter 12

PRESCRIP	Exercise 40
HOME1	Exercise 41
BUTLER	Table 12.10
SHOWTIME	Exercise 49
AUTO1	Exercise 50
MOWER	Exercise 51
HOUSING	Exercise 52
SAFETY	Computer Case
CONSUMER	Computer Case

Appendix E Answers to Even-Numbered Exercises

Chapter 1

2. a. 10
b. 4
c. Industry and Comp. vs. Shareholder Return qualitative
CEO Compensation and Sales quantitative

4. a. 10
b. *Fortune 500* Corporations, April 1994.
c. $3,756 million
d. $3,756 million

6. a, c, and d are quantitative
b and e are qualitative

8. a. Visitors to Hawaii
b. Yes; vast majority travel to Hawaii by air
c. 1 and 4 are quantitative; 2 and 3 are qualitative

10. a. 4
b. All variables are quantitative
c. Times series for 1990 to 1993

12. a. 4,436.78
b. Market was up on November 13 as compared to May 16

14. a. 56% food manufacturers; 12% HBA manufacturers; 3.6 average satisfaction.
b. 3.6
c. 12%

16. a. Percent of television sets that were tuned to a particular television show and/or total viewing audience.
b. All television sets in the United States which are available for the viewing audience. Note this would not include television sets in store displays.
c. It would be physically impossible to contact everyone in the viewing audience; even it it were possible to do so, it would be too expensive.
d. The cancellation of programs, the scheduling of programs, and advertising cost rates.

Chapter 2

2. a. .20
b. 40
c/d.

Class	Frequency	Percent Frequency
A	44	22
B	36	18
C	80	40
D	40	20
Total	200	100

4. a. Names of trucks—qualitative
b.

Truck	Frequency	Percent Frequency
C/K Pickup	13	26
Caravan	7	14
Explorer	7	14
F-Series	14	28
Ranger	9	18
Total	50	100

d. Ford F-Series pickup; Chevy C/K pickup is a close second.

6. a.

Book	Frequency	Percent Frequency
C	9	20.0
D	6	13.3
I	7	15.6
L	5	11.1
P	8	17.8
W	10	22.2
Total	45	100.0

b. W, C, P, I, D and L
c. D-13.3%, L-11.1%; Combined 24.4%

8. a.

Position	Frequency	Relative Frequency
P	17	.309
H	4	.073
1	5	.091
2	4	.073
3	2	.036
S	5	.091
L	6	.109
C	5	.091
R	7	.127
Totals	55	1.000

b. Pitcher
c. 3rd base
d. Rightfield
e. Infielders 16 to outfielders 18

10. a. Quality classifications

b.

Rating	Frequency	Relative Frequency
Poor	2	.03
Fair	4	.07
Good	12	.20
Very good	24	.40
Excellent	18	.30
Totals	60	1.00

12.

Class	Cumulative Frequency	Cumulative Relative Frequency
≤ 19	10	.20
≤ 29	24	.48
≤ 39	41	.82
≤ 49	48	.96
≤ 59	50	1.00

14. b,c.

Class	Frequency	Percent Frequency
6.0–7.9	4	20
8.0–9.9	2	10
10.0–11.9	8	40
12.0–13.9	3	15
14.0–15.9	3	15
Totals	20	100

16. a.

12 Month Return	Frequency	Relative Frequency	Percent Frequency
1–5	2	.071	7.1
6–10	5	.179	17.9
11–15	13	.464	46.4
16–20	5	.179	17.9
21–25	2	.071	7.1
26–30	1	.036	3.6
Total	28	1.000	100.0

b.

5-Year Return	Frequency	Relative Frequency	Percent Frequency
1–5	2	.071	7.1
6–10	8	.286	28.6
11–15	10	.357	35.7
16–20	5	.179	17.9
21–25	2	.071	7.1
26–30	1	.036	3.6
Totals	28	1.000	100.0

c. Very similar, with 12-Month having slightly more in the 11–15% class.

18. a. $31,000, $57,000

b.

Salary	Frequency	Relative Frequency	Percent Frequency
31–35	2	.050	5.0
36–40	4	.100	10.0
41–45	13	.325	32.5
46–50	11	.275	27.5
51–55	8	.200	20.0
56–60	2	.050	5.0
Totals	40	1.000	100.0

c. .05

d. 25%

20. a. 38%

b. 33%

c. 29%

d. 163

e. 310

22.

5	7 8
6	4 5 8
7	0 2 2 5 5 6 8
8	0 2 3 5

24.

11	6
12	0 2
13	0 6 7
14	2 2 7
15	5
16	0 2 8
17	0 2 3

26.

Stem	Leaf
−3	0
−2	
−1	1 1 7
−0	3 3 9
0	5 7
1	4
2	0 1 3 5 6
3	9
4	6 8
5	4 5
14	0

28. a.

Stem	Leaf
4	7
5	2
6	
7	
8	0 0 1
9	1 4 5 8 8
10	4
11	3 3 5 7
12	0 8
13	5 6 8 8
14	0 6 7 8 9
15	4
16	8
17	
18	
19	
20	
21	6
22	7

b.

Forecast	Frequency	Percent Frequency
4–6	2	6.67
7–9	8	26.67
10–12	7	23.33
13–15	10	33.33
16–18	1	3.33
19–21	1	3.33
22–25	1	3.33

30. Negative relationship

32. a.

		Book Value/Share				Total
		0.00–4.99	*5.00–9.99*	*10.00–14.99*	*15.00–19.99*	
Earn/Share	*0.00–0.99*	7	4	1	0	12
	1.00–1.99	2	7	1	1	11
	2.00–2.99	0	1	4	2	7
Total		9	12	6	3	30

b.

		Book Value/Share				Total
		0.00–4.99	*5.00–9.99*	*10.00–14.99*	*15.00–19.99*	
Earn/Share	*0.00–0.99*	58.3	33.3	8.4	0.0	100.0
	1.00–1.99	18.2	63.6	9.1	9.1	100.0
	2.00–2.99	0.0	14.3	57.1	28.6	100.0

34. b. Positive relationship

36. a, b.

Sport	Frequency	Relative Frequency
Baseball	7	.175
Basketball	6	.150
Football	14	.350
Ice Hockey	1	.025
Tennis	1	.025
Other	11	.275
Total	40	1.000

38.

Hourly Wage	Freq.	Rel. Freq.	Cum. Freq.	Cum. Rel. Freq.
4.00–5.99	1	.04	1	.04
6.00–7.99	3	.12	4	.16
8.00–9.99	8	.32	12	.48
10.00–11.99	6	.24	18	.72
12.00–13.99	5	.20	23	.92
14.00–15.99	2	.08	25	1.00
Totals	25	1.00		

40.

Closing Price	Freq.	Rel. Freq.	Cum. Freq.	Cum. Rel. Freq.
0–9⅞	9	.225	9	.225
10–19⅞	10	.250	19	.475
20–29⅞	5	.125	24	.600
30–39⅞	11	.275	35	.875
40–49⅞	2	.050	37	.925
50–59⅞	2	.050	39	.975
60–69⅞	0	.000	39	.975
70–79⅞	1	.025	40	1.000
Totals	40	1.000		

42. **a.**

Stem	Leaves
1	7
2	5 6 7 7
3	1 7 7 8
4	2 3 9
5	1 1 2 2 4 4 8
6	6 8
7	2 5

b.

Stem	Leaves
0	9
1	0 0 2 3 6 9
2	4 6 8 8
3	3 4 4 5 5 9
4	1 5 8 8 9
5	7 9

d. 11

e.

	Frequency	
Temperature	High Temp.	Low Temp.
0–9	0	1
10–19	1	6
20–29	4	4
30–39	4	6
40–49	3	5
50–59	8	2
60–69	2	0
70–79	2	0
	24	24

44. **a.**

		P/E Ratio					**Total**
		5–9	*10–14*	*15–19*	*20–24*	*25–29*	
Industry	*Consumer*	0	3	5	1	1	10
	Banking	4	4	2	0	0	10
	Total	4	7	7	1	1	20

b.

		P/E Ratio					Total
		5–9	*10–14*	*15–19*	*20–24*	*25–29*	
Industry	*Consumer*	0	30	50	10	10	100
	Banking	40	40	20	0	0	100

c. Consumer industry tends to have higher P/E ratios.

46. a. & b.

Year	Freq.	Fuel	Freq.
1973 or before	247	Elect.	149
1974–79	54	Nat Gas	317
1980–86	82	Oil	17
1981–91	121	Propane	7
Total	504	Other	14
		Total	504

c.

		Fuel Type				
		Elec	*Nat. Gas*	*Oil*	*Propane*	*Other*
Year Constructed	*1973 or before*	26.9	57.7	70.5	71.4	50.0
	1974–1979	16.1	8.2	11.8	28.6	0.0
	1980–1986	24.8	12.0	5.9	0.0	42.9
	1987–1991	32.2	22.1	11.8	0.0	7.1
Total		100.0	100.0	100.0	100.0	100.0

d.

		Fuel Type					Total
		Elec	*Nat. Gas*	*Oil*	*Propane*	*Other*	
Year Constructed	*1973 or before*	16.2	74.1	4.9	2.0	2.8	100.0
	1974–1979	44.5	48.1	3.7	3.7	0.0	100.0
	1980–1986	45.1	46.4	1.2	0.0	7.3	100.0
	1987–1991	39.7	57.8	1.7	0.0	0.8	100.0

e. *Observations from the column percentages crosstabulation*
For those buildings using electricity, the percentages have not changed greatly over the years. For the buildings using natural gas, the majority were constructed in 1973 or before. Most of the buildings using oil were constructed in 1973 or before. All of the buildings using propane are older.

Observations from the column percentages crosstabulation
Most of the buildings in the CG&E service area use electricity or natural gas. In the period 1973 or before most used natural gas. From 1974-1986, it is fairly evenly divided between electricity and natural gas. Since 1987 almost all new buildings are using electricity or natural gas with natural gas being the clear leader.

Chapter 3

2. 20, 22.5, 28, 29

4. a. \$103.90, \$97.50, \$75.00
b. \$59.50, \$149.50

6. a. 178
b. 178
c. Do not report a mode
d. 184

8. a. 48.33, 49; do not report a mode
b. 45, 55
c. 45, 55

10. a. 4.44, 4.25, 4.20
b. 4.10, 5.10
c. 4.15, 4.80

12. 6, 4

14. a. Mainland
115,130, 111,560
Asia
36,620, 36,695

b. Mainland
86,240 26,820 23.4%
42,970 11,400 31.1%

c. More visitors and more variation from Mainland

16. a. Range = 32, IQR = 10
b. 92.75, 9.63

18. *Dawson:* range = 2, $s = .67$
Clark: range = 8, $s = 2.58$

20. a. 10.40
b. 10.05
c. 9.50
d. 3.60
e. 6.75
f. 2.60

22. .20, 1.50, 0, −.50, −2.20

24. a. 95%
b. Almost all
c. 68%

26. a. 73.2, 13.71
b. $z = 2.54$; No
c. 16%, 2.5%

28. a. 75%, 84%, 89%
b. 95%; almost all

30. 15, 22.5, 26, 29, 34

32. a. 1.9, 10, 14.5, 20.4, 55.9
b. −5.6, 36
c. 46.1, 49.9 and 55.9 are outliers

34. a. 79.31, 78.5
b. 76.5, 80.5
c. 72, 76.5 78.5, 80.5, 90
d. Camcorders have higher variation
e. Yes; Hitachi and Mitsubishi are outliers

36. a. 9.32, 9.8
b. 6.5, 14.6
c. −5.65, 26.75
Seven mild outliers

38. b. There appears to be a linear relationship between x and y
c. $s_{xy} = 26.5$
d. $r_{xy} = .69$

40. −.91; negative relationship

42. b. 4.14
c. +.697

44. a. 3.69
b. 3.175

46. a. 2.50
b. yes

48. 10.74, 25.63, 5.06

50. a. 3.19, 3.30, 3.70
b. 2.90, 3.50
c. 2.80, .60
d. .30, .54
e. 2.00, 4.40
Two outliers

52. a. $\bar{x} = 1028.18$, median = 1000, no mode
b. Range = 510, IQR = 220
c. $s^2 = 24{,}256.36$, $s = 155.74$
d. No outliers

54. a. 32 for both
b. Public: 4.64 Auto: 1.83
c. Auto; less dispersion
d. Box plots based on the following points:

Public	25	29	32	34	41
Auto	29	31	32	33	35

56. b. $r_{xy} = .9856$; strong positive relationship

58. a. 817
b. 833

60. a. *Duke:* 87.97, 88.50
Opponent: 72.64, 71.50
b. *Duke:* 50, 21.5
Opponent: 52, 15.5

Chapter 4

2. 20 ways

4. b. (H,H,H) (H,H,T) (H,T,H) (H,T,T) (T,H,H) (T,H,T) (T,T,H) (T,T,T)
c. 1/8

6. .40, .26, .34; relative frequency method

8. a. 4; Commission Positive-Council Approves, Commission Positive-Council Disapproves
Commission Negative-Council Approves, Commission Negative–Council Disapproves

10. $P\text{ (never married)} = \frac{1106}{2038}$
$P\text{ (married)} = \frac{826}{2038}$
$P\text{ (other)} = \frac{106}{2038}$

12. No, there are 4 equally likely outcomes.
(H,T) (T,H) (T,T) (H,H)

14. a. 1/4
b. 1/2
c. 3/4

16. a. 36
c. 1/6
d. 5/18
e. No; $P(\text{odd}) = P(\text{even}) = \frac{1}{2}$
f. Classical

18. a. $P(0) = .05$
b. $P(4 \text{ or } 5) = .20$
c. $P(0, 1, \text{ or } 2) = .55$

20. a. .152
b. .340
c. .508

22. a. .40, .40, .60
b. .80, yes
c. $A^c = \{E_3, E_4, E_5\}$; $C^c = \{E_1, E_4\}$;
$P(A^c) = .60$; $P(C^c) = .40$
d. $\{E_1, E_2, E_5\}$; .60
e. .80

24. .26

26. a. $P(V) = .7197$, $P(Q) = .4498$, $P(V \cap Q) = .2699$
b. .8996
c. .1004

28. a. .698
b. .302

30. a. .67
b. .80
c. No

32. a.

	Single	Married	Total
Under 30	.55	.10	.65
30 or Over	.20	.15	.35
Total	.75	.25	1.0

b. Higher probability of under 30
c. Higher probability of single
d. .55
e. .8462
f. No

34. a.

		U.S. Car		
		Yes	*No*	Total
Foreign Car	*Yes*	.386	.226	.612
	No	.369	.019	.388
	Total	.755	.245	1.000

b. $P(U) = .755$, $P(F) = .612$
Higher probability for U.S.
c. .386
d. .981
e. .511
f. .631
g. No

36. a. .5541
c. .67

38. a. .197
b. .121
c. No

40. a. .10, .20, .09
b. .51
c. .26, .51, .23

42. a. 21
b. Yes

44. .6754

46. a. 4
b. .61, .19, .10, .10

48. **a.** .4642
b. .3458
c. .1498

50. **a.** .76
b. .24

52. **b.** .2022
c. .4618
d. .4005

54. **b.** $1 million – $2 million
c. $P(2M) = .2286$, $P(2M \mid T) = .1556$, $P(2M \mid F) = .2833$
d. No

56. **a.** .25, .40, .10
b. .25
c. B and S are independent; program appears to have no effect

58. 3.44%

60. **a.** .0625
b. .0132
c. Three

Chapter 5

2. **a.** x = time in minutes to assemble product
b. Any positive value: $x > 0$
c. Continuous

4. $x = 0, 1, 2, \ldots, 12$

6. **a.** 0, 1, 2, . . . , 20; discrete
b. 0, 1, 2, . . . ; discrete
c. 0, 1, 2, . . . , 50; discrete
d. $0 \leq x \leq 8$; continuous
e. $x > 0$; continuous

8.

x	1	2	3	4
$f(x)$	.15	.25	.40	.20

c. $f(x) \geq 0, \Sigma f(x)=1$

10. **a.** It is a proper probability distribution
b. .60

12. **a.** Yes
b. .65

14. **a.** .05
b. .70
c. .40

16. **a.** 5.20
b. 4.56, 2.14

18. **a.** $E(x) = 2.2$, same
b. $\text{Var}(x) = 1.16$, $\sigma = 1.08$

20. **a.** 166
b. –94; concern is to protect against the expense of a big accident

22. **a.** Medium: 145; large: 140
b. Medium: 2725; large: 12,400

24. **a.** $f(0) = .3487$
b. $f(2) = .1937$
c. .9298
d. .6513
e. 1
f. $\sigma^2 = .9000$, $\sigma = .9487$

26. **a.** .2301
b. .3410
c. .8784

28. **a.** Probability of a defective part must be .03 for each trial; trials must be independent
c. 2
d.

Number of defects	0	1	2
Probability	.9409	.0582	.0009

30. **a.** .2547
b. .9566
c. .8009

34. **a.** $f(x) = \dfrac{3^3 e^{-3}}{x!}$
b. .2241
c. .1494
d. .8008

36. **a.** .1952
b. .1048
c. .0183
d. .0907

38. **a.** .2001
b. 7.8
c. .9996

40. **a.** $\mu = 1$
b. .3679
c. .3679
d. .2642

42. **a.** 2.2
b. 1.16

44. **b.** $10.65 Million
c. 2.1275
d. Good. Expected profit = $1.35 million

46. **a.** .2793
b. .7762
c. .1496

48. **a.** .9510
b. .0480
c. .0490

50. .1912

52. **a.** .2240
b. .5767

Chapter 6

2. **b.** .50
c. .60
d. 15
e. 8.33

4. **b.** .50
c. .30
d. .40

6. **a.** .40
b. .64
c. .68

10. **a.** .3413
b. .4332
c. .4772
d. .4938

12. **a.** .2967
b. .4418
c. .3300
d. .5910
e. .8849
f. .2388

14. **a.** $z = 1.96$
b. $z = .61$
c. $z = 1.12$
d. $z = .44$

16. **a.** $z = 2.33$
b. $z = 1.96$
c. $z = 1.645$
d. $z = 1.28$

18. **a.** .1814
b. .9656
c. $12,816 or more

20. **a.** 50.77%
b. 15.87%
c. 23.88%

22. **a.** .7295
b. $19.49 per hour
c. .0110

24. **a.** .2266
b. .7745
c. 31.12 years

26. **a.** $\mu = 20$, $\sigma = 4$
b. yes
c. .0602
d. .4714
e. .1292

28. **a.** .7910
b. .9616
c. .1066

30. **a.** .1151
b. .2852
c. 49 or less

32. **a.** .8336
b. .0049

34. **a.** $1 - e^{-x_0/3}$
b. .4866
c. .3679
d. .8111
e. .3245

36. **b.** .6321
c. .3935
d. .0821

38. **a.** .3935
b. .5276
c. .1353

40. **b.** .20
c. 37 minutes
d. $E(x) = 35$ and $\sigma = 2.89$

42. **a.** .3174, 317.4 defects
b. .0028, 2.8 defects

44. **a.** .0918
b. 12,468 tubes

46. **a.** 47.06%
b. .0475
c. 42,480

48. **a.** .6068
b. .0146
c. .0735
d. $59,815

50. **a.** 5.16%
b. 57.87%
c. 99.55%
d. Approximately 0

52. **a.** 300
b. $\sigma^2 = 120$, $\sigma = 10.95$
c. .8008
d. .0125

54. **a.** 4 hours
b. $1/4\ e^{-x/4}$
c. .7788
d. .1353

56. **a.** 2 minutes
b. .2212
c. .3935
d. .0821

Chapter 7

2. 22, 147, 229, 289

4. **a.** Supra, Cadillac, Lincoln, Legend and Infiniti
b. 252

6. 2782, 493, 825, 1807, 289

8. 55, 126, 36, 159, 241, 99, 152, 45, 59, 258, 266, 105

10. a. Finite
b. Infinite
c. Infinite
d. Infinite
e. Finite

12. a. .50
b. .3667

14. a. .34
b. .26

16. .09

18. a. 200
b. 5
c. Normal with $E(x) = 200$ and $\sigma_{\bar{x}} = 5$
d. The probability distribution of $\bar{x}$

20. 3.54, 2.50, 2.04, 1.77
$\sigma_{\bar{x}}$ decreases as n increases

22. a. Only for $n = 30$ and $n = 40$
b. $n = 30$; normal with $E(\bar{x}) = 400$ and $\sigma_{\bar{x}} = 9.13$
$n = 40$; normal with $E(\bar{x}) = 400$ and $\sigma_{\bar{x}} = 7.91$

24. a. Normal with $E(\bar{x}) = 51{,}800$ and $\sigma_{\bar{x}} = 516.40$
b. $\sigma_{\bar{x}}$ decreases to 365.15
c. $\sigma_{\bar{x}}$ decreases as n increases

26. a. Normal with $E(\bar{x}) = 215.60$ and $\sigma_{\bar{x}} = 13.44$
b. .8639
c. .5408

28. a. 1
b. .8926

30. a. Normal with $E(\bar{x}) = 3.6$ and $\sigma_{\bar{x}} = .15$
b. .0038
c. .9924

32. a. Normal with $E(\bar{x}) = 320$ and $\sigma_{\bar{x}} = 13.69$
b. 13.69
c. .8558
d. .3557

34. a. .6156
b. 8530

36. a. .6156
b. .7814
c. .9488
d. .9942
e. High probability with larger n

38. a. Normal with $E(\bar{p}) = .71$ and $\sigma_{\bar{p}} = .0214$
b. .8384
c. .9452

40. a. .7062
b. .1469
c. .0025

42. a. Normal with $E(\bar{p}) = .15$ and $\sigma_{\bar{p}} = .0505$
b. .4448
c. .8389

44. 4324, 2875, 318, 538, 4771

46. a. Normal with $E(\bar{x}) = 30$ and $\sigma_{\bar{x}} = 1.70$
b. .7620

48. a. 67
b. 1.5
c. Normal with $E(\bar{x}) = 67$ and $\sigma_{\bar{x}} = 1.5$
d. .9082
e. .4972

50. 246

52. a. Assume population has a normal distribution
b. .9266
c. Increase n to at least 30

54. a. Normal with $E(\bar{p}) = .74$ and $\sigma_{\bar{p}} = .031$
b. .8030
c. .4778

56. .9525

Chapter 8

2. a. 30.60 to 33.40
b. 30.34 to 33.66
c. 29.82 to 34.18

4. c. 297.60 to 322.40

6. a. 12,003 to 12,333
b. 11,971 to 12,365
c. 11,909 to 12,427
d. Width increases to be more confident

8. 279 to 321

10. 5.74 to 6.94

12. a. 1.734
b. −1.321
c. 3.365
d. −1.761 and 1.761
e. −2.048 and 2.048

14. a. 15.97 to 18.53
b. 15.71 to 18.79
c. 15.14 to 19.36

16. a. 13.2
b. 7.8
c. 7.62 to 18.78
d. Wide interval; larger sample desirable

18. a. 21.15 to 23.65
b. 21.12 to 23.68
c. Intervals are essentially the same

20. 4.51 to 6.59

22. a. 9
b. 35
c. 78

24. a. 50, 89, 200
b. Only if $E = 1$ is essential

26. 385

28. 59

30. a. .6733 to .7267
b. .6682 to .7318

32. 1067

34. a. .0908
b. .0681 to .1135

36. a. .7728 to .8272
b. 1537

38. a. .9192 to .9598
b. .2036 to .2764

40. a. 85
b. 340
c. 2124
d. 8494
e. The sample size increases

42. 2.11 to 2.39

44. a. 2196.13
b. 785.31
c. 1539.49 to 2852.77

46. 9.20 to 14.80

48. 37

50. 166

52. .5165 to .6035
.2216 to .2984
A greater proportion use recognition

54. a. 1267
b. 1508

56. a. .31
b. .29 to .33
c. 8318; No, this sample size is unnecessarily large

Chapter 9

2. a. H_0: $\mu \leq 14$
H_a: $\mu > 14$

4. a. H_0: $\mu \geq 220$
H_a: $\mu < 220$

6. a. H_0: $\mu \leq 1$
H_a: $\mu > 1$
b. Claiming $\mu > 1$ when it is not true
c. Claiming $\mu \leq 1$ when it is not true

8. a. H_0: $\mu \geq 220$
H_a: $\mu < 220$
b. Claiming $\mu < 220$ when it is not true
c. Claiming $\mu \geq 220$ when it is not true

10. a. Reject H_0 if $z > 2.05$
b. 1.36
c. .0869
d. Do not reject H_0

12. a. .0344; reject H_0
b. .3264; do not reject H_0
c. .0668; do not reject H_0
d. Approximately 0; reject H_0

14. a. H_0: $\mu \leq 6.5$; H_a: $\mu > 6.5$
b. $z = 5.91$; reject H_0
c. Current cars being driven longer

16. $z = -2.74$; reject H_0
p-value = .0031

18. a. Reject H_0 if $z < -1.645$
b. $z = -1.98$; reject H_0
c. .0239

20. a. Reject H_0 if $z < -2.33$ or $z > 2.33$
b. 1.13
c. .2584
d. Do not reject H_0

22. a. .0718; do not reject H_0
b. .6528; do not reject H_0
c. .0404; reject H_0
d. Approximately 0; reject H_0
e. .3174; do not reject H_0

24. a. $z = -1.06$; do not reject H_0
b. .2892

26. $z = 6.37$; reject H_0

28. a. 71,167 to 74,433
b. Reject H_0 since 61,650 is not in the interval

30. a. 18
b. 1.41
c. Reject H_0 if $t < -2.571$ or $t > 2.571$
d. −3.47
e. Reject H_0

32. a. .01; reject H_0
b. .10; do not reject H_0
c. Between .025 and .05; reject H_0
d. Greater than .10; do not reject H_0
e. Approximately 0; reject H_0

34. $t = 1.20$; do not reject H_0

36. a. $t = -3.33$; reject H_0
b. p-value is less than .005

38. $\bar{x} = 2.4$, $s = .52$, $t = 2.43$
Reject H_0

40. a. Reject H_0 if $z < -1.96$ or $z > 1.96$
b. −1.25

c. .2112
d. Do not reject H_0

42. .1118, do not reject H_0

44. $z = -1.25$; do not reject H_0
p-value = .1056

46. **a.** $z = 1.38$; reject H_0
b. .0838

48. H_0: $p \geq .91$
H_a: $p < .91$
$z = -5.44$; reject H_0

50. $z = -1.96$; do not reject H_0

52. **a.** $z = 1.80$; do not reject H_0
b. .0718
c. 349 to 375

54. $z = -3.84$; reject H_0

56. **a.** Show $p < .50$
b. $z = -6.62$; reject H_0

58. $z = 6.81$; reject H_0

Chapter 10

2. **a.** 2.4
b. 5.27
c. .09 to 4.17

4. **a.** 3.9
b. 0.6 to 7.0

6. **a.** 125
b. 36.82 to 213.18

8. **a.** Populations normal with equal variances
b. 4.41
c. 0.71 to 3.65 ($t_{.025} = 2.042$)

10. **a.** $z = -1.53$; do not reject H_0
b. .1260

12. .84, Do not reject H_0

14. Reject H_0; change to supplier B

16. **a.** $t = 2.22$; reject H_0
b. .11 to 3.27 (thousands)

18. **a.** 3, −1, 3, 5, 3, 0, 1
b. 2
c. 2.082
d. 2
e. .07 to 3.93

20. **a.** Matched sample
b. .65 to 1.49

22. $\bar{d} = 3$, $t = 2.23$; reject H_0

24. **a.** $t = 5.88$; reject H_0
b. 1.4 to 3.0

26. **a.** MSB = 268
b. MSW = 92
c. Cannot reject H_0 since $F = 2.91 < F_{.05} = 4.26$
d.

Source of Variation	Sum of Squares	Degrees of Freedom	Mean Square	F
Between	536	2	268	2.91
Within	828	9	92	
Total	1364	11		

28. Reject H_0 since $F = 10.63 > F_{.05} = 4.26$

30. Significant difference;
$F = 7.00 > F_{.05} = 3.68$

32. Significant difference;
p-value = .015

34. Reject H_0; system B has the lower mean checkout time

36. Reject H_0; men and women do not have the same level of job satisfaction

38. **a.** $t = .42$; no significant difference
b. $F = .18$; no significant difference

40. $F = 11.65$; reject H_0 that the mean quality ratings are equal

42. p-value = 0.355; cannot reject H_0 that the means are equal

Chapter 11

2. **a.** $\bar{p} = .184$, $s_{\bar{p}_1 - \bar{p}_2} = .0354$, $z = 1.69$; Reject H_0
b. .0455

4. .04 to .14

6. $z = 1.42$; Do not reject H_0

8. **a.** $z = -3.42$; Reject H_0
b. .0637 to .2363

10. **a.** There is a significant difference
b. No, with $\bar{p}_3 = .08$, we have $n\bar{p}_3 = 2 < 5$.

12. $\chi^2 = 15.33$, $\chi^2_{.05} = 7.81473$; reject H_0

14. $\chi^2 = 99.50$, $\chi^2_{.01} = 13.28$; reject H_0
opinions have changed

16. $\chi^2 = 6.24$, $\chi^2_{.10} = 4.60517$; reject H_0

18. $\chi^2 = 8.89$, $\chi^2_{.05} = 9.48773$; do not reject H_0

20. $\chi^2 = 19.78$, $\chi^2_{.05} = 9.48773$; reject H_0

22. $\chi^2 = 6.31$, $\chi^2_{.05} = 9.48773$; do not reject H_0

24. $\chi^2 = 14.72$, $\chi^2_{.05} = 5.99147$; reject H_0

26. $\chi^2 = 37.17$, $\chi^2_{.01} = 9.21034$; reject H_0

28. $\chi^2 = 7.96$, $\chi^2_{.05} = 9.48773$; do not reject H_0

30. $\chi^2 = 102.56$, $\chi^2_{.01} = 13.2767$; reject H_0
Conclude executive opinions have changed

32. **a.** H_0: $p_1 - p_2 \leq 0$
H_a: $p_1 - p_2 > 0$
b. $z = 1.80$; reject H_0
p-value = .0359

34. .0174; reject H_0

36. $\chi^2 = 41.69$, $\chi^2_{.01} = 13.2767$; reject H_0
Attitudes differ

38. $\chi^2 = 7.44$, $\chi^2_{.05} = 9.48773$; do not reject H_0

40. $\chi^2 = 5.26$, $\chi^2_{.05} = 5.99147$; do not reject H_0

42. $\chi^2 = 59.41$, $\chi^2_{.025} = 23.3367$; reject H_0
They are related

44. $\chi^2 = 6.20$, $\chi^2_{.05} = 12.5916$; do not reject H_0

Chapter 12

2. **b.** There appears to be a linear relationship between x and y
d. $\hat{y} = 30.33 - 1.88x$
e. 19.05

4. **b.** There appears to be a linear relationship between x and y
d. $\hat{y} = 290.54 + 581.08x$
e. 2324.32

6. **b.** $\hat{y} = 25.21 + .6608x$
c. $114.42

8. **b.** $\hat{y} = 80 + 4x$
c. 116

10. **c.** $\hat{y} = -55.84 + 1.67x$
d. 44.36%
e. $\hat{y} = 36.01$; predicted value is almost the same as the observed value

12. **b.** $\hat{y} = 4.68 + .16x$
c. deductions are excessive; audit appears justified

14. **a.** SSE = 6.3325, SST = 114.80, SSR = 108.47
b. $r^2 = .945$
c. $r = -.9721$

16. **a.** SSE = 85,135, SST = 335,000, SSR = 249,865
b. $r^2 = .746$
c. $r = +.8637$

18. **a.** $\hat{y} = -54.85 + 98.80x$
b. $r^2 = .992$
c. 38.13 mgs

20. **a.** $\hat{y} = 22.66 + .26x$
b. $r^2 = .0997$
c. $r = +.3158$

22. **a.** 2.11
b. 1.453
c. .262
d. significant
$t = -7.18 < -t_{.05} = -3.182$
e. significant
$F = 51.41 > F_{.05} = 10.13$

24. **a.** Significant
$t = 3.43 > t_{.025} = 2.776$
b. Significant
$F = 11.74 > F_{.05} = 7.71$

26. They are related since $F = 238.42 > F_{.01} = 12.25$

28. $t = .74$; no significant relationship

30. **a.** .68
b. 22.53 to 26.85
c. 19.57 to 29.81

32. **a.** $1843.51 to $2224.05
b. $1586.32 to $2481.24

34. **a.** 100.12 ($100, 120)
b. 77.30 to 122.94
c. 5.58 ($5580)

36. **a.** $5046.67
b. $3815.10 to $6278.24

38. **a.** $\hat{y} = 6.1092 + .8951x$
b. $t = 6.01$; significant relationship
c. $28.49 per month

40. **b.** yes
c. $\hat{y} = -33.88 + .09253x$
d. $F = 11.02$; significant relationship

42. **a.** $\hat{y} = -7.02 + 1.59x$
b. 3.48, −2.47, −4.83, −1.60, and 5.22

44. **a.** $\hat{y} = 29.40 + 1.55x$
b. $F = 11.15$; significant relationship
d. Question assumption of a linear relationship

46. **a.** $\hat{y} = 343 + .051x$
b. 2.52, .43, −.86, .53, −.59, −.85, −.62, −.25, −.05
c. Question assumptions

48. **a.** $\hat{y} = 45.06 + 1.94x_1$; $\hat{y} = 132.36$
b. $\hat{y} = 85.22 + 4.32x_2$; $\hat{y} = 150.02$
c. $\hat{y} = -18.37 + 2.01x_1; + 4.74x_2$;
$\hat{y} = 143.18$

50. **a.** PRICE = −0.97 + 0.139 HORSEPWR
b. PRICE = 31.0 + 0.108 HORSEPWR−3.80 ZEROTO6O
c. Multiple
d. $34,068

52. **a.** $\hat{y} = -5.7 + 1.54$ STARTS + 1.81 INCOME
b. 96.96

54. **a.** $\hat{y} = -540.40 + 61.78x$
b. $2079.07
c. .5671

56. **a.** $\hat{y} = 22.173 - .1478x$
b. $F = 11.33$; significant relationship
c. .739; good fit
d. 12.296 to 17.270

58. **a.** Negative linear relationship
b. $\hat{y} = 8.10 - .344x$
c. $F = 419.67$; significant relationship
d. .711; reasonably good fit

60. **b.** −.43 for Kraft Mayonnaise
c. +.02 for Coca-Cola Classic

62. **b.** 3.19

64. **a.** $R^2 = .95, R_a^2 = .93$
b. Significant; $F = 57.84 > F_{.01} = 5.21$

66. **a.** 3.04, 3.61, 5.08
b. Both are significant

68. **b.** Significant; $F = 22.79 > F_{.05} = 5.79$
c. $R_a^2 = .861$; good fit
d. Both are significant

Chapter 13

2. **a.** 5.42
b. UCL = 6.09, LCL = 4.75

4.

	R Chart	$\bar{x}$ Chart
UCL	2.98	29.10
LCL	.22	27.90

6. 20.01, .082

8. **a.** .0470
b. UCL = .0989, LCL = −.0049 (use LCL = 0)
c. $\bar{p} = .08$; in control
d. UCL = 14.826, LCL = −0.726
Process is out of control if more than 14 defective
e. In control since 12 defective
f. *np* chart

10. $p = .02; f(0) = .6035$
$p = .06; f(0) = .2129$

12. $p_0 = .02$; producer's risk = .0599
$p_0 = .06$; producer's risk = .3396
Producer's risk decreases as the acceptance number *c* is increased

14. $n = 20, c = 3$

16. **a.** 95.4
b. UCL = 96.07, LCL = 94.73
c. No

18.

	R Chart	$\bar{x}$ Chart
UCL	4.23	6.57
LCL	0	4.27

Estimate of standard deviation = .86

20.

	R Chart	$\bar{x}$ Chart
UCL	.11	3.11
LCL	0	3.05

22. **a.** UCL = .0817, LCL = −.0017 (use LCL = 0)

24. **a.** .03
b. $\beta = .0802$

26. **a.** Producer's risk = .0064
b. Consumer's risk = .0136
c. *Advantage:* excellent control
Disadvantage: cost

Appendix F Solutions to Self-Test Exercises

Chapter 1

2. a. 10

b. 4

c. Industry and Compensation vs. Shareholder Return Rating are qualitative variables. The Compensation vs. Shareholder Return Rating variable has numeric data; however, these data values are codes that place the company into the best to worst class based on the ratio of CEO Compensation to shareholder return. CEO Compensation and Sales are quantitative variables

3. a. Average CEO compensation = \$27,580/10 = \$2758 or \$2,758,000

b. 2 of 10 are in the banking industry; 20%

c. 3 of 10 received a rating of 3; 30%

4. a. 10

b. *Fortune 500* largest U.S. industrial corporations

c. Average sales = \$37,560/10 = \$3,756 million

d. Using the sample results, estimate the average sales for the population of corporations at \$3,756 million

9. a. Quantitative

b. A time series with 13 observations

c. Volume of new equity for initial public offerings

d. The time-series is showing a decreasing trend in the most recent June to September period

Chapter 2

3. a. $360° \times 58/120 = 174°$

b. $360° \times 42/120 = 126°$

c.

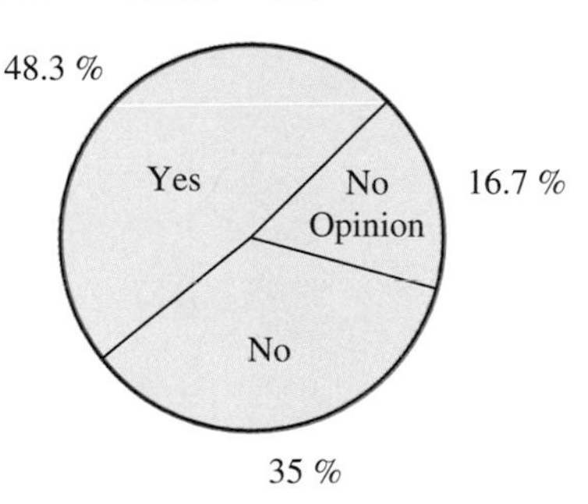

d.

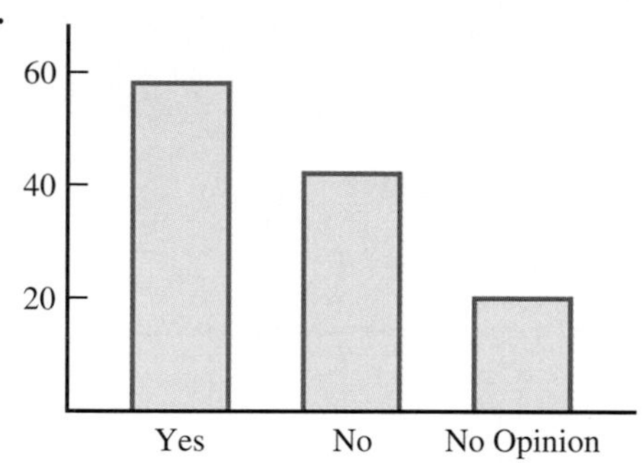

7.

Rating	Frequency	Relative Frequency
Outstanding	19	.38
Very good	13	.26
Good	10	.20
Average	6	.12
Poor	2	.04

Management should be pleased with these results: 64% of the ratings are very good to outstanding, and 84% of the ratings are good or better; comparing these ratings to previous results will show whether or not the restaurant is making improvements in its customers' ratings of food quality

12.

Class	Cumulative Frequency	Cumulative Relative Frequency
≤ 19	10	.20
≤ 29	24	.48
≤ 39	41	.82
≤ 49	48	.96
≤ 59	50	1.00

15. a, b.

Waiting Time	Frequency	Relative Frequency
0–4	4	.20
5–9	8	.40
10–14	5	.25
15–19	2	.10
20–24	1	.05
Totals	20	1.00

c, d.

Waiting Time	Cumulative Frequency	Cumulative Relative Frequency
≤ 4	4	.20
≤ 9	12	.60
≤ 14	17	.85
≤ 19	19	.95
≤ 24	20	1.00

e. $12/20 = .60$

23.

6	3
7	5 5 7
8	1 3 4 8
9	3 6
10	0 4 5
11	3

25.

9	8 9
10	2 4 6 6
11	4 5 7 8 8 9
12	2 4 5 7
13	1 2
14	4
15	1

29. a.

		y		
		1	2	**Total**
x	*A*	5	0	5
	B	11	2	13
	C	2	10	12
Total		18	12	30

b.

		y		
		1	2	**Total**
x	*A*	100.0	0.0	100.0
	B	84.6	15.4	100.0
	C	16.7	83.3	100.0

c.

		y	
		1	2
x	*A*	27.8	0.0
	B	61.1	16.7
	C	11.1	83.3
Total		100.0	100.0

d. A values are always in $y = 1$.
B values are most often in $y = 1$.
C values are most often in $y = 2$.

32. a.

		Book Value/Share				Total
		0.00–4.99	*5.00–9.99*	*10.00–14.99*	*15.00–19.99*	
Earn/Share	*0.00–0.99*	7	4	1	0	12
	1.00–1.99	2	7	1	1	11
	2.00–2.99	0	1	4	2	7
	Total	9	12	6	3	30

b.

		Book Value/Share				Total
		0.00–4.99	*5.00–9.99*	*10.00–14.99*	*15.00–19.99*	
Earn/Share	*0.00–0.99*	58.3	33.3	8.4	0.0	100.0
	1.00–1.99	18.2	63.6	9.1	9.1	100.0
	2.00–2.99	0.0	14.3	57.1	28.6	100.0

When book value is low, earnings per share tends to be low. When book value is high, earnings per share tends to be high.

Chapter 3

2. Arrange data in order: 15, 20, 25, 25, 27, 28, 30, 34.

$i = \frac{20}{100}(8) = 1.6$; round up to position 2

20th percentile = 20

$i = \frac{25}{100}(8) = 2$; use positions 2 and 3

25th percentile= $\frac{20 + 25}{2} = 22.5$

$i = \frac{65}{100}(8) = 5.2$; round up to position 6

65th percentile = 28

$i = \frac{75}{100}(8) = 6$; use positions 6 and 7

75th percentile= $\frac{28 + 30}{2} = 29$

5. a. $\bar{x} = \frac{\Sigma x_i}{n} = \frac{775}{20} = 38.75$

Mode=29 (appears three times)

b. Data in order: 22, 24, 29, 29, 29, 30, 31, 31, 32, 37, 40, 41, 44, 44, 46, 49, 50, 52, 57, 58

Median (10th and 11th positions)

$$\frac{37 + 40}{2} = 38.5$$

At home workers are slightly younger

c. $i = \frac{25}{100}(20) = 5$; use positions 5 and 6

$Q_1 = \frac{29 + 30}{2} = 29.5$

$i = \frac{75}{100}(20) = 15$; use positions 15 and 16

$Q_3 = \frac{46 + 49}{2} = 47.5$

d. $i = \frac{32}{100}(20) = 6.4$; round up to position 7

32nd percentile = 31

At least 32% of the people are 31 or younger

13. Range = 34 − 15 = 19

Arrange data in order: 15, 20, 25, 25, 27, 28, 30, 34

$i = \frac{25}{100}(8) = 2$; $Q_1 = \frac{20 + 25}{2} = 22.5$

$i = \frac{75}{100}(8) = 6;\ Q_3 = \frac{28 + 30}{2} = 29$

$\text{IQR} = Q_3 - Q_1 = 29 - 22.5 = 6.5$

$\bar{x} = \frac{\Sigma x_i}{n} = \frac{204}{8} = 25.5$

x_i	$(x_i - \bar{x})$	$(x_i - \bar{x})^2$
27	1.5	2.25
25	−.5	.25
20	−5.5	30.25
15	−10.5	110.25
30	4.5	20.25
34	8.5	72.25
28	2.5	6.25
25	−.5	.25
		242.00

$s^2 = \frac{\Sigma(x_i - \bar{x})^2}{n - 1} = \frac{242}{8 - 1} = 34.57$

$s = \sqrt{34.57} = 5.88$

19. a. Range=190 – 168=22

b. $\bar{x} = \frac{\Sigma x_i}{n} = \frac{1068}{6} = 178$

$s^2 = \frac{\Sigma(x_i - \bar{x})^2}{n - 1}$

$= \frac{4^2 + (-10)^2 + 6^2 + 12^2 + (-8)^2 + (-4)^2}{6 - 1}$

$= \frac{376}{5} = 75.2$

c. $s = \sqrt{75.2} = 8.67$

d. $\frac{s}{\bar{x}}(100) = \frac{8.67}{178}(100) = 4.87$

23. Chebyshev's theorem: *at least* $(1 - 1/k^2)$

a. $k = \frac{40 - 30}{5} = 2;\ (1 - 1/2^2) = .75$

b. $k = \frac{45 - 30}{5} = 3;\ (1 - 1/3^2) = .89$

c. $k = \frac{38 - 30}{5} = 1.6;\ (1 - 1/1.6^2) = .61$

d. $k = \frac{42 - 30}{5} = 2.4;\ (1 - 1/2.4^2) = .83$

e. $k = \frac{48 - 30}{5} = 3.6;\ (1 - 1/3.6^2) = .92$

26. a. $\bar{x} = 73.2,\ s = \sqrt{\frac{\Sigma(x_i - \bar{x})^2}{n - 1}} = 13.71$

b. $z = \frac{x_i - \bar{x}}{s} = \frac{108 - 73.2}{13.71} = 2.54$

It is a high score, but does not exceed 3 so is not an outlier

c. $z = \frac{87 - 73.2}{13.71} \approx 1$

68% within ±1, so 32%/2 = 16% 87 or more

$z = \frac{46 - 73.2}{13.71} \approx -2$

95% within ±2, so 5%/2 = 2.5% 46 or less

31. Arrange data in order: 5, 6, 8, 10, 10, 12, 15, 16, 18

$i = \frac{25}{100}(9) = 2.25$; round up to position 3

$Q_1 = 8$

Median (5th position) = 10

$i = \frac{75}{100}(9) = 6.75$; round up to position 7

$Q_3 = 15$

5-number summary: 5, 8, 10, 15, 18

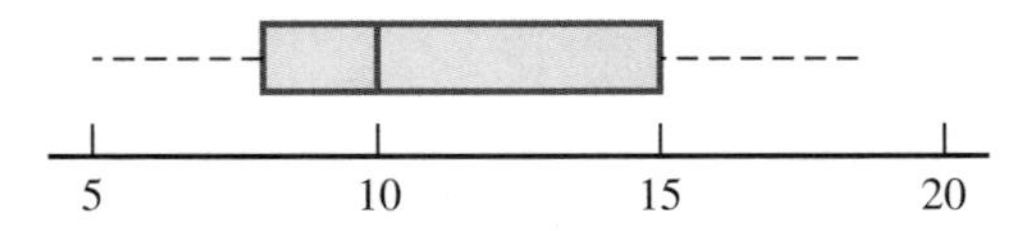

33. a. Arrange data in order low to high.

$i = \frac{25}{100}(21) = 5.25$; round up to 6th position

$Q_1 = 1872$

Median (11th position)=4019

$i = \frac{75}{100}(21) = 15.75$; round up to 16th position

$Q_3 = 8305$

5-number summary: 608, 1872, 4019, 8305, 14,138

b. $\text{IQR} = Q_3 - Q_1 = 8305 - 1872 = 6433$

1872 – 1.5 (6433) = –7777

8305 + 1.5 (6433) = 17,955

c. No; data are within limits.

d. 41,138 would be an outlier. Data value should be reviewed and corrected

e.

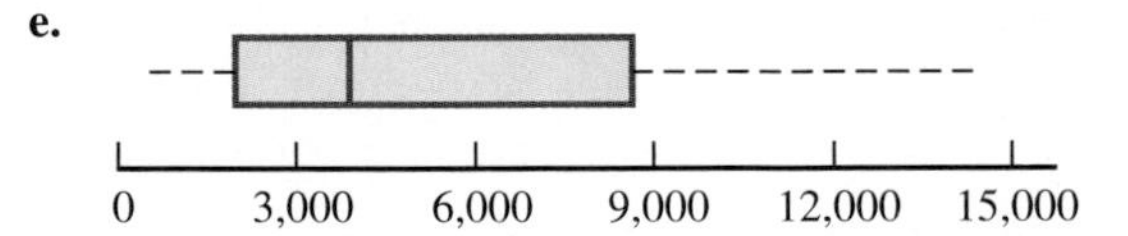

37. b. There appears to be a negative linear relationship between x and y

c.

x_i	y_i	$x_i - \bar{x}$	$y_i - \bar{y}$	$(x_i - \bar{x})(y_i - \bar{y})$
4	50	−4	4	−16
6	50	−2	4	−8
11	40	3	−6	−18
3	60	−5	14	−70
16	30	8	−16	−128
40	230	0	0	−240

$\bar{x} = 8; \bar{y} = 46$

$$s_{xy} = \frac{\Sigma(x_i - \bar{x})(y_i - \bar{y})}{n-1} = \frac{-240}{4} = -60$$

The sample covariance indicates a negative linear association between x and y

d. $$r_{xy} = \frac{s_{xy}}{s_x s_y} = \frac{-60}{(5.43)(11.40)} = -.97$$

The sample correlation coefficient of −.97 is indicative of a strong negative linear relationship

45. a.

f_i	M_i	f_iM_i
4	5	20
7	10	70
9	15	135
5	20	100
25		325

$$\bar{x} = \frac{\Sigma f_i M_i}{n} = \frac{325}{25} = 13$$

b.

f_i	M_i	$(M_i - \bar{x})$	$(M_i - \bar{x})^2$	$f_i(M_i - \bar{x})^2$
4	5	−8	64	256
7	10	−3	9	63
9	15	2	4	36
5	20	7	49	245
25				600

$$s^2 = \frac{\Sigma f_i(M_i - \bar{x})^2}{n-1} = \frac{600}{25-1} = 25$$

$$s = \sqrt{25} = 5$$

46. a.

Grade x_i	Weight w_i
4(A)	9
3(B)	15
2(C)	33
1(D)	3
0(F)	0
	60 credit hours.

$$\bar{x} = \frac{\Sigma w_i x_i}{\Sigma w_i} = \frac{9(4) + 15(3) + 33(2) + 3(1)}{9 + 15 + 33 + 3} = \frac{150}{60} = 2.5$$

b. Yes

Chapter 4

2. $$\binom{6}{3} = \frac{6!}{3!3!} = \frac{6\cdot5\cdot4\cdot3\cdot2\cdot1}{(3\cdot2\cdot1)(3\cdot2\cdot1)} = 20$$

ABC	ACE	BCD	BEF
ABD	ACF	BCE	CDE
ABE	ADE	BCF	CDF
ABF	ADF	BDE	CEF
ACD	AEF	BDF	DEF

6. $P(E_1) = .40$, $P(E_2) = .26$, $P(E_3) = .34$
The relative frequency method was used

9. $$\binom{50}{4} = \frac{50!}{4!46!} = \frac{50\cdot49\cdot48\cdot47}{4\cdot3\cdot2\cdot1} = 230{,}300$$

10. $P(\text{never married}) - \frac{1106}{2038}$

$P(\text{married}) = \frac{826}{2038}$

$P(\text{other}) = \frac{106}{2038}$

Note that the sum of the probabilities equals 1

15. a. S={ace of clubs, ace of diamonds, ace of hearts, ace of spades}
b. S={2 of clubs, 3 of clubs, . . . , 10 of clubs, J of clubs, Q of clubs, K of clubs, A of clubs}
c. There are 12; jack, queen, or king in each of the four suits
d. *For(a)*: 4/52 = 1/13 = .08
For(b): 13/52 = 1/4 = .25
For(c): 12/52 = .23

17. a. (4, 6), (4, 7), (4, 8)
b. .05 + .10 + .15 = .30
c. (2, 8), (3, 8), (4, 8)
d. .05 + .05 + .15 = .25
e. .15

23. a. $P(A) = P(E_1) + P(E_4) + P(E_6) = .05 + .25 + .10 = .40$

$P(B) = P(E_2) + P(E_4) + P(E_7) = = .20 + .25 + .05 = = .50$

$P(C) = P(E_2) + P(E_3) + P(E_5) + P(E_7)$

$= .20 + .20 + .15 + .05 = .60$

b. $A \cup B = \{E_1, E_2, E_4, E_6, E_7\}$

$P(A \cup B) = P(E_1) + P(E_2) + P(E_4) + P(E_6) + P(E_7)$

$= .05 + .20 + .25 + .10 + .05$

$= .65$

c. $A \cap B = \{E_4\}, P(A \cap B) = P(E_4) = .25$

d. Yes, they are mutually exclusive

e. $B^c = \{E_1, E_3, E_5, E_6\}$

$P(B^c) = P(E_1) + P(E_3) + P(E_5) + P(E_6)$

$= .05 + .20 + .15 + .10$

$= .50$

28. Let B = rented a car for business reasons
P = rented a car for personal reasons

a. $P(B \cup P) = P(B) + P(P) - P(B \cap P)$

$= .540 + .458 - .300$

$= .698$

b. $P(\text{Neither}) = 1 - .698 = .302$

30. a. $P(A \mid B) = \dfrac{P(A \cap B)}{P(B)} = \dfrac{.40}{.60} = .6667$

b. $P(B \mid A) = \dfrac{P(A \cap B)}{P(A)} = \dfrac{.40}{.50} = .80$

c. No, because $P(A \mid B) \neq P(A)$

33. a.

	Reason for Applying			Total
	Quality	*Cost/Convenience*	*Other*	
Full-time	.218	.204	.039	.461
Part-time	.208	.307	.024	.539
Total	.426	.511	.063	1.00

b. It is most likely a student will cite cost or convenience as the first reason (probability = .511); school quality is the first reason cited by the second largest number of students (probability = .426)

c. $P(\text{quality} \mid \text{full-time}) = .218/.461 = .473$

d. $P(\text{quality} \mid \text{part-time}) = .208/.539 = .386$

e. For independence, we must have $P(A)P(B) = P(A \cap B)$; from the table,

$P(A \cap B) = .218, P(A) = .461, P(B) = .426$
$P(A)P(B) = (.461)(.426) = .196$

Since $P(A)P(B) \neq P(A \cap B)$, the events are not independent

39. a. Yes, since $P(A_1 \cap A_2) = 0$

b. $P(A_1 \cap B) = P(A_1)P(B \mid A_1) = .40(.20) = .08$

$P(A_2 \cap B) = P(A_2)P(B \mid A_2) = .60(.05) = .03$

c. $P(B) = P(A_1 \cap B) + P(A_2 \cap B) = .08 + .03 = .11$

d. $P(A_1 \mid B) = \dfrac{.08}{.11} = .7273$

$P(A_2 \mid B) = \dfrac{.03}{.11} = .2727$

42. M = missed payment
D_1=customer defaults
D_2 = customer does not default
$P(D_1) = .05, P(D_2) = .95, P(M \mid D_2) = .2, P(M \mid D_1) = 1$

a. $P(D_1 \mid M) = \dfrac{P(D_1)P(M \mid D_1)}{P(D_1)P(M \mid D_1) + P(D_2)P(M \mid D_2)}$

$= \dfrac{(.05)(1)}{(.05)(1) + (.95)(.2)}$

$= \dfrac{.05}{.24} = .21$

b. Yes, the probability of default is greater than .20

Chapter 5

1. a. Head, Head (H, H)
Head, Tail (H, T)
Tail, Head (T, H)
Tail, Tail (T, T)

b. x = number of heads on two coin tosses

c.

Outcome	Values of x
(H, H)	2
(H, T)	1
(T, H)	1
(T, T)	0

3. Let: Y = position is offered
N = position is not offered

a. $S = \{(Y, Y, Y), (Y, Y, N), (Y, N, Y), (Y, N, N), (N, Y, Y), (N, Y, N), (N, N, Y), (N, N, N)\}$

b. Let N = number of offers made; N is a discrete random variable

c.

Experimental Outcome	(Y, Y, Y)	(Y, Y, N)	(Y, N, Y)	(Y, N, N)	(N, Y, Y)	(N, Y, N)	(N, N, Y)	(N, N, N)
Value of N	3	2	2	1	2	1	1	0

7. a. $f(x) \geq 0$ for all values of x
$\Sigma f(x) = 1$; therefore, it is a proper probability distribution

b. Probability $x = 30$ is $f(30) = .25$

c. Probability $x \leq 25$ is $f(20) + f(25) = .20 + .15 = .35$

d. Probability $x > 30$ is $f(35) = .40$

8. a.

x	$f(x)$
1	3/20 = .15
2	5/20 = .25
3	8/20 = .40
4	4/20 = .20
Total	1.00

b.

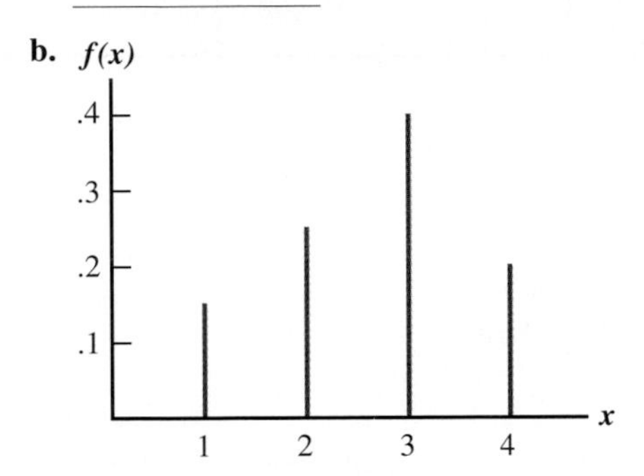

c. $f(x) \geq 0$ for $x = 1, 2, 3, 4$

$\Sigma f(x) = 1$

16. a.

y	$f(y)$	$yf(y)$
2	.20	.40
4	.30	1.20
7	.40	2.80
8	.10	.80
Totals	1.00	5.20

$E(y) = \mu = 5.20$

b.

y	$y - \mu$	$(y - \mu)^2$	$f(y)$	$(y - \mu)^2 f(y)$
2	−3.20	10.24	.20	2.048
4	−1.20	1.44	.30	.432
7	1.80	3.24	.40	1.296
8	2.80	7.84	.10	.784
			Total	4.560

$\text{Var}(y) = 4.56$

$\sigma = \sqrt{4.56} = 2.14$

18. a & b

x	$f(x)$	$xf(x)$	$(x - \mu)$	$(x - \mu)^2$	$(x - \mu)^2 f(x)$
0	.02	.00	−2.20	4.84	.0968
1	.24	.24	−1.20	1.44	.3456
2	.42	.84	−0.20	0.04	.0168
3	.20	.60	0.80	0.64	.1280
4	.08	.32	1.80	3.24	.2592
5	.04	.20	2.80	7.84	.3136
		$E(x)$ = 2.20			$\text{Var}(x)$ = 1.1600
					σ = 1.08

The expected value, $E(x) = 2.2$ of the probability distribution is the same as the average reported in the *1994 Statistical Abstract of the United States*

$\text{Var}(x) = 1.16$ television sets squared

$\sigma = \sqrt{1.16} = 1.08$ television sets

23. a.

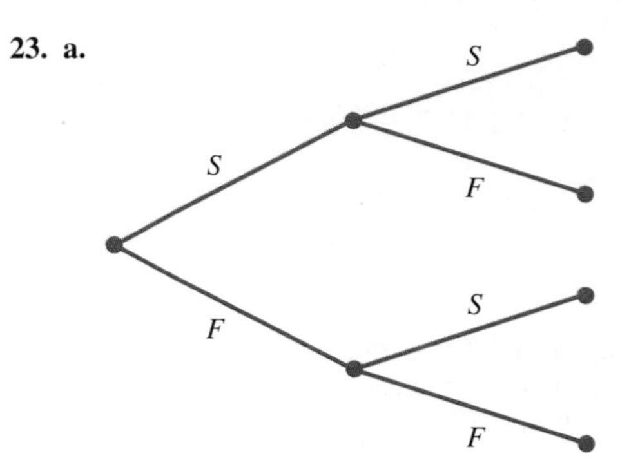

b. $f(1) = \binom{2}{1}(.4)^1(.6)^1 = \frac{2!}{1!1!}(.4)(.6) = .48$

c. $f(0) = \binom{2}{0}(.4)^0(.6)^2 = \frac{2!}{0!2!}(1)(.36) = .36$

d. $f(2) = \binom{2}{2}(.4)^2(.6)^0 = \frac{2!}{2!0!}(.16)(1) = .16$

e. $P(x \geq 1) = f(1) + f(2) = .48 + .16 = .64$

f. $E(x) = np = 2(.4) = .8$

$\text{Var}(x) = np(1 - p) = 2(.4)(.6) = .48$

$\sigma = \sqrt{.48} = .6928$

28. a. Probability of a defective part being produced must be .03 for each trial; trials must be independent

b. Let: D = defective

G = not defective

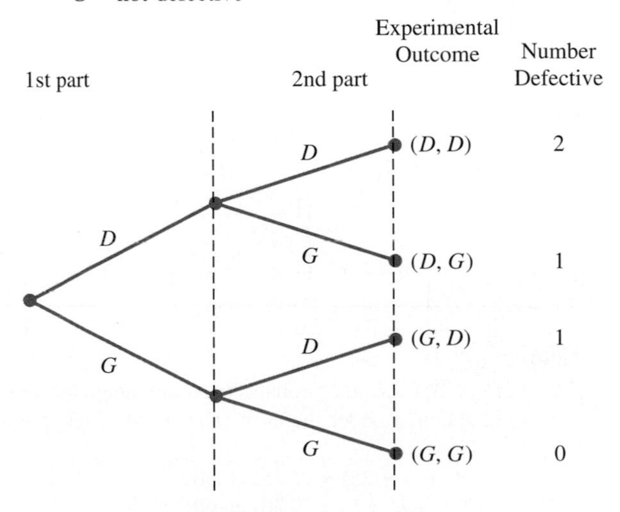

c. Two outcomes result in exactly one defect

d. $P(\text{no defects}) = (.97)(.97) = .9409$

$P(\text{1 defect}) = 2(.03)(.97) = .0582$

$P(\text{2 defects}) = (.03)(.03) = .0009$

35. **a.** $f(x) = \dfrac{2^x e^{-2}}{x!}$

b. $\mu = 6$ for 3 time periods

c. $f(x) = \dfrac{6^x e^{-6}}{x!}$

d. $f(2) = \dfrac{2^2 e^{-2}}{2!} = \dfrac{4(.1353)}{2} = .2706$

e. $f(6) = \dfrac{6^6 e^{-6}}{6!} = .1606$

f. $f(5) = \dfrac{4^5 e^{-4}}{5!} = .1563$

36. **a.** $\mu = 48(5/60) = 4$

$$f(3) = \frac{4^3 e^{-4}}{3!} = \frac{(64)(.0183)}{6} = .1952$$

b. $\mu = 48(15/60) = 12$

$$f(10) = \frac{12^{10} e^{-12}}{10!} = .1048$$

c. $\mu = 48(5/60) = 4$; one can expect 4 callers to be waiting after 5 minutes

$f(0) = \dfrac{4^0 e^{-4}}{0!} = .0183$; the probability none will be waiting after 5 minutes is .0183

d. $\mu = 48(3/60) = 2.4$

$f(0) = \dfrac{2.4^0 e^{-2.4}}{0!} = .0907$; the probability of no interruptions in 3 minutes is .0907

Chapter 6

1. **a.**

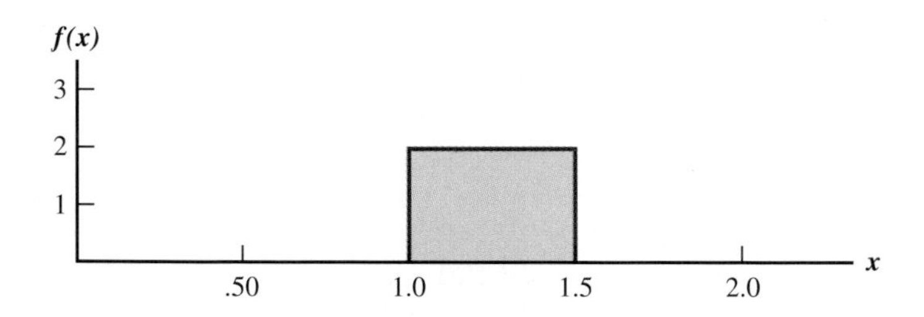

b. $P(x = 1.25) = 0$; the probability of any single point is zero since the area under the curve above any single point is zero

c. $P(1.0 \le x \le 1.25) = 2(.25) = .50$

d. $P(1.20 < x < 1.5) = 2(.30) = .60$

4. **a.**

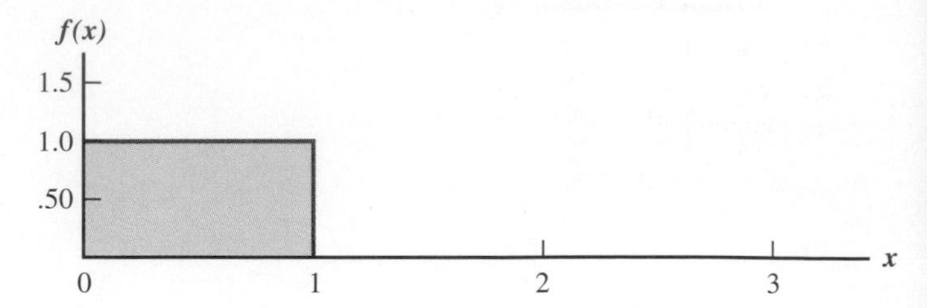

b. $P(.25 < x < .75) = 1(.50) = .50$

c. $P(x \le .30) = 1(.30) = .30$

d. $P(x > .60) = 1(.40) = .40$

13. **a.** .4761 + .1879 = .6640

b. .3888 − .1985 = .1903

c. .4599 − .3508 = .1091

15. **a.** Look in the table for an area of .5000 − .2119 = .2881; since the value we are seeking is below the mean, the z value must be negative; thus, for an area of .2881, $z = -.80$

b. Look in the table for an area of .9030/2 = .4515; $z = 1.66$

c. Look in the table for an area of .2052/2 = .1026; $z = .26$

d. Look in the table for an area of .9948 − .5000 = .4948; $z =$ 2.56

e. Look in the table for an area of .6915 − .5000 = .1915; since the value we are seeking is below the mean, the z value must be negative; thus, $z = -.50$

18. **a.** With $z = \dfrac{(12{,}000 - 10{,}000)}{2200} = .91$,

$P(x \le 12{,}000) = P(z \le .91) = .5000 + .3186 = .8186$

$P(x \ge 12{,}000) = 1 - P(x \le 12{,}000) = 1 - .8186 = = .1814$

So, the probability that a rehabilitation program will cost at least \$12,000 is .1814

b. With $z = \dfrac{(6000 - 10{,}000)}{2200} = -1.82$,

$P(x \ge 6000) = P(z \ge -1.82) = .4656 + .5000 = .9656$

The probability that a rehabilitation program will cost at least \$6000 is .9656

c. First, find the value of z that cuts off an area of .10 in the upper tail of the standard normal distribution; a value of $z = 1.28$ does this.

Now find the value of x corresponding to $z = 1.28$

$$\frac{x - 10{,}000}{2200} = 1.28$$

$$x = 10{,}000 + 1.28(2200) = 12{,}816$$

The cost range for the 10% most expensive programs is \$12,816 or more

26. a. $\mu = np = 100(.20) = 20$
$\sigma^2 = np(1 - p) = 100(.20)(.80) = 16$
$\sigma = \sqrt{16} = 4$

b. Yes, since $np = 20$ and $n(1 - p) = 80$

c. Compute $P(23.5 \leq x \leq 24.5)$

$$z = \frac{24.5 - 20}{4} = 1.13 \rightarrow \text{Area} = .3708$$

$$z = \frac{23.5 - 20}{4} = .88 \rightarrow \text{Area}=.3106$$

$P(23.5 \leq x \leq 24.5) = .3708 - .3106 = .0602$

d. Compute $P(17.5 \leq x \leq 22.5)$

$$z = \frac{17.5 - 20}{4} = -.63 \rightarrow \text{Area} = .2357$$

$$z = \frac{22.5 - 20}{4} = .63 \rightarrow \text{Area} = .2357$$

$P(17.5 \leq x \leq 22.5) = .2357 + .2357 = .4714$

e. Compute $P(x \leq 15.5)$

$$z = \frac{15.5 - 20}{4} = -1.13 \rightarrow \text{Area} = .3708$$

$P(x \leq 15.5) = .5000 - .3708 = .1292$

28. Use the normal approximation of binomial probabilities with $\mu = np = 250(.04) = 10$ and $\sigma = \sqrt{np(1 - p)} = \sqrt{250(.04)(.96)} = 3.1$

a. Compute $P(x \leq 12.5)$

$$z = \frac{12.5 - 10}{3.1} = .81 \rightarrow \text{Area} = .2910$$

$P(x \leq 12.5) = .5000 + .2910 = .7910$

b. Compute $P(x \geq 4.5)$

$$z = \frac{4.5 - 10}{3.1} = -1.77 \rightarrow \text{Area} = .4616$$

$P(x \geq 4.5) = .5000 + .4616 = .9616$

c. Compute $P(7.5 \leq x \leq 8.5)$

$$z = \frac{7.5 - 10}{3.1} = -.81 \rightarrow \text{Area} = .2910$$

$$z = \frac{8.5 - 10}{3.1} = -.48 \rightarrow \text{Area} = .1844$$

$P(7.5 \leq x \leq 8.5) = .2910 - .1844 = .1066$

34. a. $P(x \leq x_0) = 1 - e^{-x_0/3}$

b. $P(x \leq 2) = 1 - e^{-2/3} = 1 - .5134 = .4866$

c. $P(x \geq 3) = 1 - P(x \leq 3) = 1 - (1 - e^{-3/3}) = e^{-1} = .3679$

d. $P(x \leq 5) = 1 - e^{-5/3} = 1 - .1889 = .8111$

e. $P(2 \leq x \leq 5) = P(x \leq 5) - P(x \leq 2) = .8111 - .4866 = .3245$

36. a.

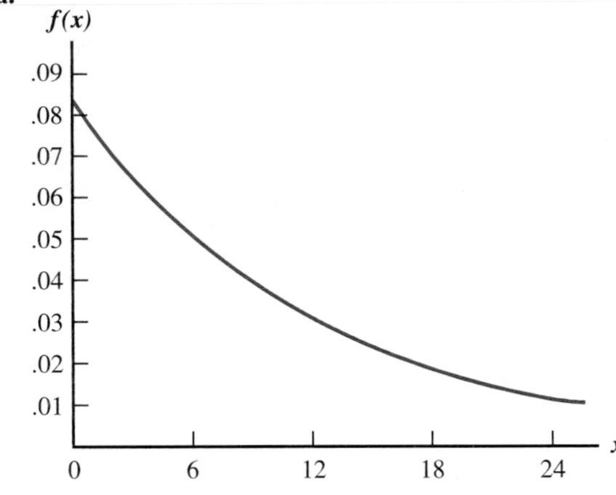

b. $P(x \leq 12) = 1 - e^{-12/12} = 1 - .3679 = .6321$

c. $P(x \leq 6) = 1 - e^{-6/12} = 1 - .6065 = .3935$

d. $P(x \geq 30) = 1 - P(x < 30)$
$= 1 - (1 - e^{-30/12})$
$= .0821$

Chapter 7

1. a. AB, AC, AD, AE, BC, BD, BE, CD, CE, DE

b. With 10 samples, each has a 1⁄10 probability

c. E and C because 8 and 0 do not apply; 5 identifies E; 7 does not apply; 5 is skipped since E is already in the sample; 3 identifies C; 2 is not needed since the sample of size 2 is complete

3. 554, 459, 147, 385, 689, 640, 113, 340, 756, 953, 401, 827

11. $\bar{x} = \frac{\Sigma x_i}{n} = \frac{54}{6} = 9$

13. $\bar{x} = \frac{\Sigma x_i}{n} = \frac{465}{5} = 93$

19. a. The sampling distribution is normal with:

$$E(\bar{x}) = \mu = 200$$

$$\sigma_{\bar{x}} = \frac{\sigma}{\sqrt{n}} = \frac{50}{\sqrt{100}} = 5$$

For ± 5, $(\bar{x} - \mu) = 5$,

$$z = \frac{\bar{x} - \mu}{\sigma_{\bar{x}}} = \frac{5}{5} = 1$$

Area $= .3413 \times 2 = .6826$

b. For ± 10, $(\bar{x} - \mu) = 10$,

$$z = \frac{\bar{x} - \mu}{\sigma_{\bar{x}}} = \frac{10}{5} = 2$$

Area $= .4772 \times 2 = .9544$

25. a.

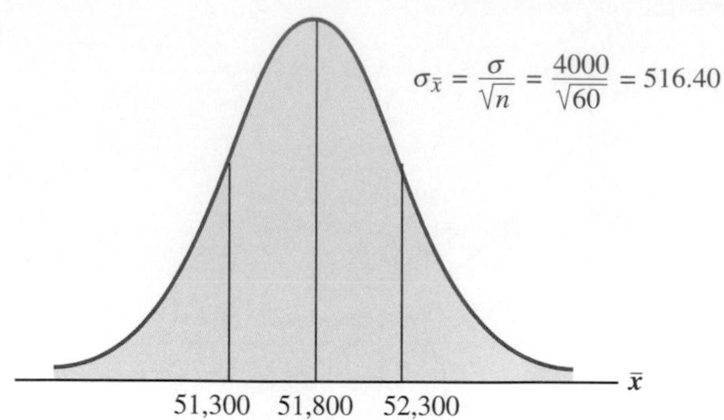

$$z = \frac{52{,}300 - 51{,}800}{516.40} = +.97$$

Area = .3340 × 2 = .6680

b. $\sigma_{\bar{x}} = \frac{\sigma}{\sqrt{n}} = \frac{4000}{\sqrt{120}} = 365.15$

$$z = \frac{52{,}300 - 51{,}800}{365.15} = +1.37$$

Area = .4147 × 2 = .8294

34. a. $E(\bar{p}) = .40$

$$\sigma_{\bar{p}} = \sqrt{\frac{p(1-p)}{n}} = \sqrt{\frac{(.40)(.60)}{200}} = .0346$$

$$z = \frac{\bar{p} - p}{\sigma_{\bar{p}}} = \frac{.03}{.0346} = .87$$

Area = .3078 × 2 = .6156

b. $z = \frac{\bar{p} - p}{\sigma_{\bar{p}}} = \frac{.05}{.0346} = 1.45$

Area = .4265 × 2 = .8530

37. a.

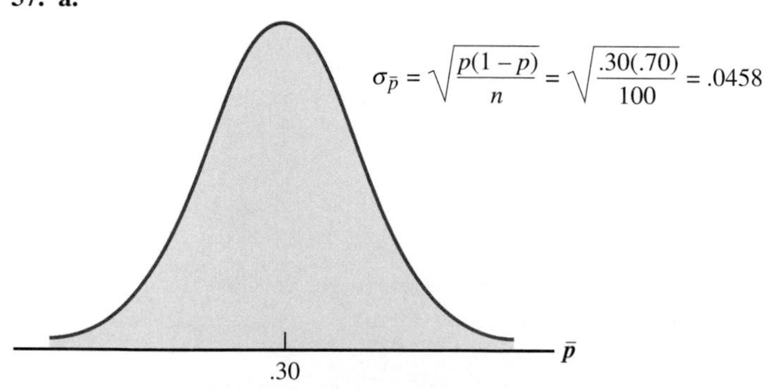

The normal distribution is appropriate because $np = 100(.30) = 30$ and $n(1 - p) = 100(.70) = 70$ are both greater than 5

b. $P(.20 \le \bar{p} \le .40) = ?$

$$z = \frac{.40 - .30}{.0458} = 2.18$$

Area = .4854 × 2 = .9708

c. $P(.25 \le \bar{p} \le .35) = ?$

$$z = \frac{.35 - .30}{.0458} = 1.09$$

Area = .3621 × 2 = .7242

Chapter 8

2. Use $\bar{x} \pm z_{\alpha/2}(\sigma/\sqrt{n})$ with the sample standard deviation s used to estimate σ

a. $32 \pm 1.645\ (6/\sqrt{50})$
32 ± 1.4; (30.6 to 33.4)

b. $32 \pm 1.96(6/\sqrt{50})$
32 ± 1.66; (30.34 to 33.66)

c. $32 \pm 2.575(6/\sqrt{50})$
32 ± 2.18; (29.82 to 34.18)

3. a. $\sigma_{\bar{x}} = \sigma/\sqrt{n} = 2.50/\sqrt{49} = .3571$
b. Sampling error less than or equal to $1.96\sigma_{\bar{x}} = .70$
c. $12.60 \pm .70$ or (11.90 to 13.30)

13. a. $\bar{x} = \frac{\Sigma x_i}{n} = \frac{80}{8} = 10$

b. $s = \sqrt{\frac{\Sigma(x_i - \bar{x})^2}{n - 1}} = \sqrt{\frac{84}{8 - 1}} = 3.46$

c. With 7 degrees of freedom, $t_{.025} = 2.365$

$$\bar{x} \pm t_{.025}\frac{s}{\sqrt{n}}$$

$$10 \pm 2.365\frac{3.464}{\sqrt{8}}$$

10 ± 2.89; (7.11 to 12.89)

15. At 90%, $80 \pm t_{.05}(s/\sqrt{n})$ with degrees of freedom=17
$t_{.05} = 1.740$
$80 \pm 1.740(10/\sqrt{18})$
80 ± 4.10; (75.90 to 84.10)

At 95%, $80 \pm 2.11(10/\sqrt{18})$ with degrees of freedom=17
$t_{.025}=2.110$
80 ± 4.97; (75.03 to 84.97)

22. a. Planning value of $\sigma = \frac{\text{Range}}{4} = \frac{36}{4} = 9$

b. $n = \frac{z_{.025}^2\sigma^2}{E^2} = \frac{(1.96)^2(9)^2}{(3)^2} = 34.6 \approx 35$

c. $n = \frac{(1.96)^2(9)^2}{(2)^2} = 77.8 \approx 78$

23. Use $n = \dfrac{z^2_{\alpha/2}\sigma^2}{E^2}$,

$$n = \frac{(1.96)^2(6.82)^2}{(1.5)^2} = 79.4 \text{ or } 80$$

$$n = \frac{(1.645)^2(6.82)^2}{(2)^2} = 31.5 \text{ or } 32$$

29. a. $\bar{p} = \dfrac{100}{400} = .25$

b. $\sqrt{\dfrac{\bar{p}(1-\bar{p})}{n}} = \sqrt{\dfrac{.25(.75)}{400}} = .0217$

c. $\bar{p} \pm z_{.025}\sqrt{\dfrac{\bar{p}(1-\bar{p})}{n}}$

$.25 \pm 1.96(.0217)$

$.25 \pm .0425$; (.2075 to .2925)

33. a. $\bar{p} = 248/400 = .62$

b. $\bar{p} \pm 1.645\sqrt{\dfrac{.62(.38)}{400}}$

$.62 \pm .04$; (.58 to .66)

37. a. $n = \dfrac{(z_{.025})^2 p(1-p)}{E^2} = \dfrac{(1.96)^2(.55)(.45)}{(.03)^2} = 1056.4 \text{ or } 1057$

b. $n = \dfrac{(1.96)^2(.55)(.45)}{(.06)^2} = 264.1 \text{ or } 265$

Chapter 9

2. a. H_0: $\mu \leq 14$
H_a: $\mu > 14$

b. No evidence that the new plan increases sales

c. The research hypothesis $\mu > 14$ is supported; the new plan increases sales.

5. a. Rejecting $H_0 : \mu \leq 8.6$ when it is true

b. Accepting H_0: $\mu \leq 8.6$ when it is false

10. a. $z = 2.05$
Reject H_0 if $z > 2.05$

b. $z = \dfrac{x - \mu}{s/\sqrt{n}} = \dfrac{16.5 - 15}{7/\sqrt{40}} = 1.36$

c. Area for $z = 1.36$ is .4131
p-value = .5000 − .4131 = .0869

d. Do not reject H_0

13. a. H_0: $\mu \geq 1056$
H_a: $\mu < 1056$

b. Reject H_0 if $z < -1.645$

$$z = \frac{\bar{x} - \mu}{s/\sqrt{n}} = \frac{910 - 1056}{1600/\sqrt{400}} = -1.83$$

Reject H_0

c. p-value = .5000 − .4664 = .0336

20. a. Reject H_0 if $z < -2.33$ or $z > 2.33$

b. $z = \dfrac{\bar{x} - \mu}{\sigma/\sqrt{n}} = \dfrac{14.2 - 15}{5/\sqrt{50}} = 1.13$

c. p-value = 2(.5000 − .3708) = .2584

d. Do not reject H_0

23. a. H_0: $\mu = 18{,}688$
H_a: $\mu \neq 18{,}688$

Reject H_0 **if** $z < -1.96$ or $z > 1.96$

$$z = \frac{\bar{x} - \mu_0}{s/\sqrt{n}} = \frac{16{,}860 - 18{,}688}{14{,}624/\sqrt{400}} = -2.50$$

Reject H_0 and conclude $\mu \neq 18{,}688$

b. p-value = 2(.5000 − .4938) = .0124

30. a. $x = \dfrac{\Sigma x_i}{n} = \dfrac{108}{6} = 18$

b. $s = \sqrt{\dfrac{\Sigma(x_i - \bar{x})}{n-1}} = \sqrt{\dfrac{10}{6-1}} = 1.41$

c. Reject H_0 if $t < -2.571$ or $t > 2.571$

d. $t = \dfrac{\bar{x} - \mu}{s/\sqrt{n}} = \dfrac{18 - 20}{1.41/\sqrt{6}} = -3.47$

e. Reject H_0 ; conclude H_a is true

33. a. $\bar{x} = \dfrac{\Sigma x_i}{n} = 270$

b. $s = \sqrt{\dfrac{\Sigma(x_i - \bar{x})^2}{n-1}} = 24.78$

c. $H_0 : \mu \leq 258$
$H_a : \mu > 258$
Reject H_0 if $t > 1.761$

$$t = \frac{\bar{x} - \mu_0}{s/\sqrt{n}} = \frac{270 - 258}{24.78/\sqrt{15}} = 1.88$$

Reject H_0 ; mean has increased

d. Between .05 and .025

40. a. Reject H_0 if $z < -1.96$ or $z > 1.96$

b. $\sigma_{\bar{p}} = \sqrt{\dfrac{.20(.80)}{400}} = .02$

$$z = \frac{\bar{p} - p}{\sigma_{\bar{p}}} = \frac{.175 - .20}{.02} = -1.25$$

c. p-value = 2(.5000 − .3944) = .2112

d. Do not reject H_0

43. $H_0 : p \geq .64$
$H_a : p < .64$
Reject H_0 if $z < -1.645$
$\bar{p} = 52/100 = .52$

$$z = \frac{.52 - .64}{\sqrt{\dfrac{.64(.36)}{100}}} = -2.5$$

Reject H_0

Chapter 10

1. a. $\bar{x}_1 - \bar{x}_2 = 13.6 - 11.6 = 2$

b. $s_{\bar{x}_1 - \bar{x}_2} = \sqrt{\dfrac{s_1^2}{n_1} + \dfrac{s_2^2}{n_2}} = \sqrt{\dfrac{(2.2)^2}{50} + \dfrac{3^2}{35}} = .595$

$2 \pm 1.645(.595)$

$2 \pm .98$ or 1.02 to 2.98

c. $2 \pm 1.96(.595)$

2 ± 1.17 or .83 to 3.17

7. a. $\bar{x}_1 - \bar{x}_2 = 15{,}700 - 14{,}500 = 1200$

b. Pooled variance

$$s^2 = \frac{7(700)^2 + 11(850)^2}{18} = 632{,}083$$

$$s_{\bar{x}_1 - \bar{x}_2} = \sqrt{632{,}083\left(\frac{1}{8} + \frac{1}{12}\right)} = 362.88$$

With 18 degrees of freedom $t_{.025} = 2.101$,

$1200 \pm 2.101(362.88)$
1200 ± 762, or 438 to 1962

c. Populations are normally distributed with equal variances

9. a. $s_{\bar{x}_1 - \bar{x}_2} = \sqrt{\dfrac{s_1^2}{n_1} + \dfrac{s_2^2}{n_2}} = \sqrt{\dfrac{(5.2)^2}{40} + \dfrac{6^2}{50}} = 1.18$

$$z = \frac{(\bar{x}_1 - \bar{x}_2) - (\mu_1 - \mu_2)}{s_{\bar{x}_1 - \bar{x}_2}} = \frac{(25.2 - 22.8)}{1.18} = 2.03$$

Reject H_0 if $z > 1.645$; therefore reject H_0; conclude H_a is true and $\mu_1 > \mu_2$

b. p-value$=.5000 - .4788 = .0212$

13. H_0: $\mu_1 - \mu_2 = 0$
H_a: $\mu_1 - \mu_2 \neq 0$

Reject H_0 if $z < -1.96$ or if $z > 1.96$

$$z = \frac{(\bar{x}_1 - \bar{x}_2) - 0}{\sqrt{\sigma_1^2/n_1 + \sigma_2^2/n_2}} = \frac{40 - 35}{\sqrt{(9)^2/36 + (10)^2/49}} = 2.41$$

Reject H_0; customers at the two stores differ in terms of mean ages

17. a. 1, 2, 0, 0, 2

b. $\bar{d} = \dfrac{\Sigma d_i}{n} = \dfrac{5}{5} = 1$

c. $s_d = \sqrt{\dfrac{\Sigma(d_i - \bar{d})^2}{n - 1}} = \sqrt{\dfrac{4}{5 - 1}} = 1$

d. With 4 degrees of freedom, $t_{.05} = 2.132$; reject H_0 if $t > 2.132$

$$t = \frac{\bar{d} - \mu_d}{s_d/\sqrt{n}} = \frac{1 - 0}{1/\sqrt{5}} = 2.24$$

Reject H_0; conclude $\mu_d > 0$

19. d = rating after – rating before
H_0: $\mu_d \leq 0$
H_a: $\mu_d > 0$

With 7 degrees of freedom, reject H_0 if $t > 1.895$; when $\bar{d} = .63$ and $s_d = 1.3025$,

$$t = \frac{\bar{d} - \mu_d}{s_d/\sqrt{n}} = \frac{.63 - 0}{1.3025/\sqrt{8}} = 1.36$$

Do not reject H_0; we cannot conclude that seeing the commercial improves the potential to purchase

25. a.

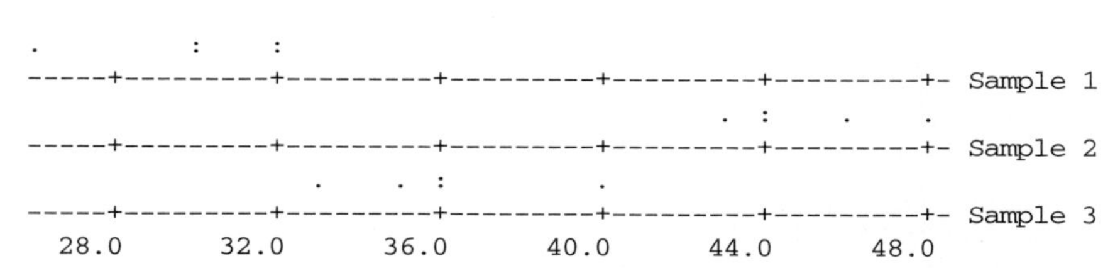

b. $\bar{\bar{x}} = (30 + 45 + 36)/3 = 37$

$$\text{SSB} = \sum_{j=1}^{k} n_j(\bar{x}_j - \bar{\bar{x}})^2$$

$$= 5(30 - 37)^2 + 5(45 - 37)^2 + 5(36 - 37)^2 = 570$$

$$\text{MSB} = \frac{\text{SSB}}{k - 1} = \frac{570}{2} = 285$$

c. $\text{SSW} = \sum_{j=1}^{k} (n_j - 1)s_j^2$

$= 4(6) + 4(4) + 4(6.5) = 66$

$$\text{MSW} = \frac{\text{SSW}}{n_T - k} = \frac{66}{15 - 3} = 5.5$$

d. $F = \dfrac{\text{MSB}}{\text{MSW}} = \dfrac{285}{5.5} = 51.82$

$F_{.05} = 3.89$ (2 degrees of freedom numerator and 12 denominator)

Since $F = 51.82 > F_{.05} = 3.89$, we reject the null hypothesis that the means of the three populations are equal

e.

Source of Variation	Sum of Squares	Degrees of Freedom	Mean Square	F
Between	570	2	285	51.82
Within	66	12	5.5	
Total	636	14		

28.

	Manufacturer 1	Manufacturer 2	Manufacturer 3
Sample mean	23	28	21
Sample variance	6.67	4.67	3.33

$\bar{\bar{x}} = (23 + 28 + 21)/3 = 24$

$$\text{SSB} = \sum_{j=1}^{k} n_j(\bar{x}_j - \bar{\bar{x}})^2$$

$= 4(23 - 24)^2 + 4(28 - 24)^2 + 4(21 - 24)^2 = 104$

$$\text{MSB} = \frac{\text{SSB}}{k - 1} = \frac{104}{2} = 52$$

$$\text{SSW} = \sum_{j=1}^{k} (n_j - 1)s_j^2$$

$= 3(6.67) + 3(4.67) + 3(3.33) = 44.01$

$$\text{MSW} = \frac{\text{SSW}}{n_T - k} = \frac{44.01}{12 - 3} = 4.89$$

$$F = \frac{\text{MSB}}{\text{MSW}} = \frac{52}{4.89} = 10.63$$

$F_{.05} = 4.26$ (2 degrees of freedom numerator and 9 denominator)

Since $F = 10.63 > F_{.05} = 4.26$, we reject the null hypothesis that the mean time needed to mix a batch of material is the same for each manufacturer

Chapter 11

2. a. $\bar{p} = \dfrac{n_1\bar{p}_1 + n_2\bar{p}_2}{n_1 + n_2} = \dfrac{200(.22) + 300(.16)}{200 + 300} = .184$

$$s_{\bar{p}_1 - \bar{p}_2} = \sqrt{(.184)(.816)\left(\frac{1}{200} + \frac{1}{300}\right)} = .0354$$

Reject H_0 if $z > 1.645$

$$z = \frac{(.22 - .16) - 0}{.0354} = 1.69$$

Reject H_0

b. p-value $= (.5000 - .4545) = .0455$

11. *Expected frequencies*: $e_1 = 200(.40) = 80$, $e_2 = 200(.40) = 80$, $e_3 = 200(.20) = 40$

Actual frequencies: $f_1 = 60, f_2 = 120, f_3 = 20$

$$\chi^2 = \frac{(60 - 80)^2}{80} + \frac{(120 - 80)^2}{80} + \frac{(20 - 40)^2}{40}$$

$$= \frac{400}{80} + \frac{1600}{80} + \frac{400}{40}$$

$$= 5 + 20 + 10 = 35$$

$\chi^2_{.01} = 9.21034$ with $k - 1 = 3 - 1 = 2$ degrees of freedom

Since $\chi^2 = 35 > 9.21034$, reject the null hypothesis; that is, the population proportions are not as stated in the null hypothesis

13. $H_0 : p_{\text{ABC}} = .29, p_{\text{CBS}} = .28, p_{\text{NBC}} = .25, p_{\text{IND}} = .18$

H_a : The proportions are not $p_{\text{ABC}} = .29, p_{\text{CBS}} = .28, p_{\text{NBC}} = .25, p_{\text{IND}} = .18$

Expected frequencies: $300(.29) = 87$, $300(.28) = 84$, $300(.25) = 75$, $300(.18) = 54$

$e_1 = 87, e_2 = 84, e_3 = 75, e_4 = 54$

Actual frequencies: $f_1 = 95, f_2 = 70, f_3 = 89, f_4 = 46$

$\chi^2_{.05} = 7.81$ (3 degrees of freedom)

$$\chi^2 = \frac{(95 - 87)^2}{87} + \frac{(70 - 84)^2}{84} + \frac{(89 - 75)^2}{75} + \frac{(46 - 54)^2}{54} = 6.87$$

Do not reject H_0; there is no significant change in the viewing audience proportions

19. H_0 : The column factor is independent of the row factor
H_a : The column factor is not independent of the row factor
Expected frequencies:

	A	B	C
P	28.5	39.9	45.6
Q	21.5	30.1	34.4

$$\chi^2 = \frac{(20-28.5)^2}{28.5} + \frac{(44-39.9)^2}{39.9} + \frac{(50-45.6)^2}{45.6} + \frac{(30-21.5)^2}{21.5} + \frac{(26-30.1)^2}{30.1} + \frac{(30-34.4)^2}{34.4} = 7.86$$

$\chi^2_{.025} = 7.37776$ with $(2-1)(3-1) = 2$ degrees of freedom

Since $\chi^2 = 7.86 > 7.37776$, reject H_0; that is, conclude that the column factor is not independent of the row factor

21. H_0 : There is no difference in shooting percentage among the teams
H_a : There is a difference in shooting percentage among the teams
Row 1 total = 629; row 2 total = 988
Column 1 total = 374; column 2 total = 341
Column 3 total = 369; column 4 total = 533
Overall total = 1617
Using these totals, we compute the expected frequencies

Expected frequencies:

	Duke	Mich.	Ind.	Cin.
Made	145.4830	132.6463	143.5380	207.3327
Missed	228.5170	208.3537	225.4620	325.6673

$$\chi^2 = \frac{(160-145.4830)^2}{145.4830} + \frac{(113-132.6463)^2}{132.6463} + \cdots + \frac{(331-325.6673)^2}{325.6673}$$

$$= 1.4486 + 2.9098 + .7625 + .1372 + .9222 + 1.8525 + .4855 + .0873$$

$$= 8.6056$$

$\chi^2_{.05} = 7.81473$ with 3 degrees of freedom

Since $\chi^2 = 8.6056 > 7.81473$, reject H_0; that is, conclude that there is a difference in 3-point shooting ability for the teams

Chapter 12

1. a.

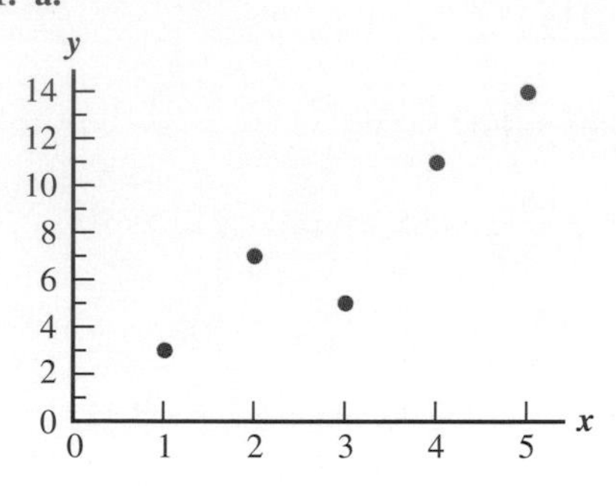

b. There appears to be a linear relationship between x and y

c. Many different straight lines can be drawn to provide a linear approximation of the relationship between x and y; in part (d) we will determine the equation of a straight line that "best" represents the relationship according to the least squares criterion

d. $\Sigma x_i = 15,\quad \Sigma y_i = 40,\quad \Sigma x_i y_i = 146,\quad \Sigma x_i^2 = 55$

$$b_1 = \frac{\Sigma x_i y_i - (\Sigma x_i \Sigma y_i)/n}{\Sigma x_i^2 - (\Sigma x_i)^2/n}$$

$$= \frac{146 - (15)(40)/5}{55 - (15)^2/5} = 2.6$$

$$b_0 = \bar{y} - b_1\bar{x}$$

$$= 8 - 2.6(3) = .2$$

$$\hat{y} = .2 + 2.6x$$

e. $\hat{y} = .2 + 2.6x = .2 + 2.6(4) = 10.6$

5. a.

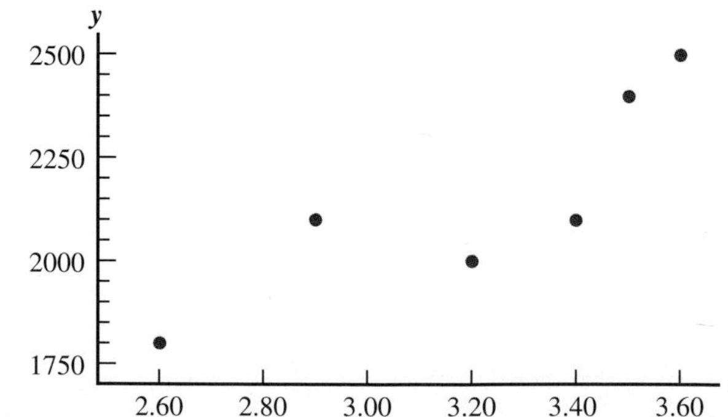

b. There appears to be a linear relationship between x and y.

c. Many different straight lines can be drawn to provide a linear approximation of the relationship between x and y; in part d we will determine the equation of a straight line that "best" represents the relationship according to the least squares criterion.

d. Summations needed to compute the slope and y-intercept are:

$\Sigma x_i = 19.2$ $\Sigma y_i = 12{,}900$ $\Sigma x_i y_i = 41{,}710$ $\Sigma x_i^2 = 62.18$

$$b_1 = \frac{\Sigma x_i y_i (\Sigma x_i \Sigma y_i)/n}{\Sigma x_i^2 - (\Sigma x_i)^2/n} = \frac{41710 - (19.2)(12900)/6}{62.18 - (19.2)^2/6} = 581.0811$$

$$b_0 = \bar{y} - b_1 \bar{x} = 2{,}150 - (581.0811)(3.2) = 290.5405$$

$$\hat{y} = 290.54 + 581.08x$$

e. $\hat{y} = 290.54 + 581.08(3) = 2033.78$

$\hat{y} = 290.54 + 581.08(3.5) = 2324.32$

13. a. $\hat{y}_i = .2 + 2.6x_i$ and $\bar{y} = 8$

x_i	y_i	$\hat{y}_i$	$y_i - \hat{y}_i$	$(y_i - \hat{y}_i)^2$	$y_i - \bar{y}$	$(y_i - \bar{y})^2$
1	3	2.8	.2	.04	−5	25
2	7	5.4	1.6	2.56	−1	1
3	5	8.0	−3.0	9.00	−3	9
4	11	10.6	.4	.16	3	9
5	14	13.2	.8	.64	6	36
				SSE=12.40		SST=80

SSR=SST − SSE=80 − 12.4=67.6

b. $r^2 = \frac{\text{SSR}}{\text{SST}} = \frac{67.6}{80} = .845$

The least squares line provided a very good fit; 84.5% of the variability in y has been explained by the least squares line

c. $\Sigma x_i = 15$, $\Sigma y_i = 40$, $\Sigma x_i y_i = 146$,

$\Sigma x_i^2 - 55$, $\Sigma y_i^2 - 400$

$$\text{SSR} = \frac{[\Sigma x_i y_i - (\Sigma x_i \Sigma y_i)/n]^2}{\Sigma x_i^2 - (\Sigma x_i)^2/n}$$

$$= \frac{[146 - (15)(40)/5]^2}{55 - (15)^2/5} = 67.6$$

$$\text{SST} = \Sigma y_i^2 - \frac{(\Sigma y_i)^2}{n}$$

$$= 400 - \frac{(40)^2}{5} = 80$$

Note that these are the same values shown in part (a)

d. $r = \sqrt{.845} = +.9192$

16. a. $\Sigma x_i = 19.2$, $\Sigma y_i = 12{,}900$, $\Sigma x_i y_i = 41{,}710$

$\Sigma x_i^2 = 62.18$, $\Sigma y_i^2 = 28{,}070{,}000$

$$\text{SST} = \Sigma y_i^2 - \frac{(\Sigma y_i)^2}{n}$$

$$= 28{,}070{,}000 - \frac{(12{,}900)^2}{6} = 335{,}000$$

$$\text{SSR} = \frac{[\Sigma x_i y_i - (\Sigma x_i \Sigma y_i)/n]^2}{\Sigma x_i^2 - (\Sigma x_i)^2/n}$$

$$= \frac{[41{,}710 - (19.2)(12{,}900)/6]^2}{62.18 - (19.2)^2/6} = 249{,}865$$

$$\text{SSE} = \text{SST} - \text{SSR} = 335{,}000 - 249{,}865$$

$$= 85{,}135$$

b. $r^2 = \frac{\text{SSR}}{\text{SST}} = \frac{249{,}865}{335{,}000} = .746$

The least squares line accounted for 74.6% of the total sum of squares

c. $r = \sqrt{.746} = +.8637$

21. a. $s^2 = \text{MSE} = \frac{\text{SSE}}{n-2} = \frac{12.4}{3} = 4.133$

b. $s = \sqrt{\text{MSE}} = \sqrt{4.133} = 2.033$

c. $\Sigma x_i = 15$, $\Sigma x_i^2 = 55$

$$s_{b_1} = \frac{s}{\sqrt{\Sigma x_i^2 - (\Sigma x_i)^2/n}}$$

$$= \frac{2.033}{\sqrt{55 - (15)^2/5}} = .643$$

d. $$t = \frac{b_1 - \beta_1}{s_{b_1}} = \frac{2.6 - 0}{.643} = 4.04$$

$t_{.025} = 3.182$ (3 degrees of freedom)

Since $t = 4.04 > t_{.05} = 3.182$, we reject H_0: $\beta_1 = 0$

e. $\text{MSR} = \frac{\text{SSR}}{1} = 67.6$

$$F = \frac{\text{MSR}}{\text{MSE}} = \frac{67.6}{4.133} = 16.36$$

$F_{.05} = 10.13$ (1 degree of freedom numerator and 3 denominator)

Since $F = 16.36 > F_{.05} = 10.13$, we reject H_0: $\beta_1 = 0$

Source of Variation	Sum of Squares	Degrees of Freedom	Mean Square	F
Regression	67.6	1	67.6	16.36
Error	12.4	3	4.133	
Total	80	4		

24. a. $s^2 = \text{MSE} = \dfrac{\text{SSE}}{n-2} = \dfrac{85{,}135.14}{4} = 21{,}283.79$

$s = \sqrt{\text{MSE}} = \sqrt{21{,}283.79} = 145.89$

$\Sigma x_i = 19.2, \Sigma x_i^2 = 62.18$

$$s_{b_1} = \frac{s}{\sqrt{\Sigma x_i^2 - (\Sigma x_i)^2/n}}$$

$$= \frac{145.89}{\sqrt{62.18 - (19.2)^2/6}} = 169.59$$

$$t = \frac{b_1 - \beta_1}{s_{b_1}} = \frac{581.08 - 0}{169.59} = 3.43$$

$t_{.025} = 2.776$ (4 degrees of freedom)

Since $t = 3.43 > t_{.025} = 2.776$, we reject H_0: $\beta_1 = 0$

b. $\text{MSR} = \dfrac{\text{SSR}}{1} = \dfrac{249{,}864.86}{1} = 249{,}864.86$

$$F = \frac{\text{MSR}}{\text{MSE}} = \frac{249{,}864.86}{21{,}283.79} = 11.74$$

$F_{.05} = 7.71$ (1 degree of freedom numerator and 4 denominator)

Since $F = 11.74 > F_{.05} = 7.71$, we reject H_0: $\beta_1 = 0$

c.

Source of Variation	Sum of Squares	Degrees of Freedom	Mean Square	F
Regression	1543.84	1	1543.84	48.17
Error	96.16	3	32.05	
Total	1640.00	4		

29. a. $s = 2.033$

$\Sigma x_i = 15, \quad \Sigma x_i^2 = 55$

$$s_{\hat{y}_p} = s\sqrt{\frac{1}{n} + \frac{(x_p - \bar{x})^2}{\Sigma x_i^2 - (\Sigma x_i)^2/n}}$$

$$= 2.033\sqrt{\frac{1}{5} + \frac{(4-3)^2}{55 - (15)^2/5}} = 1.11$$

b. $\hat{y} = .2 + 2.6x = .2 + 2.6(4) = 10.6$

$\hat{y}_p \pm t_{\alpha/2} s_{\hat{y}_p}$

$10.6 \pm 3.182(1.11)$

10.6 ± 3.53 or 7.07 to 14.13

c. $$s_{\text{ind}} = s\sqrt{1 + \frac{1}{n} + \frac{(x_p - \bar{x})^2}{\Sigma x_i^2 - (\Sigma x_i)^2/n}}$$

$$= 2.033\sqrt{1 + \frac{1}{5} + \frac{(4-3)^2}{55 - (15)^2/5}} = 2.32$$

d. $\hat{y}_p \pm t_{\alpha/2} s_{\text{ind}}$

$10.6 \pm 3.182(2.32)$

10.6 ± 7.38 or 3.22 to 17.98

32. a. $s = 145.89, \quad \Sigma x_i = 19.2, \quad \Sigma x_i^2 = 62.18$

$\hat{y} = 290.54 + 581.08x = 290.54 + 581.08(3)$

$= 2033.78$

$$s_{\hat{y}_p} = s\sqrt{\frac{1}{n} + \frac{(x_p - \bar{x})^2}{\Sigma x_i^2 - (\Sigma x_i)^2/n}}$$

$$= 145.89\sqrt{\frac{1}{6} + \frac{(3-3.2)^2}{62.18 - (19.2)^2/6}} = 68.54$$

$\hat{y}_p \pm t_{\alpha/2} s_{\hat{y}_p}$

$2033.78 \pm 2.776(68.54)$

2033.78 ± 190.27 or \$1843.51 to \$2224.05

b. $\hat{y} = 290.54 + 581.08x = 290.54 + 581.08(3)$

$= 2033.78$

$$s_{\text{ind}} = s\sqrt{1 + \frac{1}{n} + \frac{(x_p - \bar{x})^2}{\Sigma x_i^2 - (\Sigma x_i)^2/n}}$$

$$= 145.89\sqrt{1 + \frac{1}{6} + \frac{(3-3.2)^2}{62.18 - (19.2)^2/6}} = 161.19$$

$\hat{y}_p \pm t_{\alpha/2} s_{\text{ind}}$

$2033.78 \pm 2.776(161.19)$

2033.78 ± 447.46 or \$1586.32 to \$2481.24

37. a. 9

b. $\hat{y} = 20.0 + 7.21x$

c. 1.3626

d. SSE = SST − SSR = 51,984.1 − 41,587.3 = 10,396.8

MSE = 10,396.8/7 = 1485.3

$$F = \frac{\text{MSR}}{\text{MSE}} = \frac{41{,}587.3}{1485.3} = 28.0$$

$F_{.05} = 5.59$ (1 degree of freedom numerator and 7 denominator)

Since $F = 28 > F_{.05} = 5.59$, we reject H_0: $\beta_1 = 0$

e. $\hat{y} = 20.0 + 7.21(50) = 380.5$ or \$380,500

42. a. $\Sigma x_i = 70, \Sigma y_i = 76, \Sigma x_i y_i = 1264, \Sigma x_i^2 = 1106$

$$b_1 = \frac{\Sigma x_i y_i - (\Sigma x_i \Sigma y_i)/n}{\Sigma x_i^2 - (\Sigma x_i)^2/n}$$

$$= \frac{1264 - (70)(65)/5}{1106 - (70)^2/5} = 1.5873$$

$b_0 = \bar{y} - b_1\bar{x}$

$= 15.2 - 1.5873(14) = -7.0222$

$\hat{y} = -7.02 + 1.59x$

b.

x_i	y_i	$\hat{y}_i$	$y_i - \hat{y}_i$
6	6	2.52	3.48
11	8	10.47	−2.47
15	12	16.83	−4.83
18	20	21.60	−1.60
20	30	24.78	5.22

c. $y - \hat{y}$

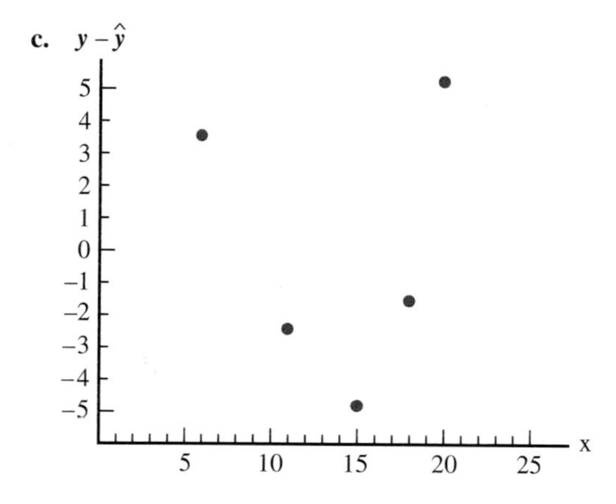

With only five observations, it is difficult to determine whether the assumptions are satisfied; however, the plot does suggest curvature in the residuals, which would indicate that the error term assumptions are not satisfied; the scatter diagram for these data also indicates that the underlying relationship between x and y may be curvilinear

44. $\Sigma x_i = 57$, $\Sigma y_i = 294$, $\Sigma x_i y_i = 2841$,

$\Sigma x_i^2 = 753$, $\Sigma y_i^2 = 13{,}350$

$$b_1 = \frac{\Sigma x_i y_i - (\Sigma x_i \Sigma y_i)/n}{\Sigma x_i^2 - (\Sigma x_i)^2/n}$$

$$= \frac{2841 - (57)(294)/7}{753 - (57)^2/7} = 1.5475$$

$$b_0 = \bar{y} - b_1\bar{x}$$

$$= 42 - (1.5475)(8.1492) = 29.3989$$

$\hat{y} = 29.40 + 1.55x$

b. SSR=691.72 and SST=1002; therefore SSE=1002 − 691.72=310.28

$$F = \frac{\text{MSR}}{\text{MSE}} = \frac{691.72}{310.28/5} = 11.15$$

$F_{.05} = 6.61$ (1 degree of freedom numerator and 5 denominator)

Since $F = 11.47 > F_{.05} = 6.61$, we reject H_0: $\beta_1 = 0$; the relationship is significant at the .05 level

c.

x_i	y_i	$\hat{y}_i$=29.40 + 1.55x_i	$y_i - \hat{y}_i$
1	19	30.95	−11.95
2	32	32.50	−.50
4	44	35.60	8.40
6	40	38.70	1.30
10	52	44.90	7.10
14	53	51.10	1.90
20	54	60.40	−6.40

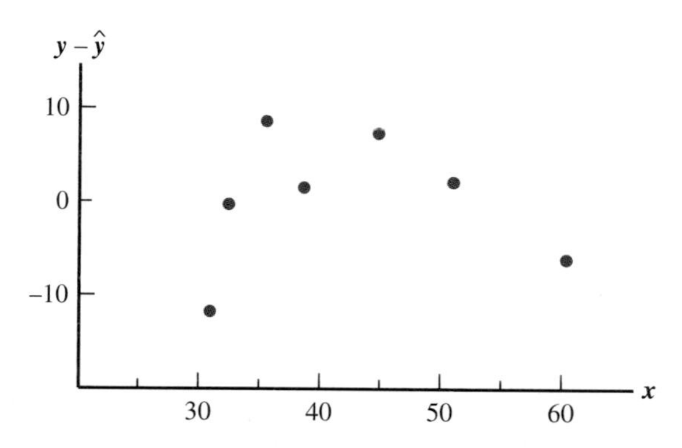

d. The residual plot leads us to question the assumption of a linear relationship between x and y; even though the relationship is significant at the $\alpha = .05$ level, it would be extremely dangerous to extrapolate beyond the range of the data (e.g., $x > 20$)

49. a. The Minitab output is shown below

```
The regression equation is
REVENUE = 88.6 + 1.60 TVADV

Predictor        Coef       Stdev     t-ratio        p
Constant       88.638       1.582       56.02    0.000
TVADV          1.6039      0.4778        3.36    0.015

s = 1.215        R-sq = 65.3%      R-sq(adj) = 59.5%

Analysis of Variance

SOURCE        DF          SS          MS          F          p
Regression     1      16.640      16.640      11.27      0.015
Error          6       8.860       1.477
Total          7      25.500
```

b. The Minitab output is shown below

```
The regression equation is
REVENUE = 83.2 + 2.29 TVADV + 1.30 NEWSADV

Predictor        Coef       Stdev    t-ratio        p
Constant       83.230       1.574      52.88    0.000
TVADV          2.2902      0.3041       7.53    0.001
NEWSADV        1.3010      0.3207       4.06    0.010

s = 0.6426     R-sq = 91.9%     R-sq(adj) = 88.7%

Analysis of Variance

SOURCE        DF        SS          MS         F          p
Regression     2    23.435      11.718     28.38      0.002
Error          5     2.065       0.413
Total          7    25.500
```

c. It is 1.60 in (a) and 2.29 in (b). In (a) the coefficient is an estimate of the change in revenue due to a one-unit change in television advertising expenditures. In (b) it represents an estimate of the change in revenue due to a one unit change in television advertising expenditures when the amount of newspaper advertising is held constant.

g. Revenue =83.2+2.29(3.5)+1.30(1.8)=93.56 or $93,560.

Chapter 13

4. *R chart:*

$$UCL = \bar{R}D_4 = 1.6(1.864) = 2.98$$

$$LCL = \bar{R}D_3 = 1.6(.136) = .22$$

$\bar{x}$ chart:

$$UCL = \bar{\bar{x}} + A_2\bar{R} = 28.5 + .373(1.6) = 29.10$$

$$LCL = \bar{\bar{x}} - A_2\bar{R} = 28.5 - .373(1.6) = 27.90$$

10. $f(x) = \dfrac{n!}{x!(n-x)!}p^x(1-p)^{n-x}$

When $p = .02$, the probability of accepting the lot is

$$f(0) = \frac{25!}{0!(25-0)!}(.02)^0(1-.02)^{25} = .6035$$

When $p = .06$, the probability of accepting the lot is

$$f(0) = \frac{25!}{0!(25-0)!}(.06)^0(1-.06)^{25} = .2129$$

Index